Taylor's Power Law

Taylor's Power Law
Order and Pattern in Nature

R. A. J. Taylor

ACADEMIC PRESS
An imprint of Elsevier

ELSEVIER

Academic Press is an imprint of Elsevier
125 London Wall, London EC2Y 5AS, United Kingdom
525 B Street, Suite 1650, San Diego, CA 92101, United States
50 Hampshire Street, 5th Floor, Cambridge, MA 02139, United States
The Boulevard, Langford Lane, Kidlington, Oxford OX5 1GB, United Kingdom

Notices
Knowledge and best practice in this field are constantly changing. As new research and
experience broaden our understanding, changes in research methods, professional practices, or
medical treatment may become necessary.

Practitioners and researchers must always rely on their own experience and knowledge in evaluating
and using any information, methods, compounds, or experiments described herein. In using
such information or methods they should be mindful of their own safety and the safety of
others, including parties for whom they have a professional responsibility.

To the fullest extent of the law, neither the Publisher nor the authors, contributors, or editors,
assume any liability for any injury and/or damage to persons or property as a matter of products
liability, negligence or otherwise, or from any use or operation of any methods, products,
instructions, or ideas contained in the material herein.

Library of Congress Cataloging-in-Publication Data
A catalog record for this book is available from the Library of Congress

British Library Cataloguing-in-Publication Data
A catalogue record for this book is available from the British Library

ISBN 978-0-12-810987-8

For information on all Academic Press publications
visit our website at https://www.elsevier.com/books-and-journals

Publisher: Charlotte Cockle
Acquisition Editor: Anna Valutkevich
Editorial Project Manager: Pat Gonzalez
Production Project Manager: Stalin Viswanathan
Cover Designer: Matthew Limbert

Typeset by SPi Global, India

Contents

Preface   xvii

1. Introduction
References   9

Part I
2. Spatial pattern
Sampling   13
Randomness   16
Poisson distribution   16
Negative binomial distribution   16
Polya-Aeppli distribution   17
Neyman's A distribution   17
Lognormal distribution   18
Inverse Gaussian distribution   19
Tweedie family of distributions   20
Origins of aggregation   21
Censuses   23
References   23

3. Measuring aggregation
Nearest neighbor   28
Indices using mean and variance   29
Variance-mean ratio   29
Negative binomial k 30
    Fitting the negative binomial   30
    Interpretation of k 31
    Negative binomial with common k 32
Morisita's I_δ patchiness index   33
Green's C_x coefficient   33
Lloyd's mean crowding   33
Iwao's patchiness index   35
Iwao's ρ index   35
Variance-mean relationship   35
    TPL as an index of aggregation   37

Adès distribution 39
Perry's spatial analyses 42
Perry and Hewitt's number of moves index 42
SADIE 43
References 45

4. Fitting TPL

The standard regression model 50
Functional regression 51
Geometric mean regression 52
Deming regression 53
Bartlett's 3-group regression 54
Methods for fitting TPL 55
Fitting split lines 57
Bias in estimating TPL 58
Comparison of models 60
Parallel-line analysis 61
Ordinary dependent regression 61
Reduced major Axis (geometric mean) regression 62
Methods used to fit TPL 64
References 64

Part II

5. Microorganisms

Free living 69
Gymnamoebae 69
Bacteria in a Siberian reservoir 70
Diatoms in Laguna di Venezia 72
Foraminifera in Delaware 72
Invasive flagellate in Sweden 73
Ciliates in the East China Sea 75
Marine viruses in California and Sweden 77
Plant hosts 78
Tobacco mosaic virus on beans 78
Verticillium dahliae in potato fields 79
Passalora fulva on tomatoes 81
Mummy berry disease of blueberries 82
Powdery mildew on apples 83
Pear scab 85
Phytophthora 86
On peppers and soybeans 86
In the air 87
Strawberry anthracnose and rain splashes 89
Animal hosts 90
Ciliates on flatworms 90

Pasteuria penetrans on *Meloidogyne arenaria* 91
Bacteria cultures 93
The human microbiome 94
Appendix: TPL estimates for microorganisms 97
References 100

6. Plants

Seedbank 104
Farmland in England 104
Field margins in Wisconsin 109
In soybean fields 110
Tree seedbank in Taiwan 112
Seedbank diversity in Catalonia 113
Invasive devil's thorn in Australia 114
Invasive ragweed in France 115
Endangered bearpoppy in Nevada 117
Grasslands 117
Grassland in Shaanxi Province 118
Tallgrass prairies in Texas 119
Rangeland in Mongolia 121
Other examples 123
Edible palm in Brazil 123
Insectivorous plants in Morocco and Iberia 124
Eelgrass in Chesapeake Bay 125
Pollination success in a Yucatan shrub 126
Commodity crops 128
Maize 128
Sugar cane 129
Wheat 130
Potatoes 131
Appendix: TPL estimates for plants 132
References 140

7. Nematodes and other worms

Nematodes 143
Extraction 144
Sampling 144
Terrestrial nematodes 148
Ecological classifications 148
Urban turfgrass in Ohio 148
Grassy pasture in Sweden 151
Oak forest in Bulgaria 151
Mountain forest in China 153
Forests in Scotland 153
Subarctic heath 153
Metabolic footprint 154

TPL stability   156
  Perrine marl soil in Florida   156
  Forest and old field   158
  Crops in Florida   158
  Cowpeas and cotton in California   158
Plant-parasitic nematodes   **160**
  Annual crops   160
    Potatoes   160
    Soybeans   164
    Broad beans   164
    Tobacco   165
    Eggplant (Aubergine)   166
    Mixed vegetables   166
  Perennial crops   167
    Citrus   167
    Clover   169
    Sugarcane   170
    Banana   170
    Pine trees   171
Entomopathogenic nematodes   **173**
  Sampling methods   173
    Steinernema feltiae and *S. glaseri*   174
    Steinernema carpocapsae and *Heterorhabditis bacteriophora* 175
    Effect of habitat   178
    Baiting effects   179
Animal-parasitic nematodes   **180**
  Invertebrate hosts   180
    Cockroaches   180
    Drosophilids   181
  Vertebrate hosts   181
    Sheep   183
    Rabbits   185
    Rats   187
    Mice   187
    Carp   188
Aquatic nematodes   **189**
Freshwater nematodes as bioindicators   190
  Rivers in Germany   190
  Highland streams in Germany   191
  Lakes in Sweden   192
  Farm ponds in Belgium   192
  Restored wetland in Georgia   194
Marine nematodes   195
  The intertidal   195
    Comparing samplers   195
    Sediment texture   196
    Predator and prey   197
  Littoral and sublittoral   197

Pollution gradient   197
Latitudinal gradient   198
Benthic   198
Depth gradient   198
The world's oceans   199
Nematodes on kelp   199
An extreme case   200
Other worms   **202**
Platyhelminths   **203**
Primary hosts—grey mullet   204
Cestodes   **205**
Secondary hosts—Shrimps   205
Primary hosts—Domestic fowl   205
Annelids   **206**
Leeches in a Cumbrian stream   206
Oligochaetes   207
Earthworms in Scotland   207
Earthworms in Colombia   208
Polychaetes   209
In the sublittoral   209
Biomass as proxy for abundance   209
A behavioral experiment   211
Ways of arranging data   211
Appendix: TPL estimates for nematodes and other worms   **213**
References   **227**

8. Insects and other arthropods

Insects   **235**
Lepidoptera   235
European corn borer   236
Winter moth   237
Gypsy moth   239
Coleoptera   242
Wireworms   242
In Washington State   243
In England and Wales   243
Colorado potato beetle   244
Japanese beetle   247
In North America   247
Larvae   247
Adults   250
In the Azores   252
Crustacea   **254**
Barnacles   254
Chthamalus species in Japan   254
Stratification in barnacle distribution   257
Settling behavior of barnacle cyprid larvae   258

A general survey   260
Consistency across space, time, and stage   261
Sampling efficiency and consistency between samplers   266
Effect of sampling method   268
Differences between trophic levels   271
    Predation   271
    Parasitism   273
    Competition   273
Effect of changes in scale   274
Crustaceans   276
Appendix: TPL estimates for arthropods   277
Appendix 8.M   280
Key to Appendix 8.M   295
References   296

9. Other invertebrates

Rotifers   305
    Lake Eufaula, Oklahoma   305
    Upper Paraña River basin, Brazil   306
    River Elbe Estuary, Germany   308
Molluscs   309
    Tellina tenuis in the Firth of Clyde, Scotland   310
    Intertidal molluscs on the Isle of Man   311
    Marine bivalves and gastropods in Denmark   313
    Terrestrial gastropods in Alberta   313
    Slugs in Northumberland   315
Echinoderms   317
    Starfish in North Wales   317
    Crinoids in São Paulo State, Brazil   318
Other Phyla   319
    Bryozoans in the Greenland Sea   319
    Hydroids in the Argentine Sea   320
    Jellyfish in Oregon-Washington coastal waters   322
Appendix: TPL estimates for other invertebrates   324
References   326

10. Vertebrates

Distance sampling   327
Fish   328
    Haddock and whiting off Massachusetts   328
    Herring and Mackerel in the Norwegian Sea   329
    Salmon in the Northeast Pacific Ocean   331
    Demersal fish in a tropical bay in Brazil   332
    Pelagic fish larvae in Portugal   334
    Pelagic fish larvae in New Jersey   335

Fish larvae entering Pamlico and Albemarle Sounds, North Carolina 336
Sea trout fry in England's Lake District 337
Adult sea trout catches in England and Wales 340
Californian commercial fisheries 342
Amphibians and reptiles **344**
Reptiles in the Florida Everglades 344
Herptiles in Arizona's Rincón Mountains 345
Frogs in an Alpine habitat in California 347
Amphibian larvae in two ephemeral ponds in Ohio 348
Eagle prey in Northern Greece 348
Birds **349**
Willow ptarmigan in Norway 349
Jays in Western USA 351
Grassland sparrows in continental USA 353
Birds in urban and nonurban environments 354
British trust for ornithology annual survey 356
Audubon Society's Christmas Bird Count 359
Mammals **363**
Cetaceans near the Azores 363
Cetaceans around the British Isles 364
Harbor porpoise in the North Sea 366
Fin whales in the Ligurian Sea 369
Herding ungulates in Kenya's Rift Valley 369
Kangaroo rat mounds in New Mexico 371
Appendix: TPL estimates for vertebrates **372**
References **378**

11. Other biological examples

General biology **383**
Rate of evolution 383
Exoenzymes in soil 385
Metabolism in a river system 387
Phosphorus in lakes 388
Physiology **389**
Metastatic cancers 389
Physiological responses to stimuli 390
Human demography **392**
United States decennial census 392
Population of Norway 395
Mortality in England and Wales 397
Movement in China 399
Human health **399**
Human immunodeficiency virus 399
Typhoid in Cambodia 401
Measles and whooping cough 402
Disease monitoring 403

Human behavior   **405**
   Stress in air traffic controllers   405
   Crime in Britain   406
   European wars   409
Socioeconomic   **410**
   Convenience store sales in Japan   410
   Size of corporations   412
   Size of cities   414
Appendix: TPL estimates for other biological examples   **416**
References   **420**

12. Nonbiological examples

Astronomical   **423**
   Cyanogen in a comet halo   423
The distribution of heavenly bodies   424
   Yale Bright Star Catalog   425
   SAO Star Catalog   426
   Principal Galaxy Catalog   426
Geophysical   **428**
   Earthquakes off the east coast of Japan   428
   Tornadoes in the continental USA   429
   Precipitation actual and simulated   431
Numerical   **433**
   Traffic through networks   433
   Foreign exchange markets   435
   Prime numbers   437
Appendix: TPL estimates for nonbiological phenomena   **440**
References   **442**

13. Counter examples

Sampling   **445**
Inadequate *NQ* or *NB*   445
   Meiobenthos in the Balearic Islands   446
   Nematodes on the Darwin Mounds   447
Inconsistent sampling   447
   Thrips in a cucumber crop   447
Trap saturation   448
   Powdery mildew on apples   448
   A modeling example   449
Catastrophic change   **450**
   Effect of pesticides   450
Ratios   **451**
Bounded ratios (%)   451
   Sex ratio   451
   Parasite prevalence   451

Unbounded ratios   452
   Commodity crops   452
Uncountable numbers   **455**
   Temperature   455
References   **456**

Part III

14. Applications of TPL

Transformations   **462**
Stabilizing variance   463
Sampling   **465**
Sampling patterns   466
   Random sampling   466
   Systematic sampling   466
   Stratified random sampling   466
Binomial sampling   466
Sequential sampling   468
Optimum sample size   469
Number of samples   470
Sampling efficiency   472
Postscript   475
Environmental assessment and monitoring   **476**
   Detecting environmental perturbations   476
   The cost of conservation   478
   Stream water quality   479
   Environmental assessment of wind farms   481
Other uses   **481**
   Testing vaccines   481
   Model calibration and validation   482
   Quality control   483
**Appendix: The delta technique for variance of a function of a
random variable**   **484**
References   **484**

15. Properties of TPL

Self-similarity   **489**
Effect of small samples   **491**
Effect of zeros   **492**
Direct effect   492
Anatomy   495
Sampling efficiency   **496**
Density-dependence   497
Trap saturation   500
References   **502**

16. Allometry and other power laws

Mathematical **503**
Pareto distribution 503
Zipf's law 504
Spectra 504
Scale-free networks 505
Diffusion-limited aggregation 506
Fractals 507
   Repetition as a source of self-similarity 507
Physical **508**
Inverse square law 508
Stefan–Boltzmann law 508
Self-organized criticality 508
Percolation 509
Meteorology 509
Hydrology 510
Geophysics 511
Biological **511**
Allometric growth 511
Dimensional relationships for flying animals 512
Fractal movement 515
Frequency of species size 515
Species-area 515
Kleiber's law of metabolism 518
Respiration 518
Self-thinning and space-filling 519
Soil fertility and crop yields 522
Binomial power law 523
Density-size and variance-size laws 524
Genetics and physiology 526
Taxonomy 526
Sociological **527**
Richardson's law of conflict 527
References **528**

17. Modeling TPL

Physical models **533**
A fractal model 535
Diffusion-limited aggregation 535
A network model 536
Statistical physics 537
A biophysical model 539
Statistical models **541**
Reformulating TPL 541
Higher moments 542

Simulated sampling 543
Sampling and feasible sets 544
A lattice model 545
Biological models **546**
Nonlinear maps 547
   Models in time 547
   Strong density dependence 548
   Effect of competition 551
Temporal TPL and stability 552
The Lewontin-Cohen model 553
A singularity in TPL 554
   Singularities in other models 556
Dispersal distance 558
Ideal free distribution 560
Agent-based models **561**
Cellular automaton 562
Arrangement in space 563
References **565**

18. Summary and synthesis

The biological evidence **571**
TPL and the pattern of sampling **572**
Physical versus biological **573**
Sources of range of means **574**
The role of the sampler **575**
Sampling effort and efficiency 576
The sampling site 577
The time of sampling 577
Transect sampling 578
Box counting 578
Analysis in two directions **579**
Temporal versus spatial 579
Orthogonal directions 579
In three dimensions 580
Intersection of TPL and Poisson line **580**
The effect of scale **581**
Super aggregation **583**
Trophic interactions **584**
Predation 584
Parasitism 585
Competition 585
Mixed-species and community TPLs **586**
Human demographics and sociology **587**
Other uses of TPL **588**
When TPL doesn't work **590**
Countable and noncountable number series 591

Models   591
Are the power laws related?   595
Is TPL universal?   596
References   599

19. Epilogue

References   **607**

Author index   609
Subject index   625

Preface

In Autumn 1964, my family took a road trip from Manhattan, Kansas, to Winnipeg, Manitoba. Sitting in the back seat of our Oldsmobile, I listened to my parents discussing an equation my father was trying to understand. It was apparently important for practical reasons and perhaps others, but was enigmatic. It didn't seem to fit into the science of the time. Five years later, I sat in a lecture theater at Imperial College, London, listening to Professor T.R.E. (Dick) Southwood lecturing on the analysis of spatial distribution of organisms and its importance in understanding population change. He too talked of an enigmatic equation that had multiple practical applications and an important function in interpreting spatial distribution, but whose origins were elusive. Four years later at Silwood Park, Professors Dick Southwood and Michael Way gave me the freedom to investigate what was by then known as Taylor's power law. I learned then that practically any rule governing organization in space iterated in time would result in a convincing power law relating spatial variance to population density. One rule I used, based on the potential energy equation of physics (the Δ-model), was particularly successful in this, but was highly dependent on initial conditions. The same input parameters applied to different starting patterns produced different power laws, apparently at random. The Δ-model has descriptive and interpretive value, as field biologists have shown, but little predictive value. One property it did have, which many other models failed to display, was its ability to generate power laws with slopes steeper than variance proportional to the mean squared. As far as Roy Taylor and I were aware in 1977, this variance-mean relationship was unique to ecology. In the last 20 years, the power law has been found to apply to data in disciplines as diverse as astronomy, geology, meteorology, criminology, sociology, economics, and computer science. Thus, it is not unique to ecology and may have nothing to do with ecology *per se*. This book grew out of my continuing interest in this enigmatic but ubiquitous equation and the desire to understand its origins. My approach is not a modeling approach, although models are important and are not ignored, but that of basic biology: the rubric of comparative anatomy. The book is organized in three parts: an introductory section deals with background and methods; a survey section describes both biological and non-biological examples; and a concluding section covers practical uses, models, and interpretation. The middle section is intended to do three things: to cover as wide a taxonomic range as possible, to show the full range of power law slopes, and to highlight similarities and differences between the biological and non-biological examples. Only in the penultimate chapter do I attempt a synthesis. Elsewhere I have tried to report accurately and to refrain from criticism. Those cited can judge whether I was successful.

Special thanks go to my teachers at Imperial College, Nadia Waloff, George Murdie, Michael Way, and Dick Southwood, and for the many conversations with my father, L.R. Taylor. In addition to my mentors, all alas now gone, I want to thank my daughter Kara who designed the book cover, and the following people who provided data or checked my use of their data: John Cardina, Malcolm Elliott, Dan Grear, Larry Madden, Adeline Murthy, Arne Peters, Wayne Polley, Peter Smits, and Xiangming Xu. Malcolm Elliott, Cathy Herms, Dan Herms, Parwinder Grewal, Sam Ma, Larry Madden, Adeline Murthy, Joe Perry, Wayne Polley, Kara Taylor, and Robert Taylor read and critiqued parts. Several coauthors also contributed data and discussion: Andy Chapple, Dick Lindquist, Mike McManus, Sunny Park, Charlie Pitts, Les Shipp, Ian Woiwod, and the late Ben Stinner. To all, I offer my sincere thanks for their assistance, support, and friendship.

This book is dedicated to my parents, Jean and Roy Taylor, and to my family, wife Kate, daughter Kara, and son Robert who not only put up with me while developing this book, but read a lot of it too. However, of the inevitable errors, I alone claim responsibility.

Wooster, Ohio
January 2019

Chapter 1

Introduction

That tigers hunt alone and wolves in packs are well-known attributes of these two predatory mammals. The likelihood of finding even one tiger these days is distressingly low, but even a century ago when they were much more abundant throughout southeast Asia and Siberia, it would have been rare to find one adult tiger in the company of another not her cub. By comparison, it was and still is rare to encounter a lone wolf. These species' spatial behavior is easily recognized as being different; but how different, and how does one measure it? I chose these two species to exemplify extremes in spatial behaviors because they are familiar archetypes. But their characteristic distributions and abundances are no different in kind to those of the myriad other species, microbes, plants, and animals vertebrate and invertebrate, with which we share this planet.

This book is primarily about the spatial distributions of populations of living things, and how their distributions are measured. It also touches on nonliving entities that appear to behave mathematically in a similar fashion. It is conceivable that the same rules that govern, or at least describe, living organisms' distributions in space and time also apply to nonliving things. If this is the case, then the mathematical construct that is central to the theme of this work may have deeper meaning than the mathematico-statistical curiosity seemed to have 55 years ago when it was first described and named.

Taylor' power law (TPL; Lincoln and Boxshall, 1982) was referred to by that name in T. R. E (Dick) Southwood's *Ecological Methods with Particular Reference to the Study of Insect Populations*, published in 1966. As far as I have been able to ascertain, this is the earliest reference to the eponymous power law relating sample variance and mean population published in *Nature* by L. R. (Roy) Taylor (Taylor, 1961, hereafter referred to as LRT61). Roy Taylor's own book on ecological experiments published the year following Southwood's references the relationship but does not name it (Lewis and Taylor, 1967).

Taylor's power law states that sample variance and mean population are related by a simple power law:

$$V = aM^b \tag{1.1a}$$

or

$$\mathrm{Log}(V) = A + b\log(M) \tag{1.1b}$$

Taylor's Power Law. https://doi.org/10.1016/B978-0-12-810987-8.00001-X

1

where V is the variance of a set of samples and M is the average value of that set; $A = \log(a)$ and b are parameters estimated from a collection of variances and means generated from multiple sets of samples taken at different times and/or places and extending over as broad a range of means as possible. I have chosen to use italics to represent calculated estimates of variance and mean as these are generally the basis for presentation and discussion of aggregation as measured by TPL. Where population or theoretical values, as opposed to estimates, are discussed, normal text for variables and Greek letters for parameters will be used:

$$V = \alpha M^{\beta}. \tag{1.2}$$

TPL data are usually analyzed by taking logarithms of the mean and variances. Common logs, $\log_{10}(\bullet)$, will normally be used, represented by $\log(\bullet)$. Natural logs, $\log_{e}(\bullet)$ when used will be denoted by $\ln(\bullet)$.

TPL is mathematically simple and its parameters provide simple metrics for describing the degree of aggregation of organisms and other entities that range from regular to random to highly clumped. But its genesis and scope are not simple. Indeed much paper has been consumed trying to account for them. This book is intended to collect together the thoughts and suggestions of the hundreds who have written and speculated about TPL. There are two main objectives: to review as wide a taxonomic range of examples and nonbiological examples of TPL and to examine the various models developed to account for it. There are also counter examples where the power law does not apply when one might expect it to. The exceptions to the law may prove to be far more illuminating than the thousands of exemplars.

Early on, in LRT61 and Southwood (1966) for example, it was proposed that a is largely a sample-size parameter while b is a population-specific, or possibly species-specific, description of aggregation. In the intervening years, these simple distinctions have been found to oversimplify the interpretation of a and b. In fact, they are often correlated making this simple dichotomy untenable. Complicating matters, there are actually three variants of the power law depending on how the samples were taken and how they are processed. These variants tell different stories that collectively shed new light on TPL.

It is no accident that the relationship between variance and mean was discovered by an agricultural and ecological entomologist. With the possible exception of plankton and other aquatic invertebrates, the most extensive data of population density and distribution are of insects, mostly those of economic—agricultural, veterinary, or medical—significance. Much of statistical theory originated with these disciplines. Ronald Fisher, the architect of experimental design, was employed at Rothamsted Experimental Station in Harpenden, England, to analyze long-term data of agricultural experiments (Fisher, 1935). While at Rothamsted, Fisher also developed methods for ecological research, including the analysis of diversity with C.B. Williams (Fisher et al., 1943) as well as spatial distribution (Bliss and Fisher, 1953). Fisher's tradition continues at Rothamsted (now Rothamsted Research) and includes the work of Roy Taylor and colleagues. While Fisher's contributions were essentially theoretical, their value lies in their application to applied entomology, plant pathology, agronomy, horticulture, agriculture and genetics, as well as population and community ecology.

The fundamental feature of TPL is the density dependence of the frequency distribution of abundance. Locally, the frequency distribution of population number per unit area may be described by the Poisson, negative binomial, Neyman's type A, or one of dozens of other statistical distributions. But at larger scale, we generally find that many statistical distributions are needed to characterize abundance. The astonishing thing is that if we plot the variance against the mean for each of the empirical frequency distributions, we find they are related by a straight line on logarithmic scales. Fig. 1.1A of four distributions of European corn borer (*Ostrinia* (=*Pyrausta*) *nubilalis*) suggests a continuum in shape of frequency distribution in which the mode shifts to the right and the right-hand tail gets longer with increasing mean density (McGuire et al., 1957 in LRT, 1965). This plot suggests a smooth change in frequency distribution with density, but a plot of *Ooencyrtus kuvanae* parasitism of gypsy moth (*Lymantria dispar*) eggs (Brown and Cameron, 1982) shows overlaps between the best fit distributions as density increases (Fig. 1.1B).

Clearly, the distribution and abundance of organisms are intimately related, and together define a population's size and structure in space. TPL implies that the shape of the frequency distribution of numbers per sample is itself density dependent. Furthermore, transitional distributions may not be well described by any known statistical frequency distribution as is evident in Fig. 1.1A.

In his book *Patterns in the Balance of Nature*, Williams (1964) makes the case that the statistical patterns of populations and communities of populations are the numerical and geographic manifestation of a natural balance. He suggested that there is a connection between the frequency distribution of abundance of species in a community and the spatial distribution of individuals in that community. Specifically, he suggested that although the numerical abundance of a species or other grouping in a community is always in a state of flux, it is possible that the overall statistical pattern remains more or less constant. The high degree of repeatability exhibited by TPL argues for a similar natural pattern. It is with that in mind that I paraphrased Williams in the subtitle *Order and Pattern in Nature*.

LRT61 (Fig. 1.2) is one of the most widely cited papers in ecology and agriculture. As of December 2018, The Web of Science (Thomson Reuters, 2018) lists nearly 2000 citations (Fig. 1.3) with 40–50 new papers per year. There are many more that refer to TPL but cite derivative or later publications. The recent growth is due in large part to the discovery of variance-mean power laws in disciplines as diverse as economics, physics, computer science, genetics, and molecular biology. This diversity was highlighted by a major review of "fluctuation scaling," as the variance-mean power law is known in physics (Eisler et al., 2008).

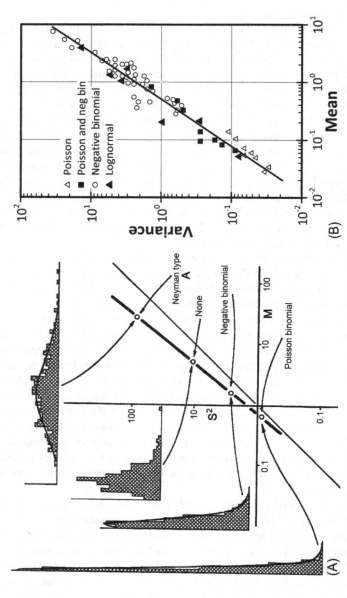

FIG. 1.1 (A) Three of four field samples of European corn borer from McGuire et al. (1957) are well fit by different frequency distribution; the fourth is not fit by a distribution. The mean-variance pairs all lie on a line with equation $V = 1.5 \, M^{1.25}$. (B) The frequency distribution of the egg parasitic wasp *Ooencyrtus kuvanae* on or near gypsy moth egg masses changes as the wasp's density increases: The original data were expressed as wasps per egg mass. The *fitted line is* $V = 2.31 \, M^{1.27}$, $r^2 = 0.94$. ((A) *From LRT (1965). Courtesy of the Council for International Congresses of Entomology. (B) Adapted from Brown and Cameron (1982).)*

Eisler et al. (2008) identified three different ways to compute TPL: spatial, temporal, and ensemble and the appropriate approach is determined by the mode of sample collection.

i. Spatial—samples at multiple locations at a point in time replicated at several points in time; each TPL point is a point in time. An example is the US decennial census in Fig. 11.5.
ii. Temporal—time series sampled with multiple contiguous windows of length Δ time units, replicated with multiple time series subjects or locations. This is generalizable to different Δs, that will generate different TPLs, for example, currency trading (Fig. 12.8). A special case is a system of samples through time at multiple locations but with a single value of Δ. An example is LRT and Woiwod (1982) in which species abundance at a site was averaged over multiple years and TPL points are sites (Fig. 15.2).
iii. The examples of aphids moths and birds (LRT and Woiwod, 1982) and European battles (Chapter 11) are amenable to both spatial and temporal analysis with fixed Δ.
iv. Ensemble—multiple simultaneous samples taken at multiple locations at a point in time—this is a common situation in ecology.
v. Other versions of ensemble also occur in which spatial and temporal samples are combined to form the variance-mean points or the TPL points are a combination of sites and times. Example of the latter is the barnacle study in Japan (Chapter 8) and the former is the rotifer study in the Elbe Estuary (Chapter 9). These ensemble TPLs I call hybrid TPLs.
vi. To these, we add community and mixed-species TPLs, what Ma (2015) calls TPL extensions. Mixed-species TPLs are plots in which each point is a species or other taxon with mean and variance of number of specimens of each taxon computed from samples taken at a single site over time or simultaneous samples from multiple sites. Community TPLs are computed from the means and variances of samples consisting of taxa (each taxon is a quadrat), again collected simultaneously at several sites or at a single site over time. Mixed-species and community TPLs of data collected at multiple sites simultaneously are, strictly speaking, ensemble TPLs, while those computed from samples taken from the same site at intervals in time are quasi-spatial. Obviously, if samples taken at different places and times are averaged, the results are hybrid mixed-species and community TPLs.

The essence of experiment is repeatability (Franklin, 1990). Most experiments in ecology and agriculture cannot be repeated precisely because the conditions can never be recreated. Moreover, experimental manipulation of most living things results in a change in their behavior—a sort of Heisenberg uncertainty. Such changes can draw any conclusions into question. Thus, statistical methods are indispensable for any enquiry involving living things. Particularly, ecology and agriculture must rely on samples, and it was the analysis of samples that lead Roy Taylor to discover that the variance of those samples scales with

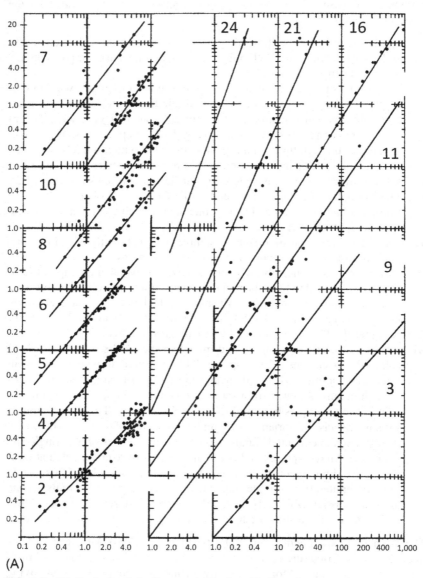

(A)

FIG. 1.2 (A) Facsimile of the figure and (B), the table from the manuscript of LRT61.

Table 1

Name	Site and sample	Range of m	s²	N	a	b	Trans-formation	Observer (refs. in brackets)
1 Shellfish on seashore, Tellina tenuis da Costa, Eulemellibranchiata : Mollusca	Sand, 63 units, various sizes	0.72-45.7	0.49-8.0	5	0.50	0.70	x⁰·⁶⁵	Holme (13)
2 European chafer larvae, Amphimallon majalis Raz., (= Melalontha melalontha L.), Coleoptera : Insecta	Pasture soil, 25 units, each 1 ft sq.	0.20-9.72	0.26-14.93	75	1.15	1.07	x⁰·⁴⁷	Burrage and Gyrisco (6)
3 Flying insects, various orders : Insecta	Open air, 16-104 units aerial density	1.9-238	1.7-606	24	1.0	1.17	x⁰·⁴²	L. R. Taylor (unpublished)
4 Wireworms, Agriotes spp. mainly obscurus, Coleoptera : Insecta	Grassland soil 20 units, 4 in. cores	0.20-4.65	0.40-17.80	2,272	2.75	1.19	x⁰·⁴¹	Yates and Finney (23)
5 Wireworms, Agriotes spp. as above	Arable land soil, 20 units, 4 in. cores	0.20-4.65	0.39-22.50	525	2.85	1.26	x⁰·³⁷	Yates and Finney (23)
6 Wireworms, Limonius spp., Coleoptera : Insecta	Arable land soil, 175 units, each 1 ft. sq.	0.39-10.89	0.58-60.42	24	2.0	1.33	x⁰·³⁴	Jones (5, 14)
7 Gall midge larvae, Jaapiella medicaginis (Rub.). Diptera : Insecta	Lucerne field Soil, 10 units, 4 in. cores	0.22-5.6	0.19-13.82	12	1.3	1.33	x⁰·³⁴	Heath (12)
8 Spruce budworm larvae, Choristoneura fumifera Iclem.), Lepidoptera : Insecta	Fir foliage, 25 units, larvae/twig	0.48-15.04	0.343-51.80	50	1	1.40	x⁰·³⁰	Waters (4, 22)
9 Virus lesions, tobacco necrosis virus	Bean leaves, 4 units, lesions/half leaf	15.38-237.56	65.13-3,265.8	120	1	1.40	x⁰·³⁰	Kleczkowski (5, 15)
10 Colorado beetle adults, Leptinotarsa decemlineata Say., Coleoptera : Insecta	Potato foliage, 2,304 counts: insects/2 ft. row	1.0-12.1	1.4-37.7	16 16 5	1	1.48	x⁰·²⁶	Beall (2, 3, 5, 19)
11 Japanese beetle larvae, Popillia japonica New., Coleoptera : Insecta	Soil, 10,000 units each 1 ft. sq.*	2.76-1,914	5.90-210,000	36	1.3	1.52	x⁰·²⁴	Fleming and Baker (10)
12 Macro-zooplankton	Water, net collection, Slide count, 10 areas	95-1,750	1,444-194,480	4	1.0	1.57	x⁰·²²	Littleford, Newcombe and Shepherd (16)
13 Macro-zooplankton	As above, 50 areas	93-1,822	169-18,769	4	0.14	1.57	x⁰·²²	Littleford, Newcombe and Shepherd (18)
14 Ticks, Ixodes ricinus L., Acarina : Arachnida	Sheep, 20-86 units ticks/sheep	0.84-85.6	2.5-2,292	10	1.0	1.66	x⁰·¹⁷	Milne (18)
15 Enchytraeid worms, mainly Fridericia bisetosa (Lev.), Enchytraeidae : Annelida	Pasture, 60-150 units, 3.6 cm cores	40.8-381.5	441-29,584	4	1.0	1.66	x⁰·¹⁷	Nielsen (20)
16 Corn borer larvae, Pyrausta nubilalis (Hubn.), Lepidoptera : Insecta	Maize stalks, 2 stages†	6.7-970	88.5-180,220	1,054	3.0	1.66	x⁰·¹⁷	Meyers and Patch (17)
17 Thrips, Thrips imagines (Bagnall), Thysanoptera : Insecta	Rose flowers, 20 units, thrips/rose	20.6-137.5	223-7,900	16	1.0	1.80	x⁰·¹⁰	Davidson and Andrewartha (8)
18 Leather-jackets, Tipula spp., Diptera : Insecta	Soil, 2 units, Nos./sq. ft.	3.10-63.50	3.5-409.0	36	0.2	1.85	x⁰·⁰⁸	Bartlett (1, 5)
18 Earthworms, all stages, Allolobophora chlorotica (Sav.), Oligochaeta : Annelida	Grassland, 4 units, 18 in. sq.	2.8-71.3	1,410	54	0.2	2.00	log x	Gerard (11)
20 Red spider mite, eggs and adults, Metatetranychus ulmi (Koch), Acarina : Arachnida	Apple leaves, 20 units, mites/leaf	8.5-216	90-55,100	162	0.4	2.19	1/x⁰·¹⁰	Daum and Dewey (7)
21 Haddock, Melanogrammus aeglefinus L., Gadidae : Pisces	Sea, 4-17 units, Nos./trawl	4.09-288	39.1-1,239,500	15	1.0	2.35	1/x⁰·¹⁸	C. C. Taylor (5, 21)
22 Earthworms, all stages Allolobophora caliginosa (Sav.), Oligochaeta : Annelida	Grassland. 4 units, 18 in. sq.	4.3-44.8	4.0-500	42	0.05	2.54	1/x⁰·²⁷	Gerard (11)
23 Symphyla, Symphylella spp., Symphyla : Myriapoda	Various soils, 60-120 units, 2 ½ in. cores	4.5-31.8	0.49-690	5	.06	2.75	1/x⁰·³⁸	Edwards (9)
24 Symphyla, Scutigerella spp., Symphyla : Myriapoda	Various soils, 60-120 units, 2 ½ in. cores	1.3-31.4	0.64-1,250	6	0.035	3.08	1/x⁰·⁵⁴	Edwards (9)

* Units combined to cover areas varying in size from 1 sq. ft. to 100 sq. ft.
† Larvae/stalk x Infested stalks/100: this introduces some bias in the variances

FIG. 1.2 (B)—CONT'D

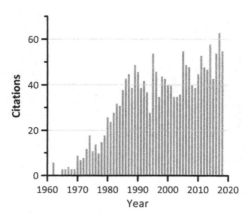

FIG. 1.3 Up to the end of 2018 there are nearly 2000 citations of LRT61 with 40–50/year over the last decade.

the mean abundance in a repeatable fashion. And completing a circle, TPL has become an integral part of designing sampling programs and analyzing sample data for both pure and applied research in many disciplines. With power law transformations, sampling plans using TPL are covered in Chapter 14.

This book is organized in three parts. Part I describes aggregation, its analysis, and the efficient fitting of power laws. Part II is a collection of case studies, the majority of which are agricultural or ecological, but which also include physiology, anthropology, sociology, and economics on the "soft" side and physics, astronomy, and computer science on the "hard" side of science. I put hard and soft in quotes because I do not consider physics any harder than ecology with its hundreds of often uncontrollable variables, although for a biologist, the mathematics is often much harder to fully understand.

Although focused on biological examples, Part II also includes case studies from the nonbiological world, included for the benefit of biologists wishing to view TPL's full scope. Furthermore, I hope that physical scientists may wish to learn of the multitude of biological examples, ecological, physiological, and sociological. For that reason, Chapters 5–11 covering taxa from microbes to man are prefaced with a short introduction to the taxon or biological field. Biologists might wish to skip the first paragraph. Chapters 12, 16, and 17 cover physical and mathematical studies: there is considerable overlap between the physical and biological realms in consideration of power laws and the models that describe or predict them.

The reader may be curious about my somewhat eclectic choice of case studies. The choice was made primarily to illustrate a use or property of TPL, secondarily to cover as wide a taxonomic and ecological range as possible, and thirdly to include examples that I found interesting because they were unusual, such as the pollination study (Fig. 6.14) or Quincy Wright's (1942) study of war

(Fig. 11.12). If, as Alexander Pope proposed in *An Essay on Man: Epistle II*, the proper study of man is man, no survey of this ubiquitous phenomenon would be complete without consideration of the population and behavioral ecology of *Homo sapiens*. A large part of Chapter 11 is devoted to the proper study and covers physiology, behavior, demography, and health, as well as some of man's more dubious achievements.

Part III covers applications and properties of TPL and other power laws, and some of the many models proposed to account for them. The penultimate chapter is an attempt at summary and synthesis of the empirical facts and the theoretical results. I have tried to treat both facts and theory impartially without injecting my own thoughts, except where I see connections. Readers wanting to know immediately what I think can turn to the short Epilog. Doing so may help understand my choice of examples, but it may also bias reading of the facts, something I have tried to avoid. The first chapter in Part III deserves special attention: Chapter 13 Counter Examples. While there is much to be learned from the case studies in which TPL provides a good fit to data, it may be that we can learn as much from the collection of examples where it does not fit, or fits poorly as from its successful applications.

References

Bliss, C.I., Fisher, R.A., 1953. Fitting the negative binomial distribution to biological data. Biometrics 9, 176–200.

Brown, M.W., Cameron, E.A., 1982. Spatial distribution of adults of *Ooencyrtus kuvanae* (Hymenoptera: Encyrtidae), an egg parasite of *Lymantria dispar* (Lepidoptera: Lymantriidae). Can. Entomol. 114, 1109–1120.

Eisler, Z., Bartos, I., Kertész, J., 2008. Fluctuation scaling in complex systems: Taylor's law and beyond. Adv. Phys. 57, 89–142.

Fisher, R.A., 1935. Design of Experiments. Oliver and Boyd, Edinburgh.

Fisher, R.A., Corbett, A.S., Williams, C.B., 1943. The relation between the number of species and the number of individuals in a random sample of an animal population. J. Anim. Ecol. 12, 42–58.

Franklin, A., 1990. Experiment Right or Wrong. Cambridge University Press, Cambridge, UK.

Lewis, T., Taylor, L.R., 1967. Introduction to Experimental Ecology. Academic Press, London.

Lincoln, R.J., Boxshall, G.A., 1982. Dictionary of Ecology, Evolution and Systematics. Cambridge University Press, Cambridge, UK.

Ma, Z., 2015. Power law analysis of the human microbiome. Mol. Ecol. 24, 5428–5445.

McGuire, J.U., Brindley, T.A., Bancroft, T.A., 1957. The distribution of European corn borer larvae *Pyrausta nubilalis* (Hbn.) in field corn. Biometrics 13, 65–78.

Southwood, T.R.E., 1966. Ecological Methods with Particular Reference to the Study of Insect Populations, first ed. Methuen, London.

Taylor, L.R., 1961. Aggregation, variance and the mean. Nature 189, 732–735.

Taylor, L.R., 1965. A natural law for the spatial disposition of insects. In: Freeman, P. (Ed.), Proceedings of the XIIth International Congress of Entomology. Royal Entomological Society, London, pp. 396–397.

Taylor, L.R., Woiwod, I.P., 1982. Comparative synoptic dynamics. I. Relationships between inter- and intra-specific spatial and temporal variance/mean population parameters. J. Anim. Ecol. 51, 879–906.

Thomson Reuters, 2018. Web of Science. http://webofknowledge.com.

Williams, C.B., 1964. Patterns in the Balance of Nature. Academic Press, London.

Wright, Q., 1942. A Study of War. The University of Chicago Press, Chicago, IL.

Part I

Chapter 2

Spatial pattern

The abundance of organisms in a community is characterized by two properties: the density or number of individuals in a defined area and the degree of separation or clumping within that area. The determination of the size of a population ordinarily requires determination of both characteristics. It might theoretically be possible to consider the total population of a species without regard to their spacing, the total number of Siberian tigers in Siberia, perhaps. As it is likely impossible to enumerate all the tigers in Siberia, if we want to know their total number we must rely on sampling smaller areas. Because the number per unit area is unlikely to be uniform across the species' range, we must consider the spatial organization of the target population in addition to the numerical attribute. To do this, we must specify the number and size of areas to be searched to make population estimates with a specified level of confidence. The use of TPL to determine of the number of samples needed for efficient population estimation is deferred to Chapter 14. In this chapter, we consider the interaction between distribution and abundance of organisms in space and the role of sampling in population estimation.

Sampling

In most instances, population estimates of both animals and plants must be obtained by sampling. Efficient and reliably informative sampling requires a stable and well-defined sampling unit (Morris, 1955). The sampling unit should be chosen to ensure that all targets have the same probability of being included in the sample and this probability does not change with time or position in the sampling universe or habitat. In practice, this means that the sampling unit should not vary either spatially or temporally and the proportion of target species vulnerable to sampling should remain constant. To achieve this ideal, the sampling unit should be easily delineated in the field and in principle should be expressible in terms of area or volume. Also, it should be cost effective, that is, the sample unit should be chosen to balance sampling variance with cost of sampling.

The ideal two-dimensional sampler is a metal frame or quadrat within which all targets are enumerable; in three dimensions, the best samplers are trapping systems (tow nets or suction traps) of known volume per unit time. The actual

Taylor's Power Law. https://doi.org/10.1016/B978-0-12-810987-8.00002-1

mechanisms for taking ecological samples are many and are described as needed in the case studies that follow this section. In the case studies, all samples are referred to as quadrats and the number of samples or quadrats is represented by *NQ*. Southwood's *Ecological Methods* (Southwood, 1966; Southwood and Henderson, 2000) is the best source for sampling and analytical methods not only for insects but also for other animal taxa. Grieg-Smith (1983) discusses sampling methods and analysis for plants. Cochran (1977) and Seber (1982) provide statistical details on the theory and analysis of sampling data.

Population density, a measure of how close together individuals are packed, is determined by a process we can visualize as a photograph of a population of animals or plants and the number in the photograph counted. Counting the individuals in this instantaneous record of the population provides information on the number of organisms per photograph (relative density) and their relative positions. Knowing the area of each photograph, we can also determine the absolute density (LRT, 1962). Recording the positions of the individuals in the photo provides information on how the individuals arrange themselves in space relative to one another. Multiple photographs taken simultaneously at several places provide a picture of the population arrangement at a larger scale. If the photographic process is continued at intervals of time, a trajectory of population density, and the corresponding spatial arrangement may be constructed. Both intra- and intergenerational population change in number and distribution may be constructed by suitable choice of interval with respect to generation time.

The population pattern in each photograph belongs in a spectrum of possibilities. Fig. 2.1 presents four patterns of 100 individuals occupying the same area captured by hypothetical photographs. The Poisson distribution (Fig. 2.1B) occupies a special place in statistical theory and also in ecological theory. It is characterized by having the mean and variance equal, $V = M$, where the mean is

$$M = \frac{1}{n}\sum_{i=1}^{n} x_i \quad \text{and} \quad V = \frac{1}{(n-1)}\sum_{i=1}^{n}(x_i - M)^2 = \frac{1}{(n-1)}\left(\sum_{i=1}^{n} x_i^2 - M\sum_{i=1}^{n} x_i\right)$$

(2.1)

is the variance of a variable x estimated from n samples. The square root of V is the standard deviation denoted as SD. Distributions with variance less than the mean have been called underdispersed, the extreme case being a grid-like pattern (Fig. 2.1A). In this extreme case, sampling reveals a mean dependent solely on the sampler's size in relation to the separation and a variance of zero as all samples contain the same number of individuals. In contrast, distributions with variance greater than the mean have been called overdispersed or contagious (Figs. 2.1C and D). The term contagious comes from epidemiology and is considered by some to be an inappropriate term for ecology and the terms over- and underdispersed are sometimes confused. Therefore we use the terms *regular*, *Poisson*, *aggregated* and now introduce the term *super-aggregated* to denote populations more tightly clustered than aggregated.

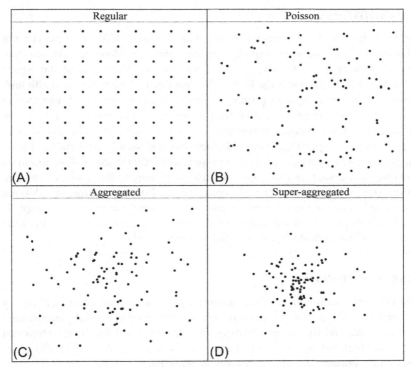

FIG. 2.1 Spatial patterns from regular (A), through Poisson (B) and aggregated (C) to super-aggregated (D) are characterized by statistical frequency distributions and TPL (Fig. 1.1).

The number of organisms in the n photos forms a frequency distribution, which may be approximated by one of a number of statistical distributions characterized by their moments. Eq. (2.1) is the first two moments; the third and fourth moments are skewness and kurtosis. Where variance is the average of the second power of deviations about the mean, skewness is the third moment about the mean, and kurtosis the fourth. Conventionally, to ensure these moments are dimensionless, a skewness coefficient, S is defined as the third moment divided by the standard deviation cubed and kurtosis, K is the fourth moment divided by the variance squared:

$$S = \frac{1}{V^{3/2}} \left[\frac{1}{n} \sum_{i=1}^{n} x_i^3 - \frac{3}{n} M \sum_{i=1}^{n} x_i^2 + 2M^3 \right] \text{ and}$$

$$K = \frac{1}{V^2} \left[\frac{1}{n} \sum_{i=1}^{n} x_i^4 - \frac{4}{n} M \sum_{i=1}^{n} x_i^3 + \frac{6}{n} M^2 \sum_{i=1}^{n} x_i^2 - 3M^4 \right]$$

(2.2)

This book is concerned primarily with the relationship between the first two moments, M and V, but the third and fourth are also important, and are used in some studies (Chapter 17).

Randomness

By definition, random events are events that lack predictability. However, not all unpredictable events are equally unpredictable. An elongated asteroid-like object recently passed through our solar system from interstellar space (Meech et al., 2017). It was the first such event that had been observed although models of planetary system formation predicted the existence of asteroids in interstellar space. Even so, neither the passage of 'Oumuamua nor any similar extraterrestrial object could be predicted, nor can the next one. The second class of randomness is specifically unpredictable but statistically predictable. These, called stochastic events, have long-term predictability subject to description by a mathematical rule, process, or statistical distribution. We also call random those events whose frequency is described by the Poisson distribution. Poisson distributed events are more properly called rare events: the classical example is von Bortkiewicz's 1898 application of the Poisson to the frequency of Prussian soldiers kicked to death by horses (Schumpeter, 1951).

Poisson distribution

The Poisson distribution occupies a central position in both ecological theory and practice. It represents a baseline against which other spatial patterns may be compared, and its tractability made it a starting point for the development of the mathematics of population dynamics (see Varley et al., 1973). The probability of r Poisson distribution events is given by

$$p(x = r) = \frac{1}{r!}\mu^r e^{-\mu} \tag{2.3}$$

where μ is the mean and also the variance of the number of events per sample:

$$V = M = \mu \tag{2.4}$$

It is a special case of the binomial distribution of alternatives, such as heads or tails in a sequence of fair coin tosses, which cannot be predicted. However, the long-term average will converge on 50% heads and 50% tails. The fact that the sequence cannot be predicted but that behavior can be predicted statistically means that this kind of randomness is a measure of uncertainty, the inverse of confidence. Although the binomial distribution is central to genetics, its role in ecology is limited.

Negative binomial distribution

Another member of the binomial family, the negative binomial distribution occupies an important place in both ecological theory and practice. Also known as the Pólya distribution, the negative binomial describes events more aggregated in time or space than the Poisson distribution. It is a Poisson distribution with parameter μ distributed as gamma distribution with parameters k and μ (Johnson et al., 1993). The probability of r negative binomial events is

$$p(x=r) = \frac{(\mu/k)^r (k+r-1)!}{(1+\mu/k)^{k+r}(k-1)!r!} \qquad (2.5)$$

where k and μ are parameters. Its mean and variance are:

$$M = \mu \quad \text{and} \quad V = \mu\left(1 + \frac{\mu}{n}\right) = \mu + \frac{\mu^2}{k} \qquad (2.6a)$$

where n is the number of samples and μ is estimated by M and an approximate value of k estimated from

$$k \approx M^2/(V-M). \qquad (2.6b)$$

Polya-Aeppli distribution

The Polya-Aeppli distribution is derived from compounding the Poisson and geometric distributions. It is the distribution that arises if individuals are arranged in Poisson distributed clusters and the number of individuals per cluster is geometrically distributed. The probability of r Polya-Aeppli distributed events is

$$p(x=0) = e^{-\mu}$$
$$p(x=r) = e^{-\mu}\lambda^r \sum_{j=1}^{r} \binom{r-1}{j-1}\left[\frac{(\mu(1-\lambda)/\lambda)^j}{j!}\right] \qquad (2.7)$$

Its mean and variance are:

$$M = \frac{\mu}{(1-\lambda)} \quad \text{and} \quad V = \frac{\mu(1+\lambda)}{(1-\lambda)^2} \qquad (2.8)$$

where μ and λ are shape parameters.

Neyman's A distribution

Another aggregated distribution is Neyman's A distribution with parameters λ and ϕ. Its probability function is:

$$p(x=0) = \frac{1}{\exp(\lambda(1-e^{-\phi}))}$$
$$p(x=r) = \sum_{i=1}^{\infty} \frac{1}{\exp(\lambda+i\phi)}\left(\frac{\lambda^i(i\phi)^r}{i!r!}\right) \qquad (2.9)$$

Its mean and variance are:

$$M = \lambda\phi \quad \text{and} \quad V = \lambda\phi(1+\phi) = M(1+\phi) \qquad (2.10)$$

Neyman's A is a compound distribution in which the parameter μ of the Poisson (Eq. 2.3) is itself distributed by a Poisson distribution with parameter λ and scale factor ϕ (Johnson et al., 1993). David and Moore (1954) called the parameter ϕ a "clumping index."

These four distributions have ecologically meaningful interpretations. The Poisson represents a sort of baseline in which the probability of an organism occupying any particular point in space is the same for all points and is independent of the occurrence of all other individuals. The Polya-Aeppli and Neyman's A distributions can be interpreted as distributions resulting from individuals clustered in space in different ways. Neyman's A is the distribution resulting from individuals randomly distributed with each producing a cluster of eggs the size of which is also Poisson distributed (Skellam, 1958). The negative binomial distribution has a number of interpretations making it a very flexible model for spatial distribution. The simplest interpretation is that the population under study is truly contagious (as defined in epidemiology): the presence of one individual increases the chance that another will be close by.

Poisson-distributed colonies with colony sizes logarithmically distributed also generate negative binomial distributions. Some population models incorporating birth, death, and immigration result in population sizes that are negative binomial distributed. The pattern of sampling or sampling plan can also produce a negative binomial as when sampling a population in order to obtain a defined number of individuals with a certain character, and a negative binomial may result from combining data obtained from several series of subsamples each of which is Poisson distributed (Pielou, 1977). However, as Pielou points out, with so many possible derivations, attributing the genesis of an actual population that is negative binomially distributed to any particular cause should be made with caution.

The Poisson, negative binomial, and Neyman's A were fitted to McGuire et al.'s (1957) data of European corn borer (*Pyrausta*, now *Ostrinia nubilalis*) by LRT (1965) who showed they belonged to a sequence (Fig. 1.1A). In a similar study Brown and Cameron (1982) used the Poisson, negative binomial, and lognormal distributions to describe the spatial distribution of *Ooencyrtus kuvanae*, an egg parasite of gypsy moth (*Lymantria dispar*) (Fig. 1.1B).

Lognormal distribution

The lognormal is a continuous distribution that generates probabilities not discrete frequencies. It is, however, extremely flexible and can, with suitable choice of parameters, generate distributions that simulate the Poisson, negative binomial, and Neyman's A. It has probability density function

$$\Phi(x) = \frac{1}{x\sigma\sqrt{2\pi}} \exp\left\{ -\frac{\ln(x-\delta)^2}{2\sigma^2} \right\} \tag{2.11}$$

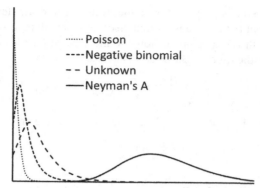

FIG. 2.2 The lognormal, a continuous distribution, is flexible enough to simulate the distributions in Fig. 1.1: Poisson distribution with lognormal parameters $\delta = 0.1$, $\sigma = 0.2$; negative binomial, $\delta = 0.4$, $\sigma = 0.25$; unknown, $\delta = 0.35$, $\sigma = 0.2$: Neyman's A, $\delta = 1.0$, $\sigma = 0.1$. For clarity, the curves are not all to the same scale.

The lognormal is the distribution of x when $\ln(x)$ is normally distributed (Johnson et al., 1994). It has mean and variance:

$$M = \exp\left(\delta + \tfrac{1}{2}\sigma^2\right) \quad \text{and} \quad V = \exp\left(2\mu + \sigma^2\right)\left\{\exp\left(\sigma^2\right) - 1\right\} \qquad (2.12)$$

Fig. 2.2 shows 3 lognormal distributions approximately the same as the Poisson, negative binomial, and Neyman's A fitted to McGuire et al.'s (1957) corn borer data (Fig. 1.1A).

A discrete version of the lognormal is the Poisson lognormal, a compound of Poisson with parameter μ distributed as a lognormal with parameters ξ and σ:

$$p(x=r) = \frac{\exp\left(r\mu + \sigma^{r(r-1)} + \tfrac{1}{2}r\sigma^2\right)}{r!\sigma\sqrt{2\pi}} \int\limits_{-\infty}^{\infty} \exp\left(-\frac{u^2}{2\sigma^2} - \exp\left(\mu + \tfrac{1}{2}\sigma^2\right)\sigma^{2r-1}e^u\right) du$$

$$(2.13)$$

It has mean and variance

$$M = \exp\left(\delta + \tfrac{1}{2}\sigma^2\right) \quad \text{and} \quad V = \exp\left(\delta + \tfrac{1}{2}\sigma^2\right) + \exp\left(2\delta + 2\sigma^2\right) - \exp\left(2\delta + \sigma^2\right)$$

$$(2.14)$$

The Poisson lognormal has an untabulated integral making fitting difficult. It was used by Bulmer (1974) with the zero term omitted to describe species abundance data.

Inverse Gaussian distribution

Both the negative binomial and Neyman's A distributions have the variance approximately proportional to the square of the mean. Similarly, the lognormal's variance is proportional to the mean squared. Its flexibility, however, is insufficient for super-aggregated distributions with their very long tails. A distribution with variance proportional to the cube of the mean and a longer right-hand tail is the inverse Gaussian distribution:

$$\Phi(x) = \frac{\lambda}{\sqrt{2\pi x^3}} \exp\left[-\lambda\left(\frac{1}{2x} - x(2\mu)^{1/2} + x(8\mu)^{1/4}\right)\right] \tag{2.15}$$

where λ is a diffusion coefficient. The mean and variance are:

$$M = \mu \quad \text{and} \quad V = \lambda\mu^3 \tag{2.16}$$

Tweedie family of distributions

The normal distribution owes its central place in statistical theory to the central limit theorem (CLT), which states if n independent variates have finite variances, then their sum will tend to be normally distributed as $n \to \infty$. The CLT is a special case of Tweedie convergence. The Tweedie convergence theorem states that all exponential dispersion models for which the variance is a power function of the mean will converge to the Tweedie model. This implies that most exponential dispersion models with finite variance have asymptotic variance-mean power functions.

The Tweedie models have power function relationship between their variance and mean:

$$V = \lambda^{1-\alpha} M^p \tag{2.17}$$

where p and α are related by

$$\alpha = \frac{(p-2)}{(p-1)} \tag{2.18}$$

There are nine Tweedie models with frequency distributions dependent on the value of p. They include the normal (Gaussian) with $p=0$, Poisson ($p=1$), Gamma ($p=2$), and inverse Gaussian ($p=3$) and distributions with fractional powers between one and infinity. There is no known Tweedie distribution with $0 < p < 1$. As the vast majority of TPL slopes are $1 \leq b \leq 3$ the gamma, Poisson-gamma ($1 < p < 2$), and positive stable ($p > 2$) distributions are of most interest. There is no closed form expression for the Poisson-gamma probability density function, but the mean and variance may be obtained from the first two derivatives of the cumulant-generating function (CGF),

$$K(s) = \lambda\kappa(\theta)\left\{\left(1 + \frac{s}{\theta}\right)^\alpha - 1\right\} \text{ where } \kappa(\theta) = \frac{\alpha-1}{\alpha}\left(\frac{\theta}{\alpha-1}\right)^\alpha \tag{2.19}$$

enabling estimation of the mean and variance in terms of the parameters:

$$M = \theta\alpha(\alpha-1)^{1-\alpha} \quad \text{and} \quad V = (\alpha-1)(\alpha\lambda\theta)^{1/(\alpha-1)} \tag{2.20}$$

Assuming TPL with parameters a and b $(1 < b < 2)$.

$$\theta = \frac{M^{1-b}}{\alpha(1-b)} \text{ and } \lambda \approx a^{b-1} \tag{2.21}$$

An example of a positive stable distribution is the positive stable Pareto distribution with density function

$$f(x) = \frac{\lambda \nu [\ln(x/\sigma)]^{\nu-1}}{x} \exp\{-\lambda[\ln(x/\sigma)]\nu\} \quad x \geq 0 \tag{2.22}$$

and cumulative density function

$$F(x) = \Pr(X \leq x) = \begin{cases} 1 - \exp\{-\lambda[\ln(x/\sigma)]^{\nu}\} & \text{if } x \geq \sigma \\ 0 & \text{if } x < \sigma \end{cases} \tag{2.23}$$

where λ and $\nu > 0$ are shape parameters and $\sigma > 0$ is a scale parameter. The mean and variance are

$$M = \lambda^{-1/\nu}\Gamma\left(1 + \frac{1}{\nu}\right) \text{ and } V = \lambda^{-1/\nu}\Gamma\left[\Gamma\left(1 + \frac{2}{\nu}\right) - \Gamma\left(1 + \frac{1}{\nu}\right)^2\right] \tag{2.24}$$

The Pareto and Zipf distributions are special cases when $\nu = 1$ and $\lambda = \nu = 1$, respectively. Both have long right-hand tails and have been used to describe the distribution of city sizes and wealth.

The Tweedie family of distributions has been used by Wayne Kendal to describe the spatial distribution of Colorado potato beetle (Kendal, 2002) and bristlecone pine tree rings (Kendal, 2017). Following his use of TPL to describe the distribution of metastases induced in laboratory models (Kendal and Frost, 1987), Kendal et al. (2000) applied Tweedie models to the distribution of metastases. Subsequently, he used Tweedie models to analyze the distribution of blood flow (Kendal, 2001), single-nucleotide polymorphisms (SNPs; Kendal, 2003), and the distribution of prime numbers (Kendal and Jørgensen, 2015). Kendal (2004) and Kendal and Jørgensen (2011a, b) have explored the relationship between TPL, Tweedie distributions, and fractals. Kendal and Jørgensen's work is examined in more detail in Chapters 11, 12, 16, and 17.

Where Tweedie models depend on nine distributions, another family of distributions, the Adès (Perry and Taylor, 1985 1988), depends on a single underlying distribution. It was developed specifically to generate TPLs to describe animal population distributions and is described in Chapter 3.

Origins of aggregation

Available evidence suggests that aggregation is the usual spatial distribution for animals and most plants. Randomness as exemplified by the Poisson distribution is comparatively rare for most common organisms; however, by definition rare species are sparsely distributed and may be indistinguishable from Poisson even if locally aggregated. At the edge of their range, common species may be rare and therefore Poisson, but at more central parts of their range they tend to be aggregated and may be locally super-aggregated (Curnutt et al., 1996, Chapter 10). Thus, the distributions described here and depicted in Fig. 1.1 may be found to apply in different parts of a species range at the same time.

The causes of aggregation are several. Members of a population may:

1. produce clusters of progeny, which may disperse with time, but are highly aggregated for a period;
2. call potential mates with colorful plumage, sound, or pheromones causing local aggregations that may be prolonged or temporary;
3. be prey that aggregate for mutual protection;
4. or predators that hunt cooperatively;
5. be attracted to food sources;
6. or habitats that confer protection.

In example 1 above aggregation is a consequence of reproduction that will be followed by dispersal from the natal site and a subsequent numerical and spatial reduction. The other processes are active responses of individuals to an external stimulus or each other. Thus, aggregation generally results from behavioral responses.

The concept of density dependence is foundational in ecology; it is the raw material of the feedback processes at the heart of the long-term persistence of most populations. Because population density is fundamentally a spatial concept, density dependence must also be a spatial phenomenon. Thus, environmental heterogeneity and variation in spatial pattern must affect population stability and resilience. It is not altogether clear whether heterogeneity is stabilizing or destabilizing. Until Stewart-Oaten and Murdoch (1990) found that spatial heterogeneity is more likely to be destabilizing than stabilizing, it was conventional wisdom that clumping is more likely to lead to stability. Yet populations are clumped over a wide range of scales and are stable over differing spatial and temporal periods.

Although purely demographic rate processes can account for moderately aggregated populations, Perry (1988) found his explicit spatial stepping-stone model could not account for the extremely high levels of aggregation frequently observed, although a chaotic model could (Perry, 1994, Chapter 17). Extreme levels of aggregation most likely emerge as a result of active congregation of a population (LRT and RAJT, 1977).

Even when a population is globally stable for long periods of time, clumps within the population range may be highly unstable (RAJT and LRT, 1979). The observed spatial pattern of a species results from interactions between the environment and the demographic activities of birth, death, and movement. These interactions are consequences of the evolved responses of organisms many of which are density dependent. Thus, more clearly for animals but also for plants, there exists a feedback between density-dependent behavior (in the broadest sense in order to include plants) and the habitat and the other species in it, organizing the spatial distribution to what we can measure quantitatively by sampling.

Perry and Taylor (1986) suggested a mechanism for host aggregation generated by avoidance behavior in predator-prey and parasite host systems. In an elegant thought experiment, Hamilton (1971) proposed that aggregation could be selected for in prey animals by each individual seeking to put another between its self and a predator. His one-dimensional case famously featured a collection of frogs on the perimeter of a pond each trying to avoid a predatory water snake by attempting to occupy a narrow gap between two other frogs. In a more realistic two-dimensional world, he showed that lions attacking a collection of cattle could in time cause the collection to condense into a herd. While Hamilton's model may not account for the initiation of aggregation, as pointed out by James et al. (2004), it would seem to reinforce aggregative behavior once initiated. Additionally, it is not always the peripheral individuals that suffer the highest mortality in fish schools (Parrish, 1989), which likely imposes some limit on aggregations.

Censuses

The majority of TPL case studies in Chapters 5–13 are of sampled data, but some systems can yield absolute censuses. Examples include the prices of commodities and currencies, which are recorded continuously by financial institutions as they change from second to second. Another system in which data are continuously monitored and recorded is the passage of packets of data through nodes of the Internet. Examples of these two systems are presented as case studies in Chapter 12. Another example, called census but actually samples with very high absolute sampling efficiency, is the periodic (usually decennial) censuses conducted by governments: an example is considered in Chapter 11.

References

Brown, M.W., Cameron, E.A., 1982. Spatial distribution of adults of *Ooencyrtus kuvanae* (Hymenoptera: Encyrtidae), an egg parasite of *Lymantria dispar* (Lepidoptera: Lymantriidae). Can. Entomol. 114, 1109–1120.

Bulmer, M.G., 1974. On fitting the Poisson lognormal distribution to species-abundance data. Biometrics 30, 101–110.

Cochran, W.G., 1977. Sampling Techniques, third ed. Wiley, New York.

Curnutt, J.L., Pimm, S.L., Maurer, B.A., 1996. Population variability of sparrows in space and time. Oikos 76, 131–140.

David, F.N., Moore, P.G., 1954. Notes on contagious distributions in plant populations. Ann. Bot. 53, 47–53.

Grieg-Smith, P., 1983. Quantitative Plant Ecology, third ed. University of California Press, Berkeley, CA.

Hamilton, W.D., 1971. Geometry for the selfish herd. J. Theor. Biol. 31, 295–311.

James, R., Bennetta, P.G., Krause, J., 2004. Geometry for mutualistic and selfish herds: the limited domain of danger. J. Theor. Biol. 228, 107–113.

Johnson, N.L., Kotz, S., Kemp, A.W., 1993. Distributions in Statistics. Discrete Distributions, second ed. Wiley, New York.

Johnson, N.L., Kotz, S., Balakrishnan, N., 1994. Distributions in Statistics. Continuous Univeriate Distributions—1, second ed. Wiley, New York.

Kendal, W.S., 2001. A stochastic model for the self-similar heterogeneity of regional organ blood flow. Proc. Natl. Acad. Sci. U. S. A. 98, 837–841.

Kendal, W.S., 2002. Spatial aggregation of the Colorado potato beetle described by an exponential dispersion model. Ecol. Model. 151, 261–269.

Kendal, W.S., 2003. An exponential dispersion model for the distribution of human single nucleotide polymorphisms. Mol. Biol. Evol. 20, 579–590.

Kendal, W.S., 2004. Taylor's ecological power law as a consequence of scale invariant exponential dispersion models. Ecol. Complex. 1, 193–209.

Kendal, W.S., 2017. $1/f$ noise and multifractality from bristlecone pine growth explained by the statistical convergence of random data. Proc. R. Soc. A: Math. Phys. Eng. Sci. 473, 20160586.

Kendal, W.S., Frost, P., 1987. Experimental metastasis: a novel application of the variance-to-mean power function. J. Natl. Cancer Inst. 79, 1113–1115.

Kendal, W.S., Jørgensen, B., 2011a. Taylor's power law and fluctuation scaling explained by a central-limit-like convergence. Phys. Rev. E. 83.

Kendal, W.S., Jørgensen, B., 2011b. Tweedie convergence: a mathematical basis for Taylor's power law, $1/f$ noise, and multifractality. Phys. Rev. E. 84.

Kendal, W.S., Jørgensen, B., 2015. A scale invariant distribution of the prime numbers. Computation 3, 528–540.

Kendal, W.S., Lagerwaard, F.J., Agboola, O., 2000. Characterization of the frequency distribution for human hematogenous metastases: evidence for clustering and a power variance function. Clin. Exp. Metastasis 18, 219–229.

McGuire, J.U., Brindley, T.A., Bancroft, T.A., 1957. The distribution of European corn borer larvae *Pyrausta nubilalis* (Hbn.) in field corn. Biometrics 13, 65–78.

Meech, K.J., Weryk, R., Micheli, M., et al., 2017. A brief visit from a red and extremely elongated interstellar asteroid. Nature 552, 378–381.

Morris, R.F., 1955. The development of sampling techniques for forest insect defoliators, with particular reference to the spruce budworm. Can. J. Zool. 33, 225–294.

Parrish, J.K., 1989. Re-examining the selfish herd: are central fish safer? Anim. Behav. 38, 1048–1053.

Perry, J.N., 1988. Some models for spatial variability of animal species. Oikos 51, 124–130.

Perry, J.N., 1994. Chaotic dynamics can generate Taylor's power law. Proc. R. Soc. B 257, 221–226.

Perry, J.N., Taylor, L.R., 1985. Adès: new ecological families of species-specific frequency distributions that describe repeated spatial samples with an intrinsic power-law variance-mean property. J. Anim. Ecol. 54, 931–953.

Perry, J.N., Taylor, L.R., 1986. Stability of real interacting populations in space and time: implications, alternatives and the negative binomial k_c. J. Anim. Ecol. 55, 1053–1068.

Perry, J.N., Taylor, L.R., 1988. Families of distributions for repeated samples of animal counts. Biometrics 4, 881–890.

Pielou, E.C., 1977. Mathematical Ecology. Wiley, New York, NY.

Schumpeter, J.A., 1951. Ladislaus von Bortkiewicz (1868–1931). In: Ten Great Economists, from Marx to Keynes. Oxford University Press, Oxford, UK, pp. 302–305.

Seber, G.A.F., 1982. The Estimation of Animal Abundance. Charles Griffen, London.

Skellam, J.G., 1958. On the derivation and applicability of Neyman's A distribution. Biometrika 45, 32–36.

Southwood, T.R.E., 1966. Ecological Methods with Particular Reference to the Study of Insect Populations, first ed. Methuen, London.

Southwood, T.R.E., Henderson, P.A., 2000. Ecological Methods, third ed. Blackwell Science, Oxford, UK.

Stewart-Oaten, A., Murdoch, W.W., 1990. Temporal consequences of spatial density dependence. J. Anim. Ecol. 59, 1027–1045.

Taylor, L.R., 1962. The absolute efficiency of insect suction traps. Ann. Appl. Biol. 50, 405–421.

Taylor, L.R., 1965. A natural law for the spatial disposition of insects. In: Freeman, P. (Ed.), Proceedings of the XIIth International Congress of Entomology. Royal Entomological Society, London, pp. 396–397.

Taylor, L.R., Taylor, R.A.J., 1977. Aggregation, migration and population mechanics. Nature 265, 415–421.

Taylor, R.A.J., Taylor, L.R., 1979. A behavioural model for the evolution of spatial dynamics. In: Anderson, R.M., Turner, B.D., Taylor, L.R. (Eds.), Population Dynamics. Blackwell Scientific Publications, Oxford, UK, pp. 1–27.

Varley, G.C., Gradwell, G.R., Hassell, M.P., 1973. Insect Population Ecology: An Analytical Approach. Blackwell Scientific Publications, Oxford, UK.

Chapter 3

Measuring aggregation

Population density is fundamentally a spatial concept. As we saw in Fig. 2.1, the same population number can have different spatial patterns. With the exception of the regular distribution (which is exceedingly rare), when we sample those patterns to obtain an estimate of population abundance, we record different numbers depending on where we sample and the size of the sampler (Fig. 3.1A and B). Calculating mean and variance (Eq. 2.1) of our sample data provides an estimate of the density and a measure of the variability of our estimate. As is clear from Figs. 2.1 and 3.1, distribution and abundance are not independent variables but intimately connected by way of the actual spatial pattern of the population. The rigor with which we design the sampling system is critically important in obtaining meaningful population estimates.

Additionally, our estimates are unlikely to be independent of the scale of sampling: not just how big and how many samplers are taken, but how large a proportion of the species range is being sampled. The examples portrayed in Fig. 2.1 are not mutually exclusive; the same species population might exhibit all but the regular pattern at different locations or the same location at different times. This suggests that populations may be clumped over a wide range of scales. Over time, the total population size may vary greatly or remain

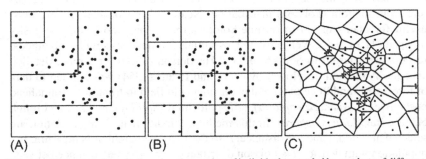

FIG. 3.1 Estimates of density based on the number of individuals recorded by quadrats of different size (A) or position (B) vary if the population is not distributed on a regular grid. The average of random samples expressed as number per unit area approximates the average density of the target population and the variance provides a measure of confidence in the estimated population. Voronoi tessellation (C) of the population shows the variation in area available for each individual.

Taylor's Power Law. https://doi.org/10.1016/B978-0-12-810987-8.00003-3

comparatively stable. Even when a population is globally stable for long periods of time, clumps within the population range may be mobile and/or highly unstable (RAJT and LRT, 1979).

Except in the case of regularity, even within a single defined area, the density experienced by each individual differs. The Voronoi tessellation in Fig.3.1C is a partitioning of the area into domains in which all points are closer to one individual of the population than to any other. These domains are approximately the inverse of density, and are called cover or mean area in plant ecology. Animal ecology does not have a specific term although they correspond to territories in territorial species. To distinguish the concept of inverse density from territory, I suggested the term population intensity (RAJT, 1992) that is rigorously defined in Chapter 17.

The key points are that results of our sampling procedures are not independent of the target population's spatial pattern and that the individual members of a population experience different densities depending on their proximity to neighbors. This is important for both practical and theoretical ecology, because a fundamental principle in ecology is density dependence. Birth, death and migration rates are frequently density dependent, the underlying reproductive, survival, and spatial behavior of individuals being influenced by their proximity to conspecifics. The differences in population intensity experienced by members of a population result in heterogeneity in spatial behavior leading to heterogeneity in spatial pattern. Thus, spatial pattern is both a cause and an effect of spatial behavior. This is important because long-term persistence of a population depends on its members adopting spatially appropriate behavior in their battle for survival.

Nearest neighbor

A measure of spatial separation, nearest-neighbor analysis, determines departure from random (Poisson) distribution. Nearest-neighbor analyses are most useful for plants or ant or termite colonies that do not move and are easily mapped. Here we cover only the essentials of nearest-neighbor methods: readers are referred to Southwood and Henderson (2000) for a more detailed account. A method developed by Clark and Evans (1954) was applied to data of forest trees and three perennial plants in an old field in Michigan; roundhead bushclover (*Lespedeza capitata*), stiff goldenrod (*Solidago rigida*), and rough blazing star (*Liatris aspera*). In this method, n individuals in a population are selected at random and the distance, r, to their closest neighbor is measure. If a population with density μ is Poisson distributed, the expected distance between nearest neighbor is

$$\bar{r}_E = \frac{1}{2\sqrt{\mu}} \tag{3.1}$$

and the actual distance is estimated from

$$\bar{r}_A = \frac{1}{n}\sum_{i=1}^{N} r_i \tag{3.2}$$

then the ratio $R = \bar{r}_A / \bar{r}_E$ is a measure of the degree of randomness. If the distribution is Poisson $R = 1.0$: if R is significantly < 1.0 it is aggregated and an $R > 1.0$ indicates regularity. The maximum value of $R = 2.149$ is obtained when the population is a perfect grid (Fig. 2.1A): and $R \to 0$ in superaggregated distributions (Fig. 2.1D). The significance of departure from Poisson is computed from

$$c = \frac{\bar{r}_A - \bar{r}_E}{SE} \tag{3.3}$$

where SE is the standard error of the mean distance in a Poisson distributed population with density μ:

$$SE = \frac{0.26136}{\sqrt{n\mu}} \tag{3.4}$$

The statistic c is the standard variate of the normal curve with values 1.96 and 2.58 representing the 5% and 1% significance levels for two-tailed tests, respectively. Clark and Evans simulated a Poisson distribution of points to test the theory before applying it to field data. The four examples were significantly more aggregated than Poisson ($p < 0.001$) and the simulation results were not significantly different from Poisson ($p > 0.45$) supporting the robustness of the method. Clark and Evans give other statistics in their Appendix. Elaborations to this basic method have been made by a number of authors and are discussed in Southwood and Henderson (2000).

Indices using mean and variance

Variance-mean ratio

As stated in the previous chapter, the Poisson distribution represents a sort of baseline as it has the convenient property that

$$V = M = \mu \tag{3.5}$$

and is mathematically tractable. Its tractability arises from the fact that the probability of $x = 0$ events occurring is $e^{-\mu}$ and the probability that $x > 0$ is therefore $1 - e^{-\mu}$, which depend solely on the mean, μ. As a consequence, the assumption of randomness, embodied in the Poisson distribution, was a convenient starting

point for early parasite-host and predator-prey models. Subsequent developments employed the zeroth term of the negative binomial distribution to simulate non-Poisson population distributions using k (Eq. 2.6b) as the aggregation
index (Varley et al., 1973; May, 1974).

Departures of the ratio of estimated variance and mean, V/M from unity indicate non-Poisson distributions: if $V/M < 1.0$, the departure is toward regularity,
if the ratio is >1.0, the data indicate aggregation. Given n independent samples
or quadrats, a significant χ^2 test with $n - 1$ degrees of freedom,

$$\chi^2 = \frac{V}{M}(n-1) \tag{3.6}$$

indicates a significant departure from Poisson (Perry and Mead, 1979). Another
dispersion index, $I = V/M - 1$, has the convenient property that the Poisson case
is now zero, more regular distributed data are negative, and aggregated data are
positive (David and Moore, 1954).

The coefficient of variation defined as,

$$CV = \frac{SD}{M}100\% \tag{3.7}$$

is a measure of heterogeneity that is occasionally used as a measure of variation
of abundance.

Negative binomial k

The reciprocal of the parameter k of the negative binomial distribution
(Eq. 2.6b) is frequently used as a measure of population aggregation. The
negative binomial may be derived from the Poisson distribution by making
the Poisson parameter (μ) vary continuously as a gamma distribution with
parameters α and β. As the variance of a negative binomial approaches the
mean, its aggregation parameter $k \to \infty$ and the distribution converges to
the Poisson. If $k \to 0$ and the frequency of zeros is ignored, the negative binomial converges to the logarithmic series, which is a good model for species
abundance (Fisher et al., 1943). The enormous flexibility endowed by k
and the limiting conditions of $k \to 0$ or $k \to \infty$ leading to distributions useful
in biology have contributed to the important position the negative binomial
occupies in ecology.

Fitting the negative binomial

Its parameter μ is estimated by the mean, M of n samples (Eq. 2.1), and the maximum likelihood (ML) estimate of parameter k is obtained by minimizing:

$$z = \sum \left(\frac{A_x}{k+x}\right) - n \cdot \ln\left(1 + \frac{M}{k}\right) \tag{3.8}$$

where n is the number of samples or quadrats and A_x the accumulated frequencies:

$$A_x = n - \sum_{i=1}^{\infty} x_i \tag{3.9}$$

The ML estimate of k is obtained by the method of false position:

$$k_{i+1} = \frac{k_{i-1} z_i - k_i z_{i-1}}{z_i - z_{i-1}} \tag{3.10}$$

where the z_i are successive values of Eq. (3.8) computed from successive estimates of k. An initial value of k is obtained from Eq. (2.6b):

$$k_0 = \frac{M^2}{(V - M)} \tag{3.11}$$

and k_1 is chosen such that z_0 and z_1 have opposite signs. The sample coefficient of variation, $CV = \sqrt{(k+M)/kM}$, thus the more aggregated the population, the larger the CV. Because k is sensitive to the number of samples, this implies the more aggregated a population the more samples are needed for reliable and efficient estimation. The goodness of fit of the negative binomial, given an estimate of k, may be made by χ^2 test of the observed frequencies and the frequencies computed using Eq. (2.5) and the solution to Eq. (3.8). Because the χ^2 test is sensitive to outliers, more precise tests employing the observed and expected variance and skewness are available (Evans, 1953).

Interpretation of k

The negative binomial may be derived a number of ways. In his derivation of the sampling theory of the negative binomial, Anscombe (1950) lists four ways in which the negative binomial distribution can arise. A negative binomial results from:

1. the distribution of the number of observations in excess of k that must be taken to obtain k individuals with a specified character when a proportion p posses the character;
2. the sampling distribution of the compound Poisson distribution with parameter μ distributed as a 2-parameter gamma distribution;
3. Poisson-distributed colonies whose sizes are distributed independently in a logarithmic distribution;
4. simple population dynamics models with constant birth and death rates of per individual and constant immigration rate.

To which Waters and Henson (1959) add a fifth:

5. a contagious distribution in which the probability of finding one individual in a sample increases the chances of finding a second.

A systematic review by Boswell and Patil (1970) lists 14 different derivations for the negative binomial. The fact the negative binomial can be derived so many ways is two edged: the variety of derivations offers confidence that the use of the negative binomial has ecological utility, but attributing an ecological meaning to any particular case may be arbitrary. However, using k to characterize distributions of different data sets is a useful way to compare the spatial distributions of different populations.

Negative binomial with common k

When the negative binomial is fitted to multiple data sets, the estimates of k may differ. A single estimate for k may be obtained by a regression method (Bliss and Owen, 1958). Two values are computed from the estimates of mean and variance,

$$x' = M - \frac{V}{n}$$
$$y' = V - M$$

$$(3.12)$$

and y' is plotted against x'. Negative values of y' are included and the regression is constrained to pass through the origin. The slope is $1/k_c$. Clearly if the regression is constrained to pass through the origin $1/k_c$ may be computed directly from

$$k_c = \frac{\sum x'}{\sum y'}.$$

$$(3.13)$$

This was called the moment estimate of k_c by Bliss and Owen who calculated k_c for several species (haddock (*Melanogrammus aeglefinus*), Colorado potato beetle (*Leptinotarsa decemlineata*), beetle larvae of the genus *Limonius* (wireworms) and tobacco necrosis virus lesions). All four cases were used in LRT61. Bliss and Owen found that $1/k_c$ was independent of the mean in two of their case studies (haddock and Colorado beetle), but $1/k_c$ was inversely proportional to M in the wireworm and virus lesion studies. In these cases, calculation of common k is probably not justified. They also provide a method for obtaining a weighted estimate of k_c that may be more reliable and a χ^2 test for systematic divergence.

In a detailed study of the behavior of k_c, LRT et al. (1979) found as a model for aggregation it had both ecological and statistical limitations. They found it behaves erratically at low densities, implying randomness where none existed and failing to model regularity. For many species, a common k_c could not be obtained for the full range of population densities recorded. As a result, $1/k$ proved to be an inconsistent measure of aggregation even when the negative binomial fit the data well. In addition, for those species with TPL slopes $1 \leq b \leq 2$ k_c is not single-valued leading to predictions of negative variance at high density.

Morisita's I_δ patchiness index

Morisita (1962) proposed an index

$$I_\delta = n\delta = n \frac{\sum_{i=1}^{N} x_i(x_i - 1)}{\sum_{i=1}^{N} x_i \left(\sum_{i=1}^{n} x_i - 1 \right)} \tag{3.14}$$

where n = the number of quadrats and x_i is the number of individuals in the ith quadrat. Morisita showed his patchiness index, I_δ, is independent of quadrat size and concluded it is superior to V/M as an index of aggregation. Assuming the underlying distribution to be negative binomial, he showed that $I_\delta \approx 1 + 1/k$ provided the number of quadrats, $n \gg I_\delta$. From this we see that if the distribution is actually Poisson, $I_\delta = 1$ and if aggregated $I_\delta > 1$.

Green's C_x coefficient

To counter the dependence of V/M ratio on density, Green (1966) proposed an index

$$C_x = \frac{(V/M) - 1}{nM - 1} \tag{3.15}$$

which he compared with several others. C_x is related to the negative binomial k by

$$C_x = \frac{1/k}{n - 1/M} \tag{3.16}$$

Using this relationship, Green developed a test for departure from randomness, giving the upper value for randomness at significance level α:

$$C_x(1-\alpha) = \frac{\chi_{1-\alpha}^2 - (n-1)}{(n-1)(nM-1)} \tag{3.17}$$

with $n - 1$ degrees of freedom. As the Poisson condition is quite rare and the majority of organisms are aggregated to some degree, Green argued that C_x was a superior description of aggregated distributions because it varies from 0 to 1 for randomness to highly aggregated and is not sensitive to either n or M.

Lloyd's mean crowding

Lloyd's (1967) mean crowding is the number of other individuals per unit area or quadrat per individual, assuming an individual does not crowd itself. It may be thought of as the intensity of interaction between individuals. Formally, it is the amount by which the variance-mean ratio minus 1 exceeds the mean density:

$$M^* = \mu + \left(\frac{\sigma^2}{\mu} - 1 \right) \tag{3.18}$$

Replacing Eq. (3.18) with sample estimates:

$$M^* \approx M + \left(\frac{V}{M} - 1\right) \tag{3.19}$$

The quantity in parentheses disappears when the distribution is Poisson and mean crowding is the mean density. For aggregated distributions, the variance mean ratio is replaced by a suitable parametric form. Assuming a negative binomial distribution, Eq. (3.19) becomes

$$M^* = M + \left(\frac{V}{M} - 1\right)\left(\frac{V}{nM} + 1\right) \tag{3.20}$$

or, if n is sufficiently large:

$$M^* \approx M + \frac{M}{k} \tag{3.21}$$

The variance is approximately:

$$\text{var}(M^*) \approx \frac{1}{n}\left(\frac{V M^*}{M^2}\right)\left(M^* + \frac{2V}{M}\right) \tag{3.22}$$

Using the mean crowding concept, Lloyd also defined an index of patchiness based on the negative binomial,

$$I_P = \left(\frac{M^*}{M}\right) = 1 + \frac{1}{k} \tag{3.23}$$

with variance

$$\text{var}(I_P) \approx \frac{V}{M^2}\sqrt{\frac{2M^*}{NM}} \tag{3.24}$$

A second approach especially useful when the population is both highly contagious (k is small) and the overall density is not too large. It uses the distribution's zeroth term:

$$k_2 \ln\left(1 + \frac{M}{k_2}\right) = \ln\left(\frac{n}{n_0}\right) \tag{3.25}$$

where n is the number of quadrats and n_0 is the number of empty quadrats. k_2 can be estimated by the method of false position used in the ML estimation of k (Eq. 3.8–3.11). It can be used to estimate mean crowding and patchiness. The standard errors for mean crowding and patchiness are given in Lloyd (1967). When n is large Lloyd's I_p is approximately equal to Morisita's I_δ (Eq. 3.14), which is independent of quadrat size; thus, Lloyd's patchiness index is also independent of quadrat size.

Lloyd includes an example using data from Hairston (1959). In a plot of M^* against M over 4 orders of magnitude of mean density M^* varies from 3 to 503. He interprets this relationship to "mean that individuals have the least tendency to wander when they are in the vicinity of some optimum number of others of their species."

Iwao's patchiness index

Iwao (1968, 1970) formalized the relationship between mean crowding and mean density with his patchiness regression model

$$M^* = M + \left(\frac{V}{M} - 1\right) = a + bM \qquad (3.26)$$

Thus, Lloyd's mean crowding index is explicitly a linear function of mean density. The intercept, a, indicates the tendency to crowding: $a > 0$ indicates crowding; $a < 0$ indicates repulsion or tendency to regularity. Iwao called this an "index of basic contagion" and suggested it is a property of the species. The slope, b, he called "density contagiousness coefficient" and interpreted as being habitat-dependent. Density-independent and density-dependent mortality are expected to influence a and b, respectively (Iwao, 1970).

Based on the parameter values, Iwao classifies spatial distributions as random or Poisson when $a = b = 1$; aggregated when $a = 0$ and $b = 1 + k$; aggregated with randomly distributed colonies when $a > 0$ and $b = 1$; aggregated with aggregated colonies when $a > 1$, $b > 0$; more regular than random when $a = 0$ and $b = 1 - 1/k$; completely uniform when $a = b = 0$ or $a = -1$ and $b \approx 1$. He provides 14 insect examples from the literature of aggregated or random distributions but no regular examples (Table 3.1).

Several of these examples were used in LRT61 and some are all reanalyzed in Chapter 8.

Iwao's ρ index

Iwao's (1972) colony size index ρ_i computes Lloyd's M^* for a sequence of quadrat sizes:

$$\rho_i = \frac{M_i^* - M_{i-1}^*}{M_i - M_{i-1}} \qquad (3.27)$$

where the subscript refers to the ith size in an ordered sequence with $i = 1$ the smallest quadrat. A plot of ρ_i ($i > 1$) against quadrat size reveals clumping of individuals or clusters and the degree of aggregation of clusters. This method is similar to one used by plant ecologists in which variance is plotted against sample area (Grieg-Smith, 1983).

Variance-mean relationship

The foregoing indices implicitly or explicitly assume the frequency distribution describing spatial pattern is negative binomial. Generally this assumption is for convenience, to illustrate the index's properties under Poisson and/or

TABLE 3.1 Examples of Iwao's (1968) mean crowding regressions applied to 14 insect data sets

Condition	Examples	Reference
More regular than random $a = 0$, $b < 1$	Azuki bean weevil (*Callosobruchus chinensis*) eggs	Iwao (1968)
	Azuki bean weevil (*Callosobruchus chinensis*) adults	Utida (1943)
	Queensland fruit fly (*Dacus tryoni*)	Monro (1967)
Random (Poisson) $a = b = 1$	Flour beetle (*Tribolium confusum*)	Naylor (1959)
Aggregated $a = 0$, $b > 1$	Spruce budworm (*Christoneura funiferana*) larvae	Bliss (1958)
	Colorado potato beetle (*Leptinotarsa decemlineata*) adults	Bliss and Owen (1958)
Aggregated with random colonies $a > 0$, $b = 1$	Cabbage white butterfly (*Pieris rapae*) eggs	Kobayashi (1966)
	Potato lady beetle (*Epilachna 28-maculata*) eggs	Iwao (1968)
Aggregated with aggregated colonies $a > 0$, $b > 1$	Brazilian bean weevil (*Zabrotes subfasciatus*)	Utida (1943)
	Brown planthopper (*Nilaparvata lugens*)	Kuno (1963)
	Green rice leafhopper (*Nephotettix cincticeps*)	Kuno (1963)
	Pea aphid (*Macrosiphum pisum*)	Forsythe and Gyrisco (1963)
	Rice stem borer (*Chilo suppressalis*)	Kanno (1962)
	Wireworms (*Limonius* spp.)	Bliss and Owen (1958)

aggregated distribution. As we see in Fig. 1.1, this assumption cannot be sustained when the mean is exceptionally high. Furthermore, as LRT and colleagues have emphasized (LRT et al., 1979; LRT, 1984; Perry and Taylor, 1986) if the negative binomial is assumed, the resulting variance-mean equations are necessarily quadratic in form, $V = a + bM + cM^2$, and for some values of M predict negative V. As negative variances are impossible, this inevitably restricts the range of population abundance over which these indices can be valid.

The distributions described in Chapter 2 form a sequence in which the variance is a function of the mean raised to an integer power (Table 3.2). The transitions between distributions in Fig. 1.1 suggest there are intermediate distributions with the variance a fractional power of the mean. A family of distribution with this property is the Adès distribution to describe data that conform to TPL (Perry and Taylor, 1985).

TPL as an index of aggregation

For any given data set based on a set of samples or quadrats, it is convenient and informative to describe the data by a single distribution. It provides information on the degree of aggregation or lack of it and permits parameter values to be compared across data sets to make inferences about the spatial patterns of a species in different places or times. However, as shown in Fig. 1.1, populations may exhibit different statistical distributions simultaneously in different parts of their range or at different times at a single location, so that a set of frequency distributions may not be fitted with equal precision or reliability. The frequency distributions shown in Fig. 1.1 are linked by having the variances predicted by a common coefficient, a, and fractional power of the mean, b (Eq. 1.1a,b).

LRT (1961, 1965, 1971, 1984) suggested that the slope of TPL was a measure of aggregation of the population sampled and that it may be species-stage specific. The specificity of the slope would therefore be similar to host specificity in parasitoids or phytophages being specific but variable (LRT et al., 1980; Elliott, 1981) and not an absolute taxonomic character equivalent to the wing venation of insects or the limb construction of vertebrates.

The discovery that the variance of spatial data frequently scales as a fractional power of the mean density (LRT61) was further investigated by LRT et al. (1978, 1979). To the two dozen examples presented in LRT61, LRT et al. (1978) added 132 new reports covering 102 species. They included data of plants, protozoans, annelids, molluscs, arthropods, echinoderms, vertebrates including *Homo sapien* obtained by sampling with quadrat counts and various traps and grab samplers. One data set was of 10 decennial censuses of the United States: this is revisited in Chapter 11.

Of the 156 data sets, only 2 sets adequately fitted the Poisson model ($V = M$) at all densities (perennial herbaceous plant *Poterium sanguisorba* and bark beetle *Ips grandicollis*) and 2 sets were significantly more regular than random ($V < M$) at all densities (fruit fly *Dacus tryoni* and bivalve mollusc *Tellina tenuis*). These 4 of 156 examples testify to the rarity of nonaggregated populations.

In this study, they also fitted Iwao's (1968) mean crowding model (Eq. 3.26) and the negative binomial with a common k (Eqs. 3.12, 3.13). Neither Iwao's regression nor k_c proved satisfactory for all examples. Depending on the fitting algorithm Iwao's model fit less well than TPL in 64% or 68% of cases, and in 20% of examples the fitted parameters predicted negative variances. In those cases where Iwao's regression performed better than TPL, the differences were

TABLE 3.2 The distributions used to describe spatial pattern all have the variance as a function of an integer power β of the mean

β	Distribution	Probability Function	Mean	Variance
1	Poisson	$p(x=r)=\frac{1}{r!}\mu^r e^{-\mu}$	μ	μ
2	Negative Binomial*	$p(x=r)=\frac{(\mu/k)^r(k+r-1)!}{(1+\mu/k)^{k+r}(k-1)!r!}$	μ	$\mu+\frac{\mu^2}{k}$
2	Neyman's A*	$p(x=r)=\sum_{i=1}^{\infty}\frac{1}{\exp(\mu+i\lambda)}\left(\frac{\mu^i(i\lambda)^r}{i!r!}\right)$	$\mu\lambda$	$\mu\lambda(1+\lambda)$
2	Poisson Lognormal*	$p(x=r)=\int_{-\infty}^{\infty}\frac{\exp\left(r\mu+\sigma^{r(r-1)}+1/2\,r\sigma^2\right)}{r!\sigma\sqrt{2\pi}}\times$ $\exp\left(-\frac{u^2}{2\sigma^2}-\exp\left(\mu+1/2\sigma^2\right)\sigma^{2r-1}e^u\right)du$	$\exp(\mu+\frac{1}{2}\sigma^2)$	$\exp\left(\mu+1/2\sigma^2\right)-$ $\exp\left(2\mu+\sigma^2\right)+$ $\exp\left(2\mu+2\sigma^2\right)$
2	Lognormal	$\Phi_X(x)=\frac{1}{x\sigma\sqrt{2\pi}}\exp\left\{-\frac{\ln(x-\delta)^2}{2\sigma^2}\right\}$	$\exp(\mu+\frac{1}{2}\sigma^2)$	$\exp\left(2\mu+\sigma^2\right)\times$ $\left\{\exp(\sigma^2)-1\right\}$
3	Inverse Gaussian	$\Phi_X(x)=\frac{\lambda}{\sqrt{2\pi x^3}}\times$ $\exp\left[-\lambda\left(\frac{1}{2x}-x(2\mu)^{1/2}+x(8\mu)^{1/4}\right)\right]$	μ	μ^3/λ

Some variance functions are polynomials (*).

generally small. A two-tailed binomial test rejected at the 1% level the hypothesis that TPL fits no better than Iwao's. Others (e.g., Allsopp and Bull, 1990; Serra et al., 2002) have found Iwao's equation to fit better than TPL, but, again the differences were usually small. Like the negative binomial's k, Iwao's regression suffers from the polynomial relationship between variance and mean, which can lead to predictions of negative variance.

In another paper, LRT et al. (1980) analyzed 97 aphid species, 263 moths, and 111 birds. These data were of samples taken simultaneously at 8-24, 21-126, and 45-210 sites over Great Britain for 11, 13, and 15 years, respectively. The points in the TPL analyses were the mean-variance pairs computed for each year (spatial TPL). The habitats sampled were mainly field and forest with some urban sites catching moths. They found that b ranged from 1.29 to 2.95 for the aphids caught in suction traps, 0.95–3.32 for moths caught in light traps, and 1.19–2.69 for birds recorded by direct observation. The analysis of bird samples revealed approximately half the TPL analyses for field and forest were statistically indistinguishable: the remainder produced parallel lines, except for eight species with divergent TPLs suggesting different spatial organization in the two habitats. They also found that in some species the points for field and forest are segregated at opposite ends of the power law line indicating similar spatial organization, but very different abundance in the two habitats. The b value of six species in the noctuid genus *Apamea* ranged from 1.20 to 3.32 showing even close relatives organize themselves in space differently.

Clearly the degree of aggregation can vary between habitats, but within similar habitats sampled in the same way RAJT et al. (1998) showed that six sets of data of western flower thrips (*Frankliniella occidentalis*) in greenhouses confirmed the stability of the estimates of slope. The practical consequence of a stable measurable spatial organization of a species in a defined habitat is that pest management or conservation plans based on density might have broad applicability.

Adès distribution

Fig. 1.1 strongly suggests that the spatial distribution of organisms is density dependent. In order to incorporate this density dependence into a rigorous statistical framework, Perry and Taylor (1985, 1988) developed the Adès family of distributions. In a system such as Fig. 1.1, Adès has two parameters (τ and r) common to all distributions and one (λ_i) unique to the distribution underlying each point on the TPL plot. Thus, to fit Adès to McGuire et al.'s (1957) European corn borer data in Fig. 1.1A, Adès has two common parameters plus one each for the individual points, for six in all.

In developing the Adès family, Perry and Taylor required that the variance-mean relationship of the individual Adès distributions should obey TPL for the entire data set of $n = NB$ points and that one parameter should be numerically equivalent to b. Thus, an Adès distribution is the underlying distribution of points on the TPL graph. They formalized this by defining a variable Y_i as a function of X_i, a random gamma variable by

$$Y_i = (\ln X_i)^{2/(2-\tau)}, \quad X_i > 1, \tau < 2, i = 1...n$$
$$Y_i = 0, \qquad\qquad X_i \leq 1 \tag{3.28}$$

where n is the number of points in a TPL plot. The probably density function for the ith point in the plot is:

$$f(y_i) = \frac{\lambda_i^\rho y_i^{-\tau/2}[1 - \tau/2]}{\Gamma(\rho)\left[1 - I(\lambda_i, \rho)\left[\exp\left(-\lambda_i \exp\left\{y_i^{[1-\tau/2]}\right\}\right)\right]\right]\left[\exp\left(y_i^{\{\tau[1-\tau/2]\}}\right)\right]}, \quad 0 < y_i < \infty \tag{3.29}$$

where $\Gamma(\rho)$ is the gamma function with $\rho = CV$ and $I(\lambda_i, \rho)$ is the incomplete gamma function:

$$\text{Prob}[Y_i = 0] = \int_0^{\lambda_i} \frac{e^{-z} z^{\rho - 1}}{\Gamma(\rho)} dz \tag{3.30}$$

The Adès distribution is sufficiently flexible to fit data conforming to different TPLs but having the same means and variances. The expected value and variance ($E(Y_i)$ and $V(Y_i)$) of the ith Adès distribution correspond to the mean and variance of the ith point on the TPL plot, and as the sample mean, M, increases, the Adès distributions change shape.

In six data sets (one of which was treated two ways) that are described well by TPL the form of the frequency distributions ranged from Poisson, asymmetrical with a high proportion of zeros to nearly symmetrical but with a long right-hand tail typical of high-density counts in the literature. The Adès family performed as well as or better than, negative binomials fitted to the same data. The mean densities ranged over four orders of magnitude with TPL slopes $1.245 \leq b \leq 3.085$ (Table 3.3). The Adès parameter τ is strongly related to TPL slope, b, and parameter ρ is weakly related to TPL intercept A ($=\log(a)$). Blackshaw and Perry (1994) found the sample-dependent parameters λ_i were related to the estimate of the true mean, μ_i, calculated from the fitted distribution by $\lambda_i = 2.74 + 1.66\exp(-0.39\mu_i) + 2.61\exp(-0.07\mu_i)$, which accounted for 99% of the variation in their data of leatherjacket (*Tipula* spp.) larvae. The λ_i are also well described by a logarithmic function of the sample mean,

$$\lambda_i = c + d \ln(M_i), \tag{3.31}$$

in four of Perry and Taylor's examples when $b < 2$, but in two examples with $b > 2$ are unrelated (Fig. 3.2). Eq. (3.31) may be preferable as it is more parsimonious, using only two parameters, and relates λ directly to the observed mean M.

As Kemp (1987) pointed out, other distributions could be used as the basis for a family conforming to TPL. She proposed the negative binomial because it has a long history of use in ecology. However, as the negative binomial has a

TABLE 3.3 TPL and Ades parameter estimates for seven insect data sets

Species	TPL		Adès		Adès: $\lambda_i = c + d \ln(M_i)$		
	a	b	ρ	τ	c	d	r
European corn borer	0.150	1.245	4.811	1.256	1.843	−0.555	−0.986
Black bean aphid	1.383	1.526	0.525	1.546	0.505	−0.109	−0.961
Black bean aphid	1.406	1.759	0.737	1.693	0.774	−0.156	−0.983
Black bean aphid eggs	0.683	1.530	2.537	1.499	1.129	−0.252	−0.906
Discestria trifolii	1.269	1.886	2.135	1.690	1.452	−0.229	−0.885
Lycophotia varia	0.047	3.085	1.749	3.036	1.476	0.064	0.336
Ellopia fasciaria	0.460	2.884	1.478	3.454	1.951	−0.300	−0.431

From *Perry and Taylor* (1985).

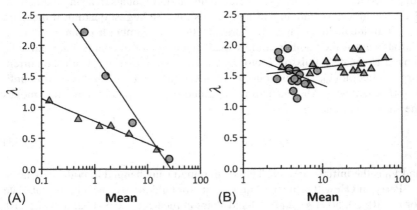

FIG. 3.2 Adès' frequency-dependent parameter λ (A) declines logarithmically with mean density (M) when $b < 2$, but (B) is independent of M when $b > 2$ (Table 3.3).

polynomial relationship between mean and variance, the gamma with $M \propto V^2$ is preferable and is better able to describe long right-hand tails. When constrained to fit TPL, the negative binomial performed similar to the Adès in 15 of 22 test cases, but "the Adès fitted noticeably better in 7 sets [that] mainly comprised long-tailed distributions with large means, typical of pest species" (Perry and Taylor, 1988).

One property of the Adès family deserves special mention. Two species with different TPL lines have the same mean and variance at the point where they cross, but do not necessarily have the same frequency distribution at that point. A characteristic distribution for data obeying TPL should have this flexibility: the Adès family has this desirable property.

Perry's spatial analyses

Perry and Hewitt's number of moves index

Joe Perry, who worked with LRT at Rothamsted for over a decade, continued the analysis of aggregation by developing methods for including spatial information in its measurement. Starting with some simple spatial models, Perry (1988) began considering ways in which spatial behavior might be incorporated into the description of spatial distribution. This was followed by Perry and Hewitt (1991) developing an analytical method for using the positional data of the samples. To begin, they defined an index of aggregation, S, that ranges between zero and one:

$$S = \frac{mtr}{mtr + mtc} \tag{3.32}$$

where mtc is the number of moves required to convert the vector of samples (quadrats) to a maximally aggregated state (all individuals in a single quadrat). This is achieved by moving all individuals into the largest quadrat: if there are x_{max} individuals in the largest quadrat and the total number is n, $mtc = n - x_{max}$. Variable mtr is the number of individuals in the vector of counts that need to be moved to create a Poisson vector by moving one individual from the current largest quadrat (x_{max}) to the current smallest quadrat (x_{min}) and repeating until $V \leq M$ is achieved. After each move x_{max} and x_{min} are updated and after i moves the sample variance is:

$$V_i = V_{i-1} - 2\frac{(x_{max} - x_{min} - 1)}{(n-1)} \tag{3.33}$$

and V_0 is the initial variance. The mean remains unchanged in both sequences.

Perry and Hewitt provide a "quick test" for randomness, which they tabulate for $n = 10, 20, 50, 100$, and 500 and Poisson means $\mu = 0.5, 1, 2, 3, 5, 7.5, 10, 17.5, 25, 35$, and 50. Simulations showed the test of S to be a more powerful test of randomness than tests based on the index of dispersion test (Eq. 3.6).

If the positions of the samples are recorded, this information may be used in an elaboration of this test. In this case, the number of horizontal and vertical moves (no diagonals) required to move individuals on a grid of quadrats is counted. The counting of mtc is straightforward, but mtr is more complex. Perry and Hewitt developed an algorithm for finding *mtr* in which they distinguished between "donor" and "receiver" samples:

1. calculate the current differences between each possible pair of neighboring samples, choose the pair with the largest difference, and move one individual from larger, donor quadrat, to the smaller, receiver quadrat;
2. if more than one donor-receiver pair have the same difference choose the pair with the fewest individuals in the receiver quadrat;
3. if this fails to select a unique pair, choose the pair whose receiver with the neighbor with the smallest count;
4. if this fails also, choose the receiver with neighbors with the smallest average count;
5. if this fails also, make a random choice among the receivers available;
6. after a move has been completed return to 1.

The cycle terminates when $V \leq M$ and the test statistic S is computed from counts of the moves as before and the same table provides the test for significance of departure from random. Simulations showed the power of this test was increased still further over the index of dispersion by inclusion of the spatial information. Perry and Hewitt used Harrington's (1987) peach-potato aphid (*Myzus persicae*) data collected from a field marked into a 3×5 grid of 10×10 m squares to test this method. Computing S for the spatial version would be tedious and time consuming by hand for any but the smallest grid, but is easily performed on a computer.

SADIE

In a series of papers, Perry continued to develop spatial analyses based on Perry and Hewitt's methods (Perry, 1995, 1998; Perry and Klukowski, 1997; Perry et al., 1999; Perry and Dixon, 2002). Called SADIE (Spatial Analysis by Distance IndicEs), the suite of methods utilizes the positions of each individual in a population. Building on the concept of moves to randomness and maximal concentration (Eq. 3.32), SADIE uses the distance of moves to regularity; the total distance moved by spatially referenced individuals to achieve as near as possible equal numbers of individuals per quadrat (uniform distribution). The process starts by assigning to each individual its own Voronoi tessellation territory (see Fig. 3.1C) within a predefined rectangle. In a series of rounds, all individuals are moved simultaneously to the centroid of its neighbors weighted by the length of the common vertex:

$$x_{new} = \frac{\sum_{i=1}^{q} l_i x_i}{\sum_{i=1}^{q} l_i}, \quad y_{new} = \frac{\sum_{i=1}^{q} l_i y_i}{\sum_{i=1}^{q} l_i} \qquad (3.34)$$

where x_i, y_i, and l_i are the coordinates and edge length of the q neighbors. At the end of each round, new Voronoi polygons are computed and the procedure repeated until the areas of the polygons are approximately the same. Individuals with an edge at the boundary are assumed to have an imaginary neighbor at the boundary. The distance moved by all individuals in the population is computed from the original and final positions, $d = \sqrt{(x_f - x_0)^2 + (y_f - y_0)^2}$ and the distance to regularity is the sum $D = \sum_{i=1}^{n} d_i$, where $n =$ the number of individuals.

Departure from randomness is measured by comparing D of the observed spatial pattern with the average distance to regularity of a series of randomly reorganized individuals, $\bar{d} = (\sum D_{rand})/n$. An index $I_a = D/\bar{d} > 1$ indicates an aggregated pattern while $I_a < 1$ indicates a regular pattern, and $I_a \approx 1$ suggests a spatially random pattern. A one-sided test of spatial randomness compares D to an ordered sequence of $S \geq 100$ values of D_{rand}. If R is the number of sets with $D_{rand} < D$, under the null hypothesis of complete randomness, $P = R/S$ is the significance level for rejecting randomness in favor of aggregation.

If the data are of spatially referenced samples with n_i individuals in the ith sample, SADIE makes equivalent calculations for all $N = \Sigma n$ individuals. The number of rounds required to place 12 aphids on a leaf into a regular distribution was 325. This represents a considerable amount of computation and for large populations it may be formidable. Fortunately, computer time is now cheap.

SADIE can also detect clusters and gaps in count data using metrics of "flow" into or out of "hot spots." Clusters and gaps are identified by dividing the area into subareas, if it is not already split up, and simultaneously comparing local density with densities elsewhere. An index of clustering (I_a) is determined for each subarea, and the overall degree of clustering into patches and gaps is assessed by a randomization test.

It is known that the spatial positioning of the samplers may interact with the population pattern as it has been shown that I_a is sensitive to the number and position of the clusters (Perry and Klukowski, 1997; Xu and Madden, 2004). Consequently, small sample sizes as well as the positions of samples can affect the significance of tests (Thomas et al., 2001). Clearly, the number of clusters detected also depends on the number of sampling locations.

Perry (1995) presented seven worked examples of SADIE over a range of spatial scales: the organisms are sycamore aphid (*Drepanosiphum platanoidis*), Japanese beetle (*Popillia japonica*) larvae, ant (*Lasius flavus*) mounds, sparrowhawk (*Accipiter nisus*) nesting territories, Japanese black pine (*Pinus thunbergii*) seedlings, redwood (*Sequoia* sp.) seedlings, and biological cells. SADIE has been used by a number of biologists to analyze population spatial patterns. Reay-Jones et al. (2016) list a number of other SADIE studies with their application to changes in aggregation over 3 years of three species of stink bugs (*Chinavia hilaris, Nezara viridula, Euschistus servus*), boll injury in cotton (*Gossypium hirsutum*), and NDVI (normalized difference vegetation index) used as a measure of plant health. Reay-Jones et al. show that not only does SADIE satisfactorily analyze spatial data of organisms but also the influence of other attributes that may influence their distribution. The basic motivation

TABLE 3.4 Summary of aggregation indices

Index	Symbol(s)	Equation(s)
Negative binomial	k and k_c	3.8–3.13
Morisita (1962) Patchiness	I_δ	3.14
Green (1966) Coefficient	C_x	3.15–3.17
Lloyd (1967) Mean crowding	M^* and I_p	3.18–3.25
Iwao (1968, 1970) M^* regression	a and b	3.26
Iwao (1972) Colony size index	ρ	3.27
LRT61 TPL	a and b	1.1
Perry and Taylor (1985) Adès	τ, r, and λ_i	3.28–3.31
Perry and Hewitt (1991) Moves index	S	3.32–3.33
Perry (1995) SADIE	I_a	3.34

of SADIE was to incorporate into the analysis of spatial pattern "a biological model for the dispersal of individuals from a source" (Perry, 1995).

The indices used to measure aggregation are summarized in Table 3.4.

References

Allsopp, P.G., Bull, R.M., 1990. Sampling distributions and sequential sampling plans for *Perkinsiella saccharicida* Kirkaldy (Hemiptera: Delphacidae) and *Tytthus* spp. (Hemiptera: Miridae) on sugarcane. J. Econ. Entomol. 83, 2284–2289.

Anscombe, F.J., 1950. Sampling theory of the negative binomial and logarithmic series distributions. Biometrika 37, 358–382.

Blackshaw, R.P., Perry, J.N., 1994. Predicting leatherjacket population frequencies in Northern Ireland. Ann. Appl. Biol. 124, 213–219.

Bliss, C.I., 1958. The analysis of insect counts as negative binomial distributions. Proc. 10th Int. Congr. Entomol. (1956) 2, 1025–1032.

Bliss, C.I., Owen, A.R.G., 1958. Negative binomial distributions with a common k. Biometrika 45, 37–58.

Boswell, M.T., Patil, G.P., 1970. Chance mechanisms generating the negative binomial distribution. In: Patil, G.P. (Ed.), Random Counts in Models and Structures. Penn State Press, University Park, PA, pp. 3–22.

Clark, P.J., Evans, F.C., 1954. Distance to nearest neighbor as a measure of spatial relationships in populations. Ecology 35, 445–453.

David, F.N., Moore, P.G., 1954. Notes on contagious distributions in plant populations. Ann. Bot. 53, 47–53.

Elliott, J.M., 1981. A quantitative study of the life cycle of the net-spinning caddis *Philopotamus montanus* (Trichoptera: Philopotamidae) in a Lake District stream. J. Anim. Ecol. 50, 867–883.

Evans, D.A., 1953. Experimental evidence concerning contagious distributions in ecology. Biometrika 40, 186–211.

Fisher, R.A., Corbett, A.S., Williams, C.B., 1943. The relation between the number of species and the number of individuals in a random sample of an animal population. J. Anim. Ecol. 12, 42–58.

Forsythe, H.Y., Gyrisco, G.G., 1963. The spatial pattern of the pea aphid in alfalfa fields. J. Econ. Entomol. 56, 104–107.

Green, R.H., 1966. Measurement of non-randomness in spatial distributions. Res. Popul. Ecol. 8, 1–7.

Grieg-Smith, P., 1983. Quantitative Plant Ecology, third ed. University of California Press, Berkeley, CA.

Hairston, N.G., 1959. Species abundance and community organization. Ecology 40, 404–416.

Harrington, R.H., 1987. Varying efficiency in a group of people sampling cabbage plants for aphids (Hemiptera: Aphididae). Bull. Entomol. Res. 77, 497–501.

Iwao, S., 1968. A new regression method for analyzing the aggregation pattern of animal populations. Res. Popul. Ecol. 10, 1–20.

Iwao, S., 1970. Analysis of spatial patterns in animal populations: progress of research in Japan. Rev. Plant Protec. Res. 3, 41–54.

Iwao, S., 1972. Application of the m^* – m method to the analysis of spatial patterns by changing the quadrat size. Res. Popul. Ecol. 14, 97–128.

Kanno, M., 1962. On the distribution pattern of the rice stem borer in a paddy-field. Jap. J. Appl. Ent. Zool. 6, 85–89 (In Japanese with English summary).

Kemp, A.W., 1987. Families of discrete distributions satisfying Taylor's power law. Biometrics 43, 693–699.

Kobayashi, S., 1966. Process generating the distribution pattern of eggs of the common cabbage butterfly, *Pieris rapae crucivora*. Res. Popul. Ecol. 8, 51–61.

Kuno, E., 1963. A comparative analysis of the distribution of nymphal populations of some leaf and planthoppers on rice plant. Res. Popul. Ecol. 5, 31–43.

Lloyd, M., 1967. Mean crowding. J. Anim. Ecol. 36, 1–30.

May, R.M., 1974. Stability and Complexity in Model Ecosystems. Princeton University Press, Princeton, NJ.

McGuire, J.U., Brindley, T.A., Bancroft, T.A., 1957. The distribution of European corn borer larvae *Pyrausta nubilalis* (Hbn.) in field corn. Biometrics 13, 65–78.

Monro, J., 1967. The exploitation and conservation of resources by populations of insects. J. Anim. Ecol. 36, 531–547.

Morisita, M., 1962. I_δ index, a measure of dispersion. Res. Popul. Ecol. 4, 1–7.

Naylor, A.F., 1959. An experimental analysis of dispersal in the flour beetle, *Tribolium confusum*. Ecology 40, 453–465.

Perry, J.N., 1988. Some models for spatial variability of animal species. Oikos 51, 124–130.

Perry, J.N., 1995. Spatial analysis by distance indices. J. Anim. Ecol. 64, 303–314.

Perry, J.N., 1998. Measures of spatial pattern for counts. Ecology 79, 1008–1017.

Perry, J.N., Dixon, P.M., 2002. A new method to measure spatial association for ecological count data. Ecoscience 9, 133–141.

Perry, J.N., Hewitt, M., 1991. A new index of aggregation for animal counts. Biometrics 47, 1505–1518.

Perry, J.N., Klukowski, Z., 1997. Spatial distributions of counts at the edge of sample areas. In: 6th Conference of the Biometric Society (Spanish Region). International Biometric Society, Washington, DC, pp. 103–108.

Perry, J.N., Mead, R., 1979. On the power of the index of dispersion test to detect spatial pattern. Biometrics 35, 613–622.

Perry, J.N., Taylor, L.R., 1985. Adès: new ecological families of species-specific frequency distributions that describe repeated spatial samples with an intrinsic power-law variance-mean property. J. Anim. Ecol. 54, 931–953.

Perry, J.N., Taylor, L.R., 1986. Stability of real interacting populations in space and time: implications, alternatives and the negative binomial k_c. J. Anim. Ecol. 55, 1053–1068.

Perry, J.N., Taylor, L.R., 1988. Families of distributions for repeated samples of animal counts. Biometrics 4, 881–890.

Perry, J.N., Winder, L., Holland, J.M., Alston, R.D., 1999. Red-blue plots for detecting clusters in count data. Ecol. Lett. 2, 106–113.

Reay-Jones, F.P.F., Greene, J.K., Bauer, P.J., 2016. Stability of spatial distributions of stink bugs, boll injury, and NDVI in cotton. Environ. Entomol. 45, 1243–1254.

Serra, G., Luciano, P., Lentini, A., Gilioli, G., 2002. Spatial distribution and sampling of *Tortrix viridana* L. egg-clusters. IOBC/wprs Bull. 25, 155–158.

Southwood, T.R.E., Henderson, P.A., 2000. Ecological Methods, third ed. Blackwell Science, Oxford, UK.

Taylor, L.R., 1961. Aggregation, variance and the mean. Nature 189, 732–735.

Taylor, L.R., 1965. A natural law for the spatial disposition of insects. In: Freeman, P. (Ed.), Proceedings of the XIIth International Congress of Entomology. Royal Entomological Society, London, UK, pp. 396–397.

Taylor, L.R., 1971. Aggregation as a species characteristic. In: Patil, G.P., Pielou, E.C., Waters, W.E. (Eds.), Statistical Ecology, Volume 1. Spatial Patterns and Statistical Distributions. Penn State Press, University Park, PA, pp. 357–377.

Taylor, L.R., 1984. Assessing and interpreting the spatial distributions of insect populations. Annu. Rev. Entomol. 29, 321–357.

Taylor, L.R., Woiwod, I.P., Perry, J.N., 1978. The density-dependence of spatial behaviour and the rarity of randomness. J. Anim. Ecol. 47, 383–406.

Taylor, L.R., Woiwod, I.P., Perry, J.N., 1979. The negative binomial as a dynamical model for aggregation, and the density dependence of k. J. Anim. Ecol. 48, 289–304.

Taylor, L.R., Woiwod, I.P., Perry, J.N., 1980. Variance and the large scale spatial stability of aphids, moths and birds. J. Anim. Ecol. 49, 831–854.

Taylor, R.A.J., 1992. Simulating populations obeying Taylor's power law. In: DeAngelis, D.L., Gross, L.J. (Eds.), Individual-Based Approaches in Ecology. Routledge, Chapman & Hall, New York, pp. 295–311.

Taylor, R.A.J., Taylor, L.R., 1979. A behavioural model for the evolution of spatial dynamics. In: Anderson, R.M., Turner, B.D., Taylor, L.R. (Eds.), Population Dynamics. Blackwell Scientific Publications, Oxford, UK, pp. 1–27.

Taylor, R.A.J., Lindquist, R.K., Shipp, J.L., 1998. Variation and consistency in spatial distribution as measured by Taylor's power law. Environ. Entomol. 27, 191–201.

Thomas, C.F.G., Parkinson, L., Griffiths, G.J.K., Fernandez Garcia, A., Marshall, E.J.P., 2001. Aggregation and temporal stability of carabid beetle distributions in field and hedgerow habitats. J. Appl. Ecol. 38, 100–116.

Utida, S., 1943. Statistical analysis of the frequency distribution of the emerging weevils on beans. Mem. Coll. Agric, Kyoto Univ. 54, 1–22.

Varley, G.C., Gradwell, G.R., Hassell, M.P., 1973. Insect Population Ecology: An Analytical Approach. Blackwell Scientific Publications, Oxford, UK.

Waters, W.E., Henson, W.R., 1959. Some sampling attributes of the negative binomial distribution with special reference to forest insects. For. Sci. 5, 397–412.

Xu, X.-M., Madden, L.V., 2004. Use of SADIE statistics to study spatial dynamics of plant disease epidemics. Plant Pathol. 53, 38–49.

Chapter 4

Fitting TPL

In order to determine the true value of the slope of any regression, it is necessary to select the proper statistical model. For TPL, we need to decide what kind of relationship exists between variance and mean: not just the mathematical form—a power function—but whether the power function is an empirical relationship we wish to use for prediction or whether it is a true functional relationship. The statistical distinction is epistemological, can be subtle, and is intimately bound up in the type and number of errors to be incorporated into the statistical model. The issue of errors is central to TPL analysis as both axes are derived from the same data and therefore subject to the same natural variation in the population and the same errors of observation. In addition, those data are nearly always arrived at by sampling, which adds further uncertainty.

A detailed exposition of the statistical issues in fitting a straight line to data is given in Madansky (1959). Madansky summarized the work of nearly a century of thought on the subject while laying out the subtleties of statistical modeling. The most recent exposition for biologists is Warton et al.'s (2006) account focusing on allometry; another accessible account is given by Ricker (1973). The following account, leading up to recommendations for fitting TPL, draws on all three papers.

Basically there are two kinds of regression model: ordinary dependent regression (ODR) and functional regression. The latter, functional regression, has a number of variants that depend critically on the assumed error structure. Two error structures are recognized: errors of measurement and errors associated with natural variability of the measured property. All biological properties vary between individuals—the raw material of evolution. In this chapter, we look at the various statistical approaches to estimating TPL parameters, their underlying assumptions, their ease of use, and possible sources of bias in estimation. We start with the simplest approach, ODR, which was used in LRT61 and by many authors since. While adequate in many situations, better methods exist and are described. TPL estimates using the several methods are compared using data of nematode distribution. It is frequently instructive to compare TPL estimates from several sources. We end with some recommendations for fitting TPL.

Taylor's Power Law. https://doi.org/10.1016/B978-0-12-810987-8.00004-5

The standard regression model

The most widely used regression model, ordinary dependent regression (ODR), has the independent variable (X-axis or abscissa) without error and the dependent variable (Y-axis or ordinate) measured or observed with constant variance of measurement and/or normally (or at least symmetrically) distributed natural variation. If there is variation in the X-axis, Berkson (1950) showed that the estimates are still unbiased if X-values are determined by the experimenter; that is to say, the experimenter has controlled, albeit imperfectly, each X-value for which Y-values have been recorded. This is the familiar regression model:

$$y_i = \alpha + \beta x_i + \varepsilon_i \quad i = 1 \ldots n \tag{4.1}$$

where n is the number of observations and ε is the error term (which incorporates natural variation and/or observational error) assumed to be symmetrically distributed with zero mean and constant standard deviation, σ. We wish to estimate α and β by fitting equation (4.1) to data. To do this, we want a solution that minimizes the residual sum of squares in the Y direction:

$$\text{SSR} = \sum_{i=1}^{n} \varepsilon_i^2 = \sum_{i=1}^{n} (y_i - \alpha - \beta x_i)^2. \tag{4.2}$$

The two techniques for model fitting, least squares and maximum likelihood, both yield the same estimates for $\hat{a}$ and $\hat{b}$:

$$\hat{b} = \frac{C_{xy}}{S_x^2} \text{ and } \hat{a} = \bar{y} - \bar{x}\hat{b} \tag{4.3}$$

where

$$S_x^2 = \sum (x_i - \bar{x})^2, \quad C_{xy} = \sum (x_i - \bar{x})(y_i - \bar{y}),$$
$$\bar{x} = \frac{1}{n}\sum x_i \text{ and } \bar{y} = \frac{1}{n}\sum y_i \tag{4.4}$$

The fitted line intercepts the abscissa at $-a/b$. The standard errors are given by

$$SE(\hat{b}) = \sqrt{\left(\frac{S_y^2}{S_x^2}\right)\frac{(1 - r^2)}{(n - 2)}} \text{ and} \tag{4.5a}$$

$$SE(\hat{a}) = \sqrt{\left(\frac{S_y^2}{S_x^2}\right)\frac{(1 - r^2)(S_x^2 + \bar{x}^2)}{(n - 2)}} \tag{4.5b}$$

where r is the correlation coefficient:

$$r = \frac{C_{xy}}{\sqrt{S_y^2 S_x^2}} \text{ and } S_y^2 = \sum (y_i - \bar{y})^2 \tag{4.6}$$

This model is suitable under most circumstances in which an equation to predict Y given X is required. According to Ricker (1973, Table 8), this approach is not suitable for the situation in which both X and Y are subject to natural variability or are sampled from a population with no objective frequency distribution resulting in an irregular or disjoint cloud of points. In this situation it is unlikely that the X-values would have been predetermined by the experimenter, but were arrived at by sampling. For this, Ricker recommends using functional regression. Functional regression is also the preferred model when there are observation errors and/or natural variation in X as well as Y.

Functional regression

If, instead of predicting one variable from another, we wish to characterize our data, to describe the relationship and perhaps compare it to another data set, the fitted line of Y given X (line A in Fig. 4.1) seems unsatisfactory. Nor does the fitted regression of X given Y (line B in Fig. 4.1). In fact, statisticians make a distinction between predictive and functional regressions. To characterize a data set such as in Fig. 4.1, we may want to use a functional regression that measures the central trend of the cloud of bivariate data points: line C in Fig. 4.1 seems to be a more intuitive fit to the data. Such a line might better represent the ideal relationship if there were no natural or measurement variation. Alternatively, if we wish to fit a known algebraic function relating two variables, say the *mass* and *volume* of a metal, in order to estimate a property

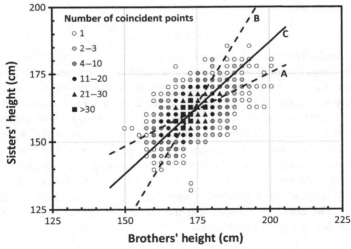

FIG. 4.1   Fitted regressions to approximately bivariate normal data of heights of sisters (Y-axis) and brothers (X-axis). A = predictive regression of sisters on brothers, B = predictive regression of brothers on sisters, C = geometric mean regression of sisters on brothers. *(Data from Pearson and Lee (1903), figure adapted from Fig. 1 in Ricker (1973).)*

known to exist, its *density*, we would want to fit the functional relationship, mass = density × volume. The question naturally arises, "is TPL a natural law, like the relationship between mass and volume?" This is a philosophical issue we defer to Chapter 18.

Geometric mean regression

In the standard regression model, the regression of Y on X minimizes the sum of squares perpendicular to the X-axis with estimates given by Eq. (4.3). One alternative to minimizing sums of squares of the Y-axis is geometric regression (GMR; Ricker, 1973). Kermack and Haldane (1950) called this reduced major axis regression. It is also known as standardized major axis regression (Warton et al., 2006). It is one of several Type II regression methods for fitting data with errors in both variables. This alternative to ODR uses the regressions of y on x and x on y, which has slope

$$\hat{d} = \frac{C_{xy}}{S_y^2} \tag{4.7a}$$

estimated by minimizing the sum of squares perpendicular to the ordinate. These estimates can be reoriented relative to the X-axis:

$$\hat{d}' = \frac{S_y^2}{C_{xy}} \tag{4.7b}$$

The intercepts are

$$\hat{c} = \bar{y} - \bar{x}\hat{d} \text{ and } \hat{c}' = \bar{y} - \bar{x}\hat{d}' \tag{4.8}$$

The standard errors for $\hat{d}$ and $\hat{c}$ are computed using Eq. (4.4a). Both the fitted regression lines (A and B in Fig. 4.1) are oriented relative to an X-axis abscissa. Looking at Fig. 4.1, we see that neither A-line nor B-line describe the data in an intuitively reasonable way. Given that the data are height measurements of a sample of brothers and sisters, we know in advance that both measurements are variable in the population: this means that neither predictive regression *describes* the data well, but each provides an equation providing a *prediction* of one variate given the other. When the two regressions are plotted together as in Fig. 4.1, the regression of X on Y is always the steeper line and the spread between the two lines decreases with increasing correlation coefficient (Eq. 4.6). The correlation coefficient has the useful property of relating the two estimates of slope to each other:

$$r = \pm\sqrt{\hat{b}\hat{d}} = \pm\sqrt{\frac{\hat{b}}{\hat{d}'}}$$

While the two regression estimates Eqs. (4.3) and (4.7a) are arrived at by minimizing the sums of squares perpendicular to the two axes, an alternative regression may be obtained by minimizing the sums of squares perpendicular to the fitted line. It can be shown that this estimate may be obtained from the geometric mean of Eqs. (4.3) and (4.7b):

$$\hat{v} = \pm\sqrt{\hat{b} \cdot \hat{d'}} = \pm\sqrt{\frac{\hat{b}}{\hat{d}}} = \pm\sqrt{\frac{S_y^2}{S_x^2}} = \pm\frac{\hat{b}}{r} \tag{4.9}$$

(Kermack and Haldane, 1950). This is the functional relationship, line C in Fig. 4.1. The intercept and standard error are calculated using

$$\hat{a} = S_y \cdot \sqrt{\frac{1-r^2}{n-2}\left(2 + \bar{x}^2\frac{1+r}{S_x^2}\right)} \text{ and } SE(\hat{a}) = S_y \cdot \sqrt{\frac{1-r^2}{n-2}\left(1 + \frac{\bar{x}^2}{S_x^2}\right)} \tag{4.10}$$

Somewhat surprisingly, the standard error of $\hat{b}$ is the same as ordinary linear regression (Eq. 4.4a) (Kermack and Haldane, 1950).

GMR has the important advantage that the best fit line is invariant under changes of scale. This is important for allometry where comparison of anatomical features measured in different units is frequently an objective. With TPL, this is a less serious issue as the units of the Y-axis are in X-axis units squared. Some users have plotted standard deviation against mean, but here we will stick with the traditional approach of plotting variance against mean.

Summarizing Ricker's (1973) Table 8, we note that the appropriate choice of statistical model is determined by the nature of the errors and whether or not the data are distributed as bivariate normal. A bivariate normal population is one with two properties, X and Y, both normally distributed (like the data of brothers' and sisters' heights) and related to one another by the correlation ρ (estimated by r). When $\rho = 0$, the two variables are independent and when $\rho = \pm 1$ they are deterministically related.

To summarize, Eq. (4.3) is used when Y only is subject to natural variability or measurement error, and X is free of error or subject only to measurement error when X-values are determined in advance by an experimenter. When X and Y are subject to what Warton et al. (2006) term equation error, or natural variation, with or without measurement error and data are sampled from a bivariate normal population Ricker recommends using Eqs. (4.3), (4.7a), (4.7b) for predictive and (4.9) for functional relationship. However, if X and Y are subject to natural variability and the data are a random sample from a population not bivariate normal (what Ricker calls "open-ended" distributions), Eq. (4.9) should be used for both predictive and functional relationships. This is the situation with most TPL data.

Deming regression

A regression model with errors in both axes more general than GMR was originally developed by Adcock (1878) and generalized by Kummell (1879). The method, however, is now known as Deming regression for the author who popularized it in his book on data analysis (Deming, 1943). Given a cloud of points relating variables Y to X, like GMR, Deming regression fits the intuitive line one would try to fit by eye. It is equivalent to the maximum likelihood estimation of a model in which the errors of both axes are assumed to be independent and normally distributed with the ratio of their variances known. Deming regression can be used when both error variances can be estimated. It is frequently used in pharmacology and clinical chemistry.

The observed values (x_i, y_i) are related to the true values (X_i, Y_i) by $y_i = Y_i + \varepsilon_i$ and $x_i = X_i + \eta_i$, where ε_i and η_i are independent errors of measurement whose variances are σ_ε^2 and σ_η^2, respectively, and whose ratio $\delta = \sigma_\eta^2 / \sigma_\varepsilon^2$ is assumed to be known. If the measurement of X and Y is by the same method, δ may equal 1, even though the errors are unknown. To estimate the parameters α and β in $Y = \alpha + \beta X$, we minimize the residual sum of squares weighted by their variances:

$$\text{SSR} = \sum \left[\frac{\varepsilon_i^2}{\sigma_\varepsilon^2} + \frac{\eta_i^2}{\sigma_\eta^2} \right] = \frac{1}{\sigma_\varepsilon^2} \sum \left[(y_i - \alpha - \beta X_i)^2 + \delta (x_i - X_i)^2 \right] \qquad (4.11)$$

All summations are for $i = 1 \ldots n$. The least squares estimators of α and β are

$$\hat{b} = \frac{1}{2C_{xy}} \left[S_y^2 - \delta S_x^2 + \sqrt{\left(S_y^2 - \delta S_x^2 \right)^2 + 4\delta C_{xy}^2} \right]$$

where $S_y^2 = \sum (y_i - \bar{y})^2$. As before $\hat{a} = \bar{y} - \hat{b}\bar{x}$. The fitted values are given by

$$\hat{x}_i = x_i + \hat{b} \frac{y_i - \hat{a} - \hat{b}x_i}{\hat{b}^2 + \delta}$$

It is easily demonstrated that if there are no X errors, $\eta_i = 0$ and $X_i = x_i$ so $\delta(x_i - X_i)^2 = 0$ and Eq. (4.11) collapses to Eq. (4.2) and the estimates for α and β are now Eq. (4.3). The special case of equal error variances minimizes the sum of squared distances from the data points perpendicular to the regression line and is the same as GMR (Eq. 4.9).

Bartlett's 3-group regression

Bartlett (1949) extended a method originally developed by Wald (1940) for fitting straight lines when both axes are subject to error. Where Wald's method divided the data into two groups, Bartlett split them into three, enabling a test for curvature. Bartlett's 3-group regression method is especially useful for fitting transformed allometric data as it permits fitting of a functional regression and is parametric while relaxing the normal regression assumptions.

The method is straightforward. The X and Y data pairs are ordered in ascending values of X and then split into three groups such that the lowest and highest X-value groups are identical in size, say N, while the middle group is $n - 1$, n, or $n + 1$ depending on the total number of data points, N. The average values of the three groups and the global averages are calculated, $\bar{x}_i$ and $\bar{y}_i$ $i=1,2,3$ or G for the global averages using Eq. (4.4). The slope can now be calculated using the means of groups 1 and 3:

$$\hat{b}_3 = \frac{\bar{y}_3 - \bar{y}_1}{\bar{x}_3 - \bar{x}_1}$$

and the intercept from the global averages:

$$\hat{a} = \bar{y}_G - \bar{x}_G \hat{b}_3$$

Several tests can be made.

1. A test for significance of regression:

$$t = \frac{1}{\sqrt{S_y^2}} \left\{ \hat{b}_3 (\bar{x}_3 - \bar{x}_1) \sqrt{\frac{n}{2}} \right\};$$

2. A test whether $\hat{b}$ differs from a reference value β:

$$t = (\bar{x}_3 - \bar{x}_1)(b - \beta) \sqrt{\frac{n}{2V_\beta^2}};$$

3. A test for curvature:

$$t = \frac{(\bar{y}_1 + \bar{y}_3 - 2\bar{y}_2) - \beta(\bar{x}_1 + \bar{x}_3 - 2\bar{x}_2)}{\sqrt{V_\beta^2 \left\{ \frac{2}{n} + \frac{4}{N-n} \right\}}} \qquad (4.12)$$

where t is Student's t with $N-3$ degrees of freedom and V_β^2 is the residual variance calculated from

$$V_\beta^2 = \frac{S_y^2 - 2\beta C_{xy} + \beta^2 S_x^2}{N-3}.$$

The S and C terms are the sums of squares and cross products in Eq. (4.4). Strictly speaking, Eq. (4.12) is not t-distributed if the estimate $\hat{b}$ is used in place of β. However, it is approximately distributed as t when $(\bar{x}_3 + \bar{x}_1 - 2\bar{x}_2) \ll (\bar{x}_3 - \bar{x}_1)$, that is, when the high and low groups are well separated.

Methods for fitting TPL

Estimation of β in the functional form of TPL, $V_i = \alpha M_i^\beta$, is complicated by the fact that the distribution of counts changes with mean density (see Fig. 1.1) and the actual form at any given mean is usually unknown. Perry (1981) tested three models for fitting TPL. The first is the same one used in LRT61:

$$\log(V_i) = \log(a) + b\log(M_i) + \varepsilon_i \quad i = 1...NB \qquad (4.13)$$

where the NB variables ε_i are independent random variables with zero mean and variances $\sigma^2/(NQ_i - 1)$ where σ^2 is assumed to be constant for all calculated values of $\log(M_i)$ and $Log(V_i)$ and NQ_i is the number of observations used to compute mean and variance. The least squares estimator $\hat{b}$ is an estimate of β. Perry's second model assumes the conditional value of V_i given M_i ($V_i \mid M_i$) is gamma distributed with $CV = k\sqrt{(NQ_i - 1)}$. In this case, if the NB points are independent of each other, $\hat{b}$ is the maximum likelihood estimate of β in

$$E(V_i \mid M_i) = aM_i^b + \varepsilon_i \quad i = 1...NB$$

where the ε_i are independent gamma distributed variables and $E(V_i \mid M_i)$ denotes the expected value of V given M.

Perry's third model uses the same form as the second, but fitting is by weighted least squares with the weights given by

$$w_i = \left\{ \frac{NQ_i^5 M_{4i}}{(NQ_i - 1)(NQ_i^2 - 3NQ_i + 3)(NQ_i^3 - NQ_i^2 + 3)} \right\} + \left\{ \frac{(3 - NQ_i^2)V_i^4}{(NQ_i^3 - NQ_i^2 + 3)} \right\}$$

where $M_{4i} = \sum_{j=1}^{NQ_i}(x_{ij} - M_i)^4/NQ_i$ for the jth sample at the ith sample location. The w_i weights are inversely proportional to the variance of V_i given M_i. He also shows the ordinary least squares approach (Model 1) is likely to be satisfactory if all NQ_i are sufficiently large and $(\log(V_i) \mid M_i) - \log(V_i)$ is constant.

Perry (1987) describes an iterative method for estimating β that uses Model 1's estimate as a starting value to provide variance stabilizing transformations for ecological and experimental data analysis. The iterative estimator is generally superior to Model 1's estimator for generating a satisfactory power transformation, has a smaller mean-squared error for $\beta \leq 2$, and is less biased for $\beta > 2$. Its use may be justified despite the extra computation required. However, evaluation by simulation exposed difficulties in simulating TPL with high values of α and β resulting in biased estimates and less effective variance stabilizing transformations. While the method may produce closer estimates of $\hat{b}$ to the population value β, it suffers from the disadvantage that standard errors are not available making it less useful for ecological interpretation and comparative analysis.

An alternative to fitting the transformed version of TPL (Eq. 4.13), it is possible to fit it using a generalized linear model (GLM; Nelder and Wedderburn, 1972) to stabilize the variance of V while fitting a predictive equation. A GLM distinguishes between the systematic or mathematical model and the random or error component(s) in the model. The systematic part can be either qualitative as with designed experiments, or quantitative as with regression analysis, or a mixture of the two. The systematic component has the dependent variable z, distributed as Gaussian, binomial, Poisson, gamma, or inverse Gaussian with parameter θ. The n independent variables, x_i, predict $Y = \sum \beta_i x$, given a set of parameters β_i to be estimated. The parameter of the distribution of the dependent variable, z, is related to the linear predictor Y by a transformation or link function $\theta = f(Y)$. The GLM fitting program GLIM (McCullagh and Nelder, 1983) and most other statistical packages recognizes link functions for the five error distributions. The Gaussian or normal distribution with identity link function resolves a GLM to the standard least squares or maximum likelihood model; the binomial is appropriate for quantal response models, such as probit analysis; the Poisson and the multinomial (which can be thought of as a set of constrained Poisson distributions) are used for analysis of contingency tables; and the gamma is used in the estimation of variance components derived from the original observations—the situation with TPL. In a comparison of four spatial models, LRT et al. (1978) used GLIM to fit TPL with gamma error distribution and log link function in a model of the form:

$$V_i = \exp(\,\ln(a) + b\,\ln(M_i)) + \varepsilon_i. \qquad (4.14)$$

In this comparison, LRT et al. concluded that TPL was fitted equally well by both statistical models (Eqs. 4.13, 4.14) to 156 examples drawn from a wide range of taxa. In his later, more detailed analysis of statistical model efficiency, Perry (1981) concluded that the model originally used in LRT61, Eq. (4.13), "performed fairly well, and should prove satisfactory for use with similar sets of animal data."

Fitting split lines

It is sometimes the case that variance-mean pairs form power law lines with $b > 1$ at high means but at lower means follow the Poisson line with $b = 1$ as in Fig. 4.2. In this instance, fitting TPL without considering the two domains will result in an underestimate of the slope at high means and overestimate at low means. If data are more variable than in Fig. 4.2, it might not be obvious there are two domains, although a test for curvature is likely to be significant. Under such circumstances, a quadratic equation, such as Iwao's patchiness index (Eq. 3.26) or a curved power law, $f(x) = ax^{b+cx}$, would fit. However, given the clear discontinuity where the steep TPL cuts the Poisson line, a "broken power law," $f(x) = ax^b$ for $x < x_T$, a threshold, and $f(x) = ax_T^{b-c}x^c$ for $x < x_T$ would be more appropriate.

Perry (1982) describes four methods for estimating "split-line" relationships. The examples he gives are all of data with a clear peak as when an aphid colony increases and then declines. Also, given are procedures for choosing the best valid fit among the alternatives and for fitting several split lines simultaneously.

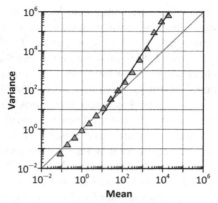

FIG. 4.2 Example of split line TPL in which the lower mean-variance pairs follow the Poisson line ($b = 1.0$) and higher mean-variance pairs obey a TPL with slope, $b \approx 1.6$. *(Data of rice wine (sake) sales from Fukunaga, et al. (2016).)*

In Perry's examples, the data were all clearly peaked although the precise position of the peak could be determined only by fitting the two regression lines. In this instance, one pair of parameters is known a priori, $a = b = 1$, reducing the fitting procedure to determining how many points to constrain to the Poisson line, in effect a 3-parameter problem, rather than 4 as in Perry's examples. The set of estimates producing the smallest residual sums of squares is the preferred model. The models given by Perry all use ODR.

Bias in estimating TPL

In considering the TPL analysis of quadrat counts of grasses at a field's edge having only $NQ = 2$, Clark and Perry (1994) investigated by simulation several factors that can bias estimation of TPL when the sample size is small. They simulated TPL with sample sizes, $NQ = 2, 3, 5, 15,$ and 30 samples per block from populations derived from the negative binomial distribution with specified relationships between V and M at each of $NB = 4$ mean densities, $M = 0.5, 1, 10,$ and 100 per sample. One thousand samples were generated at each density for all combinations of $a = 1, 2,$ and 11, and $b = 1$ and 2, resulting in 6 TPL plots. They found the most serious sources of bias in estimating $\log(a)$ and b are exclusion of samples for which $M = V = 0$ and exclusion of samples for which $V = 0$, but $M > 0$, both of which generally reduce the range of means, which sets limits on the possible values of M and V, another possible cause of bias. Other potential causes of bias they found are limitations on the possible values M and V and artificial restrictions on the minimum and/or maximum variance expressible in a sample. This can occur when the samplers, pheromone trap for example, approach saturation, which results in the mean increasing slowly while the variance stabilizes (see Chapter 13).

For highly skewed distributions, log(V) may be underestimated reducing the fitted slope. Some distributions do not have a finite variance, so if the underlying distribution were Pareto, for example, V is always underestimated. A sixth possibility given by Clark and Perry is "correlation between log(V) and log(M)." This seems contradictory as we would like to see a high correlation between log(V) and log(M) in our TPL analysis. Clark and Perry's caution is actually stating that the deviation of log(V) from the fitted line at log(M) must not be correlated with log(M). In short, the variance about the fitted line should be constant across the entire range of log(M).

In most of Clark and Perry's simulations in which power law parameters were biased, both log(a) and b were underestimated although in one example, a pseudo-Poisson case, the slope was expected to be unity and the intercept large (Fig. 7, Clark and Perry, 1994); the slope was overestimated and the intercept underestimated. The bias in parameter estimates could be summarized by quadratic response surfaces that accounted for >97% of the variation and could be used to compensate for bias in estimates made from small samples. They concluded that TPL should be used with caution if the majority of samples from which V and M are calculated have size, $NQ < 15$; LRT et al. (1988) suggest NB should also be ≥ 15.

Another possible source of bias is explored fully in Chapter 15: $\hat{b}$ appears to be sensitive to the proportion of empty quadrats contributing to a V-M pair. Eliminating zeros usually results in steeper TPL slopes. As we use samplers appropriate to the target in habitats we might reasonably expect the organism to be found, sampling outside normal habitats apparently depresses the estimate of b.

Sprugel (1983) has pointed out that back transforming any allometric equation introduces a bias in the estimation of predicted Y-values, including the estimated intercept. This is because the error term in Eq. (4.13) (ε_i) is assumed to be normally distributed on the log scale, but is lognormally distributed on the arithmetic scale. Consequently, the antilog of $\ln(\hat{a})$ yields the median, not the mean. Baskerville (1972) gives a very lucid account based on Finney's (1941) analysis of lognormally distributed variates. Finney's unbiased estimates of $\hat{a}$ and $SE[\hat{a}]$ are:

$$\hat{a}_u = \exp\left(\hat{a} + \hat{\sigma}^2/2\right) + O\left(NB^{-1}\right)$$
$$SE(\hat{a}_u) = \sqrt{\exp\left(2\hat{a} + 2\hat{\sigma}^2\right) - \exp\left(2\hat{a} + \hat{\sigma}^2\right)} + O\left(NB^{-1}\right)$$

where $\hat{\sigma}$ is the sample variance of the logarithmic equation. However, as Finney showed in an example, if the variance in the logarithmically distributed variates and the number of observations (NB) is large (i.e., if the correlation coefficient is high), the difference between corrected and uncorrected estimates is small. What is important, however, as emphasized by Sprugel, is to use natural logs: if common logs are used, recall that $\log_{10}(z) = \ln(z)/\ln(10)$.

Comparison of models

The several statistical models discussed are compared using data from Park et al. (2013). The data are of the total nematodes recovered in a survey of urban turf grass in Ohio shown in Park et al.'s Fig. 1 and are described more fully in Chapter 7. To these data were fitted five models: predictive regressions of Y on X and X on Y, geometric mean regression, a generalized linear model with log link and gamma error, and Bartlett's 3-group method. The estimates for $\hat{b}$ and $\hat{A}$ are given in Table 4.1. The standard least squares, generalized linear model, and Bartlett's 3-group estimates are remarkably close, although the standard errors differ. The small differences in $\hat{b}$ are probably significant only insofar as they lead to clear differences in $\hat{A}$. As we saw in Fig. 4.1, the two ODRs are substantially different and the GMR lies between. Perry (1981, Fig. 3) showed that GMR represents an upper limit for estimates by ODR.

These results also confirm that the standard errors of $\hat{b}$ (Eq. 4.3) and $\hat{v}$ (Eq. 4.9) are numerically identical. Despite this, unlike the standard least squares regression, $\hat{v}/SE(\hat{v})$ should not be used to test the significance of the GMR estimate. This is because the slope of the GMR is determined solely by the two standard deviations; so the variation around the fitted line, captured by the correlation coefficient, is not considered and because there is no preferred direction of slope as there is when the regression coefficient is estimated

TABLE 4.1 Comparison of regression estimates fitted to data of Park et al. (2013): predictive regression of Y on X, X on Y, geometric mean regression, generalized linear model, and Bartlett's 3-group method

Model and estimator	Estimate	Std err
Predictive regression of Y on X, $\hat{b}_{yx}$	1.989	0.164
Intercept, $\hat{A}_{yx}$	−0.429	0.363
Predictive regression of X on Y, $\hat{b}_{yx}$	2.451	0.300
Intercept, $\hat{A}_{yx}$	−1.444	0.364
Functional regression—reduced major axis (GMR), $\hat{v}$	2.208	0.164
Intercept, $\hat{A}_v$	−0.910	0.363
Generalized linear model—log link, gamma error, $\hat{b}_{GLM}$	2.074	0.175
Intercept, $\hat{A}_{GLM}$	−1.266	0.889
Bartlett's 3-group regression, $\hat{b}_3$	2.084	0.564
Intercept, $\hat{A}_3$	−0.637	

using the correlation coefficient, which has a positive or negative slope as determined by the covariance of X and Y (Eq. 4.4). A method for comparison of GMRs is given below.

Parallel-line analysis

Ordinary dependent regression

In many instances, it will be desirable to test the significance of differences between fitted lines of TPL, examples of which are in Chapters 5–12. ODR of $\log(V)$ on $\log(M)$ is simpler to use and for large samples will generally be satisfactory in most instances (Perry, 1981). Provided the correlation coefficient is sufficiently high, say $r > 0.95$, the difference between the ODR estimate of β and the GMR estimate is small and corresponding statistical tests of least squares analyses will be reliable. However, if the correlation coefficient is lower and for practical applied reasons it is important to have a closer description on the relationship between V and M, the GMR estimate should be estimated. The appropriate statistical test for differences in slope of two GMRs is a likelihood ratio test, which is approximately χ^2 distributed.

In the simpler ODR case, we can compare two TPL fits using the standard t-test:

$$t = \frac{\text{ABS}(\hat{p}_1 - \hat{p}_2)}{\sqrt{SE^2(\hat{p}_1) + SE^2(\hat{p}_2)}},$$

with $n_1 + n_2 - 4$ degrees of freedom, where p represents the parameter of interest ($\hat{A}$ or $\hat{b}$) and $n = NB$, the number of data points. This will be a two-tailed test as we are asking only if the two graphs are different. However, with larger data sets, multiple comparisons by t-test are inconvenient as well as inadvisable because the actual probability obtained by applying multiple tests is less than the nominal probability, making conclusions unreliable.

Instead of applying multiple t-tests, the appropriate approach is analysis of parallelism, a variant of covariance analysis or ANCOVA. In the analysis of parallelism of s data sets, we first fit the functional relation (Eq. 4.1) to all the data and obtain a residual sum of squares for regression, SSR. Following this, we fit a second model

$$y_{ij} = \alpha_j + \beta x_{ij} + \varepsilon_{ij} \quad i = 1 \ldots n, \ j = 1 \ldots s$$

with a single slope and s intercepts and compute the residual, SSI; then we compute the residual, SSS, for a third model with s intercepts and s slopes

TABLE 4.2 Analysis of variance table for establishing the homogeneity of regressions fitted to s data sets comprising $n = \Sigma NB_k$ for $k = 1 \dots s$ lines

Source of variation	Sums of squares	df	Mean squares	F
Regression	$SSR = \sum \left(y - \hat{a} - \hat{b}x \right)^2$	1	MSS = SSR	MSR/ RMS
Intercept	$SSI = \sum \left(y - \hat{a}_j - \hat{b}x \right)^2$	$s-1$	MSI = SSI/(s−1)	MSI/ RMS
Slope	$SSS = \sum \left(y - \hat{a}_j - \hat{b}_j x \right)^2$	$s-1$	MSS = SSS/(s−1)	MSS/ RMS
Residual	RSS = TSS − (SSR + SSI + SSS)	$n-2s$	RMS = RSS/(n−2s)	
Total	$TSS = \sum (y - \bar{y})^2$	$n-1$	V = TSS/(n−1)	

Significant F-ratios for intercept and slope indicate differences between the data sets. Summations are across the entire dataset.

$$y_{ij} = \alpha_j + \beta_j x_{ij} + \varepsilon_{ij} \quad i = 1 \dots n, \; j = 1 \dots s$$

The computations are summarized in the form of an analysis of variance table in Table 4.2. Detailed explanations of this analysis can be found in all regression texts, such as (Kutner et al., 2004). A particularly helpful account for biologists is to be found in the eighth edition of Snedecor and Cochran (1991, Ch. 14).

Reduced major Axis (geometric mean) regression

A common geometric mean regression slope may be estimated by minimizing the following equation:

$$\sum_{i=1}^{p} n_i \cdot S_{3,i}^2(\hat{b}_c) \cdot \left(\frac{1}{S_{1,i}^2(\hat{b}_c)} + \frac{1}{S_{2,i}^2(\hat{b}_c)} \right) = 0$$

where p is the number of regression lines and n_i is the number of data points in the ith line. The $S_i^2(\hat{b}_c)$ and C_i are the sums of squares and cross product obtained when fitting the common line to the data:

$$S_{1,i}^2(b_c) = G \cdot \left\{ S_{y,i}^2 - 2b_c \cdot C_{xy,i} + b_c^2 \cdot S_{x,i}^2 \right\}$$

$$S_{2,i}^2(b_c) = G \cdot \left\{ S_{y,i}^2 + 2b_c \cdot C_{xy,i} + b_c^2 \cdot S_{x,i}^2 \right\}$$

$$S_{3,i}^2(b_c) = G \cdot \left\{ S_{y,i}^2 - b_c^2 \cdot S_{x,i}^2 \right\}$$

where $G = (n_i - 1)/(n_i - 2)$ is a correction factor to be used if b_c is being estimated; it is unnecessary if all the n_i are large. The S^2 and C subscripts refer to the ith line and are the sums of squares and cross products of the p data sets.

The test for common regression line is a likelihood ratio test:

$$LR = -\sum_{i=1}^{p} \left\{ n_i - 2.5 \cdot \log \left(1 - r_i^2 \left(\hat{b}_c \right) \right) \right\},$$

which is distributed approximately as χ^2 with $p - 1$ degrees of freedom. This formidable system of equations can only be solved iteratively. A convenient starting value for b_c is the average of the GMR slopes of each individual regression, although more sophisticated approaches have been proposed (e.g., Krzanowski, 1984).

A simpler method using an F-test was suggested by Harvey and Mace (1982). The F-test with $p - 1$ and $n - 2p$ degrees of freedom compares fitting a common slope, b_c, estimated using the combined data of all p sets, to fitting each group with its own slope (b_i):

$$F_{p-1,N-2p} = \frac{(N - 2p)\sum_{i=1}^{p} (SS(b_c) - SS(b_i))}{(p - 1)\sum_{i=1}^{p} SS(b_i)}, \text{ where}$$

$$SS(b_i) = \frac{1}{2b_i}(n_i - 2)S_1^2(b_i)$$

is an estimate of the sums of squares for a line of slope b_i fitted to the ith group. Simulations by Warton et al. (2006) showed this method under some circumstances to be more prone to Type I errors than the likelihood ratio method. It has the advantage that it is more easily implemented in a spreadsheet. In many instances, an F statistic with $p - 1$ and $N - p - 1$ of degrees of freedom can be used to test for homogeneity of intercepts:

$$F_{p-1,N-p-1} = \frac{(N - p - 1)\sum_{i=1}^{p} n_i(\hat{a}_i - \hat{a}_c)^2}{(p - 1)\sum_{i=1}^{p} (n_i - 2)S_x^2(\hat{b}_c)}$$

According to Warton et al. (2006), this statistic is only reliable if the X-means and residual variances are similar for all lines. A more general method using a Wald statistic is described in Appendix XVI of Warton et al. Provided the X-means and residual variances are similar, the simpler least squares ANCOVA method gives adequate results. Provided the number of points per line is closely similar and if not, the correlation coefficients for each line are high, the F-statistic approach for both slope and intercept is satisfactory. For very different sample sizes, broadly different $\log(M)$ and $\log(V)$ average values and poor correlation coefficients (high residual variances) of the individual regression lines the more complex iterative method and Wald statistic are preferred.

Methods used to fit TPL

Unless stated otherwise, all analyses were conducted using GMR with common logs and estimates of $\hat{A} = \log(\hat{a})$ are on the $\log_{10}$ scale. In most cases in which TPL lines were compared, the simpler F-statistic approach was taken. Where previously published ODR values are cited, they were first converted to GMR whenever possible.

References

Adcock, R.J., 1878. A problem in least squares. Analyst (now Ann. Math.) 5, 53–54.

Bartlett, M.S., 1949. Fitting a straight line when both variables are subject to error. Biometrics 5, 207–212.

Baskerville, G.L., 1972. Use of logarithmic regression in the estimation of biomass. Can. J. For. Res. 2, 49–53.

Berkson, J., 1950. Are there two regressions? J. Am. Stat. Assoc. 45, 164–180.

Clark, S.J., Perry, J.N., 1994. Small sample estimation for Taylor's power law. Environ. Ecol. Stat. 1, 287–302.

Deming, W.E., 1943. Statistical Adjustment of Data. Wiley, New York.

Finney, D.J., 1941. On the distribution of a variate whose logarithm is normally distributed. J. Roy. Statist. Soc. B 7, 155–161.

Fukunaga, G., Takayasu, H., Takayasu, M., 2016. Property of fluctuations of sales quantities by product category in convenience stores. PLoS ONE 11 (6), e0157653.

Harvey, P.H., Mace, G.M., 1982. Comparisons between taxa and adaptive trends: problems of methodology. In: KCS Group (Ed.), Current Trends in Sociobiology. Cambridge University Press, Cambridge, pp. 343–361.

Kermack, K.A., Haldane, J.B.S., 1950. Organic correlation and allometry. Biometrika 37, 30–41.

Krzanowski, W.J., 1984. Principal component analysis in the presence of group structure. Appl. Stat. 33, 164–168.

Kummell, C.H., 1879. Reduction of observation equations which contain more than one observed quantity. Analyst (now Ann. Math.) 6, 97–105.

Kutner, M., Nachtsheim, C., Neter, J., 2004. Applied Linear Regression Models, fourth ed. McGraw-Hill, New York.

Madansky, A., 1959. The fitting of straight lines when both variables are subject to error. J. Am. Stat. Assoc. 54, 173–205.

McCullagh, P., Nelder, J.A., 1983. Generalized Linear Models. Chapman and Hall, London.

Nelder, J.A., Wedderburn, R.W.M., 1972. Generalised linear models. J. R. Stat. Soc. A. Stat. Soc. 135, 370–384.

Park, S.-J., Taylor, R.A.J., Grewal, P.S., 2013. Spatial organization of soil nematode communities in urban landscapes: Taylor's power law reveals life strategy characteristics. Appl. Soil Ecol. 64, 214–222.

Pearson, K., Lee, A., 1903. On the laws of inheritance in man. Biometrika 2, 357–462.

Perry, J.N., 1981. Taylor's power law for dependence of variance on mean in animal populations. Appl. Stat. 30, 254–263.

Perry, J.N., 1982. Fitting split-lines to ecological data. Ecol. Entomol. 7, 421–435.

Perry, J.N., 1987. Iterative improvement of a power transformation to stabilise variance. Appl. Stat. 36, 15–21.

Ricker, W.E., 1973. Linear regressions in fishery research. J. Fish. Res. Board Can. 30, 409–434.

Snedecor, G.W., Cochran, W.G., 1991. Statistical Methods, sixth ed. Iowa State University Press, Ames IA.

Sprugel, D.G., 1983. Corrections for bias in log-transformed allometric equations. Ecology 64, 209–210.

Taylor, L.R., Woiwod, I.P., Perry, J.N., 1978. The density-dependence of spatial behaviour and the rarity of randomness. J. Anim. Ecol. 47, 383–406.

Taylor, L.R., Perry, J.N., Woiwod, I.P., Taylor, R.A.J., 1988. Specificity of the spatial power-law exponent in ecology and agriculture. Nature 332, 721–722.

Wald, A., 1940. The fitting of straight lines if both variables are subject to error. Ann. Math. Stat. 11, 284–300.

Warton, D.I., Wright, I.J., Falster, D.S., Westoby, M., 2006. Bivariate line-fitting methods for allometry. Biol. Rev. 81, 259–291.

Part II

Chapter 5

Microorganisms

This chapter deals with single-celled organisms, viruses and bacteria, plus oomycete and ascomycete fungi. Although the fungi are multicelled for most of their lifecycle, they have single celled spores that are the infective stages. All four groups include economically significant plant and animal pathogens.

Free living

GYMNAMOEBAE

Amoebae are among the most abundant and diverse microbes in soil. Their taxonomy is poorly understood and identification of most species is difficult. The gymnamoebae or naked amoebae form a single functional group despite their diversity and taxonomic heterogeneity. All amoebae are essentially aquatic; those species in soil inhabit the soil water between soil particles and are secondarily terrestrial.

Soil was sampled by Anderson (2000) for gymnamoebae monthly from 1995 to 1998 near a freshwater pond on Columbia University's Palisades, NY campus. Cylindrical $0.8\,cm^2$ soil cores were taken to a depth of $2\,cm$ and the soil mixed with filtered pond water. 72 2-mL subsamples of filtrate per soil sample were cultured on agar for 14 days after which the cultures were examined and the gymnamoebae counted. Anderson identified most of the cultured gymnamoebae to genus and separated them into four morphotypes based on their shape and style of locomotion (fan- or discoidal-shaped, eruptive locomotion, noneruptive motion, or locomotion by protruding subpseudopodia). The number of each morphotype was converted to density (#/g soil) and the average computed from 72 replicates for 43 sampling occasions. Density of gymnamoebae ranged 156–5838/g of soil with an overall mean of 1600 ± 190/g ($\pm$SE). Anderson examined several environmental variables but only precipitation correlated significantly with abundance of gymnamoebae ($P < 0.02$). The El Niño winter of 1997–98 was unusually mild and moist with density of gymnamoebae approximately four times that of the other more normal winters. They also tended to be larger in the El Niño winter.

A spatial TPL analysis of morphotype densities given in Anderson's Table 1 in Fig. 5.1 (Appendix 5.A) with morphotype as the quadrat ($NQ = 4$, $NB = 43$) found gymnamoebae to be superaggregated with $b = 2.44 \pm 0.121$. The

Taylor's Power Law. https://doi.org/10.1016/B978-0-12-810987-8.00005-7

69

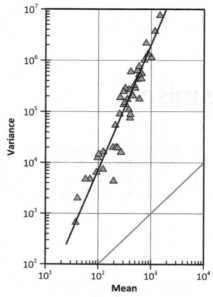

FIG. 5.1 Spatial community TPL ($NQ = 4$, $NB = 43$) morphotypes of gymnamoebae in soil near a freshwater pond are superaggregated. *(Data from Table 1 in Anderson (2000).)*

numerical difference in morphotype abundance apparently increases through time with the most abundant morphotype (fan- or discoidal-shaped) growing relatively faster than the others and the least abundant (noneruptive motion) increasing most slowly.

BACTERIA IN A SIBERIAN RESERVOIR

Cyanobacteria are blue-green prokaryote photosynthetic bacteria. They are found in both fresh and salt water and in moist soils. They may be free living or symbiotic with fungi as lichens, unicellular, filamentous, or laminate in form and many have resistant spores called akinetes to survive hostile conditions. Phosphorus runoff from fields into water bodies stimulates cyanobacterial blooms that are toxic, endangering other aquatic species as well as humans. Akinetes resting in littoral sediments are thought to be the main source of blooms exploiting agricultural runoff. The factors influencing their distribution and abundance are therefore important for predicting and managing harmful bacterial blooms.

Kravchuk et al. (2011) examined akinete distribution and abundance in littoral sediments in Bugach reservoir near Krasnoyarsk, Siberia, with a view to predicting the abundance of akinetes in water bodies to help prevent the cyanobacterial blooms. Three sediment cores were taken from the middle of the reservoir in June 2004 and June 2006. A variable number of subsamples from the top 1 cm and a 1 cm slice at 5 cm were incubated in flasks containing filtered reservoir water for 27 days in 2004. In 2006, the top 1 cm and a slice from 10 cm were incubated for 17 days but at a range of concentrations. In both years, phytoplankton were identified to species or genus. 39 species were identified in 2004 and 27 in 2007: a total of 22 taxa of chlorophyta, 11 taxa of cyanophyta, 8 taxa of bacillariophyta, and 3 other taxa were identified and enumerated. The biomass (μg/L) was estimated from the number and volume of each taxon. By an order of magnitude, the noxious cyanobacterium *Anabaena flos-aquae* was the most abundant species derived from akinetes stored in the sediment. Kravchuk et al.'s Table 3 lists the mean and SE biomass for 44 species of phytoplankton.

Counting each flask as a quadrat ($NQ = 12$), the standard errors in Kravchuk et al.'s Table 3 were converted to variance for a community TPL analysis of both years individually and combined. Although the methods differed and the abundance of phytoplankton in the 2 years differed and not all species were detected in both years, the TPLs of biomass of the community of phytoplankton did not differ between the years (Fig. 5.2, Appendix 5.B). The close similarity of the 2 years' TPL indicates very good agreement between the two methods used to estimate biomass. Two points, both of another cyanobacterium, *Synechocystis salina*, appear to be outliers. Relative to the rest of the phytoplankton community, *S. salina* is noticeably less variable for its mean.

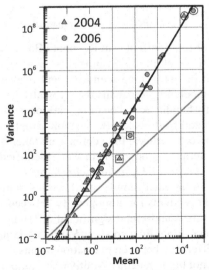

FIG. 5.2 Freshwater plankton mixed-species TPL ($NQ = 12$, $NB = 66$) of experiments conducted in 2004 and 2006 do not differ in slope or intercept. Each point is a species found in samples from a reservoir in Siberia. Circled points are of the noxious cyanobacterium *Anabaena flos-aquae* and the squared outliers are of cyanobacterium *Synechocystis salina*. (*Data from Table 3 in Kravchuk et al. (2011).*)

DIATOMS IN LAGUNA DI VENEZIA

The Venice lagoon to the east and southwest of Venice is $\sim$550km^2 and exchanges $\sim$60% of its water with the Adriatic every tide. Except where canals are maintained for shipping the depth is $\sim$1 m. Facca and Sfriso (2007) investigated the benthic diatom abundance in the Laguna di Venezia as part of a study of trophic food webs in shallow coastal waters. In 2003, they took two 80cm^2 sediment cores every month at six sites in the Venice lagoon. The surface layer of the sediment samples was subsampled and the subsamples diluted with synthetic seawater, fixed and the diatom cells identified, and counted in settling chambers. The abundance of diatoms in the lagoon showed little seasonal variation and variation at the six sample sites were not synchronous. Nor were they apparently correlated with temperature or light variation.

Facca and Sfriso provide monthly means and SDs of diatom abundance (millions/cc) at each sample site. An hybrid ensemble TPL of diatom abundance (millions/cc) at the $NQ = 6$ sites over $NB = 12$ months is entirely below the Poisson line (Fig. 5.3; Appendix 5.C). The top and right-hand axes of Fig. 5.3 show the same data expressed as number/μL above the hatched Poisson line. The pattern, slope, and correlation coefficient are unchanged by the change in scale. Both A and $SE(A)$ change as the scatter of points is moved up and to the right (Appendix 5.C) causing the means of log(M) and log(V) to increase. The variances of log(M) and log(V) are invariant under scaling and so do not change.

FORAMINIFERA IN DELAWARE

Buzas' (1970) study of foraminiferans in Rehoboth Bay, Delaware, was one of the data sets used by LRT et al. (1978) in their survey of 102 species' spatial distributions. The set includes three identified and one unidentified species. Foraminiferans are amoeboid protozoans that typically have an external shell or test. They are mostly benthic found on or in seafloor sediment although some are found in brackish and fresh water. Buzas took samples at each vertex of a 4×4 grid of 10m between vertices. Five cores were taken at each station (vertex) and the foraminiferans in the top 1 cm were identified and counted. The number of *Ammonia beccarii*, *Elphidium clavatum*, *E. tisburyensis*, and an unknown *Elphidium* sp. in each sample at all sample stations is given in his Table 1. Buzas tested his counts for homogeneity between stations by analysis of variance of log numbers and found significant differences between stations for all but the unknown *Elphidium* species. Multivariate analysis of variance found a significant difference between the station means confirming the multi-species population is not homogeneous. Multivariate canonical analysis identified the heterogeneity was due to a single station's very high values.

Means and variances per station ($NQ = 5$) for each species at each station ($NB = 16$) were used to compute ensemble TPLs for each species. Two species were superaggregated with $2.50 < b < 2.83$, one is aggregated ($b \approx 2$) and the comparatively rare *Elphidium* sp. is moderately aggregated ($b = 1.66 \pm 0.22$)

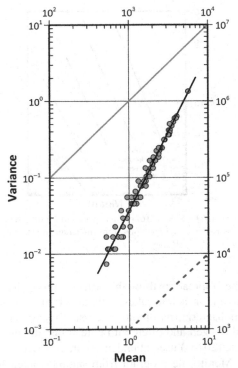

FIG. 5.3 Hybrid temporal TPL ($NQ = 6$, $NB = 72$) of diatoms in the Laguna di Venezia do not change in appearance, as the sample units are changed from millions/cc (lower and left axes, solid Poisson line) to number/μL (upper and right axes, hatched Poisson line). As the means and variances increase with rescaling, the intercept, A and $SE(A)$ increase as the points move up and right in the V-M domain, but r, b and $SE(b)$ are unchanged. *(Data from Fig. 2 in Facca and Sfriso (2007).)*

over a density range of ~1.5 orders of magnitude (Fig. 5.4; Appendix 5.D). The TPL results suggest that the log transformation may not have been adequate for this species: with $b = 1.66$, a transformation of $z = x^{(2-b)/2} = x^{0.17}$, where x is the count at a station, would have been more appropriate (Healy and Taylor, 1962; Chapter 14). One station stands out as a having a very high mean and variance, establishing the very strong aggregation of all but *Elphidium* sp. We note that with $NQ = 5$ being less than the recommended 15 samples, these results may be subject to bias (Clark and Perry, 1994).

INVASIVE FLAGELLATE IN SWEDEN

Flagellates are protozoans with one or a small number of long whip-like hairs called flagella that are used for locomotion. Lifestyles include autotrophic, heterotrophic, parasitic, and symbiotic. *Gonyostomum semen* is an invasive

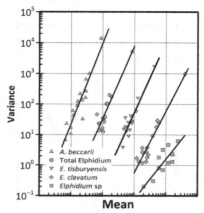

FIG. 5.4 Ensemble TPLs ($NQ = 5$, $NB = 16$) of four species of foraminifera in seaflore sediment. The four species show a range of aggregations from moderate to superaggregated. *(Data from Table 1 in Buzas (1970).)*

flagellate that has been spreading through northern Europe. It is a phytoplankton responsible for blooms in acidic lakes, although it also occurs in higher pH lakes. Unlike cyanobacteria, its blooms are usually associated with the high levels of dissolved oxygen and often with cold water streams and lakes.

Trigal and Ruete (2016a) used abundance data of *G. semen* obtained by the Swedish National Monitoring program from samples taken from 76 lakes in Sweden in 1997–2010. Their purpose was to investigate the influence of environmental stochasticity on the persistence and establishment of *G. semen* with the hope of anticipating the risk of the organism becoming established and developing useful management measures.

The study lakes were mostly shallow with average depths between 1.6 and 12.6 m. Phytoplankton samples were collected from each lake once each summer using 3-cm-diameter tube sampler. One sample was taken in lakes $<1 \mathrm{km}^2$ and five samples were taken in larger lakes to account for spatial variability. Samples were taken at a depth of 0–4 m. All samples were processed separately and plankton identified and *G. semen* density estimated as #/L.

They found *G. semen* population dynamics differed between 50 lakes of the southern lowlands and 26 lakes of the boreal forest of the Fennoscandian region to the north. Blooms were larger and more frequent in the south and smaller and less regular in the north. The difference was largely due to climatic differences and larger production of resistant cysts in the longer warmer summers in the south.

Data of *G. semen* abundance in 76 lakes over 14 years were retrieved from the Dryad Digital Repository (Trigal and Ruete, 2016b). The density data were converted from #/L to #/10cc and analyzed both spatially and temporally (Fig. 5.5, Appendix 5.E). The temporal plot of $NB = 76$ lakes ($NQ = 14$) extends

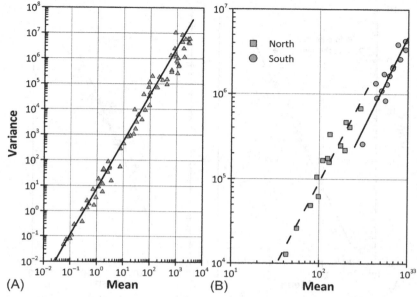

FIG. 5.5 Temporal (A) and spatial (B) TPLs of the abundance (#/10cc) of the invasive flagellate *Gonyostomum semen* in 76 lakes in Sweden over 14 years. The degree of aggregation in northern and southern lakes differ slightly, but not significantly. *(Data from the Supplement (Trigal and Ruete (2016b) to Trigal and Ruete (2016a).)*

over 5 orders of magnitude with a slope of $b = 1.76 \pm 0.026$ is moderately aggregated: the spatial plot ($NQ = 76$, $NB = 14$) is superaggregated with $b = 2.39 \pm 0.21$ but short, extending over <1 order of magnitude. Separating the north and south lakes we see the strong influence of the southern lakes on the spatial plot (Fig. 5.5B). The northern lakes' regression has a slightly lower slope and slightly higher correlation than the southern lakes but the slopes are not significantly different ($P > 0.45$).

CILIATES IN THE EAST CHINA SEA

Ciliates are protozoans with short hair-like organelles called cilia that undulate to provide locomotion. Cilia are very similar to flagella but are much shorter and generally cover a substantial proportion of the cell membrane.

The Three Gorges dam on the Yangtze River became operational and began filling in July 2003 changing the flow of freshwater into the East China Sea. Not only did the dam reduce the flow of freshwater, but it probably altered the flow of nutrients into the sea. Tsai et al. (2011) had sampled the East China Sea offshore from the Yangtze delta in July 1998 and did so again in June and August 2003 and again in 2004, 2005, and 2007. Following the damming of the

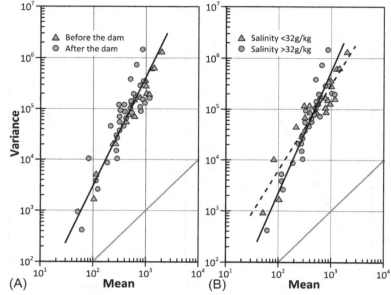

FIG. 5.6   (A) There was no discernable change in spatial TPL ($NQ = 17$–30, $NB = 53$) of hetero-trophic ciliates in the East China Sea before and after the Three Gorges dam went online in July 2003. (B) A strong difference in spatial distribution is evident above and below the 32 g/kg isohaline, which moved inshore after July 2013. *(Data from Fig. 4 in Tsai et al. (2011).)*

Yangtze, freshwater discharge declined causing the isohalines to move inshore. Tsai et al. examined the distribution and abundance of heterotrophic ciliates to determine if the marine environment was being altered by the change in fresh-water discharge. They sampled microorganisms at five depths in two zones dif-fering in their salt concentration (above and below 32 g/kg water).

Depending on the year, 17–30 samples were taken on 4–8 transects perpen-dicular to the shore at 5 depths to 100 m. Heterotrophic ciliates were identified and counted in each sample and are given in Tsai et al.'s Fig. 4 as mean (1000 cells/m^2) and SD of ciliate density at five depths on seven sampling occa-sions from July 1998 to July 2007 in zones above and below the 32 g/kg isoha-line. A number of depths yielded no ciliates, resulting in $NB = 53$ points, of which 17 were pre-dam (Fig. 5.6A). Although there are fewer points before the dam went online, the range of means before and after is closely similar (Appendix 5.F) and the spatial distributions are not significantly different ($P > 0.25$). However, the spatial distributions above and below the 32 g/kg iso-haline (Fig. 5.6B) are significantly different ($P < 0.002$) indicating a much higher level of aggregation at higher salt concentrations ($b = 1.81$ below and $b = 2.57$ above the 32 g/kg isohaline). We see here how different partitions of the same data can result in different TPLs.

MARINE VIRUSES IN CALIFORNIA AND SWEDEN

Seawater samples were taken at a range of depths on transects off Santa Monica and La Jola, California, in 1990 and 1991 and at three sites in the Gulf of Bothnia, Sweden, in 1991 by Cochlan et al. (1993). They examined the vertical and horizontal distributions of viruses in relation to bacterial abundance and chlorophyll-*a*. They found higher concentrations of viruses in the upper 50 m of the water column and the largest component of the community were bacteriophages <60 nm, which correlated strongly with the abundance of bacteria.

Subsamples were centrifuged onto specimen grids for transmission electron microscope (TEM) imaging. Nonfilamentous viruses were counted in five size classes. The smallest class (<30 nm) was thought to have been underestimated because viruses this small could not be distinguished clearly at TEM resolution. Cochlan et al.'s Tables 2 and 3 give the mean density in millions per ml of seawater in the five size classes of samples from California and Sweden, respectively.

The mean and variance of the density per size class ($NQ=5$) was computed for 32 California samples taken at depths 0–900 m and 13 samples from Sweden at depths 0–210 m. Ensemble TPL of $NB=45$ observations over a range of depths with means over 2 orders of magnitude has a slope of $b=2.0$ (Fig. 5.7; Appendix 5.G). The data from Sweden have a narrower range of

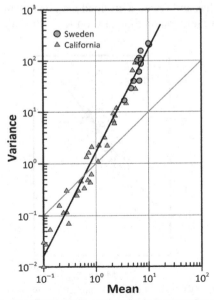

FIG. 5.7 The ensemble TPL ($NQ=5$, $NB=45$) of marine viruses from the Southern California bight and the Gulf of Bothnia, Sweden, forms a single line. *(Data from Tables 2 and 3 in Cochlan et al. (1993).)*

means but extend the line and increase the correlation. Taken alone Sweden's TPL is significantly steeper than California's ($P < 0.05$). However, its narrower range and $NB < 15$ may have introduced bias, while its excellent fit with and extension to the California data argue for a single TPL using quadrats based on size class.

Plant hosts

TOBACCO MOSAIC VIRUS ON BEANS

The number of tobacco mosaic virus (*Tobamovirus* sp.) lesions on bean leaves was one of the studies in LRT61. The data used were taken from Bliss and Owen (1958) who used Kleczkowski's (1949) data as an example of estimating common k_c of the negative binomial. Kleczkowski's original data of individual lesion counts, presented in his Table 1, are of the number of lesions on the top 4 half-leaves of 120 young bean (*Phaseolus vulgaris*) plants 4 days following inoculation. Bliss and Owen grouped the lesion counts into 17 class intervals from 5 to 275 with 3 to 15 plants per class ($NQ = 12–60$ half-leaves). Their Table 5 gives the means for each group and an estimate of $1/k$ from which variance may be computed using Eq. (2.6b). Fitting TPL using GMR, the estimate of $b = 1.367 \pm 0.081$ (triangles in Fig. 5.8; Appendix 5.H) is very close to LRT61's ODR estimate of $b = 1.40$.

Fitting TPL to Bliss and Owen's grouped data results in a correlation coefficient $r = 0.97$. However, conducting TPL on Kleczkowski's raw data (dots in

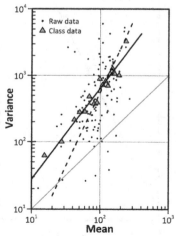

FIG. 5.8 Ensemble TPL of Kleczkowski's (1949) tobacco mosaic virus lesion data arranged in $NB = 17$, $NQ = 12–60$ class intervals (*triangles*) from Table 5 in Bliss and Owen (1958). Fitting a TPL to lesions per half-leaf ($NQ = 4$) on $NB = 120$ plants (*black dots*) forms a cloud of points with much greater variability about the fitted (*dashed*) line, and a steeper regression.

Fig. 5.8) results in a fit with $r = 0.51$ of average number of the lesions on $NQ = 4$ half-leaves and $NB = 120$ plants (Fig. 5.8). Computing means and variances on total lesions per leaf ($NQ = 10$ half-leaves) is no better ($r = 0.44$). Both fits to the raw data result in slopes substantially greater than Bliss and Owen's grouped data in part as a result of the low correlation coefficient, the likely bias due to the $NQ < 15$.

VERTICILLIUM DAHLIAE IN POTATO FIELDS

Verticillium wilt is a plant disease caused by a complex of ascomycete fungi in the genus *Verticillium*. Symptoms include chlorosis and necrosis of leaves, discoloration in stems and roots, and wilting on warm, sunny days. Severely diseased plants may be stunted or die. The complex has a wide host range, including both annual and perennial plants. *Verticillium dahliae* causes economic losses in crops throughout the temperate zones. Its spores (microsclerotia) can persist in soil for many years in the absence of a host, making management difficult; consequently, it is one of the most studied fungal pathogens.

In two studies, Wheeler et al. (1994, 2000) sampled potato fields in Ohio to establish the spatial distribution of *V. dahlia* for efficient sampling programs for the disease. In the 2000 study, each of 6 commercial potato fields were divided into contiguous quadrat sizes of 10, 50, 100, 250, 1000, and $4000\,m^2$ with the smaller quadrats nested within the larger ones (2 fields were too small to receive $4000\,m^2$). Twenty quadrats of each size were randomly selected for soil core sampling. Within each selected quadrat, 20 $5\,cm^2$ by $18\,cm$ cylindrical cores were taken and combined, resulting in 100 or 120 samples per field. A 1000-mL subsample was taken and returned to the laboratory where dilution plate assays were conducted and the number of colonies after 13 days' incubation counted to estimate the number of spores/cc of soil. The means and variances of *V. dahlia* colonies per quadrat size per field are given in Wheeler et al.'s Table 1. Wheeler et al. fitted ensemble spatial TPLs for each quadrat size separately by ODR: analysis of parallelism showed the slopes to differ. The TPL analysis is repeated here using GMR (Appendix 5.I). The slopes declined from 2.34 at $10\,m^2$ to 1.55 at $500\,m^2$ and increased again to 3.42 at $4000\,m^2$, suggesting aggregation changes with sample spacing.

Note that only the spacing of the samples varied. Although the fields were divided into quadrats of different sizes and the samples taken over different areas, the *number* of samples and *volume* of soil assayed were the same for all quadrat sizes and fields. With only $NB = 6$ or 4 points over very restricted ranges, such a conclusion can only be tentative. An ensemble TPL combining all 34 field-quadrat points results in a TPL extending over 2 orders of magnitude of means with a slope of $b = 2.48 \pm 0.15$ (Fig. 5.9; Appendix 5.I).

In a parallel study, Wheeler et al.'s Table 4 shows the number of colonies in assays of 7 different fields divided into 2000 or $4000\,m^2$ quadrats from which

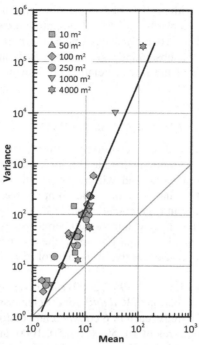

FIG. 5.9 *Verticillium dahliae* microsclerotia/cc of soil in potato fields are highly aggregated when samples in all quadrat sizes are combined in a single ensemble TPL (*NQ* = 20, *NB* = 34). *(Data from Table 1 in Wheeler et al. (2000).)*

cores were taken in spring (April/May) and autumn (September/October). A composite sample of 20 soil cores taken from each of 20 quadrats per field was assayed for *V. dahlia* as before and the spatial distributions in spring and autumn compared.

The purpose of Wheeler et al.'s studies was to use TPL to determine the number of samples required to estimate *V. dahlia* density with a defined precision for pest management sampling (see Chapter 14). They found aggregation in spring was higher than in autumn, requiring four to five times the number of samples in spring than autumn. Counting the spring points on the Poisson line underestimates the spring TPL slope (Appendix 5.I), suggesting the difference between spring and autumn may be even greater.

Comparison of the 2000 and 1994 studies shows the slope $b = 1.745 \pm 0.250$ for the autumn samples in the 2000 study are almost identical to the value obtained in their 1994 study ($b = 1.755 \pm 0.196$). The intercepts differ as might be expected from two studies conducted slightly differently and in different locations. In the 1994 study, 2 potato fields were sampled per year from 1987 to 1991 along transects split into $NQ = 22$–107 contiguous plots (depending on field size). From plot 10–12 2.5 cm² by 20–30 cm cylindrical soil cores were taken prior to planting. The cores from each plot were composited and assayed for *V. dahlia* microsclerotia. Colonies were counted following incubation and the preplant density of *V. dahlia* expressed as microsclerotia/ cc of soil. It is not obvious why the spring and autumn TPLs in the 2000 study are so different unless they are due to the small number of points and limited range of means.

PASSALORA FULVA ON TOMATOES

Kawaguchi and Suenaga-Kanetani (2014) analyzed the distribution of green-house tomato plants infected by tomato leaf mold (*Passalora fulva*) by TPL and Iwao's model (Eq. 3.26). Their purpose was to distinguish between 2 models of leaf mold population growth: increasing population by adding new colonies of fixed size; or by increasing the size of a fixed number of colonies. They examined tomato plants in 17 commercial greenhouses at intervals from June to September in 2009 and 2010. The greenhouses, consisting of 3 pairs of rows, were divided into 9 quadrats of ∼36 plants each. On each visit, the total number of infected plants and newly infected plants per quadrat were recorded separately. Means and variances for 50 sets of observations are reported in supplementary Table S2.

Ensemble analysis of the supplementary data shows the TPLs for total number and newly diseased plants (Fig. 5.10, Appendix 5.J) are not significantly different in either intercept ($P > 0.80$) or slope ($P > 0.75$). Kawaguchi and

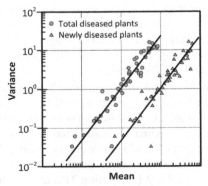

FIG. 5.10 Hybrid spatial TPLs ($NQ = 9$, $NB = 50$) of the number of tomato plants newly infected and total infected with tomato leaf mold in 17 commercial greenhouses in Japan. The distributions are almost identical and moderately aggregated ($b = 1.3$). Nearly half the points in both plots have $M \leq V$, but there is no indication of a change in slope at $V = M$. Newly diseased plants are offset by 1 cycle for clarity. *(Data from supplementary table S2 in Kawaguchi and Suenaga-Kanetani (2014).)*

Suenaga-Kanetani's analysis by Iwao's regression method found the regressions to be almost identical also (total diseased plants: $M^* = 0.104 + 1.101 M$, $r^2 = 0.974$ and newly diseased plants $M^* = 0.144 + 1.091 M$, $r^2 = 0.953$). Both the Iwao-Lloyd model and TPL fit the data equally well and both models showed an increase in aggregation with density. Approximately 40% of both total and newly diseased plants had $M \geq V$ with no evidence of curvature in the TPLs for total or newly diseased plants. Kawaguchi and Suenaga-Kanetani concluded on the basis of Iwao's regressions that newly infected plants form new independent foci during outbreaks—the colony increase model.

MUMMY BERRY DISEASE OF BLUEBERRIES

Blueberry fruit infected with mummy berry disease caused by the ascomycete fungus *Monilinia vaccinii-corymbosi* bring a substantial reduction in economic return to the producer as a very small number in a load of $\sim$200 L would cause it to be rejected. An efficient sampling procedure with high confidence of detecting infected fruit is therefore desirable.

Copes et al. (2001) developed a sequential sampling plan using Iwao's (1968) regression using Lloyd's mean crowding (Eq. 3.18) to describe the heterogeneity of mummies per load at a packing facility. They took 20–100 550cc samples of fruit from 23 loads in June–July of 1997 and 1998. The fruit were examined and the number of infected fruit recorded. The number of fruit per sample varied with fruit size, but averaged 270 fruit/550cc sample. Copes et al. computed mean crowding (Eq. 3.19) from the mean and variance of number of infected fruit per sample and computed Iwao's patchiness index by regressing M^* on the mean (Eq. 3.26). The parameters a and b of Iwao's index were used to specify the heterogeneity (clustering) of mummies in the derivation of upper and lower stop limits. The sequential sampling plan determined whether the incidence of infected fruit is above or below the threshold of 1.25 infected fruit per sample of 250 fruit at a significance level of $P < 0.05$. Because fruit size determines the number of fruit per sample, they also performed a sensitivity analysis for different sizes of berries.

Copes et al. give their data of fruit per sample, mean and variance of mummies per sample as well as aggregation statistics based on negative binomial k and M^* in their Table 1. They note that in about half their samples the mean and variance are approximately equal, that is, Poisson, over a range of densities. Also in eight cases, k could not be computed. Fig. 5.11 shows the spatial TPL of Copes et al.'s data with half the points following the Poisson line and half with a slope of $b = 1.54 \pm 0.152$. Fitting TPL to all points lowers the estimated slope to $b = 1.29 \pm 0.052$ (Appendix 5.K). The parameters describing aggregation are important ingredients for designing sampling plans; underestimating the degree of aggregation will result in increased risk of rejecting an otherwise acceptable batch (Type 1 error).

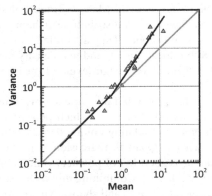

FIG. 5.11 Spatial TPL of mummy berry of blueberries, (caused by *Monilinia vaccinii-corymbosi*) is Poisson up to about 0.6 mummy berries per sample. Above $M = 0.6$, mummy berries are more aggregated with $b = 1.43$ ($NQ = 20–100$, $NB = 23$). Failure to allow for the change in slope at the Poisson line underestimates the degree of aggregation. *(Data from Table 1 in Copes et al. (2001).)*

POWDERY MILDEW ON APPLES

Powdery mildew is a disease affecting a wide range of plants caused by asco-mycete fungi in the order Erysiphales. A common disease agent of agricultural crops, particularly pome fruit, is *Podosphaera leucotricha*. Symptoms include white powdery spots on the leaves and stems, particularly the lower parts. The spots grow larger and more numerous with time as more asexual spores are produced and spread over the plant. Disease severity is typically quantified in two ways: incidence and intensity. Intensity in this case is equivalent to density and is expressed as the number of lesions, colonies, or other observable symptom per plant part (flower, fruit, leaf, shoot, etc.), while incidence is expressed as the proportion of plant parts supporting symptoms. Generally, it is easier to assess incidence while density estimates provide more information for use in pest management decision making.

Xu and Madden (2002) sampled apple (*Malus pumila*) orchards for powdery mildew in Kent, England, annually in 1994–97. Their objective was to examine relationships between incidence and severity at two scales (shoot and leaf) and develop a simple and robust method for predicting intensity using only inci-dence on leaves or shoots. In May of each season, 10 trees in 2 orchards were selected and 4 shoots per tree tagged for sampling. Samples were taken approx-imately three times per week from late May to mid-August. On each sample date, the top four leaves on each tagged shoot were examined for powdery mil-dew. The samples consisted of number of powdery mildew colonies per leaf on 4 leaves per shoot, 4 shoots per tree, and 10 trees for a total of 320 leaves per orchard on each sample date: they took 22, 25, 14, and 17 samples per year, 1994–97. From these data, incidence was determined from the proportion of leaves and shoots with one or more colonies.

To characterize their data, Xu and Madden fitted Poisson, negative binomial, Polya-Aeppli, and Neyman's A distributions (Chapter 2) to the frequency
data of mildew on leaves and shoots. They also fitted TPL to characterize variability in density at both scales. They found that density data from their study
were well fit by TPL except at very high levels of mean density as occurred
in 1997.

Omitting the 1997 data, they found the variance-mean ratio of colony density was more or less constant over time, which permitted them to establish
regressions with high precision relating colony density and disease incidence
on both shoots and leaves. Quantal regression using the complementary
log-log transformation gave good fits of powdery mildew incidence on lesion
density for both shoots and leaves. Combining this with a strong binomial relationship between incidence on leaves and shoots, Xu and Madden were able to
predict colonies per leaf from disease incidence per shoot. Using the leaf
incidence-density relationships they developed, it is possible to make practical
disease management decisions based on powdery mildew incidence on shoots.

Drs. Xiangming Xu and Larry Madden have generously made their data
available for reanalysis by GMR. Ignoring the 1997 data and fitting TPL to
the first 3 years' data results in an excellent fit with the variance almost equal
to the mean at all densities. The TPL estimates of $b = 1.08 \pm 0.027$ ($r = 0.98$) for
leaves (Fig. 5.12A, Appendix 5.L), although very nearly Poisson are significantly different from $b = 1.0$ ($P < 0.01$). Even this small difference would not
justify an assumption of $V = M$ as Xu and Madden discovered. Assuming the
colonies per leaf to be Poisson, their models overestimated disease incidence.
The slightly aggregated pattern of the colonies per leaf detected by TPL was

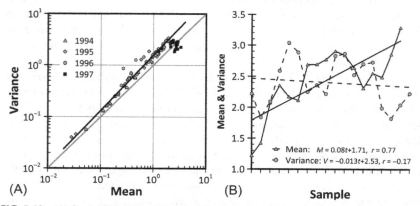

FIG. 5.12   (A) Spatial TPL ($NQ = 320$, $NB = 59$) of powdery mildew density on apple tree leaves,
excluding the 1997 data. (B) Mean and variance of the 1997 data show the mean continues to
increase through the season, while the variance initially tracks the mean and then oscillates and
declines slightly as available space on the leaves disappears. *(Data used in their 2002 paper generously made available by Drs. Xiangming Xu and Larry Madden.)*

confirmed by the distribution analysis that showed the aggregated distributions fit better than the Poisson.

The anomalous results of 1997 deserve special attention. In 1997, powdery mildew infected the orchards very early and spread rapidly: incidence reached 80% by early June and remained high, reaching 100% in August. As a result, the number of colonies per leaf ranged from ~ 1.2 in early June to 3.3 in August, almost double the number in previous years. Xu and Madden attributed the lack of fit of TPL for 1997 to a limitation of the leaves to sustain more colonies. This is certainly the case. It is equivalent to trap saturation (see Chapter 15) in which the ability of the sampler, in this case the leaf, to accept any more targets (colonies) declines with increasing mean density (Fig. 5.12B). As the mean increases beyond a critical value, the difference between samples declines as more samplers become saturated and the variance eventually drops to zero when all samplers are saturated. If the maximum number of colonies per leaf is constant or known, Hughes and Madden's (1992) binomial power law (Chapter 16) could be used to describe the heterogeneity.

PEAR SCAB

Pear scab caused by the ascomycete fungus *Venturia nashicola* creates black lesions on pear shoots, leaves, and fruit. It is a serious economic problem in all temperate and subtropical regions where pears are grown. Ascospores overwinter on infected leaves and are released over several months starting in spring, resulting in new infections. Good orchard hygiene practices can reduce yield losses, but when management is required, understanding the relationship between incidence and density is required for efficient prediction and effective management.

Following the methods of Xu and Madden (2002), Li et al. (2007) counted scab lesions on pear (*Pyrus ussuriensis*) shoots, leaves, and fruits caused by *V. nashicola* on 10–20 trees in 3 orchards in 2 regions of China at intervals of 7–14 days in 2002–04. Their data consisted of 58 datasets of lesions on leaves and 48 on fruit. In addition, they used data of scab density on 1-year old plants in two greenhouse experiments under controlled environments (212 datasets). Altogether, they used 318 datasets from 6 experiments to develop a procedure for estimating pear scab density from incidence data. Like Xu and Madden, Li et al. fit the Poisson, negative binominal, Polya-Aeppli, and Neyman's A distributions to their data. All but eight failed to fit the Poisson distribution; the negative binomial fit the most datasets, but a number of datasets with extremely large number of lesions per leaf or fruit were too skewed to the right to fit any of the distributions. Consequently, they based their models on complementary loglog transform regressions. From regressions of scab incidence on scab density, parameter estimates for leaves and fruit were obtained and also of incidence on shoots and leaves. There were small but insignificant differences in model parameters at the two locations, permitting a single model to predict scab density from incidence on shoots.

Li et al.'s TPL analysis by ODR showed moderate aggregation in the number of lesions per leaf ($b = 1.47 \pm 0.016$) over 8 orders of magnitude of mean (Appendix 5.M). The data included the field sample and greenhouse data of lesions/leaf and field data of lesions/fruit. They used analysis of parallelism to compare field and greenhouse, the two regions, fruit and leaves, and 3 years. Although they found some heterogeneity in the TPL parameters, the improvement in fit of separate lines added only 2% to the explained variance. Combining all the data from the six experiments accounted for 95% of variance in $\log(V)$. They noted a slight curvature in their V-M plot, but this is due to the absence of variances less than the means between 3×10^{-5} and 0.028: the points in this range follow the Poisson line precisely. Given the large number of points above $M = 0.028$ and the high correlation ($r = 0.98$), bias in the estimation of A and b is immeasurable.

The practical upshot of the steeper TPL and longer right-hand tails of pear scab distribution compared to apple powdery mildew was the necessity of using the complementary log-log regression to relate incidence of scab on leaves and shoots instead of the binomial model used by Xu and Madden. With this difference, Li et al.'s model for predicting pear scab density could be used in making practical disease management decisions.

PHYTOPHTHORA

Phytophthora blight and root rot are diseases of plants in the Cucurbitaceae, Fabaceae, and Solanaceae families caused by oomycete fungi in the genus *Phytophthora*. The diseases occur commonly in poorly drained fields where standing water persists for several days. The blight discolors the roots and causes seedlings to topple over. In beans, foliage becomes water-soaked with necrosis of stem and pods and in peppers stems display dark, water-soaked areas that girdle the plant resulting in death. Hyphae grow on fruit, which may become mummified. *Phytophthora* spores can remain in the soil for several years.

On peppers and soybeans

Phytophthora blight of peppers (*Capsicum annuum*) and root rot of soybeans (*Glycine* max) are caused by *P. capsici* and *P. sojae*, respectively. *P. capsici* and *P. sojae* oospores are the fungus' overwintering structure. They germinate with low frequency on selective media resulting in poor density estimates from cultures which hampers disease management decision making. Miller et al. (1997) used an enzyme-linked immunosorbent assay (ELISA) for these species to determine the heterogeneity of populations fields with a view to developing sampling programs for these diseases in pepper and soybean.

An extensive survey of 64 soybean fields was made with 6 soil samples taken per field in 1990–1991. Four soybean fields and four pepper fields were intensively sampled on an 8×8 grid per field in 1991. The crops were also assessed for incidence of phytophthora soybean root rot or pepper blight.

Soil cores were taken with a golf course cup cutter and organic matter, including *Phytophthora* life stages, were extracted by flotation from 50 g subsamples. The extracts were subjected to ELISA analysis and the number of *Phytophthora* antigen units (PAU) determined where 1 PAU is roughly equivalent to a single oospore. Two ELISA tests were conducted on each 1990 sample but only one in 1991, making for 636 tests in all. The mean PAU values for fields in which root rot or blight had been moderate to severe were higher than in fields in which disease incidence had been low or not observed.

Miller et al. analyzed the spatial distribution of PAUs by TPL ($NQ = 6$, $NB = 106$) and the intensively sampled fields also by spatial autocorrelation. Spatial autocorrelation coefficients were not significant, indicating there was no relationship between PAU values in the intensively sampled fields at the scale of sampling. TPL showed the distribution of PAUs (oospores) to be highly aggregated at both sampling scales. Their Tables 1 and 2 give the mean and variance ($NQ = 64$) of ELISA results for intensively sampled soybean and pepper fields, respectively. Their Fig. 1 shows a TPL plot of PAUs from the extensively sampled soybean fields in 1990 and 1991. Miller et al.'s data have been reanalyzed by GMR (Appendix 5.N) and the extensive and intensive data plotted together (Fig. 5.13); the fitted line is based only on the extensive data.

Based on TPL, they estimated the optimum sample size to estimate the mean density would in most cases be 20. The GMR estimate of $b = 2.23$ instead of the ODR estimate of $b = 1.97$ suggests a higher number might be required.

In the air

Phytophthora spores and sporangia containing them disperse on the wind. Fall et al. (2015) sampled sporangia of *Phytophthora infestans*, the causal agent of potato blight in the air in potato production areas located near Florenceville and Grand Falls, New Brunswick, in 2010–12. They used rotating-arm spore samplers at 3-m agl on the edges of eight fields at each location and operated for 4.5 h 3 days per week from June to September; $\sim$50 days sampling. The effective air-sampling rate was 21 L/min with a calculated sporangia sampling efficiency of 37%. Sporangia captured on the samplers were counted under a microscope and the number of sporangia/sampler was converted to time-averaged sporangia/m^3. They used these data to test whether a network of samplers could give early warning of potato blight risk.

The daily mean densities varied widely over the course of the summer. Maximum aerial densities recorded at Florenceville were 32.87 and 28.71 sporangia/m^3 in 2010 and 2011, respectively. At Grand Falls, the maxima were 61.37 and 24.65 sporangia/m^3 in 2010 and 2011, respectively. Samples taken in 2012 were too low to analyze. The earliest occurrences in the traps occurred a week before the first cases of late blight were detected in both years and at both locations.

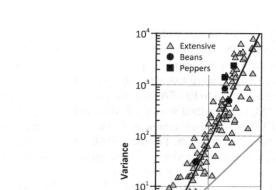

FIG. 5.13   Ensemble TPL ($NQ=6$, $NB=106$) of abundance of *Phytophthora sojae* is highly aggregated in extensively sampled soybean fields in Ohio as determined by enzyme-linked immunosorbent assay, although a small proportion of fields had variance less than the mean. Intensively sampled soils for *P. sojae* and *P. capsici* in four soybean and four pepper fields fall within the cloud of extensively sampled field data with two pepper fields on the Poisson line. *(Data from Fig. 1 and Tables 2 and 3 in Miller et al. (1997).)*

If the aerial density were homogeneous over the region, only a single sampler would be needed. To determine the spatial heterogeneity of atmospheric sporangia, Fall et al. applied TPL to the daily trap counts in both years. They found aerial density to be highly aggregated with $b=2.16\pm0.052$ and $A=-0.048$ by ODR: the GMR estimate of $b=2.25$ (Appendix 5.O).

Analysis of parallelism showed no difference between years in either slope or intercept ($P>0.15$). With $b \geq 2$, a network would therefore be required. There was a weak correlation between the aerial density and weather patterns suggesting that the earliest sporangia in the traps were due to long-distance transport from an earlier, more southerly crop. As the season progressed, the increased aerial density resulted from locally produced sporangia. This result, plus the high spatial heterogeneity, supported the conclusion that a spore-sampling network "may be a suitable approach for early detection of incoming inoculum."

These results with *P. infestans* and Miller et al.'s of *P. sojae* and *capsici* show that at both field scale and at regional scale, *Phytophthora* spores are very highly aggregated spatially. Fall et al.'s Figs. 2–4 showing the aerial density trajectories suggest that temporal heterogeneity is also high at the regional scale.

STRAWBERRY ANTHRACNOSE AND RAIN SPLASHES

Aerial dispersal of plant pathogens accounts for dispersal at all scales. At very local scales, rain splash can also spread pathogen spores. Madden (1993) conducted experiments with simulated rain to quantify the local dispersal of spores of *Colletotrichum acutatum*, the causal agent of strawberry anthracnose.

In the first experiment, Madden placed sampling plates with a culture medium for *C. acutatum* under rainshields at 20 and 60 cm from infected strawberries and exposed the fruit to simulated rainfall of 15 and 30 mm/h. Plates were exposed for 1 min every 5 min for a total of 10 exposures. The plates were incubated and the number of colonies counted, providing 10 spore densities at each distance and rainfall rate. The experiment was replicated four times. Colonies per plate in the first experiment ranged from 0 to 4000, with 95% <1000. At each distance and rain intensity, the number of colonies increased over time to a maximum around 15–25 min, and then declined. The number of colonies was considerably higher at 20 cm than at 60 cm.

The mean, variance, and Lloyd's index of patchiness (M^*) were calculated for each distance, time, rain intensity, and replicate. Lloyd's index confirmed high variability and clustering of colonies, and hence spores, resulting from splash dispersal with M^* increasing with distance from the source. The number of colonies differed significantly between the rain intensities at 60 cm, but not at 20 cm. Aggregation was confirmed by TPL with a significant slope of $b = 1.67 \pm 0.05$ by ODR.

Intercepts differed between rain intensities, but slopes depended on distance between source and sampler. The intercepts were $\ln(a) = 0.83 \pm 0.27$ and $\ln(a) = 0.71 \pm 0.33$ at 30 and 15 mm/h, respectively; the slopes were $b = 1.65 \pm 0.05$ at 20 cm but 1.77 ± 0.06 at 60 cm indicating aggregation increasing with distance (Appendix 5.P). Although slopes and intercepts were statistically significant, the differences were small, justifying the use of the common curve, $V = 1.55 M^{1.67}$. Madden used the TPL results to derive an exact power transformation of $z = x^{0.2}$ to stabilize the variance for analysis of variance (Chapter 14). Both distance and rain intensity significantly affected the number of colonies per dish.

In the second experiment, six infected fruit were arranged in a rectangle and six sampling plates were placed 30 cm from two strawberries. Plates were again exposed for 1 min every 5 min and rain simulated at 30 mm/h. This experiment was replicated three times. Fig. 5.14 shows the TPL for this experiment with a GMR estimate of $b = 1.75 \pm 0.109$. Madden concluded that the physical process of splash dispersal produces aggregation similar to that produced by other ecological processes.

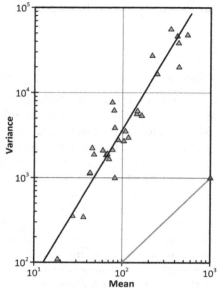

FIG. 5.14 Spatial TPL ($NQ = 6$, $NB = 30$) of Madden's second rain splash experiment in which sampling plates were exposed to splashes from infected plants during simulated rain. *(Adapted from Fig. 4 in Madden (1993).)*

Animal hosts

CILIATES ON FLATWORMS

Another example from LRT et al. (1978) is Reynoldson's (1950) account of ciliate abundance on planarian flatworms. As parasites are often hard to detect and/or count, their abundance is often given as percentage of hosts infected. The ciliate *Urceolaria mitra* lives on the flat worm *Polycelis tenuis*. They attach principally to the flat, dorsal surface of their planarian hosts and are apparently nondamaging, so are technically epizoic rather than parasitic. The fact that ciliates attached to planarians are easily counted (under a microscope) makes this system particularly useful in studying the population dynamics of two species systems. Although nonparasitic, *U. mitra* are not found free living except during dispersal.

Starting in December 1946, Reynoldson took approximately weekly samples of planarians in a small pond on the campus of Bangor University in North Wales. He found *U. mitra* infesting *P. tenuis* but also occasionally on related planarians *P. nigra* and *Dugesia lugubris*. The populations of both *U. mitra* and host *P. tenuis* were quite stable during the period the pond was sampled. The maximum number of *U. mitra* observed on a single worm was 85: the majority of planarians examined had ≤20 *U. mitra* with only 23 (~2.4%) uninfested. Reynoldson noted that the number of *U. mitra* per worm was apparently not limited by size of host as the largest *P. tenuis* could have accommodated

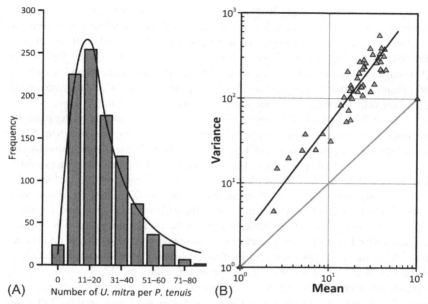

FIG. 5.15 (A) The frequency distribution of ciliates per planarian is approximately lognormal but lacks a long right-hand tail. (B) Spatial TPL ($NQ = 20$, $NB = 50$) ciliates *Urceolaria mitra* epizoic on planarian *Polycelis tenuis* are moderately aggregated ($b = 1.4$). *(Data from Table 2 in Reynoldson (1950).)*

more ciliates. The distribution of *U. mitra* per planarian is well described by the lognormal distribution at low and intermediate frequencies, but it overestimates the number of ciliates at the high end, indicating the actual distribution is less skewed than expected of a lognormal variate (Fig. 5.15A).

Of the 950 *P. tenuis* collected, 20 per sample date were examined for *U. mitra*. The number and SE of *U. mitra* per host are given in Reynoldson's Table 2. A spatial TPL analysis of *U. mitra* on *P. tenuis* (Fig. 5.15B; Appendix 5.Q) shows *U. mitra* are moderately aggregated on *P. tenuis* over 1.5 orders of magnitude of mean with $b = 1.396 \pm 0.075$. Reynoldson found wide variation in *U. mitra* numbers on worms of equal size, and no relation between the size of the worm and the ciliate population except when the *U. mitra* population peaked in June and July (when they were also found on *D. lugubris*). Younger planarians generally had fewer ciliates, but even on the smallest hosts there was apparently room to spare.

PASTEURIA PENETRANS ON MELOIDOGYNE ARENARIA

Pasteuria penetrans is a bacterial parasite of nematodes that has emerged as a useful biological control agent (BCA) for root-knot and cyst-forming nematodes, particularly those in the family Heteroderidae. Natural populations of *P. penetrans* have a near-worldwide distribution; augmentation as a biological pesticide has met with success in a number of crops. Root-knot nematodes in the genus *Meloidogyne* are among the most economically important plant-parasitic nematodes. They have a worldwide distribution and once established in a field are difficult to control. Juvenile *Meloidogyne* nematodes hatch from eggs and have a short free-living (J2) stage in the soil after which they invade the rhizosphere and attach to host plants. The host forms galls around developing juveniles where they complete their lifecycle. Females lay hundreds of eggs over a 2–3 month life span.

A study by Chen and Dickson (1997) developed binomial sampling plans to estimate *P. penetrans* endospore attachment to J2 *Meloidogyne arenaria*. They used data from previous studies (Chen et al., 1996a, b) to estimate the maximum number of individuals per sample that may be treated as zero (tally thresholds) for binomial sampling. Their data compared two methods for estimating *P. penetrans* endospores on *M. arenaria* on tomato (*Solanum lycopersicum*) roots. Data from a field experiment testing the efficacy of *P. penetrans* against *M. arenaria* in a peanut (*Arachis hypogaea*) field are also included. The 3 datasets comprised 70 and 33 estimates of *P. penetrans* on J2 *M. arenaria* on tomato roots by centrifugal flotation and incubation bioassays, respectively, and 111 estimates of the mean number of endospores attached per J2 obtained by centrifugal flotation of field samples of *M. arenaria* on peanuts following application of *P. penetrans* in efficacy trials. The datasets were analyzed by TPL and the parameters used to evaluate the utility and accuracy of tallies up to 10 endospores/J2. Several tallies were sufficiently robust to estimate the mean number of endospores attached per J2 while reducing the sampling effort.

Chen et al. used ODR to estimate TPL: their three datasets were reanalyzed by GMR (Appendix 5.R) and are plotted in Fig. 5.16. TPLs of the field

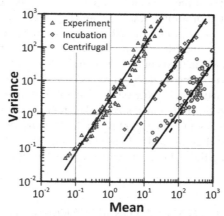

FIG. 5.16 TPLs (*NQ* variable, *NB* = 29–92) comparing three bioassays estimating *Pasteuria penetrans* endospores/J2 *Meloidogyne arenaria* nematodes. The bioassays are offset by 1 and 2 cycles for clarity. The hatched line is TPL fitted to Centrifugal bioassay without the Poisson points. (*Data from the text and Fig. 1 in Chen and Dickson (1997).*)

data and the incubation bioassay are not significantly different ($P > 0.4$). The centrifugal bioassay is marginally different from the others ($P < 0.08$) possibly due to the fact that lowest seven means lie close to the Poisson line. Ignoring the Poisson means the estimate of slope is $b = 1.74$ and not significantly different from the field experiment ($P > 0.35$) or the incubation assay ($P > 0.26$). In this instance with a lot of variation around the fitted line, the impact of Poisson points is not obvious. However, as a discontinuity at the Poisson line (Fig. 5.11) can adversely affect inferences, caution should be exercised when drawing inferences from TPL with points close to Poisson.

BACTERIA CULTURES

In an experiment to test whether competition changed spatial distribution, Ramsayer et al. (2012) compared TPLs for two species of bacteria cultured individually and in competition. *Serratia marcescens* is a human pathogen in the family Enterobacteriaceae. It has frequently been implicated in hospital-acquired infections of the gastrointestinal, respiratory, and urinary systems. Unlike *S. marcescens*, *Pseudomonas fluorescens* is not an animal pathogen but it does produce enzymes responsible for spoiling milk. It is a flagellate typically found in water and the water film of soil. Ramsayer et al. set up eight replicate sets of cultures of *S. marcescens* and *P. fluorescens* individually and together at eight concentrations of a nutrient culture medium. The media were seeded with populations of *S. marcescens* and *P. fluorescens* in proportion to the concentration of the culture media. After 24 h of incubation, the cultures were serially diluted and samples plated for 24–36 h when the number of new colonies was counted. Population densities of the original cultures were calculated and the mean and variance computed for each dilution, and ensemble TPLs ($NQ = 8$ replicates, $NB = 8$ cultures) computed for *S. marcescens* and *P. fluorescens* individually and together.

Ramsayer et al. found that the populations competing for nutrients were smaller than the solo populations, indicating competition had a negative impact on population growth for both species. They give estimates for TPL slope and 95%CI from which SE(b) and r were calculated (Appendix 5.S). The regressions over 3 orders of magnitude of mean are highly significant, with all correlation coefficients, $r > 0.97$. When cultured alone, the two species' TPLs were not significantly different ($P > 0.15$). The TPLs for both species in competition are slightly, but not significantly, greater than the TPLs for the species alone: $P > 0.25$ and $P > 0.8$ for *P. fluorescens* and *S. marcescens*, respectively.

Some theoretical studies have predicted that interspecific competition should reduce the slope of temporal TPL (e.g., Kilpatrick and Ives, 2003). Ramsayer et al. averaged over an ensemble of sites with different carrying capacities, rather than over a collection of times. This experiment may not simulate temporal TPL exactly but it is the first, and apparently only experimental, test of TPL's sensitivity to competition. The absence of a response under these conditions suggests competition per se may not affect TPL directly, although it is possible that potential competitors avoid competition by adopting spatial distributions that produce different ensemble TPLs (see Park et al., 2013, Chapter 7).

THE HUMAN MICROBIOME

In Ma's (2015) study of the human microbiome, TPL was applied to one of the largest datasets in terms of number of samples and species, the vast US National Institutes of Health (NIH) Microbiome Project database (http://hmp.dacc.org; HMP Consortium, 2012; Gajer et al., 2012). In the context of microbial ecology, "species" is often replaced by "operational taxonomic unit" (OTU). An OTU is a sequence of DNA or RNA strands representative of similar genetic sequences collected from the environment, but whose taxonomic status may be unknown. The concept of microbial species is a rather complex concept because species are hard to define using traditional species concepts of animal and plants. The term OTU was coined to represent a mathematical entity obtained by clustering various samples (from individuals) based on certain features. To a certain extent OTUs are subjective, because they are based on genetic sequences having an arbitrary degree of statistical similarity. In the case of OTUs based on 16s-rRNA sequences, a cutoff of 97% similarity in sequence reads is considered equivalent to species; Ma used the 97% similarity threshold. Environmental studies using OTUs include studies in water, soil, and plant or animal tissues, especially the gut.

An important objective for Ma was to establish whether TPL applies to the genetic traces of microbes in humans. Not only did he apply TPL to identifiable species and OTUs, he also applied it to collections of OTUs ordered as communities or multispecies sets. While multispecies and community analyses had been made before (see Polley et al. (2007) in Chapter 6 and Park et al. (2013) in Chapter 7), Ma's important contribution was to describe and formalize what he called power law extensions (PLEs) and test them rigorously. Type I and II PLEs characterize community spatial and temporal aggregation (heterogeneity and stability), and Types III and IV characterize the heterogeneity and stability of mixed-species populations. His results are summarized in Appendix 5.T.

Ma's first test, based on data of Gajer et al. (2012), was of 40 taxa, most identified to genus, in $15 \leq NB \leq 32$ subjects (3 taxa were <10) of the human vaginal microbiome with $NQ \approx 30$ observations/subject taken over a 3-month period. All 40 temporal TPLs were highly significant with $0.951 \leq r \leq 0.998$ and $1.4 \leq b \leq 2.0$ (Fig. 5.17A). Having established a baseline, Ma tested his PLEs with the NIH data.

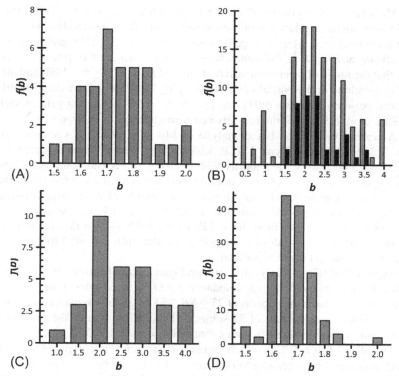

FIG. 5.17 Frequency distributions of b for TPL and PLE analyses of data from the NIH Human Microbiome Project. (A) TPL slopes for populations of 40 microbe genera from the human vaginal microbiome range between $1.5 < b < 2.0$. (B) Type I PLE of $NQ = 9$ location-based samples of 140 subjects, of which 43 had $NB > 14$ points. Both the complete set and the sub-set with $NB > 14$ points (black bars) show a range of $1.0 > b > 3.5$ with the median value of $b = 2.15$. (C) Type II PLE slopes for community temporal stability of the human vaginal microbiome (D) Type III PLE for mixed-species population spatial aggregation with location-based samples of 9 locations of oral microbiome. *(Data from Table 3 and Supplementary Tables S2–S3 in Ma (2015).)*

Type I PLE is the TPL of mean and variance of individuals per species of NQ species found at a particular site in NB subjects. Type I PLE is therefore an ensemble spatial TPL measuring community heterogeneity at a specific site in the body. There are two ways to compute Type I PLEs: subject-based sampling in which each individual subject is a sample unit; and location-based sampling where samples are taken in several sites within a broader category of locations, as for example several sites within the buccal cavity. TPLs applied to subject-based and location-based samples are ensemble spatial TPLs.

Ma's Table 2 gives the results for 18 subject-based samples ($87 \leq NB \leq 190$) of 18 sites, including skin, buccal cavity, respiratory tract, gastrointestinal tract, and urogenital area. Slopes range between $1.74 \leq b \leq 2.16$; all regressions are significant at $P < 10^{-3}$. The comparatively narrow range of slopes belies the fact that the range of b between different areas of skin ($1.74 \leq b \leq 2.08$) and different locations in the buccal cavity ($1.76 \leq b \leq 2.16$) are greater than the difference between skin ($b = 1.91$) and mouth ($b = 1.92$). This is a pattern seen by Park et al. (2013) within and between nematode guilds (Chapter 7).

An extensive location-based analysis of 140 subjects in Ma's Supplementary Table S3 shows Type I PLE with $NQ = 9$ buccal samples taken on $3 \leq NB \leq 25$ occasions, the majority with $9 \leq NB \leq 18$ occasions/subject. Thus, this is a spatial TPL measuring community heterogeneity at a specific site in the body. This location-based analysis shows estimates of b ranging between $0.5 \leq b \leq 4.5$, of which 26% were not significant and 61% were significant at $P < 0.001$. Only 30% of these have TPLs with $NB > 14$, but the distribution of the smaller set is almost the same as the full set of 140 subjects (Fig. 5.17B) and 91% are significant.

Type II PLE (Fig. 5.17C) is the temporal equivalent of fitting TPL to a time series of mean and variance of abundance per OTU in samples from NQ subjects. The example in the paper is of $25 \leq NB \leq 33$ temporally separated samples of the vaginal microbiomes of 32 subjects. This PLE measures the microbial community temporal stability with estimated $b = 2.25 \pm 0.362$.

Type III PLE is a TPL in which variance-mean pairs are computed for each OTU from their frequencies in NQ subjects and TPL is plotted for NB species or OTUs. The equivalent temporal plot is computed for NB mixed-species abundances over time where mean is computed from the abundances of NB OTUs averaged over a time series. In this case, b measures the stability of mixed-species population. Supplement Table S4 has data of 146 Type III PLE for $NQ = 9$ locations of oral microbiome containing $954 \geq NB \geq 7812$ OTUs from 146 subjects. Type III slopes range $1.46 \leq b \leq 1.99$ (Fig. 5.17D) with $r > 0.93$. The test of Type IV PLE for temporal aggregation (stability) of mixed-species population was conducted with the human vaginal microbiome consisting of $NQ \approx 90$ samples of $NB = 32$–179 vaginal biome OTUs from 32 subjects. Type IV slopes ranged from $1.51 \leq b \leq 1.79$ with all $r > 0.97$.

Ma's results fitting four PLEs with NIH data were almost all highly statistically significant—the exceptions were all cases in which $NQ < 10$, confirming the validity of mixed-species and community TPLs as well as population TPLs. Types III (spatial) and IV (temporal) TPLs, the mixed-species analyses all have $b \leq 2$, and the Types I (spatial) and II (temporal) have broader distributions with half the cases having $b > 2$. The Type II analysis suffered from very low $NQ = 3$ compared to $NQ > 9$ for all other tests. The TPL and Types I and III are all temporal TPLs, whereas Types II and IV are ensemble spatial TPLs. In view of the small number of quadrats, it is possible these results are biased (Clark and Perry, 1994), a conclusion supported by the very low correlation of some analyses.

Formally naming and defining these extensions to TPL as well as applying it to microbiome are important contributions. Ma also suggests that TPL's application to spatial heterogeneity and temporal stability at both population and community levels offers support for the conclusion that populations obeying TPL are nonstationary and like other power laws TPL possesses the properties of scale invariance and universality. We return to the question of universality in Chapter 19.

Appendix: TPL estimates for microorganisms

	NB	Range of means	r	A	SE[A]	b	SE[b]
Free-living							
A Spatial TPL ($NQ = 4$) of gymnamoeba in a freshwater pond in Palisades, New York. Data from Table 1 in Anderson (2000)							
Gymnamoeba	43	37.5 - 1460	0.945	−1.018	0.305	2.441	0.121
B Mixed-species TPLs ($NQ = 12$) of biomass of phytoplankton akinates in reservoir sediments. Data from Table 3 in Kravchuk et al. (2011)							
2004	39	0.03 - 14,908	0.989	0.763	0.059	1.840	0.043
2006	27	0.10 - 38,642	0.987	0.817	0.086	1.746	0.053
Combined	44	0.03 - 38,642	0.989	0.774	0.049	1.803	0.033
C Hybrid spatial TPL ($NQ = 2$) of diatoms in the Venetian lagoon. Data from Fig. 2 in Facca and Sfriso (2007)							
Diatoms (millions/cc)	72	0.515 - 5.63	0.986	−1.433	0.010	2.060	0.041
(#/μL)	72	515 - 5626	0.986	−1.614	0.013	2.060	0.041
D Ensemble TPLs ($NQ = 5$) of foraminifera in Rehoboth Bay, Delaware. Data from Table 1 in Buzas (1970)							
Ammonia beccarii	16	6.6 - 87.8	0.929	−1.523	0.339	2.828	0.262
Elphidium tisburyensis	16	4.4 - 42.2	0.921	−0.750	0.224	2.497	0.243
Elphidium clavatum	16	1.2 - 39.2	0.957	−0.293	0.093	2.109	0.153
Elphidium sp.	14	0.2 - 2.8	0.871	0.253	0.073	1.660	0.218
Total Elphidium	16	6.6 - 83.8	0.926	−1.049	0.268	2.558	0.241
E Spatial and temporal TPLs of invasive flagellate (*Gonyostomum semen*) in Swedish lakes. Data from Trigal and Ruete (2016a,b)							
Temporal ($NQ = 14$)	76	0.029 - 3078	0.992	0.840	0.053	1.761	0.026
Spatial ($NQ = 76$)	14	217 - 676	0.946	−0.269	0.547	2.392	0.206
South ($NQ = 50$)	14	307 - 933	0.925	−0.312	0.664	2.346	0.239
North ($NQ = 26$)	14	42.3 - 291	0.958	0.715	0.341	2.122	0.162
F Spatial TPLs ($NQ = 17–30$) of heterotrophic ciliates in the East China Sea before and after the Three Gorges dam went online. Data from Fig. 4 in Tsai et al. (2011)							
Heterotrophic ciliates (thousand/m^3)	53	51.2 - 1990	0.919	−0.802	0.307	2.142	0.116
Predam	16	104 - 1990	0.982	−0.683	0.269	2.053	0.096
Postdam	37	51.2 - 1171	0.894	−1.065	0.433	2.266	0.167
Salinity <32g/kg	25	51.2 - 1990	0.930	0.081	0.365	1.813	0.133
Salinity >32g/kg	28	61.4 - 1161	0.929	−1.881	0.462	2.567	0.179

Continued

	NB	Range of means	r	A	SE[A]	b	SE[b]

G Ensemble TPLs ($NQ = 5$) of marine viruses in the Southern California bight and the Gulf of Bothnia. Data from Tables 2 and 3 in Cochlan et al. (1993)

	NB	Range of means	r	A	SE[A]	b	SE[b]
Viruses (millions/mL)	45	0.10 - 10.4	0.986	0.222	0.031	2.013	0.050
California	32	0.1 - 5.64	0.976	0.205	0.040	1.955	0.075
Sweden	13	3.48 - 10.44	0.907	−0.271	0.255	2.651	0.309

Plant hosts

H Tobacco mosaic virus lesions on bean leaves. Data from Table 1 in Kleczkowski (1949) and Table 5 in Bliss and Owen (1958)

	NB	Range of means	r	A	SE[A]	b	SE[b]
Grouped in classes ($NQ = 12$–60)	17	15.4 - 238	0.970	0.076	0.158	1.367	0.081
Total lesions per leaf ($NQ = 10$)	12	277 - 507	0.441	−2.934	1.933	2.863	0.742
Average lesions per leaf ($NQ = 4$)	120	12.5 - 270	0.506	−1.688	0.345	2.218	0.175

I Ensemble TPLs ($NQ = 22$–107) of *Verticillium dahliae* in potato fields in Ohio. Data from Table 2 in Wheeler et al. (1994) and Table 1 in Wheeler et al. (2000)

	NB	Range of means	r	A	SE[A]	b	SE[b]
Wheeler et al. (2000) Quadrat $= 10\,m^2$ (Wheeler, 2000)	6	1.90 - 9.80	0.821	−0.104	0.424	2.238	0.521
50 m^2	6	1.50 - 14.1	0.970	0.087	0.185	2.142	0.211
100 m^2	6	1.60 - 11.8	0.970	0.025	0.179	2.106	0.208
250 m^2	6	1.80 - 11.5	0.950	0.296	0.153	1.551	0.197
1000 m^2	6	2.30 - 35.9	0.975	−0.714	0.263	2.861	0.262
4000 m^2	4	7.20 - 120	0.996	−1.802	0.202	3.421	0.145
Quadrat samples – combined	34	1.50 - 121	0.935	−0.343	0.142	2.478	0.151
Spring with Poisson samples	7	1.00 - 31.0	0.957	−0.473	0.231	2.279	0.249
Spring minus Poisson samples	4	5.00 - 31.0	0.978	−1.138	0.340	2.907	0.301
Autumn samples	7	2.00 - 79.0	0.926	−0.236	0.359	1.745	0.250
Wheeler et al. (1994)	10	0.30 - 111	0.936	0.450	0.199	1.755	0.196

J Hybrid ensemble TPLs ($NQ = 9$) of leaf mold (*Passalora fulva*) on greenhouse tomato plants Kawaguchi and Suenaga-Kanetani (2014)

	NB	Range of means	r	A	SE[A]	b	SE[b]
Total diseased plants	50	0.056 - 8.40	0.950	0.030	0.034	1.339	0.059
Newly diseased plants	50	0.056 - 6.28	0.940	0.046	0.035	1.318	0.064

K Spatial TPLs ($NQ = 20$–100) of *Monilinia vaccinii-corymbosi*, the causal agent of mummy berry of blueberries. The variance per sample is equal to the mean up to about 0.6 mummy berries per sample: above $M = 0.6$ the variation in mummy berry abundance increases with number per sample. Data from Table 1 in Copes et al. (2001)

	NB	Range of means	r	A	SE[A]	b	SE[b]
Poisson	11	0.05 - 0.73	0.956	0.135	0.073	1.127	0.099
Aggregated	12	1.10 - 12.4	0.940	0.112	0.085	1.540	0.152
Combined	23	0.05 - 12.4	0.981	0.231	0.034	1.290	0.052

L Spatial TPLs ($NQ = 320$) of powdery mildew on leaves of apple trees in Kent orchards, 1994–1996. Original data used in their study (Xu and Madden, 2002) generously made available by Xiangming Xu and Larry Madden

	NB	Range of means	r	A	SE[A]	b	SE[b]
1994	21	0.053 - 1.90	0.996	0.212	0.012	1.128	0.021
1995	24	0.028 - 1.21	0.987	0.187	0.016	1.027	0.034
1996	14	0.25 - 1.59	0.931	0.181	0.027	0.822	0.080
1994–1996	59	0.028 - 1.90	0.983	0.193	0.012	1.055	0.026

Continued

	NB	Range of means	r	A	SE[A]	b	SE[b]
M	Hybrid spatial TPL ($NQ > 60$) fit by ODR of pear scab (*Venturia nashicola*) lesions on fruits and leaves in field sampling and greenhouse experiment. Data from the text and range of means digitized from Fig. 2 in Li et al. (2007)						
Pear scab on fruits and leaves	318	$\sim 3 \times 10^{-5}$ - ~ 2623	0.982	0.722	0.014	1.465	0.016
N	Ensemble TPLs ($NQ = 6$) of *Phytophthora sojae* and *P. capsici* in soybean and pepper field soils as determined by enzyme-linked immunosorbent assay (ELISA). Data from Fig. 1 and Tables 2 and 3 in Miller et al. (1997)						
Extensive – ODR	106	1.00 - 84.0	0.883	−0.09	0.12	1.97	0.10
Extensive – GMR	106	1.00 - 84.0	0.883	−0.371	0.121	2.227	0.101
Intensive	8	4.32 - 29.7	0.960	−1.198	0.343	3.122	0.311
Beans (*P. sojae*)	4	5.72 - 24.4	0.974	−0.234	0.282	2.261	0.255
Peppers (*P. capsici*)	4	4.32 - 29.7	0.987	−1.749	0.322	3.596	0.287
O	Spatial TPLs ($NQ = 16$) of aerial density of *Phytophthora infestans* sporangia caught with rotating-arm spore samplers on 16 potato fields in New Brunswick, Canada, in 2010 and 2011. Data from the text in Fall et al. (2015)						
Aerial density – ODR	~ 100	1.00 - 61.4	0.959	−0.048	-	2.160	0.050
GMR	~ 100	1.00 - 61.4	0.959	-	-	2.225	0.050
P	Ensemble TPLs ($NQ = 10$) of rain splash experiments with *Colletotrichum acutatum* the causal agent of strawberry anthracnose. Data from the text and Figs. 2–4 in Madden (1993)						
Experiment 1 – Combined fit (ODR)	160	0.34 - ~ 4000	0.938	0.19	0.12	1.67	0.05
15 mm rain/h @ 20 cm	40	-	0.935	−0.05	0.12	1.65	0.05
30 mm rain/h @ 20 cm	40	-	0.935	0.31	0.14	1.65	0.05
15 mm rain/h @ 60 cm	40	-	0.920	−0.05	0.12	1.77	0.06
30 mm rain/h @ 60 cm	40	-	0.920	0.31	0.14	1.77	0.06
Experiment 1 – Combined fit (GMR)	160	0.34 - ~ 4000	0.938	0.191	0.133	1.780	0.05
Experiment 2 (GMR)	30	18.5 - 545	0.940	0.068	0.222	1.750	0.109

Animal hosts

	NB	Range of means	r	A	SE[A]	b	SE[b]
Q	Spatial TPL ($NQ = 20$) of ciliate protozoan *Urceolaria mitra* epizoic on planarian worm *Polycelis tenuis*. Data from Table 1 in Reynoldson (1950)						
Ciliate epizoic on flatworms	50	2.40 - 43.8	0.925	0.309	0.099	1.396	0.075
R	TPL ($NQ > 10$) of *Pasteuria penetrans* endospores on 2nd stage (J2) juvenile *Meloidogyne arenaria* from 2 studies. Data from the text and Fig. 1 in Chen and Dickson (1997)						
Field experiment	111	0.050 - 29.4	0.948	0.407	0.042	1.622	0.027
Incubation bioassay	33	0.264 - 57.6	0.985	0.126	0.049	1.593	0.023
Centrifugal bioassay	70	0.176 - 20.9	0.877	0.167	0.048	1.479	0.058
Minus Poisson points	63	0.666 - 20.9	0.800	−0.059	0.074	1.739	0.127
S	Ensemble TPL ($NQ = 8$) of competition experiments with bacterial populations, replicated $NQ = 8$ times. Data from the text in Ramsayer et al. (2012)						
P. fluorescens – alone	8	6.9×10^7 - 4.2×10^{10}	0.969	-	-	2.104	0.217
In competition	8	3.8×10^7 - 4.2×10^{10}	0.974	-	-	1.784	0.168
S. marcescens – alone	8	3.3×10^7 - 1.1×10^{10}	0.968	-	-	1.677	0.177
In competition	8	1.6×10^7 - 1.5×10^{10}	0.977	-	-	1.734	0.155

Continued

	NB	Range of means	r	A	SE[A]	b	SE[b]
Human microbiome							

T   Results of various TPLs and 4 types of power law extension (PLE) applied by ODR to parts of the human microbiome. N is the number of examples in the tables. Original data from US NIH Microbiome Project, TPL and PLE estimates from Supplementary Tables S2–S4 and Tables 2–5 in Ma (2015)

	NB	Range of means	r	A	SE[A]	b	SE[b]
Temporal TPL (NB = 7–32 subjects)							
TPL (vaginal) genera	40	1.41 - 2.00	0.984	2.052	0.182	1.715	0.063
$NQ \approx 30$							
Subject-based samples (NQ = 7–30)							
Type I PLE (locations)	18	1.73 - 2.16	0.928	5.633	0.103	1.926	0.063
NB = 87–190							
Type III PLE (locations)	18	1.44 - 1.69	0.949	2.465	0.012	1.545	0.003
NB > 2900 Location-based samples							
Type I PLE (oral)	140	0.01 - 6.98	0.853	4.128	0.286	2.533	0.534
NQ = 9, NB = 3–25							
NB = 15–25	39	1.47 - 3.36	0.858	4.463	0.213	2.145	0.327
Type II PLE (vaginal)	32	0.52 - 3.78	0.754	2.782	1.257	2.250	0.362
NQ = 3, NB = 25–33							
Type III PLE (oral)	146	1.46 - 1.99	0.978	1.389	0.010	1.657	0.006
NQ = 9, NB > 950							
Type IV PLE (vaginal)	32	1.51 - 1.79	0.986	2.093	0.091	1.672	0.034
$NQ \approx 90$, NB = 32–179							

References

Anderson, O.R., 2000. Abundance of terrestrial Gymnamoebae at a Northeastern U.S. site: a four year study, including El Nino winter of 1997–1998. J. Eukaryot. Microbiol. 47, 148–155.

Bliss, C.I., Owen, A.R.G., 1958. Negative binomial distributions with a common k. Biometrika 45, 37–58.

Buzas, M.A., 1970. Spatial homogeneity: statistical analyses of unispecies and multispecies populations of foraminifera. Ecology 51, 874–879.

Chen, Z.X., Dickson, D.W., 1997. Estimating incidence of attachment of *Pasteuria penetrans* endospores to *Meloidogyne* spp. with tally thresholds. J. Nematol. 29, 289–295.

Chen, Z.X., Dickson, D.W., Hewlett, T.E., 1996a. Quantification of endospore concentration of *Pasteuria penetrans* in tomato root material. J. Nematol. 28, 50–55.

Chen, Z.X., Dickson, D.W., McSorley, R., Mitchell, D.J., Hewlett, T.E., 1996b. Suppression of *Meloidogyne arenaria* race 1 by soil application of endospores of *Pasteuria penetrans*. J. Nematol. 28, 159–168.

Clark, S.J., Perry, J.N., 1994. Small sample estimation for Taylor's power law. Environ. Ecol. Stat. 1, 287–302.

Cochlan, W.P., Wikner, J., Steward, G.F., Smith, D.C., Azam, F., 1993. Spatial distribution of viruses, bacteria and chlorophyll *a* in neritic, oceanic and estuarine environments. Mar. Ecol. Prog. Ser. 92, 77–87.

Copes, W.E., Scherm, H., Ware, G.O., 2001. Sequential sampling to assess the incidence of infection by *Monilinia vaccinii-corymbosi* in mechanically harvested rabbiteye blueberry fruit. Phytopathology 91, 348–353.

Facca, C., Sfriso, A., 2007. Epipelic diatom spatial and temporal distribution and relationship with the main environmental parameters in coastal waters. Estuar. Coast. Shelf Sci. 75, 35–49.

Fall, M.L., van der Heyden, H., Brodeurb, L., Leclerc, Y., Moreau, G., Carisse, O., 2015. Spatiotemporal variation in airborne sporangia of *Phytophthora infestans*: characterization and initiatives toward improving potato late blight risk estimation. Plant Pathol. 64, 178–190.

Gajer, P., Brotman, R.M., Bai, G., et al., 2012. Temporal dynamics of the human vaginal microbiota. Sci. Transl. Med. 4. 132ra52.

Healy, M.J.R., Taylor, L.R., 1962. Tables for power-law transformations. Biometrika 49, 557–559.

HMP Consortium, 2012. A framework for human microbiome research. Nature 486, 215–221.

Hughes, G., Madden, L.V., 1992. Aggregation and incidence of disease. Plant Pathol. 41, 657–660.

Iwao, S., 1968. A new regression method for analyzing the aggregation pattern of animal populations. Res. Popul. Ecol. 10, 1–20.

Kawaguchi, A., Suenaga-Kanetani, H., 2014. Spatiotemporal distribution of tomato plants naturally infected with leaf mold in commercial greenhouses. J. Gen. Plant Pathol. 80, 430–434.

Kilpatrick, A.M., Ives, A.R., 2003. Species interactions can explain Taylor's power law for ecological time series. Nature 422, 65–68.

Kleczkowski, A., 1949. The transformation of local lesion counts for statistical analysis. Ann. Appl. Biol. 36, 139–152.

Kravchuk, E.S., Ivanova, E.A., Gladyshev, M.I., 2011. Spatial distribution of resting stages (akinetes) of the cyanobacteria *Anabaena flos-aquae* in sediments and its influence on pelagic populations. Mar. Freshw. Res. 62, 450–461.

Li, B.-H., Yang, J.-R., Li, B.-D., Xu, X.-M., 2007. Incidence-density relationship of pear scab (*Venturia nashicola*) on fruits and leaves. Plant Pathol. 56, 120–127.

Ma, Z., 2015. Power law analysis of the human microbiome. Mol. Ecol. 24, 5428–5445.

Madden, L.V., 1993. Aggregation of *Colletotrichum acutatum* in response to simulated rain episodes. J. Phytopathol. 138, 145–156.

Miller, S.A., Madden, L.V., Schmitthenner, A.R., 1997. Distribution of *Phytophthora* spp. in field soils determined by immunoassay. Phytopathology 87, 101–107.

Park, S.-J., Taylor, R.A.J., Grewal, P.S., 2013. Spatial organization of soil nematode communities in urban landscapes: Taylor's power law reveals life strategy characteristics. Appl. Soil Ecol. 64, 214–222.

Polley, H.W., Wilsey, B.J., Derner, J.D., 2007. Dominant species constrain effects of species diversity on temporal variability in biomass production of tallgrass prairie. Oikos 116, 2044–2052.

Ramsayer, J., Fellous, S., Cohen, J.E., Hochberg, M.E., 2012. Taylor's Law holds in experimental bacterial populations but competition does not influence the slope. Biol. Lett. 8, 316–319.

Reynoldson, T.B., 1950. Natural population fluctuations of *Urceolaria mitra* (Protozoa, Peritricha) epizoic on flatworms. J. Anim. Ecol. 19, 106–118.

Taylor, L.R., Woiwod, I.P., Perry, J.N., 1978. The density-dependence of spatial behaviour and the rarity of randomness. J. Anim. Ecol. 47, 383–406.

Trigal, C., Ruete, A., 2016a. Asynchronous changes in abundance over large scales are explained by demographic variation rather than environmental stochasticity in an invasive flagellate. J. Ecol. 104, 947–956.

Trigal, C., Ruete, A., 2016b. Data from Dryad Digital Repository, https://doi.org/10.5061/dryad.685vq.

Tsai, A.-Y., Gong, G.-C., Chiang, K.-P., Chao, C.-F., Guo, H.-R., 2011. Long-term (1998–2007) trends on the spatial distribution of heterotrophic ciliates in the East China Sea in summer: effects of the Three Gorges Dam construction. J. Oceanogr. 67, 725–737.

Wheeler, T.A., Madden, L.V., Riedel, R.M., Rowe, R.C., 1994. Distribution and yield-loss relations of *Verticillium dahliae, Pratylenchus penetrans, P. scribneri, P crenatus* and *Meloidogynae hapla* in commercial potato fields. Phytopathology 84, 843–852.

Wheeler, T.A., Madden, L.V., Rowe, R.C., Riedel, R.M., 2000. Effect of quadrat size and time of year for sampling of *Verticillium dahliae* and lesion nematodes in potato fields. Plant Dis. 84, 961–966.

Xu, X.-M., Madden, L.V., 2002. Incidence and density relationships powdery mildew on apple. Phytopathology 92, 1005–1014.

Chapter 6

Plants

Whereas most animal species can be individually enumerated, many plants are not easily enumerated because of their habit of vegetatively spreading by rhizome or root sprouts. Bunch grasses like species in the genus *Festuca* form clumps with many stems or tillers that may be identifiable as individual clumps of several square meters. And bamboos and many grasses like the biofuels crop *Miscanthus giganteus* spread by rhizome. An extreme case is the quaking aspen (*Populus tremuloides*), which grows as clonal communities with a common root network comprising thousands of stems that extend for scores of hectares. Because of their continual renewal, some root networks may be tens of thousands of years old (Mitton and Grant, 1996). Individual clones may only be recognizable in autumn when the groves change color at different times. The difficulty of identifying individual plants has led to methods for population estimation that differ from animal methods.

Plant ecologists recognize two broad characteristics of plant population description: cover and density. The latter fits our expected definition of density as number per unit area, but the enumerated objects may be whole or parts of plants. Estimation may be by quadrat or line transect. Individual annual plants can usually be identified and counted, but for perennial grasses and quaking aspen that spread by rhizome or expand by putting out tillers the concept of number per unit area is less useful. For those species, the concept of cover, the proportion of ground covered by a species, is more often used. Estimation of cover is by point quadrats in which the presence or absence of a species is recorded vertically below pins arranged in an array. The percentage of pins pointing to a species is the percentage cover. A similar procedure using line transects is also used to estimate cover. Percent cover may also be expressed as a frequency, although this statistic is not independent of the size of the sampler.

These two concepts of density and cover are related in that the reciprocal of density is area per plant, sometimes called mean area, which is directly related to cover. However, neither of these estimates of abundance considers the spatial distribution or distance between individuals or clumps. Peter Greig-Smith was one of the first plant ecologists to seriously tackle the analysis of plant spatial distribution. His 1983 book, *Quantitative Plant Ecology*, is still one of the best sources for the numerical analysis of plant populations and communities.

Taylor's Power Law. https://doi.org/10.1016/B978-0-12-810987-8.00006-9

Because of the difficulty of identifying and enumerating individuals, the development of quantitative plant ecology followed a different path to animal ecology. As a result, the botanical literature is not as rich in examples of TPL as the zoological literature; nor are there as many data suitable for TPL analysis. However, one characteristic of plant cover that has proven useful for TPL analysis is biomass per unit area. For this, the aerial parts are excised, identified, dried, and the dry weight used as a proxy for abundance. Recently, there has been a marked increase in the use of TPL to examine the distribution of plants. Particularly in rangelands both wild and managed, TPL has been used to analyze the distribution and abundance of forage species. And in Texas, TPL was applied to plant dry weight in relict and restored tallgrass prairies. In addition, seeds in the soil seedbank are clearly enumerable individually and there are a number of data sets suitable for TPL analysis.

One class of plants that are well suited to TPL analysis are annual commodity crops cultivated in rows for which there are reliable historical records covering large areas. The US Department of Agriculture (USDA) has published surveys of agricultural production since 1866. These data are freely available for downloading from several sites supported by the USDA.

Seedbank

The collection of dormant plant seeds in the soil is collectively called the seedbank. This is the basis for recovery of plant communities following range and forest fires. The diversity of ecosystems and rate of recovery after damage or destruction are directly related to the abundance and diversity of its seedbank.

The longevity of seeds in the soil ranges from a few months (e.g., the cereal weed corncockle, *Agrostemma githago*) to many hundreds of years. Lambsquarters (*Chenopodium album*) is an important weed pest whose seeds are viable in the soil for 40 years and possibly much longer. An experiment, started in 1879 in which seeds were buried in bottles and recovered at intervals, documented the longevity of 21 species' seeds, three of which were viable after 100 years (Telewski and Zeevaart, 2002). Carbon-dated seeds of lotus (*Nelumbo nucifera*) recovered from a pond were found to be still viable after 1200 years (Black et al., 2006).

The seed bank is the source of many pest weeds in field crop. Because of its importance in agriculture, seed banks in arable fields have been investigated quantitatively by methods familiar to animal ecologists rendering some data suitable for TPL analysis. In arid environments, the absence of a seedbank leads directly to desertification. The presence of a seedbank can lead to spectacular blooms following the occasional precipitation events.

FARMLAND IN ENGLAND

In an analysis of the distribution of weedy plants in arable fields at two farms in England, Marshall (1989) identified four types of plant distribution: plants limited to the hedgerow margins of the fields; plants occasionally found in the margins but abundant in the field; plants in the margins and at decreasing density into the crop; and plants in the margins and in the field but only close to the edge. The majority of species in the margins did not occur in the crop area, but about 30% of species were found in both the margins and at varying distances into the crop. The significance of this research was the finding that the field margins apparently served as a source for several but not all important weeds in field crops.

In April 1984, Marshall sampled 12 cereal fields near Basingstoke in southern England by taking 40 3.1 cm^2 soil cores at the hedgerow and at 6 distances into the field up to 50 m from the hedgerow. Seeds were separated and allowed to germinate in sterilized soil. Six grass species were identified: the data of their abundance ($\pm SE$) given in Marshall's Table 2 yield $NB = 32$ points for a mixed-species TPL analysis of the grass community's seed bank (Fig. 6.1; Appendix 6.A). At the lowest abundances, the distributions were indistinguishable from Poisson, but the most abundant seeds were highly aggregated. Overall, the community of seeds in the seedbank was strongly aggregated with $b = 1.86 \pm 0.05$ over 4 orders of magnitude. This slope is slightly lower than that of broad-leaved species ($b = 1.94 \pm 0.04$).

In his second study at Boxworth, near Cambridge, Marshall sampled a field for seeds in January 1984 and again in November for plants. The seed bank was sampled with a soil core in the hedgerow and the field and the plants were

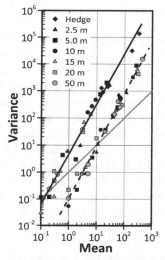

FIG. 6.1 Left: Six viable grass species' seeds in $NQ = 12$ arable fields near Basingstoke in January 1984. Points ($NB = 32$) are a species at distances up to 50 m into the fields including the hedgerow. Right: Broad-leaved plants in the same fields in April ($NB = 46$), excluding the hedgerow; shifted right one cycle for clarity. Both seeds and plants are strongly aggregated with the plant community's TPL slightly steeper than the seedbank's. *(Data from Tables 2 and 4 in Marshall (1989).)*

counted at 10 points in 11 transects to 5 m into the field. His Table 6 gives the data for both broad-leaved plants and grasses. The slopes of the Boxworth TPLs are both lower than those at Basingstoke, and again the plant community TPL is slightly steeper than the seed bank's ($b = 1.59 \pm 0.03$ vs. $b = 1.42 \pm 0.08$, respectively).

Not surprisingly, at both sites the density of hedgerow species declined exponentially with distance into the field providing a range of means for TPL analysis. More surprisingly perhaps, the TPLs for seeds and plants do not segregate out by distance. The communities of grass seeds and broad-leaved plants at Basingstoke conform to steep TPL curves but with the species-distance points distributed across the density range of 3 to 4 orders of magnitude. The pattern of TPL points of 28 grass and weedy plants at Boxworth is very similar (Appendix 6.A).

In annual surveys of grass weeds at Boxworth, Marshall (1985, 1988, 1989) collected data of grass, weed, and seed distributions in 6 fields from 1984 to 1988. These data with some other published and unpublished plant and seed bank data were analyzed by Clark et al. (1996) to determine if the generality of TPL found in animal distribution and abundance data applied also to plant data, and to examine the effect of scale on TPL. In all, they analyzed 69 data sets in 6 groups, including data from France and Spain. The latter set of winter wild oat (*Avena sterilis ludoviciana*) panicles was collected in 20 randomly placed 1 m^2 quadrats at 20 sites near Madrid (unpublished data of L. Navarrete). In Hautes Landes, France, 300 18 cm^2 soil cores were taken 2 m apart in a 12 by 25 grid in 2 areas of a field in 1983 and 1985. The published data (Chauvel et al., 1989) provided 4 variance-mean pairs for crab grass seeds (*Digitaria sanguinalis*) (Appendix 6.B).

Ten years data of black grass (*Alopecurus myosuroides*) in 16 cereal fields at Neville's Farm near Wantage in central England (Wilson and Brain, 1991) were analyzed to examine how scale affected TPL. Clark et al. calculated TPL at three scales: at the smallest scale, TPL was calculated from data for each year and field; two larger scales pooled samples before computing TPL. At the smallest scale, the black grass data of 6–10 fields surveyed at rate of 6–7 points/ha over a period of 10 years. As the field sizes varied, the number of sample points per field ranged from $NQ = 28$–120 and the number of TPL points per year ranged from $NB = 6$–10. Combining all years together in a single TPL regression resulted in $NB = 83$ points (Scale 1). Analysis of parallelism found that the 10 years were not significantly different in either slope or intercept; a single regression described the distribution at Scale 1. TPL was also computed for each year from total count for each field ($NQ = 6$–10) yielding a point for each year ($NB = 10$; Scale 2). Fig. 6.2 (Appendix 6.C) shows Scale 2 ($b = 2.00 \pm 0.187$) to be significantly steeper ($P < 0.05$) than Scale 1 ($b = 1.59 \pm 0.038$).

In the annual surveys, Marshall recorded the abundance of 4 grasses: black grass, meadow brome (*Bromus commutatus*), barren brome (*B. sterilis*), and

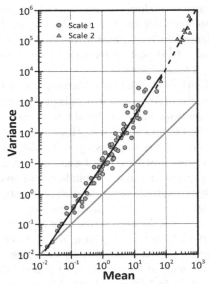

FIG. 6.2 Black grass at Neville's Farm near Wantage is slightly more aggregated at the Scale 2 than Scale 1. Each of $NB = 83$ points at Scale 1 is a field-year ($NQ = 28$–120) of panicles/m^2; each Scale 2 point is 1 year ($NB = 10$) with mean and variance computed from the total panicles/m^2 for each field ($NQ = 6$–10). *(Adapted from Fig. 2 in Clark et al. (1996).)*

common couchgrass (*Elymus repens*). These data Clark et al. analyzed in three studies. In the first, barren brome, meadow brome, and couchgrass distributions were analyzed at two scales. The small scale was analyzed twice: panicles/m^2 in $NB = 23$ areas of 0.46 ha with $13 \leq NQ \leq 17$ samples and in $NB = 12$ areas of 0.92 ha with $16 \leq NQ \leq 33$ samples. At the larger scale, 13–17 individual samples were aggregated within the 0.46-ha areas yielding $NB = 6$ points comprising $3 \leq NQ \leq 4$ samples for each grass species. In this study, the slopes computed for the three species followed no consistent pattern with respect to scale (Appendix 6.D).

Another study analyzed meadow brome and black grass samples at three scales in four fields over four years. Two 0.25 m^2 quadrat counts of grass panicles were taken at 10×11 grid points within each field (an 11th row sampled within the hedgerow was not included). At the smallest scale, Scale 1, means, and variances were calculated from the pairs of samples at each point ($NQ = 2$) at $22 \leq NB \leq 75$ nonzero samples. At Scale 2, counts of each pair of samples were combined and the survey area divided into 10 subareas, each containing 10 grid points. This yielded $6 \leq NB \leq 10$ data points each with means and variances based on $NQ = 10$ sample points. At the largest scale (Scale 3), the samples in each subarea were combined and variance-mean pairs computed for each field-year yielding $NB = 10$ for meadow brome and $NB = 7$ for black grass, both with $NQ = 10$. Comparison of slopes across the three scales showed different patterns in relation to scale for the two species (Appendix 6.E). The slope for meadow brome increased from Scale 1 to 2 to 3 in field TP in 1985. For black grass in the same field-year, b increased from Scale 1 to 2 and then decreased from Scale 2 to 3.

In a third study, Clark et al. used data Marshall had taken of four $0.25\,\text{m}^2$ samples of meadow brome in $NQ = 8$ or 9 sample points and $5 \leq NB \leq 12$ points per field-year. In this analysis, they found that b for meadow brome was highly variable $(1.40 \leq b \leq 2.41)$ in space and time (Fig. 6.3; Appendix 6.F).

Clark et al. analyzed these data using weighted least squares (WLS; Chapter 4) and the exception of Neville's Farm and Hautes Landes, both NB and $NQ < 15$. To compensate for the small number of samples, Clark et al. used Clark and Perry's (1994) quadratic response surface equations to adjust for the likely bias of parameter estimates. Unadjusted, the slopes were $1.40 \leq b \leq 2.41$ and adjusted they were $1.32 < b < 2.61$. Adjusting b increased the slope in all but five data sets: two analyses were unchanged and 3 decreased b. Fifty sets increased by an average of 6% but 6 increased by an average 50% (barren brome and black grass in four fields at Boxworth in 1985 and 1987) at Scale 1. The same fields and years at Scale 2 increased only 6%, in common with the other analyses.

Overall, after adjustment, the TPL estimates of $1.32 \leq b \leq 2.61$ and $-0.85 \leq A \leq 1.58$ were within the range of values estimated in previous multispecies studies of plant and animal populations (LRT61; LRT et al., 1978, 1980). Parameter estimates, varied with sample size and spatial sample scale, but unpredictably, leading Clark et al. to conclude that "changes in parameter estimates may be unpredictable when spatial scale is altered."

Clark et al.'s analysis provides sufficient data to compare the results of WLS fitting with small sample adjustment for bias with GMR regression without bias

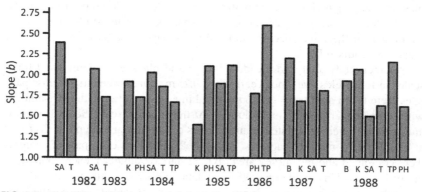

FIG. 6.3 Variability of TPL in space and time adjusted b for meadow brome (panicles/m^2) varied from 1.32 to 2.61 in 6 fields (B, T, SA, K, TP, and PH) at Boxworth 1982–1988. *(Data from Table 1 in Clark et al. (1996).)*

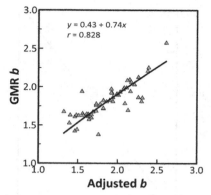

FIG. 6.4 Comparison of GMR estimates of b with WLS estimates adjusted for small sample bias shows they are well correlated with GMR slightly underestimating relative to adjusted WLS.

adjustment (Fig. 6.4). An extreme outlier is not shown: estimates of b for barren brome at Scale 2 in the first study were $b_{unadjusted} = 1.73$, $b_{adjusted} = 2.04$ and $b_{GMR} = 3.54$. Overall (excluding the barren brome outlier), the GMR estimates of b correlated well with the WLS estimates ($r = 0.83$), ranging from 16% underestimate to 27% overestimate, averaging 2.1% underestimate, which is significantly different from zero at $p < 10\%$.

FIELD MARGINS IN WISCONSIN

In a study of weed abundance in field margins, Sosnoskie et al. (2007) surveyed the boundaries of 63 fields in 6 Wisconsin counties. The counties differed in topography, geology, and land use. Field margins ranging in length from 1.35–2.04 km were separated into 307 segments and classified into 7 habitat types: buffer strips of annual or perennial grasses planted on erodible land; field boundaries with annual and perennial grass and broadleaf species; fence rows with annual and perennial broadleaf and grass species; fence rows with woody shrubs and trees; forest boundaries dominated by trees; within-field edges with uncultivated area (e.g., drainage ditches); and road ditches between field and road to carry runoff from the road.

The surveys were conducted by walking the perimeter of each field, identifying the margin segments and recording the first occurrence of each agricultural weed species within each field margin segment. The width the survey area was limited to a visual estimate of a transect extending from 0.5 m into the crop to 0.5 m into the peripheral habitat. The frequency of occurrence of each of 46 species was estimated from $F_{ij} = 100 * n_{ij}/m_i$, where n_{ij} = number of margin segments of type i that species j occurred and m_i = the total number of field segments recorded for margin type i. The frequency of occurrence of each species in the seven habitat classes is given in Sosnowskie et al.'s Table 1.

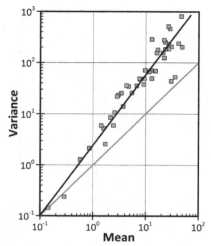

FIG. 6.5   Distribution of *NB* = 46 weed species in *NQ* = 7 habitats forming 307 margin segments of 63 fields in Wisconsin. *(Data from Table 1 in Sosnoskie et al. (2007).)*

In addition to recording the species diversity in relation to habitat type, Sosnowskie et al. also examined diversity in relation to three geophysical variables: elevation variability, field perimeter length, and field shape. These nonbiological variables proved to be good predictors of field species richness, their influence varying between habitat margin types. In all margin types weed diversity increased with field perimeter length, with forest margin having the least and buffer strips the most influence. They conclude that farmers should pay attention to weed management of the ecotones at the field edges as well as the fields themselves.

Sosnowskie et al.'s Table 1 provides material for a mixed-species TPL analysis of the frequency of occurrence of a community of *NB* = 46 weed species using the *NQ* = 7 habitat types as samples. The community of weed species averaged over 7 habitats in these 63 fields obeys TPL very well ($b = 1.378 \pm 0.063$) with a few exceptions (Fig. 6.5; Appendix 6.G). Two species (foxtail barley (*Hoedeum jubatum*) and field sandbur (*Cenchrus incertus*)) were uncommon with $M < 1.0$ and distributions indistinguishable from Poisson and two species with $M > 10$ were closer to the Poisson line than the fitted TPL line. The latter two species, ladiesthumb (*Polygonium persicaria*) and wild carrot or Queen Anne's lace (*Daucus carota*), are naturalized and locally invasive in North America. Their lower aggregation relative to the rest of the community reflects the wide distribution of these invasive weeds.

IN SOYBEAN FIELDS

In a study of the spatial distribution of weeds in soybeans, Cardina et al. (1996) marked out two soybean fields into 3 × 6 m grids: 146 grid squares in a plowed field and 162 in a no-till field. As part of the study, 10 38.5 cm^2 soil cores were taken in each of five randomly selected grids in each field in early April 1990–93. Seeds were removed by sieving, identified, and counted. The fields were planted to soybeans 1 month following the soil samples, and no post emergence weed control was applied to the fields. Seedlings were identified and counted periodically in the fields from April to July in permanent 25 × 40 cm quadrats placed adjacent to the hole left by the soil samples. The seedlings were removed following counting.

Cardina et al.'s Table 2 gives the average and SE of the number of lambsquarters (*Chenopodium album*) and annual grass (giant foxtail (*Setaria faberi*), large crabgrass (*Digitaria sanguinalis*), and fall panicum (*Panicum dichotomiflorum*)) seeds and seedlings per square meter in both fields in each of the 4 years. The abundance of lambsquarters seeds and seedlings was generally higher in the plowed field than the no-till field; the reverse was the case with the grasses. Fig. 6.6 (Appendix 6.H) shows TPL analyses for lambsquarters seeds and seedlings in plowed and no-till fields.

Plowed and no-till management had no effect on the spatial distribution of either seedbank or seedlings. The slopes of seedbank and seedlings differed in both taxa. The TPL slope of lambsquarters seedbank was steeper than the seedlings ($P < 0.02$), and seedlings slope was almost significantly different from Poisson ($P < 0.06$). The slopes of grass and seedbank were not significantly different ($P > 0.20$).

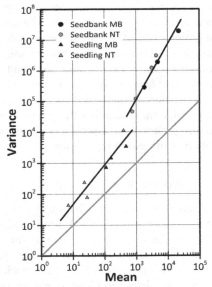

FIG. 6.6 Tillage practice did not affect the spatial TPLs ($NQ = 50$, $NB = 8$) of lambsquarters (or grasses, not shown) seedbanks and seedlings in mold-board plowed and no-till soybean fields. *(Data from Table 2 in Cardina et al. (1996).)*

The results of a study by Mulugeta and Boerboom (1999) for lambsquarters seedbank in soybeans agree broadly with Cardina et al.'s lambsquarters results. Their study's main purpose was to establish if row spacing influenced weed distribution and abundance sufficiently to require different management strategies. The seedbank was sampled with $6.6\,cm^2$ soil cores in plots with row spacings of 18 and 76 cm. Seeds were separated from the cores, identified and the density of viable seeds recorded ($\#/m^2$) for each plot. Seedlings in 16 randomly placed $625\,cm^2$ quadrats were identified, counted and removed at intervals; 3, 6, 8, 10, and 13 weeks after planting. The experiment was conducted, in 1996 and 1997.

In addition to lambsquarters, giant foxtail and velvetleaf (*Abutilon theophrasti*) were sufficiently abundant to analyze by TPL. Lambsquarters accounted for 60%–75% of viable seeds and ~22% of seedlings. Mulugeta and Boerboom's TPL analysis of the distribution of these three species used ODR (Appendix 6.I). Foxtail and lambsquarters seeds in the seedbank were aggregated, while velvetleaf was not significantly different from Poisson. Mulugeta and Boerboom also calculated the negative binomial k and Lloyd's patchiness index (Chapter 3) and found these indices broadly agreed with TPL. However, Lloyd's patchiness index declined exponentially with increasing density making it less suitable as a measure of aggregation.

Estimates of b and A for foxtail and lambsquarters, but not velvetleaf seedlings differed significantly between row spacings and years. The inconsistency in estimates may result from bias due to a narrow range of means. The range of densities of velvetleaf seeds was 140–$305\,seeds/m^2$, a range insufficient for an efficient TPL fit (Chapter 4; Clark and Perry, 1994). The number of samples per plot, $NQ = 16$ quadrat samples and $NQ = 15$ soil core samples per plot, both with $NB = 8$ regression points, are almost identical, but the soil cores of seeds produced more consistent results across the two treatments and years than the seedlings in quadrats. The $625\,cm^2$ quadrats may have been too small to sample the plants efficiently. Another potential source of bias was the practice of spreading the sampling over 13 weeks and removing the plants while combining the counts. Removing seedlings altered intraspecific competition, which may have biased sampling by altering the sampling efficiency.

TREE SEEDBANK IN TAIWAN

Lin et al. (2011) documented the seed distribution of trees in a 10-ha tropical forest permanent plot in Taiwan and examined factors influencing their distribution. An array of 18 seed traps was checked weekly for 1 year from August 2006 and the seeds collected and identified. Of 27 tree species collected, 11 were sufficiently abundant for analysis. All 11 tree species were aggregated according to the indices of Morisita and Green (Chapter 3).

Using the total number of seeds collected and values for Green's index in Lin et al.'s Table 1, means and variances of seeds per seed trap ($NQ = 72$) were

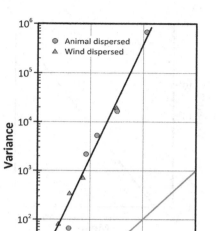

FIG. 6.7 Mixed-species TPL ($NQ = 72$, $NB = 11$) of the community of tree seeds in a tropical forest is strongly aggregated ($b=2.32\pm0.114$). The distribution is not influenced by the mode of dispersal, by wind, or animal where each point is a tree species. *(Data from Table 1 in Lin et al. (2011).)*

computed for each species. In addition, the table identifies the mode of seed dispersal: by wind or animal. Both modes of dispersal are identified in a mixed-species TPL of the tree community (Fig. 6.7; Appendix 6.J).

All the seeds were collected in the same sampler providing absolute density estimates. The abundance of seeds depends on the abundance of the parents and their reproductive output, both of which can be expected to differ between species and provide a range of means. Perhaps the most surprising feature of this example is the comparison of the mode of dispersal, wind or animal. Even though the wind-borne seed species are fewer ($NB=4$) than the animal dispersed seeds' ($NB=7$) and cover a narrower range of abundance, the wind-borne and animal-dispersed seeds are surprisingly consistent.

SEEDBANK DIVERSITY IN CATALONIA

Izquierdo et al. (2009) examined how the spatial distribution of weed seedbank diversity was affected by weed control. They examined the changing spatial distribution of weed seeds in an 8-ha winter wheat field (*Triticum aestivum*) field in western Catalonia from 2001 to 2003. The field was regularly treated with herbicides to control grass and broadleaf species, except only grass herbicides were administered in 2002 and 2003. 16-cm^2 soil cores were taken at 10-m intervals

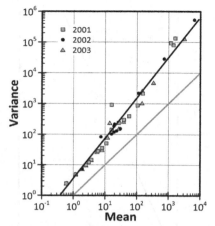

FIG. 6.8   There is no difference between years in the mixed-species TPL ($NQ = 254$, NB = 68) of a community of 30 weed species in the seedbank of a field in Catalonia. Each point in this graph is a different weed species recovered in sampling the seedbank by soil core. *(Data from Table 1 in Izquierdo et al. (2009).)*

on a 150×150 m grid in the wheat field in January of each year. Seeds were allowed to germinate in a greenhouse, identified, and the density per square meter estimated for each species at 254 sample points (2 points were skipped). The distribution of weed seed diversity within the 2.25-ha area was mapped for each year. Izquierdo et al. found that the spatial distributions of Shannon diversity and evenness became increasingly patchy over time. Both grass and broadleaf weed patches moved and varied in size from year to year. In general patches of broadleaf weeds decreased in response to herbicide application, but the absence of a grass herbicide application in the first year enabled grass patches to expand contributing to increased patchiness. Izquierdo et al.'s Table 1 gives the density and SE of seeds (#/m^2) for 30 weed species. Despite the year-to-year variation in diversity and spatial distribution, the 3 years' mixed-species TPLs do not differ significantly and are best described by a single line (Fig. 6.8; Appendix 6.K).

INVASIVE DEVIL'S THORN IN AUSTRALIA

Devil's thorn (*Emex australis*) is a native of southern Africa that colonizes disturbed areas and has become widespread in southern Australia where it is an economically important weed. Weiss (1981) sampled for seeds and plants in permanent quadrats established at two arable sites in New South Wales (NSW) and an uncultivated site in South Australia (SA). At each site, 6 0.5 m^2 quadrats were placed randomly within a 100-m^2 area chosen for their similarity of devil's thorn density. The positions of all plants within the quadrats

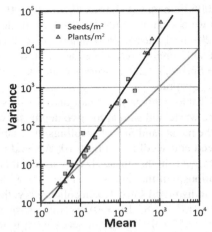

FIG. 6.9 Spatial TPL ($NQ = 18$, $NB = 26$) of devil's thorn plants and seeds produced by those plants are virtually identical. *(Data from Table 3 in Weiss (1981).)*

were recorded at 3-week intervals at the NSW sites and every 6 weeks at the SA site during the growing season for 2 years from July 1975. In 1977, the seed production of each plant was also recorded. Weis found a significant negative density-dependent relationship between seed production and plant density. Nearest-neighbor analysis suggested the spatial distribution of plants was random.

Weis' Table 3 gives the mean and standard deviation of plants and seeds per square meter at each site on four or six sampling occasions. Spatial TPL analysis confirms both seed and plant distributions at about half of the sites were indistinguishable from Poisson (Fig. 6.9; Appendix 6.L). The TPLs for seed production and plants are not significantly different ($P > 0.2$), suggesting limited seed dispersal.

INVASIVE RAGWEED IN FRANCE

The invasive common ragweed (*Ambrosia artemisiifolia*), introduced to Europe from North America, is highly competitive and a major cause of crop loss on both continents. Although it propagates mainly by rhizome, ragweed also sets seed, which can remain in the seedbank for several years. Its wind-borne pollen is highly allergenic and the cause of respiratory distress in a large fraction of the human population. Fumanal et al. (2008) sampled ragweed seeds and seedlings at nine locations in France, four fields, two conservation set-asides, and three wastelands, to determine the spatial and vertical distributions, the relationship between seedbank and plant abundance, the proportion of viable seeds, and the effect of habitat disturbance on germination rate.

In early Spring 2006 Fumanal et al. took 10, 20, or 100 $20\,cm^2$ soil cores at the nine sites. Cores were separated into depths 0–5 cm, the maximum germination depth and 5–20 cm, the usual plow depth for field crops. Seeds sieved from a subsample of soil cores were counted as dead, live, dormant, and nondormant. Later in the Spring, seedlings were sampled at each site using 10 randomly placed $0.25\,m^2$ quadrats.

Fumanal et al.'s Table 2 gives the mean and SE of seeds/m^2 ($NQ = 20$, except $NQ = 10$ at one wasteland site) at the two depths for each seed category. Their Table 5 gives the means and SEs of seedlings/m^2 enabling comparison of ensemble TPLs of seed and seedlings (Fig. 6.10; Appendix 6.M).

The density of live seeds in the surface layer was less than in the lower layer for field-crop populations, but the reverse was the case with the set-asides and wasteland with their undisturbed soils. Taken collectively, the distribution of all seeds in the surface layer ($b = 1.82 \pm 0.300$) was more strongly aggregated than in the lower layer ($b = 1.55 \pm 0.171$). Dead seeds were more aggregated with $b = 1.62 \pm 0.111$, compared to $b = 1.42 \pm 0.126$ for living seeds and $b = 1.31 \pm 0.143$ for seedlings, suggesting density-dependent seed mortality. Dormant and dead seeds' TPLs were almost the same ($b \approx 1.6$) and greater than nondormant and live seeds and seedling TPL ($b \approx 1.4$), also supporting a conclusion of density-dependent seed mortality and/or seedling recruitment. This conclusion is further supported by the fact the seedbank density was highest

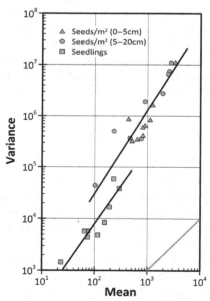

FIG. 6.10 Ensemble TPLs of common ragweed seeds in soil cores and seedlings (#/m^2) taken at $NB = 9$ ($NQ = 10$–100) locations in France. *(Data from Tables 2 and 5 in Fumanal et al. (2008).)*

and the seedling density lowest in the conservation set-asides with seedling recruitment about 10% of recruitment in the fields and the wasteland, possibly due to competition from other plants in the set-asides.

ENDANGERED BEARPOPPY IN NEVADA

The Las Vegas bearpoppy (*Arctomecon californica*) is a rare, endangered, short-lived, herbaceous perennial endemic to the Mojave Desert of southeastern California and southern Nevada. The bearpoppy depends on a long-lived seed-bank in order to persist in the driest desert in North America. Growth occurs following rare precipitation events sufficient to stimulate germination, which result in occasional spectacular flowering of the desert. The year-to-year variation in bearpoppy populations is therefore extreme. The distribution and abundance of bearpoppy plants and seeds was studied by Megill et al. (2011) in an area near Las Vegas, Nevada. Megill et al.'s study was set up in a 100-ha area fragmented by urban expansion and uncontrolled and unpermitted landuse. Seven randomly placed 1-ha sites were established and surveyed during the winter and spring of 2005–06. Each survey site consisted of 2178 contiguous $4.5\,m^2$ quadrats in which plants were counted in 954 randomly chosen quadrats. The seedbank was sampled with $64\,cm^2$ soil core in $954 \leq NQ \leq 2141$ quadrats and the soil cores separated into 0–2 and 2–4 cm depths from which bearpoppy seeds were sieved. There was no difference in abundance at the two depths. Seeds and plants were found together only in areas with high densities of both plants and seeds. Megill et al. also surveyed other plants in the community and found that bearpoppys are associated with only two of nine other species. In general, bearpoppy plants occurred in open areas with low vegetative cover.

Bearpoppy seeds were very rare at the study location: only 114, of which only one-third were viable, were recovered in a total of 9170 cores. Megill et al.'s analysis does not distinguish between viable and nonviable seeds. They fitted each station's sample data to the Poisson distribution and were able to reject the Poisson in two of seven stations and concluded that at five stations the distribution was random implying that the variance and mean were equal. Given the small number of seeds recovered, this is a reasonable conclusion. Using the bootstrapping resampling method, Megill et al. computed 95% CI for both depths at all sites (their Fig. 2). Ensemble TPL analysis of density $(\#/m^2)$ has a gradient of $b = 1.52 \pm 0.130$ (Appendix 6.N), confirming Megill et al.'s conclusion bearpoppy is aggregated at the most populated sites but shows only two points on the Poisson line. The overlap of site-depth points in the plot confirms the absence of a difference in abundance at the two depths sampled.

Grasslands

Shiyomi et al. (2001) proposed that in a mixed species temporal TPL, the vertical deviation from Poisson line, δ_i is a spatial heterogeneity index measuring the heterogeneity of species in relation to the community. The larger the value of δ_i, the greater the heterogeneity of species i. The fitted TPL line represents the level of spatial heterogeneity for the community as a whole and the vertical deviation, ε_i, of a species' point from the fitted line measures the difference in heterogeneity between that species and the other members of the community. If $\varepsilon_i < 0$, the spatial heterogeneity of the ith species is less than that of the community as a whole and if $\varepsilon_i > 0$, that species' heterogeneity is higher than community's. This proposition has been tested in several dryland grass biomes.

GRASSLAND IN SHAANXI PROVINCE

In China's Shaanxi Province, Guan et al. (2016) surveyed the plant compositions of a grassland and a periurban weed patch in June 2013 and May 2014, respectively. The grassland site was about 600 m higher and $\sim$80 km further north than the weed community, although the sites experienced similar climates. All plants were identified and classified to species in 100 randomly placed 0.25 m^2 quadrats in each community. They recorded the number of tillers of grasses (Gramineae) and sedges (Cyperaceae) and the number of individual plants of other species. The mean and variance of number of stems per quadrat were computed for each species. Of the 30 species in the grassland and 32 in the weed patch only 3 were common to both habitats.

Their main objective was to determine if mixed-species temporal TPL effectively quantifies the spatial heterogeneity of plant abundance. Their Table 1 gives the mean number per quadrat, M_i, and δ_i ($= \log(V_i) - \log(M_i)$) from which variance can be calculated. Mixed-species TPLs of grassland and weed communities showed the weed patch to be slightly more aggregated than the grassland (grassland $b = 1.41 \pm 0.061$, weed patch $b = 1.57 \pm 0.054$) and the difference significant at $P = 0.054$ (Fig. 6.11; Appendix 6.O).

A second objective was to determine if TPL could be used to measure community stability and by inference sustainability. Their results show positive correlation of δ_i and $\log(M_i)$: $r = 0.69$ for the grassland and $r = 0.89$ for the weed community. Because species with both high δ_i and M_i are important for the stability and succession of communities, these species are of particular interest for maintaining rangeland sustainability. However, high absolute values of ε_i contribute to community instability, so the ideal species are those with high δ_i and low ε_i. By defining a community-wide spatial heterogeneity index, $\delta_c = (\sum m_i \delta_i)/\sum m_i$, a weighted average of the δ_i summed over all species, $i = 1 \ldots S$, Guan et al. identified those species with the greatest influence on the community spatial heterogeneity and thus the most important for stability. Many species had high m_i and δ_i, but two in the grassland and three in the weed patch were particularly high and therefore important for maintaining habitat stability, tending to support Grime's (1998) hypothesis that dominant species may be more important than diversity for stability.

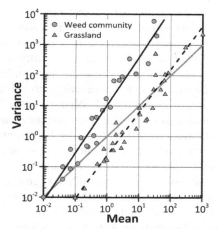

FIG. 6.11 Mixed-species temporal TPLs of plant communities in a grassland ($NQ = 100$, $NB = 32$) and a peri-urban setting ($NQ = 100$, $NB = 30$) show the grassland community to be both more variable and more aggregated. Grassland offset by one cycle for clarity. *(Data from Table 1 in Guan et al. (2016).)*

TALLGRASS PRAIRIES IN TEXAS

In a study to establish if community stability was governed by plant diversity or the dynamics of dominant species, Polley et al. (2007) compared interannual variation of above-ground biomass in paired restored and remnant tallgrass prairies at two locations in central Texas. Restored and remnant prairies were 4.6 ha and 9.8, respectively, at Temple and 0.5 and 1.6 ha at Riesel 58 km NE of Temple. The restored tallgrass prairies had been seeded from the remnant prairies in fields previously grown to maize 9 years prior to sampling at Riesel and 20 years at Temple. All four prairies had been managed similarly and were not grazed by domestic animals.

Polley et al. established 12 randomly located 5 m radius circular permanent plots in each prairie and took samples twice a year for 5 years from June 2001. Above-ground biomass was clipped to 2 cm height in one randomly chose 0.5 m^2 subplot in each of the 12 circular plots. Sampling was timed to correspond with periods of peak biomass for early-season (June) and late-season (November) species. Live tissue from the June harvest was separated and identified to species: plant tissue from the November harvest was separated into C4 (mostly perennial grasses) and C3 (mostly perennial forbs) plants. The number of species recovered at the remnant prairies was higher (60 species at Riesel, 69 at Temple) than at the restored (48 at Riesel and 47 at Temple) prairies.

The biggest influence on plant production in Central Texas is the seasonal pattern of rainfall (~900 mm/year concentrated in spring and autumn). Polley et al. found that the biomass response to rain depended as much on characteristics of dominant species, notably little bluestem grass (*Schizachyrium scoparium*) as on differences in diversity, even though diversity of the remnant prairies was double the restored prairies. Using the variation in above-ground dry weight as measured by mixed species temporal TPLs as proxies for community stability, they found little difference in stability of the four prairies' communities. They concluded that in these tallgrass prairie ecosystems the dominant species contributed more to community stability than species diversity, a result also consistent with Grime's (1998) hypothesis. This insight differs from the older hypothesis that more diverse communities are inherently more stable. Pimentel (1961) studying insect outbreaks, for example, concluded that community stability increased with the number of interspecific interactions, which, in turn, is directly related to species diversity.

The lead author of this study, Dr. Wayne Polley, generously made his raw data for June available to me to reanalyze the species abundances at the four sites. The data comprise the dry weight (g/m^2) of a variable number of species in each of 12 subplots in 5 years at the 4 sites. For each site and species, means and variances were averaged over $NQ = 5$ years. Results of the reanalysis using GMR are slightly different from Polley et al.'s ODR estimates. The patterns observed by Polley et al. are however preserved. Fig. 6.12 shows the mixed-species temporal TPL for the remnant prairie at Temple with each point the dry weight of a single species in one of the 12 subplots averaged over 5 years.

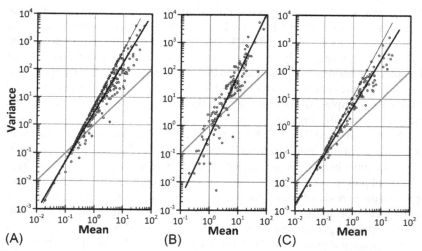

FIG. 6.12 Mixed-species temporal TPLs of the remnant prairie community at Temple, Texas. (A) ($NQ = 5$, $NB = 315$) Censoring of points above $V = 5M^2$ (light line) is clearly visible in this and the other three plots. The *heavy line* is fitted by GMR. Each point is the dry weight of a species in one of the 12 plots established in the prairies. (B) The upper cutoff is not visible in the TPL of the remnant prairie at Temple computed by averaging only plots with nonzero counts ($2 < NQ < 12$, $NB = 130$). (C) Including plots with zero counts ($2 < NQ < 12$, $NB = 132$) in the TPL of the remnant prairie at Temple exposes the cutoff. *((Data courtesy of Dr. Wayne Polley.))*

The other three prairies look very similar and produced similar statistics (Appendix 6.P, Case 1), with the four slopes not differing significantly between sites and age of prairie. While the intercepts do not differ between remnant and restored sites at either Reisel or Temple, they do differ significantly ($P \ll 0.001$) between Reisel and Temple, reflecting the difference in plant cover at the two locations. Thus, except for slight differences in the positions of the TPL lines, the four plots are indistinguishable.

Fig. 6.12A shows an abrupt limit to points on the graph at the line $V = 5M^2$ identified by the lighter line. All four regressions have this cutoff. Tokeshi (1995) defined the zone in which a TPL could exist: an area in the first quadrant of the variance-mean domain bounded by $V = M$ and $V = nV^2$, where $n =$ the number of sample units ($NQ = 5$ years). Polley et al. computed the mean and variance of each species' dry weights for all plots with at least one nonzero year. Averaged over $NQ = 5$ years for all species, the number of points is $NB = 209-315$ for the four analyses. An alternative approach is to average over $NQ = 12$ plots and plot TPL using all species for up to 5 years. Counting only nonzero quadrats produces TPLs ($NB = 67-130$) with substantially higher slopes (Fig. 6.12B; Appendix 6.P, Case 2). Because only nonzero counts are included in the plot database, averages were computed over a variable number of plots, $NQ = 2-12$ ($NQ = 1$ is excluded because variance is undefined). Excluding zero plots or quadrats is a source of bias (RAJT et al., 2017) that inflates the estimate of b. The cutoff is not visible in Fig. 6.12B because n is variable.

Recasting the analysis by computing variance and mean from all the plots, including zeros ($NQ = 12$) for each species, results in TPLs with higher intercepts and slopes similar to those obtained with $NQ = 5$ (Appendix 6.P, Case 3). As with the first approach averaging over years, this plot also has a cutoff at $V = nM^2$, but here the number of plots, $n = 12$ (Fig. 6.12C).

RANGELAND IN MONGOLIA

TPL provides a powerful quantitative method for measuring stability, and other authors since Polley et al. have used TPL in studies of community structure and stability. The temperate steppe in Inner Mongolia has been subjected to grazing for millennia, but with increasing human and domestic animal populations, the sustainability of traditional practices is being questioned. Although conservation or best management practices are known to affect community structure and productivity, the impact on community stability is less certain. In a survey of four steppe areas subject to different management practices, Zhang et al. (2016) collected plant abundance data over 3 years. The four sites formed a continuum in disturbance from minimal to extreme: one area had been free of grazing for 50 years, another was enclosed and not grazed or mowed for about a decade; one was mowed annually for about a decade and a fourth had been grazed continuously for 50 years.

Each August from 2011 to 2013, Zhang et al. took three 1-m^2 quadrat samples from three permanent plots at each site; a total of 27 samples. Plants were clipped in each quadrat, identified to species, and oven-dried; species abundance was expressed as above-ground dry weight (g/m^2). The species compositions of the four sites were similar, but the relative and absolute abundance as measured by above ground dry weight differed markedly between sites. As a result, both the community diversity and stability differed significantly. Needlegrass (*Stipa grandis*) was the single dominant species in the long-term ungrazed site, whereas Chinese rye grass (*Leymus chinensis*) dominated the long-term grazed site and was one of several species dominant in the short-term sites. The long-term ungrazed site had lower species richness (with 8.4 species/plot) than the other sites, which averaged 11.2 species/plot.

Zhang et al. computed the mean and variance of each species' dry weight for each permanent plot ($NQ = 9$) and used ODR to compute TPL for the plant communities at each site ($NB = 75\text{--}100$); their results are given in Appendix 6.Q. They found the mixed-species temporal TPL slopes were significantly greater than 1 at all four sites, but b was significantly lower ($P < 0.001$) in the long-term graze-free site than the others, indicating a more stable plant community in the minimally disturbed site. The long-term reservation grassland with minimal human disturbance featured a single dominant species, whereas the long-term free-grazing grassland, the mowed grassland, and the enclosed grassland (i.e., with more disturbances) featured higher species richness and multiple dominant species.

There are two sources of variance of a community: the variance of each species' abundance and the covariance between species. High diversity sites with many species must necessarily have smaller differences in abundance between species resulting in lower covariances than sites with a single dominant species and lower overall diversity. The long-term ungrazed site dominated by needlegrass had a lower TPL slope and greater stability than the more disturbed sites with higher species richness, whereas the mowed and enclosed grasslands had higher diversity.

The expected variance for the most abundant species at each site was estimated from TPL and the expected and observed CVs were compared. Most comparisons were not significantly different, but the observed CVs of one species in the mowed site, *Cleistogenes squarrosa*, and *C. squarrosa* and *Chenopodium album* in the short-term graze-free site were significantly higher than expected indicating their populations were substantially less stable than the average community members. Neither species was the most abundant at these sites.

An earlier study by Tsuiki et al. (2005) using the Hughes and Madden (1992) binomial power law (Chapter 16) reached similar conclusions in a study of grazing pressure at three sites in the semiarid steppe of China's Heilongjiang Province. Zhang et al.'s results also agree broadly with Polley et al. (2007) that at least in the short term species stability is maintained by dominant species rather than high diversity.

Simulation experiments by Tredennick et al. (2017) of five semiarid grass-lands found that the synchrony of per capita growth rates of community members was higher when environmental variation was present, while interspecific interactions and demographic stochasticity had little effect on synchrony. Simulations based on empirical data from four long-term grassland biodiversity experiments support Pimentel's (1961) position that species richness stabilizes communities by increasing community biomass and reducing the strength of demographic stochasticity (de Mazancourt et al., 2013). One of these long-term sites was in semiarid Central Texas; the other sites had higher rainfall. It is possible that ecologically similar species will exhibit synchronous dynamics, but even species belonging to the same functional group may respond differently to the same environment by arranging themselves differently in space (Park et al., 2013). In this context, Tredennick et al. (2017) suggest that subtle differences among dominant species may ultimately determine ecosystem stability.

All three case studies were conducted in semiarid grassland where growth is limited by sparse seasonal rains likely to synchronize plant growth and reproduction. Thus, the empirical data support Grime's (1998) hypothesis that ecosystem function, including stability, is most strongly influenced by the dominant species while the modeling studies, which included nonarid biomes, support a more nuanced conclusion.

Other examples

EDIBLE PALM IN BRAZIL

In a project to determine if density-dependent mortality or the community drift model accounts for population dynamics of tropical forest trees, Silva Matos et al. (1999) tested models of the edible palm *Euterpe edulis* population dynamics against field data. *Euterpe edulis* is a subcanopy palm that grows in Brazil's Mata Atlântica rainforest and has been so intensively harvested for heart of palm that it is now in decline. It prefers aquic soils and is often found in swampy areas. Silva Matos et al.'s field data were obtained in the Municipal Reserve of Santa Genebra in São Paulo State. In a swampy area of semideciduous forest in the reserve where *E. edulis* is abundant, a 1-ha area was marked in an array of 400 25 m^2 plots. 100 plots were randomly selected and within each plot plants within a 1-m^2 quadrat and with ≤ 3 leaves were tagged. All plants with >3 leaves were tagged in the 25-m^2 plots. The plots were monitored for 3 years from January 1991 and plants were classified in seven size classes and their transitions between classes recorded.

Plants with ≤ 3 leaves (seedlings) experienced the highest mortality and the largest size class (reproductives) the lowest mortality. The probabilities of mortality and transition between size classes and the recruitment rate of seedlings per reproductive were calculated from the field data and used to parameterize a matrix population model (Caswell, 1989). The matrix model was simulated with and without density-dependent mortality. They found that the matrix model incorporating density dependence predicted size distributions and densities matching the field data. However, density-dependent mortality varied substantially between size classes, being strongest in the seedling class. By incorporating seed dispersal into the model, Silva matos et al. also examined the spatial dynamics of *E. edulis* and found that predictions of densities are sensitive to the precise spatial dynamics of the population.

Silva Matos et al.'s Table 1 gives each age class' density ($\#/25\,m^2$) and *CV*: ensemble TPL of the *E. edulis* segregated by size has a highly significant slope of $b = 1.94 \pm 0.080$ that is not significantly different from 2.0 (Appendix 6.R). In this example, the age classes provide the range of means for TPL analysis; the points derived from a variable number of plants in $NQ = 100\text{--}400$ plots that change status from year to year.

INSECTIVOROUS PLANTS IN MOROCCO AND IBERIA

Insectivorous plants are among the most specialized plants: they are essentially sedentary predators that attract and trap insects and other arthropods to obtain nutrients. They are typically found in nutrient poor habitats: thin soils, acid bogs, rocky terrain, and xeric habitats. There are nearly 600 species described plus about 300 specializing in protozoans. *Drosophyllum lusitanicum* is the only insectivorous plant in the northern hemisphere adapted to xeric habitats. Garrido et al. (2003) surveyed and analyzed the environmental and spatial factors affecting populations in Morocco, Portugal, and Spain to document *Drosophyllum* plant communities in this range with the view to evaluating its conservation status.

Garrido et al. selected 32 sites (5 in Morocco, 7 in Portugal, and 20 in Spain) with known populations and established 200-m^2 plots within which all *Drosophyllum* plants were counted. *Drosophyllum* densities ranged between 0.5 and 4.13 plants/m^2. Cover by 53 species of woody plants and bare ground were estimated on a 25-m transect.

Drosophyllum populations varied in size, density, and age structure with populations becoming older and sparser toward the northwestern part of the range. Competition with woody plants was an important determinant of population vigor. To quantify the variation in *Drosophyllum* communities, they were classified using TWINSPAN (Hill et al., 1975; McCune and Mefford, 1999), a program for classifying species sample data that produces an ordered two-way table of their occurrence using a hierarchical classification system. Samples are successively divided into categories, and species are categorized using the sample classification. Garrido et al.'s TWINSPAN analysis identified 4 *Drosophyllum* communities: all Moroccan sites and those on Iberia with highly disturbed sites on low elevation high alkaline soils close to the coast; evergreen plantations on acid soils on the Atlantic coast; oligotrophic heathlands; and high elevation inland sites. In addition to the TWINSPAN classification, two other sources of variation were apparent: latitude and altitude. Garrido et al.'s

Table 1 gives the latitude, altitude, and *Drosophyllum* population density (#/ 200 m^2) at the 32 sites. Table 2 gives the mean population density and standard deviation of the TWINSPAN scores and the number of sites per classification.

Latitude, altitude, the TWINSPAN score offer three ways to segregate the data for TPL analysis. The 4 TWINSPAN scores were composed of $NQ = 4$–15 sites and ranking the sites by latitude and altitude into $NB = 8$, $NQ = 4$ site groupings for TPL analysis. Not surprisingly, Appendix 6.S. shows the three approaches produce virtually identical TPL regressions as both altitude and latitude were identified by TWINSPAN as key variables in its classification.

EELGRASS IN CHESAPEAKE BAY

Eelgrass (*Vallisneria americana*) is a fresh water plant that tolerates salt and is frequently found in estuarine submersed aquatic vegetation (SAV) beds. SAV beds provide ecological services by providing food for waterfowl and habitat for fish and aquatic invertebrates and by enhancing water quality. Eelgrass is a dominant, possibly keystone, species in the Chesapeake Bay on the U.S. eastern seaboard. It is a deep-rooted plant that grows under water but flowers above the surface. However, eutrophication and sedimentation have reduced the number and total area of SAV beds and the population of eelgrass.

Using maps of SAV patches in the tidal region of the Bay developed by the Virginia Institute of Marine Science, a program of The College of William & Mary (http://www.vims.edu), Lloyd et al. (2016) developed coverage maps representing patches of SAV that are within the salinity and depth limits of eelgrass for the years 1984–2010. These data showed that 2128 SAV patches totaling 250 km^2 likely contained eelgrass in patches between 0.09 and 4834 ha in extent (median area: 0.27 ha). Temporal changes in patch distribution provide insight into the colonization dynamics and thus the long-term sustainability of SAV patches in the Bay.

Seeds released from fruit settle within tens of meters of the parent and detached reproductive shoots are known to be transported a few kilometers on currents. Fruit and seeds carried by waterfowl probably disperse further, but by an unknown amount. Thus, connectivity by dispersal between SAV patches probably extends over many kilometers. Lloyd et al. examined the connectivity between patches at a range of critical dispersal distances from 100 m to the distance at which all patches are connected. At each critical distance, the connectivity of SAV patches was estimated using graph theoretic methods. At the lowest critical distance, each patch is isolated and at the highest, all patches are networked together into a single group or component. At each successive critical distance, the number of patches in a networked group decreases and the average number of patches per group increases. They found that all patches were connected at critical distances between 236 and 251 km, but critical distances <10 km were considered most important as eelgrass dispersal mechanisms were likely limited to this range.

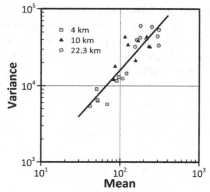

FIG. 6.13   Ensemble TPL ($NQ = 7$–26, $NB = 27$) of submersed aquatic vegetation patches containing eelgrass in networks connected at three critical distances (4, 10, and 22.3 km) in Chesapeake Bay, U.S.A. The average number of patches per network increased with critical distance, providing a range of means for analysis. *(Data from Table 2 in Lloyd et al. (2016).)*

Lloyd et al.'s Table 2 gives the number of networked patch groups, the mean number of patches, and SDs estimated for three critical distances (4 km, 10 km, and 22.3 km) for 9 years. These values form three populations of patches that are linked together in a network. The method employed to find the number of groups at any particular critical distance is equivalent to sampling the population of SAVs expected to have eelgrass in quadrats proportional to the critical dispersal distance. As such, these data are amenable to TPL analysis. Fig. 6.13 (Appendix 6.T) shows the 4-km critical distance set extending over an order of magnitude, the 10-km range is narrower, and the 22.3-km is restricted to the highest means. Although the fitted estimates differ, the three sets lie on a common line with $b = 1.31 \pm 0.113$, indicating the eelgrass beds are moderately aggregated in space. That these points are derived from network critical distances suggests TPL is sensitive to the degree of connectedness of elements in a network.

POLLINATION SUCCESS IN A YUCATAN SHRUB

Called in Spanish chaya silvestre (wild chaya) and known to the Maya as tsah, *Cnidoscolus souzae* is endemic to Mexico's Yucatan Peninsula. Although the leaves are edible, it bears abundant stiff, sharp, stinging hairs said to be more painful than the Holarctic stinging nettle (*Urtica dioica*). Tsah's small white insect pollinated male and female flowers bloom in July–October and remain open for 1 day. Because of the short period during which pollination can occur, tsah are ideal for a field study to document the spatial variation in pollination success. Pollen deposition and pollen tube formation are important determinants of reproductive success in flowering plants. Arceo-Gómez et al.'s

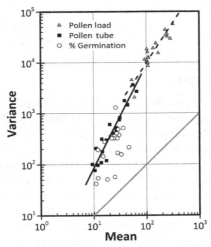

FIG. 6.14 Spatial distribution of reproductive success in a tropical shrub, *Cnidoscolus souzae*. Spatial TPLs ($36 \leq NQ \leq 69$, $NB = 19$) of per flower pollen load on the stigma, pollen tubes to the base of the style and germination rate (pollen that succeeded in entering the style). *(Data from Table 1 in Arceo-Gómez et al. (2016).)*

(2016) study documents how spatial variation is partitioned in populations, plants, and flowers. Variation in pollen load and pollen tube number per flower was mostly explained by within-individual and between-population (spatial) variance. Both within-plant flower differences and spatial differences in habitat can play an important role in determining pollination success.

Arceo-Gómez et al. sampled 19 tsah populations at 0.5–5.0 km intervals on a ~50 km SE-NW transect on the Yucatan Peninsular. At each site, 3 wilted female flowers were collected from 12 to 23 plants over a 3-day period and returned to the lab where microscopic examination and dissection revealed the pollen load on the stigma, the number of pollen grains that succeeded in entering the style and started a pollen tube (pollen germination), and the number of pollen tubules reaching the base of the style where the ovaries are located. Their Table 1 gives the mean, SE and CV of pollen load and successful pollen tubes per flower and the germination rate at all 19 sites. On the basis of these data, Arceo-Gómez et al. concluded that pollen quantity and quality did not limit tsah reproductive success in its Yucatan range.

TPLs of pollen load on the stamen, successful tubes to the base of the style, and germination rate (pollen entering the style) are shown in Fig. 6.14 (Appendix 6.U). It is noteworthy that germination rate, a bounded proportion (Chapter 13), produces a poor diffuse TPL ($r = 0.53$) compared to pollen load and tube success, which differ markedly in slope. The difference in spatial distributions of pollen load ($b = 1.57 \pm 0.143$) and pollination success ($b = 2.04 \pm 0.150$) suggests the number of pollen grains on the stamen that succeed in reaching the base of the style grows more slowly than the number of pollen grains landing on the stamen, resulting in a density-dependent effect on pollen tube formation revealed by the difference in TPL slopes.

Commodity crops

The US Department of Agriculture (USDA) started keeping records of crop production in 1866, 4 years after its foundation in May 1862. The earliest records included the commodity crops barley (*Hordeum vulgare*), cotton (*Gossypium* spp.), maize (*Zea mays*), oats (*Avena sativa*), potatoes (*Solanum tuberosum*), rye (*Secale cereale*), tobacco (*Nicotiana tabacum*), and wheat (*Triticum* spp.). Since then, minor crops and livestock have been added to the annual surveys. The surveys include area harvested as well as total production by state and county; other statistics such as area irrigated are also recorded. The number of states has grown from 36 in 1866 to 48 contiguous states today, plus Alaska and Hawaii. For confidentiality reasons some data are not available at the county level, but state production data for hundreds of agricultural products are downloadable at 〈http://www.nass.usda.gov/Data_and_Statistics and http://agcensus. mannlib.cornell.edu〉. Although the states vary substantially in size, these political units provide a useful quadrat for analyzing the spatial distribution of farmland devoted to a range of crops as well as their annual productivity. Because of the disparity in size of the quadrats one might expect TPL not to apply, but this is not the case. In fact TPL not only applies to these data, but can also identify major changes in farming habits in response to changing technology and economics. Interestingly, while harvested quantity and area harvested obey TPL, yield, harvest per unit area, often does not. Here we examine harvest and area but defer yield to Chapter 13.

A summary of TPL analyses of 16 commodity crops is presented in Appendices 6.V and 6.W and four are described in detail.

MAIZE

Maize or corn (*Zea mays*) is one of the largest crops in the United States. It is grown commercially in nearly every state with half the agricultural area of the Midwestern states from Ohio to Kansas and Texas to North Dakota devoted to this crop. Currently, annual production is generally $>350 \times 10^6$ metric tons per year on $>350,000 \, \text{km}^2$, averaging ~ 10 metric tons/ha. In 1866, 35 states reported corn production totaling 530,000 tons on $3470 \, \text{km}^2$. Since then, the number of states has ranged from 35 to 48 with 41 states reporting production for the last 50 years. Much of the increase in total production is clearly the result of increased area harvested, but average yield has also increased substantially, due to improved varieties and agricultural methods, including the "Green Revolution" of the 1950s. The Second World War saw the development of new pest control chemicals and better matching of germplasm to inorganic fertilizers.

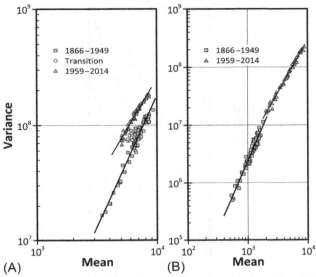

FIG. 6.15 Spatial TPLs ($NQ = 34-48$, $NB = 84$ and 56) of (A) area of maize production (km^2) and (B) total production (1000 metric tons) show transitions in the 1950s coincident with the transition to modern chemically intensive farming methods introduced by the Green Revolution.

This led to a change in agricultural practices as well as yields. Fig. 6.15. shows the TPLs for area harvested and total production of corn from 1866 to 2014. The plot of area shows two distinct variance-mean relationships: 1866–1946 and 1959–2014, with a transition period in the 1950s with no TPL as agriculture realigned in response to the new cultivars, fertilizers, and pest control methods. It was shortly after this period, in 1967, that the number of states reporting commercial corn production dropped from 48 to 41. The companion figure of corn harvested shows a concurrent reduction in the variance-mean slope as production became more consolidated in a smaller number of states.

SUGAR CANE

Sugar cane is grown in only four states, up from two from 1866 to 1933 and three from 1934 to 1972. With so few "quadrats," it is surprising that TPLs are evident. In fact, there are four separate TPLs for area harvested corresponding to the three periods of sugar cane cultivation with the middle period separated into two distinct variance-mean relationships. The effect of number of quadrats on the TPL is clearly evident here with different intercepts but similar slopes for the four periods. It is not immediately obvious why 1934–72 is discontinuous at 1961 as the same three states (Florida, Louisiana, and Hawaii[1])

1. Unlike the other commodities, Hawaii is included in this analysis.

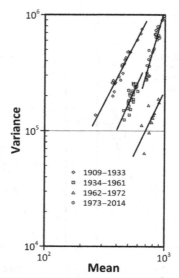

FIG. 6.16 Spatial TPLs of sugar cane area of production (km²) break down into four distinct periods with slightly different slopes reflect the increasing number of states (i.e., quadrats) producing sugar from $NQ = 2$ to 4 states ($NB = 15$, 28, 11, and 41).

grew sugar cane. Examination of the data shows a dramatic increase in area growing sugar cane in Florida. Thus, while the slope remained the same, more than doubling the area harvested in one state substantially reduced the variability. In 1973, commercial sugar cane production started in Texas, resulting in a larger area harvested and slightly steeper regression. Fig. 6.16 (Appendix 6.W) shows the sugar cane area harvested clearly segregated into four relationships. The total production (metric tons) produces a single regression with comparatively low correlation.

WHEAT

Wheat has been recorded since 1866 and grown in about 41 states since 1880, and during that time total harvest has grown from about 330 K tons to 1400 K tons per year. A large portion of that gain was made in the 1950s: in 1945 the yield was 1.3 T/ha, by 1965 the yield had risen to almost 2 T/ha and now it stands at almost 4 T/ha. Despite these dramatic gains in yield, the TPL for total harvest is remarkably consistent (Fig. 6.17B). By contrast, the TPL for area harvested shows a discontinuity between 1901 and 1909 (Fig. 6.17A). A possible explanation for the transition is that prior to 1909, wheat reporting did not distinguish between spring and winter sown wheat. Since then the area harvested, the total

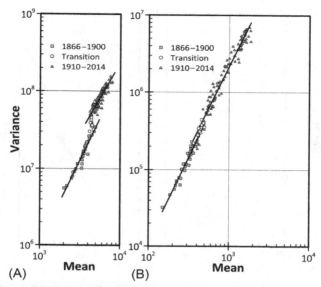

FIG. 6.17 Spatial TPLs ($NQ = 31$–43, $NB = 35$ and 105) of area of wheat production (A) separates into two distinct TPLs with a transition coinciding with an increase in winter wheat planting in the first decade of the 20th century and introduction of separate reporting for winter and spring wheat. Interestingly, the total production (B) does not show a transition.

harvest and the average yield have remained very similar for both crops on an annual basis (Appendices 6.V and 6.W). The reason for the discontinuity in the area TPL could therefore be due to a methodological change in data collection that started in 1901, rather than a real change in the spatial aggregation of wheat production.

POTATOES

Like sugar cane, the area of potatoes harvested segregates into several distinct variance–mean relationships, and like wheat and corn, there is a range of years that do not fit but appear to be transitional. There are two periods of transition in area harvested (1901–09 and 1946–58), but only one for total harvested (1946–58). The separation of 1866–1900 and 1910–46 may reflect a methodological change in data collection, but the transition of both area harvested and quantity harvested (Fig. 6.18) to the latest period certainly reflects the changes in land use following the Green Revolution as we saw with corn. Unlike corn, however, quantity harvested became more not less aggregated as the productivity increased from the 1960s.

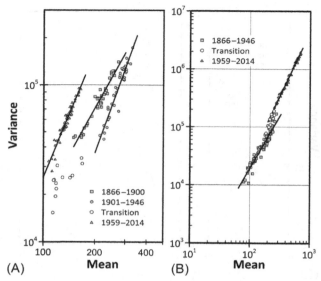

FIG. 6.18 Area of potatoes (A) segregates out into three TPLs ($NQ = 29$–48, $NB = 35, 46$, and 56) with a cloud of transitional years comprising two periods (1901–09 and 1946–58). Total potato production (B) has two distinct variance mean relationships separated by a single transitional period during the Green Revolution. The slope is slightly steeper following the transition reflecting the smaller area devoted to potatoes but a continued increase in productivity.

In this chapter, we have seen a variety of qualitatively different data collected in different ways well described by TPL. The use of TPL to distinguish between hypotheses is notable, as is the consistency of result obtained by quadrats of different size (e.g., U.S. states), a property to be seen again in Chapter 11. The breakdown of TPL and transitions between periods governed by different rules is also noteworthy.

Appendix: TPL estimates for plants

	NB	Range of means	r	A	SE[A]	b	SE[b]
Seedbank							
A Seedbank and weeds in fields at Basingstoke and Boxworth in England. Data from Tables 2, 4, 6, and 8 in Marshall (1989)							
Basingstoke – Grasses in cereal fields	32	0.1 - 337	0.987	0.666	0.052	1.855	0.052
Dicots in April	46	0.1 - 44.5	0.990	0.951	0.032	1.940	0.040
Boxworth – Weed seeds in January	48	25 - 1,324	0.919	1.670	0.165	1.423	0.081
Weed plants in November	110	0.4 - 964	0.974	0.789	0.030	1.589	0.034

Continued

	NB	Range of means	r	A	SE[A]	b	SE[b]
B Winter wild oats in fields near Madrid, Spain and crab grass seeds in 2 fields at Hautes Landes, France, 1983 and 1985 (Chauvel et al., 1989). In this and the next 4 examples, estimation of b and A was by weighted least squares (Chapter IV). The Range entry is the range of log(M). Data from Table 1 in Clark et al. (1996)							
Winter wild oat (panicles/m^2)	20	1.59	0.921	0.46	0.082	1.52	0.144
Crab grass seed (#/m^2)	4	0.97	0.982	0.50	0.068	1.46	0.140
C Black grass in fields at Neville's Farm, Wantage, England, 1977–1986. Data from Table 1 in Clark et al. (1996)							
Black grass – 1977 (panicles/m^2)	9	5.98	0.941	0.86	0.156	1.72	0.206
1978	10	2.95	0.983	0.89	0.068	1.50	0.088
1979	7	2.32	0.975	1.14	0.121	1.58	0.135
1980	6	2.46	0.974	1.23	0.136	1.72	0.162
1981	9	2.85	0.980	1.04	0.097	1.71	0.115
1982	8	1.69	0.975	0.93	0.077	1.63	0.131
1983	7	2.94	0.981	1.06	0.097	1.49	0.112
1984	9	2.04	0.990	1.11	0.071	1.61	0.078
1985	10	2.22	0.398	1.99	0.101	1.56	1.137
1986	8	2.53	0.975	1.07	0.091	1.58	0.128
All years – Scale 1	83	3.4	0.977	1.03	0.030	1.59	0.038
Scale 2	10	0.96	0.959	0.34	0.448	2.00	0.187
D Three grass species at different scales in fields at Boxworth, England. Data from Table 1 in Clark et al. (1996)							
Scale 1, 0.46 ha survey area							
Barren brome	23	1.64	0.960	0.71	0.066	1.75	0.107
Meadow brome	11	1.17	0.949	0.97	0.079	1.87	0.130
Couchgrass	19	1.78	0.903	0.83	0.045	1.61	0.176
Scale 1, 0.96 ha survey area							
Barren brome	12	1.33	0.938	0.76	0.122	1.85	0.198
Meadow brome	8	0.68	0.881	1.09	0.204	1.71	0.324
Couchgrass	11	1.30	0.969	0.92	0.060	1.56	0.120
Scale 2, 0.46 ha survey area							
Barren brome	6	0.43	0.489	0.13	2.280	1.73	1.260
Meadow brome	6	0.80	0.692	0.69	0.0250	1.16	0.494
Couchgrass	6	1.37	0.953	0.39	0.268	1.85	0.240
E Two grass species at different scales in 4 fields, 1984–1987 at Boxworth, England. Data from Table 1 in Clark et al. (1996)							
Scale 1							
Meadow brome – K, 1985	22	1.48	0.877	0.20	0.079	1.63	0.190

Continued

	$N\beta$	Range of means	r	A	SE[A]	b	SE[b]
T, 1987	75	2.13	0.740	0.16	0.091	1.02	0.107
TP, 1985	54	1.62	0.773	0.26	0.093	1.38	0.154
Black grass – EC, 1985	35	1.94	0.889	0.23	0.122	1.61	0.140
EC, 1987	44	1.45	0.843	0.18	0.093	1.57	0.151
T, 1987	66	1.71	0.844	0.13	0.071	1.43	0.112
Scale 2							
Meadow brome – K, 1984	8	1.54	0.962	0.61	0.09	1.55	0.156
K, 1985	8	1.03	0.978	0.61	0.063	2.08	0.157
T, 1984	10	1.79	0.963	0.35	0.076	1.37	0.122
T, 1986	6	1.44	0.990	0.46	0.063	1.70	0.098
T, 1987	10	1.59	0.921	0.36	0.222	1.62	0.217
TP, 1984	8	1.63	0.988	0.79	0.062	1.87	0.103
TP, 1985	10	1.66	0.928	0.65	0.131	1.51	0.191
TP, 1986	7	1.59	0.963	0.22	0.084	1.39	0.146
Black grass – EC, 1984	9	2.14	0.930	0.79	0.205	1.42	0.187
EC, 1985	8	1.21	0.945	0.65	0.175	1.72	0.211
EC, 1987	10	1.12	0.949	0.57	0.092	1.54	0.162
T, 1986	10	1.03	0.877	0.62	0.108	1.43	0.248
T, 1987	10	2.19	0.990	0.50	0.06	1.58	0.071
Scale 3							
Meadow brome	10	1.98	0.969	0.36	0.189	1.85	0.148
Black grass	7	2.88	0.973	0.69	0.215	1.58	0.141

F Consistency of TPL slope of meadow brome in 6 fields, 1982–1988, at Boxworth, England. Data from Table 1 in Clark et al. (1996)

Meadow brome – B, 1987 (panicles/m^2)	6	2.57	0.995	0.75	0.071	2.02	0.084
B, 1988	8	2.48	0.995	0.68	0.054	1.80	0.064
K, 1984	8	1.32	0.931	0.62	0.563	1.79	0.248
K, 1985	7	1.12	0.859	0.73	0.258	1.40	0.316
K, 1987	8	2.04	0.961	0.69	0.132	1.61	0.163
K, 1988	8	1.59	0.971	0.36	0.137	1.93	0.168
PH, 1984	7	1.78	0.968	0.66	0.103	1.63	0.160
PH, 1985	8	1.16	0.986	0.24	0.097	1.96	0.116
PH, 1986	7	1.06	0.929	0.44	0.101	1.65	0.249
PH, 1988	8	1.09	0.932	0.45	0.082	1.53	0.211
SA, 1982	7	1.65	0.968	0.60	0.104	2.18	0.213
SA, 1983	7	2.00	0.971	0.71	0.126	1.91	0.179
SA, 1954	6	1.28	0.972	0.68	0.099	1.88	0.185
SA, 1855	7	1.19	0.970	0.62	0.062	1.77	0.169
SA, 1987	8	1.69	0.988	0.54	0.062	2.18	0.123
SA, 1988	8	1.23	0.912	0.74	0.158	1.48	0.236
T, 1982	11	1.2	0.961	0.56	0.161	1.79	0.156
T, 1983	11	1.48	0.911	0.58	0.112	1.62	0.221

Continued

	NB	Range of means	r	A	SE[A]	b	SE[b]
T, 1984	10	1.91	0.957	0.56	0.088	1.73	0.166
T, 1987	12	3.12	0.987	0.62	0.077	1.70	0.081
T, 1988	12	2.79	0.985	0.56	0.072	1.55	0.078
TP, 1984	7	1.48	0.964	0.61	0.081	1.58	0.164
TP, 1985	7	0.96	0.940	0.37	0.139	1.96	0.268
TP, 1986	6	0.72	0.936	0.37	0.106	2.41	0.371
TP, 1988	5	0.60	0.974	0.31	0.059	2.01	0.208

G Ensemble TPL of weeds in $NQ = 7$ habitats forming the margins of fields in Wisconsin. Data from Table 1 in Sosnoskie et al. (2007)

	NB	Range of means	r	A	SE[A]	b	SE[b]
Weeds in field margins	46	0.143 - 48.6	0.951	0.39	0.068	1.38	0.063

H Spatial TPLs ($NQ = 5$) of seedbank and seedlings of lambsquarters and annual grasses in soybean plots. Data from Table 1 in Cardina et al. (1996)

	NB	Range of means	r	A	SE[A]	b	SE[b]
Lambsquarters – Seedbank	7	760 - 22,391	0.981	−0.392	0.466	1.814	0.133
Seedlings	7	7 - 476	0.952	0.401	0.294	1.270	0.147
Annual grasses – Seedbank	7	209 - 12,662	0.972	0.618	0.419	1.464	0.129
Seedlings	8	21 - 1,052	0.992	−0.776	0.163	1.653	0.072

I Spatial TPLs of giant foxtail, lambsquarters, and velvetleaf seeds ($NQ = 15$) and plants ($NQ = 16$) in soybeans planted at 18 cm or 76 cm row spacing. ODR estimates from Tables 3 and 5 in Mulugeta and Boerboom (1999)

	NB	Range of means	r	A	SE[A]	b	SE[b]
Seeds – Foxtail, 18-cm rows – 1996 - 1997	8	-	0.97	−0.55	0.28	1.50	0.14
	8	-	0.89	0.56	1.80	1.96	0.39
76-cm rows – 1996 - 1997	8	-	0.96	−0.54	0.59	1.46	0.15
	8	-	0.96	0.43	0.73	1.23	0.14
Lambsquarters, 18-cm rows – 1996 - 1997	8	-	0.84	−0.21	1.53	2.14	0.55
	8	-	0.87	−0.11	5.24	2.05	0.46
76-cm rows – 1996 - 1997	8	-	0.62	−0.17	2.85	2.12	1.07
	8	-	0.94	−0.62	5.52	1.80	0.30
Velvetleaf, 18-cm rows – 1996 - 1997	8	-	0.99	−0.57	1.02	0.76	0.02
	8	-	0.96	0.54	0.35	0.86	0.10
76-cm rows – 1996 - 1997	8	-	0.93	−0.89	1.23	1.13	0.18
	8	-	0.94	0.50	0.52	1.03	0.16
Seedlings - Foxtail, 18-cm rows – 1996 - 1997	8	-	0.98	−0.03	0.74	1.29	0.11
	8	-	0.89	−0.34	2.94	1.55	0.32

Continued

	NB	Range of means	r	A	SE[A]	b	SE[b]
76-cm rows – 1996 - 1997	8	-	0.98	−0.82	1.99	1.60	0.13
	8	-	0.93	0.26	1.98	2.22	0.37
Lambsquarters, 18-cm rows – 1996 - 1997	8	-	0.84	−0.09	2.95	2.00	0.51
	8	-	0.68	0.14	2.04	1.02	0.45
76-cm rows – 1996 - 1997	8	-	0.85	0.31	0.69	0.77	0.19
	8	-	0.80	−0.10	2.35	1.32	0.40
Velvetleaf, 18-cm rows – 1996 - 1997	8	-	0.73	0.15	1.23	0.87	0.33
	8	-	0.85	0.14	0.62	0.85	0.21
76-cm rows – 1996 - 1997	8	-	0.75	0.05	1.52	1.08	0.38
	8	-	0.73	0.17	0.99	0.84	0.32

J Mixed-species TPL (NQ = 72) of seeds of 11 tropical tree species in Taiwan dispersed by animals and wind Data from Table 1 in Lin et al. (2011)

	NB	Range of means	r	A	SE[A]	b	SE[b]
Animal dispersed seeds/m²	7	1.43 - 115	0.985	0.92	0.177	2.36	0.151
Wind dispersed seeds/m²	4	2.51 - 31.2	0.995	1.11	0.100	2.12	0.103
All tree seeds/m²	11	1.43 - 115	0.986	0.95	0.124	2.32	0.114

K Mixed-species TPLs (NQ = 254) of weeds in a wheat field in Catalonia did not change from year to year. Data from Table 1 in Izquierdo et al. (2009)

	NB	Range of means	r	A	SE[A]	b	SE[b]
2001	23	0.60 - 1,608	0.986	0.54	0.072	1.37	0.047
2002	22	0.60 - 6,405	0.994	0.58	0.040	1.31	0.029
2003	23	0.60 - 3,135	0.994	0.58	0.036	1.25	0.028
Combined	68	0.60 - 6,405	0.991	0.56	0.030	1.33	0.022

L Spatial TPLs (NQ = 18) of plants and seeds of devil's thorn (*Emex australis*), a weed introduced to Australia from southern Africa. Data from Table 3 in Weiss (1981)

	NB	Range of means	r	A	SE[A]	b	SE[b]
Devil's thorn – plants/m²	14	4.1 - 520	0.976	−0.18	0.132	1.44	0.085
seeds/m²	12	2.7 - 1112	0.992	−0.39	0.105	1.60	0.058
combined	26	2.7 - 1112	0.986	−0.33	0.084	1.55	0.050

M Ensemble TPLs (NQ = 10-20) of ragweed in France. Data from Tables 2 and 5 in Fumanal et al. (2008)

	NB	Range of means	r	A	SE[A]	b	SE[b]
Dormant seeds/m² (0–5 cm)	5	26.0 - 134	0.947	1.97	0.371	1.46	0.210
Dormant seeds/m² (5–20 cm)	5	27.0 - 255	0.972	1.67	0.323	1.68	0.177

Continued

	NB	Range of means	r	A	SE[A]	b	SE[b]
Dormant seeds/m^2 (0–20 cm)	10	26.0 - 255	0.956	1.77	0.265	1.60	0.148
Nondormant seeds/m^2 (0–5 cm)	9	306 - 2627	0.861	0.91	0.831	1.73	0.293
Nondormant seeds/m^2 (5–20 cm)	9	26.0 - 2343	0.942	2.21	0.417	1.32	0.148
Nondormant seeds/m^2 (0–20 cm)	18	26.0 - 2627	0.914	1.96	0.374	1.38	0.132
Live seeds/m^2 (0–5 cm)	9	332 - 2761	0.860	0.91	0.834	1.72	0.292
Live seeds/m^2 (5–20 cm)	9	26.0 - 2343	0.960	2.12	0.359	1.36	0.126
Live seeds/m^2 (0–20 cm)	18	26.0 - 2761	0.927	1.85	0.358	1.42	0.126
Dead seeds/m^2 (0–5 cm)	9	26.0 - 713	0.962	1.79	0.309	1.54	0.141
Dead seeds/m^2 (5–20 cm)	9	26.0 - 1045	0.958	1.34	0.372	1.71	0.163
Dead seeds/m^2 (0–20 cm)	18	26.0 - 1045	0.957	1.58	0.247	1.62	0.111
Total seeds/m^2 (0–5 cm)	9	433 - 3324	0.869	0.57	0.890	1.82	0.300
Total seeds/m^2 (5–20 cm)	9	102 - 2761	0.944	1.55	0.509	1.55	0.171
Total seeds/m^2 (0–20 cm)	18	102 - 3324	0.912	1.24	0.467	1.63	0.157
Seedlings/m^2	9	23.0 - 292	0.945	1.55	0.295	1.31	0.143

N Ensemble TPL (*NQ* > 950) of endangered desert plant, Las Vegas bearpoppy (*Arctomecon californica*), in Nevada is the same at 0–2 cm and 2–4 cm depth. Data digitized from Fig. 2 in Megill et al. (2011)

	NB	Range of means	r	A	SE[A]	b	SE[b]
Seeds/m^2 at 0–2 cm	7	0.218 - 3.26	0.981	2.35	0.054	1.49	0.110
Seeds/m^2 at 2–4 cm	5	0.218 - 2.08	0.915	2.45	0.119	1.53	0.276
Seeds/m^2 combined depths	12	0.218 - 3.26	0.955	2.39	0.058	1.52	0.130

Grasslands

O Mixed-species TPL (*NQ* = 100) comparison of a grassland and weed community in Shaanxi Province, China. Data from Table 1 in Guan et al. (2016)

	NB	Range of means	r	A	SE[A]	b	SE[b]
Grassland	32	0.01 - 102	0.970	0.78	0.068	1.41	0.061
Weed community	30	0.01 - 35.4	0.982	1.02	0.068	1.57	0.054

Continued

	NB	Range of means	r	A	SE[A]	b	SE[b]
P	Mixed-species TPLs showing community stability in 4 tallgrass prairies in Central Texas. Data courtesy of Dr. Wayne Polley.						
Case 1: Averaging years (NQ=5)							
Temple remnant prairie	315	0.020 - 53.18	0.981	0.46	0.016	1.80	0.020
Temple restored prairie	209	0.020 - 50.28	0.982	0.47	0.021	1.79	0.023
Reisel remnant prairie	297	0.020 - 41.44	0.985	0.36	0.017	1.77	0.017
Reisel restored prairie	227	0.020 - 40.58	0.977	0.33	0.025	1.72	0.024
Case 2: Averaging plots without zeros (NQ=2-12)							
Temple remnant prairie	130	0.20 - 90.8	0.908	−0.36	0.068	2.23	0.082
Temple restored prairie	67	0.20 - 46.5	0.901	−0.28	0.113	2.19	0.116
Reisel remnant prairie	103	0.13 - 40.4	0.910	−0.26	0.074	2.19	0.090
Reisel restored prairie	72	0.15 - 39.3	0.936	−0.20	0.061	2.13	0.088
Case 3: Averaging plots with zeros (NQ=12)							
Temple remnant prairie	132	0.0083 - 22.7	0.979	0.71	0.024	1.74	0.031
Temple restored prairie	58	0.0083 - 45.0	0.980	0.65	0.045	1.70	0.044
Reisel remnant prairie	125	0.0083 - 23.5	0.987	0.61	0.026	1.78	0.026
Reisel restored prairie	77	0.0083 - 36.0	0.973	0.49	0.048	1.64	0.043
Q	Community stability in a series of sites in Inner Mongolia as measured by TPL is greatest in the least disturbed site (no grazing for 50 years) and greatest in the most disturbed site (continuous grazing for 50 years). Each point (NQ = 9) is a steppe plant species, many common to all sites. ODR estimates from Fig. 3 in Zhang et al. (2016)						
Long-term exclusion	~90	0.0033 - 146	0.964	0.13	-	1.66	0.066
Long-term grazing	~100	0.0096 - 77.2	0.990	0.52	-	1.83	0.040
Short-term exclusion	~100	0.0032 - 70.5	0.964	0.34	-	1.75	0.070
Short-term mowing	~100	0.0033 - 85.1	0.969	0.34	-	1.77	0.065

Other examples

	NB	Range of means	r	A	SE[A]	b	SE[b]
R	Ensemble TPL (NQ=100) of edible palm tree density (#/25 m^2) constructed from 7 age classes. Data from Table 1 in Silva Matos et al. (1999)						
Edible palm trees	7	0.31 - 74.5	0.994	0.72	0.071	1.94	0.080

	NB	Range of means	r	A	SE[A]	b	SE[b]
S							

	NB	Range of means	r	A	SE[A]	b	SE[b]
TWINSPAN	4	23.0 - 461	0.976	−1.39	0.611	2.50	0.273
Latitude	8	19.8 - 399	0.965	−1.77	0.567	2.71	0.251
Altitude	8	26.0 - 419	0.930	−1.41	0.733	2.51	0.327

S Distribution of insectivorous plant, *Drosophyllum lusitanicum*, in Morocco, Spain, and Portugal. TPL (*NQ*=4-15) regressions of abundance at sample sites segregated by TWINSPAN classification, latitude, and altitude are very similar in part because altitude and latitude were key variables identified by TWINSPAN in its classification. Data from Tables 1 and 2 in Garrido et al. (2003)

T Spatial TPLs (*NQ*=7-26) of number of submersed aquatic vegetation patches containing eelgrass at three critical distances (scales) estimated from maps of Chesapeake Bay. Data from Table 2 in Lloyd et al. (2016)

	NB	Range of means	r	A	SE[A]	b	SE[b]
4 km	9	41.5 - 120	0.871	2.07	0.308	1.01	0.166
10 km	9	83.1 - 239	0.639	2.09	0.607	1.09	0.279
22.3 km	9	154 - 309	0.207	2.78	0.609	0.80	0.259
Combined	27	41.5 - 309	0.893	1.58	0.242	1.31	0.113

U TPL of per flower (36 < *NQ* < 69) pollination success of *Cnidoscolus souzae*, a shrub endemic to the Yucatan Peninsula. Data from Table 1 in Arceo-Gómez et al. (2016)

	NB	Range of means	r	A	SE[A]	b	SE[b]
Pollen load on style/flower	19	52.4 - 317	0.917	0.85	0.308	1.57	0.143
Tubes to base of style/flower	19	9.40 - 64.8	0.948	−0.11	0.203	2.04	0.150
% of Tube Formation	19	11.0 - 46.0	0.526	−0.78	0.612	2.30	0.448

Commodity crops

V Spatial TPLs (NQ=4-48) of area of commodity crops harvested (km^2) in the continental U.S., 1866-2014. Data from USDA annual surveys.

	NB	Range of means	r	A	SE[A]	b	SE[b]
Barley	149	145 - 1,730	0.988	0.569	0.074	1.990	0.026
Canola	18	467 - 1574	0.673	1.212	0.912	1.751	0.305
Corn (maize) grain	149	3471 - 9349	0.780	−1.718	0.497	2.529	0.130
Cotton bales	149	951 - 10,619	0.812	−0.747	0.394	2.264	0.108
Oats	149	111 - 3839	0.994	−0.353	0.064	2.211	0.020
Peanuts	106	156 - 1087	0.975	0.071	0.118	2.027	0.043
Potatoes	149	107 - 329	0.795	1.507	0.166	1.469	0.073
Rice	120	182 - 2558	0.889	1.046	0.204	1.609	0.067
Rye	149	60.9 - 855	0.935	−1.008	0.176	2.556	0.074
Grain sorghum	86	693 - 3463	0.909	0.347	0.307	2.099	0.094
Soybeans	91	76.3 - 10,843	0.989	1.226	0.094	1.737	0.027
Sugarbeet	91	106 - 515	0.976	0.422	0.102	1.871	0.042
Cane sugar (includes Hawaii)	87	264 - 952	0.312	−0.192	0.574	2.027	0.206
Sunflower oil	39	512 - 5252	0.950	0.678	0.292	1.858	0.093
Tobacco	149	92.5 - 452	0.987	−0.349	0.073	2.355	0.031
Winter wheat	107	2784 - 5772	0.879	−0.982	0.400	2.415	0.111
Spring wheat	96	1461 - 7185	0.922	2.027	0.232	1.617	0.064
Total wheat	107	2011 - 7680	0.960	−3.507	0.302	3.038	0.082

Continued

	NB	Range of means	r	A	SE[A]	b	SE[b]
W	Spatial TPLs (NQ=4-48) of quantity of commodity crops harvested (1000 metric tons) in the continental U.S., 1866-2014						
Barley	149	23.4 - 591	0.993	0.466	0.045	2.018	0.020
Canola	18	467 - 1574	0.673	1.212	0.912	1.751	0.305
Corn (maize) grain	149	530 - 8807	0.994	−0.181	0.066	2.194	0.020
Cotton bales	149	41.5 - 306	0.895	−2.418	0.251	3.160	0.116
Oats	149	111 - 3839	0.994	−0.353	0.064	2.211	0.020
Peanuts	106	13.4 - 306	0.995	0.192	0.038	1.991	0.020
Potatoes	149	79.8 - 699	0.987	−0.910	0.080	2.542	0.033
Rice	120	17.7 - 1838	0.919	−0.049	0.181	1.929	0.069
Rye	149	10.5 - 80.9	0.886	−0.155	0.125	2.358	0.090
Grain sorghum	86	693 - 3463	0.909	0.347	0.307	2.099	0.094
Soybeans	91	6.03 - 3482	0.994	1.022	0.057	1.777	0.020
Sugarbeet	91	283 - 3196	0.984	0.965	0.097	1.730	0.032
Cane sugar (includes Hawaii)	87	850 - 8237	0.692	2.077	0.379	1.357	0.105
Sunflower oil	39	512 - 5252	0.950	0.678	0.292	1.858	0.093
Tobacco	149	6.16 - 59.1	0.989	0.071	0.042	2.282	0.028
Winter wheat	107	271 - 1391	0.966	−0.124	0.154	2.192	0.055
Spring wheat	96	99.0 - 1786	0.982	1.004	0.099	1.841	0.036
Total wheat	107	149 - 1048	0.987	−1.042	0.105	2.486	0.039

References

Arceo-Gómez, G., Alonso, C., Abdala-Roberts, L., Parra-Tabla, V., 2016. Patterns and sources of variation in pollen deposition and pollen tube formation in flowers of the endemic monoecious shrub *Cnidoscolus souzae* (Euphorbiaceae). Plant Biol. 18, 594–600.

Black, M., Bewley, J.D., Halmer, P. (Eds.), 2006. The Encyclopedia of Seeds: Science, Technology and Uses. CABI Publishing, Wallingford, UK.

Cardina, J., Sparrow, D.H., McCoy, E.L., 1996. Spatial relationships between seedbank and seedling populations of common lambsquarters (*Chenopodium album*) and annual grasses. Weed Sci. 44, 298–308.

Caswell, H., 1989. Matrix Population Models. Sinauer, Sunderland, MA.

Chauvel, B., Gasquez, J., Darmency, H., 1989. Changes of weed seed bank parameters according to species, time and environment. Weed Res. 29, 213–219.

Clark, S.J., Perry, J.N., 1994. Small sample estimation for Taylor's power law. Environ. Ecol. Stat. 1, 287–302.

Clark, S.J., Perry, J.N., Marshall, E.J.P., 1996. Estimating Taylor's power law parameters for weeds and the effect of spatial scale. Weed Res. 36, 405–417.

de Mazancourt, C., Isbell, F., Larocque, A., et al., 2013. Predicting ecosystem stability from community composition and biodiversity. Ecol. Lett. 16, 617–625.

Fumanal, B., Gaudot, I., Bretagnolle, F., 2008. Seed-bank dynamics in the invasive plant, *Ambrosia artemisiifolia* L. Seed Sci. Res. 18, 101–114.

Garrido, B., Hampe, A., Maranon, T., Arroyo, J., 2003. Regional differences in land use affect population performance of the threatened insectivorous plant *Drosophyllum lusitanicum* (Droseraceae). Divers. Distrib. 9, 335–350.

Grieg-Smith, P., 1983. Quantitative Plant Ecology, 3rd ed. University of California Press, Berkeley, CA.

Grime, J.P., 1998. Benefits of plant diversity in ecosystems: immediate, filter and founder effects. J. Ecol. 86, 902–910.

Guan, Q., Chen, J., Wei, Z., Wang, Y., Shiyom, M., Yang, Y., 2016. Analyzing the spatial heterogeneity of number of plant individuals in grassland community by using power law model. Ecol. Model. 320, 316–321.

Hill, M.O., Bunce, R.G.H., Shaw, M.W., 1975. Indicator species analysis, a divisive polythetic method of classification, and its application to a survey of native pinewoods in Scotland. J. Ecol. 63, 597–613.

Hughes, G., Madden, L.V., 1992. Aggregation and incidence of disease. Plant Pathol. 41, 657–660.

Izquierdo, J., Blanco-Moreno, J.M., Chamorro, L., González-Andójar, J.L., Sans, F.X., 2009. Spatial distribution of weed diversity within a cereal field. Agron. Sustain. Dev. 29, 491–496.

Lin, Y.-C., Lin, P.-J., Wang, H.-H., Sun, I.-F., 2011. Seed distribution of eleven tree species in a tropical forest in Taiwan. Bot. Stud. 52, 327–336.

Lloyd, M.W., Widmeyer, P.A., Neel, M.C., 2016. Temporal variability in potential connectivity of *Vallisneria americana* in the Chesapeake Bay. Landsc. Ecol. 31, 2307–2321.

Marshall, E.J.P., 1985. Weed distributions associated with cereal field edges—some preliminary observations. Asp. Appl. Biol. 9, 49–58.

Marshall, E.J.P., 1988. Field scale estimates of grass weed populations in arable land. Weed Res. 28, 191–198.

Marshall, E.J.P., 1989. Distribution patterns of plants associated with arable field edges. J. Appl. Ecol. 26, 247–257.

McCune, B., Mefford, M.J., 1999. PC-ORD. Multivariate Analyses of Ecological Data, Version 4. MjM Software Design, Gleneden Beach, OR.

Megill, L., Walker, L.R., Vanier, C., Johnson, D., 2011. Seed bank dynamics and habitat indicators of *Arctomecon californica*, a rare plant in a fragmented desert environment. West N Am. Naturalist 71, 195–205.

Mitton, J.B., Grant, M.C., 1996. Genetic variation and the natural history of quaking aspen. Bioscience 46, 25–31.

Mulugeta, D., Boerboom, C.M., 1999. Seasonal abundance and spatial pattern of *Setaria faberi Chenopodium album*, and *Abutilon theophrasti* in reduced-tillage soybeans. Weed Sci. 47, 95–106.

Park, S.-J., Taylor, R.A.J., Grewal, P.S., 2013. Spatial organization of soil nematode communities in urban landscapes: Taylor's power law reveals life strategy characteristics. Appl. Soil Ecol. 64, 214–222.

Pimentel, D., 1961. Species diversity and insect population outbreaks. Ann. Entomol. Soc. Am. 54, 76–86.

Polley, H.W., Wilsey, B.J., Derner, J.D., 2007. Dominant species constrain effects of species diversity on temporal variability in biomass production of tallgrass prairie. Oikos 116, 2044–2052.

Shiyomi, M., Takahashi, S., Yoshimura, J., Yasuda, T., Tsutsumi, M., Tsuiki, M., Hori, Y., 2001. Spatial heterogeneity in a grassland community: use of power law. Ecol. Res. 16, 487–495.

Silva Matos, D.M., Freckleton, R.P., Watkinson, A.R., 1999. The role of density dependence in the population dynamics of a tropical palm. Ecology 80, 2635–2650.

Sosnoskie, L.M., Luschei, E.C., Fanning, M.A., 2007. Field margin weed-species diversity in relation to landscape attributes and adjacent land use. Weed Sci. 55, 129–136.

Taylor, L.R., Woiwod, I.P., Perry, J.N., 1978. The density-dependence of spatial behaviour and the rarity of randomness. J. Anim. Ecol. 47, 383–406.

Taylor, L.R., Woiwod, I.P., Perry, J.N., 1980. Variance and the large scale spatial stability of aphids, moths and birds. J. Anim. Ecol. 49, 831–854.

Taylor, R.A.J., Park, S.-J., Grewal, P.S., 2017. Nematode spatial distribution and the frequency of zeros in samples. Nematology 19, 263–270.

Telewski, F.W., Zeevaart, J., 2002. The 120th year of the Beal seed viability study. Amer. J. Bot. 89, 1285–1288.

Tokeshi, M., 1995. On the mathematical basis of the variance-mean power relationship. Res. Popul. Ecol. 37, 43–48.

Tredennick, A.T., de Mazancourt, C., Loreau, M., Adler, P.B., 2017. Environmental responses, not species interactions, determine synchrony of dominant species in semiarid grasslands. Ecology 98, 971–981.

Tsuiki, M., Wang, Y.-S., Yiruhan, Y., Tsutsumi, M., Shiyomi, M., 2005. Analysis of grassland vegetation of the southwest Heilongjiang steppe (China) using the power law. J. Integr. Plant Biol. 47, 917–926.

Weiss, P.W., 1981. Spatial distribution and dynamics of populations of the introduced annual *Emex australis* in south-eastern Australia. J. Appl. Ecol. 18, 849–864.

Wilson, B.J., Brain, P., 1991. Weed monitoring on a whole farm patchiness and the stability of distribution of *Alopecurus myosuroides* over a ten year period. In: Integrated Weed Management in Cereals, Proceedings of the 7th European Weed Research Society Symposium, Helsinki. pp. 45–51.

Zhang, J., Huang, Y., Chen, H., Gong, J., Qi, Y., Yang, F., Li, E., 2016. Effects of grassland management on the community structure, aboveground biomass and stability of a temperate steppe in Inner Mongolia, China. J. Arid. Land 8, 422–433.

Chapter 7

Nematodes and other worms

Taxonomists recognize nine phyla of worms of which Annelida, Platyhelminthes, and Nematoda are the three most abundant and diverse. The platyhelminths, or flatworms, comprising two major classes, Cestoda (tapeworms) and Trematoda (flukes), are endoparasitic. The annelids include nonparasitic earthworms and ectoparasites, like the leach. By far the largest phylum, in terms of number of species, individuals, and mass are the nematodes. With their enormous range of habitats and life histories, quantitative assessments of worm abundance therefore present an ideal opportunity to investigate the properties and extent of TPL. In this chapter, I present a survey of articles on worms either presenting TPL results or including data sufficient for an *ex post* analysis.

The nematodes constitute much the larger part because of their abundance, diversity, and importance in human and veterinary health, in agriculture, and certain environmental investigations. The nematode section is divided into two habitat groups, terrestrial and aquatic. A third section covers the annelids and platyhelminths. Some of the articles cited are of surveys with several taxonomic groups and/or habitats surveyed; data of other taxa are presented in the appropriate taxonomic chapters.

Nematodes

The nematodes or round worms are divided into four classes comprising 26 orders. Probably 80% of all metazoans are nematodes; they inhabit all habitats and regions of the world, including Antarctica. It has been suggested that if all solid materials except nematode worms were to be eliminated the planet would, as USDA nematologist Cobb (1915) put it, still be "visible as a ghostly outline." Billions of nematodes may live in a hectare of field or forest soil or in freshwater, marine littoral, or benthos mud. Over 25,000 species have been described and there are likely double that number undescribed. Nematodes have exploited all lifestyles but primary producer: free-living nematodes are fungal feeders, bacteria feeders, and predators of other meiofauna including nematodes; plant-parasitic nematodes (primarily on or in plant roots) are important agricultural pests; nematodes are parasitic on all metazoans studied and are both a medical and veterinary pest; entomopathogenic nematodes (EPNs) are specialists

Taylor's Power Law. https://doi.org/10.1016/B978-0-12-810987-8.00007-0

that, like insect parasitoids, invade and develop inside an insect host. EPNs have a special place in agricultural and horticultural pest control. We start with general consideration of analyzing soil inhabiting nematodes.

Extraction

A number of methods and variants have been employed to extract nematodes from soil. Boag (1974) describes a sieving and decanting technique based on a method of Cobb (1918) that is followed by a Baermann funnel extraction. The extraction efficiency of the Baermann funnel has been assessed by Ruess (1995) who compared it with direct microscopic examination. In general, the proportion extracted increased with extraction time and with depth in the profile, but even at 160 h, extraction was not complete. Furthermore, extraction efficiency was not the same for all trophic groups: efficiency increased with time for plant and fungal feeders, but decreased with time for bacteria feeders. Efficiency estimates by other authors ranged from ~50% (Griffiths et al., 1991) to >90% (Yeates, 1972), the range probably due to differences in soils (Harrison and Green, 1976) and the quantity of soil used: Sohlenius (1980) recommended 10 cc. Ruess concluded that the Baermann funnel is useful and much less labor intensive than microscopic examination, but "less reliable in providing results of community compositions" because of the differences between trophic groups in extraction efficiency. For the purpose of evaluating spatial and temporal distribution by TPL, the widely used Baermann method is quite adequate unless its efficiency should prove to be density dependent (see Chapter 15).

Sampling

The number of nematodes recorded in soil samples depends critically on the sampling method used and the efficiency with which individuals are extracted from the samples. Soil nematode samples are generally taken using an auger or corer of some kind. In a comparison of soil samplers, Boag and Brown (1985) took multiple samples with nine devices to obtain 200 g of soil with each tool. Samples were taken from two grass fields near Dundee, Scotland, and the nematodes extracted using a modified Cobb (1918) method.

Remarkably, Boag & Brown's data of two taxa sampled with different samplers generate a convincing ensemble TPL analysis. Each point in Fig. 7.1A (Appendix 7.A) is one species obtained by a different sampler fit by a single line, with one possible outlier (screw auger, circled). Analysis of parallelism for the two species shows them not to be significantly different in intercept ($P > 0.75$) or slope ($P > 0.10$). Apparently the different efficiencies of the sampling tools provided the range of means necessary to establish a TPL. In this instance, TPL appears to be measuring a property of sampling.

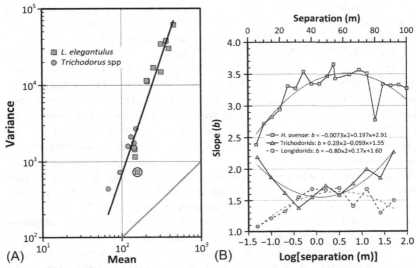

FIG. 7.1 (A) Ensemble TPL of *Longidorus elongatus* (*squares*) and trichodorid (*triangles*) nematodes per 200 g soil extracted from samples taken with $NB = 9$ ($NQ = 10$) different sampling tools. Each point is a different tool. Both nematodes conform to the same line, with the exception of *L. elongatus* collected with a screw auger (*circled*). (B) Ensemble *b* of *L. elongatus* (*circles*) and trichodorid (*triangles*) over a range of scales appear to be the reverse of each other as indicated by the quadratic equations. Ensemble TPL of *Heterodera avenae* (*squares*) increased with sample separation to a maximum at ~50 m and is also described by a quadratic equation. (*(A) Data from Table 1 in Boag and Brown (1985). (B) Data from Table 2 in Boag et al. (1987) and from Fig. 5 in Webster and Boag (1992).)*

Boag and Brown (1985) found the aggregation of *Longidorus elongatus* to be approximately Poisson when samples were taken close together (~5 cm), but aggregation increased to $b > 2.4$ at separations over 10 m. In a more detailed study, Boag et al. (1987) found that when *L. elongatus* and five other trichodorids were sampled with a tulip planter they produced very different TPL slopes over scales ranging from <5 cm to >50 m, the patterns mirroring each other (Fig. 7.1B). Clearly TPL is not independent of scale of sampling, at least at scales up to ~50 m, but the scale dependency may be population specific. A survey by Boag and Topham (1984) found the change in TPL slope of *L. elongatus* was probably influenced by soil type. The nematodes were approximately Poisson distributed in samples taken close together with slope increasing with distance apart to a maximum of $b = 2.0$. Webster and Boag (1992) examined the distributions of cysts of the nematodes *Globodera rostochirnsis* and *Heterodera avenae* in fields in eastern Scotland. Analysis showed that the greatest spatial variation occurred within 5–50 m. Both species were strongly aggregated with *H. avenae* occurring in larger patches. Like the trichodorids, TPL slope of *H. avenae* increased to a maximum at ~50 m and then declined, but unlike the trichodorids, *H. avenae* was extremely aggregated with all values of $b > 2.0$.

In the three-dimensional world of soil, distribution and abundance may change with depth in the profile as well as in areal extent. In this three-dimensional world, a population pattern can be projected onto the vertical and horizontal planes. Boag et al. (1987) give the density (#/200g soil) of six species at five depths (0–9, 10–19, 20–29, 30–39, and 40–49 cm). Fig. 7.2 (Appendix 7.A) shows four species' ensemble TPLs projected onto the horizontal (A) and vertical (B) planes. The TPL of the horizontal projection does not distinguish between the four species—all but two points (both *leptocephalus*) lie on a common line. By contrast, the samples projected onto the vertical plane show clear differences between the species. More obvious than on the horizontal projection, *L. leptocephalus* is much less abundant than the other three species when viewed in the vertical plane. All four species have slopes $2.0 < b < 2.5$ and are not significantly different from each other; only *Trichodorus primitivus* slope is significantly >2.0 ($P < 0.05$). Notably, the overall variability measured by A differs between species, with *L. elegantus* the most variable at all densities in the vertical plane.

Sampling an unmanaged pasture for the plant-parasitic nematode *Criconemella sphaerocephalus*, Wheeler et al. (1987) obtained different values of b with samples of different soil core volumes. Combinations of two soil core sizes and volumes of soil removed provided a range of different volumes (quadrat

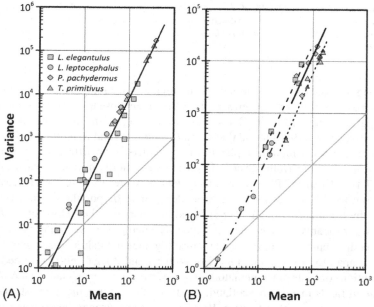

FIG. 7.2 (A) Ensemble horizontal TPL ($NQ = 5$, $NB = 33$) of four species integrated over five depths fit a single line. (B) Vertical ensemble TPL of the same four species at $NB = 5$ ($NQ = 4$–16) depths separate out with common slope. *(Data from Table 1 in Boag et al. (1987).)*

sizes of 0.5, 1, 2, 4, and $8\,\text{m}^2$) for analysis. Of 19 combinations of sample date and quadrat volume, twelve were well described by the negative binomial distribution. On four sampling occasions with four or five different quadrat volumes, slopes ranged from $b = 1.44 \pm 0.099$ to $b = 2.55 \pm 0.277$ (Appendix 7.B). Combining the 19 combinations of quadrats and sampling occasions results in an ensemble TPL with slope of $b = 1.96 \pm 0.162$ ($r = 0.94$). This kind of sampling with a range of quadrat sizes is called box counting and is discussed later and in Chapters 8 and 16.

Differences in variation projected onto the vertical plane suggest that the spatial distribution of a population changes with depth in the profile. A study by Somerfield et al. (2007) provides data of species abundance at three depths in marine sediment. They took 15 188.4 cc sediment cores at low water spring tide on a semiexposed sandflat on St Martin's Island in the Scilly Isles (40 km SW of England) and recovered 152 species of free-living nematodes. The cores were split into three 62.8 cc slices 0–5, 5–10, and 10–15 cm below the surface. Of the 152 species recorded, 91 were non-Poisson in at least one depth and their total number and variance/mean ratio were reported by Somerfield et al.

Fig. 7.3 (Appendix 7.C) shows the multispecies ensemble TPLs at the three depths with the deeper slices offset by one and two log cycles for clarity. Collectively, the community of nematodes at St Martin's mud flat was moderately aggregated with a common slope of $b = 1.37 \pm 0.023$. Somerfield et al. reported a positive relationship between V/M and M leading them to suggest that "a species' tendency to aggregate is relatively constant and not markedly influenced by the depth at which it occurs." Analysis of parallelism shows the three slices confirmed that the three TPLs are statistically indistinguishable ($P < 0.001$). Thus, the community-level aggregation did not differ at depths to 15 cm. In all three plots, a row of Poisson points is clearly distinguishable as a straight line of points diverging from the fitted line. About a dozen species that were sufficiently scarce as to be indistinguishable from random at one depth were nonrandom at another depth, suggesting that within species aggregation varied with depth as suggested by Boag et al.'s (1987) results.

The evidence suggests that TPL for a species, if not a community, may vary spatially. In addition to the spatial disposition of the target taxon, sampling and extraction methods, as well as the scale of sampling, can affect TPL slope. These examples also highlight the fact that for organisms living in a 3-dimentional world there are several ways to structure a TPL analysis: abundance of a taxon and/or community projected onto a vertical or horizontal surface or within a slice or layer. Boag et al.'s (1987) data produced vertical and horizontal TPLs (Appendix 7.A) that were not statistically distinguishable.

The rest of this chapter presents sections on soil, freshwater and marine nematodes, plant-parasitic, insect-parasitic (entomopathogenic), and animal-parasitic nematodes. A final section examines data of other round worms (annelids) and flatworms (platyhelminths). In each section, several case studies are presented in detail and some in lesser detail.

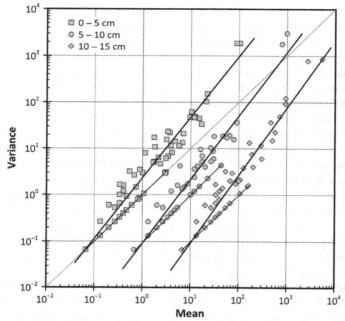

FIG. 7.3 Mixed-species ensemble TPLs ($NQ = 15$, $NB = 74$, 72, and 69) of nematodes sampled from intertidal sediments at three depths are not different. The plot of 5–10 cm and 10–15 cm depth offset by 1 and 2 cycles for clarity. The faint lines link points on the Poisson line. *(Data from Table 1 and Fig. 5 in Somerfield et al. (2007).)*

Terrestrial nematodes

Ecological classifications

URBAN TURFGRASS IN OHIO

Using data of a large-scale longitudinal study of a soil nematode community, Park et al. (2013) analyzed abundance of 28 genera to examine the relationships between nematode life history characteristics and TPL. Nematodes are classified by life history group according to whether they are small with short life-spans and high reproductive rates (colonizers) or larger and longer lived with lower reproductive rates (persisters). This *cp*-classification (Bongers, 1990) is roughly equivalent to population ecology's $r - K$ continuum. Nematodes are also classified by trophic group and functional guild, a combination of trophic group and *cp*-class.

Park et al. (2010) took ten 7-cm^2 soil cores from 18 lawns in Ohio in July and October 2007 for a total of 360 samples. The soil cores were treated identically: extracted by modified Baermann funnel (Flegg and Hooper, 1970) from 10 g of soil and identified to genus, *cp*-class, and trophic group (plant feeding, fungal feeding, bacterial feeding, omnivore, and predator). >64,000 individuals were identified to genus. Of the 47 genera identified, 28 were considered sufficiently abundant for TPL analysis. Means and variances were calculated for all nematodes combined, for each cp-class, trophic group, and functional guild and for each genus separately (Fig. 7.4) with $NQ = 10$ samples and a maximum of $NB = 36$ points. Data points with zero means or infinite variances were excluded and TPL parameters were estimated by ODR to facilitate analyses of parallelism. The tests of homogeneity of power law lines within and between genera, cp-class, trophic group, and functional guild found all categorizations of the data were heterogeneous. Importantly, fitting each of the three life-history classifications to the variance-mean data was a significant improvement in fit ($P < 10^{-8}$) over fitting 28 individual regressions at the level of genus. ODR estimates of TPL slope for the 28 genera ranged $1.21 \leq b \leq 2.34$; GMR estimates are slightly higher ($1.39 \leq b \leq 2.56$, Appendix 7.D). GMR estimates of b for cp-class, trophic group, and functional guild are also higher (Appendix 7.E).

Overall, nematodes tend to be more highly aggregated at the functional guild level than at the genus level with the six functional guilds having a narrower range and higher average of b values than the genera taken separately. Nematodes in lower cp-classes were more aggregated than those in higher cp-classes. Bacterial- and plant-feeding groups were more highly aggregated than omnivorous and predatory nematodes. Significantly, b segregates across these two life-history classifications: Table 7.1 shows GMR and ODR estimates of b

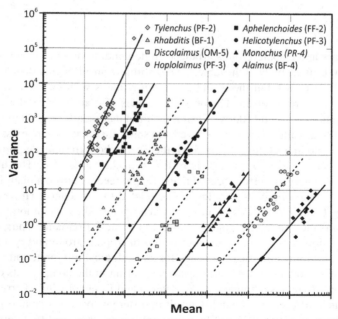

FIG. 7.4 Ensemble TPLs ($NQ = 10$, NB $= 17-36$) of nematodes representative of eight functional guilds with $1.39 \leq b \leq 2.56$. *(Data from Park et al. (2013).)*

TABLE 7.1 GMR (ODR in parentheses) estimates of TPL slopes for functional group separated into trophic group and cp-class, segregate along life-history axes

Trophic group		BF	PF	FF	PR	OM
	b	2.23 (2.08)	2.22 (1.97)	1.93 (1.73)	1.61 (1.53)	1.68 (1.44)
cp-Class	b	*Functional groups*				
1	2.13 (1.97)	**2.13 (1.97)**	–	–	–	–
2	2.23 (1.94)	**2.09 (1.91)**	**2.70 (2.36)**	**1.93 (1.73)**	*	–
5	1.84 (1.575)	*	**1.94 (1.73)**	*	*1.45 (1.49)*	**1.67 (1.60)**
3	1.70 (1.51)	1.66 (1.72)	**1.67 (1.51)**	*1.72 (1.53)*	*1.45 (1.36)*	–
4	1.68 (1.39)	**1.39 (1.28)**	**1.82 (1.60)**	*1.61 (1.43)*	**1.63 (1.53)**	**1.68 (1.42)**

*– Functional group not found in nature. * Functional group not found in this study. Bold: Observed values of b. Italic: b estimated from marginal means.*
From Table 4 in Park et al. (2013).

for cp-class and trophic groups arranged in descending order with b for functional guilds in the body of the table. GMR estimates are all higher than the original ODR estimates, but the essential pattern remains with the exception that cp-1 and cp-2 are reversed. The observed values of b are bold, and estimates for low abundance guilds derived using the marginal means are in italics. Several functional guilds were not observed and some are not known to exist. There is a clear trend in declining b from top left corner to bottom right supporting the premise that life history strategy is reflected in the value of b. The trend in decreasing aggregation from bacterial and fungal feeders through plant feeders to predators and omnivores could also be related to abundance. However, by removing *Tylenchus*, the single most abundant genus, the correlation between b and abundance disappears, but the pattern in Table 7.1 remains. Thus, degree of aggregation is at least in part a reflection of trophic group and cp or $r - K$ strategy, providing empirical evidence for an ecological interpretation of TPL.

Each functional guild contains species differing in the components of their life-history strategies (Bongers, 1990; Ettema and Bongers, 1993; Ferris et al., 1996). Such differences inevitably result in different specific spatial organizations because of differences in life history traits such as fecundity, tolerance to environmental stresses, and interspecific competition. The range of b for individual genera within the same functional guild suggests that genera within a *cp*-class or trophic group organize themselves differently in space, thereby carving out different spatial niches based on their degree of aggregation, a conclusion also reached by McSorley et al. (1985) and Ettema et al. (1998). In common with others (e.g., Boag et al., 1994; Elliott, 1986, 2002), Park et al. concluded that *b* measures properties of spatial behavior that are sensitive to a species' lifestyle or life history strategy. This implies that the aggregation pattern adopted by two taxa with otherwise similar demographic and feeding styles can reduce their niche overlap by exploiting space differently. Such differences in spatial organization are quantifiable by TPL.

GRASSY PASTURE IN SWEDEN

A study by Viketoft (2007) examined the nematode communities in a seminatural grassy pasture near Uppsala, Sweden. The communities were sampled in soil under monospecific patches of white clover (*Trifolium repens*), a legume, and sheep's fescue (*Festuca ovina*), a grass, and from soil under patches of mixed vegetation. Pairs of 42-cc samples were taken by auger under 10 patches of clover and fescue and 20 patches of mixed vegetation. Nematodes extracted by a modified Baermann funnel method were identified to genus with some to family or species, and assigned to a trophic group. The bacterial feeders *Acrobeloides* dominated all patch types, while *Eucephalobus* and *Panagrolaimus* were more common under clover, and fescue supported more *Alaimus*. Plant feeders were found in ~40% of cores, with *Helicotylenchus*, *Paratylenchus*, and *Tylenchus* found in all three patch types. Omnivores were less common and predators were scarce under all patch types.

Viketoft provides data for a mixed-species ensemble TPL of five trophic groups. The differences in abundance between habitats provided the range of means to fit TPL. Fitting TPL to all the species means and variances produces a moderately aggregated mixed-species ensemble graph (Fig. 7.5; Appendix 7.F) that clearly shows the segregation of trophic groups by abundance. Like Park et al.'s (2013) analysis, Viketoft's results also show TPL segregated by trophic group, although the sequence is not the same. In Viketoft's pasture, the bacteria feeders were less aggregated than either fungus or plant feeders: as 25% of Viketoft's samples were from under clover with its community of nitrogen-fixing bacteria, the lower aggregation may reflect a higher relative density and more even distribution of bacteria than Park et al's Ohio lawns.

OAK FOREST IN BULGARIA

A longitudinal study of nematodes in a 60–80-year-old oak (*Quercus dalechampii*) forest in Bulgaria tested the utility of nematodes as potential indicators of environmental quality. Lazarova et al. (2004) sampled nine microhabitats (soil from oak and grass rhizospheres, litter from oak and grass, moss growing on soil, stone and tree trunk, decaying wood, and oak bark). The number and

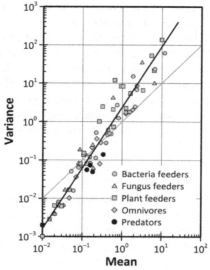

FIG. 7.5 Mixed-species ensemble TPL ($NQ = 40$, $NB = 5$–29) of nematodes (mostly genus, but some species and families) in a seminatural grass pasture in Sweden and labeled by trophic groups are fit by a single regression. *(Data from Table 2 in Viketoft (2007).)*

volume of samples taken varied by microhabitat, but all data were converted to number/100 g substrate. 79 genera were collected and classified into trophic groups and *cp*-classes. Different microhabitats supported different nematode communities with only fungal-feeding *Plectus* and bacteria-feeding *Aphelenchoides* found in all microhabitats. The community structure reflected microhabitat differences in primary production and decomposition processes: plant feeders dominated the soil and mosses growing on soil while bacterial and/or fungal feeders dominated the litter and decaying wood microhabitats. Soils had more diverse communities while the communities in litter, decaying wood, moss, and bark had considerably lower diversity and were dominated by a few taxa, predominately low *cp*-class taxa with few persisters (*cp*-class >2).

Ensemble TPL analyses of Lazarova et al's data of $NQ = 3$–15 samples for trophic group and *cp*-group as well as total nematodes were made in all microhabitats (Appendix 7.G) with each point a microhabitat. They also segregate by trophic group and *cp*-class although the pattern differs from both Park et al. (2013) and Viketoft (2007). The three sets of TPL analyses differ in part because they are analyzing different ensembles of nematode abundance data, not just different communities. They also differ in the manner of their collection and treatment of the data. TPLs of Park et al's ecological classes are pure ensemble analyses based on single-taxon population density in the same habitat. Lazarova et al's TPLs are based on density estimates from different habitats, while Viketoft's TPLs are mixed-species analyses.

MOUNTAIN FOREST IN CHINA

The diversity of soil nematodes in four forest types (mixed coniferous-broadleaf, spruce-fir, spruce, and Ermans birch (*Betula ermanii*) forest) at different elevations in the Changbai Mountains of NE China was investigated by Zhang et al. (2012). In each forest type, six 28-cm^2 soil cores were taken at four locations and separated into the O and A horizons. Nematodes were extracted from 100 g of fresh soil by a cotton-wool filtration method and identified to genus. Sixty-two genera were identified and classified as bacterivores, fungivores, plant-parasitic, omnivores, or carnivores. Nematode abundance ranged from 355/100 g of dry soil in the spruce forest A horizon to 3367/100 g in the O horizon of the coniferous-broadleaf forest. Forest type and horizon accounted for most of the variation in abundance. Fungivores dominated the samples, accounting for 45%–63% of total nematodes.

A mixed-species ensemble TPL of the trophic groups is fit by a single regression extending over three orders of magnitude of mean (Appendix 7.H).

As with Viketoft's (2007) plot (Fig. 7.5), the four trophic groups separate along the TPL line by abundance. Plant-parasitic nematodes cover a wide range of densities, while fungivores are most abundant but over a narrow range of means. Bacterivores, omnivores, and predators occupy comparatively narrow ranges of means.

FORESTS IN SCOTLAND

Boag (1974) surveyed forests in Northumberland and Durham in England and much of Scotland for soil nematodes. Wherever possible, samples were taken in paired coniferous and deciduous forests in the same 10-km^2 square. Nematodes were extracted by sieving and decanting 200 g of soil taken from the top 20 cm using an auger or small spade. Twelve plant parasitic nematodes were identified to genus. Boag's Table 2 gives the average nematode density (#/200 g soil) for six categories of deciduous and five coniferous forests. Abundance varied between forest types, but only *Criconemoides* showed a consistent preference for deciduous forest. Taking the mean and variance of forest type for each genus and genus for each forest class creates TPLs in two planes: mixed-species ensemble and community ensemble (Fig. 7.6; Appendix 7.I). Unlike most spatial and temporal TPLs computed from the same data, these orthogonal mixed-species TPLs do not differ in slope ($P > 0.65$), although intercept and correlation do reflect the difference in range and abundance.

SUBARCTIC HEATH

In an investigation of the toxic effect of plant defensive compounds on nematode diversity, abundance, and community structure, Ruess et al. (1998) collected soil samples and leaves from two dwarf shrub heaths and mountain birch from a subarctic dwarf shrub heath in a fellfield in Swedish Lapland. To investigate the nematode community, six 5-cm-thick soil samples were

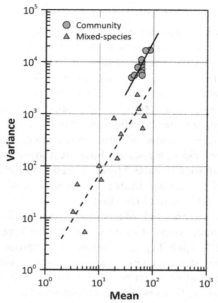

FIG. 7.6 Mixed-species and community ensemble TPLs of 12 nematode genera taken in samples from 11 different forest types. Averaging each genus across forest type (mixed-species; $NQ = 11$, $NB = 12$) results in a wider range of means than averaging over genus (community; $NQ = 12$, $NB = 11$). Despite differences in integration, the slopes do not differ. *(Data from Table 2 in Boag (1974).)*

taken from the side of a pit 30 cm deep. Five 25 g subsamples were taken from each layer and nematodes were extracted and classified as bacterial, fungal and plant feeders, obligate plant parasites, omnivores, and predators arranged by layer. A range of densities (#/g soil) varying by depth in the profile provide for a mixed-species ensemble TPL (Appendix 7.J). The trophic groups fit a single TPL regression. With the exception of the very rare obligate plant parasitic species, the trophic groups form a continuum from rare predators and omnivores to abundant bacterivores and fungivores.

METABOLIC FOOTPRINT

Occupying all trophic roles but primary producer, nematodes are important constituents of soil food webs. Because they are well-documented indicators of ecosystem condition, their metabolic footprints provide metrics for ecosystem functions and services in the soil. In addition, morphometrics are published for many species enabling estimation of their volume and approximate biomass from their simple tubular body plan. Ferris (2010) collected morphometric data of 1368 species and computed means and variances of biomass estimates for 77 families. 43 estimates were based on <15 species, 17 estimates were based on 15–26 species, the remainder ranged to up to 178 species. Using biomass as a proxy for metabolic footprint, Ferris was able to assess the ecosystem services performed by each trophic group. Ferris' Table 1 gives the biomass and SD from which mean and variance provide for a mixed-species TPL of biomass separated into *cp*-groups (Fig. 7.7; Appendix 7.K). Similar to the previous examples, biomass varies by *cp*-group and obeys TPL over a wide range of means.

The foregoing examples, starting with Park et al.'s (2013) analysis of genera and ecological classes, illustrates TPLs ubiquity in describing the distribution and abundance of ecological classes in addition to taxonomic groupings. Although the examples do not all agree with Park et al's observation of a spectrum of slopes from low *cp*-class bacterivores to high *cp*-class omnivores, they all show mean abundance for ecological classes that transition from rare to abundant while obeying TPL. That the variance-mean ratio increases with mean for ecological groupings reflecting life history strategies suggests that V/M is sensitive to demographics and spatial behavior, just as the taxonomic analyses of Park et al. (2013) suggested.

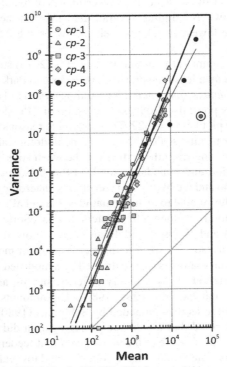

FIG. 7.7   Mixed-species TPL ($NQ = 2–178, NB = 76$) of nematode biomass separated by *cp*-group, fit a single regression (*bold line*). The lighter lines are fit to the individual *cp*-groups. Three points are outliers; two points on the Poisson line are included in the analyses, but the circled point (Longidoridae), computed from only three species, is omitted. *(Data from Table 1 in Ferris (2010).)*

TPL stability

PERRINE MARL SOIL IN FLORIDA

McSorley et al. (1985) used TPL to compare the spatial distributions of free-living nematodes with plant-parasitic species. Three fallow fields in Florida were divided into six 0.4-ha plots, from which twelve 3.1-cm^2 soil cores were collected in a regular grid. Nematodes were extracted from 9.6 cc of soil from each core, identified to species or genus and counted. Ten nematode genera and six species were identified. By pooling the core data in different ways, four ensemble TPL analyses were made at four spatial scales. For each of the 16 taxa, means and variances were computed from 6 cores from 36 plots, 12 cores from each of 18 plots, 24 cores from 9 plots, and 36 cores from 6 plots. In each analysis, the area represented increased with core number from 0.2 to 1.2 ha.

All but five taxa produced significant TPLs at all scales and three taxa had $b < 1.0$ at one or more scales (Fig. 7.8). The majority of taxa had a small spread between the scales. The slope did not vary much with plot size and spanned a comparatively narrow range for most of the taxa. The 0.4-ha plot generally had the highest values of b, which decreased with plot sizes >0.4 ha while A tended to decrease. The poor fit and generally low estimates of b in the 1.2-ha plots are likely due to the low $NB = 6$ plots. Overall, the slopes at the four scales were well correlated, the lowest correlation being between the 0.2 and 1.2 ha plots (Table 7.2).

Ten genera are common to this and Park et al.'s (2013) study and estimates of b for four genera in at least one plot size were close to Park et al.'s estimates: *Acrobeloides, Alaimus, Aphelenchoides,* and *Rhabditis.* However, overall, agreement was very poor; the strongest correlation of TPL slopes was between Park and the 0.2-ha plot with $r < 0.22$. There are three reasons these studies produced so dissimilar results: scale differences, methodological differences, and real differences resulting from the differences between soils. It is possible that nematode distribution and abundance might differ between the loamy well-drained soils of Ohio and the hydric and sometimes anaerobic Perrine marl soil, which may be less hospitable to many nematodes. Park et al's replicate samples were taken much closer together (a few meters) than McSorley et al's, which as Boag et al. (1987) and Webster and Boag (1992) showed (Fig. 7.1B), can change TPL slope in a species-specific manner. The major methodological difference was Park et al's samples were extracted by a modified Baermann funnel method and McSorley et al. by a highly efficient sieving and centrifugation method (McSorley and Parrado, 1981, 1982). The amounts of soil used for extraction were $\sim$10 cc as recommended by Sohlenius (1980).

McSorley et al. noted that the degree of aggregation did not appear to be related to the trophic preference of the nematode. Plant feeders, fungal feeders, and bacterial feeders were found to have both high and low values of b. As Park et al. (2013) found, it is the ensembles of ecological classifications that segregate, not the individual members.

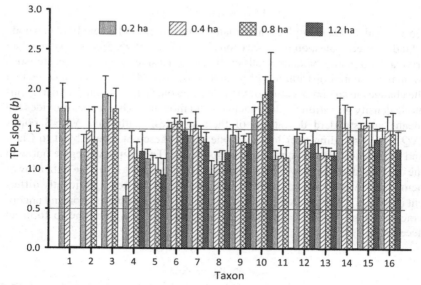

FIG. 7.8 Estimates of (ODR) *b* with SE for ensemble TPLs of 16 nematode taxa at 4 spatial scales vary between species but are mostly consistent across scales. An exception is *Aphelenchus* at the 0.2-ha scale, which approaches regularity. *(Data from Tables 1 and 2 in McSorley et al. (1985).)* Key to taxa:

1	Acrobeloides	2	Alaimus	3	Aphelenchoide
4	Aphelenchus	5	Cephalobus	6	Criconemella onoensis
7	Diphtherophora	8	Eucephalobus	9	Helicotylenchus dihystera
10	Meloidogyne incognita	11	Panagrolaimus	12	Psilenchus hilarulus
13	Quinisulcius acutus	14	Rhabditis	15	Rotylenchulus reniformis
16	Tylenchus				

TABLE 7.2 Correlation coefficients between *b* at different plot sizes

	0.2 ha	*0.4 ha*	*0.8 ha*
0.4 ha	0.794		
0.8 ha	0.770	0.881	
1.2 ha	0.600	0.777	0.928

Data from Tables 1 and 2 in McSorley et al. (1985).

FOREST AND OLD FIELD

Gorres et al. (1998) sampled soil nematodes in a forest and an old field in Rhode Island, United States, on four occasions in 1994 and 1995. Samples were taken from a 5×5 matrix of square cells each 3.4 m on a side. They used geostatistics to analyze spatial distribution of nematode density (#/g soil) and documented the changes in spatial structure of the forest and old field through the year. Nematode density was strongly dependent on soil moisture. Their paper provides the means and standard deviations of their samples. With only $NB = 8$ points ($NQ = 25$) extending over less than one order, these data are subject to Clark and Perry's (1994) cautions about sample sizes. The points segregate out with the old field densities lower than the forest's (Appendix 7.L) in line with a lower nematode diversity of the old field. The two slopes are not significantly different ($P > 0.25$). What is significant is that, despite the low correlation and unconvincingly high slope of the old field plot, the two sets combine to form an acceptable TPL.

CROPS IN FLORIDA

McSorley and Dickson (1991) combined three studies to examine the stability of TPL of plant parasitic nematodes (PPNs) in several crops in Florida. Samples in these studies were taken over a 2-year period from October 1986 from fallow ground, maize, soybeans, rye, or hairy vetch. Three replicate soil cores were taken from plots divided into 12 sections. The 12 cores from each replicate were combined, nematodes were extracted from a 100-cc portion by a sieving-centrifugation method, and PPNs of interest were identified and counted. Means and variances of $NQ = 3$ samples of four species (*Belonolaimus longicaudatus*, *Criconemella sphaerocephala*, *Pratylenchus brachyurus*, and *Meloidogyne incognita*) expressed as #/100 cc were analyzed by ensemble TPL with $NB = 8$–32 depending on crop (Appendix 7.M). In most cases, McSorley & Dickson found that TPLs of nematode density in the same crop to differ little between sampling occasions. Analysis of parallelism showed TPLs of nematodes in the five crops sampled on different dates differed in the intercept (at $P < 0.05$) in seven of 20 crop-species comparisons and slope in three cases. When all crops and dates were combined, only *C. sphaerocephala* was heterogeneous in both slope and intercept, and *B. longicaudatus* in intercept only. The strong agreement in TPLs over 18 sampling dates and five crops led McSorley & Dickson to examine the precision estimates obtained using general and specific sampling plans. They concluded that TPL was sufficiently consistent that sampling plans for some PPNs based on only SEM for single plots would likely be adequate when TPL parameters were unavailable.

COWPEAS AND COTTON IN CALIFORNIA

In a project to determine whether sample programs based on TPL were transferable from one field to another, Ferris et al. (1990) sampled a field in Tulare

County California for nematodes on a 25×25 grid. A single 4-cm^2 core was taken from each cell on 11 occasions from September 1982 to October 1983 during which time cowpeas were followed by cotton. Six plant-parasitic nematode species were extracted from the 6875 samples and their numbers entered in a database. By randomly sampling the database with sample sizes from one to six cores, Ferris et al. established the stability of b and validated a model for a as a function of sample number, n ($=NQ$):

$$V = (cn^d)M^b \text{ or}$$
$$\log(V) = \log(c) + d\log(n) + b\log(M) \tag{7.1}$$

where c and d are empirical constants in which $d < 0.0$. The six species produced two curves for $a = cn^d$ (Fig. 7.9 and Table 7.3). The model (Eq. 7.1) was tested for *Paratrichodrus minor* at four sites in California's Central Valley and three

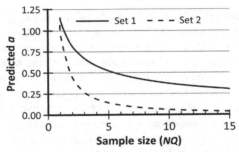

FIG. 7.9 Predicted intercept a using Eq. (7.1). Set 1 is typical of results for *Meloidogyne incognita* and *Xiphinema americanum*; Set 2 is typical of *M. javanica, Helicotylenchus dihystera, Paratrichodorus minor*, and *Tylenchorhynchus cylindricus*. (*Data from Ferris et al. (1990).*)

TABLE 7.3 Values of c, d, and b, and standard errors (SE) relating variance, mean, and sample size (Eq. 7.1) for six plant-parasitic nematodes in cowpea and cotton

Species	c	SE(c)	d	SE(d)	b	SE(b)
Helicotylenchus dihystera	0.94	0.19	−1.19	0.43	1.94	0.04
Meloidogyne javanica	1.06	0.09	−1.43	0.35	2.19	0.12
M. incognita	1.10	*	−0.49	*	2.07	0.04
Paratrichodorus minor	1.24	0.59	−1.20	0.76	1.81	0.39
Tylenchorhynchus cylindricus	1.01	*	−1.20	*	2.08	0.08
Xiphinema americanum	1.20	*	−0.49	*	2.01	0.04

Data from Table 2 in Ferris et al. (1990). * = SE not given.

sites for *M. incognita*. In every case, they found TPL to be repeatable and robust. Table 7.3 gives the estimates for *b*, *c*, and *d*. Simulations showed that increasing the volume of the sample unit from 100 to 1000 g soil shifts the regression along the abscissa but leaves *b* unchanged, while both SE(*b*) and *a* increased with sample volume.

Ferris et al. (1990) concluded that when parameter values for TPL in similar cropping systems in a large geographical area are pooled estimates of *a* and *b* "appear sufficiently robust to describe variance-mean relationships for the same species in various locations."

Plant-parasitic nematodes

Annual crops

POTATOES

Plant-parasitic nematodes feed on all parts of the plant, from root hairs to flowers and fruit using specialized organs, the size and shape of which define their mode of feeding. All are technically ectoparasites and many of the most serious agricultural pests are soil inhabiting, attacking the roots and root hairs. Some form egg-bearing cysts that can remain dormant for long periods until a suitable host grows nearby. Golden eelworms or potato cyst nematodes (PCNs), principally *Globodera* (=*Heterodera*) *rostochiensis*, native to the Andes where potatoes originated, are a major cause of crop loss and are very difficult to control because of their cysts' long-term viability. They live on the roots of potatoes, tomatoes, and other Solanaceae. PCN cysts can remain dormant for 20 years with up to 400 eggs, which hatch in response to host plant root exudates. The larvae migrate to the root tips where they feed and mature. After mating, females encyst on root and potato surfaces. At very high population densities, PCN causes root damage leading to early senescence and up to 60% yield reduction.

In the 1950s, the UK's National Agricultural Advisory Service (ADAS) adopted a soil-sampling technique for PCN. The methodology developed was based on statistical analyses by statistician Anscombe (1950). He used the negative binomial distribution to characterize samples taken from fields distributed throughout England to estimate sampling errors and develop ADAS' soil sampling plan. The regional field samples were taken with a 5-cm^2 auger in a stratified random sample of 50 cores spread widely in most fields. Cysts were separated from soil by flotation and sieving, counted and cyst density expressed as #/kg soil. Ensemble TPL of PCN has slope only slightly greater than Poisson (Fig. 7.10; Appendix 7.N): $b = 1.19 \pm 0.128$ and $A = 2.61 \pm 0.352$ ($r = 0.941$). Expressed as #/g soil, the slope and correlation are unchanged, but the intercept is $A = 0.18 \pm 0.040$.

To test the generality of a model of PNC infestation in a variety of soils and cropping sequences, Schomaker and Been (1999) sampled 37 fields in the

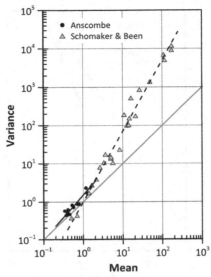

FIG. 7.10   Ensemble TPL of PCN cysts (#/g soil) in samples taken from $NB = 12$ fields ($NQ = 23–120$) in England (solid line). Ensemble TPL ($NQ = 20$, $NB = 24$) of PCN cysts (#/kg soil) in samples of an area of $480\,m^2$ of a field in the Netherlands (hatched line). *(Data from Table 1 in Anscombe (1950) and from Fig. 1 in Schomaker and Been (1999).)*

Netherlands known from prior sampling to have locally abundant foci of *G. rostochiensis* and *Globodera pallida* cysts. A field in the province of Friesland was intensively sampled with 1.5 kg of soil taken from an array of 480 numbers of 1-m^2 cells. Data of this array arranged as $NB = 24$ blocks of $NQ = 20$ quadrats produce an aggregated ensemble TPL of PCN cysts with $b = 1.85 \pm 0.059$ (Fig. 7.10, Appendix 7.O). In Anscombe's survey, PCN cysts were sampled over a very large scale, their distribution virtually the same as Poisson, but at the scale of a few hundred square meters, the variance is roughly proportional to the mean squared.

To determine how management practices influence spatial distribution and temporal stability of nematodes, Morgan et al. (2002) investigated the population dynamics of the root lesion nematode *Pratylenchus penetrans*. Nematodes extracted from soil samples taken in fumigated and unfumigated potato fields in central Wisconsin have different but adjoining TPLs.

The fields were grown in a 3-year rotation following crops that included corn, soybean and vegetables, and a winter cereal cover crop planted following harvest of the main crop. Eighty five to 134 four soil samples were taken per crop-year in late June at georeferenced sampling grid points in each field. Morgan et al. present *P. penetrans* density data (#/100 cc) for 15 crop-years suitable for spatial TPL analysis. Combining fumigated with nonfumigated samples gives a range of means over three orders of magnitude (Fig. 7.11; Appendix 7.P).

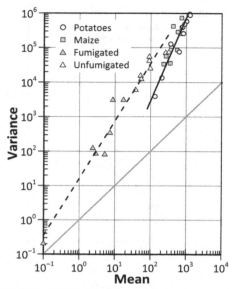

FIG. 7.11 Spatial TPL ($NQ = 85$-134, $NB = 15$) of *Pratylenchus penetrans* in fields in 3-year potato rotations. Fumigated fields have higher densities and lower TPL slopes than the fumigated fields. Combining fumigated and unfumigated crop-years provides a three order of magnitude range in mean. Spatial TPLs ($NQ = 9$, $NB = 20$) of *P. scribneri* density (#/100 cc) at 5 depths under maize and potatoes do not differ in superaggregation in the vertical profile. *(Data from Table 1 in Morgan et al. (2002) and from Fig. 2 in MacGuidwin and Stanger (1991).)*

The unfumigated fields had the greatest abundance of *P. penetrans* over a fairly narrow range of means and lower TPL slope ($b = 1.19 \pm 0.222$) than the nematodes in the fumigated fields ($b = 1.94 \pm 0.222$). The difference is due largely to the two highest means, which occurred in pepper and pea crops, not in potatoes. Combining fumigated and unfumigated densities extends the TPL by an order of magnitude and improves the correlation ($r = 0.978$).

MacGuidwin and Stanger (1991) monitored the vertical distribution of *Pratylenchus scribneri* under irrigated corn and potatoes in field plots in central Wisconsin. Six soil cores taken with a 5-cm^2 corer on nine dates were separated into five 7.5-cm lengths. The density (#/100 cc) at all layers increased synchronously through the season to a peak and then declined by harvest suggesting changes in *P. scribneri* numbers were due to reproduction rather than vertical movement. Spatial TPLs of density computed over the five depths are steep (Fig. 7.11, Appendix 7.Q) for both crops. Neither intercepts nor slopes differ between crops ($P > 0.65$), the TPLs are best described by a single regression that is significantly steeper than $b = 2.0$ ($P \ll 0.001$) tending to confirm MacGuidwin & Stanger's conclusion that vertical movement was minor under both crops.

Wheeler et al. (1994) fitted TPLs to data of *Meloidogyne hapla*, *Pratylenchus penetrans*, *P. scribneri*, and *P. crenatus* obtained from samples from commercial potato fields in Ohio. They used the TPL estimates to develop sampling plans for potato nematodes. Samples of 10–12 5 cm² soil cores were taken from $NQ = 22–107$ plots before planting in a total of 10 fields over 5 years. Spatial TPL show *M. hapla* density (#/100 cc) was moderately aggregated ($b = 1.30 \pm 0.111$) (Appendix 7.R). The three *Pratylenchus* species were both individually ($1.14 \leq b \leq 1.48$) and collectively ($b = 1.27 \pm 0.048$) were also moderately aggregated. This result differs markedly from Park et al.'s (2013) ensemble estimate for *Pratylenchus* of $b = 1.74 \pm 0.119$. Often ensemble and spatial TPLs are similar; in this case, the difference is likely due to the relative scales of the studies.

In a later study, Wheeler et al. (2000) divided six commercial potato fields into contiguous quadrat sizes of 10, 50, 100, 250, 1000, and 4000 m² with smaller quadrats nested within the larger ones. Twenty quadrats of each size were randomly selected for soil core sampling. Within each selected quadrat, 20 5-cm² soil cores were taken, in all 100 or 120 samples per field. TPL slope declined from 10 m² quadrats to 100 m² above which it increased but with a dip at 2000-m² quadrat (Fig. 7.12; Appendix 7.R), suggesting minimum variation may occur at the 100-m² scale. Wheeler et al. concluded there is no discernable

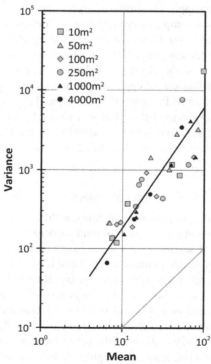

FIG. 7.12 Spatial TPL ($NQ = 20$, $NB = 34$) of *Pratylenchus* spp. in sampled at scales from 10 to 4000 m² in potato fields in Ohio. (*Data from Table 2 in Wheeler et al. (2000).*)

systematic relationship discernable between quadrat size and variance. The comparatively small number of points per quadrat size, the high SE(b), and the five quadrat sizes fitting a single TPL ($b = 1.53 \pm 0.124$) confirm their conclusion.

Seven additional fields were divided into 2000- or 4000-m^2 grids, and each grid was sampled in both autumn and spring for *Pratylenchus* spp. and *M. hapla*. Both taxa show a decrease in slope from autumn to spring (Appendix 7.R), although the differences are not significant owing to very high SE(b). The overall population density of *M. hapla* declined dramatically from autumn to spring. Over the same period, *Pratylenchus* spp's population density increased almost twofold, so the decline in slope was probably not related to the change in density.

SOYBEANS

Another cyst nematode, the soybean cyst nematode (SCN; *Heterodera glycines*), is also a serious pest. In a pair of papers, Avendano et al. (2003, 2004) described their investigation to determine if SCN's spatial and temporal dynamics were sufficiently stable in space and time to facilitate site-specific management in SCN-infested fields. They surveyed two fields in Michigan for SCN cysts in a nested sampling design with distances reduced in geometric progression in order to apply geostatistical techniques to analyze spatial distribution. 157 and 109 soil samples were taken in two fields, at planting and again at harvest in 1999 and 2000. Cysts were extracted by elutriation from single-core soil samples and density (eggs/100 cc soil) was estimated from cyst density (#/100 cc) and number of eggs/cyst. Avendano et al. provide data for spatial TPLs relating the distributions of egg and cyst density and eggs/cyst. Although significantly different from Poisson ($P < 0.05$), the TPL slope of eggs/cyst is only slightly more aggregated than Poisson ($b = 1.17 \pm 0.083$). As a result, the distribution of eggs/100 cc is almost identical to the distribution of cysts/cc, although the abundance is 1½ to 2 orders of magnitude greater (Fig. 7.13; Appendix 7.S).

BROAD BEANS

Ditylenchus dipsaci is a migratory endoparasite that enters hosts through stomata or plant wounds and creates galls or malformations in plant growth allowing for the entrance of secondary pathogens such as fungi and bacteria. Most generations are passed inside plants with eggs and larvae overwintering in dried infected hosts. High fecundity (females can lay 200–500 eggs) and extreme polyphagy (400–500 known hosts worldwide) make this species a serious agricultural and horticultural pest. Control is by fumigation and field sanitation.

Infected broad beans (*Vicia faba*), which tend to be smaller and distorted, are unmarketable. Green (1979) estimated the infestation rate of broad beans in 17 samples of $NQ = 43$–94 beans by submerging seeds individually in small dishes of water and counting *D. dipsaci* emerging after 4 days. Of the 963 broad beans

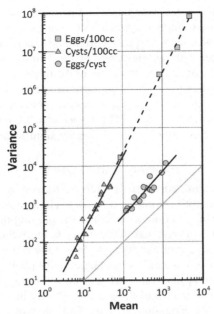

FIG. 7.13 Spatial TPLs ($NQ = 110-160$) of soybean cyst nematode (SCN) cysts ($NB = 20$), eggs ($NB = 4$), and eggs/cyst ($NB = 20$). The distribution of eggs/cyst is pseudo-Poisson, acting as a simple multiplier having little impact on the TPL of eggs/100cc of soil. *(Data from Table 1 in Avendano et al. (2003) and Table 2 in Avendano et al. (2004).)*

examined, 48 were host to 1–12 nematodes. With only 5% of broad beans infested, the distribution of nematodes/bean was highly skewed. Green computed an ensemble TPL by ODR. Repeating his analysis of $NB = 17$ samples by GMR, the distribution of nematodes/bean is highly aggregated with $b = 1.80 \pm 0.13$ (Appendix 7.T). Green concluded that the high aggregation of *D. dipsaci* with nematodes concentrated in a small number of beans would result in a number of scattered foci giving the nematode the best chance of a finding a favorable environment in the field and increasing the likelihood of finding a new host.

TOBACCO

Samples for nematodes in tobacco (*Nicotiana tabacum*) fields in North Carolina provide data for TPL analysis of three species, *M. incognita*, *Tylenchorhynchus claytoni*, and *Helicotylenchus dihystera*. Noe and Campbell (1985) took soil samples in a grid pattern in seven fields: four fields with 1-m² quadrats and three with 9-m² quadrats. The 1-m² quadrats were arranged on an 8×8 grid and the 9-m² quadrats were arranged on either a 10×10 grid or an 8×8 grid. Four fields were sampled in October and March and three fields in September. In each quadrat, ten 5-cm^2 soil cores were taken and nematodes extracted, identified, and counted. NOE and Campbell applied several spatial analyses to determine cluster size. They concluded that spatial correlation was greater in autumn than spring and was greater with samples taken by the 1-m^2 quadrats than the 9-m^2 quadrats. Spatial TPLs of density (#/500 cc) of the three species differ in slopes of $1.38 \leq b \leq 2.41$; however, only the two extreme species are significantly different ($P < 0.05$). Plotted together, the three species appear, with a single outlier (*T. claytoni*), to conform to a single TPL (Appendix 7.U). Combining apparently different TPLs can produce a single highly significant regression.

EGGPLANT (AUBERGINE)

The bacterium *Pasteuria penetrans*, an obligate parasite of nematodes, forms endospores and is nonmotile requiring the host to come in contact with the bacterial endospores for infection to occur. Since the phasing out of the soil fumigant methyl bromide, interest has focused on several *Pasteuria* species as biopesticides to protect a number of crops. In a study of the efficacy and sustainability of *P. penetrans* as a biocontrol tactic for the plant-parasitic nematode *Meloidogyne javanica* in Senegal, Dabiré et al. (2007) established 10 microcosms of tomato plants in each of three soils differing in texture (sand, sandy clay, and clay) and two irrigation regimes. Estimates of bacterial endospore leaching and nematode survival in the different soils led Dabiré et al. to conclude that the best balance of spore percolation and nematode infection occurred in sandy-clay soil. Their results of nematode density (#/plant) and bacterium density (#/g of soil) provide data for a comparison of the ensemble TPLs of bacterium and nematode host. Despite differences in substrate (soil or plant roots), data of both species produce convincing super-aggregated TPLs with $b \approx 2.35$ (Appendix 7.V). Neither the intercepts nor the slopes differ significantly ($P > 0.45$), but only the nematode's slope is significantly greater than $b = 2.0$ ($P < 0.02$). The agreement in slopes confirms that in this experiment at least, the distribution and abundance of *P. penetrans* depended heavily on its host's spatial distribution—as one would expect of nonmotile endospores.

MIXED VEGETABLES

The cotton root-knot nematode *M. incognita* has a worldwide distribution and numerous hosts. It attacks the roots of plants deforming root cells and stimulating the formation of galls. It is another nematode for which *P. penetrans* may be used as a biocontrol agent. In a field experiment in Martinique, soil samples for *M. incognita* juveniles were taken every 3 weeks for 1 year by Ciancio and Quénéhervé (2000). The experiment with five replicates of six crop rotations consisted of combinations of resistant and susceptible tomato, okra, cabbage beans, peppers, and eggplant interspersed with fallow periods. On average, 20%–30% of juvenile nematodes were found with *P. penetrans* endospores attached and susceptible crops had higher nematode densities. Pooling data from all treatments, Cianco & Quénéhervé found the frequency of density classes of juvenile *M. incognita* followed the Gutenberg-Richter power law, a negative exponential law describing relationship between the magnitude and total number of earthquakes (see Chapters 12 and 16). Cianco & Quénéhervé provide data for a temporal TPL of *M. incognita* juveniles in which the treatments provide a range of means over one order of magnitude (Appendix 7.W). As with *M. javanica*, *M. incognita* was strongly aggregated ($b = 2.20 \pm 0.123$), but is significantly >2.0 at only $P < 0.10$, but with $NQ = 85$ and $r = 0.994$, the slope is probably not overestimated. Significantly, TPL of *M. incognita* infected with *P. penetrans* is more regularly distributed than Poisson ($b = 0.53 \pm 0.110$) over a similar range of means, but a lower overall abundance. The near-regular distribution probably reflects the application process.

Perennial crops

CITRUS

The citrus nematode *Tylenchulus semipenetrans* is the only member of its family that is economically important. It is a semiendoparasite with a fairly narrow host range that includes olive, grape, and citrus in which it causes a slow decline with increasing crop loss. It has a worldwide distribution and can live in a wide variety of soils. Other nematodes found on citrus include *Helicotylenchus pseudorobustus* and some species of *Criconemella*. Abd-Elgawad (1992) applied TPL to population data of *T. semipenetrans*, *H. pseudorobustus*, and *Criconemella* spp. obtained from stratified random samples taken from 2.1-ha citrus blocks in Egypt's Nile delta region. In a second systematic survey, *T. semipenetrans* only was extracted. In the stratified random set, samples were taken from beneath a single tree in each block and nematodes extracted and counted. Ensemble TPL was computed from up to $NB = 35$ means and variances computed from $NQ = 7$ samples. In the second survey, one sample was collected from beneath the canopy of a single tree. Samples from $NQ = 7$ trees were obtained from each of 24 adjacent blocks and nematodes extracted. Ensemble TPLs were computed for *T. semipenetrans* in $NB = 24, 12, 8,$ and 6, representing citrus blocks of 2.1, 4.2, 6.3, and 8.4 ha. Abd-Elgawad's (ODR) results are in Appendix 7.X.

While either NQ and/or NB is <15 in all regressions, Abd-Elgawad's results show remarkable consistency between the stratified random and four systematic analyses ($1.89 \le b \le 1.98$) for *T. semipenetrans*. The other results, *H. pseudorobustus* and *Criconemella* spp., are similar and not significantly different from *T. semipenetrans* ($P > 0.20$). Assuming validity of Park et al.'s (2013) hypothesis that differences in b of potential competitors indicate avoidance of competition, the close similarity of the three TPL results would suggest these species do not compete directly or if they do that *T. semipenetrans* is dominant. Abd-Elgawad's data confirm *T. semipenetrans* was much more abundant than either of the other taxa, with the criconemellids least abundant. Their lifestyles are also somewhat different. *H. pseudorobustus* is thought to be largely partheno-genic as males are rarely found. Like *T. semipenetrans*, it is semiendoparasite but may be facultatively completely endoparasitic. It is not considered an eco-nomically important pest of any crop, including citrus. *Criconemella* are not usually a problem on citrus and were comparatively rare, found in only 11 of 35 blocks with nonzero mean and variance. Thus, direct competition is probably not extreme. Abd-Elgawad used the TPL estimates to calculate optimum sam-ple sizes to achieve predetermined levels of sampling error for the nematodes. He also used his TPL results to compute exact transformations for further analysis.

In Florida, Duncan et al. (1989) conducted an extensive survey of 50 citrus orchards and an intensive survey of 80 trees in a single orchard for *T. semipe-netrans*. In the extensive survey, soil samples at 8 locations within the drip line of 20 randomly selected trees were taken by shovel and composited. Juvenile and male nematodes were extracted from 1 L subsamples. In the intensive sur-vey, sixteen 4.2-cm^2 soil cores were taken from under each tree and composited. Nematode males and juveniles were extracted by Baermann funnel from 60-cc subsamples. Spatial TPL was computed from the intensive survey with $NB = 18$ sample events ($NQ = 80$). The extensive survey yielded the means and variances from $NB = 41$ orchards used in an ensemble TPL ($NQ = 20$) (Appendix 7.Y). Comparison of the spatial and ensemble (Duncan et al. called the TPLs geo-graphic and temporal, respectively) resulted in parallel TPLs with different intercepts and slopes not significantly different ($P > 0.70$). Duncan et al. used these results to determine optimum sample sizes for estimating and managing *T. semipenetrans* populations.

Dividing the extensive (=geographic) survey into orchards with large ($NB = 19$ orchards with $\geq 90\%$ of samples positive for nematodes) and small ($NB = 16$ orchards with $\leq 65\%$ positives) patches of *T. semipenetrans* resulted in markedly different TPL slopes. It is noteworthy that Duncan et al's estimates of slope, except for the small patch survey, are much lower than Abd-Elgawad's ($1.50 \leq b \leq 1.59$ versus $1.89 \leq b \leq 1.98$). Although not significantly different from the large patch TPL ($P > 0.15$), the influence of the small patch size orchards on calculated optimum sample size was substantial. Excluding small patch orchards reduced the number of samples required for efficient population estimation of *T. semipenetrans* by 17%. The differences in slope estimate and hence the optimum sample size may be due to differences in extraction effi-ciency. As noted, the various extraction methods are not equally efficient, and there is a possibility that some have density-dependent efficiency, which can bias, usually decrease, the estimate of b. Duncan et al. do not give the den-sities in the large and small patches, but given the criteria for selecting the patches, it's likely that the overall density was higher in the large patches than the small one. If the efficiency of extraction by Baermann funnel is density dependent, it could account for the lower b obtained for the large patch set of orchards.

A later study (Duncan et al., 1993) estimated the population density of T. *semipenetrans* in an orchard of mature grapefruit trees in central Florida to identify factors affecting population dynamics. The density of nematode females on roots and juveniles and males in soil was correlated with root mass density and plant carbohydrate concentration and inversely with soil moisture and root lignin content. Samples taken weekly for 27 months provide data for spatial TPLs of females and juveniles plus males. Means and variances were calculated from pairs of samples ($NQ = 2$) producing regressions with $NB = 103$ weeks. The spatial estimate of b for juveniles and males was identical to the earlier ensemble estimate for small patch size and similar to Abd-Elgawad's ensemble estimates. The estimate of b for females was lower than for juveniles and males. While differences in aggregation likely reflect the different habitats sampled, females in or on plant roots and males and juveniles in the soil and the small number of quadrats ($NQ = 2$) could also contribute to the difference in estimates of b.

Duncan et al. (1994) examined sampling efficiency by varying NQ and NB using data from the previous spatial studies and a data set from Egypt. Again, the objective was to develop and evaluate sampling plans for females on the roots and males and juveniles in the soil. A number of analyses were performed with TPL results given of four data sets of juvenile and male nematodes from Florida and one from Egypt. These estimates are of two policies for compositing cores into samples and five different combinations of NQ and NB. The data from Egypt comprising $NB = 36$ sets of $NQ = 2$ samples of 15 composited cores were analyzed as a spatial TPL. The Florida datasets were analyzed as ensemble TPLs (Appendix 7.Y). The estimates are not significantly different ($P > 0.20$), although they range from $b = 1.48–1.93$. The lack of significance can be attributed to the high SE(b)s, which, in turn, are due to either low NB or low NQ. The two spatial analyses with $NQ = 2$ had correlations of $r \approx 0.50$, which is unusually low. The three ensemble analyses with higher values for NQ all had correlations $r \geq 0.98$, but the low NB resulted in comparatively high SE(b).

CLOVER

The phytonematode populations of the rhizosphere in 41 clover fields in Egypt were sampled by Abd-Elgawad and Hasabo (1995) to obtain data for nematode TPL analysis, sampling plans, and population analysis. Three 28-cm^2 soil cores were taken with a hand trowel in five contiguous 800-m^2 areas of each field. The three cores were composited and nematodes extracted by sieving and centrifugation from 100 g of soil were identified and counted. All stages of twelve taxa were identified, six to genus and six to species, and ensemble TPLs computed by ODR for each taxon with the number of nonzero points, $NB = 5–39$: TPL slopes ranged $1.22 \leq b \leq 1.87$ (Appendix 7.Z). In addition to plant-parasitic nematodes, *Aphelenchus avenae* and *Tylenchus* spp., both fungus feeders, were also analyzed and were not different from the plant-parasitic species. Abd-Elgawad & Hasabo used the TPL estimates to compute necessary sample sizes to achieve predetermined levels of sampling error. They also computed exact transformations $z = (1 - b/2)$ and compared the results of analyses with data transformed by $y = \log(x)$, $y = \sqrt{x}$ and $y = x^z$ and found the exact transformation best stabilized the variance for analysis of the nematode population data.

SUGARCANE

Allsopp (1990) sampled nematodes during nematicide trials in sugarcane fields in Queensland, Australia. In each of 35 treatment plots, five 840-cc soil samples were taken adjacent to randomly chosen sugarcane roots and composited. Nematodes were extracted from 500 g of soil and 40-60 g root and abundance recorded as #/200 cc of soil or #/100 g fresh weight of roots. Allsopp gives ensemble TPL estimates of six taxonomic groups: root-lesion, root-knot, spiral, stubby-root, stunt, and ring nematodes (Appendix 7.A1). A mixed-species ensemble TPL of the seven groups in soil and three groups (root-lesion, root-knot, and spiral) in the roots is fit by a single ensemble TPL ($b = 1.95 \pm 0.112$, $r = 0.987$). The three root points extend the TPL by one order of magnitude; omitting them reduces both the correlation coefficient ($r = 0.959$) and the slope slightly ($b = 1.79 \pm 0.228$): the differences are not significant ($P > 0.50$). Evidently #/2 cc of soil represents the same density of nematodes as #/g of root. In addition to TPL, Allsopp applied the Poisson distribution, negative binomial distribution, and Iwao's regression model to his data. He concluded that TPL "gave [a] better fit compared with the other models" and used the TPL estimates to compute sample sizes for fixed levels of precision and fixed-precision-level stop lines for sequential sampling for all species.

BANANA

Ten banana crops in Queensland and ten in New South Wales, Australia, were sampled over a period of years for *Radopholus similis* and *Pratylenchus goodeyi* by Stanton et al. (2001). They analyzed the nematode population density (#/100 g root) and a disease index by fitting the negative binomial and spatial TPL. Northern Queensland is tropical and New South Wales is subtropical. TPL gave the better fit and was used to determine fixed-precision stop lines for sequential sampling for both the disease index and for nematode populations. There was no difference in TPLs of nematode density between tropical and subtropical sites (Appendix 7.B1), but the TPL of the damage index for the tropical sites was significantly less than the subtropical sites ($P < 0.01$). Apparently, the damage index variance grows more slowly in the tropics suggesting it is more predictable than in the subtropics, possibly because damage/nematode is greater on account of the higher temperatures and/or longer growing season.

PINE TREES

The pinewood nematodes *Bursaphelenchus xylophilus* and *B. mucronatus* are phoretic on cerambycid (longhorn) beetles *Monochamus carolinensis* and *M. saltuarius*, respectively. In addition to the direct damage to trees by the beetles, the nematodes cause pine wilt disease, which can be fatal. Warren and Linit (1992) asked if the presence of *M. carolinensis* was necessary for *B. xylophilus* infestation, other than for transport. They compare the distribution and abundance of *B. xylophilus* in Scots pine (*Pinus sylvestris*) bolts infested with and without *M. carolinensis*. Blue-stain fungus was introduced to paired bolts from each of seven healthy pine trees to enhance nematode population growth and nematodes were introduced 1 week later. One bolt of each pair was exposed to beetles for 3 days to permit oviposition. The seven pairs of pine bolts were incubated for 60–70 days until beetles began to emerge. Each bolt was sampled for nematodes and nematode counts analyzed by TPL and Green's coefficient (Chapter 3). The nematode distributions were aggregated in both sets of bolts (Appendix 7.C1), with super-aggregated TPL slopes not significantly different ($P > 0.18$) in beetle-infested and uninfested bolts (common $b = 3.10 \pm 0.389$). However, the pattern of density within the bolts differed with the radial distribution more heterogeneously in the noninfested bolts than the infested.

Many *Bursaphelenchus* species are obligate fungus feeders, while *B. xylophilus* and *B. mucronatus* feed on both pine wood and fungus growing in the pine. The relationship between *B. xylophilus* and the fungi cohabiting Japanese black pine (*Pinus thunbergii*) was examined by Sriwati et al. (2007) who isolated 18 species of fungus from black pine trees. In a laboratory experiment, they investigated the relationship between the 18 fungi and *B. xylophilus*. Isolates of each fungus were cultured and then inoculated with 300 *B. xylophilus*. After inoculation, nematodes were extracted from the fungus cultures and counted. Sriwati et al. provide the log of population density (#/plate) and SD for the 18 fungi. The grey mold *Botrytis cinerea*, a preferred food of the *B. xylophilus* culture was included as a control. Nematodes were counted 5, 10, and 15 days post inoculation. The association between fungus and nematode population growth ranged from negative through neutral to very positive, providing a range of population means suitable for TPL analysis. Fig. 7.14 (Appendix 7.D1) shows the ensemble TPL identifying the three time frames over seven orders of magnitude of mean. The three *B. xylophilus* treatments fed *B. cinerea* (circled) are all below the best fit line, indicating lower variation on the preferred food source. The three time periods do not differ in either intercept or slope ($P > 0.55$). As the nematode populations grew on the fungus cultures, the variance mean relationship remained constant. The longest time period spanned the entire seven orders of magnitude of mean and the 5 day period spanned 1.5 orders. This experiment resulted in significantly lower slopes than Warren & Linit's experiment probably as a result of the differences in setup and sampling procedures.

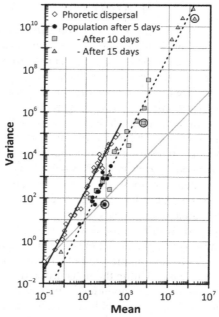

FIG. 7.14   Mixed-species ensemble TPL ($NQ = 5, NB = 36$) of *Bursaphelenchus xylophilus* grow-ing on 19 fungus cultures recorded at 5, 10, and 15 days post inoculation. Each point is nematode density (#/plate) on a different fungus culture with the control fungus *Botrytis cinerea* (circled). Temporal TPL ($NQ = 2$–14, $NB = 37$) of phoretic *B. mucronatus* released from longhorn beetle *Monochamus saltuarius* over a 70 day period. *(Data from Table 2 in Sriwati et al. (2007) and from Table 1 in Jikumaru et al. (2001).)*

The related pinewood nematode *B. mucronatus* carried by the longhorn bee-tle *M. saltuarius* was investigated by Jikumaru and Togashi (2001). They recov-ered 43 emerging adult *M. saltuarius* from dead Japanese red pine (*Pinus densiflora*), which were transferred to individual containers with red pine twigs. Every 5 days, the beetles were transferred to new containers and nematodes extracted and counted: nematodes washed from the old containers were also counted. At the end of the experiment, the nematodes still on the beetles and washed from the containers were added to the nematodes recovered from the twigs to estimate the total nematode load and from that the transfer rate for each beetle. Transfer rate was divided into nematode departure efficiency and nem-atode transmission efficiency, both of which were highly variable. Jikumaru & Togashi record the number of nematodes recovered from the twigs exposed to each beetle during each 5-day period up to 70 days. Of the 43 beetles, four died in the first 10 days and failed to transmit any nematodes and two were unin-fected. The mean and variance of nematodes recovered per 5-day exposure period from $NB = 37$ beetles provide for a temporal TPL with $NQ = 2$–14 (Fig. 7.14; Appendix 7.E1). Two factors contributed to the range of means over nearly four orders of magnitude: the beetle survival, which spanned 2–120 days with half the beetles surviving 50 of 70 days and the nematode transmission efficiency, which ranged from 0% to 50% and averaged 11%. Both variables are skewed and their combination contributed to the TPL's steep slope ($b = 1.85 \pm 0.039$), which is not significantly different ($P > 0.35$) from the estimate obtained from Sriwati et al.'s (2007) experiment with *B. xylophilus*.

Entomopathogenic nematodes

Entomopathogenic nematodes (EPNs) in the families Steinernematidae and Heterorhabditidae are ubiquitous obligate parasites of soil insects. They are found naturally in soils throughout the world and are now applied as biological control agents for a range of insect crop pests. Their economic utility has stimulated research in EPN genetics, physiology, and ecology as well as that of their symbiotic enterobacteria that facilitate the exploitation of insect hosts. Invasion of the insect host is effected by infective juveniles (IJs) that release the bacteria, which partially digest the host biomass ingested by the nematodes. It is the enterobacteria that are largely responsible for the host's death. EPNs reproduce sexually and/or asexually for several generations before exiting the cadaver as IJs. Exiting IJs may number 100s of 1000s of individuals, which disperse (Bal et al., 2014) to seek new hosts using several hunting strategies (Grewal et al., 1994). They do not feed, but only complete their development after successfully invading a new insect host. Stuart et al. (2006) review the biotic and abiotic factors that influence EPN distribution and abundance and the sampling methods used to estimate their spatial and temporal population dynamics.

Sampling methods

The procedures for estimating EPN numbers in samples differ from those used for other nematode groups. The usual method is by baiting developed originally by Bedding and Akhurst (1975) or a variant. Fan and Hominick (1991) assessed the efficiency of *Galleria* baiting for rhabditids in sand and soil. A measured volume of soil is placed in a container to which is added one or more last instar wax moth (*Galleria mellonella*) larvae. The sample is examined periodically, usually every 3 days and kept moist as needed. Dead *Galleria* are replaced and dissected, and the nematodes in each cadaver identified and counted. Several rounds of exposure, removal, and dissection of *Galleria* cadavers are made until no further *Galleria* mortality has occurred. This process can take a month or more. Koppenhöfer et al. (1998) did as many as eight baiting rounds over a 3-week period. Stuart and Gaugler (1994) developed a variant in which the bioassays were examined twice a week and dead *Galleria* were replaced and transferred to damp filter paper in a Petrie dish (White trap; White, 1927) where IJs emerging after about a week could be identified and counted, saving time-consuming dissections. Even so, they reported as many as 14 rounds of baiting during an 11-week period in one bioassay. While less work is involved in this indirect method of population estimation than the methods used for nonentomopathogenic nematodes in soil or sediment, it can be very time consuming and may require the use of a large number of *Galleria* larvae. Additionally, many rounds of baiting may exceed the life span of some IJs, notably heterorhabditids.

To reduce the time involved in estimating EPN density, I developed a short cut, published in Stuart et al. (2006), that reduces the time investment from 3 to 6 weeks to about 3 days. The number of EPNs in a sample may be estimated by making the prey number and prey killed a function of the parasite number using the Holling (1959) Type III predator-prey model and solving for parasite number. The technique is demonstrated using data of Koppenhöfer et al. (1998). To my knowledge, we are the only researchers to have used this method.

STEINERNEMA FELTIAE AND S. GLASERI

Spridonov and Voronov (1995) reported on the small-scale distribution of *Steinernema feltiae* infective juveniles (IJs) found in a fallow field on sandy soil in southern Estonia. Soil cores were taken with a 215-cc auger at 5-cm intervals along eight parallel transects. Nematodes were extracted from 386 soil samples using a modified Baermann funnel technique. Each transect was tested for aggregation using the variance-mean ratio and only in one transect were the IJs randomly distributed: all others were aggregated, two of them extremely aggregated. Spatial autocorrelation between transects showed spatial correlation to decline rapidly even between closely adjacent transects.

Spiridonov & Voronov provide the mean and variance of *S. feltiae* IJs per 215-cc sample for each transect ($NQ = 47$–52). An ensemble TPL analysis of their data was originally published in RAJT (1999) with data of *S. feltiae* and of *S. glaseri*. As part of his PhD research, Arne Peters (1994) sampled natural populations of *S. feltiae* at $NB = 7$ locations in Germany. At each site, $NQ = 2$ soil cores were taken and the nematodes extracted, identified, and counted. Dr. Peters kindly made available to me the means and variances per core of *S. feltiae* for an ensemble TPL. In a series of experiments to investigate the persistence and vertical migration of *S. glaseri*, Dr. Peter Smits extracted nematodes from soil samples taken with an 80-cm^2 corer, which were separated into four, five, or six 2.5-cm layers. Samples were taken on several occasions over a year following inoculation of the soil by *S. glaseri* IJs with the objective of tracking the vertical movement of the IJs in the soil. The number of sampling occasions differed between the five experiments conducted at different locations all on sandy soil. Dr. Smits kindly made available to me the original sample counts, which I used to perform a spatial TPL analysis on the total number of nematodes recovered from each soil column. The means and variances were calculated ($NQ = 5$–29) for each time period in each experiment, giving a total of $NB = 23$ data points. I am grateful for the generosity of Drs. Peters and Smits for allowing me to use their unpublished data. Results of the three TPL analyses

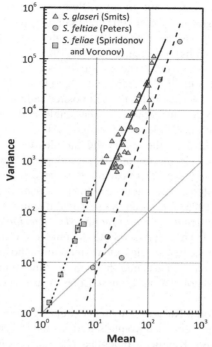

FIG. 7.15 Small-scale distribution of *Steinernema feltiae* infective juveniles in a fallow field in Estonia ($NQ = 47$–52, $NB = 8$) and $NB = 7$ sites ($NQ = 2$) in Germany; *S. glaseri* IJs following inundative release in five experiments ($NQ = 5$–29, $NB = 29$). *(Data from Table 1 in Spiridonov and Voronov (1995) data provided by Dr. Arne Peters and Dr. Peter Smits and used in RAJT (1999).)*

are presented in Fig. 7.15 (Appendix 7.F1). Visually, the *S. glaseri* slope is less than either *S. feltiae* slope, but the three slopes are not significantly different from each other ($P > 0.13$), but all are significantly steeper than $b = 2.0$ ($P < 0.05$) indicating extreme aggregation in these EPNs.

STEINERNEMA CARPOCAPSAE AND HETERORHABDITIS BACTERIOPHORA

Duncan et al. (1996) used TPL to estimate optimum sample sizes for detecting EPNs following inundative release. They established 46-m^2 experimental arenas centered on two grapefruit trees in a citrus grove in Florida. *Heterorhabditis bacteriophora* suspensions were applied to the experimental arenas using a hand sprayer at a rate of 2×10^6 IJs/m^2. One hundred 100-cc soil cores were taken in each arena in a stratified random arrangement 1 h, 2 and 7 days following application. The cores were split into two layers, each 15 cm deep. To check for movement, four additional sets of 15 samples were taken 1 m from each edge of the arena on each sampling occasion and composited. A second set of samples was taken from one arena on day 7. Nematodes were extracted from 12 sets (2 trees × 2 depths × 3 dates) of 100 single samples to compute variance and mean for TPL. In addition, there were 12 composited samples from the periphery and 8 (4 sets × 2 depths) composited samples from the arena to examine how combining samples affects TPL. The spatial TPLs for single and composited samples differ significantly in slope ($P < 0.02$) with $b = 1.78 \pm 0.062$ for the single samples and $b = 1.54 \pm 0.069$ for the composited samples. The intercepts are not different ($P > 0.60$), but combining the single and composited samples reduces the correlation coefficient (Appendix 7.G1). Substantially more nematodes were recovered from the composite samples than the single samples, which had both higher correlation and steeper slope. As sampling procedure, aside from compositing, was identical for both sets, the difference between TPLs is likely due to density dependence in the Baermann extraction efficiency.

Campbell et al. (1995) measured the seasonal dynamics of natural populations of EPNs *Steinernema carpocapsae* and *H. bacteriophora* in turfgrass in central New Jersey. Their purpose was to document and quantify the natural seasonal dynamics of the two species and their common hosts, larvae of the Japanese beetle (*Popillia japonica*) and other soil- and turf-inhabiting arthropods, to determine if natural populations of these EPNs have a regulatory effect on the arthropods.

Four grass plots on a Rutgers University property were sampled biweekly from May 1992, 19 dates in all. Two plots were set to tall fescue (*Festuca arundinacea*) and two were Kentucky bluegrass (*Poa pratensis*). The plots were divided into 1000 300 × 760 cm sections. Four 2.5-cm^2 soil cores were taken in each of 12 randomly selected sections, composited and EPNs extracted by *Galleria* baiting.

S. carpocapsae were recovered in about four times as many samples as *H. bacteriophora*, although *H. bacteriophora* was about four times as abundant where it was found than *S. carpocapsae*, indicating quite different spatial distributions of the two EPNs, possibly related to their host-finding strategies of active hunting and passive ambushing, respectively. Locations with *H. bacteriophora* had significantly lower Japanese beetle densities compared to sections without nematodes, while *S. carpocapsae* did not have a measurable impact on Japanese beetle or the other arthropod species.

Campbell et al. give the proportion of positive sections, the average number of nematodes recovered from each section, and the number of nematodes per cm^2 in each section for both species and grasses. Very few *H. bacteriophora* were recovered from the tall fescue plot, insufficient for TPL analysis but the other three sets are adequate (Fig. 7.16A; Appendix 7.H1).

In a later study, Campbell et al. (1998) established eight transects of 30 sections each on tall fescue, Kentucky bluegrass or on mixed species lawns. Two transects were close to the first site and the others were established at three other sites in New Jersey. Samples were taken from transect sections on one, two, or three occasions between August 1993 and June 1994. Using the same sampling

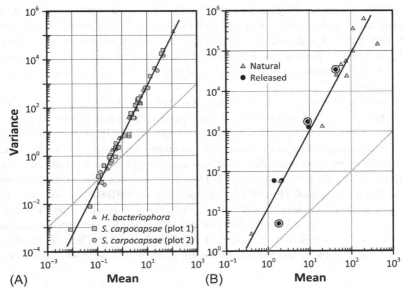

FIG. 7.16   (A) Spatial TPLs ($NQ=12$, $NB=51$) of *Heterorhabditis bacteriophora* under tall fescue and *Steinernema carpocapsae* under two plots of Kentucky bluegrass coincide almost perfectly
with a slope of $b=2.07\pm0.039$. (B) Spatial TPLs ($NQ=30$, $NB=18$) of *H. bacteriophora* recovered immediately after and 2 months following application on to turfgrass plots, and natural
populations in other plots. Circled points are of the release plots the day after application. *((A) Data
from Figs. 2–4 in Campbell et al. (1995). (B) Data from Campbell et al. (1998).)*

methods, Campbell et al. estimated the number of EPNs per section from the
number of nematode-infected larvae per section using an equation developed
by Koppenhöfer et al. (1998) predicting the density of nematodes (y) from
the number of cadavers (x):

$$\log(y) = -0.25 + 2.08\log(x) \tag{7.2}$$

Three transects were determined to be free of *H. bacteriophora*. To these
transects, inundative releases of *H. bacteriophora* were made at a rate of
2.5×10^9 IJs/ha using a watering can. The day following application in August
and again in October, the transects were sampled for nematodes and population
density in the sections estimated as described. Six other transects were sampled
1–3 times for natural populations of *H. bacteriophora*.

Campbell et al. tested for aggregation using Green's index (Eq. 3.15) and
concluded that *H. bacteriophora* populations, both natural and released, were
aggregated in most sample sets. Fig. 7.16B (Appendix 7.H1) shows the spatial
TPLs of natural and released EPNs are best fit by a single line. Neither the intercepts nor the slopes of this ($A=1.08\pm0.174$, $b=1.93\pm0.116$) and the previous
($A=0.84\pm0.033$, $b=2.07\pm0.039$) experiment are significantly different

$(P > 0.17)$. It's noteworthy that samples from two of the three release plots taken the day after application (circled) as well as those taken 2 months later fit the line very well. Despite the nearly uniform application of nematodes, within 24 h, the released populations had adopted the same TPL as the native populations. Also, despite the population decline over 2 months, the spatial distribution was unchanged.

To test their efficacy in controlling citrus root weevil (*Diaprepes abbreviatus*), Duncan and McCoy (1996) applied commercial formulations of the nematodes *S. riobravis* and *H. bacteriophora* at a rate of 2×10^6 IJs per tree to two trees in an orange grove in Florida. They monitored the nematodes at 1 h, 2 and 7 days following application by taking four samples of 6 cores beneath each tree split into three layers (0–1, 1–3, and 3–15 cm). Nematodes were extracted from 10 cc of soil and spatial TPLs of #/sample and density (#/100 cc of soil) were fit by ODR for both species (Appendix 7.I1). Although not significantly different, converting #/sample to #/100 cc reduced the TPL slope for both species as the conversion factors differed by depth ($\times 1$, $\times 0.5$, and $\times 0.083$). This is equivalent to variable efficiency and emphasizes the importance of standardized sampling.

EFFECT OF HABITAT

Campos-Herrera et al. (2008) surveyed natural EPN populations in the La Rioja region of northern Spain with the objective of studying EPN diversity, occurrence, and life history characteristics under different agricultural management systems: conventional and organic, annual and perennial crops. Samples were taken on eight occasions from May 2003 to February 2005 in three to five permanent plots established at four agricultural sites and a natural area. On each occasion, five scoops of soil were taken and composited from five points in each plot. Nematode extraction was by *Galleria* baiting. Seven EPN species and strains were identified using morphological and molecular characteristics, including five strains of *S. feltiae*, which was widely distributed and two strains of *S. carpocapsae* found only under perennial crops. Campos-Herrera et al. provide EPN density at the five sites using Koppenhöfer et al.'s (1998) equation (eqn 7.2). Density ranged from 0.01 IJs/100 cc of soil at the organic annual crop site to 21.2 IJs/100 cc in the natural area: no EPNs were detected at the conventional annual crop site. With $NB = 4$ points ($NQ = 40$) for a hybrid TPL analysis (Appendix 7.J1), the regression passes through all four points and extends over three orders of magnitude with slope $b = 2.45 \pm 0.093$ ($r = 0.999$), significantly steeper than $b = 2.0$ ($P < 0.05$). The use here of the habitats to provide the range of means is not unique, but the tight regression is. The fact that the order of abundance of EPNs is conventional annual (with none detected), organic annual, conventional perennial, organic perennial, and natural area is interesting from an agricultural perspective, but the same variance-mean relationship extending so precisely over four distinct habitat types is very surprising.

Efron et al. (2001) investigated the spatial distribution of EPNs and their host insects in a citrus orchard in Israel's arid Negev region. The spatial patterns were studied in plots differing on their shade level: 25%–50%, 50%–75%, and 75%–100%. 130-cc soil samples were taken beneath the canopies of 20 trees in each plot monthly from November 1994 to September 1995. *Galleria* baiting was used and the number of EPN-killed larvae was taken as an indicator of EPN abundance on each sampling occasion. Cadavers were dissected to identify EPNs present, which were not counted. A single spatial TPL ($NQ = 20$, $NB = 27$) of the monthly counts of EPN-killed larvae per sample fits all three plots better than three separate regressions. However, the TPL slope of the 50%–75% shade plot is significantly steeper than the 25%–50% plot ($P < 0.05$), suggesting that EPNs select preferentially the soil temperature under moderate shade (Appendix 7.K1).

In a project to examine competition and coexistence of EPNs, Koppenhöfer and Kaya (1996) exposed larvae of the masked chafer (*Cyclocephala hirta*) and larvae of black cutworm (*Agrotis ipsilon*) to five combinations of *S. carpocapsae*, *S. riobravis* and *S. glaseri*. Each combination was replicated 12 times with a single larva of each species placed in a container of autoclaved soil seeded with perennial rye grass. The containers were checked daily for 10 days and dead larvae were checked for IJs, the species easily identified by their relative sizes. Koppenhöfer & Kaya provide the means and SDs of IJs recovered from the insect larvae and the number of larvae killed by the EPNs. Ensemble TPLs of these data have slopes significantly different ($P < 0.05$) with the masked chafer slope not different from $b = 2.0$ ($P > 0.50$) and the black cutworm slope not different from Poisson ($P > 0.55$) (Appendix 7.L1). As samplers, the two insects have very different sampling efficiencies. The methodology was the same for both bait species, the numbers of individual and combinations of EPNs was the same, and recoveries comparable, thus differences are due to the relative attractiveness and/or susceptibility of the hosts. Sampling efficiency of the black cutworm is density dependent, relative to masked chafer. The use of baits to estimate EPN abundance is exactly equivalent to the use of pheromone and light traps in sampling for insects as they all rely on the target's response to the trap (Chapter 15).

Abd-Elgawad (2014) used a *Galleria*-baiting bioassay to assess the population of mostly *H. bacteriophora*, in a mango orchard in Egypt. He found IJs in 152 baits in 29 of 30 soil samples with up to 13 rounds of *Galleria* baiting. Means and variances were computed from the number of *Galleria* infected in each of the 29 positive soil samples. An ensemble TPL by ODR ($b = 0.90 \pm 0.099$) was not different from Poisson ($P > 0.30$), suggesting that the distribution of heterorhabditids is indistinguishable from random. Examining the individual variance-mean ratios, Abd-Elgawad concluded that despite the apparent Poisson TPL, EPNs in the mango orchard were indeed aggregated.

The GMR estimate of $b = 1.20 \pm 0.099$ is significantly different from $b = 1.0$ at $P < 0.05$. The TPL has a low correlation coefficient ($r = 0.75$), owing in part to the comparatively narrow range of means of *Galleria* infection. Abd-Elgawad's conclusion that heterorhabditids are not randomly distributed, supported by the GMR regression, illustrates the importance of the statistical model used for TPL (and other allometric relationships). Taking more samples and estimating the nematode population at each site would improve the fit and likely increase the correlation coefficient and also the ODR estimate of slope.

Animal-parasitic nematodes

There is probably at least one nematode parasite of all large metazoans and some are host to whole communities. In this section, we examine the spatial and temporal distributions of nematode parasites of both vertebrates and invertebrates. In general, animal-parasitic nematodes impose a chronic health burden on hosts without killing them, although extreme nematode burdens can lead to death. The important exception is the EPNs that kill their hosts.

Parasitologists recognize three statistics to describe the structure of parasite populations: prevalence, intensity, and abundance. Relative to other, free-living populations, quantifying parasite populations is complicated by the fact that the host population also has a structure and the distribution and abundance of the parasite population is not independent of the population distribution and abundance of the host. Hence, the three basic statistics with different biological interpretations that nevertheless overlap somewhat. As with other organisms, parasites exhibit a range of spatial distributions from Poisson to highly aggregated, both at the population level and within their hosts.

The proportion of hosts in a sample that is infected with the parasite of interest is called prevalence. This is distinct from intensity, which is the number of parasites found in the infected hosts, ignoring parasite-free hosts; it may be expressed as a mean or median when characterizing the frequency distribution of parasite load. The third statistic is abundance, which is defined as the mean number of parasites found in all hosts in a sample including uninfected hosts. Intensity TPL plots produces steeper gradients than do abundance plots. This phenomenon is examined here and explored further in Chapter 15.

Invertebrate hosts

COCKROACHES

Four species of nematode (*Hammerschmidtiella diesingi*, *Leidynema appendiculatum*, *Thelastoma periplaneticola* and *T. bulhoesi*) in the hindgut of 328 lab-reared American cockroaches (*Periplaneta americana*) were recorded by Adamson and Noble (1992) over a 3.5-year period. Adamson & Noble provide the intensities (#/infected host) of males and females of *H. diesingi*, *L. appendiculatum*, and the *Thelastoma* species combined in adult male and female and two age classes of immature cockroaches. The means and variances of intensity averaged over time, representing temporal stability, form significantly different ($P < 0.02$) temporal TPLs for male and female nematodes (Appendix 7.M1). As intensity is computed only from hosts positive for the parasite ($NQ = 6$–48), the TPLs exclude zero counts. Knowing the total number of samples ($NQ = 328$), it is simple to recalculate means and variances including the zeros because Σx and Σx^2 are unchanged, while the sample size (NQ) is increased. TPLs for male and female nematodes with zeros included results in almost identical slopes ($P > 0.75$) and marginally different intercepts ($P < 0.055$) and different from intensity. The practice of using intensity as a measure of parasite population abundance can give a very misleading impression of the temporal (or spatial) distribution of the parasites. In the case of temporal TPL, as a measure of stability, the male results in particular illustrate the possibility of misinterpretation.

DROSOPHILIDS

The nematode *Howardula aoronymphium* is parasitic on mycophagous drosophilids, entering the fly larvae while they feed on mushrooms of a dozen genera. *Drosophila* spp. emerging from field-collected mushrooms were identified and dissected by Jaenike (1994) to determine the numbers of adult *H. aoronymphium* they carried. He used the negative binomial k as a measure of the nematode's aggregation and suggested the different levels of exposure of the larvae to the parasites in different mushroom species and sites resulted in the observed aggregation. Mean intensity of parasitism in flies emerging from individual mushrooms is highly correlated between species at a site (Jaenike and James, 1991). Jaenike gives data of nematodes/fly (not intensity) enabling ensemble TPLs of nematodes/fly for site and mushroom species. The regressions are $b = 1.24 \pm 0.089$ for the site comparison and $b = 1.58 \pm 0.166$ for mushrooms (Appendix 7.N1). They differ only at $P < 0.10$ and reflect the different ways the original data were aggregated confirming the aggregation measured by k: both slopes are steeper than $b = 1.0$ ($P < 0.001$).

Vertebrate hosts

In an exhaustive search of the literature on patterns of parasite load and aggregation of macroparasites in or on vertebrate hosts, Shaw and Dobson (1995) collected 269 examples of parasitism of which 104 are nematodes: the remainder are platyhelminths, acanthocephalids, and arthropods. The survey covered fish, herptile, bird, and mammal hosts and provided quantitative analyses of mean parasite burden and estimates of the degree of aggregation of parasite burdens between hosts by negative binomial k. Mean parasite burden was found to be log-normally distributed across all examples, and the variance-mean ratios of all but two parasite infections were substantially above one, indicating aggregation. There are two examples of the nematode, *Rictularia coloradensis*, in mouse: one has $V/M = 1.12$, close to Poisson, and the other has $V/M = 4.71$.

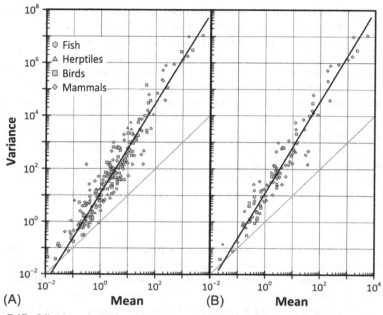

FIG. 7.17 Mixed-species TPLs of parasite load segregated by vertebrate host—fish, herptiles, birds, and mammals. (A) All endoparasites ($NB = 227$). (B) Nematodes ($NB = 104$). *(Data from Appendix A in Shaw and Dobson (1995).)*

This suggests that a TPL of *R. coloradensis* in mice would be aggregated with the lower value close to its intersection with the Poisson line.

Mixed-species TPLs of the complete endoparasite collection (Fig. 7.17A; Appendix 7.O1) and the nematode subset (Fig. 7.17B) cover almost six orders of magnitude of mean and have closely similar slopes of $b = 1.69 \pm 0.034$ and $b = 1.73 \pm 0.046$, respectively. Separating the nematode set into bird and mammal hosts, the slopes or intercepts are not different ($P > 0.13$) and the estimates of TPL for all hosts, which includes one fish and seven herptiles, are intermediate ($A = 1.094 \pm 0.055$ and $b = 1.725 \pm 0.046$).

Some of Shaw & Dobson's nematode data were included in Morand and Guégan's (2000) compilation of 828 populations of 326 species of adult nematodes in the gut of 66 mammal species. Their database contains 104 means and variances of nematode abundance (#/host) from which they computed an (ODR) ensemble TPL with $b = 1.79 \pm 0.039$ and $A = 3.270 \pm 0.170$ (Appendix 7.P1). Morand & Guégan's estimates differ significantly ($P < 0.02$) from the ODR estimates for nematodes using Shaw & Dobson's data of nematodes in mammals ($b = 1.64 \pm 0.057$ and $A = 1.20 \pm 0.073$). A possible reason for the discrepancy is that Morand & Guégan's database extends over nine orders of magnitude of mean with two extremely high means and variances that were not included in Shaw & Dobson's database: these may have had a strong influence on the regression.

A motivation for Morand & Guégan's study was to try to determine why parasite loads are almost always aggregated and sometimes highly aggregated. They found that the prevalence of nematodes in mammal hosts showed a U-shaped distribution in which there are many hosts with few or no parasites and also many with vary large parasite burdens. The inverse Gaussian (Eq. 2.15) and one form of the beta distribution are U-shaped, the former has the variance proportional to the mean cubed. They concluded that "nematode parasite species might adjust their spatial distribution and burden in mammal hosts for simple epidemiological reasons," suggesting that a parasite population's spatial distribution is at least partly under the control of its members.

SHEEP

Stear et al. (1998) conducted a survey of 514 six-month-old, Scottish Blackface lambs from a single farm in Scotland. They found differences between years in the prevalence and mean intensity of infection by five nematode species and two genera. Most lambs had relatively few nematodes, but a small proportion had many, indicating skewed distributions. One species, *Ostertagia circumcincta*, was particularly abundant and found in most lambs. Stear et al. fitted the negative binomial distribution to the abundance data but found it fit only *O. circumcincta*. They give the means and variances of intensity of infection by the seven taxa for each of 4 years. Mixed-species TPL for the seven taxa fit a common line with moderate slope of $b = 1.43 \pm 0.057$ (Appendix 7.Q1). Stear et al. do not give the prevalence of infection for all taxa and years, but they note that *O. circumcincta* was found in 513 of 514 lambs, so for this species intensity and abundance are essentially identical. With the exception of *Nematodirus* spp., the slopes of the individual taxa are steeper than the mixed-species collection. The fact that we get a very good mixed-species TPL from a collection of species with different slopes and intercepts is curious, made more so by the fact that the data are of intensity with an unknown (presumably variable) number of zero counts. In a later study, Stear et al. (2006) examined fecal egg counts of nematode-infected Scottish Blackface lambs from four farms. In this study, it is probable that eggs/lamb and eggs/infected lamb are identical because *O. circumcincta* was found in almost all lambs in the other study. The lambs were sampled at regular intervals for 4–5 months in 4 years (generations). The spatial TPL plot of eggs/lamb has slope of $b = 1.34 \pm 0.097$ (Appendix 7. Q1), which is not different from the mixed-species TPL of intensity ($P > 0.20$).

Boag et al. (1989) used TPL to investigate the spatiotemporal distribution of gastrointestinal nematode larvae in a sheep paddock in northeast England. Grass samples cut to ground level were taken on August and in September from a grid of 144 125-cm^2 plots. Nematode larvae on the grass were counted and the densities recorded as #/125 cm^2. TPL was computed for both sample occasions and for samples at a range of separations. Except for a dip at 20 m in August, the slope remained fairly constant at separations of 5–30 m. In addition, aggregation

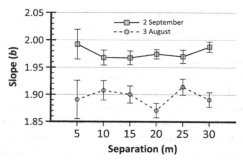

FIG. 7.18 TPL slope of gastrointestinal nematodes of sheep varied spatially with distance of separation of samples and temporally over the course of 1 month in sheep pastures. *(Data from Figs. 4 and 5 in Boag et al. (1989).)*

increased by ~4% from August to September (Fig. 7.18). The relative stability of b over 30 m and 30 days was sufficient for use in devising an efficient pasture sampling strategy. The 1989 paper did not distinguish between nematode species, but a later paper (Boag et al., 1992) identified 15 gastrointestinal nematodes at necropsies of 160 lambs. Ensemble TPLs (Appendix 7.R1) of 12 of 15 species were not significantly different from $b = 2.0$ and are close to the overall average of the TPLs of the paddock grass samples.

Gaba et al. (2005) compared the fit of the negative binomial distribution with the Gaussian, lognormal, exponential, & Weibull distributions to gastrointestinal nematode distributions in 11 sheep host populations taken from the literature. They used the Akaike Information Criterion (AIC; Pan, 2001) to distinguish between the fits and found the "Weibull distribution was clearly more appropriate over a very wide range of degrees of aggregation." The Weibull is more flexible in fitting the heavily infected hosts and less sensitive to sample size than the other distributions, including the negative binomial. Gaba et al. did not fit a TPL to their data, but provided the means and variances for 11 data sets.

Eight of the eleven data sets fit an ensemble TPL; three sets do not (solid points in Fig. 7.19). These three points from Gruner et al. (1994) are of nematodes recovered from 30 lambs 8 weeks after different procedures for artificially infecting them with 7000 *Teladorsagia circumcincta* infective larvae per animal. The AIC selected these three, along with six other sets, as best fit by the Weibull and almost as well by the Gaussian: they were the sets with the lowest V/M ratios. The averages, a little less than half the applied rate, combined with the substantially lower variances than expected from the other naturally occurring infestations suggest the populations of nematodes in these three sets had not yet stabilized. A fourth point (circled) is also of artificially infected lambs but fits the TPL. Despite receiving different treatments and quantities of nematodes, the means of the four artificially infected lambs are similar while the variances are very different. It is important that the statistical populations

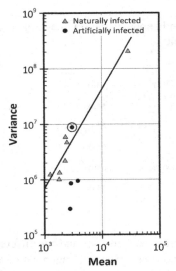

FIG. 7.19 Ensemble TPL ($NQ = 17–74$, $NB = 11$) of intestinal nematode *Teladorsagia circum-cincta* in lambs naturally and artificially infested. The circled point fits the pattern of naturally infested lambs, but the other 3 points of artificially infested lambs 8 weeks after infection have exceptionally low variances. *(Data from Table 1 in Gaba et al. (2005).)*

from which TPL points are drawn are the same (Fig. 13.2): in this case, the source of nematode infection specifies the statistical population and has an impact on the measured variance-mean relationship.

RABBITS

Another of Dr. Brian Boag's TPL studies of nematode distribution examined the distributions of three gut nematodes and two cestodes of the wild rabbit in Scotland (Boag et al., 2001). 2963 rabbits were collected by shooting at a site in Perthshire between January 1977 and December 1999. They were classified as kittens, juveniles, or adults, by sex, and by myxomatosis status. Myxomatosis is a fatal viral disease of rabbits causing skin tumors and blindness that is spread by direct contact and by insects. Boag et al. used bootstrapping (Good, 2010) to estimate means and SEs of *Graphidium strigosum*, *Passalurus ambiguus*, and *Trichostrongylus retortaeformis* found in the rabbit intestines. Table 7.4 shows the range of *b* values obtained for TPLs of parasite abundance computed for each month and each year. Both the annual and monthly slopes varied considerably. Boag et al. attributed the instability of *b* to variations in weather conditions acting upon infectious stages. As *P. ambiguous* females lay eggs in the perianal skin where larvae develop, the incidence of host self-infection is likely to be high because of rabbits' habit of copraphagy. Although the range of variation in *b* for *P. ambiguous* abundance is high, it is substantially narrower than

TABLE 7.4 Monthly and annual variation in aggregation of nematode parasites of wild rabbit

		Monthly b	Annual b
Graphidium strigosum	Min	1.63	1.20
	Max	3.27	2.78
Passalurus ambiguus	Min	1.73	1.70
	Max	2.23	2.54
Trichostrongylus retortaeformis	Min	1.83	1.09
	Max	3.18	3.21

the other two species. There were also sex and age differences in parasite aggregations. *P. ambiguous* and *T. retortaeformis* aggregation was lower in adult than juvenile rabbits, while *G. strigosum* aggregation tended to increase with age. The biggest and most consistent variation in parasite aggregation was due to myxomatosis status: b was substantially lower in myxomatosis-infected rabbits than infection-free rabbits, approximately Poisson in three instances (Fig. 7.20). Of all the factors investigated, myxomatosis had the most consistent impact, lowering the degree of aggregation for all parasites in all age groups. Boag et al. suggest that this may be due to variation in host immunity to nematode infection and the decreases in aggregation could be explained by the breakdown in immunity of rabbits infected with myxomatosis.

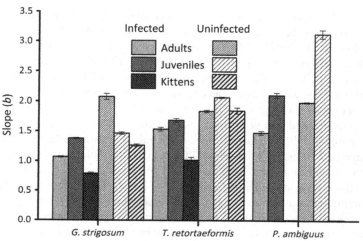

FIG. 7.20 Infection by myxamatosis reduces the aggregation of three gut nematodes in adult, juvenile, and kitten rabbits. *(Data from Table 3 in Boag et al. (2001).)*

RATS

Strongyloides ratti, an intestinal nematode parasite of rats, does not promote host self-infection. Eggs and larvae are passed in the hosts' faeces and subsequently develop outside their host where they may continue free living for several generations. The number of *S. ratti* per host is usually aggregated and Harvey et al. (1999) questioned if heterogeneity in the distribution of infective stages within the environment contributed to heterogeneity in their rat hosts. In an experiment in which laboratory rats were injected with 3rd instar *S. ratti* larvae at rates of 10, 100 or 500 larvae/rat, Harvey et al. counted the number of infectives in faeces at 5, 10, and 15 days post treatment. They give the number of pellets, the mean and variance of infective stages per pellet, and the goodness of fit of the Poisson and negative binomial distributions. The Poisson failed to fit any of the nine results, while no distributions of counts were different from the negative binomial. Ensemble TPL of their data ($NQ = 32–48$, $NB = 9$) has a slope of $b = 1.52 \pm 0.180$ (Appendix 7.S1). The TPL is of abundance as all pellets, including those without infective stages, are included in the analysis. This result is close to that of Shaw and Dobson's (1995) mixed-species TPL of $1.5 < b < 1.7$ and its moderate value is lower than that of self-infection.

MICE

Grear and Hudson (2011), noting that the aggregation of infection intensity of some nematode taxa is much higher than others, examined the factors affecting the unexpectedly high TPL slopes of species in the genus *Syphacia*. Female *Syphacia* can deposit eggs on the perianal region of the host, allowing for host-self-infection during grooming. This reproductive strategy, they suggested, may cause a positive feedback as hosts with higher intensities are subject to higher self-infection rates. The intensity of infection in hosts already infected would increase faster than the rate of infection in uninfected hosts, resulting in highly aggregated intensity distributions compared to other nematodes. Comparing *Syphacia* spp. in *Apodemus* and *Peromyscus* mice species with a non-self-infecting species, *Heligmosomoides polygyrus* in *A. flavicollis* they found that prevalence of *H. polygyrus* in *A. flavicollis* was significantly higher than *Syphacia* spp. in either *A. flavicollis* or *P. leucopus*, tending to support the feedback hypothesis. Detailed comparisons of the intensity relationships between host and nematode were equivocal. However, an ensemble TPL (Fig. 7.21; Appendix 7.T1) of parasite intensity of the two nematodes, $b = 2.80 \pm 0.208$ for *Syphacia* spp. and $b = 1.91 \pm 0.180$ for *H. polygyrus*, shows a significant difference in slopes ($P < 0.002$). As these data are of intensity (i.e., zeros excluded), the slopes are inflated. The difference between abundance slopes would be less, though still probably significant as Grear & Hudson give the overall prevalence of *Syphacia* and *H. polygyrus* as 10% and 58%, respectively. A simulation model developed to examine the role of the behavioral ecology of the mouse-nematode interaction also supported the hypothesis that the higher aggregation of *Syphacia* was due to host-self-infection.

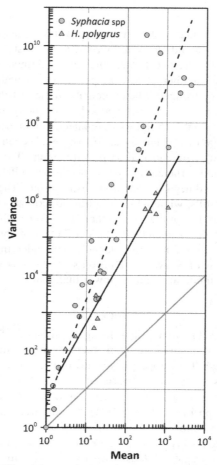

FIG. 7.21   Ensemble TPLs of the parasite intensity of *Syphacia* spp. (*NB* = 23) and *Heligmoso-moides polygyrus* (*NB* = 11) in *Apodemus* and *Peromyscus* mice differ as a result of differences in nematode reproductive strategy. (*Adapted from Fig. 1 in Grear and Hudson (2011).*)

CARP

Camallanus cotti is an intestinal parasite of the Chinese hooksnout carp (*Opsariichthys bidens*). Its seasonal population dynamics in the carp was studied by Wu et al. (2007) in the Danjiangkou Reservoir in central China. They found a weak positive correlation between abundance and fish length and a significant seasonal pattern of mean abundance of *C. cotti* (#/host). Prevalence also varied seasonally with higher levels of infection in summer than in winter. Wu et al. give the prevalence and mean monthly abundance with SD from which intensity and its SD can be calculated. Fig. 7.22 (Appendix 7.U1) shows the

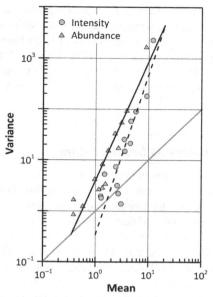

FIG. 7.22 Spatial TPLs ($NQ = 50$, $NB = 14$) of *Camallanus cotti* burden in the hooksnout carp (*Opsariichthys bidens*). The *solid* line is abundance and the hatched line is intensity. The lines intersect at $M = 20.5$, $V = 4649$. (*Data from Fig. 2 in Wu et al. (2007).*)

spatial TPLs for mean monthly abundance of *C. cotti* (#/host) and mean monthly intensity (#/infested host), showing the divergence of the two TPLs from the common point of $M = 20.5$, $V = 4649$. The slopes are significantly different ($P < 0.05$).[1]

Much of parasitism work focuses on prevalence and intensity, often ignoring abundance. The very different TPLs for abundance and intensity (unless the number of zero counts is very small) clearly indicate different levels of dependence of mean on variance and therefore aggregation, with consequences for statistical analysis and interpretation.

Aquatic nematodes

In addition to the various corers used to take terrestrial samples, aquatic samples are also taken by various grab and box samplers. Southwood and Henderson (2000) devote a chapter to sampling aquatic habitats. Most are for sampling the water, but the various grab samplers are for taking sediment samples. Box samplers are designed to take a large volume of sediment, from 200–2500 cm^2 and up to 50 cm deep. The Day sampler, widely used for marine sediment

---

1. Statistical comparisons are not strictly appropriate as intensity and abundance are not independent.

sampling, takes a $1000\,cm^2$ by 15 cm (15 L) sample. Multicore samplers comprising arrays of corers collect replicate samples over a small area. Except in the intertidal and sublittoral, samplers are deployed from boats or by divers. As aquatic sediments are waterlogged, extraction methods differ from terrestrial soil. Centrifuguation (Pfannkuche and Thiel, 1988), flotation (Caveness and Jensen, 1955) and elutriation (Uhlig et al., 1973) are commonly used and may be combined with Baermann funnel.

Freshwater nematodes as bioindicators

Bioindicators are taxa or groups of taxa that respond recognizably to changes in the condition of their environment. Changes in the abundance and/or behavior of bioindicators can signify deterioration or improvement in habitat condition. Nematodes occupy multiple positions in food webs as primary, secondary, and tertiary consumers and collectively can provide a comprehensive picture of the functional status of a habitat. The role of nematodes in the food web can be inferred from their morphology, enabling whole communities to be characterized. The relative frequency of *cp*-classes in a community, for example, is one such character. The data used by Park et al. (2013) to examine the spatial distribution of soil nematodes were collected to examine the potential for use of nematodes to indicate the health of urban soils (Park et al., 2010). Nematode communities have been extensively used to characterize the health of freshwater and marine habitats.

RIVERS IN GERMANY

Heininger et al. (2007) tested the hypothesis nematodes might be useful indicators for biomonitoring freshwater systems. In a study of nematode community structure in relation to physicochemical properties of river sediments, they sampled a gradient of anthropogenic contamination by heavy metals and organic pollutants at eight sites on three rivers in Germany: one site on the Rhine, two on the River Oder, and five on the River Elbe. At the time of sampling, two sites on the Elbe were subject to continuous pollution inputs and six sites were free of contemporaneous pollutant input. At each site, 3–6 50-cc subsamples were collected from the uppermost 2–3 cm of sediment. Nematodes were extracted, counted, identified, and assigned to a feeding guild: deposit feeders (mainly bacterivores); epistrate feeders (on algae); suction feeders (on plants and fungi); and chewers (omnivores and predators). A total of 168 nematode species were recovered in 62 samples. Nematode abundance ranged from 98 to 958/100 cc of sediment. The more heavily contaminated sites had relatively high densities of omnivores and predators, while sites with low-to-medium contamination were dominated by bacterivores or suction feeders. These differences broadly support Heininger et al's hypothesis.

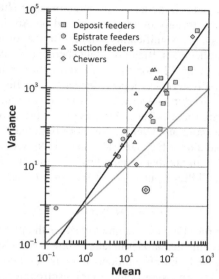

FIG. 7.23 Ensemble TPL ($NQ = 10, NB = 31$) of nematode feeding guilds from eight sites on three rivers in Germany. The *circled* outlier is chewers at the least polluted reach of the River Elbe. *(Data from Table 5 in Heininger et al. (2007).)*

Heininger et al's Table 5 provides data of total nematodes, number of genera, and numbers of the four feeding guilds for the eight sites. A mixed-species ensemble TPL of Heininger et al's results (Fig. 7.23, Appendix 7.V1) shows the four guilds and eight sites fit a common regression with a single outlier (circled), chewers at the least polluted site on the Elbe at which site epistrate feeders were absent. This point has a much lower variance for its mean than expected of the omnivores and predators that generally were more abundant in the more polluted sites: based on only two samples this point is probably biased.

HIGHLAND STREAMS IN GERMANY

To determine the extent to which pollution affected nematode community composition, Beier and Traunspurger (2003) sampled the streambed surface of two small streams in southwest Germany. The streams differ in geomorphology, velocity, water level, and key chemical variables, including the organic content of the sediment. Both streams are located on mainly limestone and sandstone and differ in their anthropogenic influence. They were expected to exhibit community differences related to pollution levels as one (the Körsch) passes through intensely used farmland and industrial areas, while the other (the Krähenbach) rises in a nature reserve and passes mainly through grazed pastures and woodland.

A total of 13,249 nematodes were recovered with the mean population abundance (#/10 cm^2) of the Körsch nearly 10 times that of the Krähenbach.

The latter site displayed a bimodal temporal abundance (July and February), while the Körsch was strongly unimodal centered on April. The mean abundance of nematodes was significantly higher in the organically enriched Körsch than in the Krähenbach. Bieier & Traunsberger provide data for a spatial TPL ($NQ = 4$, $NB = 11$ or 12) for the three most numerous families in the two streams, the Monhysteridae (deposit feeders), Tobrilidae (chewers), and Tylenchidae (suction feeders) (Appendix 7.W1). In general, all three families were highly aggregated and differed little between the two sites, except the Monhysteridae were more abundant at the more polluted Körsch site and the Krähenbach produced a poor but steep TPL that is likely biased by the narrow range of means and small number of samples. The consistency and wider range of means of the other TPLs argues against significant bias in most cases.

LAKES IN SWEDEN

Ristau and Traunspurger (2011) examined whether the proportion of omnivorous and bacterial-feeding nematodes in lake sediment are associated with nutrient levels. They sampled littoral sediments in eight southern Swedish lakes of different nutrient levels: an oligotrophic, four mesotrophic, and three eutrophic lakes. Four replicate 15-cc sediment samples were taken from 5 to 8 sample sites per lake. Samples were taken in April when the water columns were well mixed and again in September when the water column was more settled, conditions that might affect the nematode community. 11,710 individual nematodes in 157 species were extracted and identified to species. Their abundance ranged from 48 nematodes per $10\,cm^2$ in a mesotrophic lake to 369 nematodes/$10\,cm^2$ in one of the eutrophic lakes. Bacteria feeders, algae feeders, and omnivores comprised nearly equal parts of the total nematode community; suction feeders and predators were less common with the density of predators declining with increasing eutrophication. Overall nematode abundance, biomass, and diversity appeared unaffected by degree of eutrophication, although the oligotrophic and mesotrophic lakes supported a larger number of species than the eutrophic lakes.

Ristau & Traunspurger provide data for TPL analysis of total nematodes in spring and autumn and in the eutrophic and noneutrophic lakes. Differences between trophic status and seasons provided a range of means for a very steep TPL to emerge ($b = 2.69 \pm 0.308$). Fig. 7.24 (Appendix 7.X1) shows the nematode community TPL with points segregated by season and by trophic status to be extremely aggregated with a broader range of means in Autumn than Spring but community trophic status and season conforming to the same TPL.

FARM PONDS IN BELGIUM

In a study of nematode diversity and population response to eutrophication, Bert et al. (2007) sampled 14 ponds across a range of agricultural intensities on farms in Belgium. Agricultural intensification ranged from nature reserve to extensive

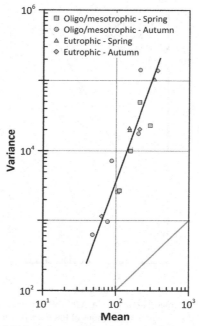

FIG. 7.24   Hybrid spatial TPL ($NQ = 8$, $NB = 16$) of nematode in samples (#/100cc) collected from three eutrophic and five healthy lakes in southern Sweden. *(Data from Table 2 in Ristau and Traunspurger (2011).)*

arable crops to intensively grazed cattle. In August and September, eight 255-cc sediment samples were taken in each pond and composited. Nematodes were extracted from 150-cc subsamples, identified to species, and abundance expressed as #/10 cm^2. Seventeen genera of free-living benthic nematodes were identified with deposit feeders and chewers dominating all communities but one in which suction feeders were predominant. *Tobrilus gracilis* and *Eumonhystera filiformis* comprised 58% of the total nematofauna. Although phosphorus, which is associated with intensification, is usually an important variable structuring nematode communities, this study failed to find a clear relationship between environmental variables such as phosphorus and the nematode assemblages across the intensification gradient.

Bert et al. provide data for ensemble TPLs for total nematodes and nematodes in the genera *Dorylaimus*, *Eumonhystera*, *Monhystera*, *Mononchus*, and *Tobrilus*, in addition to mixed-species TPLs for the five regions and three agricultural intensification rankings. Fig. 7.25 (Appendix 7.Y1) shows the overall ensemble TPL with the three intensity classes marked. At this level of organization, the nematode genera form a strongly aggregated community with $b = 1.71 \pm 0.077$ and $r = 0.976$. Separating this regression into component parts, the five ensemble TPLs of genera range from near Poisson (the *cp*-4 predator

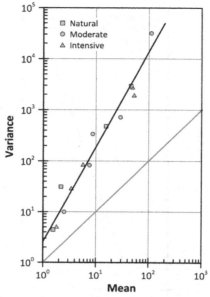

FIG. 7.25 Mixed-species ensemble TPL ($NQ = 14$, $NB = 25$) of nematodes in ponds in a range of agricultural intensifications in five agricultural regions of Belgium. *(Data from Table 3 in Ber et al. (2007).)*

Mononchus, $b = 1.03 \pm 0.199$) to very aggregated (*Monhystera*, a *cp*-2 bacterivore; $b = 1.63 \pm 0.142$), ignoring the poor fit of *Eumonhystera* whose estimate of $b = 2.59$ is not reliable. With the exception of *Eumonhystera*, the individual ensemble analyses have high correlations despite the low $NQ = 6$–10 and $NB = 5$. The mixed-species TPLs for ponds in areas ranked as natural, extensive, and intensive do not differ, confirming Bert et al's conclusion that landuse had little effect on the pond nematode communities.

RESTORED WETLAND IN GEORGIA

Ettema et al. (1998) sampled bacterivorous nematode populations in the sediment of a restored riparian wetland in the Coastal Plain of Georgia, United States. Prior to 1985, the area was forested, it was then cleared and used as wet pasture until 1991 when it was revegetated with riparian tree species. At the time of sampling, the trees were 2.5–3.5 years old and the understory vegetation comprised a mixture of native wetland grasses and rushes. Soil samples were taken at 24 georeferenced locations four times from November 1993 to the following August with an extra 12 points sampled in August. The distance between samples ranged from 4.5 to 106.5 m. Two 57-cc soil cores were taken at each location, composited, and nematodes extracted.

Geostatistical analysis showed that a substantial portion of the variation in nematode populations and soil resources was spatially dependent. The distance over which populations were spatially dependent varied from 15 m for *Chronogaster*, to 67 m for *Acrobeloides*. As a result of temporal variation, populations rose and fell locally. Two numerically dominant taxa, *Acrobeloides* and *Prismatolaimus*, had wider spatial ranges than the less abundant bacterivores. Many bacterivorous taxa were spatially separated with populations aggregating in different parts of the wetland: where *Acrobeloides* were abundant, *Chronogaster* were less abundant and vice versa. Unlike earlier studies, this study examined distribution and abundance in the four seasons. Individual spatial TPLs ($NQ = 23$–36, $NB = 4$) of the abundance of a superfamily and seven genera are very variable with slopes of $1.36 \leq b \leq 3.07$. An ensemble TPL comprising all the data ($NB = 32$) has an intermediate slope of $b = 1.76 \pm 0.129$ (Appendix 7.Z1). The genera *Chronogaster* and *Acrobeloides* have minimal spatial overlap and minimal overlap in density, consistent with Park et al.'s (2013) proposition that space may be partitioned by species to avoid competition.

Marine nematodes

Free-living nematodes are the most abundant component of the meiofauna in most coastal and oceanic sediments. They are also frequently associated with intertidal and subtidal algae.

The intertidal

COMPARING SAMPLERS

In one of the first studies to correlate nematode density at the genus level to food density, Moens et al. (1999) examined the relationship between nematode abundance and microphytobenthos abundance at a site on the Molenplaat, an intertidal mudflat in the Westerschelde Estuary in SW Netherlands. The sediment at this site is a medium sand with relatively low silt fraction.

Sampling the Molenplaat, Moens et al. took a set of 15 samples within a 33-m^2 area using a standard 10-cm^2 cylindrical corer and another set of 15 with a 1.25-cm^2 "microcorer" adapted from a syringe. Meiofauna were extracted from the top 2 cm of the cores and the nematodes were identified to genus and counted. Nematode densities in both sample sets were expressed as #/10 cm^2 of sediment surface. At the time of sampling, nematodes accounted for 98% of the total meiofauna in the samples. Genera *Tripyloides*, *Viscosia*, and *Eleutherolaimus* comprised nearly two-thirds of total nematode numbers: ciliate feeders and facultative predators were the dominant feeding types.

Total nematodes, individual feeding types, and the 10 most abundant genera were randomly distributed at both standard and microcore scales according to Green's (1966) index (Eq. 3.15). Moens et al. give the mean number per 10 cm^2 and SDs for total nematodes, 28 genera, and seven feeding guilds. A mixed-species

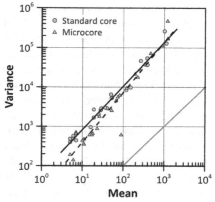

FIG. 7.26 Mixed-species ensemble TPLs of nematode genera (#/10 cm²) in samples collected from sediments in the Westerschelde estuary, Netherlands, using meiocore ($NQ = 12$, $NB = 23$) and microcore ($NQ = 14$, $NB = 24$) samplers. Different lines suggest density-dependent efficiency or scale differences in the two sample sets. *(Data from Table 1 in Moen et al. (1999).)*

ensemble analysis (Fig. 7.26, Appendix 7.A2) of the two samplers shows them to differ significantly in intercept ($P < 0.005$) and marginally in slope ($P < 0.055$). Both are significantly steeper than Poisson ($P < 0.005$). TPL analysis of feeding type showed a significant difference in slope between the two samplers ($P < 0.05$). The divergence between the TPLs based on the two samplers suggests that the sampling efficiency of the microcore sampler may be density dependent (Chapter 15) with respect to the standard sampler. Alternatively, because they are very different in area, there may be a scale effect, with the samplers differing with respect to the natural spacing of the nematodes.

SEDIMENT TEXTURE

The spatial heterogeneity of the nematode community of another intertidal flat in the Westerschelde estuary was investigated by Steyaert et al. (2003). Three sites were selected on the basis of their sediment characteristics: Site 1 with intermediate sediment characteristics; Site 2, a fine sediment site; and Site 3 with sandy sediment.

Five 10-cm² cores separated by 10 m were taken at each site. The samples were divided into 12 horizontal slices varying in thickness from 0.5 cm at the surface to 5 cm at the bottom of the profile. Nematodes in each slice were identified to species and counted. Steyaert et al. noted a twofold difference in total species in the vertical profiles as well as differences between replicates at each site. At the sandy site, nematode abundance was generally lower and fluctuated with depth. Species richness differed little between the three sites when integrated over the entire profile. However, they noted distinctly different distribution patterns of species at the three sites with most nematode species in the finest sediment confined to the surface layers with a particularly steep decline in abundance down the profile. Appendix 7.B2 has results of spatial TPLs in which the points are abundance by depth. The TPLs ($NQ = 5$, $NB = 12$) of density (#/ 10 cm^2) of total nematodes and two species common to all sites are not significantly different ($P > 0.15$) on account of the small NQ. However, the slopes for all nematodes and both *Theristus blandicor* and *Viscosi viscosia* are lowest at the intermediate sediment Site 1 and highest at the sandy sediment Site 3, indicating different aggregations with substrate texture.

PREDATOR AND PREY

Gallucci et al. (2005) assessed the distribution of the predatory nematode *Enoploides longispiculosus* and of its prey nematodes in the Paulina intertidal flat in the Westerchelde estuary. Three 20-cc replicate samples were taken at 18 stations during low tide on the Paulina. The predator and its potential prey species were extracted and counted. *E. longispiculosus* was present at 16 stations at average densities of 1.56–189/10 cm^2. Average densities of the potential prey species ranged from 19.5–2435/10 cm^2 and were significantly ($P < 0.001$) negatively correlated with *E. longispiculosus* density. Experiments suggest the sediment characteristics strongly affect predation efficiency, affecting predator-prey dynamics. Ensemble TPLs of the predator and its potential prey species (Appendix 7.C2) are not different in either intercept or slope ($P > 0.30$). Their common TPL has a wider range of means (and higher correlation) but is dominated by the much more abundant prey species. Their similar spatial distribution suggests the predatory nematodes track their prey efficiently as required by the Ideal Free Distribution hypothesis (Fretwell and Lucas (1969).

Littoral and sublittoral

POLLUTION GRADIENT

Neilson et al. (1993) surveyed the nematode populations in sediments of the Tay Estuary in eastern Scotland, near to a shortfall sewage outlet to provide a range of means for TPL analysis. Replicate sediment samples of 15-cm^2 cores were taken at 10 stations in 2 transects from the sewage outlet. Extracted nematodes were identified to species where possible; juveniles were identified only to genus. Means and variances were computed from $NQ = 4$ samples at $NB = 10$ distances from the outlet and ensemble TPLs calculated to test the efficacy of Healy and Taylor's (1962) exact TPL transformation and compare it with log, square root, and two additional fixed powers ($z = 1/3$ and 1/4). Their ensemble TPLs for the 15 most abundant species, eight genera, and six families are in Appendix 7.D2. Species slopes ranged $0.98 \leq b \leq 2.25$ with average $b = 1.63 \pm 0.095$ and the genera ranged $1.10 \leq b \leq 2.06$, average $b = 1.70 \pm 0.085$. The increase in b with increased data aggregation is expected and the narrower range at the genus level than the species is in line with Park et al.'s. (2013) experience.

Nielson found the exact transformations gave the best results for genus and family and ranked second for species. For two species, the square root transformation performed better than the exact transformation and for 12 of 15 species the data transformed by the exact method were approximately normally distributed. Neilson et al. concluded that the "transformation derived from Taylor's power law was frequently the best." Overall, Neilson et al. concluded that the cube root transformation was the best single choice: this is the transformation corresponding to $b = 1.33$, substantially lower than the average values of slope for species and genus.

LATITUDINAL GRADIENT

Between January and March 2011, Hua et al. (2016) sampled meiofauna at nine sandy beaches across a latitudinal gradient (18° to 40°N) along China's coast. One site they classified as tropical, five as subtropical, and three as temperate. Their purpose was to examine the effect of latitude on meiofauna abundance and diversity. Secondarily, they examined anthropogenic influences. At all but one beach, two transects were established >50 m apart perpendicular to the waterline. Four transects were established at the exception, adjacent to the city of Sanya: two strongly influenced by anthropogenic activity and two by weaker anthropogenic activity. Three sample stations, at high, middle, and low tidal zones, were taken on each of the 20 transects. Three 4.5-cm^2 cores were taken at each station and were cut into five 4-cm slices. Samples were sieved and all meiofauna were hand sorted into 18 major taxa.

Nematodes accounted for 39%–93% of the meiofauna recovered and were least abundant in the subtropical sites. The abundance and dominance of nematodes increased with latitude suggesting a preference for cooler waters. Overall meiofauna abundance was highest in the low tidal zone at the tropical beach and in middle tidal zone at the temperate beaches. In general, abundance and richness of meiofauna, especially in the upper beach zones, were lowest at the tourist beaches. The vertical profiles were not reported. Hua et al. give the mean and SDs of nematodes permitting an ensemble TPL of nematodes on a latitudinal gradient (Appendix 7.E2). With abundance varying over 22° of latitude and three orders of magnitude, temporal TPL slope ($b = 1.22 \pm 0.177$) of total nematodes is not significantly different from Poisson ($P > 0.25$) indicating totally random temporal variation in nematode distribution across latitude.

Benthic

DEPTH GRADIENT

Moodley et al. (2000) recorded the abundance of nematodes and other metazoan meiobenthic taxa at four contrasting sites in the Adriatic Sea. Samples were taken at a shallow site (Station 11 at 15-m depth) located close to the Po River delta, at two sites on the continental shelf (Stations 2 and 4 at 19 and 89 m, respectively), and a fourth (Station 5) in the central basin at the base of the continental slope (1039 m). Sediment samples were taken at each station with boxcore samplers and two 10-cm^2 corer samples taken. Cores were cut into ten 1-cm slices, the nematodes extracted and identified to genus. Station 11, which supported the highest densities, also had the lowest genus diversity. Except at the deepest site, nematodes were more abundant well below the oxygen penetration depth than above it.

Moodley et al. list the 10 most abundant nematode genera giving the mean density (#/10 cm^2) and SD in the top 5 cm and 5–10 cm depth for the four stations. Despite major differences in the densities between sites and depths, all eight points fit a single highly aggregated ensemble TPL over two orders of magnitude of mean ($b = 1.80 \pm 0.105$: Appendix 7.F2). Also included in Appendix 7.F2 are the TPLs for the vertical profiles at the individual stations. With only two quadrats per layer, the results are questionable: three of four sites show moderate-to-high aggregation over 2–3 orders of magnitude, but one continental shelf site is Poisson ($b = 0.91 \pm 0.321$) with poor correlation contrasting sharply with the other sites ($b = 2.05 \pm 0.160$). Despite the small number of samples, nematodes seem to be highly aggregated on the vertical profile.

THE WORLD'S OCEANS

Mokievskii et al. (2007) developed and analyzed a database of 665 records of the abundance of deep sea free-living nematodes and other oceanic meiofauna at depths to almost 10 km. They divided the world's oceans into 14 zones and separated the records into 4 depth intervals; 20–400, 401–600, 601–3000, and >3000 m, corresponding to upper and lower continental shelf, the slope, and abyssal. The number of records declines exponentially with depth reducing the confidence in the estimates with increasing depth. Mokievski et al. estimated their database included >90% of available information for depths >1000 m and about 80% of published studies from 100 to 1000 m, the lower part of the shelf and upper part of the slope. Abundance records for different samplers were adjusted for differences in efficiency and standardized to #/10 cm^2.

The proportion of nematodes in meiofauna communities increased with the depth even as total numbers declined. Nematode biomass (μg C/10 cm^2) was estimated in 167 samples. Mokievskii et al. provide means and SDs for estimates of nematode abundance and biomass in the four depth ranges. Ensemble TPLs for abundance ($b = 1.86 \pm 0.370$, $r = 0.963$) and biomass ($b = 1.88 \pm 0.265$, $r = 0.981$) do not differ in either slope ($P > 0.55$) or intercept ($P > 0.65$) (Fig. 7.27; Appendix 7.G2) and are best fit by a common regression ($b = 1.76 \pm 0.113$, $r = 0.988$) with a much greater range of means. It's curious that TPLs for density and biomass coincide so well.

NEMATODES ON KELP

Free-living nematodes on the kelp, *Macrocystis integrifolia*, were studied by Trotter and Webster (1983) in the relatively sheltered waters of Dodger Channel off the west coast of Vancouver Island, British Columbia, Canada. Three blades

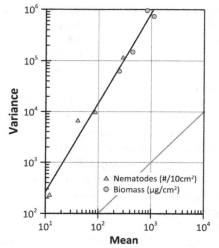

FIG. 7.27 Ensemble TPL $(NB=8)$ of nematode density $(\#/10\,cm^2; NQ=53-252)$ and biomass $(\mu g/10\,cm^2; NQ=13-62)$ in the ocean's deep-water zones. *(Data from Figs. 1 and 2 in Mokievskii et al. (2007).)*

from twelve *M. integrifolia* plants were collected at 2-m intervals along a transect ranging in depth from 2 to 14 m through the kelp bed and perpendicular to the shore: one from 1 m above the holdfast, one from 1 m from the distal end of the plant, and one from the midpoint. The transect was sampled monthly from July 1978 to July 1979, except December 1978, and then bimonthly until November 1979 for a total of $NB=14$ sampling occasions. The blades were bagged before they were severed from the main stem. The contents of each bag were washed off the blades and sieved to extract nematodes.

Nine nematode species in six families were recovered in the samples. Three species, *Monhystera disjuncta*, *M. refringens*, and *Prochromadorella neapolitana*, accounted for >90% of the total nematode population on kelp. Nematodes occurred in their largest numbers on the lower kelp blades with significantly fewer on the middle and upper blades, and distributed evenly along the length of the upper surface of the blades. The total number of nematodes increased with mean depth of water.

Trotter & Webster give the density and SD of the three dominant nematode species per 500 cm^2 of kelp blade for each sample date. Spatial TPL analyses $(NQ=36, NB=14)$ for three dominant species are given in Appendix 7.H2. All three species are moderately to highly aggregated on kelp and are fit by a single regression (Fig. 7.28).

AN EXTREME CASE

We end the section on aquatic nematodes with an extreme example. Soetaert et al. (1991) estimated nematode diversity on a transect over a range of depths

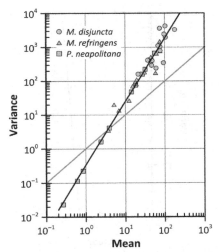

FIG. 7.28 Mixed-species spatial TPL ($NQ = 36$, $NB = 14$) of three marine nematodes (#/500 cm^2 blade) on kelp *Macrocystis integrifolia* samples collected from Berkley Sound, British Columbia. *(Data from Fig. 2 in Trotter and Webster (1983).)*

off the coast of Corsica in the Mediterranean Sea. Samples were taken at six stations at depths ranging from 160 to 1220 m by box core from which two 10-cm^2 cores were taken and cut into two 0.5-cm-thick slices in the top 1-cm and four 1-cm-thick slices below. Nematodes were extracted from the 72 total samples and identified to species. Total recoveries ranged from 535 nematodes/ 10 cm^2 in 201 species at the 1220 m station to 1037/10 cm^2 nematodes in 226 species from the 160-m site, although overall abundance did not correlate with depth. This example with $NQ = 12$ and $NB = 6$ has the extraordinary TPL slope of $b = 13.18 \pm 2.265$, $r = 0.94$ (Fig. 7.29, Appendix 7.I2).

A model of Cohen (2014) (Chapter 17) predicts a singularity in TPL in which b becomes sharply positive, goes to infinity, and rebounds from minus infinity through steeply negative values to its customary positive value. If the model has validity and there is indeed a TPL singularity, it is possible this example constitutes evidence for it.

However, we know from the simulation studies of Clark and Perry (1994) that TPLs with NQ and $NB < 15$ are unlikely to be unbiased and the narrow range of means lends support to skepticism. The vertical profile and its different thicknesses of slices would act as a variance multiplier and increase the slope, but it's remarkable that it would simultaneously introduce such a large bias without spreading the points further from the fitted line with its high correlation coefficient of $r = 0.94$. Furthermore, neither the mean nor the variance correlate well with depth, which would be another obvious variance multiplier. Like the TPL of total number of nematodes, the equivalent TPL of number of species, whose numbers are correspondingly lower, also has a very steep slope but

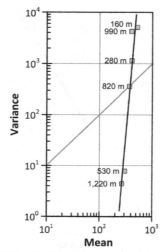

FIG. 7.29 Ensemble TPL ($NQ = 12$, $NB = 5$) of nematodes in Mediterranean sediment. The extremely steep regression shows similar overall densities but extreme variation with depth. *(Data from Table 1 in Soetaert et al. (1991).)*

poorer correlation ($b = 12.51 \pm 5.612$, $r = 0.441$): its slope is not significantly different from the abundance TPL slope ($P > 0.90$).

The preceding case studies cover a wide range of species, lifestyles, and habitats, in which it has been possible to fit convincing TPLs. With the very large number of nematode studies with replicated samples over wide ranges of means in scores of distinctly different habitats, it is unfortunate that so many base their estimates of variance on so few samples, often only two or three. Even when more samples are taken, they are frequently composited deliberately to minimize variance in the mistaken view that variability needs to be eliminated in order to obtain reliable estimates of mean. The fact that variance scales with mean makes estimates of the mean from composited samples less reliable not more. Park et al.'s (2013) study was consciously designed to provide data meeting the requirements for fitting reliable TPLs for a range of genera in order to expose the real differences in their abundance.

Other worms

In addition to the nematodes there are, depending on taxonomist, three other worm phyla: Annelida, Plathyhelminthes, and Acanthocephala. The annelids are segmented worms that include the familiar earthworm (oligochaets) and ragworms (polychaetes); the platyhelminths are unsegmented worms that include the predatory planarians and parasitic flat worms, including tapeworms (cestodes) and flukes (trematodes); and the acanthocephalans or thorny-headed worms comprise a small and exclusively parasitic taxon. Many of the parasitic worms have complex lifecycles that include two or more hosts, a definitive host that is usually a vertebrate, and one or more secondary hosts that are usually invertebrates, especially arthropods but also other worms.

Platyhelminths

Shaw and Dobson's (1995) survey of vertebrate parasites included three other taxa: the phylum Acanthocephala and the cestode and digenean flatworms in the phylum Platyhelminthes. A multispecies TPL of these three taxa results in almost identical estimates to the nematodes (Fig. 7.17). There is little difference in the mixed-species ensemble TPLs for worms segregated by host (Fig. 7.30A) or by worm taxon (Fig. 7.30B, Appendix 7.J2). Evidently at this level of organization, endoparasitism has a fairly characteristic level of aggregation ($b \approx 1.7$) regardless of host or parasite. Five points lie close to the Poisson line and only one point that could classify as an outlier, the tapeworm (cestode) *Taenia macrocystis* in coyotes. As the span of variance at any given mean abundance is quite narrow (on log scales), there must be constraints on variability. Shaw & Dobson suggest that parasites that enter hosts passively (e.g., nematodes on forage waiting to be consumed by sheep) tend to exhibit higher levels of mean

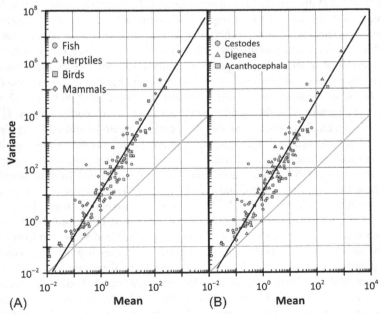

FIG. 7.30 Mixed-species ensemble TPLs ($NB = 125$) of Plathyhelminthes and Acanthocephala parasite abundance identified by (A) host and (B) parasite taxa. *(Data from Appendix 1 in Shaw and Dobson (1995).)*

burden. Such parasites typically have very high fecundity leading to large worm burdens of infected hosts, but low probability of infection, leading to very high levels of variability. Cestodes, for example that are taken in actively by their primary host while consuming their secondary host prey, tend to have lower burdens in part because the parasite's larger size imposes limits on the number a host can accommodate. The factors balancing the scaling relationship over such a wide range of densities, hosts, and parasites must be fundamental to the parasitic life-history strategy.

PRIMARY HOSTS—GREY MULLET

Sarabeev et al. (2017) amassed a collection of data for endohelminths, monogeneans, adult and larval digeneans parasitic on grey mullets, *Liza haematocheilus*, and *Mugil cephalus*. The parasite loads of native and introduced populations of *L. haematocheilus* in the Sea of Japan and the Sea of Azov, respectively, were compared with the parasites of native host populations and of parasites of *M. cephalus* in the Mediterranean, Azov, and Japan Seas. The total mean abundance and aggregation of the helminth communities varied substantially between the different populations of grey mullet. Native *L. haematocheilus* supported helminth populations 15 times greater than introduced individuals and the helminth communities of *L. haematocheilus* were less aggregated in the invasive host population compared with native populations of the same species for both community and mixed-species abundance data.

Mixed-species and community TPLs were used to analyze the distribution and abundance of the helminth parasite communities for five parasite taxa in two the host species in five seas. A total of 695 specimens of *L. haematocheilus* and 560 of *M. cephalus* were used with host sample sizes of $NQ = 13-62$ and $NB = 2-14$ locations in the various seas. Sarabeev et al. computed 38 TPLs of hosts from the Sea of Azov and the Mediterranean with 14 and 10 TPL points, respectively (Appendix 7.K2). Supplementary Tables provide more detailed analyses with estimates of b, SE(b), r^2, and NB for the helminth mixed-species and community TPLs broken down into six groups defined by host species, sampling locality, sea, season, and year of capture. The number of TPL points ranges from $10 \leq NB \leq 19$ and $12 \leq NB \leq 54$ for the mixed-species and community TPLs of total parasites, respectively: the number of TPL points for the subsets are smaller. The mixed-species slopes are all similar to Shaw and Dobson's (1995) value of $b \approx 1.7$: the composite average of 42 cases is $b = 1.74 \pm 0.08$ for total parasites and the values for the four subsets are also close to $b \approx 1.7$. The estimates of slope for the community TPLs are significantly higher with a composite average for total parasites of $b = 2.02 \pm 0.15$: the subsets are similarly steeper than the mixed-species subsets. These detailed analyses, are further support for Shaw & Dobson's more general finding that a mixed-species TPL slope of $b \approx 1.7$ is characteristic of parasite aggregations.

Cestodes

SECONDARY HOSTS—SHRIMPS

The definitive hosts of the tapeworm *Hymenolepis tenerrima* are ducks and other water fowl. It has an intermediate host in the freshwater shrimp *Herpetocypris reptans*. Evans (1983) describes the occurrence of *H. tenerrima* in *H. reptans* for a 12-month period obtained by monthly net and substrate sampling taken at four preselected sites around the margin of a ~2 ha shallow pond in West Sussex, England. *H. reptans* were extracted from the samples and examined by microscope for the larval stages (cysticercoids) of *H. tenerrima*. Parasite prevalence was minimal during winter as hosts died and maximal in summer. Adult shrimp harbored a maximum of three cysticercoids/host, their presence greatly reducing host survival and fecundity. Evans provides mean and variance data for parasite and host, respectively. Spatial TPLs reveal that host population was highly aggregated ($b = 2.12 \pm 0.30$), while the parasite's abundance distribution was indistinguishable from Poisson ($b = 0.98 \pm 0.075$). Intensity (number per infected host), however, is extremely aggregated ($b = 6.46 \pm 1.554$) owing to the narrow range of means and wide range of variances resulting from the exclusion of many zero infections (Fig. 7.31, Appendix 7.L2).

PRIMARY HOSTS—DOMESTIC FOWL

Hodasi (1969) compared the distribution and abundance of trematode and cestode parasites in 108 native and 108 introduced fowl in Ghana. The native fowl were free running and unconfined birds bought in native markets, while the

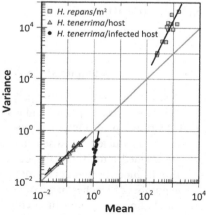

FIG. 7.31 Spatial TPLs ($NQ = 82$–294, $NB = 12$) of abundance of the tapeworm *Hymenolepis tenerrima* and its freshwater shrimp secondary host, *Herpetocypris reptans*. Also shown is the TPL for parasite intensity with its extreme TPL slope. *(Data from Fig. 1 and Table 1 in Evans (1983).)*

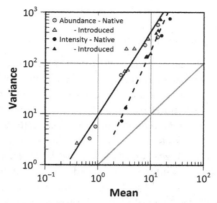

FIG. 7.32 Mixed-species TPLs ($NQ = 108$, $NB = 12$) of platyhelminth abundance (*solid line*) and intensity (*hatched line*) in native and introduced fowl in Ghana. Despite differences in infestation rates in introduced and native birds, both fit single TPL lines. *(Data from Tables 1 and 2 in Hodasi (1969).)*

introduced birds were free range but confined on an experiment station; thus, their ecologies differed with the native birds possibly more exposed to parasites than the introduced birds. The infestation rates for introduced and native birds differed between the parasite taxa, but the overall infestation was greater in the native birds: 35,807 worms were collected from the native birds and 23,527 from the introduced birds. Hodasi's Tables 1 and 2 provide prevalence, intensity, and abundance data for seven species of platyhelminths found in both classes of bird. Mixed species TPLs of abundance did not differ between native and introduced birds ($P > 0.95$) (Fig. 7.32; Appendix 7.M2) despite the large differences in parasite infestation rates. Intensity of parasitism in all birds is significantly steeper than abundance ($P < 0.0001$).

Annelids

LEECHES IN A CUMBRIAN STREAM

Elliott (2004) sampled two stream reaches ~300 m apart, for the leech *Erpobdella octoculata*, a deep section with macrophytes and a shallow stony section and examined the spatial distribution of adults and cocoons in reaches with major habitat differences. He conducted censuses in June and August for 25 years. Both reaches were divided into 10 equal sections and three samples were taken from each section in June for young and mature leeches, and again in August for cocoons. The number of leeches in cocoons was also determined. Abundance of all stages, including the young in cocoons, was expressed as number/900 cm^2. Elliott's interest in the leeches' population dynamics included survival, which he described by Ricker curves and the changes in spatial distribution through several stages. He fitted spatial TPLs to the leech abundance data (Appendix 7.N2) by ODR: the difference between ODR and GMR estimates is negligible given the extremely high correlations $(r \geq 0.99)$. With $NQ = 30$ and $NB = 23$ or 24, combined with samples that are essentially censuses, the fits for all stages are exceptional. There were no differences in spatial distribution between reaches, despite their differences. Nor were there differences between young and mature leeches in June, which were separated numerically but not spatially. All four TPLs of age and reach coincided almost perfectly. However, cocoons were more strongly aggregated than either young or mature leeches, indicating increased aggregation by mature leeches for mating and egg laying.

Elliott's Ricker curves showed the survival of the two populations differed substantially, indicating that the population dynamics differed between the two sites—a high incidence of cannibalism was observed at the stony site. Although both reaches were subject to the same weather conditions, oscillations in abundance were strongly synchronized only in extreme years. Although leeches were generally more abundant at the deeper site, the ranges of density of leeches and cocoons over 25 years were similar for both sites. Elliott's results show clearly that although subpopulations in a metapopulation may have different densities and density fluctuations, their spatial dynamics can be remarkably consistent across space.

Oligochaetes

EARTHWORMS IN SCOTLAND

A stratified random sample of 200 permanent pasture and arable fields in Scotland were surveyed for the planarian predator *Artioposthia triangulata*, which feeds on earthworms. Five samples 2500 cm^2 in area were taken in each field by formaldehyde extraction. Two applications were made, the first to extract earthworms and the second for planarians. Thirteen species of earthworm were recorded, the most common being *Aporrectodea longa*, *A. caliginosa*, and *Lumbricus terrestris*, which with eight others were recorded in all areas surveyed. Data from this survey were used by Boag et al. (1994) to assess the spatial distribution of earthworms by TPL. Blackshaw (1990) had already established that the distribution of *A. triangulate* was indistinguishable from Poisson (Appendix 7.O2), and Boag et al. combined all species in a sample to perform an earthworm community ensemble TPL analysis. To maximize the data for analysis, all means and variances were relocated by addition of 0.001 before taking logs. Following Perry (1981), they computed both ODR regressions, b_{yx} and $1/b_{xy}$ to bracket the "true" value of the TPL relationship (see Fig. 4.1), but did not compute the GMR. Appendix 7.O2 gives the GMR estimates computed from Boag et al's ODR results.

Although most species were more abundant in, and more total earthworms were recovered from, the permanent pastures than the tilled arable fields

(Boag et al., 1997), the differences in TPL between the two field types are small and not significant, as are the differences between adults and juveniles. Boag et al's estimates of slope are somewhat less than those obtained by LRT et al. (1978) for *A. calignosa* and *Allolobophora chlorotica* ($b \approx 1.6$), a difference possibly due to combining 13 species. It should also be borne in mind that including zero means and variances by relocating them by 0.001 may place undue emphasis on the virtual point $\log(M) = \log(V) = -3.0$ if there were many zero means and/or indeterminate variances, although in this instance their effect would have been to raise the slope not reduce it. The question of zeros is examined more fully in Chapter 15.

EARTHWORMS IN COLOMBIA

Jiménez et al. (2001) compared the spatial distributions of earthworms in a Colombian savannah and an established pasture by spatial TPL. At both sites, $40 \times 40 \times 15\,\text{cm}$ (24 L) blocks of soil were taken close to the nodes of an 8×8 array 70 m on a side and the earthworms identified, counted, and returned to the soil. The samples were taken three times between November 1993 and June 1995. Spatial TPLs for six species were computed with all species aggregated in both savannah and pasture habitats. TPLs for species in savannah and pasture were similar and ranged from $b = 1.39$ for *Aymara* sp. to $b = 1.89$ for *Andiodrilus* sp. Jiménez concluded that "land use had no significant effect on the spatial distribution of earthworms." Mixed-species spatial TPLs of worms at the two sites (Fig. 7.33, Appendix 7.P2) were moderately aggregated ($b \approx 1.28$) and not different in either slope or intercept ($P > 0.70$).

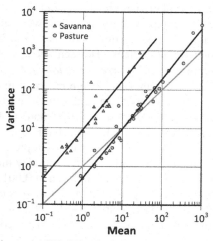

FIG. 7.33   Mixed-species spatial TPL ($NQ = 54$–81) of earthworms in savanna ($NB = 29$) and pasture ($NB = 34$) habitats are closely similar in intercept and slope. Pasture is offset by one cycle for clarity. *(Data from Tables 1 and 2 in Jiménez et al. (2001).)*

Polychaetes

IN THE SUBLITTORAL

Deudero and Vincx's (2000) survey of the sublittoral fauna of the sea sediments around the Balearic Islands included members of the polychaete and oligo-chaete classes. Ensemble TPLs of the oligochaete and polychaete assemblages are not significantly different ($P > 0.30$), although the polychaetes were clearly more numerous, the two populations overlap and both are moderately aggregated with common $b = 1.50 \pm 0.102$ (Appendix 7.Q2).

Following a hypoxic event in the Mariager Fjord in Denmark, Hansen et al. (2002) initiated a sampling program to document the recolonization by molluscs, arthropods, and polychaetes. Polychaete larvae were among the earliest to reinvade and their settlement of 30–$40 \times 10^3/\text{m}^2/\text{day}$ was monitored at roughly weekly intervals over a three-month period at three stations in the fjord. Hansen et al. gave the mean and SD of counts of polychaete settling on commercial settlement strips per station. Spatial TPL of the polychaete juveniles showed them to be highly aggregated with $b = 1.74 \pm 0.111$ (Appendix 7.R2).

BIOMASS AS PROXY FOR ABUNDANCE

Sampling the meiofauna in the Beaufort Sea in the Arctic Ocean, Bessière et al. (2007) took subsamples from box cores of the sediment at 23 stations. They give the density and biomass with SDs of polychaetes recovered from 16 stations. Ensemble TPLs of abundance ($\#/10\,\text{cm}^2$) and biomass ($\mu g\,C/10\,\text{cm}^2$) result in slopes and intercepts that are not significantly different ($P > 0.35$, Fig. 7.34, Appendix 7.S2). However, the fitted estimates of slope suggest variance of biomass may grow faster than variance of density. This might be expected, given that the slope of $\log(V)$ on $\log(\text{biomass})$ is expected to be more negative than that of $\log(M)$ on $\log(\text{biomass})$ (Lagrue et al., 2015).

Schratzberger et al. (2008) compared the distribution and abundance of benthic invertebrate assemblages in sediments in the Celtic Deep and the NW Irish Sea between Great Britain and Ireland. The sites differed in depth, tidal stress, temperature, sediment composition, and organic carbon. Five 15-L samples were collected from the muddy sea floor at each of nine stations at both locations. Twenty nine of the forty five samples were filtered for polychaetes. Species diversity and similarity of assemblage composition was higher in the Celtic Deep than in the shallower Irish Sea location and of the 88 species recorded half were common to both sites. Schratzberger et al. give the average density ($\#/\text{sample}$) and biomass (mg/sample) with SEs. As with the previous example, ensemble TPLs of density and biomass are not significantly different with the biomass slope slightly steeper than density's (Fig. 7.35, Appendix 7.T2). In situations where it is impractical to count specimens, biomass can be used as a proxy for abundance.

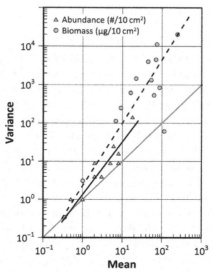

FIG. 7.34   Ensemble TPLs ($NQ = 3$) of density ($NB = 16$) and biomass ($NB = 14$) of polychaetes in Arctic Ocean sediment. *(Data from Tables 3 and 4 in Bessière et al. (2007).)*

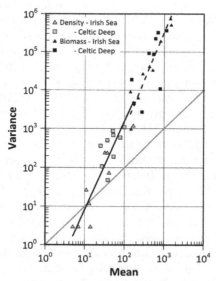

FIG. 7.35   Ensemble TPLs ($NQ = 5$, $NB = 18$) of polychaete density (#/sample) and biomass (mg/sample; hatched line) in sediment in the sea between Great Britain and Ireland. *(Data from Tables 3 and 4 in Schratzberger et al. (2008).)*

A BEHAVIORAL EXPERIMENT

The polychaete worm *Capitella capitata* burrows in the littoral and sublittoral mud or sand sediment, forming visible tubes. It is essentially sessile and can produce benthic larvae rapidly to exploit favorable habitats or planktonic larvae to find new homes. In a laboratory experiment by Bonsdorff and Pearson (1997), eight laboratory-reared populations of *C. capitata* were subjected to predation by two shrimp (*Crangon crangon*) for 3 days during which time the density of visible worm tubes declined from $\sim$150/100 cm^2 to near zero. The shrimp killed some worms and damaged more by taking bites from them causing them to retreat into the sediment. Following removal of the shrimp predators, the density of tubes recovered to about 50% in $\sim$4 days and only recovered to $\sim$100% when enclosures were removed allowing immigration to the experimental area. Bonsdorf & Pearson gave means and SEs of number of visible tubes during the course of the experiment. A spatial TPL of tube number shows the density of tubes is moderately aggregated ($b = 1.41 \pm 0.083$) (Appendix 7.U2). What is interesting is that TPL is recording the numerical expression of a behavioral response to predation and the spatial response of immigration.

WAYS OF ARRANGING DATA

In a class exercise on a 1971 field trip to the Marine Biological Station at Millport, Cumbrae, Scotland, the lugworm *Arenicola marina* was surveyed on a sandy shore at low tide. The worm casts in a 1-m^2 quadrat were counted. The quadrat was then increased in size to 4 m^2, and the casts counted. Quadrats of 9, 16, 25, and 36 m^2 followed. Multiple teams of students conducted 90 sets of counts following the tide to low water. There are four ways to compute an ensemble TPL from these data with $NQ = 90$ and $NB = 6$. First, take the sample mean and variance for each successive quadrat size, ignoring the fact that each contains the counts from the previous sizes; second, repeat but with the counts expressed as #/m^2 by dividing each quadrat count by the quadrat area; third, compute means and variances from each sample less the number in the smaller sizes for effective sample sizes of 1, 3, 5, 7, 9, and 11 m^2; and fourth, repeat with #/m^2.

Distinguishing between counts per quadrat (overlapping samples) and counts per sample (nonoverlapping samples), we find that a plot suggests that all four approaches coincide (Fig. 7.36), although TPL for #/m^2 computed from overlapping quadrats is steeper than the other plots (Appendix 7.V2). Calibrating by quadrat area typically affects both intercept and slope when the quadrats are different sizes: if all are the same size, it affects only the intercept. This effect, though not significant ($P > 0.20$), is apparent in the conversion of overlapping samples to density, and to a lesser extent the nonoverlapping samples

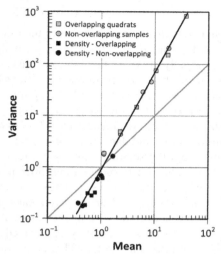

FIG. 7.36   Ensemble TPL ($NQ = 90$, $NB = 24$) of *Arenicola marina* at Millport, Isle of Cumbrae in Scotland. Plotted data were processed four different ways, but fit a common regression. *(Data from the author's field notebook and course report.)*

converted to density. With $NB = 6$ in all cases, the Clark and Perry (1994) cautions apply, but what is intriguing about these results is the lack of a discernible difference between the overlapping and nonoverlapping quadrat TPLs, which have identical correlation coefficients and $SE(b)$s. Apparently, the counts common to both sets dominate. A possible explanation is that there is a correlation between the size of sample and the number of additional worm casts at each quadrat size. This is entirely possible as the number of exposed casts/m^2 increased as the tide receded and the quadrats increased in size. That such a correlation should exactly compensate for the changing area is the intriguing result. As Ma (2015) has emphasized with his TPL Extensions, the type of basic data and the way TPL is computed are critically important in interpreting TPL. Sampling with increasing quadrat size is called box counting, which can be used to estimate fractal dimensions. It does not make any difference apparently whether the boxes overlap or not in the estimation of TPL. The box-counting method can be used to estimate a fractal dimension (Chapters 16). Whether it does so in this case is not certain, but it seems like a suitable point to end the survey of nematodes and other worms.

Appendix: TPL estimates for nematodes and other worms

	NB	Range of means	r	A	SE[A]	b	SE[b]

Soil nematodes

A Ensemble TPLs of population density (#/200 g soil) of *Longidorus elongatus* and 5 other trichodorid nematodes sampled (*NQ* = 10) with 9 sampling tools; each point in Fig. 7.1A is a different tool. Data from Table 1 in Boag and Brown (1985)

Ensemble TPLs of densities of 4 species in 3 dimensions projected onto horizontal and vertical planes. The densities projected onto the horizontal plane form a single TPL (Fig. 7.2A) but the vertical projections do not, although the slopes are similar. All slopes are *b* > 2.0 *(NQ* = 4−16). Data from Table 1 in Boag et al. (1987)

	NB	Range of means	r	A	SE[A]	b	SE[b]
Boag and Brown (1985)							
Longidorus elongatus	9	157 - 455	0.882	−4.877	1.620	3.701	0.660
Trichodorus spp.	9	67 - 151	0.871	−1.148	0.794	2.043	0.379
Combined	18	67 - 455	0.947	−3.255	0.557	3.044	0.244
Boag et al. (1987)							
Longidorus elongates – vertical	5	13.4 - 61.4	0.989	−0.167	0.294	2.248	0.192
horizontal	16	1.60 - 153	0.932	−0.493	0.273	2.055	0.199
L. leptocephalus – vertical	5	1.77 - 17.3	0.991	−0.401	0.153	2.180	0.164
horizontal	4	4.67 - 32.0	0.999	0.101	0.085	1.960	0.075
Paratrichodorus pachydermus - vertical	5	49.1 - 120	0.817	−0.486	1.462	2.298	0.765
horizontal	7	4.67 - 393	1.000	0.023	0.037	2.004	0.019
Trichodorus primitivus – vertical	5	31.7 - 152	0.989	−1.179	0.422	2.472	0.212
horizontal	6	41.7 - 353	0.999	0.070	0.090	1.978	0.042
Combined – vertical	20	1.77 - 152	0.978	−0.350	0.178	2.191	0.108
horizontal	33	1.60 - 393	0.963	−0.464	0.173	2.183	0.105

B Spatial TPLs of *Criconemella sphaerocephalus* in an unmanaged pasture. Data from Tables 2 and 3 in Wheeler et al. (1987)

	NB	Range of means	r	A	SE[A]	b	SE[b]
Sampling occasion A	5	-	0.97	1.67		2.55	0.277
B	5	-	0.97	8.28		1.65	0.166
C	5	-	0.63	4.34		1.89	0.840
D	4	-	0.99	18.69		1.44	0.099
Combined	19	1.50 - 248	0.940	0.582	0.217	1.955	0.162

C Ensemble multispecies TPLs of the horizontal distribution of nematode density (#/100 cc) in *NQ* = 15 samples at 3 depths collected from a sand flat on St. Martin's Island in the Scilly Isles, UK. Data from Table 1 in Somerfield et al. (2007)

	NB	Range of means	r	A	SE[A]	b	SE[b]
0–5 cm	74	0.07 - 104	0.969	0.368	0.029	1.310	0.038
5–10 cm	72	0.07 - 98.7	0.975	0.331	0.027	1.372	0.037
10–15 cm	69	0.07 - 53.9	0.970	0.421	0.030	1.457	0.043
Combined	215	0.07 - 104	0.971	0.366	0.017	1.371	0.023

Terrestrial nematodes

D Ensemble TPLs (*NQ* = 10) of soil nematodes sampled from 18 lawns on 2 occasions in 2008. Data from Park et al. (2013)

	NB	Range of means	r	A	SE[A]	b	SE[b]
Acrobeles (BF-2)	22	0.10 - 4.10	0.944	0.532	0.055	1.457	0.107
Acrobeloides (BF-2)	36	2.70 - 45.6	0.904	0.199	0.137	1.673	0.122
Alaimus (BF-4)	17	0.10 - 2.10	0.924	0.389	0.064	1.387	0.137

Continued

	NB	Range of means	r	A	SE[A]	b	SE[b]
Aphelenchoides (FF-2)	36	4.10 - 77.0	0.907	−0.122	0.199	1.964	0.142
Aphelenchus (FF-2)	36	1.60 - 28.3	0.884	0.067	0.132	1.882	0.151
Cephalobus (BF-2)	36	1.20 - 51.0	0.926	0.127	0.119	1.845	0.119
Criconemoides (PF-3)	29	0.10 - 4.00	0.972	0.374	0.039	1.467	0.066
Diplogaster (BF-1)	34	0.10 - 6.90	0.935	0.371	0.042	1.570	0.098
Discolaimus (OM-5)	19	0.10 - 3.70	0.957	0.508	0.075	1.670	0.118
Dorylaimus (OM-4)	36	0.80 - 10.1	0.767	0.210	0.104	1.582	0.174
Eucephalobus (BF-2)	30	0.10 - 17.2	0.975	0.679	0.047	1.775	0.075
Eudorylaimus (OM-4)	36	0.40 - 12.4	0.917	0.292	0.048	1.609	0.110
Filenchus (PF-2)	36	0.90 - 35.7	0.778	0.111	0.194	1.947	0.210
Helicotylenchus (PF-3)	35	0.10 - 43.5	0.962	0.453	0.084	1.759	0.084
Heterodera (PF-3)	23	0.10 - 10.8	0.965	0.635	0.058	1.800	0.103
Hoplolaimus (PF-3)	32	0.10 - 7.44	0.961	0.378	0.042	1.519	0.076
Monohystera (BF-1)	34	0.10 - 14.1	0.950	0.429	0.048	1.718	0.094
Mononchus (PR-4)	35	0.10 - 3.70	0.938	0.377	0.046	1.627	0.098
Panagrolaimus (BF-1)	35	0.40 - 49.3	0.947	0.310	0.096	1.762	0.098
Parathylenchus (PF-2)	31	0.40 - 22.8	0.944	0.540	0.075	1.850	0.113
Plectus (BF-2)	36	0.20 - 17.5	0.937	0.193	0.073	1.647	0.099
Pratylenchus (PF-3)	36	0.20 - 51.7	0.923	0.298	0.119	1.735	0.114
Rhabditis (BF-1)	36	0.20 - 34.2	0.940	0.167	0.100	1.792	0.105
Trichodorus (PF-4)	11	0.10 - 2.70	0.891	0.437	0.114	1.433	0.217
Tylenchorhynchus (PF-3)	30	0.22 - 5.80	0.931	0.369	0.050	1.735	0.119
Tylenchus (PF-2)	36	2.60 - 159	0.918	−0.731	0.233	2.557	0.174
Wilsonema (BF-2)	31	0.10 - 10.4	0.971	0.486	0.036	1.616	0.072
Xiphinema (PF-5)	21	0.10 - 2.10	0.939	0.635	0.074	1.841	0.145

E Ensemble TPLs (*NQ* = 10) of soil nematode ecological classes sampled from 18 lawns on 2 occasions in 2008. Data from Park et al. (2013)

	NB	Range of means	r	A	SE[A]	b	SE[b]
Total nematodes	36	33.0 - 415	0.901	−0.910	0.363	2.208	0.164
Colonizer-persister class							
cp-1	36	1.00 - 115	0.956	−0.345	0.160	2.131	0.108
cp-2	36	26.9 - 275	0.880	−0.797	0.366	2.229	0.182
cp-3	36	1.60 - 82.1	0.895	0.259	0.181	1.701	0.130
cp-4	36	1.80 - 18.5	0.802	0.007	0.153	1.681	0.172
cp-5	15	0.10 - 1.80	0.940	0.586	0.084	1.851	0.175
Trophic group							
Bacteria feeder (BF)	36	6.00 - 152	0.933	−0.740	0.240	2.225	0.138
Fungus feeder (FF)	36	6.90 - 89.4	0.896	−0.120	0.223	1.930	0.147
Plant feeder (PF)	36	8.10 - 202	0.888	−0.690	0.314	2.222	0.175
Omnivore (OM)	36	1.60 - 18.9	0.856	0.048	0.124	1.678	0.149
Predator (PR)	35	0.10 - 3.70	0.938	0.377	0.046	1.627	0.098
Functional guild							
BF-1	36	1.00 - 72.3	0.953	-0.342	0.149	2.127	0.111
BF-2	36	5.00 - 82.6	0.913	−0.434	0.221	2.090	0.146
BF-4	17	0.10 - 2.10	0.924	0.389	0.064	1.387	0.137
FF-2	36	6.90 - 89.4	0.896	−0.120	0.223	1.930	0.147
PF-2	36	6.60 - 168	0.885	−1.228	0.334	2.695	0.215
PF-3	36	1.50 - 82.1	0.905	0.306	0.170	1.671	0.122
PF-4	11	0.10 - 2.70	0.891	0.437	0.114	1.433	0.217
PF-5	19	0.10 - 3.70	0.957	0.508	0.075	1.670	0.118
OM-4	36	1.60 - 18.5	0.837	0.044	0.128	1.680	0.158
OM-5	19	0.10 - 3.70	0.957	0.508	0.075	1.670	0.118
PR-4	35	0.10 - 3.70	0.938	0.377	0.046	1.627	0.098

	NB	Range of means	r	A	SE[A]	b	SE[b]
F							

F Ensemble TPLs ($NQ = 40$) of nematode trophic group population density (#/100 g soil) in samples collected from a seminatural grassland near Uppsala, Sweden. Data from Table 2 in Viketoft (2007)

	NB	Range of means	r	A	SE[A]	b	SE[b]
Bacteria feeders	29	0.010 - 11.7	0.987	0.263	0.048	1.479	0.046
Fungus feeders	11	0.015 - 6.50	0.960	0.479	0.177	1.606	0.149
Plant feeders	26	0.010 - 9.9	0.977	0.574	0.085	1.679	0.073
Omnivores	12	0.010 - 1.83	0.970	0.162	0.125	1.467	0.113
Predators	5	0.010 - 0.34	0.990	-0.249	0.109	1.222	0.098
Combined	83	0.010 - 11.7	0.972	0.378	0.046	1.583	0.041

G Ensemble TPLs ($NQ = 5-15$) of nematode population density (#/100 g) in samples from 9 microhabitats in an oak forest in Bulgaria. Data from Table 3 in Lazarova et al. (2004)

	NB	Range of means	r	A	SE[A]	b	SE[b]
Number of genera	9	3.00 - 27.4	0.857	-0.890	0.634	1.250	0.284
Total nematodes	9	103 - 21,130	0.983	-0.460	1.079	2.017	0.143
Trophic group							
Bacteria feeders	9	4.00 - 10,440	0.982	0.825	1.297	1.869	0.137
Fungus feeders	9	7.00 - 9790	0.976	0.533	0.937	1.827	0.155
Plant feeders	6	16.0 - 347	0.955	1.525	0.676	2.032	0.315
Omnivores	7	0.25 - 780	0.981	1.180	0.907	1.632	0.146
Predators	5	6.00 - 2160	0.967	2.152	0.567	2.299	0.348
Colonizer-persister class							
cp-1	9	1.00 - 3490	0.990	-0.716	1.598	1.687	0.091
cp-2	9	8.00 - 16,740	0.992	1.714	0.439	1.933	0.090
cp-3	7	9.00 - 479	0.974	0.303	0.628	2.292	0.238
cp-4	8	1.00 - 80.0	0.932	-0.606	0.929	1.561	0.247
cp-5	6	6.00 - 2340	0.987	2.007	0.740	2.102	0.684

H Ensemble TPL ($NQ = 6$) of nematode population density (#/100 g soil) in samples collected from 2 horizons in 4 woodlands in China. Data from Table 1 in Zhang et al. (2012)

	NB	Range of means	r	A	SE[A]	b	SE[b]
Total nematodes	8	355 - 3367	0.899	-0.388	0.953	1.761	0.315
Bacterivores	8	105 - 728	0.837	0.555	0.776	1.452	0.325
Fungivores	8	161 - 2010	0.920	-1.539	0.978	2.217	0.354
Plant parasitic	8	2.00 - 450	0.972	0.802	0.233	1.478	0.142
Omnivores and predators	8	47.0 - 592	0.886	-0.470	0.772	1.898	0.360
Combined	32	2.00 - 2010	0.945	0.572	0.199	1.467	0.087

I Mixed-species ensemble TPLs ($NQ = 5-84$) of nematode population density (#/200 g soil) in samples taken from soils adjacent to trees at sites in Scotland and N England. Statistics calculated from 12 genera at 11 tree species. Sample data from Tables 1 and 2 in Boag (1974)

	NB	Range of means	r	A	SE[A]	b	SE[b]
Mixed species – averaged by tree	12	3.27 - 67.4	0.888	0.073	0.340	1.764	0.257
Averaged by nematode	11	38.9 - 85.7	0.794	0.492	0.696	1.950	0.395

J Ensemble TPLs ($NQ = 5$) of nematode density (#/100 g soil) in samples from a dwarf shrub heath at Abisko, Swedish Lapland. Data from Table 1 in Ruess et al. (1998)

	NB	Range of means	r	A	SE[A]	b	SE[b]
Bacterial feeders	5	20.6 - 6076	0.958	0.205	0.714	1.606	0.267
Fungal feeders	5	15.8 - 3083	0.959	-0.086	0.708	1.691	0.277
Plant feeders	5	42.7 - 3302	0.919	-1.218	1.249	2.094	0.477
Obligate plant parasites	4	7.20 - 20.0	0.533	-0.354	1.779	2.657	1.589
Omnivores	5	0.90 - 284	0.996	0.557	0.148	1.700	0.089
Predators	3	1.02 - 155	0.994	0.822	0.261	1.714	0.181
Combined	27	0.90 - 6076	0.945	0.604	0.214	1.509	0.099

K Ensemble TPL ($NQ = 2-178$) of estimates of nematode biomass from morphometric data in the literature for the 5 colonizer-persister groups. NB is the number of genera in each *cp*-group. Data from Table 2 in Ferris (2010)

	NB	Range of means	r	A	SE[A]	b	SE[b]
cp-1	13	130 - 7440	0.804	-2.694	1.483	2.754	0.494
cp-2	14	30.0 - 9390	0.971	-1.810	0.476	2.614	0.179
cp-3	26	90.0 - 7120	0.917	-2.838	0.627	2.856	0.233

Continued

	NB	Range of means	r	A	SE[A]	b	SE[b]
cp-4	16	250 - 8540	0.955	−1.646	0.638	2.449	0.194
cp-5	7	1250 - 44,270	0.937	−0.268	1.227	2.045	0.320
Combined	76	30.0 - 44,270	0.939	−1.903	0.301	2.527	0.101

L Ensemble TPLs ($NQ = 65$) plot of nematodes recovered from an old field and a forest in Rhode Island, USA. Nematodes in the forest site were more abundant than the old field site with a narrower range of means. Combined, the two sites form a continuum to produce a single TPL. Data from Table 2 in Gorres et al. (1998)

Forest	4	2.24 - 9.16	0.939	0.273	0.305	1.599	0.388
Old field	4	1.89 - 3.64	0.435	−0.673	0.768	2.794	1.779
Habitats combined	8	1.89 - 9.16	0.885	−0.405	0.287	2.381	0.453

M Ensemble TPLs ($NQ = 3$) of estimates of *Belonolaimus longicaudatus* density (#/100 cc soil) in samples taken from maize and soybean fields in Florida, USA. Data from Fig. 1 in McSorley and Dickson (1991)

Maize − August 1987	9	0.32 - 15.00	0.978	0.514	0.182	1.366	0.110
Maize − August 1988	8	3.67 - 63.25	0.907	−1.707	1.080	1.946	0.368
Soybeans − October 1987	8	0.32 - 18.97	0.985	0.317	0.205	1.517	0.107
Soybeans − October 1988	13	0.32 - 32.72	0.980	0.166	0.156	1.075	0.065

Plant-parasitic nematodes

N Ensemble TPLs ($NQ = 50$) of potato cyst nematode *Globodera rostochiensis* cysts in fields in England expressed as cysts/g soil and per kg of soil are weakly aggregated and differ only in intercept. Data from Table 1 in Anscombe (1950)

Cysts/g soil	12	0.34 - 1.23	0.941	0.180	0.040	1.190	0.128
Cysts/kg soil	12	340 - 1230	0.941	2.609	0.352	1.190	0.128

O Ensemble TPL ($NQ = 20$) of potato cyst nematodes *Globodera rostochiensis* and *G. pallida* at small scale: #/m² in 480 m² section of field. Data from Fig. 1 in Schomaker and Been (1999)

Globodera rostochiensis and *G. pallida*	24	0.50 - 174	0.989	−0.004	0.071	1.853	0.059

P Spatial TPLs ($NQ = 90–138$) of root lesion nematode *Pratylenchus penetrans* (#/100 cc soil) in fumigated and unfumigated fields across 3 rotations including 1 year in potatoes. Each point is 1 year. Data from Morgan et al. (2002)

Fumigated	7	0.10 - 17.3	0.966	1.133	0.189	1.935	0.222
Unfumigated	8	37.8 - 324	0.889	2.106	0.445	1.189	0.222
Combined	15	0.10 - 324	0.978	1.210	0.153	1.667	0.097

Q Spatial within-season TPLs of *Criconemella sphaerocephalus* (#/100 cc soil) in corn and potato fields from $NQ = 5$ depths. Data from Fig. 2 in MacGuidwin and Stanger (1991)

Corn	10	134 - 1259	0.969	−1.458	0.590	2.402	0.211
Potato	10	232 - 743	0.819	−1.888	1.436	2.652	0.538
Combined	20	134 - 1259	0.915	−1.342	0.625	2.403	0.229

R Ensemble and spatial TPLs ($NQ = 22–107$) of *Pratylenchus* spp and *Meloidogyne hapla* (#/sample) in potato fields in Ohio: effect of season and scale of sampling. Data from Table 2 in Wheeler et al. (1994) and Tables 2 and 4 in Wheeler et al. (2000)

Wheeler et al. (1994)							
Meloidogyne hapla	7	1.00 - 395	0.982	1.567	0.191	1.296	0.111
Pratylenchus penetrans	7	1.00 - 37.0	0.985	0.900	0.119	1.479	0.114
P. scribneri	5	9.00 - 110	0.982	1.189	0.227	1.316	0.142
P. crenatus	7	3.00 - 167	0.974	1.479	0.172	1.144	0.117
Pratylenchus spp. combined	10	10.0 - 371	0.994	1.301	0.089	1.273	0.048

	NB	Range of means	r	A	SE[A]	b	SE[b]
Wheeler et al. (2000)							
Pratylenchus spp – 10 m²	6	7.50 - 96.5	0.927	0.498	0.458	1.721	0.322
50 m²	6	6.80 - 83.3	0.957	1.351	0.243	1.183	0.171
100 m²	6	8.50 - 74.2	0.883	1.246	0.332	1.083	0.254
250 m²	6	14.5 - 63.9	0.668	0.498	0.934	1.721	0.640
1000 m²	6	10.5 - 78.1	0.948	0.686	0.350	1.474	0.235
4000 m²	4	6.50 - 52.6	0.997	0.232	0.144	1.893	0.111
Combined	34	6.50 - 96.5	0.888	0.736	0.174	1.527	0.124
Autumn	7	53.0 - 107	0.641	-1.117	1.577	2.410	0.827
Spring	7	40.0 - 220	0.857	0.596	0.688	1.528	0.352
Meloidogyne hapla – Autumn	6	3.00 - 183	0.979	0.369	0.269	1.983	0.202
Spring	7	1.00 - 18.0	0.904	0.917	0.288	1.594	0.305

S   Spatial TPLs (*NQ* = 110 or 160) of soybean cyst nematode cysts/100 cc, eggs/100 cc are almost identical, separated by the essentially random factor of eggs/cyst. Data from Table 1 in Avendano et al. (2003) and Table 2 in Avendano et al. (2004)

	NB	Range of means	r	A	SE[A]	b	SE[b]
Cysts/100 cc	20	4.00 - 48.0	0.965	0.268	0.151	2.012	0.125
Eggs/cyst	20	11.8 - 139	0.953	1.531	0.140	1.167	0.083
Eggs/100 cc	4	86.6 - 4939	0.998	0.248	0.289	2.061	0.095

T   Ensemble TPL (*NQ* = 43–94) of stem nematode *Ditylenchus dipsaci* on broad bean (*Vicia faba*) seeds (#/seed). Data from Table 1 in Green (1979)

	NB	Range of means	r	A	SE[A]	b	SE[b]
Ditylenchus dipsaci	17	0.02 - 4.20	0.960	1.461	0.117	1.796	0.130

U   Hybrid ensemble TPLs (*NQ* = 64 or 100) of 3 nematode species in 7 North Carolina tobacco fields. Data from Table 1 in Noe and Campbell (1985)

	NB	Range of means	r	A	SE[A]	b	SE[b]
Helicotylenchus dihystera	7	31.0 - 143	0.942	1.561	0.386	1.379	0.207
Meloidogyne incognita	11	39.0 - 8.153	0.957	0.898	0.495	1.712	0.165
Tylenchorhynchus claytoni	11	45.0 - 457	0.933	−0.575	0.601	2.211	0.265
Combined	29	31.0 - 8153	0.959	0.538	0.245	1.810	0.098

V   Ensemble TPLs (*NQ* = 10) of parasitic bacterium *Pasteuria penetrans* and nematode host *Meloidogyne javanica* are parallel and highly aggregated. Data from Tables 6 and 7 in Dabiré et al. (2007)

	NB	Range of means	r	A	SE[A]	b	SE[b]
M. javanica – Juveniles/plant	10	1.00 - 83,100	0.991	−1.115	0.389	2.277	0.109
P. penetrans – Spores/g soil	11	12,200 - 858,696	0.927	−2.294	1.554	2.350	0.294

W   Ensemble TPLs (*NQ* = 5) of J2 *Meloidogyne incognita* and nematodes infected with *Pasteuria penetrans* endospores. Abundance (#/100 g soil) averaged over multiple nematode-resistant and susceptible crops in rotation. Data from Table 1 in Ciancio and Quénéhervé (2000)

	NB	Range of means	r	A	SE[A]	b	SE[b]
Meloidogyne incognita – Total	6	60.0 - 396	0.994	−0.951	0.282	2.197	0.123
Infected	6	12.2 - 115	0.910	−0.270	0.191	0.531	0.110

X   Ensemble TPLs (*NQ* = 7) of nematodes in citrus orchards sampled in a stratified random pattern or systematically at a range of spatial scales. Data from Table 1 in Abd-Elgawad (1992)

	NB	Range of means	r	A	SE[A]	b	SE[b]
Stratified random sampling							
Tylenchulus semipenetrans	35	-	0.924	−0.081	-	1.950	0.141
Helicotylenchus pseudorobustus	26	-	0.812	−0.268	-	2.050	0.301
Criconomella spp.	11	-	0.981	−0.444	-	2.180	0.144
Systematic sampling							
T. semipenetrans – 2.1-ha sites	24	-	0.964	0.070	-	1.950	0.115
4.2-ha sites	12	-	0.976	0.228	-	1.980	0.141
6.3-ha sites	8	-	0.976	0.599	-	1.900	0.172
8.4-ha sites	6	-	0.987	0.693	-	1.890	0.154

Continued

	NB	Range of means	r	A	SE[A]	b	SE[b]
Y TPLs of *Tylenchulus semipenetrans* (#/100 cc soil) in samples (temporal $NQ = 80$, geographic $NQ = 20$) taken from citrus groves in Florida. Data from Table 1 in Duncan et al. (1989). Comparison of TPLs ($NQ = 2$) derived from different scales of sampling in orchard surveys in Florida and Egypt. Data from Duncan et al. (1994)							
Duncan et al. (1989)							
Temporal	18	-	0.960	1.138	-	1.50	0.109
Geographic – Total	41	-	0.987	0.964	-	1.57	0.041
Large patch	16	-	0.971	0.799	-	1.59	0.104
Small patch	19	-	0.995	0.951	-	1.83	0.045
Duncan et al. (1994)							
Florida – 9 sets of 50 single core samples	9	-	0.980	0.856	-	1.770	0.137
4 sets of 20 single core samples	4	-	0.985	0.701	-	1.750	0.218
4 sets of 12 15-core samples	4	-	0.980	0.599	-	1.480	0.214
103 sets of 2 15-core samples	103	-	0.529	−1.155	-	1.830	0.292
Egypt – 36 sets of 2 15-core samples	36	-	0.480	−0.921	-	1.610	0.505
Z Ensemble TPLs ($NQ = 5$) of nematodes recovered from soil samples in Egyptian barseem clover fields. Data from Table 1 in Abd-Elgawad and Hasabo (1995)							
Aphelenchus avenae	6	0.40 - 7.60	0.992	0.641	-	1.695	0.012
Criconomella spp.	6	0.40 - 4.20	0.955	0.488	-	1.218	0.036
Ditylenchus spp.	8	0.20 - 2.80	0.966	0.431	-	1.476	0.026
Helicotylenchus dihystera	11	0.20 - 22.4	0.996	0.670	-	1.869	0.003
Heterodera trifolii	26	0.40 - 36.0	0.923	0.437	-	1.445	0.015
Hoplolaimus spp.	7	0.40 - 2.40	0.946	0.530	-	1.590	0.059
Meloidogyne spp.	22	0.40 - 42.0	0.959	0.542	-	1.729	0.013
Pratylenchus brachyurus	28	1.00 - 44.4	0.956	0.576	-	1.398	0.007
Rotylenchulus reniformis	5	0.40 - 6.80	0.990	0.500	-	1.509	0.015
Trichodorus spp.	9	0.20 - 21.0	0.918	0.736	-	1.327	0.047
Tylenchorhynchus clarus	39	1.60 - 377	0.877	−0.029	-	1.816	0.027
Tylenchus spp.	32	0.20 - 13.4	0.942	0.486	-	1.428	0.009
A1 Mixed-species ensemble TPL ($NQ = 5$) of nematodes in sugar cane. Density in soil samples expressed as #/200 cc soil, root samples as #/100 g root. Data from Tables 1 and 3 in Allsopp (1990)							
Soil samples							
Root lesion	35	* - 5250	0.938	1.54	-	1.64	0.105
Root knot	30	* - 8100	0.980	1.95	-	1.65	0.064
Spiral	27	* - 1375	0.889	2.62	-	1.37	0.141
Stubby root	33	* - 350	0.781	3.41	-	1.1	0.166
Stunt	22	* - 320	0.854	3.11	-	1.18	0.160
Ring	23	* - 350	0.943	2.56	-	1.45	0.111
Root samples							
Root lesion	35	* - 80,360	0.949	1.55	-	1.76	0.102
Root knot	33	* - 129,320	0.943	2.03	-	1.76	0.111
Spiral	16	* - 2126	0.980	3.4	-	1.42	0.077
Combined – soil samples	7	5.00 - 669	0.959	0.215	0.451	1.790	0.228
Soil and root samples	10	5.00 - 9876	0.987	−0.061	0.282	1.954	0.112
B1 Spatial TPLs of a disease index and *Radopholus similis* and *Pratylenchus goodeyi* density on banana in tropical and subtropical climates ($NQ = 10$ and 20, respectively). Data from Stanton et al. (2001)							

	NB	Range of means	r	A	SE[A]	b	SE[b]
Disease Index – Tropical	44	-	0.788	1.114	*	1.042	0.126
Subtropical	51	-	0.936	0.635	*	1.449	0.078
Nematodes	91	-	0.95	1.256	*	1.723	0.060

C1 Ensemble TPLs ($NQ = 30-59$) of pinewood nematode *Bursaphelenchus xylophilus* in Scots pine bolts infested and uninfested by beetle *Monochamus carolinensis*. Data from Table 3 in Warren and Linit (1992)

Beetle-infested	7	21.0 - 147	0.980	−0.678	0.430	2.863	0.257
Noninfested	7	25.9 - 78.9	0.853	−2.486	1.467	3.710	0.865
Combined	14	21.0 - 147	0.901	−1.267	0.656	3.103	0.389

D1 Ensemble TPLs ($NQ = 5$) of pinewood nematode *Bursaphelenchus xylophilus* in Japanese black pine infected with 19 species of fungus at 5, 10, and 15 days post infection. Points are pinewood nematode density (#/g of wood) associated with a different fungus. Data from Table 2 in Sriwati et al. (2007)

After 5 days	13	0.59 - 167	0.942	−0.712	0.303	1.850	0.187
After 10 days	12	17.6 - 10,519	0.958	−0.682	0.474	1.834	0.166
After 15 days	11	0.69 - 1,698,243	0.990	−1.004	0.367	1.949	0.090
Combined	36	0.59 - 1,698,243	0.987	−0.852	0.159	1.910	0.054

E1 Temporal TPL ($NQ = 2-14$) of pinewood nematode *Bursaphelenchus mucronatus* transmitted by vector beetle *Monochamus saltuarius*. Data are of the total number of nematodes recovered from 43 beetles over a 70-day period. Data from Table 1 in Jikumaru and Togashi (2001)

Bursaphelenchus mucronatus	37	0.11 - 350	0.992	0.471	0.056	1.852	0.039

Entomopathogenic nematodes

F1 Ensemble TPLs (NQ variable) of *Steinernema glaseri* and *S. feltiae* are extremely aggregated. Results originally published in Taylor (1999) from data of Spridonov and Voronov (1995), Peters (1994), and Dr. Peter Smits.

S. glaseri (Spiridonov and Voronov)	29	13.6 - 121	0.917	−0.274	0.309	2.461	0.189
S. feltiae (Peters)	7	9.00 - 372	0.937	−2.404	0.859	3.147	0.491
S. feltiae (Smits)	8	1.36 - 7.38	0.983	−0.257	0.142	2.900	0.218

G1 Hybrid ensemble TPL (NQ variable) of *Heterorhabditis bacteriophora* sampled with single and composited cores (#/100 cc soil) following inundative release beneath grapefruit trees in a Florida citrus grove. Adapted from Fig. 1 in Duncan et al. (1996)

H. bacteriophora – single samples	11	1.13 - 268	0.994	0.740	0.086	1.777	0.062
Composite samples	17	0.55 - 625	0.985	0.678	0.099	1.544	0.069
Combined	28	0.55 - 625	0.979	0.737	0.089	1.639	0.065

H1 Spatial TPLs ($NQ = 30$) of *Heterorhabditis bacteriophora* and *Steinernema carpocapse* under tall fescue and Kentucky bluegrass ($NQ = 12$). Data from Figs. 2–4 in Campbell et al. (1995). Spatial TPLs of naturally occurring and released *H. bacteriophora*. Data from Campbell et al. (1998)

Campbell et al. (1995)							
H. bacteriophora – Tall fescue	10	0.26 - 113	0.994	0.899	0.061	2.057	0.078
S. carpocapse – Kentucky bluegrass	22	0.008 - 44.0	0.990	0.839	0.059	2.070	0.066
Kentucky bluegrass	19	0.13 - 45.5	0.992	0.806	0.054	2.078	0.065

Continued

	NB	Range of means	r	A	SE[A]	b	SE[b]
Combined	51	0.0083 - 113	0.991	0.839	0.033	2.074	0.039
Campbell et al. (1998)							
H. bacteriophora – Natural	12	0.40 - 433	0.975	1.155	0.228	1.866	0.132
Released	6	1.40 - 42.9	0.947	0.810	0.341	2.389	0.384
Combined	18	0.40 - 433	0.970	1.084	0.174	1.931	0.116

I1 Spatial TPLs ($NQ = 4$) of *Heterorhabditis bacteriophora* and *Steinernema riobravis* applied to soil in a citrus orchard and sampled (#/sample and #/100 cc soil) at 0, 7, and 14 days post-treatment. Data from Fig. 4 in Duncan and McCoy (1996)

	NB	Range of means	r	A	SE[A]	b	SE[b]
H. bacteriophora – #/sample	9	1.46 - 149	0.889	−0.135	0.465	1.896	0.328
#/100 cc	9	7.74 - 787	0.888	0.541	0.425	1.478	0.257
S. riobravis – #/sample	9	1.22 - 124	0.947	−0.782	0.388	2.144	0.261
#/100 cc	9	6.45 - 658	0.911	−0.469	0.511	1.927	0.301

J1 Temporal TPL ($NQ = 120{-}200$) of nematodes (#/100 cc) in 4 habitats ranging from a natural area to conventional annual cropland in Spain. Data from Table 2 in Campos-Herrera et al. (2008)

	NB	Range of means	r	A	SE[A]	b	SE[b]
Total EPNs	4	0.01 - 21.2	0.999	0.303	0.123	2.448	0.093

K1 Temporal TPL ($NQ = 20$) of nematode-killed *Galleria* larvae by EPNs () recovered from a citrus grove in Israel. Data from Fig. 1 in Efron et al. (2001)

	NB	Range of means	r	A	SE[A]	b	SE[b]
25%–50% cover	9	0.12 - 4.00	0.949	0.471	0.092	1.378	0.164
50%–75% cover	9	0.10 - 0.77	0.900	1.023	0.235	2.277	0.376
75%–100% cover	9	0.22 - 3.03	0.964	0.740	0.066	1.623	0.164
Combined	27	0.10 - 4.00	0.939	0.625	0.059	1.604	0.110

L1 Ensemble TPLs ($NQ = 5{-}12$) of IJs recovered from white grubs (principally Japanese beetles) and vlack cutworms, exposed to combinations of *Steinernema carpocapsae*, *S. glaseri*, and *S. riobravis*. Data from Table 4 in Koppenhöfer and Kaya (1996)

	NB	Range of means	r	A	SE[A]	b	SE[b]
White grubs	5	3.30 - 201	0.974	−0.707	0.477	2.149	0.284
Black cutworm	7	14.8 - 615	0.811	1.274	0.585	1.175	0.307

Animal-parasitic nematodes
Invertebrate hosts

M1 Temporal TPLs ($NQ = 3{-}27$) of male and female nematodes. Intensity (#/noninfected host) and density (#/host) produce very different graphs. Intensity has no hosts with zero parasites and therefore produces steeper TPLs that differ significantly between male and female. Density (includes zeros) results in parallel TPLs for male and female nematodes/host. Data from Tables 8 and 9 in Adamson and Noble (1992)

	NB	Range of means	r	A	SE[A]	b	SE[b]
Intensity – male nematodes	9	1.05 - 31.7	0.961	−1.992	0.231	2.845	0.299
Female nematodes	12	1.45 - 53.0	0.981	−1.702	0.098	1.975	0.120
Density – male nematodes/host	16	0.0091 - 2.13	0.984	1.203	0.106	1.800	0.086
Female nematodes/host	12	0.076 - 6.63	0.994	0.973	0.040	1.766	0.063

N1 Ensemble TPLs ($NQ = 86{-}2339$ flies) of *Howardula aoronymphium* parasitic on mycophagous drosophilds caught at 2 sites and on 3 mushroom species. Points are nematodes/host on each of 4 *Drosophila* species. Data from Tables 3 and 6 in Jaenike (1994)

	NB	Range of means	r	A	SE[A]	b	SE[b]
H. aoronymphium – by site	7	0.004 - 0.28	0.987	0.405	0.121	1.241	0.089
By mushroom species	12	0.02 - 0.44	0.943	0.737	0.180	1.579	0.166

Vertebrate hosts

O1 Mixed-species TPLs (NQ variable) of all endoparasites and nematode parasites per host on vertebrate hosts. Data from Appendix A in Shaw and Dobson (1995)

	NB	Range of means	r	A	SE[A]	b	SE[b]
All endoparasites	229	0.012 - 5234	0.954	1.102	0.036	1.687	0.034
Nematodes	104	0.023 - 5234	0.963	1.094	0.055	1.725	0.046

	NB	Range of means	r	A	SE[A]	b	SE[b]
P1 Mixed-species TPLs (*NQ* variable) of nematode parasites per host on mammal hosts. Data from Morand and Guegan (2000)							
Worm burden (#/host)	104	0.010 - 2.14E+07	0.980	3.270	0.170	0.515	0.023
Q1 Spatial TPLs (*NQ* > 100) of nematodes recovered at necropsy from Scottish Blackface lamb. Data from Tables 2 and 3 in Stear et al. (1998). Spatial TPLs (*NQ* > 100) of faecal egg counts, eggs/lamb sampled monthly, and faecal egg count averaged per year. Data from Fig. 3 and Table 1 in Stear et al. (2006)							
Stear et al. (1998)							
Cooperia spp.	4	83.9 - 4369	0.997	1.197	0.267	1.707	0.095
Ostertagia. circumcincta	4	2301 - 12,936	0.996	0.337	0.435	1.862	0.117
Nematodirus spp.	4	120 - 456	0.572	2.111	1.814	1.302	0.755
Teladorsagia vitrinus	4	116 - 1022	0.932	1.751	0.983	1.527	0.392
Combined	21	0.90 - 12,936	0.985	1.960	0.150	1.426	0.057
Stear et al. (2006)							
Faecal egg count – Eggs/lamb	42	6.01 - 950	0.890	1.789	0.228	1.339	0.097
eggs/g faeces	4	87.0 - 1767	0.973	0.926	0.708	1.660	0.269
R1 Ensemble TPLs (*NQ* = 160) of nematodes recovered from lambs by necropsy, separated into species that do and do not overwinter in the pasture. Data from Table 1 in Boag et al. (1992)							
Overwintering nematodes							
Nematodirus battus	40	-	0.948	-	-	2.152	0.110
N. filicollis	44	-	0.966	-	-	2.029	0.080
Teladorsagia circumcincta	46	-	0.973	-	-	2.044	0.070
T. pinnata	47	-	0.981	-	-	2.070	0.060
T. trifurcate	41	-	0.987	-	-	1.975	0.050
Chabertia ovina	45	-	0.980	-	-	1.980	0.060
Trichuris ovis	44	-	0.679	-	-	1.943	0.220
Strongyloides papillosus	42	-	0.971	-	-	1.864	0.070
Nonoverwintering nematodes							
Haemonchus contortus	29	-	0.971	-	-	2.194	0.100
Trichostrongylus axei	29	-	0.982	-	-	1.925	0.070
T. colubriformis	26	-	0.993	-	-	2.045	0.050
T. vitrinus	31	-	0.983	-	-	1.800	0.060
Bunostomum trigonocephalum	27	-	0.981	-	-	2.100	0.080
Cooperia curticei	14	-	0.997	-	-	1.985	0.040
Oesophagostomum venulosum	29	-	0.961	-	-	1.519	0.080
S1 Ensemble TPL (*NQ* = 33–48) of *Strongyloides ratti* injected at 3 rates into laboratory rats and recovered in faeces and counted at 5, 10, and 15 days post treatment (#/pellet). Data from Table 1 in Harvey et al. (1999)							
Strongyloides ratti	9	4.20 - 669	0.950	1.247	0.294	1.523	0.180
T1 Ensemble TPLs (*NQ* variable) of the parasite intensity of *Syphacia* spp and *Heligmosomoldes polygyrus* in mice. Data from Fig. 1 in Grear and Hudson (2011)							
Syphacia spp.	23	1.00 - 4044	0.941	0.559	0.405	2.804	0.208
H. polygyrus	11	2.97 - 1102	0.959	0.823	0.380	1.908	0.180
U1 Spatial TPLs (*NQ* = 13–30) of abundance and intensity of *Camallanus cotti* in Chinese hooksnout carp. Data from Fig. 2 in Wu et al. (2007)							
Abundance	14	0.38 - 8.85	0.936	0.603	0.097	2.338	0.237
Intensity	14	1.25 - 12.0	0.890	−0.470	0.245	3.156	0.415

Continued

	NB	Range of means	r	A	SE[A]	b	SE[b]

Freshwater nematodes

V1 Ensemble TPLs ($NQ = 2-8$) of nematode populations (#/100 cc sediment) in samples collected from 8 contaminated rivers in northern Europe. Sample data from, Table 5 in Heininger et al. (2007).

	NB	Range of means	r	A	SE[A]	b	SE[b]
Total nematodes	8	98.0 - 958	0.910	−3.839	2.605	2.600	0.485
Genera	8	7.00 - 19.3	0.831	−6.683	2.564	3.907	1.068
Deposit feeders	8	44.0 - 516	0.873	−1.924	2.051	1.855	0.423
Epistrate feeders	7	0.18 - 8.69	0.931	1.543	0.321	1.075	0.189
Suction feeders	8	5.18 - 50.6	0.927	−1.132	1.157	2.325	0.385
Chewers	8	11.9 - 397	0.719	−1.427	2.771	1.814	0.715

W1 Spatial TPLs ($NQ = 4$) of nematode population density (#/10 cm^2 sediment) in samples collected from 2 rivers in southwest Germany. Data from Fig. 5 in Beier and Traunspurger (2003)

	NB	Range of means	r	A	SE[A]	b	SE[b]
Krahenbach River							
Monohysteridae (deposit feeders)	11	11.4 - 52.4	0.557	−2.082	3.466	2.178	1.083
Tobrilidae (chewers)	11	3.38 - 33.2	0.848	−0.052	0.785	1.671	0.348
Tylenchidae (suction feeders)	11	1.08 - 6.71	0.875	0.433	0.301	1.483	0.273
Korsch River							
Monohysteridae	12	7.24 - 1609	0.925	0.253	1.077	1.730	0.225
Tobrilidae	11	0.68 - 141	0.968	0.654	0.357	1.636	0.141
Tylenchidae	12	4.58 - 164	0.924	0.037	0.793	1.718	0.225

X1 Hybrid ensemble TPLs ($NQ = 8$) of nematode populations (#/100 cc sediment) in samples collected from 3 eutrophic and 5 healthy lakes in southern Sweden. Data from Table 2 in Ristau and Traunspurger (2011)

	NB	Range of means	r	A	SE[A]	b	SE[b]
Spring	8	105 - 334	0.873	−4.236	3.178	2.691	0.614
Autumn	8	48.0 - 369	0.938	−4.441	2.065	2.791	0.422
Oligo/mesotrophic	10	48.0 - 293	0.873	−4.081	2.589	2.674	0.527
Eutrophic	6	64.0 - 369	0.980	−3.572	1.392	2.612	0.265
Combined	16	48.0 - 369	0.919	−4.109	1.552	2.693	0.308

Y1 Ensemble TPLs ($NQ = 6$) of nematodes population density (#/10 cm^2 sediment) in samples collected from ponds in 5 regions of Belgium. Sample data from Bert et al. (2007), Table 3

	NB	Range of means	r	A	SE[A]	b	SE[b]
All data combined	25	0.10 - 160	0.976	0.338	0.086	1.709	0.077
Total nematodes	5	63.9 - 248	0.726	0.207	4.295	1.609	0.880
Tobrilus spp. (inc. juveniles)	5	3.95 - 160	0.908	0.775	1.584	1.584	0.423
Dorylaimus spp.	5	0.10 - 7.58	0.982	0.710	0.321	1.631	0.182
Mononchus spp.	5	0.60 - 4.65	0.942	1.262	0.183	0.967	0.199
Monhystera spp. (inc. juveniles)	5	0.28 - 11.5	0.992	0.967	0.247	1.910	0.142
Eumonhystera spp. (inc. juveniles)	5	19.3 - 45.7	0.716	0.407	3.592	1.854	1.045
By region							
Blankaart	5	0.17 - 20.8	0.995	0.484	0.205	1.739	0.104
Knokke	5	1.75 - 48.7	0.969	0.748	0.670	1.698	0.248
Geraardsbergen	5	0.60 - 160	0.994	1.338	0.346	1.750	0.109
Zottegem	5	0.10 - 22.5	0.994	0.927	0.740	1.757	0.570
Temse	5	1.20 - 85.7	0.963	1.630	0.520	1.131	0.184
By Landuse type							
Natural areas	4	1.56 - 45.6	0.981	1.359	0.588	1.753	0.246
Extensive arable	5	2.01 - 88.9	0.978	0.585	0.731	2.015	0.249
Intensive livestock grazing	5	1.81 - 52.6	0.991	1.032	0.356	1.731	0.133

	NB	Range of means	r	A	SE[A]	b	SE[b]

Z1 Spatial TPLs ($NQ = 23-36$) of bacterivorous nematode populations (#/10 cm² sediment) in samples collected from a restored riparian wetland soils sampled in 4 seasons in Georgia, USA. Data from Table 1 in Ettema et al. (1998)

Combined	32	1.10 - 24.2	0.928	0.528	0.263	1.761	0.129
Acrobeloides	4	9.40 - 24.2	0.930	0.760	1.391	1.730	0.484
Chronogaster	4	7.07 - 12.1	0.816	−3.375	2.035	2.292	1.150
Eumonhystera	4	1.31 - 6.40	0.986	−1.229	4.804	1.357	0.163
Heterocephalobus	4	2.83 - 6.40	0.815	−0.146	2.491	2.244	1.127
Monhystrella	4	1.10 - 9.76	0.999	0.843	0.093	1.687	0.062
Prismatolaimus	4	8.72 - 16.6	0.936	−0.308	1.845	3.070	0.814
Rhabditinae	4	2.95 - 5.56	0.899	0.006	1.204	2.380	0.818
Rhabdolaimus	4	8.85 - 11.2	0.649	0.426	0.213	2.563	2.124

Marine nematodes

A2 Mixed-species ensemble TPL of nematode genera (#/10 cm²) in samples collected from sediments in the Westerschelde Estuary, the Netherlands, using meiocore ($NQ = 12$) and a microcore ($NQ = 14$) samplers. Data from Table 1 in Moens et al. (1999)

Meiocore	23	5.00 - 1124	0.988	4.160	0.157	1.109	0.037
Microcore	24	5.00 - 906	0.949	3.283	0.334	1.206	0.085

B2 Ensemble TPL ($NQ = 5$) of nematodes sampled from the Molenplaat intertidal flat in the Westerschelde estuary in SW Netherlands. Sample data digitized and expressed as #/10 cc. Data from Figs. 3–6 in Steyaert et al. (2003)

Total nematodes – Site 1 (intermediate)	11	104 - 656	0.786	0.067	0.873	1.772	0.365
Site 2 (fine)	11	4.00 - 1024	0.932	−0.146	0.351	1.820	0.209
Site 3 (sandy)	12	80.0 - 280	0.576	−1.701	1.462	2.521	0.652
Combined	35	4.00 - 1024	0.913	−0.203	0.279	1.851	0.132
Theristus blandicor – Site 1	12	2.38 - 395	0.958	0.587	0.270	1.628	0.147
Site 2	12	0.16 - 1.87	0.889	0.374	0.081	1.751	0.254
Site 3	10	1.17 - 183	0.946	0.101	0.383	1.788	0.204
Viscosi viscosia – Site 1	11	0.73 - 134	0.829	0.556	0.241	1.430	0.267
Site 2	11	0.48 - 97.7	0.954	0.151	0.177	1.663	0.166
Site 3	9	1.32 - 12.3	0.817	0.378	0.302	1.869	0.407

C2 Ensemble TPLs ($NQ = 3$) of the predatory nematode *E. longispiculosus* and its prey nematodes (#/10 cm² sediment) in samples from Paulina intertidal flat of the Westerschelde Estuary in SW Netherlands. Data from Fig. 2 in Gallucci et al. (2005)

E. longispiculosus	16	1.56 - 189	0.911	0.475	0.196	1.212	0.134
Prey nematodes	18	19.5 - 2435	0.787	0.101	0.746	1.659	0.256
Combined	34	1.56 - 2435	0.947	0.025	0.219	1.651	0.094

D2 Ensemble TPLs ($NQ = 4$) of Marine nematode populations (#/10 cm²) in samples from the Tay Estuary in Scotland. Data from Tables 1 and 2 in Neilson et al. (1993)

Species							
Anoplostoma viviparum	39	-	0.601	−2.02	2.50	2.06	0.45
Calomicrolaimus honestus	25	-	0.720	−0.57	1.25	2.09	0.42
Chromadorina sp. A	33	-	0.728	1.30	0.83	1.36	0.23
Chromadorita tentabunda	37	-	0.689	−1.65	1.82	2.25	0.40
Daptonema procerum	22	-	0.743	1.61	1.08	1.49	0.30
D. setosum	21	-	0.775	2.18	1.02	1.39	0.26
D. tenuispiculum	25	-	0.872	1.99	0.56	1.28	0.15
Daptonema sp. A	34	-	0.540	2.79	1.10	0.98	0.27
Desmolaimus zeelandicus	39	-	0.458	−1.21	3.73	1.91	0.61
Dichromadora cephalota	35	-	0.649	0.91	1.32	1.57	0.32
D. geophila	34	-	0.739	0.89	0.85	1.55	0.25

Continued

	NB	Range of means	r	A	SE[A]	b	SE[b]
Leptolaimus papilliger	39	-	0.508	−1.56	3.22	1.90	0.53
Paracanthonchus caecus	36	-	0.623	−0.56	1.90	1.95	0.42
Paracanthonchus sp. A	23	-	0.857	1.71	0.64	1.45	0.19
Ptycholaimellus ponticus	24	-	0.835	2.19	0.73	1.28	0.18
Genus							
Anoplostoma	39	-	0.601	−2.02	2.5	2.06	0.45
Chromadorina	33	-	0.728	1.3	0.83	1.36	0.23
Chromadorita	39	-	0.641	−0.2	1.74	1.83	0.36
Daptonema	39	-	0.326	2.21	2.89	1.11	0.53
Desmolaimus	39	-	0.458	−1.21	3.73	1.91	0.61
Dichromadora	39	-	0.351	0.27	3.74	1.64	0.72
Leptolaimus	39	-	0.559	−2.23	3.02	2.01	0.49
Paracanthonchus	38	-	0.652	0.45	1.52	1.65	0.32

E2 Ensemble TPL ($NQ = 6$) of nematode populations (#/10 cm^2 sediment) in samples collected from sediments at 9 tropical (18°N) to temperate (40°N) beaches in China. Sample data from Hua et al. (2016), Table 2

Total nematodes	9	87.7 - 2253	0.934	3.333	1.129	1.221	0.177

F2 Ensemble TPLs ($NQ = 2$) of nematode populations (#/10 cm^2 sediment) in samples ($NQ = 10$ depths) collected from sediment at 4 sites in the Adriatic Sea; also TPLs of the vertical profile at each station ($NQ = 2$). Data from Fig. 2 in Moodley et al. (2000)

Total nematodes	8	106 - 4791	0.990	−2.595	0.707	1.780	0.105
Station 11	10	57.7 - 904	0.866	−0.669	1.664	1.434	0.292
Station 2	10	3.70 - 437	0.706	1.918	1.304	0.905	0.321
Station 4	10	14.8 - 151	0.977	0.291	0.609	2.052	0.160
Station 5	10	0.62 - 45.7	0.985	1.253	0.251	1.912	0.118

G2 Ensemble TPLs Estimates of nematodes in the ocean's deep-water zones (abundance: $NQ = 53-252$, biomass: $NQ = 13-62$). Data from Tables 1 and 2 in Mokievskii et al. (2007)

Nematodes (#/10 cm^2)	4	247 - 1116	0.963	0.783	2.351	1.858	0.370
Biomass (µg C/10 cm^2)	4	12.0 - 28.9	0.981	1.125	1.126	1.877	0.265
Combined	8	12.0 - 1116	0.988	0.620	0.265	1.775	0.113

H2 Spatial TPLs ($NQ = 36$) of nematode populations (#/500 cm^2 blade) on kelp *Macrocystis integrifolia* samples collected from Berkley Sound, British Columbia. Data from Fig. 2 in Trotter and Webster (1983)

Monhystera disjuncta	14	21.1 - 179	0.769	−0.052	1.650	1.610	0.387
M. refringens	14	3.73 - 77.3	0.961	−0.440	0.443	1.655	0.137
Prochromadorella neapolitana	14	0.26 - 56.6	0.988	−1.221	0.228	1.998	0.089

I2 Ensemble TPLs ($NQ = 5$) of nematode populations (#/10 cm^2 sediment) in samples collected from sediments on a transect of the Mediterranean deep. Data from Soetaert et al. (1991), Table 1

Total nematodes	6	268 - 519	0.939	−31.439	5.822	13.180	2.265
Number of species	6	101 - 148	0.441	−24.036	11.624	12.506	5.612

Other worms

J2 Mixed species TPLs (NQ variable) of cestode, digenean, and acanthocephalan parasites in vertebrate hosts. Data from Appendix A in Shaw and Dobson (1995)

Cestodes	60	0.012 - 68.0	0.928	1.151	0.069	1.538	0.075
Digenea	46	0.09 - 865	0.952	1.079	0.081	1.770	0.082
Acanthocephala	19	0.20 - 272	0.961	0.775	0.118	1.754	0.118

	NB	Range of means	r	A	SE[A]	b	SE[b]
K2 Mixed-species and community TPLs ($NQ = 13-62$) of parasites of gray mullets, *Liza haematocheilus* native to the Sea of Japan and introduced to the Azov Sea, and *Mugil cephalus* native to both the Mediterranean and Sea of Japan. Data from supplementary Table S2 in Sarabeev et al. (2017)							
L. haematocheilus in the Sea of Japan							
Mixed-specie – All	8	-	0.999	-	-	1.792	0.040
Endohelminths	8	-	0.999	-	-	1.852	0.040
Monogeneans	8	-	0.980	-	-	1.704	0.140
Adult digeneans	8	-	0.984	-	-	1.921	0.140
Community – All	8	-	0.986	-	-	2.181	0.150
Endohelminths	8	-	0.997	-	-	2.066	0.070
Monogeneans	8	-	0.991	-	-	1.806	0.100
Adult digeneans	8	-	0.997	-	-	1.976	0.060
Larval digeneans	4	-	1.000	-	-	2.000	0.004
L. haematocheilus in the Azov Sea							
Mixed-specie – All	14	-	0.983	-	-	1.719	0.090
Endohelminths	14	-	0.977	-	-	1.781	0.110
Monogeneans	13	-	0.870	-	-	1.816	0.270
Adult digeneans	14	-	0.928	-	-	1.853	0.200
Larval digeneans	8	-	0.918	-	-	1.786	0.290
Community – All	14	-	0.989	-	-	1.911	0.080
Endohelminths	14	-	0.989	-	-	1.911	0.080
Monogeneans	14	-	0.979	-	-	1.879	0.110
Adult digeneans	14	-	0.967	-	-	1.903	0.140
Larval digeneans	11	-	0.942	-	-	1.975	0.220
M. cephalus in the Mediterranean							
Mixed-specie – All	10	-	0.995	-	-	1.759	0.060
Endohelminths	10	-	0.995	-	-	1.759	0.060
Monogeneans	5	-	0.957	-	-	1.787	0.300
Adult digeneans	10	-	0.957	-	-	1.944	0.200
Larval digeneans	5	-	0.985	-	-	1.878	0.190
Community – All	10	-	0.987	-	-	2.128	0.120
Endohelminths	10	-	0.989	-	-	2.103	0.110
Monogeneans	10	-	0.974	-	-	1.889	0.150
Adult digeneans	10	-	0.993	-	-	1.934	0.080
Larval digeneans	10	-	0.976	-	-	2.059	0.160
M. cephalus in the Azov Sea							
Mixed-specie – All	8	-	0.994	-	-	1.751	0.080
Endohelminths	8	-	0.994	-	-	1.751	0.080
Monogeneans	7	-	0.973	-	-	1.829	0.190
Adult digeneans	8	-	0.990	-	-	1.727	0.100
Community – All	8	-	0.990	-	-	2.051	0.120
Endohelminths	8	-	0.988	-	-	2.065	0.130
Monogeneans	8	-	0.989	-	-	2.002	0.120
Adult digeneans	8	-	0.989	-	-	2.002	0.120
Larval digeneans	4	-	1.000	-	-	2.010	0.020
L2 Spatial TPLs of intensity ($NQ = 4-44$) and abundance ($NQ = 82-281$) of cestode parasite *Hymenolepis tenerrima* and its host *Herpetocypris reptans* ($\#/m^2$) Data from Fig. 1 and Table 2 in Evans (1983)							
H. tenerrima intensity	11	1.00 - 1.50	0.691	−1.249	0.146	6.455	1.554
H. tenerrima/host	12	0.022 - 0.33	0.970	0.058	0.077	0.976	0.075
H. reptans/m^2	12	238 - 1575	0.894	−2.046	0.842	2.117	0.300

Continued

	NB	Range of means	r	A	SE[A]	b	SE[b]
M2 Mixed-species ensemble TPLs ($NQ = 108$) of abundance and intensity of parasites in native and introduced fowl to Ghana. Data from Tables 1 and 2 in Hodasi (1969)							
Abundance in native birds	6	0.7 - 13.5	0.990	0.851	0.091	1.646	0.118
In introduced birds	6	0.4 - 8.8	0.972	1.097	0.122	1.649	0.195
Abundance	12	0.40 - 13.5	0.970	0.977	0.088	1.642	0.126
Intensity	12	2.81 - 22.8	0.986	-0.107	0.125	2.357	0.126
N2 Spatial TPLs ($NQ = 30$) of life stages of the leech *Erpobdella octoculata* in 2 reaches of a stream in N England. Data from Table 1 in Elliott (2004)							
Deep reach – Young leeches	23	-	1.00	−0.013	0.024	1.510	0.030
Mature leeches	24	-	0.99	0.013	0.015	1.480	0.040
Young + mature leeches	47	-	1.00	0.004	0.009	1.500	0.020
Cocoons	24	-	1.00	−0.337	0.018	1.820	0.020
Stony reach – Young leeches	23	-	0.99	-0.027	0.022	1.520	0.050
Mature leeches	24	-	0.99	0.025	0.012	1.470	0.040
Young + mature leeches	47	-	0.99	0.017	0.014	1.480	0.030
Cocoons	24	-	1.00	-0.292	0.017	1.800	0.020
O2 Ensemble TPLs ($NQ = 5$) of earthworms (principally *Aporrectodea caliginosa*, *A. longa* and *Lumbricus terrestris*) adults and juveniles in pasture and arable fields in Scotland. Data from Table 1 in Boag et al. (1994). TPL data ($NQ = 20$) Data of planarian *Artioposthia triangulate* from Blackshaw (1990)							
A. triangulata	8	∼0.05 - ∼1.26	0.96	0.034	-	1.071	*
Earthworms							
Arable – Adults	68	-	0.858	-	-	1.380	0.100
Juveniles	68	-	0.921	-	-	1.430	0.070
Total	68	-	0.925	-	-	1.486	0.070
Pasture – Adults	132	-	0.806	-	-	1.570	0.080
Juveniles	132	-	0.875	-	-	1.466	0.060
Total	132	-	0.867	-	-	1.530	0.070
Combined – Adults	200	-	0.834	-	-	1.486	0.060
Juveniles	200	-	0.907	-	-	1.443	0.050
Total	200	-	0.899	-	-	1.501	0.050
P2 Mixed-species spatial TPLs ($NQ = 64$) of earthworms from soil in pastures and savannah in Colombia. Data from Tables 1 and 2 in Jiménez et al. (2001)							
Pasture	34	0.09 - 102.5	0.981	0.983	0.035	1.291	0.044
Savannah	29			0.957	0.069	1.272	0.097
Q2 Ensemble TPLs ($NQ = 5$) of indicates polychaetes and oligochaetes in sediments around the Balearic Islands are similarly distributed similar although polychaetes were more numerous. Data from Table 2 in Deudero and Vincx (2000)							
Oligochaetes	5	0.80 - 26.4	0.973	1.167	0.177	1.313	0.175
Polychaetes	7	1.60 - 226	0.979	0.940	0.200	1.560	0.142
Combined	12	0.80 - 226	0.976	1.015	0.129	1.495	0.102
R2 Spatial TPL ($NQ = 3$) of polychaetes settling on settlement strips at 3 stations during the recovery from a hypoxia event in Mariager Fjord, Denmark. Data from Table 1 in Hansen et al. (2002)							
Polychaetes	25	170 - 702,657	0.952	−0.009	0.494	1.739	0.111
S2 Ensemble TPLs of ($NQ = 3$) abundance (#/10 cm^2) and biomass (μg C/10 cm^2) of polychaetes in the Arctic Ocean. Data from Tables 3 and 4 in Bessière et al. (2007)							
Abundance	16	0.50 - 18.0	0.930	0.126	0.077	1.383	0.135
Biomass	14	0.35 - 245	0.864	0.359	0.377	1.646	0.240

	NB	Range of means	r	A	SE[A]	b	SE[b]

T2 Ensemble TPLs ($NQ = 3$ or 5) of abundance (#/sample) and biomass (mg/sample) of polychaetes at 2 locations in the Irish Sea. Data from Tables 3 and 4 in Schratzberger et al. (2008)

	NB	Range of means	r	A	SE[A]	b	SE[b]
Celtic Deep – Polychaetes	9	25.0 - 99.0	0.567	−1.275	1.194	2.285	0.712
Biomass	9	150 - 1129	0.682	−2.424	1.963	2.672	0.739
NW Irish Sea – Polychaetes	9	5.00 - 166	0.915	−1.188	0.435	2.073	0.316
Biomass	9	140 - 1489	0.951	−1.726	0.746	2.394	0.281
Combined – Polychaetes	18	5.00 - 166	0.890	−1.364	0.400	2.281	0.261
Biomass	18	140 - 1489	0.841	−1.995	0.898	2.503	0.338

U2 Spatial TPL ($NQ = 8$) of the behavioral response of a population of polychaetes (*Capitella capitata*) that removed their tubes when subjected to attack by shrimps and restored them after the shrimps were removed and immigration was permitted. Data from Fig. 1 in Bonsdorff and Pearson (1997)

	NB	Range of means	r	A	SE[A]	b	SE[b]
C. capitata	21	2.82 - 142	0.967	0.641	0.143	1.409	0.083

V2 Four ensemble TPLs ($NQ = 90$) with means and variances computed 4 different ways. *Arenicola* worm casts on a sandy shore at Millport, Isle of Cumbrae, Scotland

	NB	Range of means	r	A	SE[A]	b	SE[b]
Arenicola/overlapping quadrat	6	1.13 - 38.0	0.997	0.101	0.060	1.729	0.062
Arenicola/nonoverlapping sample	6	1.12 - 18.1	0.997	0.133	0.045	1.689	0.062
Arenicola/m² (quadrat)	6	0.50 - 1.13	0.886	−0.060	0.105	2.501	0.580
Arenicola/m² (sample)	6	0.37 - 1.65	0.944	−0.063	0.071	1.778	0.292

References

Abd-Elgawad, M.M., 1992. Spatial distribution of the phytonematode community in Egyptian citrus groves. Fundam. Appl. Nematol. 15, 367–373.

Abd-Elgawad, M.M., 2014. Spatial patterns of *Tuta absoluta* and heterorhabditid nematodes. Russ. J. Nematol. 22, 89–100.

Abd-Elgawad, M.M., Hasabo, S.A., 1995. Spatial distribution of the phytonematode community in Egyptian berseem clover (*Trifolium alexandrinum* L.) fields. Fundam. Appl. Nematol. 18, 329–334.

Adamson, M.L., Noble, S., 1992. Structure of the pinworm (Oxyurida, Nematoda) guild in the hindgut of the American cockroach, *Periplaneta americana*. Parasitology 104, 497–507.

Allsopp, P.G., 1990. Sequential sampling plans for nematodes affecting sugar cane in Queensland. Aust. J. Agric. Res. 41, 351–358.

Anscombe, F.J., 1950. Soil sampling for potato root eelworm cysts. Ann. Appl. Biol. 37, 286–295.

Avendano, F., Schabenberger, O., Pierce, F.J., Melakeberhan, H., 2003. Geostatistical analysis of field spatial distribution patterns of soybean cyst nematode. Agron. J. 95, 936–948.

Avendano, F., Pierce, F.J., Melakeberhan, H., 2004. The relationship between soybean cyst nematode seasonal population dynamics and soil texture. Nematology 6, 511–525.

Bal, H.K., Taylor, R.A.J., Grewal, P.S., 2014. Ambush foraging entomopathogenic nematodes employ 'sprinters' for long-distance dispersal in the absence of hosts. J. Parasitol. 100, 422–432.

Bedding, R.A., Akhurst, R.J., 1975. A simple technique for rhe detecrion of insect parasitic rhabdirid nematodes in soil. Nematologica 2 (I), 109–110.

Beier, S., Traunspurger, W., 2003. Temporal dynamics of meiofauna communities in two small submountain carbonate streams with different grain size. Hydrobiologia 498, 107–131.

Bert, W., Messiaen, M., Hendrickx, F., Manhout, J., de Bie, T., Borgonie, G., 2007. Nematode communities of small farmland ponds. Hydrobiologia 583, 91–105.

Bessière, A., Nozais, C., Brugel, S., Demers, S., Desrosiers, G., 2007. Metazoan meiofauna dynamics and pelagic–benthic coupling in the Southeastern Beaufort Sea, Arctic Ocean. Polar Biol. 30, 1123–1135.

Blackshaw, R.P., 1990. Studies on *Artioposthia trangulata* (Dendy) (Tricladidae: Terricola), a predator of earthworms. Ann. Appl. Biol. 116, 169–176.

Boag, B., 1974. Nematodes associated with forest and woodland trees in Scotland. Ann. Appl. Biol. 77, 41–50.

Boag, B., Brown, D.J.F., 1985. Soil sampling for virus-vector nematodes. Asp. Appl. Biol. 10, 183–189.

Boag, B., Topham, P.B., 1984. Aggregation of plant-parasitic nematodes and Taylor's power law. Nematologica 30, 348–357.

Boag, B., Brown, D.J.F., Topham, P.B., 1987. Vertical and horizontal distribution of virus-vector nematodes and implications for sampling procedures. Nematology 33, 83–96.

Boag, B., Topham, P.B., Webster, R., 1989. Spatial distribution on pasture of infective larvae of the gastro-intestinal nematode parasites of sheep. Int. J. Parasitol. 19, 681–685.

Boag, B., Hackett, C.A., Topham, P.B., 1992. The use of Taylor's power law to describe the aggregated distribution of gastro-intestinal nematodes in sheep. Int. J. Parasitol. 22, 267–270.

Boag, B., Legg, R.K., Neilson, R., Palmer, L.F., Hackett, C.A., 1994. The use of Taylor's power law to describe the aggregated distribution of earthworms in permanent pasture and arable soil in Scotland. Pedobiologia 38, 303–306.

Boag, B., Palmer, L.F., Neilson, R., Legg, R., Chambers, S.J., 1997. Distribution, prevalence and intensity of earthworm populations in arable land and grassland in Scotland. Ann. Appl. Biol. 130, 153–165.

Boag, B., Lello, J., Fenton, A., Tompkins, D.M., Hudson, P.J., 2001. Patterns of parasite aggregation in the wild European rabbit (*Oryctolagus cuniculus*). Int. J. Parasitol. 31, 1421–1428.

Bongers, T., 1990. The maturity index: an ecological measure of environmental disturbance based on nematode species composition. Oecologia 83, 14–19.

Bonsdorff, E., Pearson, T.H., 1997. The relative impact of physical disturbance and predation by *Crangon crangon* on population density in *Capitella capitata*: an experimental study. Ophelia 46, 1–10.

Campbell, J.F., Lewis, E., Yoder, F., Gaugler, R., 1995. Entomopathogenic nematode (Heterorhabditidae and Steinernematidae) seasonal population dynamics and impact on insect populations in turfgrass. Biol. Control 5, 598–606.

Campbell, J.F., Orza, G., Yoder, F., Lewis, E., Gaugler, R., 1998. Spatial and temporal distribution of endemic and released entomopathogenic nematode populations in turfgrass. Entomol. Exp. Appl. 86, 1–11.

Campos-Herrera, R., Gómez-Ros, J.M., Escuer, M., Cuadra, L., Barrios, L., Gutiérrez, C., 2008. Diversity, occurrence, and life characteristics of natural entomopathogenic nematode populations from La Rioja (Northern Spain) under different agricultural management and their relationships with soil factors. Soil Biol. Biochem. 40, 1474–1484.

Caveness, F.E., Jensen, H.J., 1955. Modification of the centrifugal-flotation technique for the isolation and concentration of nematodes and their eggs from soil and plant tissues. Proc. Helminthol. Soc. Wash. 22, 87–89.

Ciancio, A., Quénéhervé, P., 2000. Population dynamics of *Meloidogyne incognita* and infestation levels by *Pasteuria penetrans* in a naturally infested field in Martinique. Nematropica 30, 77–86.

Clark, S.J., Perry, J.N., 1994. Small sample estimation for Taylor's power law. Environ. Ecol. Stat. 1, 287–302.

Cobb, N.A., 1915. Nematodes and Their Relationships. Yearbook of Agriculture for 1914. Government Printing Office, Washington, DC.

Cobb, N.A., 1918. Estimating the Nema Population of the Soil. Agricultural Technical Circular. USDA, Washington, DC.

Cohen, J.E., 2014. Taylor's law and abrupt biotic change in a smoothly changing environment. Theor. Ecol. 7, 77–86.

Dabiré, R.K., Ndiaye, S., Mounport, D., Matielle, T., 2007. Relationships between abiotic soil factors and epidemiology of the biocontrol bacterium *Pasteuria penetrans* in a root-knot nematode *Meloidogyne javanica*-infested Weld. Biol. Control 40, 22–29.

Deudero, S., Vincx, M., 2000. Sublittoral meiobenthic assemblages from disturbed and non-disturbed sediments in the Balearics. Sci. Mar. 64, 285–293.

Duncan, L.W., McCoy, C.W., 1996. Vertical distribution in soil, persistence, and efficacy against citrus root weevil (Coleoptera: Curculionidae) of two species of entomogenous nematodes (Rhabditida: Steinernematidae; Heterorhabditidae). Environ. Entomol. 25, 174–178.

Duncan, L.W., Ferguson, J.J., Dun, R.A., Noling, J.W., 1989. Application of Taylor's power law to sample statistics of *Tylenchulus semipenetrans* in Florida citrus. J. Nematol. Suppl. 21, 707–711.

Duncan, L.W., Graham, J.H., Timmer, L.W., 1993. Seasonal patterns associated with *Tylenchulus semipenetrans* and *Phytophthora parasitica* in the citrus rhizospher. Phytopathology 83, 573–581.

Duncan, L.W., El-Morshedy, M.M., McSorley, R., 1994. Sampling citrus fibrous roots and *Tylenchulus semipenetrans*. J. Nematol. 26, 442–451.

Duncan, L.W., McCoy, C.W., Terranova, A.C., 1996. Estimating sample size and persistence of entomogenous nematodes in sandy soils and their efficacy against the larvae of *Diaprepes abbreviatus* in Florida. J. Nematol. 28, 56–67.

Efron, D., Nestel, D., Glazer, I., 2001. Spatial analysis of entomopathogenic nematodes and insect hosts in a citrus grove in a semi-arid region in Israel. Environ. Entomol. 30, 254–261.

Elliott, J.M., 1986. Spatial distribution and behavioural movements of migratory trout *Salmo trutta* in a Lake District stream. J. Anim. Ecol. 55, 907–922.

Elliott, J.M., 2002. A quantitative study of day-night changes in the spatial distribution of insects in a stony stream. J. Anim. Ecol. 71, 112–122.

Elliott, J.M., 2004. Contrasting dynamics in two subpopulations of a leech metapopulation over 25 year-classes in a small stream. J. Anim. Ecol. 73, 272–282.

Ettema, C.H., Bongers, T., 1993. Characterization of nematode colonization and succession in disturbed soil using the maturity index. Biol. Fertil. Soils 16, 79–85.

Ettema, C.H., Coleman, D.C., Velidis, G., Lowrance, R., Rathbun, S.L., 1998. Spatiotemporal distributions of bacterivorous nematodes and soil resources in a restored riparian wetland. Ecology 79, 2721–2734.

Evans, N.A., 1983. The population biology of *Hymenolepis tenerrima* (Linstow 1882) Fuhrmann 1906 (Cestoda, Hymenolepididae) in its intermediate host *Herpetocypris reptans* (Ostracoda). Parasitol. Res. 69, 105–111.

Fan, X., Hominick, W.M., 1991. Efficiency of the *Galleria* (wax moth) baiting technique for recovering infective stages of entomopathogenic rhabditids (Steinernematidae and Heterorhabditidae) from sand and soil. Rev. Nematol. 14, 381–387.

Ferris, H., 2010. Form and function: metabolic footprints of nematodes in the soil food web. Eur. J. Soil Biol. 46, 97–104.

Ferris, H., Mullens, T.A., Foord, K.E., 1990. Stability and characteristics of spatial description parameters for nematode populations. J. Nematol. 22, 427–439.

Ferris, H., Eyre, M., Venette, R.C., Lau, S.S., 1996. Population energetics of bacterialfeeding nematodes: stage-specific development and fecundity rates. Soil Biol. Biochem. 28, 217–280.

Flegg, J.M., Hooper, D.J., 1970. Extraction of free-living stages from soil. In: Southey, J.F. (Ed.), Laboratory Methods for Work With Plant and Soil Nematodes. HMSO, London, pp. 5–22.

Fretwell, S.D., Lucas, H.L., 1969. On territorial behavior and other factors influencing habitat distribution in birds. I. Theoretical development. Acta Biotheor. 19, 16–36.

Gaba, S., Ginot, V., Cabaret, J., 2005. Modelling macroparasite aggregation using a nematode-sheep system: the Weibull distribution as an alternative to the negative binomial distribution? Parasitology 131, 393–401.

Gallucci, F., Steyaert, M., Moens, T., 2005. Can field distributions of marine predacious nematodes be explained by sediment constraints on their foraging success? Mar. Ecol. Prog. Ser. 304, 167–178.

Good, P.I., 2010. Permutation, Parametric, and Bootstrap Tests of Hypotheses. Springer-Verlag, New York.

Gorres, J.H., Dichiaro, M.J., Lyons, J.B., Amador, J.A., 1998. Spatial and temporal patterns of soil biological activity in a forest and an old field. Soil Biol. Biochem. 30, 219–230.

Grear, D.A., Hudson, P., 2011. The dynamics of macroparasite host-self-infection: a study of the patterns and processes of pinworm (Oxyuridae) aggregation. Parasitology 138, 619–627.

Green, C.D., 1979. Aggregated distribution of *Ditylenchus dipsaci* on broad bean seeds. Ann. Appl. Biol. 92, 271–274.

Green, R.H., 1966. Measurement of non-randomness in spatial distributions. Res. Popul. Ecol. 8, 1–7.

Grewal, P.S., Lewis, E.E., Gaugler, R., Campbell, J.F., 1994. Host finding behavior as a predictor of foraging strategy in entomopathogenic nematodes. Parasitology 108, 207–215.

Griffiths, B.S., Young, I.M., Boag, B., 1991. Nematodes associated with the rhizosphere of barley (*Hordeum vulgare*). Pedobiologia 35, 265–272.

Gruner, L., Mandonnet, N., Bouix, J., Vu Tien Khang, J., Cabaret, T.J., Hoste, H., Kerboeuf, D., Barnouin, J., 1994. Worm population characteristics and pathological changes in lambs after a single or trickle infection with *Teladorsagia circumcincta*. Int. J. Parasitol. 24, 347–356.

Hansen, B.W., Stenalt, E., Petersen, J.K., Ellegaard, C., 2002. Invertebrate re-colonisation in Mariager Fjord (Denmark) after severe hypoxia. I. Zooplankton and settlement. Ophelia 56 (3), 197–213.

Harrison, J.M., Green, C.D., 1976. Comparison of centrifugal and other methods for standardization of extraction of nematodes from soil. Ann. Appl. Biol. 82, 299–308.

Harvey, S.C., Paterson, S., Viney, M.E., 1999. Heterogeneity in the distribution of *Strongyloides ratti* infective stages among the faecal pellets of rats. Parasitology 119, 227–235.

Healy, M.J.R., Taylor, L.R., 1962. Tables for power-law transformations. Biometrika 49, 557–559.

Heininger, P., Höss, S., Claus, E., Pelzer, J., Traunspurger, W., 2007. Nematode communities in contaminated river sediments. Environ. Pollut. 146, 64–76.

Hodasi, J.K.M., 1969. Comparative studies on the helminth fauna of native and introduced domestic fowls in Ghana. J. Helminthol. 43, 35–52.

Holling, C.S., 1959. Some characteristics of simple types of predation and parasitism. Can. Entomol. 91, 385–398.

Hua, E., Zhang, Z., Zhou, H., Mu, F., Li, J., Zhang, T., Cong, B., Liu, X., 2016. Meiofauna distribution in intertidal sandy beaches along China shoreline (18°–40°N). J. Ocean Univ. China 15, 19–27.

Jaenike, J., 1994. Aggregations of nematode parasites within *Drosophila*: proximate causes. Parasitology 108, 569–577.

Jaenike, J., James, A.C., 1991. Aggregation and the coexistence of mycophagous drosophila. J. Anim. Ecol. 60, 913–928.

Jikumaru, S., Togashi, K., 2001. Transmission of *Bursaphelenchus mucronatus* (Nematoda: Aphelenchoididae) through feeding wounds by *Monochamus saltuarius* (Coleoptera: Cerambycidae). Nematology 3, 325–333.

Jiménez, J.J., Rossi, J.-P., Lavelle, P., 2001. Spatial distribution of earthworms in acid-soil savannas of the eastern plains of Colombia. Appl. Soil Ecol. 17, 267–278.

Koppenhöfer, A.M., Kaya, H.K., 1996. Coexistence of two steinernematid nematode species (Rhabditida: Steinernematidae) in the presence of two host species. Appl. Soil Ecol. 4, 221–230.

Koppenhöfer, A.M., Campbell, J.F., Kaya, H.K., Gaugler, R., 1998. Estimation of entomopathogenic nematode population density in soil by correlation between bait insect mortality and nematode penetration. Fundam. Appl. Nematol. 21, 95–102.

Lagrue, C., Poulin, R., Cohen, J.E., 2015. Parasitism alters three power laws of scaling in a metazoan community: Taylor's law, density-mass allometry, and variance-mass allometry. Proc. Natl. Acad. Sci. U. S. A. 112, 1791–1796.

Lazarova, S.S., de Goede, R.G.M., Peneva, V.K., Bongers, T., 2004. Spatial patterns of variation in the composition and structure of nematode communities in relation to different microhabitats: a case study of *Quercus dalechampii* Ten. forest. Soil Biol. Biochem. 36, 701–712.

Ma, Z., 2015. Power law analysis of the human microbiome. Mol. Ecol. 24, 5428–5445.

MacGuidwin, A.E., Stanger, B.A., 1991. Changes in vertical distribution of *Pratylenchus scribneri* under potato and corn. J. Nematol. 23, 73–81.

McSorley, R., Dickson, D.W., 1991. Determining consistency of spatial dispersion of nematodes in small plots. J. Nematol. 23, 65–72.

McSorley, R., Parrado, J.L., 1981. Effect of sieve size on nematode extraction efficiency. Nematropica 11, 165–174.

McSorley, R., Parrado, J.L., 1982. Relationship between 2 nematode extraction techniques on 2 Florida soils. Proc. Soil Crop Sci. Soc. Fla. 41, 30–36.

McSorley, R., Dankers, W.H., Parrado, J.L., Reynolds, J.S., 1985. Spatial distribution of the nematode community in perrine marl soils. Nematropica 15, 77–92.

Moens, T., Van Gansbeke, D., Vincx, M., 1999. Linking estuarine nematodes to their suspected food. A case study from the Westerschelde Estuary (south-west Netherlands). J. Mar. Biol. Assoc. U. K. 79, 1017–1027.

Mokievskii, V.O., Udalov, A.A., Azovskii, A.I., 2007. Quantitative distribution of meiobenthos in deep-water zones of the World Ocean. Mar. Biol. 47, 797–813.

Moodley, L., Chen, G., Heip, C., Vincx, M., 2000. Vertical distribution of meiofauna in sediments from contrasting sites in the Adriatic Sea: clues to the role of abiotic versus biotic control. Ophelia 53, 203–212.

Morand, S., Guégan, J.F., 2000. Distribution and abundance of parasite nematodes: ecological specialisation, phylogenetic constraint or simply epidemiology? Oikos 88, 563–573.

Morgan, G.D., MacGuidwin, A.E., Zhu, J., Binning, L.K., 2002. Population dynamics and distribution of root lesion nematode (*Pratylenchus penetrans*) over a three-year potato crop rotation. Agron. J. 94, 1146–1155.

Neilson, R., Boag, B., Hackett, C.A., 1993. Observations on the use of Taylor's power law to describe the horizontal spatial distribution of marine nematodes in an intertidal estuarine environment. Russ. J. Nematol. 1, 55–64.

Noe, J.P., Campbell, C.L., 1985. Spatial pattern analysis of plant-parasitic nematodes. J. Nematol. 17, 86–93.

Pan, W., 2001. Akaike's information criterion in generalized estimating equations. Biometrics 57, 120–125.

Park, S.-J., Cheng, Z., Gardener, B.B.G., Grewal, P.S., 2010. Are nematodes effective bioindicators of soil conditions and processes along distance from roads and age of development in urban areas? J. Environ. Indic. 5, 28–47.

Park, S.-J., Taylor, R.A.J., Grewal, P.S., 2013. Spatial organization of soil nematode communities in urban landscapes: Taylor's power law reveals life strategy characteristics. Appl. Soil Ecol. 64, 214–222.

Perry, J.N., 1981. Taylor's power law for dependence of variance on mean in animal populations. Appl. Stat. 30, 254–263.

Peters, A., 1994. Interaktionen zwischen den Pathogenitätsmechanismen entonopathogener Nematoden und den Abwehrmechanismen von Schnakenlarven (*Tipula* spp.) sowie Möglichkeiten zur Virulenzsteigerung der Nematoden durch Selektion [Interactions between the pathogenicity mechanisms of entomopathogenic nematodes and the defense mechanisms of leatherjackets (*Tipula* spp.) as well as opportunities for increasing the virulence of nematodes by selection] (PhD Thesis). University of Kiel, Germany.

Pfannkuche, O., Thiel, H., 1988. Sample processing. In: Higgins, R.P., Thiel, H. (Eds.), Introduction to the Study of Meiofauna. Smithsonian Institution Press, Washington, DC, pp. 134–145.

Ristau, K., Traunspurger, W., 2011. Relation between nematode communities and trophic state in southern Swedish lakes. Hydrobiologia 663, 121–133.

Ruess, L., 1995. Studies on the nematode fauna of an acid forest soil—spatial-distribution and extraction. Nematologica 41, 229–239.

Ruess, L., Michelsen, A., Schmidt, I.K., Jonasson, S., Dighton, J., 1998. Soil nematode fauna of a subarctic heath: potential nematicidal action of plant leaf extracts. Appl. Soil Ecol. 7, 111–124.

Sarabeev, V., Balbuena, J.A., Morand, S., 2017. Testing the enemy release hypothesis: abundance and distribution patterns of helminth communities in grey mullets (Teleostei: Mugilidae) reveal the success of invasive species. Int. J. Parasitol. 47, 687–696.

Schomaker, C.H., Been, T.H., 1999. A model for infestation foci of potato cyst nematodes *Globodera rostochiensis* and *G. pallida*. Phytopathology 89, 583–590.

Schratzberger, M., Maxwell, T.A.D., Warr, K., Ellis, J.R., Rogers, S.I., 2008. Spatial variability of infaunal nematode and polychaete assemblages in two muddy subtidal habitats. Mar. Biol. 153, 621–642.

Shaw, D.J., Dobson, A.P., 1995. Patterns of macroparasite abundance and aggregation in wildlife populations: a quantitative review. Parasitology 111, S111–S133.

Soetaert, K., Heip, C., Vincx, M., 1991. Diversity of nematode assemblages along a Mediterranean deep sea transect. Mar. Ecol. Prog. Ser. 75, 275–282.

Sohlenius, B., 1980. Abundance, biomass and contribution to energy flow by soil nematodes in terrestrial ecosystems. Oikos 34, 186–194.

Somerfield, P.J., Dashfield, S.L., Warwick, R.M., 2007. Three-dimensional spatial structure: nematodes in a sandy tidal flat. Mar. Ecol. Prog. Ser. 336, 177–186.

Southwood, T.R.E., Henderson, P.A., 2000. Ecological Methods, third ed. Blackwell Science, Oxford, UK.

Spiridonov, S.E., Voronov, D.A., 1995. Small scale distribution of *Steinernema feltiae* juveniles in cultivated soil. In: Griffin, C.T., Gwynn, R.L., Masson, J.P. (Eds.), Ecology and Transmission Strategies of Entomopathogenic Nematodes. COST 819. European Commission, Luxembourg, pp. 36–41.

Sriwati, R., Takemoto, S., Futai, K., 2007. Cohabitation of the pine wood nematode, *Bursaphelenchus xylophilus*, and fungal species in pine trees inoculated with *B. xylophilus*. Nematology 9, 77–86.

Stanton, J.M., Pattison, A.B., Kopittke, R.A., 2001. A sampling strategy to assess banana crops for damage by *Radopholus similis* and *Pratylenchus goodeyi*. Aust. J. Exp. Agric. 41, 675–679.

Stear, M.J., Bairden, K., Bishop, S.C., Gettinby, G., McKellar, Q.A., Park, M., Strain, S., Wallace, D.S., 1998. The processes influencing the distribution of parasitic nematodes among naturally infected lamb. Parasitology 117, 165–171.

Stear, M.J., Abuagob, O., Benothman, M., Bishop, S.C., Innocent, G., Kerr, A., Mitchell, S., 2006. Variation among faecal egg counts following natural nematode infection in Scottish Blackface lambs. Parasitology 132, 275–280.

Steyaert, M., Vanaverbeke, J., Vanreusel, A., Barranguet, C., Lucas, C., Vincx, M., 2003. The importance of fine-scale, vertical profiles in characterising nematode community structure. Estuar. Coast. Shelf Sci. 58, 353–366.

Stuart, R.J., Gaugler, R., 1994. Patchiness in populations of entomopathogenic nematodes. J. Invertebr. Pathol. 64, 39–45.

Stuart, R.J., Barbercheck, M.E., Grewal, P.S., Taylor, R.A.J., Hoy, C.W., 2006. Population biology of entomopathogenic nematodes: concepts, issues, and models. Biol. Control 38, 80–102.

Taylor, L.R., Woiwod, I.P., Perry, J.N., 1978. The density-dependence of spatial behaviour and the rarity of randomness. J. Anim. Ecol. 47, 383–406.

Taylor, R.A.J., 1999. Sampling entomopathogenic nematodes and measuring their spatial distribution. In: Gwynn, R.L., Smits, P.H., Griffin, C., Ehlers, R.-U., Boemare, N., Masson, J.-P. (Eds.), Application and Persistence of Entomopathogenic Nematodes (EUR 18873 EN). European Commission, Brussels, pp. 43–60.

Trotter, D., Webster, J.M., 1983. Distribution and abundance of marine nematodes on the kelp *Macrocystis integrifolia*. Mar. Biol. 78, 39–43.

Uhlig, G., Thiel, H., Gray, J.S., 1973. The quantitative separation of meiofauna. A comparison of methods. Helgoländer Meeresun. 25, 173–195.

Viketoft, M., 2007. Plant induced spatial distribution of nematodes in a semi-natural grassland. Nematology 9, 131–142.

Warren, J.E., Linit, M.J., 1992. Within-wood spatial dispersion of the pinewood nematode, *Bursaphelenchus xylophilus*. J. Nematol. 24, 489–494.

Webster, R., Boag, B., 1992. Geostatistical analysis of cyst nematodes in soil. J. Soil Sci. 43, 583–595.

Wheeler, T.A., Kenerley, C.M., Jeger, M.J., Starr, J.L., 1987. Effect of quadrat and core sizes on determining the spatial pattern of *Criconemella sphaerocephalus*. J. Nematol. 19, 413–419.

Wheeler, T.A., Madden, L.V., Riedel, R.M., Rowe, R.C., 1994. Distribution and yield-loss relations of *Verticillium dahliae*, *Pratylenchus penetrans*, *P. scribneri*, *P crenatus* and *Meloidogynae hapla* in commercial potato fields. Phytopathology 84, 843–852.

Wheeler, T.A., Madden, L.V., Rowe, R.C., Riedel, R.M., 2000. Effects of quadrat size and time of year for sampling of *Verticillium dahliae* and lesion nematodes in potato fields. Plant Dis. 84, 961–966.

White, G.F., 1927. A method for obtaining infective nematode larvae from culture. Science 66, 302–303.

Wu, S.G., Wang, G.T., Xi, B.W., Gao, D., Nie, P., 2007. Population dynamics and maturation cycle of *Camallanus cotti* (Nematoda: Camallanidae) in the Chinese hooksnout carp *Opsariichthys bidens* (Osteichthyes: Cyprinidae) from a reservoir in China. Vet. Parasitol. 147, 125–131.

Yeates, G.W., 1972. Nematoda of a Danish beech forest. I. Methods and general analysis. Oikos 23, 178–189.

Zhang, M., Liang, W.-J., Zhang, X.-K., 2012. Soil nematode abundance and diversity in different forest types at Changbai Mountain, China. Zool. Stud. 51, 619–626.

Chapter 8

Insects and other arthropods

The Arthropoda is an invertebrate phylum of species with a segmented body, an exoskeleton, and multiple pairs of jointed legs, some of which may be highly modified. It is the most highly species-rich phylum with over a million described species that include the insects, arachnids, and crustaceans. They are highly diverse in morphology and lifestyle with most undergoing some form of metamorphosis. Most species are mobile as adults, but adult barnacles, for example, are sessile. Arthropods range in size from tenths of a millimeter to over a meter and in weight from a few milligrams to tens of kilograms. Many of the smaller species are parasites and some of the larger species, especially crustaceans, are important food sources.

Insects

The insects comprise the largest class of arthropods with more than a million described species and probably only <20% described. They are primarily terrestrial and found on every continent and also in fresh and salt water. All species undergo metamorphosis developing through several distinct stages. The Lepidoptera and Hymenoptera are important pollinators and probably coevolved with flowering plants. Many species are agriculturally and medically important, and consequently there are abundant data of these insects.

This chapter is in two parts: seven detailed case studies of classic and new data followed by a general survey of examples drawn mostly from agricultural entomology. The second part discusses with reference to examples, issues arising from sampling and the methods used, differences and similarities of TPLs in space, time and by life stage, the influence of predation, parasitism, and competition, and the effect of scale on TPL parameters.

Lepidoptera

Lepidoptera is the insect order that includes butterflies and moths. There are >180,000 described butterflies and moths that account for ~10% of all species. Butterflies and moths have four life stages: Egg, larva, pupa, and adult. Most adults are winged and some are long-distance migrants. Some species are important pollinators, while other butterflies and many more moth species are agricultural or forestry pests. The adults' high fecundity, often producing

Taylor's Power Law. https://doi.org/10.1016/B978-0-12-810987-8.00008-2

hundreds to thousands of eggs per female, contribute to the pest status of the larval stage. As a result of their agricultural importance, the quantitative ecology of many butterflies and moths is well studied.

EUROPEAN CORN BORER

The European corn borer (ECB; *Ostrinia* (= *Pyrausta*) *nubilalis*) is a moth pest of cereal crops, particularly maize (*Zea mays*). It is an introduced species in North America where it has invaded most of the United States and Canada east of the Rocky Mountains. The larvae feed by tunneling in the aerial parts of the maize plant, including the ear, and can drastically reduce yields if uncontrolled. Originally mostly univoltine, it is now multivoltine in much of its range.

In the autumns of 1931 and 1932, Meyers and Patch (1937) surveyed 66 counties in Indiana, Michigan, Ohio, Pennsylvania, and New York for ECB to determine sampling variability within and between fields. In the 1930s, ECB was mostly univoltine in these states. In each state, a county was selected and 20–25 randomly selected maize fields were surveyed. Within field sampling consisted of four sets of 25 consecutive plants examined for the presence of ECB. Following determination of infestation rate, 5–10 infested plants were dissected and the ECB counted. The product of infestation rate and number/plant is an estimate of the field mean of ECB per 100 plants. Meyers and Patch present equations for the variance and a worked example for the variance of a product. They provide means and variances for percentage of infested plants, the number of ECB per infested plant, and the number of ECB per 100 plants for 20 fields in Wood County, Ohio, in the Autumn 1932 survey.

TPL analysis of all three variables is shown in Fig. 8.1 (Appendix 8.A). The most obvious feature is the poor fit for the percentage of infested plants ($r = 0.61$), while the log means of both borers/infested plant and borers/100 plants are convincingly dependent on log variance. The number of borers per infected plant has a very narrow range and fairly good correlation coefficient, but the number of borers per 100 plants, the statistic assumed to be representative of the field population, extends over almost two orders of magnitude of mean and has a correlation coefficient of $r = 0.97$. The slope $b = 2.07 \pm 0.291$ for borers/infested plant indicates a high degree of aggregation, but ignores the plants with no borers. The field estimate takes the uninfested plants into account and consequently has a lower slope of $b = 1.78 \pm 0.109$ (Chapter 15). An ensemble TPL of ECB data of 18 grouped frequencies in 1054 fields surveyed in 66 counties in Michigan, Ohio, Pennsylvania, and New York conducted in 1931 and 1932 produces substantially the same estimates as the single county data (Fig. 8.1A; Appendix 8.A). This dataset from Meyers and Patch was used in LRT61. LRT et al. (1978) found it to be significantly curved due to the smaller increases in variance at the high means. Densities of ~ 10 borers per plant may be approaching the maximum number/plant.

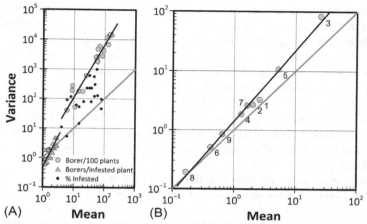

FIG. 8.1   Ensemble TPLs of European corn borer (ECB), *Ostrinia nubilalis*. (A) Wood County, Ohio, in Autumn 1932, borers per infested plant and borers per 100 plants but not percentage of infested plants produce convincing TPLs. (B) Borers/plant in 9 (numbered) Iowa fields. *(A: Data from Table 2 in Meyers and Patch (1937). B: Data from the appendix in McGuire et al. (1957).)*

LRT (1965, 1984) chose four of nine of McGuire et al.'s (1957) distributions of ECB larvae in maize plants to illustrate the continuity of change in frequency distribution of abundance with density (Fig. 1.1A). Three of the 4 distributions were good fits to McGuire et al.'s (1957) data of ECB in field corn: Poisson, negative binomial, and Neyman's A, but the third in the series, between the negative binomial and Neyman's A distributions was fit by no then-known distribution.[1]

In summer 1952, McGuire et al. dissected corn plants from a field in Northwest Iowa. Three additional fields were sampled the following summer. McGuire et al. gave the frequency distributions of the nine sample plots in the Appendix to their article. These distributions range from extremely left skewed to nearly symmetrical, but with a long right-hand tail. Ensemble analysis of $NB = 9$ distributions with $NQ = 311$–3205 samples (Fig. 8.1B) results in an ensemble TPL with slope $b = 1.20 \pm 0.036$ ($r = 0.996$), a result closely similar to LRT's (1965) ODR estimate of $b = 1.25$: the GMR estimates are given in Appendix 8.A.

WINTER MOTH

Winter moth, *Operophtera brumata*, is a native of Europe, the Caucasus, and Turkey. It is a temperate species active throughout the autumn and on warmer nights in winter. Only the males fly, and as females have vestigial wings and cannot fly, dispersal is by young larvae "ballooning" on silk threads. Winter

---

1. This distribution is fit by the Adès and Tweedie families of distributions (Chapters 2 and 3).

238 PART | II

moth is an invasive defoliator of hardwoods in North America and was first confirmed in Nova Scotia, Canada, in the early 1930s. It is now well established in the Maritime Provinces, New England, and the Pacific Northwest.

In 1954, Embree (1961, 1965) established nine study plots near Bridgewater, Nova Scotia, and sampled for winter moth life stages regularly for 9 years to develop a life table for the moth. Two other sites were sampled in 1957 and 1958. Ten trees were monitored at 8 sites, 20 trees at one site, and eight trees each at the two extra sites. All the sites were typical of hardwood forests in the area, with northern red oak (*Quercus rubra*) dominant. To develop a lifetable, an estimate is needed of all individuals entering each stage and by difference establishing the survival of each stage.

Emergence traps were used to estimate the total number of adult males and females. Embree estimated the number of females per tree by trapping them as they ascended the trees to call for mates and oviposit. The number of eggs per tree was determined by estimating the average number of eggs per female per tree and from an empirical calibration curve relating egg load to abdomen width of each female caught ascending the trap trees. Larval density was estimated from larvae on 12 leaf clusters per tree collected at random from each sample tree and the number of leaf clusters per tree estimated from branch diameter. Prepupae dropping from the crown to pupate in the soil were caught in traps beneath the sample trees. Survival of pupae was estimated by experiment. With these data, Embree was able to create a lifetable for each year of his survey. He gave the mean and variances of the number of 1st instars/leaf cluster/tree and the number of females trapped/tree. Also included are data of mean and variance of eggs/in^2 of lichen and prepupae falling onto, and all adults emerging from 1ft^2 of soil, respectively. In most cases, the sample size was 10 trees.

Appendix 8.B has estimates of spatial TPLs of Embree's data of all adult moths, female moths, eggs, 1st instar larvae, and prepupae: adults and prepupae are illustrated in Fig. 8.2. Adult moths recovered as they emerged from their pupation sites were very aggregated with $b = 2.39$, while females climbing trees to call, mate and oviposit, eggs laid, 1st instar larvae on leaves and prepupae falling to pupate have progressively lower slopes. Throughout the lifecycle of winter moth, the degree of aggregation as measured by b declined. As the emerging adults had a very high level of aggregation but the prepupal stage was only slightly aggregated, the question arises of whether the pupae become aggregated by active congregation of the prepupae. Embree determined the pupal survival by experiment and not by sampling. One reason he did this was the soil around the sample trees was very stony, which may indicate available pupation sites were limited, resulting in the prepupae congregating in areas with sufficient soil to burrow into. In any event, the progressive reduction in aggregation through the lifecycle suggests spatial rearrangement in and near the trees by a dispersive process.

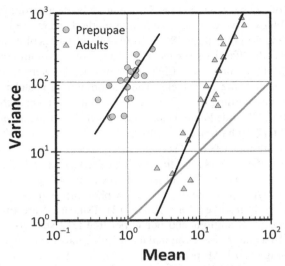

FIG. 8.2 Spatial TPLs of winter moth adults emerging from the soil and prepupae preparing to enter the soil. Density estimates of #/ft² averaged over 10 trees in forests in Nova Scotia. *(Data from Figs. 25 and 26 in Embree (1961).)*

GYPSY MOTH

Gypsy moth *Lymantria dispar* is a Palaearctic species that disperses primarily by silking first instar larvae. Adult females from the western Palaearctic are flightless; females in the eastern Palaearctic are capable of limited flight. Specimens originating from France were accidentally released from a home in Medford, Massachusetts, in 1868. Attempts to contain this exotic moth were unsuccessful, with the result that its range has expanded down the Appalachian Mountain system, reaching south central Pennsylvania by the late 1970s. It will feed on almost all deciduous hardwood foliage, on many shrubs and even evergreen foliage. In the 1970s and 1980s, it frequently caused near-complete defoliation over large areas of the Appalachian Mountains, including Central and South Central Pennsylvania where there were few natural enemies to control it. More recently, the fungus *Entomophaga maimaiga* has emerged as a potent mortality factor. Gypsy moth has been the subject of intensive research for most of the time since its release at Medford.

Three projects to quantify gypsy moth population density were conducted in Pennsylvania between 1979 and 1987. The first dataset, collected in southern Pennsylvania during the summer and autumn of 1979, was part of a project to detect new, low-density gypsy moth infestations at and ahead of the leading edge (Embody and Bachlor, 1980). Data consisted of pheromone trap catches at 2174 sites spaced on a 2-km grid across 13 counties. Egg mass counts were taken in the following autumn at a subset of 101 2 × 2 km² blocks chosen to cover a range of catches from 25 to 843 male moth by the pheromone trap at the center of the block. Two egg mass-sampling methods were used: fixed-radius 0.01-ha plots and 5-min walk. An 8×8 array of fixed-radius plots was surveyed at the 40 lowest adult density sites and a 7×7 array of fixed-radius surveys were made in the 61 remaining sites. Eighteen 5-min walks were distributed randomly within each of the 101 blocks. The pheromone trap catches are divided into $NB = 67$ sets according to the 7.5' USGS quad map on which they fall, with $30 \leq NQ \leq 40$ sites per quad map. Data comprise $NB = 101$ means and variances of $NQ = 49$ or 64 0.01 ha fixed-area plot and $NQ = 18$ 5-min walks.

A survey in autumn 1984 and spring 1985 was conducted at 14 sites in central Pennsylvania 5–10 years after the leading edge had passed and populations were close to their maxima. The objective was to relate egg mass and subsequent pupal densities with adult catches in pheromone traps (discussion of the adult catches is reserved for Chapter 13). Data used are of egg masses per tree counted at 21 stations and pupae per tree at 12 stations in each of 14 36-ha blocks. Egg masses were counted using binoculars in variable radius plots established using a 20 BAF prism (Bell, 2002) to select trees for inspection. Twenty-five trees at 12 stations distributed between the variable radius plots were selected and banded with burlap to provide pupation sites. In June, when the larvae had pupated, the burlap bands were examined and the pupae counted and sexed. Data comprise mean and variance of male and female pupae per tree at $NB = 168$ stations ($NQ = 25$ trees/station) and mean and variance of egg masses per tree at $NB = 294$ stations ($3 \leq NQ \leq 9$ trees/station).

An experiment to calibrate the efficiency of USDA pheromone traps was conducted in summer 1987 (RAJT et al., 1991; RAJT, 2018). The experiment was conducted in Huntingdon County, Pennsylvania, an area of predominantly oak forest typical of that part of the Appalachians. The stand was close to the peak of the population wave front with high larval density having caused almost complete defoliation. The data used are the daily mean and variance of $NQ = 15$ pheromone trap counts over a $NB = 27$ day period commencing 2 days after the first detected moth flight. The primary TPL data therefore comprise a temporal sequence of spatially separated pheromone traps sampling flying male gypsy moths.

Comparison of the egg mass TPLs (Fig. 8.3A, Appendix 8.C) provides information on the relative efficiency of the methods. The three TPLs of the egg mass sampling methods differ in both slope and intercept suggesting that the methods are not equally efficient. As the variable radius sample is expressed per tree and the others per station, differences in intercept are to be expected. The differences in slope suggest differences in efficiency of the three methods, with the variable radius prism point most efficient and with the smallest intercept the most precise. It might be expected that the fixed-radius method would be the preferred method, but its slope is very significantly lower than the prism point slope ($P < 0.001$) suggesting density-dependent decline in efficiency of the fixed-radius method (RAJT, 2018). With the highest intercept and the lowest

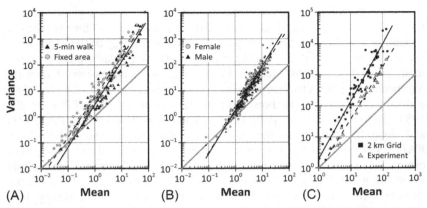

FIG. 8.3 Ensemble TPLs of gypsy moth eggmasses (A), pupae (B), and adult male moths (C). *((A) Data from unpublished field work of the author. (B) Data from Embody and Batchlor (1980). (C) Data from RAJT et al. (1991) and RAJT (2018).)*

slope the fixed-radius method is the least efficient method with the highest relative variance and lowest precision. All trees in the fixed- and variable-area methods are examined with binoculars imposing a significant "handling time" of 20–30 min for two people to examine all the trees at a station. By comparison, the 5-min walk covers a larger area but includes only egg masses visible to the observer' naked eye. Although the 5-min walk would seem to be the least controlled method and therefore the least efficient, its slightly lower slope $(P = 0.051)$ suggests density-dependent efficiency relative to the prism point method. However, its high speed suggests it may be the most cost effective, if not the most precise sampling procedure.

Neither the slopes nor the intercepts of the pupae TPLs (Fig. 8.3B) differ significantly $(P > 0.15)$ justifying a common regression and the conclusion that the burlap banding method for estimating pupal density is equally efficient for males and females. Although the range of means of males is slightly higher than females, the 8248 males and 9756 females pupated under burlap indicate a significant 54–46 female bias. The sex ratio of eggs is 50% female (Campbell, 1963, 1967), but the proportion of female pupae is known to decline from ~70% females at low density to ~40% females at very high density (Myers et al., 1998). A sex ratio of 54% females therefore indicates a moderate-to-high population density.

The number of male moths captured in the 1987 trap efficiency experiment was recorded as #/trap/day, while the 1979 leading-edge survey catches were expressed as #/trap/season as the traps had been deployed before first flight and were retrieved after the last flight in order to catch the full flight period of 4–5 weeks. Thus, the experiment is a classic spatial TPL, while the survey is an ensemble TPL (Fig. 3C), which are expected to have similar TPL slopes but different intercepts resulting from the differences in time integration and/or the overall population levels (Eisler et al., 2008). The overall average number/ trap in the survey was approximately double the number/trap in the experiment. Assuming a flight period of ~30 days, the number/trap/day of the experiment was ~15 times the survey number/trap/day, reflecting a substantially higher moth population in 1987 than in 1979. Given the suppression of density-dependent sampling efficiency due to competition at the trap (RAJT, 2018), a divergence between the power law lines is expected. The TPLs for the 1985 experiment and the 1979 survey differ in both intercept ($P \ll 0.001$) and slope ($P < 0.03$) with the 1979 survey data having the higher slope. RAJT (2018) showed that after correcting for density-dependent trapping efficiency, the experiment's true TPL slope is $b = 1.77$, which parallel line analysis shows is not different from the survey catch converted to density ($P > 0.35$): the intercepts differ at $P < 0.001$. The longer service interval of the survey traps apparently did not contribute significantly to catch suppression due to trap saturation and/or odor as the numbers were substantially <1000/trap that Elkinton (1987) identified as a threshold for significant catch suppression.

Unlike the winter moth, Gypsy moth's three life stages exhibit about the same degree of aggregation as measured by TPL, although the adult survey catch has the steepest slope and the adult experiment corrected for density-dependent efficiency is also steeper. The relationship between trap catch efficiency and the estimation of TPL is discussed in more detail in Chapter 15.

Coleoptera

The order Coleoptera contains the beetles, insects with the forewings hardened into wing cases or elytra. There are more than 400,000 described species of beetle, making the Coleoptera the single most species rich and diverse order of any metazoan phylum. They inhabit every habitat, but the oceans and poles. They are herbivores, omnivores, detritivores, and carnivores as larvae and/or adults. At least one family is parasitic on other beetles. The predatory beetles, for example ladybird beetles, are frequently used as biological control agents against plant-sucking insects, such as aphids, whiteflies, and thrips. Many beetle larvae and/or adults are agricultural or forestry pests: these species' ecologies are well studied.

WIREWORMS

Wireworms are the soil-inhabiting larvae of click beetles or Elateridae. Most wireworms are saprophagous, some are predatory, and some species, *Agriotes* and *Limonius* for example, can be serious agricultural pests. As the larvae typically spend several years in the soil, they may be controlled by plowing and crop rotation. Adults are mostly nocturnal and are rarely economically important.

In Washington State

Jones (1937) reported a project to find a time- and labor-saving system of soil sampling for wireworms in the genus *Limonius* that would give acceptably accurate results. The preferred method for estimating wireworm populations in soil consisted of sample units at randomly chosen points in a field. Data were obtained with three sizes of quadrat: $1 ft^2$, $\frac{1}{4} ft^2$, and $\frac{1}{16} ft^2$. Also investigated was the optimal number of samples: 25, 50, or 100 per field.

Jones' interest was in determining the smallest number of sampling units needed to provide reliable population estimates. Tests of field samples comparing the three quadrats' sampling errors, expressed as CV, found sampling error to be stable at 50 samples per field at low population levels but fewer samples were adequate at intermediate and high levels of population. The CV of the 1-ft^2 quadrat was consistently smaller than the smaller quadrats'. The consistency of results showed the 1-ft^2 quadrat to give a better estimate of the wireworm population than the smaller units with the $\frac{1}{4} ft^2$ more accurate than the $\frac{1}{16} ft^2$ quadrat.

The article gives the means and SEs for combinations of quadrat size and number of samples taken in seven and 25 fields. For the smaller quadrats, the means are smaller but can be standardized to $1 ft^2$ by multiplying by 4 or 16, as appropriate.

Ensemble TPLs of the data of the three sample sizes show heterogeneity in both intercept and slope, with the smallest ($NQ = 25$) and largest ($NQ = 100$) sample sizes producing steeper slopes than the intermediate ($NQ = 50$) (Appendix 8.D). Converting the 2 smaller quadrat sizes to #/ft^2 results in a single TPL (Fig. 8.4A). Analyzing Jones' data as #/quadrat and #/m^2 results in two different ensemble TPLs (Fig. 8.3B). For comparison, data were also analyzed as #/ft^2, which differs from #/m^2 only in intercept (Appendix 8.D). The results of TPL analysis can depend critically on how data are recorded and aggregated for analysis. Numbers per quadrat can give convincing TPLs even when the quadrats differ in size. Conversion to an absolute density estimate can alter the slope as well as the intercept.

In England and Wales

Wireworms, principally *Agriotes obscurus*, were the target of the Wireworm Survey of England and Wales established in 1939 (Finney, 1941, 1946). As in the United States, the Survey's purpose was to characterize wireworm spatial distribution in order to develop an efficient sampling scheme for protecting vulnerable grasslands. The Wireworm Survey established 13 districts in which 473 fields were surveyed from Autumn 1939 to Spring 1940. In most fields, 20 samples were taken with 2 randomly located in each tenth of the field to account for spatial variation. The recommended quadrat size was 8×8 inches, but in some areas quadrats of 6×6 inches or 3-in.-diameter cores were taken. To establish a rigorous sampling method, Yates and Finney (1942) analyzed the 1940–41 data to determine the sampling error and best quadrat size. They also examined the

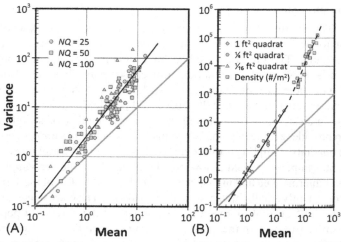

FIG. 8.4 Ensemble TPLs of wireworms sampled with different number and size of quadrat. (A) Converting the quadrat counts to #/ft² brings all 3 sample sizes together into a single ensemble plot (*NB* =). (B) *Solid line* fitted to all 3 quadrat sizes. *Hatched line* shows conversion to #/m² reduces the range of means and increases the slope. *(Data from Tables 2 and 3 in Jones (1937).)*

geographical distribution of populations and compared the populations in arable fields and pastures.

In the 1940–41 field season, 525 arable and 2272 grassland fields were sampled using 4-in. diameter cores. Yates and Finney grouped the field data in bins of mostly 100/acre increments and compared the distribution with the Poisson. The Poisson line fit the data only at the very lowest densities, elsewhere it greatly underestimated the SE. Yates and Finney give means and SEs of the field data, which were converted to mean and variance of wireworms/m² and plotted in Fig. 8.5 (Appendix 8.D). This very extensive survey resulted in spatial TPLs with slopes of $b = 1.18 \pm 0.015$ and 1.26 ± 0.034 for arable and grass fields, respectively. Conversion of 1000s/acre to #/m² affected only the intercepts. Combining the arable and grass field data resulted in an intermediate value slope and an increase in the intercept. Correlation declined slightly reflecting the small but significant difference in arable and grass field intercepts ($P \ll 0.001$) and slopes ($P < 0.05$). Ordinarily rescaling data moves the TPL plot in the variance-mean domain, but relative positions are maintained (see Chapter 15). In this example, conversion from 1000s/acre to #/m² reversed the order of the intercepts: $A_{arable} > A_{grass}$ for 1000s/acre, but $A_{grass} > A_{arable}$ for #/m². Presumably the result of rounding error, the differences are small for the original scale but larger for the transformed scale.

COLORADO POTATO BEETLE

The Colorado potato beetle (CPB; *Leptinotarsa decemlineata*) is native to North America and has invaded Europe and Asia. Its principal food plants

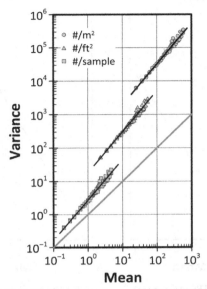

FIG. 8.5 New figure with 3 sample area measurements—Spatial TPL of wireworms sampled in 525 arable and 2272 grass fields in England and Wales in winter 1940–41. Data were binned at 100/acre intervals and converted to #/m². *(Data from Fig. 1 in Yates and Finney (1942).)*

are in the family Solanaceae, which includes potato, *Solanum tuberosum*. Both larvae and adults feed on foliage. Its high fecundity and very rapid development enables it to transition from egg to adult in as little as 3 weeks. Consequently, CPB can be a very serious pest on potatoes and other *Solanum* species such as eggplant (aubergine), peppers, and tomatoes.

Beall (1939) estimated the number of the CPB in a heavily infested field near Chatham, Ontario, on 14 August 1935. He marked out the field into a 48×48 grid of 2ft^2 units each containing two potato plants. Beall's objective was to determine the most efficient sampling plan for determining the population density of CPB. His analyses showed the population density to vary over the field with areas of high and low densities, which led him to recommend a stratified sampling scheme. Beall was also concerned with finding a transformation that would make these and other entomological data suitable for analysis of variance and used the good fit he obtained for the polynomial $V = M + KM^2$, where $1/K = k$ the negative binomial parameter, to justify the transformation $z = K^{-\frac{1}{2}}\sinh^{-1}[(Kx)^{-\frac{1}{2}}]$, which reduces to $z = x^{\frac{1}{2}}$ in the Poisson case ($K = 0$).

Beall gives the number of adult beetles in each of the 2304 sample units: data are divided into 16 aggregate units of 12×12 sample units each. Taking the mean and variance of the $NQ = 144$ units in each of $NB = 16$ aggregate units results in a TPL of slope $b = 1.60 \pm 0.132$ (Fig. 8.6A; Appendix 8.E). In a later study, Beale (1942) plotted the CPB variance on mean on linear scales and

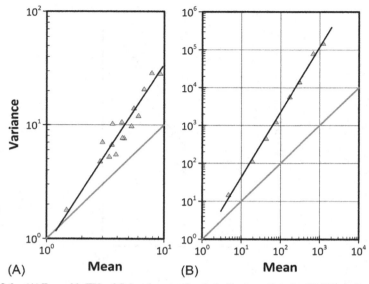

FIG. 8.6   (A) Ensemble TPL of Colorado potato beetle in Chatham, Ontario. (B) TPL derived from the sums of successive subdivisions of the 48 × 48 matrix of CPB counts: from the bottom left, single unit, 2 × 2, 3 × 3, 4 × 4, 6 × 6, 8 × 8, 12 × 12, and 16 × 16 sample units combined to compute mean and variance. *(Data from Table 6 in Beall (1939).)*

noted the strong dependence of V on M, but did not plot the data on log scales, so did not discover the polynomial was unnecessary. One of the other examples Beale used was data of European corn borer (see later).

These data were analyzed in LRT61 and LRT et al. (1978). They were also analyzed by Kendal (1995) who noted TPL's scale invariance, which suggested to him a statistical model based on diffusion-limited aggregation (DLA; Chapter 16). Kendal's test of the DLA model involved aggregating the 2304 quadrats into successively larger quadrats and computing mean and variance of each new quadrat size (see the *Arenicola* case study in Chapter 7). Kendal's largest quadrat was 9×9 units. His experiment is repeated here but with a different set of quadrats: 2304 quadrats each of a single sample unit, 576 quadrats of 2×2 units, 256 3×3 units, 144 4×4 units, 64 6×6 units, 36 8×8 units, 16 12×12 units, and 9 16×16 units. The means and variances range from $M = 4.74$, $V = 15.0$ for $NQ = 2304$ quadrats to $M = 1213$, $V = 153,732$ for $NQ = 9$ quadrats. This TPL has slope $b = 1.72 \pm 0.035$ (Fig. 8.6B; Appendix 8.E), which is not significantly different from the ensemble estimate $(P > 0.30)$.

Kendal also tested his DLA model with data of the distribution of houses in a 12-km^2 area of Toyama Prefecture, Japan (Matui, 1932), obtaining a similar result. He notes that TPLs comprising increasing quadrat sizes possess the property of self-affinity, which "implies that specific requirements are imposed on the forces that cause individuals to aggregate." Furthermore, self-affinity is a characteristic of some fractals and that therefore "b can be interpreted as a fractal dimension." In a later work, Kendal (2002) showed how the CPB data fit the compound Poisson-negative binomial distribution, one of the scale-invariant exponential dispersion models constituting the Tweedie family of distributions (Chapter 2). These data were also used by Bliss and Owen (1958) to illustrate estimation of the negative binomial constant k (Chapter 3).

JAPANESE BEETLE

In North America

The Japanese beetle (JB; *Popillia japonica*) is a scarab beetle introduced in New Jersey from Japan sometime before 1916. It has since spread to most of the country east of the Mississippi River and into Canada: populations are also established in Pacific coast areas. The beetles were also introduced into the Azores in the 1970s and more recently into Italy. As an invasive exotic, it has fewer natural enemies than in Japan and is a serious pest both as larva and adult. The larvae are soil dwellers feeding on grass roots making them a serious pest of turfgrass. They remain in the soil for $\sim$11 months and the adults are active for $\sim$1 month in the summer. Adults skeletonize the foliage of hundreds of plant species, reducing or eliminating photosynthetic area. They may also feed on fruit when present. Since the 1920s, JB population dynamics and control methods have received considerable attention. Two major studies in the 1930s produced extensive data of the distribution and abundance of Japanese beetle larvae. More recently, adult monitoring programs have also accumulated large datasets.

Larvae

In 1917, the US Department of Agriculture (USDA) set up a Japanese Beetle Laboratory at the presumed point of entry at Riverton, NJ. The lab made regular collections of JB larvae in and around Riverton and from 1927 to 1934 collections were made at four sites in New Jersey and four in Pennsylvania; all eight sites are within 20 km of Riverton on the south bank of the Delaware River opposite Philadelphia. The collections over 7 years comprising 1-ft^2 samples of sod and soil made at approximately 3-week intervals were used by Fox (1937) to document the seasonal trends in abundance. His article include data of the mean and SE computed from 1272 to 1834 1 ft^2 samples and $NB = 18$ sampling occasions. A spatial TPL analysis of these data has a slope of $b = 1.39 \pm 0.061$ (Fig. 8.7, Appendix 8.F). A subset from 1930 to 1934 of 601–1246 samples ($NB = 18$) has a steeper slope of $b = 1.77 \pm 0.121$. Given the large number of samples, the steeper slope probably reflects a real increase in aggregation in the later years.

The practice of taking 1-ft^2 samples of sod and soil for JB larval surveys had been in use at Riverton almost from the inception of the USDA JB Lab. A very intensive survey of JB larvae conducted by Fleming and Baker (1936) was

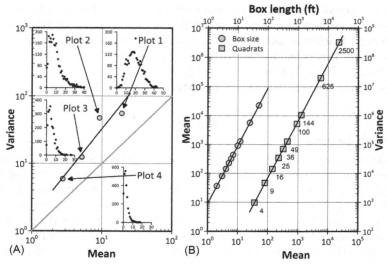

FIG. 8.7   (A) Ensemble TPL of Japanese beetle larvae in 4—2500ft² plots with the frequency distributions of #/ft². (B) Right: Box counting ensemble TPL of number of larvae in 4 to 2500 ft² quadrats. Left: mean against box length has the same slope. *(Original Fleming & Baker (1936) data given to LRT by Dr. Chester Bliss.)*

intended to verify this practice or develop a better one. Their study in Jacobstown, New Jersey, subsequently received attention by statisticians such as Chester Bliss, who made the original data available to LRT and which are used here. In their article, Fleming and Baker described an intensive sampling experiment in 4 grass plots 50 × 50 ft. (15.24 × 15.24 m) in which they took 2500 1ft² (0.093m²) soil samples. Their Fig. 2 shows the approximate positions of 23,034 JB larvae in Plot 2. By aggregating the 1-ft² samples into increasingly larger areas and estimating the sampling error, Fleming and Bake concluded that the 1-ft² sampling units were most accurate for estimating the population. As the size of the sampling unit was increased, the error became progressively larger and correlated with the population density. They recommended samples consist of a minimum of 25 1-ft² uniformly distributed samples in areas <2500 ft², and in larger areas sample units should be spaced <10 ft. apart.

When Bliss (1941) analyzed the Jacobstown data, he found significant departure from the Poisson distribution in three of the four plots (Fig. 8.7A). Interestingly, he plotted variance against the mean (his Figs. 3 and 4) for each of the four plots with larval numbers aggregated into 4 × 4 sample units and found the result to be close enough to Poisson to justify using the square root transformation for analysis of variance. In a series of "dummy experiments," Bliss simulated insecticide trials to test the utility of different experimental designs. He concluded that replication and randomization lead to unbiased estimates of treatment effects when treatment effects could not be distinguished from spatial heterogeneity in unreplicated plots. Furthermore, he showed that sampling each plot both before and after treatment could reduce both plot and sample variation and increase the efficiency of efficacy trials by 30%–60%. His simulations assumed that density but not spatial distribution was affected by insecticide treatment. For weakly aggregated populations like JB larvae, this assumption is valid, but this may not be so for strongly aggregated populations (Trumble, 1985; RAJT, 1987; Chapter 15).

Fig. 8.7A shows the frequency distributions of number of JB larvae per 1-ft^2 sample for each of the 4 plots with the corresponding ensemble TPL. The lowest density plot, Plot 4, the distribution is clearly very skewed with many zero samples, while Plot 1 with the highest density the distribution has no zero samples and a longer right-hand tail; Plots 2 and 3 are intermediate. Averaging at this level results in a moderately aggregated ensemble TPL ($b = 1.31 \pm 0.285$). Aggregating samples, as Bliss did, produces lower TPL slopes (Appendix 8.F). Combining the 1-ft^2 samples into progressively larger units (2, 3, 4, 5, 6, 7, 10, 12, 25, and 50 ft on a side) produces TPLs with $1.116 \le b \le 1.438$. Generally as NQ decreases and NB increases, the correlation declines and the slope increases. All slopes but the 25×25 sample unit with $NB = 16$ are significantly greater than Poisson. The $NB = 4$ and $NQ = 4$ cases do not meet the criteria for unbiased fitting of TPL recommended by Clark and Perry (1994): the other cases all meet these criteria. The results are clearly influenced by the choice of sampling frame.

The variance mean plots for the box-counting approach (Fig. 8.7B) shows the dependence of the mean on the box size (2–50 ft on a side). The regressions of variance on mean with different quadrat sizes (right) and mean on box side length (left) are in perfect agreement ($b = 1.972$ and 2.003, respectively), steeper than the TPLs derived from fixed box sizes. The box-counting regression is usually expressed as number of boxes required to cover the spatial pattern against box length, which is proportional to the inverse of average number per box: the usual box-counting regression would therefore have a slope of $\alpha = -2$.

In a recent study, Jordan et al. (2012) conducted field studies in Virginia to predict spring infestation levels of white grubs, a complex of beetle larvae dominated by JB. In an experiment to validate a sampling method and to assess an insecticide seed treatment of maize, Jordan et al. took soil samples at 15 randomly placed points in 14, 18, and 12 fields in three successive years first in autumn and then in the following spring from 2005 to 2008. The samples were smaller than the 1-ft^2 samples taken by Fox and Fleming and Baker: 20.3×20.3 cm (8×8 inches). The autumn and spring population densities were strongly correlated indicating that autumn sampling could be used to predict spring infestations. They used TPL to characterize the spatial distribution and develop sampling plans. Jordan et al's ODR TPL estimates for each of the 3 years are in Appendix 8.G. Their estimates converted to GMR have average $b = 1.27$, a value very similar to those obtained with Fleming and Baker's data.

Jordan et al. used their TPL estimates to determine the optimum sample sizes in autumn to predict the spring density. Only six autumn samples per field were needed for 95% confidence of predicting the spring density $\pm 25\%$ of the true value: for $\pm 15\%$ of the true value at 95% confidence, 15 samples per field would be required. See Chapter 14 for the use of TPL to develop sampling plans.

Adults

Adult female JB emit a pheromone to attract a mate. Tumlinson et al. (1977) analyzed and synthesized the sex attractant, which they called Japonilure. Subsequently, Ladd et al. (1981) combined the sex attractant with several other lures collectively known as PEG. In a study to determine optimum sample sizes for management decision making, Allsopp et al. (1992) tested Japonilure and PEG separately and together in a randomized complete block design with three treatments and eight replicates. Ellisco traps were set at 1.1 m above the ground for 12 weeks in 1985 and 14 weeks in 1986. Results showed the combination of PEG and Japonilure to be most attractive to both males and females and Japonilure alone to be least attractive.

Prior to analysis of the catches, Allsopp et al. tested the spatial distribution using Iwao's patchiness regression (Chapter 3) and a quadratic model (Routledge and Swartz, 1991), both of which fit their data poorly. Neither the Poisson distribution nor the negative binomial adequately described the results. They found spatial TPL ($NQ = 8$, $NB = 26$) to be a better fit to their JB catches with all three treatments and for both sexes. As the treatments with PEG attract both male and female JB, the beetles were sexed: against expectation Japonilure also caught females. Allsopp et al's TPL results (reproduced in Appendix 8.H) were used to find the optimum transformations for analysis. They found that Healy and Taylor's (1962) exact transformation (Chapter 14) to completely remove the dependence of the variance on the mean for all three treatments, including the small sample of females caught by the Japonilure traps. The highest correlation between variance and mean after transformation was the females caught with Japonilure, $r = .256$, which is not significantly different from zero. Results were also used to estimate the number of traps required to estimate male and female populations with 10% and 25% precision.

The objective of trapping Japanese beetles is to monitor the population to make pest management decisions with the best empirical support. Another objective that has been proposed, but for which support is equivocal, is to use pheromone traps to attract beetles away from food plants and to disrupt mating. The population of JB can become very high locally requiring standard attractant traps to be changed at least daily. Traps with large collection containers would eliminate the need to empty traps once or more a day when JB are numerous. A study by Alm et al. (1994) tested the dual-lure system used by Allsopp et al. in several traps designed to maximize the catch for control purposes.

Alm et al. modified commercially available "Trécé Catch-Can" traps by attaching large 121-L containers to collect the trapped beetles, thereby reducing the need for daily servicing of the standard trap with its 1.2-L container. Alm et al. noted that they had found that the odor of dead and decaying beetles may reduce trap efficiency. One modification had the lures attached to the inside of an open 121-L container 15 cm above 68 L of water in which the beetles were caught. Another focus of Alm et al.'s study was on the configuration of the opening to the trap and how that might be manipulated to increase the efficiency of the trap. A third was to compare the trap efficiency with respect to height above the ground.

Alm et al.'s article includes means and SEs for four modified Trécé traps plus the standard Trécé trap. The five traps and four replicates were arranged in a randomized complete block design that was rearranged weekly. The traps on a golf course in Rhode Island were serviced twice weekly in a 10-week experiment in 1992.

The standard and large container traps produced almost identical TPLs although the residual variance around the fitted line is much higher for the standard trap than the large container trap (Fig. 8.8, Appendix 8.I). The large container trap captured significantly more beetles than the open-top trap containing water due almost certainly to the fact that the open-top water trap

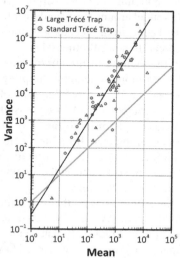

FIG. 8.8 Spatial TPLs of Japanese beetle adults caught in standard and large Trécé traps are not different, but the results of the standard trap are more variable because of sampling efficiency effects. *(Data from Tables 1&2 in Alm et al. (1994).)*

retained fewer beetles than the closed traps. Alm et al. concluded that the odor of dead beetles near the lures probably did not reduce a trap's efficiency basing their conclusion solely on the numbers caught. Results in the Appendix 8.I show that the water traps had significantly steeper spatial TPLs than either the standard or large container traps very likely as a result of density-dependent trapping efficiency of the standard trap (RAJT, 2018). The difference in slopes despite the differences in absolute numbers suggests the absence of odors from dead beetles did increase the sampling efficiency: it was the open top that reduced the number caught and the trapping efficiency.

With the exception of the open top water trap, the modifications to the openings did not materially change the samples, but in the comparison of efficiency with height Alm et al. found that significantly more beetles were captured by traps with collection openings at 13 cm compared with openings at 90 cm above the ground because males fly low while searching for newly enclosed females. Spatial TPLs of these two traps produced closely similar results, but difference in slopes compared to the standard Trécé traps in the main experiment remain to be explained. With both NQ and $NB < 15$, bias cannot be ruled out, although low NQ generally depresses both A and b (Clark and Perry, 1994). As all the experiments had $NQ = 4$ replicates, any bias would likely be in the same direction for all treatments.

IN THE AZORES

The Japanese beetle was introduced into the Azores in the early 1970s. The first detections were made in the vicinity of Lajes Airport, Terceira Island. Despite control efforts, the beetle spread around the island over the next 20 years. Quarantine efforts to keep the beetle from spreading to the other Azorean islands have not prevented its spread to several other islands.

Random sampling in the vicinity of the airport was started soon after the beetle's discovery. In 1982 a more systematic sampling program centered on the airport was started, and in 1984 a regular grid of 1×2 km, offset by 1 km on alternate rows, was established. Over the ensuing years, the grid was extended round the island, completing coverage by 1999 with 168 traps. In 2000, a number of traps were eliminated, bringing the number to 101. The traps were serviced weekly, on a rotating basis. A combination of pheromone and floral lure was changed every 6 weeks during the summer. In the Azores, adult beetles are typically present for ~20 weeks.

From 1996 to 2000, the author assisted the operators of the survey on Terceira, Serviço de Desenvolvimento Agrário da Terceira,[2] in data management and analysis and population modeling for JB management. I continued to cooperate with service personnel for several more years. They generously made the data from 1986 to 2008 available to me. The population spread from Lajes to the

---

2. Terceira Agricultural Service.

west and more slowly to the south. With the exception of the higher elevations and the extinct volcano Santa Bárbara, the beetle had colonized the entire island by 2000. Several biocontrol agents have been introduced and are having some effect, but populations are still very high and continue to threaten the island's agriculture.

Spatial (Fig. 8.9A, Appendix J) and temporal (Fig. 8.9B) TPLs have been computed on the annual totals, which represent most of the period of the range

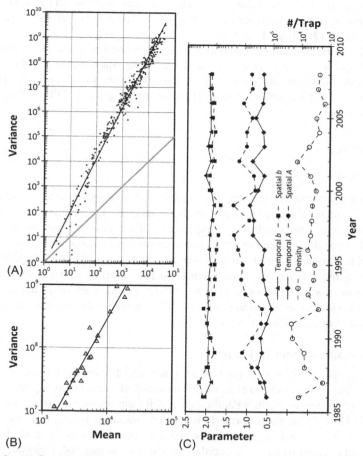

FIG. 8.9 (A) Spatial and (B) temporal TPLs of annual catches of Japanese beetle on Terceira Island, Azores, 1986–2008. (C) Time series of the parameters for intrayear spatial TPLs of JB on Terceira. The slopes are more consistent from year to year than the intercepts and the temporal parameters less variable than the spatial. The average number caught per trap in each year has varied over a factor of 10, but has generally been in decline since the 2002. *(Data courtesy of the Terceira Agricultural Service.)*

expansion around the island. In addition, spatial and temporal intrayear analyses have also been computed (summarized in Fig. 8.9C). Both spatial and temporal annual TPLs indicate strong aggregation throughout the period of population range expansion and the subsequent period of gradual decline as introduced controls have begun to regulate the population. The annual temporal TPL shows increased variation around the fitted line at densities close to the Poisson line suggesting that in some low-density areas, traps may have been influenced by hot spots. The variation in intrayear temporal TPL slope has been remarkably stable throughout the invasion and stabilization periods, while spatial b has seen some spikes. The intercepts were much more variable with spatial A more variable than temporal A. The number of weeks ranged from $NQ = 17-23$ and the number of traps $NQ = 97-316$ over the $NB = 23$ years, so small sample bias does not account for the variation. As both spatial and temporal b have become more stable since the end of the range expansion, it seems likely that TPL is sensitive to the more fluid population distribution during range expansion, resulting in greater TPL within-year variation.

Crustacea

Barnacles

Barnacles are crustaceans that have mobile larval stages and a sessile adult stage. Fertilized eggs hatch into free-living nauplius larvae that pass through several larval instars before molting into cyprid larvae. Unlike the nauplia, cyprid larvae do not feed but settle onto a substrate and metamorphose into sessile adult barnacles. Adult barnacles are 6- or 8-sided pyramid-shaped filter feeders using feathery cirri that extend into the water through the operculum, a diamond-shaped trapdoor at the apex of the pyramid. They range in size from a few millimeters to >5 cm in diameter and are found in the intertidal to littoral zones, and occasionally to depths >500 m.

Chthamalus species in Japan

To study the within-season population dynamics of intertidal barnacles in the genus *Chthamalus*, Fukaya et al. (2013) established a set of census plots at sites along ~1800 km of Japan's Pacific coast. Adjacent regions were separated by 263–513 km. At the most northerly region on Hokkaido Island, *C. dalli* dominated while *C. challengeri* was dominant in other regions. The study was designed to examine barnacle population processes at national, regional, shore, and plot scales. In each of 5 regions, 3–5 semiexposed rocky shores separated by 2.7–17 km were chosen. At each shore site, 1–5 census plots 50 cm wide by 100 cm high were established on rock walls with the midpoint at the mean tide level. In all, 88 census plots were set up and surveyed in spring and autumn from autumn 2002 for periods from 5½-9 years.

Cover of *Chthamalus* was measured by point sampling the 5000-cm^2 survey plots divided into a 10×20 grid and the presence or absence of a barnacle at each of the 200 nodes recorded. Cover estimates were used in a population dynamic model to estimate density-dependent and density-independent processes as well as population parameters including growth rate and density. Output from the model was analyzed for each region by hybrid TPL with each point a season-plot extending over 1.5 orders of magnitude of mean abundance.

Fukaya et al. found seasonal and spatial-scale differences in population size and growth rate at all spatial scales. Also the seasonality of rates and processes varied between sites at plot to region spatial scales. These results provided strong evidence that environmental variability can affect population dynamics at multiple spatial scales. But despite the differences in abundance, growth rates and population processes between regions and sites, spatiotemporal differences in the variability of population size were best described by a single temporal TPL: $\log(V) = 0.825 + 1.501\log(M)$. Using ODR, Fukaya et al. tested TPL for parallelism in a model using Region, Season, and the interaction Region*Season for both intercept and slope. Models were ranked using the Akaike Information Criterion (AIC), which is frequently used to reduce terms in generalized linear models (Pan, 2001). The second best model selected was TPL with common slope and region-specific intercepts. The large difference in latitude, and therefore timing of phenology may account for differences in intercept and therefore overall variation.

In a later study, Fukaya et al. (2014) applied TPLs to their barnacle data to test the competing hypotheses that temporal variability of populations is larger in the center of the range compared to the margins ($V_{center} > V_{periphery}$) or vice versa. In the former case, temporal variation in the center implies that overcompensatory density dependence destabilizes population dynamics. The reverse case implies that the edges are more susceptible to environmental fluctuations than the center. In this study, they used the cover data to estimate mean and variance with their population dynamics model for the 5 shore sites on Hokkaido where *C. dalli* dominated. The data comprise spring and autumn censuses at 25 plots for 10.5 years spanning the elevational range of 50 cm above and 50 cm below the midtide point. The plots were divided into subplots above and below the midtide mark comprising four elevation levels. The vertical range of upper, middle-upper, middle-lower, and lower plots apparently covered most of the elevational range of *C. dalli*. The lower and upper subplots had lower minimum means than the middle plots giving them a substantially greater range of means and variances: the two midtide subplots extended over $\sim$2.5 orders of magnitude, while the highest and lowest subplots extended over $\sim$3.5 orders. The midtide subplots also displayed high population means with very small range of variance as may occur when all available space is occupied (Figs. 5.12, 13.3, and 15.6). These clumps of points, similar to those recorded by Xu and Madden (2002), may have reduced the measured TPL slopes. Overall, TPL

TABLE 8.1 Heterogeneity in temporal TPL for barnacle *C. dalli* at 25 plot-sites on Hokkaido, Japan

Factor	Estimate	SE	t
Intercepts			
Upper height	0.26	0.123	2.10
Mid-upper height	0.45	0.152	2.94
Slopes			
Upper height, autumn	1.66	0.049	33.6
Mid-upper height	−0.20	0.063	3.21
Lower height	0.18	0.052	3.44
Upper height, spring	0.08	0.038	2.15

Only effects significant at P < 0.05 are listed.

fit all four heights very well with analysis of parallelism showing some heterogeneity in both intercept and mean. Using AIC to select models with the most support showed $\log(M)$, Height, and Height*$\log(M)$ were the most important predictors of $\log(V)$ followed by the interaction Height*Season*$\log(M)$ (Table 8.1). Surprisingly, the intercepts were less variable than the *s*lopes, which ranged $1.63 \le b \le 1.79$. The other important result was that $b < 2.0$ for all heights and seasons, supporting an inverse relationship between relative population variability and mean population size.

The mean population size was largest at middle-upper height and decreased toward edges of the elevational range, but the relative variability of population size was smallest at middle-upper and increased to the edges. The lower subplot had the largest relative variability. Fukaya et al. attributed this pattern to an increase in the magnitude of stochastic fluctuations of growth rates at the extremes, a result supporting the $V_{center} < V_{periphery}$ temporal variance hypothesis. However, they caution that population abundance can be highest at one edge of the range, as may occur with a geographic discontinuity, and decrease toward another edge. If this is the situation, the $V_{center} < V_{periphery}$ hypothesis cannot hold. The case of *C. dalli* indicates that populations are not uniformly high even when located near the center of the range. This emphasizes that population structure is not monolithic in space time, but continuously variable with local extinctions and new colonizations as proposed by LRT and RAJT's (1977) fern stele metaphor. Thus, Fukaya et al. concluded that a full understanding of the population dynamics of a species over its range requires the spatial structure to be considered.

STRATIFICATION IN BARNACLE DISTRIBUTION

Adult *Balanus* (= *Semibalanus*) *balanoides* barnacles that grow up to ~15mm in diameter attach to rocks and other solid substrates. It is native to the northern oceans' intertidal zones and is frequently the dominant species of both sheltered and exposed rocky shores. Although capable of living in the sublittoral zone, *B. balanoides* is usually confined to the intertidal by competition and predation. It is common throughout the British Isles, except the southwest.

On a marine biology field course in September 1971 at Millport, Cumbrae Island, Scotland, the author conducted a survey of adult *B. balanoides* on a rocky shore adjacent to Kames Bay. Over a two-day period, two areas of the shore were surveyed: a comparatively steep rocky shore, rising ~1 m for each 10 m and a shallower shore rising ~1 m for about 25 m. In both areas, the barnacles in 20 0.25-m^2 quadrats on 20 parallel transects perpendicular to the low water mark were counted. The transects were 1–2 m apart on both shores. On the steeper shore, the quadrats were placed 0.5 m apart and on the shallower shore they were 1.5 m apart. The barnacles were small, 4-6 mm in diameter, and in places very densely packed. In the densely packed quadrats, 10 5 × 5cm (10% of the area) subsamples were taken. This meant that densities over about 300/quadrat were subsampled, about 30% of the total of 400 quadrats/site. At both sites, *B. balanoides* density declined from the midtide mark to the low- and high-tide levels as Fukaya et al. found with *C. dalli*.

Ensemble TPLs were computed by averaging along (perpendicular to the sea), across (parallel to the sea) the transects and resultants composed of blocks of 5 × 5 quadrats (Fig. 8.10, Appendix 8.K). The overall density of barnacles was ~39% higher on the steeper shore and the range of means for both shores was greater when averaged perpendicular to the sea than parallel to the sea. Averaging along the transects resulted in steeper TPLs than across the transects but only at the steeper shore were the parallel and perpendicular slopes significantly different.[3] Thus, barnacles were more aggregated on the steeper shore than the shallow shore but only in the direction perpendicular to the sea. The two resultant analyses were virtually identical in slope and intercept although the range of means differed. The TPLs for the shallower shore are very similar to the resultants for both shores, suggesting that the higher slope of the perpendicular analysis on the steep shore is a function of the gradient, which is almost absent from the shallow shore and obscured by averaging square blocks on the steep shore. Only the perpendicular intercept in the steeper shore is significantly different from the others.

TPL slope is sometimes correlated with abundance and in this study the perpendicular TPLs appear to conform to this, but the TPLs parallel to the are

---

3. Statistical comparisons are not strictly valid as the samples are not independent. However, the *t*-test probability for perpendicular versus parallel slopes on the shallower shore is $P > 0.80$ compared to $P < 0.02$ for the steeper shore, indicating real differences but with unknown statistical significance. A similar result was obtained for the intercepts.

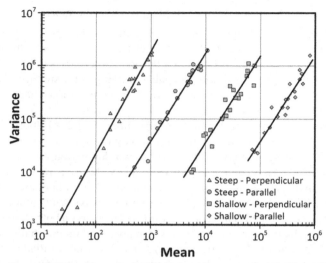

FIG. 8.10 Ensemble TPLs of barnacle *Balanus balanoides* on steep and shallow rocky shores on Cumbrae Island, Scotland. $NQ = 20$ quadrats on $NB = 20$ transects analyzed perpendicular to and parallel to the sea. Successive plots are shifted right by one cycle for clarity. Results of a survey made by the author on a field course in September 1971.

almost identical despite the difference in abundance and the steeper shore's parallel and perpendicular TPLs have substantially different slopes but exactly the same abundance. Thus, in this study, TPL slope would seem to be most influenced by the interaction of geographic variation and cyprid settling behavior.

SETTLING BEHAVIOR OF BARNACLE CYPRID LARVAE

The barnacle *Elminius modestus* is a native of Australia that was introduced to Britain during the Second World War probably on the hulls of ships. It is common in southern England and Wales where it competes with *B. balanoides* on nonexposed shores. Knight-Jones and Stevenson (1950) investigated the barnacle's gregariousness with experiments on the settling behavior of *E. modestus* cyprid larvae.

Glass plates with microscope slides attached were exposed in pairs, one with a clean slide and the other with barnacles already settled. The plates were exposed for 2–3 days below Low Water Spring Tide at Burnham-on-Sea in southwest England on 10 occasions between 6 August and 6 September 1949. On recovering the plates, the newly settled barnacles were counted in 10 concentric rectangles centered on the slide. The average density of barnacles in each rectangle was computed and expressed as #/10 cm^2 for the slide and 10 rectangles. They give the density of settled barnacles on the slides and in the rectangles for both "barnacled" and "not barnacled" slides for all repetitions.

The density declined exponentially from the slide to the outermost rectangle (Fig. 8.11). However, in both treatments, the density of cyprid larvae was higher in the rectangle immediately adjacent to the slide, suggesting that the presence of the slide may have influenced settling. Knight-Jones and Stevenson suggested the corner between the plate and the edge of the slide was especially attractive. Ignoring rectangle I, the barnacled treatment is still exponential but the other treatment is not.

Means and variances computed across the 10 repetitions and 11 areas provide for temporal and spatial TPLs, respectively (Appendix 8.L). While care should be exercised in interpreting these results with both NQ and $NB < 15$, the high correlation coefficients, especially the temporal TPL, suggest the differences in barnacled and not barnacled distributions are real. Furthermore, any bias due to lower than ideal NQ and NB is likely to act in the same direction by reducing the slopes (Clark and Perry, 1994). If the relative values of barnacled and not barnacled TPLs are representative, the divergence of temporal TPL for barnacled $(b > 2.0)$ and not barnacled $(b < 2.0)$ is important. A real divergence indicates that the not-barnacled slides support an inverse relationship between relative population variability and mean population size, while the barnacled slides support a direct relationship between variability and size. Thus, settling behavior differed radically in the two treatments. The relationship between population variability and mean population size must switch when the settled nauplia reach a critical density, a density not reached in the 2–3 days of the experiment.

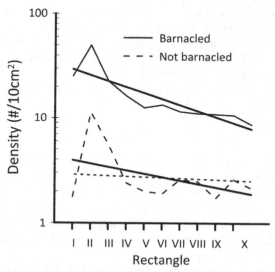

FIG. 8.11   Density declined exponentially with distance from the barnacled slide (upper) but not from the not-barnacled slide (lower) if rectangle I is ignored in that treatment *(dotted line)*. The rate of settlement on barnacled plates was higher than not banacled, but the same TPL fits both sets. *(Data from Table 3 in Knight-Jones and Stevenson (1950).)*

If the edge of the slides was especially attractive to *E. modestus* because either vertical surfaces or corners are preferred settlement areas as suggested by Knight-Jones and Stevenson, the presence of more vertical surface would make the steeper rocky shore at Millport the more attractive settlement site for *B. balanoides* and account for the greater degree of aggregation in the perpendicular direction.

A general survey

This section reviews some of the hundreds of examples illustrating how TPL has been used in agricultural, medical, and veterinary entomology. Some examples of aquatic and marine arthropods are also included. As a much-abbreviated survey of the enormous entomological literature, its purpose is to explore the consistency and variability of TPL and extend the scope of the issues explored in the detailed case studies.

This discussion refers to entries in Appendix 8.M, which includes both TPL estimates from the literature and some new analyses. The Appendix is organized taxonomically by principal organism and where appropriate with associated predators, parasites, or competitors. The habitat, host, or prey is given along with the type of TPL, and whether the study was to develop a sampling plan. The variables NQ, NB, r, A, $SE(A)$, b, and $SE(b)$ are given. Blanks are of variables not specified in the original work and that could not be deduced from other information. Sometimes when NB is not given it can be counted in a plot or deduced from the statistical analyses. NQ can sometimes be deduced from text, but with multilevel sampling or replicated experiments it is not always possible to determine how or if samples were aggregated before computing mean and variance. Where NQ alone is given, the total number of sample units was specified but the distribution of NQ and NB was not.

Because NB, r, b, and $SE(b)$ are connected by the equation.

$$SE(b) = \sqrt{\frac{(1 - r^2)b^2}{(NB - 2)r^2}}$$

if three are known, the fourth can be calculated. Calculating $SE(A)$ from the text or table is rarely possible as it requires additional information, the residual mean squares and the average of $\log(M)$. Estimates are reported as given in the original reports, which mostly used ODR of the logs of mean and variance (Eqs. 4.1–4.10). For consistency, new TPL analyses also used ODR. Citations numbered in the text refer to the entry in Appendix 8.M and to the key to References at the end of the appendix.

There are multiple factors influencing the estimated values of TPL parameters in addition to the intrinsic spatial distribution of the target species. The sampler and its operation are the most obvious factors confounding TPL estimates, but the scale of the survey and the substrate (host plant or animal) and its condition, management, and habitat can also influence A and b. Given the number of possible confounding factors, it is surprising how little variation there is in TPL's parameters for a given species. With that in mind, this section examines a variety of cases and compares estimates over space and time, between life stages, on different hosts, and in the presence of predators, parasites, and competitors.

Consistency across space, time, and stage

The hundreds of examples of TPL in the agriculture literature is a reflection of its discovery in a primarily agricultural environment—Rothamsted Experimental Station. The TPL entomological literature also includes basic ecological research as well as investigations of TPL's properties. However, the majority of papers citing LRT61 used TPL to develop sampling schemes for insect pest management. Several kinds of sampling plan use TPL: fixed precision, sequential, double sample (Chapter 14). These plans are only as useful as TPL's stability in space and time, and the consistency of the combination of environment and sampler.

The consistency of TPL over a range of taxa, environments, plant and animal hosts or prey, and between stages and scales has been questioned. These criteria are linked and in many cases inseparable. For example, most data of parasite distribution and abundance are expressed as numbers per host. Thus, the host animal is the sampling unit. Similarly, plants or plant parts are often the sampling units. Expressing insect abundance in terms of whole plants or plant parts, introduces the question of scale, and whether scale and environment can be separated.

Understanding the sampling mechanism's properties and its relationship to crop damage or host health is crucial for effective pest management decision making. Sampling with an objective sampler is important for repeatability, although samplers, such as light traps, colored sticky traps, and pheromone traps, are convenient and objective, their effectiveness depends on the target's response to stimuli, which can change with age or season or time of day, with its visibility or fetch, and with changing environmental conditions. Most importantly, the targets' responsiveness may also depend on how many others are competing for access to the trap (RAJT, 2018).

Although visual sampling can be effective, it must be sustained for long enough to obtain a usefully large sample. The consistency between the several visual samples of gypsy moth egg masses on different occasions and by different teams is extraordinary. However, operator variation can be practically significant even when it is not statistically different. Quadrat samples of red-legged earth mite (*Halotydeus destructor*) on soil between rows of oilseed rape (canola) by three operators resulted in two almost identical TPLs, but that of the 3rd operator was sufficiently different that sampling plans based on the 1st or 2nd operator were different from those of the 3rd operator [76].

Other sources of variation include the positioning of the sampler as RAJT et al. (1998) found with western flower thrips (WFT; *Frankliniella occidentalis*) caught on yellow sticky cards [1]. Samples of WFT taken within the canopy of a cucumber crop produced no TPL ($r^2 \approx 0.01$) or resulted in TPLs markedly different from those taken above the canopy. However, WFT caught on sticky cards above a variety of crops in greenhouses in Ohio, British Columbia, and Ontario revealed consistent TPLs over a range of plant architectures; cucumbers, peppers, and a variety of flowers [1]. The table developed for optimum sample sizes for sampling WFT in greenhouse crops using yellow sticky cards also showed how the number of samples required for a fixed level of precision can be manipulated by varying the sampling efficiency or size of sampler and thus the intercept A (Chapter 15). This study concluded that TPL sampling the same reproductive population under different conditions of stage, phenology, or habitat can result in different power law estimates, as had been reported by some authors. The condition or management of the host plant may influence TPL. Cutting grass and forage crops affect survival and can increase mortality, thereby changing TPL slope. WFT on alfalfa cut every other month were significantly more aggregated ($P < 0.01$) than those on alfalfa cut monthly [4].

Another study of WFT distribution, of larvae and adults on greenhouse cucumbers included samples of the predatory mite *Amblyseius cucumeris* [2]. The distribution of WFT larvae and adults differed with the larvae's slope markedly greater than the adults', and the mite's slope lower still ($b_{larvae} > b_{adult} > b_{predator}$). In addition, within the plant, both larval and adult WFT slopes were lower in the middle part than either the apex or basal areas. The pattern was different for the mite whose slope was steepest in the basal area and shallowest in the apex. The mites were more strongly aggregated low on the vine and apparently preferred to hunt WFT there.

WFT and *Pezothrips kellyanus*, a new citrus pest, in 22 citrus groves in Valencia, Spain, were similarly distributed despite being potential competitors [3]. All stages of both species, whether on flowers and fruitlets, were aggregated with the immature thrips of both species showing significantly higher aggregation ($b_{larvae} > b_{adult}$; $P < 0.05$).

A comparison of the distribution and abundance of onion thrips (*Thrips tabaci*) on eight soybean cultivars showed variation in TPL between cultivars [6]. Although the densities (#/leaf) are all low, the ranges of means extend over two orders of magnitude for all cultivars and with the exception of the two shallowest slopes, the correlations are high. Of the eight cultivars, onion thrips' slope is significantly greater than Poisson in four cases ($P < 0.02$). Of the remainder, three slopes are not significantly different from $b = 1.0$ ($P > 0.25$) and one is marginal < 1.0 ($P < 0.086$).

On weeds in mango orchards, the larvae of multiple thrips species were more strongly aggregated than adults ($b_{larvae} > b_{adult}$) in both wet and dry seasons and with and without insecticide [7]. Comparisons between wet and dry season and between pesticide and no pesticide suggested the environmental conditions had little influence on TPL.

The spatial distribution of response of plum aphid (*Brachycaudus heli-chrysi*) on two sunflower cultivars in France showed no difference in either A or b [9]. A third analysis including data from three other cultivars was also closely similar. In this case with $NQ = 25–320$ samples, $NB = 17–53$, and very high correlation coefficients, there is high confidence of no cultivar effect on plum aphid TPL. In contrast, pea aphid (*Acyrthosiphon pisum*) on two varieties of alfalfa in Oklahoma and one in Wisconsin sampled on 32–120 stems per date over a 2–3 year period produced very different TPL slopes and intercepts [10]. With high correlations and large numbers of sample units and TPL points, the distributions of pea aphid on alfalfa in Wisconsin and Oklahoma were clearly different. Similarly, pea aphids on Indian pea (*Lathyrus sativus*) at two locations in Ethiopia over a three-year period resulted in heterogeneity of estimates [11]. In this case, the small number of sample units ($NQ = 5$) and TPL points ($NB = 12–13$) contributed to the heterogeneity and probably to uncommonly high slope estimates ($2.04 \leq b \leq 4.18$). TPLs of the caterpillars of moths *Helicoverpa zea* and *Spodoptera frugiperda* on sorghum in Oklahoma and two locations in Texas were not different in A or b, nor were they when combined with data from Kansas [53].

The potato leafhopper (PLH; *Empoasca fabae*) is known to reproduce on over 200 plants and feed on many more (Lamp et al., 1994). Ensemble TPL intercepts and slopes of PLH on 24 clones of red maple differed significantly between months from a high of $b = 1.46 \pm 0.061$ in June to a low in August of $b = 0.99 \pm 0.028$ [29]. Notably, TPL slopes differed markedly between clones, from a low of $b = 0.92 \pm 0.001$ to a high of $b = 1.50 \pm 0.080$. Although the sample sizes were small, all but two of the correlation coefficients were $r > 0.95$, suggesting real differences in aggregation on the different clones. The difference in distribution between clones is presumably related to differences in acceptability of the clones as food sources and the changes with season may also be in response to changes in host chemistry. Habitat differences via the host plant may also influence TPL as is evident in Ranchman's tiger moth (*Platyprepia virginalis*) on lupines in wet and dry habitats [52] where they differed in A but not b, indicating greater variability in the dry sites.

Aggregation also can differ between some host plants and not others. TPLs of the maize ear borer (*Mussidia nigrivenella*) on maize compared to four alternate wild hosts were significantly different when considered collectively [46]. However, the heterogeneity in slopes was entirely due to one alternate host, the baobab tree. The slope of borers on the baobab was significantly steeper than maize ($P < 0.002$), which was not significantly different from the other hosts ($P > 0.80$). TPLs of *M. nigrivenella* on several crops in Benin [47] did not differ in slope and only one, of borers on jackbean, a cover crop, differed in intercept, denoting increased overall varibility

Aggregation differences between different parts of a host plant also occur. Aggregation of the mealybug *Phenacoccus peruvianus* on bougainvillea [16] increased significantly from number/leaf to number/twig ($P < 0.001$), although neither is significantly different from the intermediate slope of number/bract $(P > 0.40)$. The distribution of silverleaf whitefly (*Bemisia tabaci*; also called sweet potato whitefly) on river tamarind in India [18] and field tomatoes in Florida [17] did not differ between locations on the plant. The tamarind study found no difference between years either, but the TPL slopes of the Florida tomatoes differed in the order $b_{eggs} > b_{nymphs} > b_{pupae}$ with pupae significantly different from the other stages $(P < 0.01)$. The most aggregated stage of the greenhouse whitefly (*Trialeurodes vaporariorum*) on green beans was adult $(b_{adult} > b_{eggs} > b_{larvae} > b_{pupae})$ [19] with only the slopes of 3rd instar larvae and pupae not significantly different.

Aggregation of potato psyllid (*Bactericera cockerelli*), vector of psyllid yellow disease of tomato was generally low and declined with stage [23], but a second study found nymphs to be significantly more aggregated on bell peppers [22]. Similarly, nymphs of the lemon psyllid (*Diaphorina citri*) on lemon trees were more strongly aggregated than adults, and also eggs $(b_{nymph} > b_{adult} > b_{eggs})$ [21]. On emerging from egg masses, Colorado potato beetle (CPB; *Leptinotarsa decemlineata*) neonates on potatoes are very highly aggregated and the stages become progressively less aggregated as they develop $(b_{small\ larvae} > b_{large\ larvae} > b_{adults} > b_{egg\ masses})$ [38].

Species with b declining systematically from eggs to adults include the leafhopper *Empoasca kraemeri* on common beans [31], the carob moth (*Ectomyelois ceratoniae*) on dates [48], and the glassy-winged sharpshooter (*Homalodisca vitripennis*) on orange trees [32, 33]. The kudzu bug (*Megacopta cribraria*) is an invasive pest first seen in Georgia in 2009 that has spread throughout the southeast US. A native of Asia, it feeds on many legumes including soybeans, which it frequently infests in very high numbers. For TPLs of the kudzu bug on soybeans in 2012 and 2013 the order of aggregation was $b_{eggs} > b_{nymphs} > b_{adults}$. But in 2011, the adult slope was extremely high $(b_{adults} = 3.27 \pm 0.115)$ and the order reversed $(b_{adults} > b_{nymphs} > b_{eggs})$ [36,37]. This was also the order of the pear psyllid (*Cacopsylla pyri*) on pear trees where the adults were significantly $(P < 0.0001)$ more aggregated than nymphs [25]. Another species with $b_{adults} > b_{nymph} > b_{eggs}$ is the alfalfa weevil (*Hypera postica*) on alfalfa [41].

Neither the Oriental red mite (*Eutetranychus orientalis*) on ornamentals [74] nor the southwestern corn borer (*Diatraea grandiosella*) on maize [45] showed any differences between stages. The latter species' slopes were close to Poisson for all stages. Another species with Poisson distributed eggs but not significantly less aggregated than larvae is the citrus leaf miner (*Phyllocnistis citrella*) [50].

Although there are exceptions, the general pattern is for decreasing aggregation from egg to pupae, with adults similar to pupae or aggregated more like eggs. Exceptionally egg masses tend to be distributed more closely to random than aggregated, but neonates must start out highly aggregated. An interesting exception to this common pattern is the prepupal larvae of the caddis flies

Philopotamus montanus and *Silo pallipes,* which actively aggregate prior to pupation ($b_{larvae} < b_{pupae}$) [55,56].

There may also be differences in distribution between the sexes, although as with gypsy moth pupae, the distributions of male and female citrus red mite (*Panonychus citri*) on citrus do not differ significantly ($P = 0.23$) [81]. The presence of males attempting to mate with ovipositing fruit flies (*Drosophila melanogaster*) apparently changes the spatial distribution of eggs [58] in much the same way that male gypsy moths interfere with each other as they approach a pheromone source (RAJT et al., 1991; RAJT, 2018).

Stability of TPL in time and space is important for the portability of sampling plans. The temporal TPLs of eggs of two species of mosquito *Aedes aegypti* [61] and *A. albopictus* [62,63] on two continents are remarkably similar. Neither TPL parameter of eggs of *A. aegypti*, vector of yellow fever and *A. albopictus*, vector of chikungunya sampled in ovitraps in Rio de Janeiro, Brazil, Split, Croatia, and northern Italy differ significantly. Also similar is the spatial TPL of *A. albopictus* adults in Split. Temporal TPLs of adult female *A. aegypti* are closely similar to the egg distributions. Samples of the invasive *Culex tritaeniorhynchus* in rice paddies in Greece are also very similar over 2 years, but a year with the lowest counts is not different from Poisson [64]. *C. tritaeniorhynchus* is a vector of dengue and Japanese encephalitis.

In a search of the literature, Jones (1990) concluded that a single value for *b* could be used for mite pests *Panonychus citri and ulmi* and the predator *Tetranychus* spp. (mostly *T. urticae*) [82]. Confirming this conclusion with a separate validation set, he concluded that for these species "variation within a study is often greater than that observed between studies." Ensemble TPLs of citrus red mite (*Panonychus citri*) on lemon in California were not different between four sites nor between seasons at one site [81]. The distribution of 2nd instar larvae of oleander scale (*Aspidiotus nerii*) on jojoba trees in two consecutive years did not vary ($P > 0.80$) despite an order of magnitude difference in abundance [15].

Changing environmental conditions, such as weather from year to year, can upset TPLs repeatability. The cotton aphid (*Aphis gossypii*) on cotton in the Central African Republic did not differ in slope in three of 4 years [12]. In the third year, the aphid population developed later but grew more rapidly than the other years possibly influencing the distribution resulting in a higher *b* than the other years ($P > 0.10$). The sugarcane weevil borer (*Acrotomopus atropunctellus*) also shows heterogeneity in *b* between years, with 1 year not different from Poisson [94], but the cacao mirid bug (*Sahlbergella singularis*) on cacao trees in Cameroon *A* and *b* are virtually identical in two consecutive years [34]. In a study of the connection between dispersal distance and spatial distribution of the stable fly (*Stomoxys calcitrans*), an important pest of livestock, *b* did not differ in three of 4 years [60]. In 2005, however, $b > 2$ was significantly different ($P < 0.05$) from the other 3 years. Male gypsy moths at the leading edge of the invasions of Wisconsin and West Virginia between 1996 and 2006 were highly aggregated in space resulting from long distance dispersal founding local populations [54].

Separate from changes in distribution with stage, distributions may change within year, as seen with PLH on red maple. A study of the chironomid larvae living on bulrushes in a stream in England compared the ensemble TPLs of eight species in spring and autumn [59]. Of eight species, two showed no difference in TPL parameters in spring and autumn, four showed differences in intercept and one, *Cricofopus fuscus*, showed significant differences in both *A* and *b*; one congener *C. bicinctus* differed only in *A* and *C. sylvestris* was not different in either parameter. When a group of related insects in the same habitat display a variety of TPL behaviors, small differences in TPL may reflect small niche and/or behavior differences that permit their coexistence.

Spatial distribution may change from day to night as organisms' activity cycles. The distributions of 12 of 21 common species of Ephemeroptera, Plecoptera and Trichoptera in a stony stream examined at midday and midnight at intervals over 4 years changed between day and night [68]. Of the 12 species, four TPLs were reduced significantly at night (all case-building Trichoptera larvae) and aggregation increased at night in eight species, one of which, *Isoperla grammatica*, is a nocturnal predator.

Clearly the time of day populations are sampled can make a difference, although in most cases where data are used for decision making, sampling lasts several days, so diurnal differences will not matter. Other examples show that distribution can change by season: PLH on maples [29] and chironomid larvae on bulrushes [59], for example. Differences in distribution between populations on host plants with different architectures [46] are not surprising, nor differences between parts of plants. Systematic changes in distribution through the lifecycle reflect the life history strategy of each stage: often aggregation declines with stage from egg to adult, but there are also many exceptions. Perhaps the most surprising similarity is that of distributions of several mosquito species in several different countries [61–66], although perhaps the distribution of eggs in ovitraps has more to do with the traps than the trapped as the traps are set to attract females.

Sampling efficiency and consistency between samplers

Because of the importance of obtaining reliable quantitative population estimates for making pest management decisions, the reliability and repeatability of samplers has received considerable attention, most prominently with work on light traps and suction traps at Rothamsted Experimental Station in England (Johnson, 1950; Williams, 1951; LRT, 1962a,b). The emphasis at Rothamsted on rigorously calibrating sampling methods for aerial insects contributed to the discovery of TPL. One of the largest bodies of work using TPL is by among agricultural entomologists comparing sampling methods to determine the most useful ones for pest management decision making. Central to developing decision-making programs for assessing pest populations is characterizing the relationship between mean and variance for samplers and for identifying the most cost-effective samplers.

We can distinguish between two kinds of sampling methods: those that produce absolute and those producing relative density estimates. The latter can be subdivided into those that can be expressed as number per unit habitat, per unit effort, or per trap. The last may be further divided into attractant and passive samplers. Relative samplers include sweep nets, beating sticks with trays or buckets, to catch falling insects. Attractant traps, including light traps, sticky traps, and food- or pheromone-baited traps, are useful relative samplers that usually require less effort.

Quadrat counts, suction traps, and airplane tow nets estimate absolute density because the area of quadrat or volume of air being filtered can be measured and the extraction efficiency determined. Knowing the volume of air being filtered by a suction trap or tow net and its extraction efficiency, the absolute aerial density of a population can be estimated. A terrestrial version of the aerial suction trap, the D-vac was developed specifically to sample insects from ground vegetation (Dietrick et al., 1959) and has become a standard sampling tool. However, the D-vac's efficiency can be habitat- and stage dependent (Duffey, 1980) and, unlike aerial suction traps, density dependent (Dewer et al., 1982). The A-vac, adapted from a domestic leaf-blower, is neither calibrated nor standardized. It is less powerful than the D-vac suggesting its efficiency may be lower and more variable, but as it is comparatively inexpensive and more portable it has seen increased use recently. Visual sampling is frequently expressed in absolute terms, but the efficiency is often low and unknown, effectively making visual methods relative per unit habitat per unit effort. Provided sufficient time is permitted for them, visual methods can have acceptably high efficiency for sessile or slow-moving targets. For easily dislodged and some mobile targets, beating trays or similar devices are effective. Although attractant traps rely on the behavioral response of the targets, they are often cost-effective samplers and easily calibrated to crop condition or damage. Behavioral traps present special problems as the target's response to light, food, or pheromone is often unknown and may change with habitat, age, or density.

LRT (1962a) showed comparison of catches by a relative estimation trap with simultaneous catches by an absolute estimation trap will permit a relative trap to be calibrated. For most pest management purposes, a correlation between a trap's catch and the risk to the crop may be sufficient, but for population estimation, sampler calibration will be required.

TPL has a role to play in the comparison of sampling methods as two samplers may catch different fractions of the population, but if their TPLs are similar, the samplers' relative efficiency remains constant as population density changes. TPL is sensitive to variations in traps' relative efficiency, which is reflected in differing TPL slopes. It was determined very early on that the parameter most influenced by sampling method is the intercept A (Southwood, 1966). Subsequently, it was realized that A and b may be correlated (Clark and Perry, 1994) and that therefore b can also be influenced by the sampling scheme. Heterogeneity of slopes ($b_1 \neq b_2$) indicates changing distribution of one population relative to another as density changes, and this applies to the same population sampled by different devices. RAJT (2018) has shown that sampling efficiency may change with density and that this impacts both A and b. Furthermore, if sampling efficiency is density dependent, as when a target's response to an attractant changes with density, A and b must be correlated. This implies that if they are uncorrelated, sampling efficiency is likely to be density independent.

Effect of sampling method

Measuring the distribution and abundance of parasites presents problems as the host's distribution and abundance must also be considered. Ticks that leave one host to wait for the next one may be sampled like any other free-living species. Milne (1943) sampled sheep tick (*Ixodes ricinus*) on sheep and using a blanket dragged across grazing land [70]. The life system of the sheep tick, an ectoparasite of large mammals such as sheep and cattle, is typical of many ticks. It will also feed on humans to whom it can transmit Lyme disease and viral encephalitis. Sheep ticks live for 2–4 years and have different hosts for each of three life stages. Upon hatching larvae feed on small mammals, such as mice and shrews, and molt into nymphs that seek larger hosts such as rabbits and birds. After molting to adults, sheep ticks seek large mammals on which to feed and mate. Blood-engorged mated females drop off to lay several thousand eggs.

Milne gives 10 frequency distributions of ticks per sheep. Ensemble TPL of ticks/sheep shows ticks to be moderately aggregated on sheep ($b = 1.46 \pm 0.114$). He also gives the number of nymphs caught on 6 45.7-m drags of bracken beds on 33 occasions in Spring 1941. The sample data fit neither the Poisson nor the Neyman's A distributions, but a spatial TPL of $NB = 33$ days and $NQ = 6$ blanket drags suggests 2 distinct populations, one random and one aggregated (Fig. 8.12). Five of the Poisson points occurred in the last 2 weeks of the season when nymph numbers were in decline. The appearance of two populations is likely due to mortality reducing the TPL slope late in the season.

The blanket drag sampler has also been used for Rocky Mountain wood tick (*Dermacentor andersoni*), a major vector of anaplasmosis that causes severe anemia in cattle. It occurs in rangeland in western North America, including Alberta, Oregon, and Washington, where 222 site-date combinations at 13 locations were sampled in 2008–10 [71]. An ensemble TPL analysis made of means and variances computed from samples of $NQ = 86$–250 10 m^2 drags ($NB = 222$) resulted in a slope of $b = 1.33 \pm 0.02$ indicating moderate aggregation. Using this estimate of b, the number of samples needed for efficient estimation was underestimated, while Iwao's method overestimated the required sample size. Using the negative binomial with common k for the non-Poisson points, Rochon et al. (2012) found required sample sizes closest to empirically calculated sample sizes. 85 (38%) of site dates were indistinguishable from Poisson but included in Rochon et al's TPL analysis, causing b to be underestimated by >5%. A revised

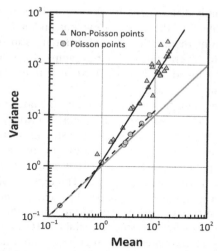

FIG. 8.12   Spatial TPL of sheep ticks sampled by blanket drags show moderate aggregation. The comparatively high density points suggest two populations were sampled: an aggregated population early and and a random population later accounting for 5 of the 7 Poisson points. *(Data from Table 6 in Milne (1943).)*

estimate of $b \approx 1.4$ increases the required sample size and brings the slope closer to Milne's spatial estimate for blanket dragged sheep tick nymphs of $b \approx 1.58$.

The glassy-winged sharpshooter (*Homalodisca vitripennis*), a vector of Pierce's disease, discolors citrus fruit reducing marketability. Castle and Naranjo (2008) found differences in spatial TPL between adult and nymph glassy-winged sharpshooters on citrus in Arizona and also between samplers [32]. A-vac, D-vac, and two dislodging methods (beating pole and bucket and pole and net) were evaluated for estimating relative densities of nymphs and adults on citrus trees. The methods were not equally efficient but all showed more males caught, suggesting a male-biased sex ratio. The A-vac was least efficient and the D-vac most efficient adult sampler, with the dislodging methods intermediate. The trend in adult spatial TPL slopes reflected the trend in sampling efficiency and with minor differences so did the nymphs' TPLs. For both adults and nymphs, the D-vac produced the steepest slope and the A-vac the shallowest; the difference is significant ($P < 0.01$) for adults but not nymphs ($P > 0.18$). The differences in efficiencies of the A-vac and D-vac samplers reflect the differences in suction created by the two devices.

The pear psyllid *Cacopsylla pyri* is a major pest of pears. Adults and nymph pear psyllids and *Pilophorus gallicus* a mirid predator were sampled ~160 times in 4 pear orchards in southern Spain in 2007–10 [25]. Sampling for adult and nymph psyllids was by counting psyllids on 60 shoot samples/orchard and one leaf/shoot ($NQ = 240$); sampling for adult psyllids and mirid predators was by beating the trunks of 50 trees/orchard ($NQ = 200$) and catching insects falling into nets. Spatial TPL slopes for shoot samples versus beating net for adult pear psyllids are significantly different, as are slopes for shoot and leaf sampling for nymphs. The difference in TPL slopes of nymphs on leaves and twigs may be due to differences in method, although real differences in spatial distribution seem more likely. However, with such large sample sizes and $NB \approx 160$, the two sampling methods for adults clearly led to two different TPLs, which relates to differences in sampling efficiency.

The efficiency and precision of five sampling methods for the Asian citrus psyllid (*Diaphorina citri*) were assessed in Florida citrus groves in 2009–13 [20]. Asian citrus psyllid is the vector of the devastating citrus greening disease. In increasing amount of effort (seconds/tree), the methods were stem tap (2 sample selection policies), sweep net, visual, sticky trap, and A-vac. TPL was used to evaluate the methods' precision. The reported slopes ranged from $b = 1.20 \pm 0.037$ (sweep net) to $b = 1.84 \pm 0.046$ (suction trap) and again tended to increase with the amount of effort required to obtain the samples.

Several visual methods for assessing rice stinkbug (*Oebalus pugnax*) populations in California were compared with 10 sweeps with a net [35]. The visual methods consistently resulted in lower TPL slopes than the sweep net ($b = 1.41 \pm 0.072$). A "sweep stick" was used to gently disturb adult *O. pugnax* from rice panicles; adults seen flying adjacent to the stick within 38 cm of the end were counted in each of five sweeps. This method resulted in progressively higher estimates of b (all <1.0) as effort increased from 1 to 5 sweeps. Another similar visual method parted the panicles with a stick and counted all adults seen in a full $180°$ sweep. This latter had a value ($b = 1.18 \pm 0.97$) intermediate between the sweep stick and sweep net. Clearly, in this study, the value of b was strongly influenced by the degree of effort applied, which determined the magnitude of the catch.

Thirty-one fields of timothy in California were sampled for *Anaphothrips obscurus* on several occasions in 2006–08 [5]. Ten tillers in each field were visually inspected and thrips counted. Simultaneously and close by, an individual tiller was sampled using a method in which the tiller was rapped 10 times inside a cup. The beat cup method resulted in generally lower means than the direct method. Means and variances ($NQ = 10$) of thrips/tiller were computed for each field-year and TPL computed for each method. Slope was consistently higher for the direct method than the beat cup method for each year separately and all years combined. The consistently higher slope of the direct observation method suggests the beat cup method, which generally resulted in lower counts, may have lower efficiency at higher densities.

The most abundant insects in grain storage and processing facilities are weevils in the genera *Tribolium* and *Sitophilus*. Two traps were evaluated in a rice-processing facility [39], one baited with a food-based oil and the other baited with a synthetic male aggregation pheromone for *Sitophilus* spp. Both traps attracted *Tribolium castaneum* in addition to *S. oryzae* and *S. zeamais*. The differences in TPL estimates for the two traps and three species were sufficiently small that a common fixed-precision sampling plan for weevils could be developed and validated. It is noteworthy that estimates of b were higher for *Sitophilus* caught in traps with the food-based lure than the aggregation lure, but the reverse was the case with *T. castaneum*. It is possible these small differences ($P > 0.20$) were due to density-dependent effects.

The sugarcane weevil borer (*Acrotomopus atropunctellus*) has become an important pest of sugarcane in northwest Argentina [42]. A TPL analysis of five levels of sampling effort for weevils, 2, 4, 6, 8, and 10 min spent at 1 m of cane furrow, showed TPL slope tended to increase with increasing effort as the range of means increased; simultaneously the correlation coefficient tended to decrease.

Other examples of differences between sampling method include a mixed-species TPL of leafhoppers on weeds near potatoes [30] and potato leafhopper on green beans [28]. In the mixed-species case, *b* is greater for sweep netting than sticky cards and in the latter case *b* is greater for visual than D-vac sampling for both nymph and adult potato leafhoppers. In both cases, the differences are plausibly due to differences in effort, and with the mixed-species comparison density-dependent sampling efficiency may play a role as sticky cards are behavioral samplers.

The number of samples can be critical. As LRT et al. (1978) pointed out, randomness is quite rare. Thus, if a series of samples is not significantly different from Poisson at all densities, it may denote insufficiently large samples to detect the acceleration of variance with mean. It often occurs that samples taken with minimal effort are indistinguishable from Poisson. Consequently, as effort increases, so too does sampling efficiency and with it TPL slope. As a result, it is usually not possible to determine the true spatial distribution of organisms unless the sampling efficiency is high. Sampling for longer or aggregating samples can overcome inefficiencies in sampling method. Another, less obvious source of variation in TPL concerns whether or not one counts the sample units (*NQ*) with zero counts. Excluding empty quadrats from calculation of mean and variance can inflate the TPL slope (RAJT et al., 2017). TPLs of the invasive fire ant (*Pseudacteon tricuspis*) in Louisiana [65] with and without zeros are significantly different in both *A* and *b* ($P < 0.005$).

Differences between trophic levels

Park et al. (2013) proposed that *b* is at least partly influenced by life history strategy. In particular, they found a trend in decreasing aggregation as measured by *b* from bacterial- and fungal-feeding nematodes through plant feeders to predators and omnivores. The trend was independent of abundance, which also declines with increasing trophic level. Much of the information on spatial distribution of predator and prey comes from the biological control literature as predatory mites, thrips, and mirid bugs are frequently used in greenhouses: parasitoids are also used.

Predation

Biological control of greenhouse pests has been practiced for decades and generalist predators are favored biocontrol agents. Visual sampling for western flower thrips larvae and adults and the predatory mite *Amblyseius cucumeris* on cucumbers in greenhouses in Alberta found spatial TPL slopes of western flower thrips larvae/leaf to be steeper than the TPL for adults and both were steeper than the mite [2]. The distributions of both predator and prey differed between locations on the cucumber vine, although the pattern of $b_{prey} > b_{predator}$ was consistent. Also on cucumbers, the population densities of the predatory mite (*Phytoseiulus persimilis*) and its red spider mite (*Tetranychus urticae*) prey in greenhouses in Denmark displayed distinct predator-prey oscillations [78]. Although not significantly different, spatial TPLs of the number of predator and prey mites/3 leaves also resulted in $b_{prey} > b_{predator}$. Ensemble TPLs produce similar results. TPL slopes of all stages of the predatory mite *Amblyseius fallacis* were less than slopes of prey European red mites (*Panonychus ulmi*) on apple trees in Michigan orchards [80]. The red mite was more than ten times as abundant as the predator and on the assumption $NB = 64$, the number of orchards surveyed, the difference between b_{prey} and $b_{predator}$ is highly significant ($P \ll 0.001$).

Eutetranychus orientalis can cause total defoliation of the ornamental tree *Melia azedarac*. A study in Spain identified two predatory mites (*Euseius scutalis* and *E. stipulates*), and a more abundant predatory thrips *Scolothrips longicornis* on *M. azedarach* trees in a plantation [74]. Predation by the thrips was observed and its density tracked *E. orientalis*. Distribution of all stages of *E. orientalis* combined was aggregated ($b = 1.62 \pm 0.061$) with immatures and males highly aggregated. Both predators' distributions were significantly less aggregated than all stages of *E. orientalis* ($b_{prey} > b_{predator}$) and from each other ($P < 0.005$) with the predatory mite's not different from Poisson ($P = 0.7$). The difference in b of the two predators shows very different spatial distributions, possibly avoiding competition.

Both the polyphagous mirid *Macrolophus pygmaeus* and the predatory mite *Phytoseiulus macropilis* are used against the carmine spider mite *Tetranychus urticae*, a serious pest of greenhouse tomatoes. Being smaller, the predatory mite is potentially a prey item for the larger mirid, as well as a competitor [79]. In a cage experiment, carmine spider mite exposed to the predatory mite and mirid bug separately and together and with no predators was more aggregated in the presence of the predators than alone. The predatory mite was also strongly aggregated, but less so than the prey, while the mirid's distribution was indistinguishable from Poisson. There was no evidence of predation of the smaller predatory mite by the mirid, but the significant difference in slopes ($P < 0.05$) suggests they may have avoided competition by adopting different spatial distributions.

The spatial TPLs of the mirid predator *Pilophorus gallicus* and its psyllid prey *Cacopsylla pyri* also resulted in $b_{prey} > b_{predator}$ [25]. However, the TPL slopes for adult psyllids and mirids caught by beating net were not significantly different, while the other visual methods for psyllid adults and nymphs resulted in significantly steeper TPLs. The phytophagous mites *Cenopalpus irani* and *Bryobia rubrioculus* and a predatory mite *Zetzellia pourmirzai* in a sprayed orchard in Iran were similarly distributed with closely similar b [24]. However, in this study, in which the population density of *C. irani* per leaf was 10 × that of the other phytophage and $b_{predator} > b_{prey}$. The predatory mite (*Zetziella mali*) was intermediate in aggregation between two competing phytophagous mites (*Eutetranychus frost* and *Tydeus longisetosus*), although no slopes were significantly different ($P > 0.20$), in an unsprayed apple orchard in Iran.

Parasitism

With some possible exceptions, the biocontrol data support the pattern proposed by Park et al. (2013) that $b_{prey} > b_{predator}$. Insect parasitoids are similar to predators in that they consume the host. The hymenopteran parasitoid *Agriotypus armatus* parasitizes pupae of the caddis fly *Silo pallipes* [57]. Elliott (1983) found that as *S. pallipes* aggregation increases with density so does the aggregation of *A. armatus* parasitism with respect to the host density, but at a lower rate. The slope of ensemble TPL of parasitized *S. pallipes* pupae ($b = 1.39 \pm 0.108$) is significantly less ($P < 0.01$) than that of unparasitized pupae ($b = 1.85 \pm 0.102$). Thus, in this case $b_{host} > b_{parasite}$.

Because parasite distribution and abundance are usually expressed as number/host, determining a parasite's distribution independent of its host can be difficult. The Varroa mite (*Varroa destructor*) is a leading cause of honey bee colony death. A survey of 31 apiaries in which 24–57 colonies were sampled and ~300 adult bees examined for mites [73]. Clearly at the scale of sampling the aggregation of bees is extremely high, but without an estimate of b for adult bees it's hard to know if $b_{host} > b_{parasite} = 1.24 \pm 0.050$, but it seems highly likely.

The slope of gypsy moth egg mass temporal TPL ranged $1.54 \leq b \leq 1.77$ depending on sampling method, a range substantially above Brown and Cameron's (1982) estimate of $b = 1.27 \pm 0.04$ for the distribution of *Ooencyrtus kuvanae* parasites per egg mass [66]. The survey in Pennsylvania counted *O. kuvanae* on the same 1775 egg masses over a period of months in a range of gypsy moth population densities. Despite the difference in sampling program, again $b_{host} > b_{parasite}$. Parasitized eggs of the invasive leafhopper (*Siphonia rufofascia*) in Hawaii were significantly less aggregated than all eggs ($P \ll 0.001$) on a variety of hosts [27]. A similar result was obtained in a study of lemon psyllid (*Diaphorina citri*) and potential natural enemies in an orchard in Iran [21]. In addition to egg, nymph, and adult lemon psyllids, a predatory ladybird beetle (*Menochilus sexmaculatus*) and two parasitoid wasps (*Marietta leopardine* and *Tamarixia radiata*) were found in sufficient numbers for spatial TPL analysis. Both parasitoids were less aggregated than their psyllid nymph hosts and one, *M. leopardine*, was Poisson distributed as was the ladybird beetle, a generalist predator.

Competition

Like predation and parasitism, competition is frequently a negative effect for one participant and may be evident in TPL ($b_1 \neq b_2$). In four apple orchards in Spain, two with one species of earwig and two with two congeners, competition increased the slope of *Forficula auricularia* significantly ($P < 0.005$) from $b \approx 1.46$ to $b = 1.73$ [67]. The competing species *F. pubescens* was not different from Poisson ($P > 0.17$) with virtually identical slopes in the orchards with and without *F*.

auricularia. A study in France compared the distributions of three competing predators in apple orchards [72]. Among 11 phytoseiids found in 173 orchards, *Amblyseius andersoni, Kampimodromus aberrans,* and *Typhlodromus pyri* dominated. Ensemble TPLs of the three phytoseiids were similar, ranging $1.23 \leq b \leq 1.31$ and not significantly different ($P > 0.30$). While *K. aberrans* was largely restricted to the Mediterranean region, both *A. andersoni* and *T. pyri* were widespread and competed in at least 12 orchards. Although their abundances in those orchards differed, their spatial distributions were practically identical ($b_1 = b_2$).

TPLs of bird cherry-oat aphid (*Rhopalosiphum padi*) and greenbug (*Schizaphis graminum*) on winter wheat were not significantly different ($P > 0.50$), but sampling plans derived from the parameters recommended different number of samples [8]. What is surprising is that these two aphids in direct competition on the same plant are almost identically distributed, reminding us that lack of statistical significance does not necessarily denote an absence of biologically meaningful significance.

Both sugarcane delphacids (*Perkinsiella saccharicida*) and predatory mirids (*Tytthus* spp) are more aggregated as nymphs than adults ($b_{nymph} > b_{adult}$) and the delphacids are slightly more strongly aggregated than the predator ($b_{prey} > b_{predator}$) [26]. Similarly for three-stored product pests with densities ranging over four orders of magnitude, red flour beetle (*Tribolium castaneum*), lesser grain beetle (*Rhyzopertha dominica*), and rusty grain beetle (*Cryptolestes ferrugineus*), both intercept and slope of ensemble TPLs are closely similar suggesting that at least at the scale of sampling these beetles do not compete for space ($b_1 = b_2$) [40]. Hybrid TPLs of competing cassava green mite (*Mononychellus progresivus*) and cotton red mite (*Oligonychus gossypii*) on cassava were significantly different ($P < 0.07$) [77].

In general, predators and parasites have different, usually lower aggregation as measured by TPL b, than their prey or hosts. By contrast, competition usually does not influence b as Ramsayer et al. (2012) determined experimentally with competing bacterial colonies. A proposition by Hamilton (1971) that predation could lead to aggregation of prey is well supported by the entomological literature on TPL.

Effect of changes in scale

A number of studies of the effect of scale on TPL slope seem to support an increase in b with scale. For example, the TPL slope of green oak tortrix (*Tortrix viridana*) on oak increased from eggmasses/branch/tree within plots to eggmasses/branch within trees to eggmasses/branch/tree within plots [51]. The differences in b are small and not statistically significant, but the trend is clear. Similarly, TPL slope of red scale (*Aonidiella aurantii*) on citrus in Australia is higher for scales/tree/block than for scales/fruit/tree [13].

In contrast, a study of the tomato leafminer (*Tuta absoluta*) on tomatoes in Egypt shows clearly significant ($P < 0.05$) decreases in b in the order leafminers/leaflet, leafminers/leaf, and leafminers/plant with the latter indistinguishable from Poisson [44]. Similarly, a study of six species of leafminer on apple trees in

Spain support a decline in b with scale [49]. For five of the six species, b decreased from mines/leaf to mines/leaf/tree to mines/tree. The differences are small, except for *Callisto denticulella* mines/leaf ($b = 1.88 \pm 0.054$). With the exception of *C. denticulella*, the slopes are not significantly different from Poisson. Oddly, the pattern common to all species individually is not repeated when all six species are combined: the order then is mines/tree, mines/leaf, mines/leaf/tree.

There is also strong evidence for no change in scale. Hamid et al. (1999) found no difference in b and minor differences in A in a study of the weevil *Gymnetron pascuorum* on ribwort plantain spanning three scales and two years at two sites [43]. In this study, the number of weevils/seed head within plants, weevils/seed head within areas, and weevils/plant within area was compared. In the first two examples, the common metric weevils/seed head differ over two scales, with $b = 1.61$. Weevils/seed head within areas and weevils/plant within areas compare the same metrics at a larger scale. On the basis of these observations, scale, whether at the resolution of seed head, plant, or area, had no impact on b. A more complex picture emerges with the distribution of *Mecinus pyraster*, another weevil on the same host, but occupying a different niche. The slope of weevils/plant in two areas and 2 years are identical and similar to *G. pascuorum* with $b = 1.66$. The slopes of weevils/seed head within areas are also identical, but with $b = 0.56$, the distribution is near regular instead of highly aggregated. The results for weevils/seed head within areas in 2 years at one site were close to Poisson, and the other site were close to regular ($b = 0.25$). A noticeable difference between the two species is the range of means for TPL estimation: 2.5 orders of magnitude for *G. pascuorum* and <1 order for *M. pyraster* with most points < 1 weevil/seed head and 1–10 weevils/plant. Clark et al. (1996) had concluded that changes in scale could lead to unpredictable responses in TPL parameters, and this appears to be the case with *M. pyraster* on plantain, a situation exacerbated when the targets were scarce and densities covered a narrow range. Thus, the conclusion of increase in b with scale for *M. pyraster* should be regarded as tentative.

The strongest case for an increase in TPL slope with scale comes from a study in the Azores [69]. The Azorean archipelago in the North Atlantic, 1450 km from the European mainland, is a chain of 9 volcanic islands, plus some islets. As an isolated island chain, the Azores is an ideal location to test ecological hypotheses such as the connection between TPL and the abundance-occupancy relationship. The Azorean islands have a rich endemic community of arthropods as well as many native and introduced species. Gaston et al. (2006) took pitfall samples and canopy samples from the endemic Azores juniper tree (*Juniperus brevifolia*) on transects in reserves or protected areas on seven islands in 1999 and 2000 and a second survey on Terceira in 2003. The arthropods caught were identified to species and classified as endemic, native but not endemic, and introduced.

The motivation for this study was to examine empirical evidence for He and Gaston's (2003) abundance-variance-occupancy model that combines abundance-occupancy with TPL into a trivariate relationship in which occupancy, p, is a function of mean and variance of abundance:

$$p = 1 - \left(\frac{M}{V}\right)^{M^2/(V-M)} \tag{8.1}$$

Their results of arthropod assemblages of differing endemicity and three spatial scales fit the model extremely well. Regardless of the specific mechanisms generating TPL and abundance-occupancy, Gaston et al. conclude the distribution of species appears to be "entirely determined by abundance, occupancy and variance."

Overall soil arthropods showed increasing spatial variance as measured by b increasing from reserve to archipelago. And on average b was steepest for introduced species and shallowest for native species. Within this broad result, the pattern of scale's influence on b was not consistent with native species' b lowest at the island scale. The range of $1.45 \leq b \leq 1.71$ is not great but represents the range in b for introduced species from reserve to archipelago. As all the TPLs have high correlations ($0.964 \leq r \leq 0.989$), some of these small differences are statistically significant.

The soil arthropod pattern was not repeated with arthropods in the juniper canopy. The spatial variance of canopy arthropods was highest at the reserve scale and lowest at the island scale with archipelago between, and overall the native species had the highest b and endemics the lowest. Thus, the largest range of $1.46 \leq b \leq 1.70$ is from endemic to native at the reserve scale. The differences in pattern of scale and species status between the soil and canopy TPLs again support Clark et al.'s (1996) conclusion with plants that the relationship between scale and TPL slope is unpredictable.

Crustaceans

We end with some examples of crustacean distributions. The freshwater ostracod *Ankylocythere sinuosa* is commensal on the invasive crayfish *Procambarus clarkii* in Spanish rivers [83]. It is highly aggregated on its host with mean abundance ranging as high as 300/crayfish. Nauplius larvae of the copepod salmon louse (*Lepeophtheirus salmonis*) settle on their host and are sessile before molting to the mobile preadult copipodit stage. TPLs of the sessile stage of salmon louse were more aggregated on sea trout than the mobile chalimus that are indistinguishable from Poisson [84]. Hybrid spatial TPLs of monthly samples of the related copepod parasite *L. pectoralis* on plaice were not different in two consecutive years [85]. Cirripeds in a Danish fjord recovering from a hypoxia event were moderately aggregated [86]. An alien species, Sally Lightfoot crab (*Percnon gibbesi*) is very highly aggregated on rocky bottoms to 4 m below the surface in areas around Malta [87]. A multispecies-stage TPL of five stages plus adult females of two species of littoral copepods (*Calanus* spp.) were moderately aggregated [88]. While the adult females were not common, the abundance of juveniles generally increased with stage, so that the later stages were more strongly aggregated than the younger copepodites. This is in sharp contrast to most insects in which aggregation generally declines with stage.

Appendix: TPL estimates for arthropods

	NB	Range of means	r	A	SE[A]	b	SE[b]
A	Ensemble TPLs of European corn borer (*Ostrinia nubilalis*) in maize plants. Table 2 in Meyers and Patch (1937) of ECB in Wood County, Ohio, in 1932 gives the means and variances of borers/100 plants based on data of % infested plants and borers/infested plant. Their Table 3 gives the means and variance of 18 grouped frequencies in 1054 fields surveyed in 66 counties in Michigan, Ohio, Pennsylvania, and New York conducted in 1931 and 1932. The more variable single-county result is substantially the same as the combined data. Data from the Appendix in McGuire et al. (1957) of ECB per plant in 9 fields show a range of distributions from Poisson to Neyman's A with a much lower slope than Meyers and Patch's data						

	NB	Range of means	r	A	SE[A]	b	SE[b]
Meyers and Patch (1937)							
% of Infested plants	20	4.00 - 80.0	0.614	−0.027	0.393	1.466	0.273
Borers/infested plant	20	1.10 - 2.80	0.804	−0.313	0.081	2.073	0.291
Wood County − Borers/ 100 plants	20	6.00 - 168	0.966	0.253	0.186	1.783	0.109
All counties − Borers/100 plants	18	6.70 - 970	0.998	0.540	0.059	1.632	0.028
McGuire et al. (1957)	9	0.16 - 25.6	0.996	0.147	0.026	1.197	0.036
Taylor (1965) GMR estimates	4	0.65 - 25.6	1.000	0.143	0.015	1.252	0.019

	NB	Range of means	r	A	SE[A]	b	SE[b]
B	Spatial TPLs of winter moth (*Operophtera brumata*) life stages in Nova Scotia. Data from Figs. 24–26 in Embree (1961) and Figs. 10 and 11 in Embree (1965)						

	NB	Range of means	r	A	SE[A]	b	SE[b]
Emerging adults	19	2.56 - 43.0	0.912	−0.877	0.278	2.390	0.237
Climbing females	34	2.56 - 80.3	0.906	−0.135	0.167	1.903	0.142
Eggs on lichen	11	2.33 - 24.0	0.880	0.035	0.285	1.829	0.289
1st instar larvae on leaf clusters	49	0.033 - 9.10	0.904	−0.074	0.049	1.377	0.086
Falling prepupae	18	0.25 - 2.24	0.610	2.002	0.057	1.363	0.270

	NB	Range of means	r	A	SE[A]	b	SE[b]
C	Ensemble TPLs of egg masses, pupae, and adult males in 2 surveys (1979 and 1985) and a spatial TPL of adult male (1987) Gypsy moth (*Lymantria dispar*) in Pennsylvania. 1979 data from Embody and Bachlor (1980); 1985 data collected by the author; 1987 data from Taylor (2018)						

	NB	Range of means	r	A	SE[A]	b	SE[b]
Egg masses							
5-min walk (1979)	89	0.056 - 78.8	0.961	0.457	0.040	1.644	0.049
Fixed-radius plot (1979)	100	0.020 - 48.2	0.973	0.746	0.030	1.543	0.036
Prism point plot (1985)	245	0.167 - 62.0	0.912	0.157	0.042	1.774	0.047
Pupae							
Male (1985)	168	0.08 - 25.8	0.952	0.211	0.023	1.604	0.038
Female (1985)	168	0.08 - 19.7	0.940	0.245	0.028	1.683	0.045
Combined (1985)	336	0.08 - 25.8	0.944	0.227	0.018	1.648	0.030
Adult male moths							
2-km grid (1979)	67	1.02 - 141	0.959	0.288	0.088	1.855	0.065
Experiment (1987)	27	2.33 - 129	0.990	−0.110	0.067	1.685	0.046

	NB	Range of means	r	A	SE[A]	b	SE[b]
D	Ensemble TPLs of wireworms (Elateridae) sampled with different sized and number of quadrats Ensemble TPLs of wireworm survey in pasture and arable fields in England and Wales. Correlation coefficient, r, b, SE(b) do not change with conversion of data from #/ sample to absolute density estimates nor with the area units used; not only A but also SE(A) change with scale. Data from Yates and Finney (1942)						

	NB	Range of means	r	A	SE[A]	b	SE[b]
Jones (1937)							
$NQ = 25$	41	0.12 - 14.2	0.982	0.181	0.026	1.453	0.044
$NQ = 50$	42	0.08 - 9.46	0.989	0.233	0.019	1.319	0.031
$NQ = 100$	46	0.05 - 10.8	0.982	0.276	0.024	1.380	0.040
Combined (#/ft^2)	129	0.20 - 14.2	0.922	0.405	0.027	1.354	0.047
#/quadrat − ^{1}/16 ft^2	7	0.36 - 1.52	0.987	0.300	0.030	1.902	0.138
¼ ft^2	7	0.96 - 5.56	0.966	0.091	0.102	2.136	0.248
1 ft^2	7	5.40 - 19.0	0.975	−0.441	0.218	2.319	0.228
combined	21	0.36 - 1.09	0.984	0.218	0.043	1.693	0.070
#/ft^2	21	3.84 - 24.3	0.927	−0.347	0.214	2.456	0.212
#/m^2	21	41.3 - 262	0.927	-0.818	0.430	2.456	0.212
Yates and Finney (1942)							
Grass field − #/sample	35	0.20 - 4.68	0.997	0.431	0.006	1.182	0.015
1000/acre	35	100 - 2,350	0.997	2.640	0.043	1.182	0.015
#/m^2	35	24.7 - 581	0.997	2.143	0.034	1.182	0.015
Arable field − #/sample	35	0.20 - 4.68	0.988	0.440	0.013	1.264	0.034
1000/acre	35	100 - 2350	0.998	2.427	0.098	1.264	0.034
#/m^2	35	24.7 - 581	0.988	1.981	0.078	1.264	0.034
Combined - #/m^2	70	24.7 - 581	0.992	2.059	0.044	1.224	0.019

E TPLs of samples of Colorado potato beetle (*Leptinotarsa decemlineata*) in a potato field created 2 ways: 1. ensemble TPL of $NB = 16$ points of $NQ = 144$ samples in a 48 × 48 grid of sample points; 2. TPL derived from aggregation of 2034 grid points into 576 2 × 2 units, 256 3 × 3 units, 144 4 × 4 units, 64 6 × 6 units, 36 8 × 8 units, 16 12 × 12 units, and 16 × 16 units. Data from Table 6 in Beall (1939)

	NB	Range of means	r	A	SE[A]	b	SE[b]
Ensemble TPL of 2304 units	16	1.48 - 9.24	0.951	−0.058	0.088	1.595	0.132
Aggregated sample units	8	4.74 - 1213	0.999	−0.080	0.075	1.719	0.035

F Hybrid spatial TPLs of Japanese beetle (*Popillia japonica*) larvae in grass plots at 8 locations in New Jersey and Pennsylvania 1927–1934 and a subset 1930–1934. Data from Table 1 in Fox (1937). Ensemble TPLs of JB larvae in 4 grass plots in Jacobstown, NJ. The per plot TPL ($NB = 4$, $NQ = 2500$) slope is not significantly different from 2.0, but TPLs with $NB > 15$ points have lower and similar slopes. Copies of the original data sheets provided to LRT by Chester Bliss

	NB	Range of means	r	A	SE[A]	b	SE[b]
New Jersey and Pennsylvania							
1927–1934	18	0.10 - 16.9	0.984	0.388	0.061	1.393	0.061
1930–1934	18	0.20 - 14.8	0.962	0.245	0.117	1.771	0.121
Jacobstown, NJ							
$NB = 4$, $NQ = 2500$	4	2.77 - 19.1	0.974	1.629	0.287	1.974	0.317
$NB = 16$, $NQ = 625$	16	2.37 - 24.8	0.939	0.276	0.092	1.116	0.102
$NB = 100$, $NQ = 100$	100	1.20 - 27.7	0.926	0.089	0.041	1.206	0.046
$NB = 196$, $NQ = 49$	196	0.90 - 28.7	0.915	0.024	0.031	1.219	0.035
$NB = 400$, $NQ = 25$	400	0.56 - 31.8	0.921	−0.022	0.022	1.139	0.022
$NB = 576$, $NQ = 16$	576	0.38 - 31.5	0.891	−0.086	0.021	1.245	0.024
$NB = 2474$, $NQ = 4$	2,474	0.25 - 35.5	0.713	−0.431	0.018	1.439	0.020
Box counting	10	36.3 - 22,676	1.000	−0.096	0.014	1.972	0.005
Mean × box length	10	2.00 - 50.0	1.000	0.951	0.003	2.003	0.003

G Ensemble TPLs of white grub (predominately Japanese beetle) in soil samples taken in autumn and the following spring from corn fields in Virginia. ODR estimates from Jordan et al. (2012)

	NB	Range of means	r	A	SE[A]	b	SE[b]
Year 1 − Autumn '05 − Spring '06	28	-	0.88	0.125	-	1.12	0.119

Continued

	NB	Range of means	r	A	SE[A]	b	SE[b]
Year 2 – Autumn '06 – Spring '07	36	-	0.86	0.128	-	1.07	0.109
Year 3 – Autumn '07 – Spring '08	24	-	0.97	0.222	-	1.26	0.067

H Ensemble TPLs ($NQ = 8$) of adult Japanese beetles caught in traps with 3 different lure combinations. ODR estimates from Allsopp et al. (1992)

	NB	Range of means	r	A	SE[A]	b	SE[b]
Japonilure – Male	26	0.75 - 927	0.970	0.126	0.169	1.58	0.087
Female	26	0.13 - 66.8	0.990	0.517	0.052	1.68	0.044
PEG – Male	26	1.00 - 1206	0.975	−0.169	0.169	1.71	0.078
Female	26	1.00 - 1905	0.975	−0.143	0.200	1.71	0.083
Japonilure + PEG – Male	26	2.00 - 2828	0.975	−0.421	0.213	1.82	0.082
Female	26	1.00 - 3960	0.990	−0.208	0.143	1.74	0.054

I Spatial TPLs ($NQ = 4$) of adult Japanese beetles caught in modified Trécé Catch-Can traps and spatial TPLs of JB caught with traps with openings at 2 heights. Data from Tables 1–3 in Alm et al. (1994)

	NB	Range of means	r	A	SE[A]	b	SE[b]
Trécé standard trap	30	1.00 - 4674	0.908	−0.316	0.358	1.698	0.135
Trécé trap with large container	30	5.50 - 12,533	0.923	−0.745	0.373	1.770	0.129
Combined	60	1.00 - 12,533	0.913	−0.477	0.256	1.716	0.092
Open-top trap with water	20	6.80 - 294	0.941	−1.002	0.324	2.019	0.161
Standard Trécé trap at 13 cm	13	65.3 - 2792	0.922	−2.405	0.823	2.363	0.276
Standard Trécé trap at 90 cm	13	10.3 - 1256	0.976	−1.579	0.394	2.247	0.147

J Spatial and temporal TPLs of annual Japanese beetle samples invading Terceira Island, Azores, 1986–2008, during the spread around the island from the presumed port of entry, Lajes Airport. Data courtesy of the Terceira Agricultural Service

	NB	Range of means	r	A	SE[A]	b	SE[b]
Spatial	23	1536 - 20,205	0.969	0.927	0.373	1.886	0.102
Temporal	346	2.00 - 42,597	0.985	−0.080	0.064	2.091	0.019

K Ensemble TPLs ($NQ = 20$) of barnacles (*Balanus balanoides*) on steep and shallow rocky shores: 20 transects of 20 0.25-m² quadrats analyzed along transects perpendicular to and parallel to the low water mark. Results of a survey made by the author on a field course on Cumbrae Island, Scotland, in September 1971

	NB	Range of means	r	A	SE[A]	b	SE[b]
Steep rocky shore – Perpendicular	20	24.2 - 1034	0.979	0.374	0.240	1.973	0.096
Parallel to LWM	20	50.5 - 1072	0.983	1.172	0.185	1.695	0.073
Resultant ($NQ = 25$)	16	68.1 - 1101	0.973	1.194	0.264	1.677	0.104
Shallow rocky shore – Perpendicular	20	56.2 - 782	0.953	1.090	0.297	1.715	0.122
Parallel to LWM	20	71.6 - 805	0.960	1.186	0.271	1.680	0.111
Resultant ($NQ = 25$)	16	22.0 - 786	0.974	1.177	0.246	1.678	0.102

L Temporal TPLs ($NQ = 10$) of introduced barnacle *Elminius modestus* on the south coast of England. Density (#/10 cm²) of settled barnacle cyprid larvae on glass plates with microscope slides and without pre-existing settled cyprid larvae. Data from Table 3 in Knight-Jones and Stevenson (1950)

	NB	Range of means	r	A	SE[A]	b	SE[b]
Spatial – Barnacled	9	1.26 - 36.0	0.881	0.117	0.370	1.773	0.317
Not barnacled	10	0.17 - 8.72	0.900	0.260	0.152	1.630	0.251
Combined	19	0.17 - 36.0	0.934	0.239	0.133	1.670	0.145
Temporal – Barnacled	11	8.70 - 50.0	0.975	0.040	0.166	1.860	0.138
Not barnacled	11	1.72 - 11.4	0.957	0.072	0.104	2.174	0.211
Combined	22	1.72 - 50.0	0.983	0.212	0.066	1.751	0.072

Appendix 8.M

Thysanoptera (thrips)

Ref.	Latin name (common name)	Habitat/host/prey	Type	NQ	NB	r	A	SE(A)	b	SE(b)	Comments (√=sampling plan)
1	Frankliniella occidentalis (western flower thrips)	Greenhouse crops	S	10	12	0.954	0.240	0.403	1.940	0.021	√ Yellow sticky cards – greenhouse 1 in Ohio (impatiens and other flowers)
			S	10	14	0.985	−0.320	0.213	1.900	0.100	Greenhouse 2 (mixed pot plants)
			S	10	14	0.980	−0.110	0.208	2.170	0.122	Greenhouse 3 (mixed pot plants)
			S	10	11	0.975	0.160	0.365	1.770	0.133	Greenhouse 4 (chrysanthemums)
			S	10	7	0.985	0.580	0.193	1.650	0.119	Greenhouse 5 (mixed pot plants)
			S	10	12	0.990	0.540	0.164	1.660	0.066	Greenhouse 6 (chrysanthemums)
			S	10	70	0.964	0.150	0.350	1.850	0.061	Combined
			S		16	0.922	−0.560	0.490	1.990	0.222	Sweet peppers (Ontario)
			S		81	0.980	0.440	0.311	1.780	0.038	Cucumbers and peppers (British Columbia)
			S		30	0.943	0.080	0.437	1.750	0.123	Cucumbers (British Columbia)
			S		44	0.911	−0.130	0.478	1.950	0.233	Cucumbers (Ohio)
			S		241	0.954	0.080	0.410	1.840	0.039	All data combined
2	Frankliniella occidentalis (western flower thrips)	Cucumber	S	50	68	0.980	0.588		1.730	0.040	Larvae – apical
			S	50	64	0.975	0.677		1.620	0.050	– Middle
			S	50	64	0.980	0.759		1.710	0.040	– Basal
			S	50	66	0.933	0.392		1.380	0.070	Adults – apical
			S	50	64	0.959	0.441		1.310	0.050	-middle
			S	50	63	0.970	0.536		1.560	0.050	-basal
	Amblyseius cucumeris (predatory mite)	Western flower thrips	S	50	49	0.938	0.401		1.290	0.070	A. cucumeris – apical
			S	50	43	0.995	0.396		1.310	0.070	– Middle
			S	50	43	0.943	0.501		1.480	0.090	– Basal
3	Frankliniella occidentalis (western flower thrips)	citrus	S	50	160	0.978	0.320		1.190	0.020	Adult in flower+fruitlet
			S	50	119	0.976	0.664		1.400	0.029	Nymphs in flower+fruitlet
	Pezothrips kellyanus (Kelly's citrus thrips)	Citrus	S	50	55	0.988	0.246		1.160	0.025	Adult in flower
			S	50	26	0.931	0.600		1.380	0.110	Nymphs in flower

#	Species	Host	S/E	n1	n2						√	Notes
4	*Frankliniella occidentalis* (western flower thrips)	Alfalfa	S		20	0.950	0.130	0.060	1.090	0.060		Alfalfa cut monthly
			S		21	0.930	0.380	0.060	1.370	0.080		Alfalfa cut bimonthly
5	*Anaphothrips obscurus*	Timothy	E	10	31	0.889	0.086		1.460	0.140		2006 – direct observations
			E	10	31	0.825	0.033		0.920	0.117		– Beat cup
			E	10	26	0.959	0.143		1.330	0.080		2007 – direct observations
			E	10	26	0.970	0.124		1.300	0.067		– Beat cup
			E	10	27	0.964	0.117		1.400	0.077		2008 – direct observations
			E	10	27	0.849	0.104		1.210	0.151		– Beat cup
			E	10	84	0.943	0.272		1.350	0.052		Combined – direct observations
			E	10	84	0.943	0.170		1.190	0.046		– Beat cup
6	*Thrips tabaci* (onion thrips)	Soybeans	S	43	17	0.976	−0.658		1.200	0.060		Sahar
			S	41	17	0.990	−1.046		1.180	0.040		Tellar – good range of means
			S	40	16	0.980	−0.854		1.160	0.060		Variety – Williams
			S	42	15	0.992	−1.046		1.130	0.030		Zane
			S	46	17	0.917	−0.886		1.110	0.130		L17
			S	47	17	0.891	−1.222		0.870	0.110		Sari
			S	45	14	0.672	−0.638		0.860	0.250		Ks3494
			S	44	17	0.735	−0.523		0.430	0.310		Dpx
7	Multiple thrips species	Weeds in mango orchard	S	70	12	0.894	−1.770		2.260	0.357	√	Adults – dry season, treated
			S	70	12	0.917	−1.650		2.330	0.322		Larvae
			S	70	7	0.959	−0.400		1.540	0.203		Adults – wet season, treated
			S	70	7	0.975	−0.710		2.090	0.214		Larvae
			S	70	11	0.742	−0.680		1.750	0.528		Adults – dry season, untreated
			S	70	11	0.917	−1.910		2.230	0.324		Larvae
			S	70	7	0.964	0.730		1.080	0.133		Adults – wet season, untreated
			S	70	7	0.938	0.070		1.370	0.226		Larvae

Hemiptera (Sternorrhyncha) Aphids

#	Species	Host	S/E	n1	n2						√	Notes
8	*Schizaphis graminum* (greenbug)	Wheat	E	100	159	0.978	1.755	0.059	1.390	0.024	√	Competitors
	Rhopalosiphum padi (bird cherry-oat aphid)		E	100	134	0.990	1.607	0.044	1.376	0.017		

Continued

Ref.	Latin name (common name)	Habitat/host/prey	Type	NQ	NB	r	A	SE(A)	b	SE(b)	Comments (√=sampling plan)
9	*Brachycaudus helichrysi* (plum aphid)	Sunflower	E-S	25–320	18	0.975	0.662	0.125	1.524	0.044	Cultivar – Mirasol √
			E-S	25–321	17	0.985	0.525	0.103	1.532	0.078	– Viki
			E-S	25–322	53	0.980	0.583	0.064	1.528	0.068	-Mirasol, Viki, and 3 others
10	*Acyrthosiphon pisum* (pea aphid)	Alfalfa	S	60–120	35	0.900	0.529	1.529	1.116	0.097	Oklahoma √
			S	32–64	33	0.970	0.451	0.097	1.536	0.066	Wisconsin
			S	32–120	68	0.970	0.502	1.527	1.438	0.049	Combined
11	*Acyrthosiphon pisum* (pea aphid)	Indian pea	S	5	13	0.883	−1.100		3.160	0.506	Wondata, Ethiopia, 2009-2010 √
			S	5	13	0.938	−2.580		4.180	0.465	– 2010–2011
			S	5	12	0.980	−0.680		3.060	0.198	– Woreta 2009–2010
			S	5	12	0.849	−0.100		2.040	0.402	– 2010–2011
12	*Aphis gossypii* (cotton aphid)	Cotton	S	12	22	0.953	1.574	0.084	1.294	0.038	1981
			S	12	21	0.975	1.715	0.064	1.251	0.122	1982
			S	12	22	0.939	0.952	0.099	1.466	0.050	1983
			S	12	20	0.898	1.552	0.149	1.312	0.067	1984

Coccoidea (scale insects)

Ref.	Latin name (common name)	Habitat/host/prey	Type	NQ	NB	r	A	SE(A)	b	SE(b)	Comments (√=sampling plan)
13	*Aonidiella aurantii* (red scale)	Citrus	S		1340	0.975			1.680	0.133	Scales/fruit/tree
			S		24	0.964			1.750	0.102	Scales/tree/block
14	*Aulacaspis tubercularis* (white mango scale)	Mango	S	80	12	0.987	0.169	0.059	1.144	0.058	Colonies (males)/leaf
			S	80	13	0.978	0.087	0.045	1.088	0.069	Females/leaf
15	*Aspidiotus nerii* (oleander scale)	Jojoba	S	56	13	0.966	−0.610	0.378	1.976	0.159	Larvae – 1981 and 1982
			S	40	12	0.972	−0.965	0.306	2.125	0.163	– 1984
			S	40–56	25	0.978	−0.826	0.197	2.060	0.092	– Combined
16	*Phenacoccus peruvianus* (mealybug)	Bougainvillea	S	50	238	0.980	2.100		1.620	0.022	Total/leaf √
			S	10	207	0.980	1.810		1.760	0.025	Total/bract
			S	20	140	0.985	1.700		1.790	0.027	Total/twig

Aleyrodidae (whiteflies)

Ref.	Latin name (common name)	Habitat/host/prey	Type	NQ	NB	r	A	SE(A)	b	SE(b)	Comments (√=sampling plan)
17	*Bemisia tabaci* (silverleaf whitefly)	Tomato	E		36	0.900	−0.219	0.345	2.059	0.167	Eggs/3 leaflets √
			E		36	0.883	0.347	0.341	1.777	0.158	Nymphs/3 leaflets
			E		35	0.959	0.572	0.094	1.308	0.068	Pupae/3 leaflets
18	*Bemisia tabaci* (silverleaf whitefly)	River tamarind	S		12	0.873	−1.207	0.401	2.269		Upper plant
			S		12	0.911	−1.109	0.316	2.206		Middle
			S		12	0.925	−1.578	0.329	2.530		Lower

No.	Species	Host	Type	Range	n	R²					Method	
19	*Trialeurodes vaporariorum* (greenhouse whitefly)	Green bean	E	74–311	20	0.991	0.464	0.041	1.262	0.023	Adults/quadrat	
			E	127–210	20	0.986	0.358	0.037	1.150	0.021	Eggs/quadrat	
			E	124–202	20	0.994	0.109	0.026	1.061	0.013	3rd instars/quadrat	
			E	127–202	20	0.994	0.088	0.026	1.052	0.014	Pupae/quadrat	

Psyllidae (plant lice)

No.	Species	Host	Type	Range	n	R²					Method	
20	*Diaphorina citri* (Asian citrus psyllid)	Citrus	S	10–20	9	0.995	−0.073	0.106	1.198	0.037	Sweep net	√
			S	10–20	1071	0.975	0.517	0.027	1.279	0.009	Stem tap – multiple trees/sample site	
21	*Diaphorina citri* (lemon psyllid)	Lemon	S	10–20	175	0.970	0.174	0.021	1.291	0.025	Stem tap – single trees/sample site	
			S	10–20	1202	0.954	1.081	0.032	1.302	0.011	Visual	
			S	10–20	191	0.975	0.413	0.021	1.544	0.025	Sticky trap	
			S	10–20	99	0.970	0.303	0.044	1.840	0.046	Suction trap	
			S		24	0.918	0.755	0.142	1.447	0.131	Nymph	
			S		23	0.927	0.757	0.103	1.287	0.105	Adult	
			S		11	0.898	0.813	0.239	1.144	0.178	Egg	
	Tamarixia radiata	Lemon psyllid	S		15	0.957	0.769	0.093	1.320	0.172	Parasitoid wasp	
	Marietta leopardina	Lemon psyllid	S			0.980	0.057	0.081	1.036	0.079	Parasitoid wasp	
	Menochilus sexmaculatus	Lemon psyllid	S		12	0.923	−0.193	0.091	0.800	0.111	Predatory ladybird beetle	
22	*Bactericera cockerelli* (potato psyllid)	Bell pepper	S	8–18	20	0.943	1.950		1.630	0.135	Nymphs	√
23	*Bactericera cockerelli* (potato psyllid)	Tomato	E	39–84	15	0.976	0.927	0.048	1.382	0.086	Eggs	√
			E	39–84	15	0.953	0.590	0.053	1.155	0.101	Nymphs	
			E	39–84	14	0.936	0.065	0.073	1.054	0.114	Adults (excluding an outlier)	
			E	39–84	15	0.883	0.443	0.147	1.613	0.238	Adults (including an outlier)	
24	*Bryobia rubrioculus*	Apple	S	130	12	0.992	0.502	0.045	1.130	0.044	Apple mites in a sprayed orchard	√
	Cenopalpus irani	Mites	S	130	13	0.976	0.364	0.066	1.126	0.076		
	Zetzellia pourmirzai	Mites	S	130	11	0.987	0.511	0.067	1.225	0.065	Predatory mite	
25	*Cacopsylla pyri* (pear psyllid)	Pear	S	240	~160	0.982	1.706	0.029	1.507	0.014	Adults/shoot	√
			S	240	~160	0.975	0.719	0.044	1.206	0.017	Nymphs/shoot	
			S	240	~160	0.987	1.510	0.045	1.335	0.015	Nymphs/leaf	
	Pilophorus gallicus	Pear psyllid	S	200	~160	0.970	0.409	0.030	1.136	0.016	Beating net – adults	
			S	200	~160	0.970	0.429	0.057	1.121	0.026	Beating net – predator	

Continued

Hemiptera (Auchenorryncha)

Planthoppers and leafhoppers

Ref. name	*Latin name* (common name)	Habitat/host/prey	Type	NQ	NB	r	A	SE(A)	b	SE(b)	Comments (√=sampling plan)
26	*Perkinsiella saccharicida* (sugarcane delphacid)	Sugarcane	E		>300	0.975	1.030	0.050	1.230	0.010	√ Prey – egg masses/stalk
			E		>300	0.975	0.670	0.040	1.260	0.020	– Nymphs/stalk
			E		>300	0.975	0.180	0.030	1.110	0.010	– Adults/stalk
	Tytthus spp. (predatory mirids)	Sugarcane delphacid	E		>100	0.980	0.580	0.070	1.240	0.020	Predator – nymphs/stalk
			E		>100	0.938	0.160	0.100	1.100	0.050	– Adults/stalk
27	*Sophonia rufofascia* (two spotted leafhopper)	Assorted plants	S	164–4153	22	0.950	1.837	0.122	2.125	0.142	Total eggs/m^2 leaf
			S		21	0.886	0.126	0.107	1.139	0.143	Parasitized eggs/m^2 leaf
28	*Empoasca fabae* (potato leafhopper)	Green beans	S	16	11	0.688	−2.654	2.419	3.179	1.116	Visual – nymphs
			S	16	12	0.730	0.382	1.056	1.783	0.527	– Adults
			S	16	16	0.919	0.084	0.240	1.667	0.191	D-vac –nymphs
			S	16	8	0.681	0.203	0.086	0.866	0.380	– Adults
29	*Empoasca fabae* (potato leafhopper)	Red maples	E	24	40	0.959	0.300	0.095	1.150	0.056	May
			E	24	45	0.812	0.060	0.089	1.460	0.061	June
			E	24	42	0.889	0.640	0.033	1.310	0.068	July
			E	24	27	0.990	−0.030	0.059	0.990	0.028	August
			S	4	5	1.000	−0.240	0.002	0.920	0.001	Clone 56028
			S	4	5	0.990	0.890	0.142	1.500	0.080	Clone 55896
30	Leafhoppers	Potato	M-S		16	0.840	1.014	0.235	1.411	0.228	Leafhoppers/sticky card
			M-S		34	0.880	0.658	0.112	1.645	0.152	Leafhoppers/100 sweeps
31	*Empoasca kraemeri*	Green beans	E	379–431	31	0.944	−1.728	0.098	3.006	0.196	Nymphs/beating tray
			E	379–431	31	0.953	−1.988	0.054	2.685	0.159	Adults/beating tray
32	*Homalodisca vitripennis* (glassy-winged sharpshooter)	Orange	S	20	16	0.980	0.494	0.078	1.430	0.078	A-vac – nymphs
			S	20	24	0.990	0.152	0.035	1.160	0.035	√ – adults
			S	20	14	0.964	0.515	0.124	1.570	0.124	Beating pole and net – nymphs
			S	20	25	0.990	0.173	0.040	1.340	0.040	– Adults
			S	20	17	0.980	0.558	0.083	1.580	0.083	Beating pole and bucket – nymphs
			S	20	25	0.985	0.294	0.047	1.280	0.047	– Adults
			S	20	15	0.980	0.654	0.090	1.590	0.090	D-vac – nymphs
			S	20	24	0.980	0.270	0.059	1.360	0.059	– Adults

No.	Species	Commodity		Sample size	n							Life stage / method
33	Homalodisca vitripennis (glassy-winged sharpshooter)	Oranges	S	17		0.985	1.285	0.140	1.583	0.077	√	Nymphs
			S	25		0.985	0.679	0.098	1.282	0.049		Adults
			S	25		0.985	0.717	0.230	1.425	0.115		Both
Hemiptera (Heteroptera)												
True bugs												
34	Sahlbergella singularis (cacao mirid bug)	Cacao	S	51			1.110	0.324	1.500	0.064	√	2003
			S	61			1.120	0.713	1.500	0.025		2004
35	Oebalus pugnax (rice stink bug)	Rice	S	69		0.922	-0.080		1.411	0.072	√	Sweep net – 10 sweeps
			S	49		0.872	0.029		1.178	0.097		180° stick sweep
			S	69		0.707	0.150		0.791			38 cm stick sweep – 1 pass
			S	69		0.800	0.249		0.888	0.081		– 2 passes
			S	69		0.877	0.233		0.980	0.065		– 3 passes
			S	69		0.877	0.274		0.996	0.067		– 4 passes
			S	69		0.787	0.331		0.982			– 5 passes
36	Megacopta cribraria (kudzu bug)	Soybeans	S	100		0.985	-3.523		3.270	0.115	√	Sweep net – 2011 – adults
			S	100		0.952	1.143		1.390	0.091		– Nymphs
			S	100		0.803	0.377		1.310	0.229		– Egg mass
			S	104		0.901	0.844		1.680	0.186		– 2012 – adults
			S	104		0.989	0.389		1.730	0.060		– Nymphs
			S	104		0.876	0.410		1.760	0.223		– Egg mass
37	Megacopta cribraria (kudzu bug)	Soybeans	S	25–58		0.501	1.279		0.901	0.318	√	2012/13 – sweep net – adults
			S	25–58		0.996	0.328		1.615	0.028		–Nymphs
			S	25–58		0.746	1.459		0.871	0.159		– beating tray – adults
			S	25–58		0.985	0.473		1.620	0.058		– Nymphs
Coleoptera												
38	Leptinotarsa decemlineata (Colorado potato beetle)	Potato	S	50–80	313	0.973	1.010		1.600	0.004	√	Small larvae
			S	50–80	321	0.963	0.570		1.400	0.004		Large larvae
			S	50–80	378	0.978	0.180		1.110	0.002		Adults
			S	50–80	384	0.977	0.170		1.090	0.002		Egg masses
39	Sitophilus oryzae and S. zeamais	Stored products	S	25	35	0.975	0.674		1.560	0.031	√	Food-based oil lure
			S	25	36	0.889	0.545		1.480	0.066		Sitophilure aggregation pheromone

Continued

Ref.	Latin name (common name)	Habitat/host/prey	Type	NQ	NB	r	A	SE(A)	b	SE(b)	Comments (√=sampling plan)
	Tribolium castaneum (red flour beetle)		S	25	71	0.980	0.609	0.067	1.440	0.020	Both traps
			S	25	48	0.970	0.500	0.037	1.400	0.071	Food-based oil lure
			S	25	34	0.927	0.625	0.097	1.510	0.112	Sitophilure aggregation pheromone
40	*Tribolium castaneum* (red flour beetle)	Stored products	S	25	82	0.959	0.528	0.039	1.400	0.051	Both traps
			E	21	24	0.971	1.861	0.071	1.743	0.091	√ Samples from 4 grain storage bins with 4 orders of magnitude
	Rhyzopertha dominica (lesser grain borer)		E	21	22	0.983	1.824	0.089	1.701	0.072	range
	Cryptolestes ferrugineus (rusty grain beetle)		E	21	24	0.975	1.862	0.120	1.655	0.080	of densities
	Combined		E	21	70	0.980	1.847	0.051	1.685	0.042	
41	*Hypera postica* (alfalfa weevil)	Alfalfa	S	65	20	0.959	0.515	0.089	1.355	0.089	√ Larvae/quadrat
			S	65	19	0.953	0.412	0.090	1.300	0.090	Pupae/quadrat
			S	65	9	0.988	0.491	0.068	1.416	0.068	Adults/quadrat
42	*Acrotomopus atropunctellus* (sugarcane weevil borer)	Sugarcane	S		18	0.854	0.185		1.250	0.157	√ 2011–2012
			S		20	0.949	0.179		1.220	0.095	2012–2113
			S		19	0.975	0.053		1.050	0.061	2013–2014
			S		56	0.938	0.140	0.029	1.180	0.060	Combined
			S	30	44	0.970	0.143	0.037	1.124	0.042	Examination time = 2 min
			S	30	44	0.954	0.170	0.033	1.164	0.055	ET = 4 min
			S	30	44	0.959	0.167	0.029	1.167	0.053	ET = 6 min
			S	30	44	0.933	0.149	0.033	1.177	0.068	ET = 8 min
			S	30	44	0.938	0.146	0.029	1.167	0.060	ET = 10 min
43	*Gymnetron pascuorum*	Ribwort plantain	S				0.320		1.610	0.035	Site 1 – #/seedhead within plants
			S				0.690		1.610	0.035	– #/seedhead within areas
			S				0.500		1.610	0.035	– #/plant within area
			S				0.500		1.610	0.035	Site 2 – #/seedhead within plants
			S				0.630		1.610	0.035	– #/seedhead within areas
			S				0.810		1.610	0.035	– #/plant within area

No.	Species	Host	E/S		n						√	Notes
	Mecinus pyraster		S			-0.170		0.560	0.067			Site 1 1993 – #/seedhead within plants
			S			-0.010		0.930	0.120			– #/seedhead within areas
			S			0.280		1.660	0.170			– #/plant within area
			S			-0.260		0.560	0.067			Site 1 1994 – #/seedhead within plants
			S			-0.030		1.010	0.220			– #/seedhead within areas
			S			0.420		1.660	0.170			– #/plant within area
			S			-0.170		0.560	0.067			Site 2 – #/seedhead within plants
			S			-0.170		0.250	0.283			– #/seedhead within areas
			S			0.220		1.660	0.170			– #/plant within area

***Lepidoptera* "Microlepidoptera"**

No.	Species	Host	E/S		n						√	Notes
44	*Tuta absoluta* (tomato leafminer)	Tomato	E		229	0.851	0.761		1.654	0.068		#/leaflet
			E		55	0.866	0.788		1.315	0.104		#/leaf
			E		12	0.648	1.987		0.973	0.362		#/plant
45	*Diatraea grandiosella* (southwestern corn borer)	maize	S		49	0.980	0.587	0.222	1.066	0.333	√	Eggs
			S		93	0.975	0.255	0.024	1.099	0.026		Small larvae
			S		98	0.980	0.115	0.018	1.025	0.021		Medium larvae
			S		57	0.980	0.040	0.024	0.984	0.026		Nondiapause large larvae
			S		40	0.980	-0.090	0.018	0.908	0.038		Prediapause large larvae
			S		69	0.990	0.077	0.021	1.028	0.019		Pupae
			S		406	0.954	0.204	0.015	1.090	0.017		Total
46	*Mussidia nigrivenella* (maize ear borer)	Maize	E	20	140	0.954	0.430		1.360	0.036	√	On maize
			S	20	~150	0.917	0.452		1.360	0.049		On *Parkia biglobosa* (leguminous tree)
			S	20	~174	0.781	-0.078		1.380	0.084		On *Gardenia sokotensis* (shrub)
			S	10	60	0.894	0.573		1.380	0.091		On *Ximenia americana* (shrub)
			S	10	52	0.938	0.669		1.670	0.087		On *Adansonia digitata* (baobab tree)

Continued

Ref.	Latin name (common name)	Habitat/host/prey	Type	NQ	NB	r	A	SE(A)	b	SE(b)	Comments (√=sampling plan)
47	Mussidia nigrivenella (maize ear borer)	Maize	S	200	4	0.566	0.810	0.060	1.120	1.050	Main season
		Jack bean	S	200	7	0.480	0.890	0.140	1.350	1.130	Cover crop
		Maize	S	200	4	0.624	0.840	0.050	1.210	1.040	Minor season
		Jack bean	S	200	7	0.510	1.790	0.310	1.570	1.180	Cover crop
48	Ectomyelois ceratoniae (carob moth)	Dates	S	600	21	0.843	−0.935		1.883	0.276	Eggs/fruit on the ground
			S	600	19	0.980	0.638		1.322	0.065	Larvae/fruit on the ground
			S	600	21	0.883	0.086		1.619	0.197	Eggs/abscised fruit
			S	600	21	0.877	0.449		1.201	0.151	Larvae/abscised fruit
49	Phyllonorycter blancardella	Apple	S	10	1131	0.964	0.139		1.141	0.014	√ Mines/leaf
			S	50	66	0.983	0.096		1.102	0.037	Mines/leaf/tree
			S	50	66	0.997	0.129		1.062	0.013	Mines/tree
	P. corylifoliella		S	10	411	0.950	0.134		1.132	0.027	Mines/leaf
			S	50	57	0.988	0.063		1.028	0.031	Mines/leaf/tree
			S	50	57	0.995	0.068		1.024	0.019	Mines/tree
	Leucoptera scitella		S	10	437	0.957	0.253		1.254	0.027	Mines/leaf
			S	50	41	0.993	0.279		1.176	0.03	Mines/leaf/tree
			S	50	41	0.992	0.244		1.082	0.031	Mines/tree
	Stigmella malella		S	10	492	0.973	0.193		1.195	0.019	Mines/leaf
			S	50	59	0.988	0.079		1.039	0.031	Mines/leaf/tree
			S	50	59	0.994	0.046		1.005	0.020	Mines/tree
	Lyonetia clerkella		S	10	451	0.961	0.166		1.163	0.023	Mines/leaf
			S	50	64	0.992	0.172		1.132	0.026	Mines/leaf/tree
			S	50	64	0.996	0.123		1.049	0.017	Mines/tree
	Callisto denticulella		S	10	54	0.989	0.883		1.883	0.054	Mines/leaf
			S	50	25	0.975	0.323		1.175	0.081	Mines/leaf/tree
			S	50	25	0.977	0.520		1.183	0.072	Mines/tree
	All species combined		S	10	1906	0.961	0.068		1.073	0.010	Mines/leaf
			S	50	66	0.975	0.110		1.104	0.045	Mines/leaf/tree
			S	50	66	0.995	0.099		1.041	0.018	Mines/tree – trees selected at random

#	Species	Host	Type	Range	n	R²					Description	√
50	*Phyllocnistis citrella* (citrus lime leafminer)		S	90	90	0.872	0.310		1.120	0.067	Eggs/leaf	√
			S	90	90	0.906	0.380		1.210	0.060	Larvae/leaf	
51	*Tortrix viridana* (green oak tortrix)	oak	E	4	503	0.800	0.082		1.389	0.047	Egg masses/branch within trees	√
			E	8	64	0.877	−0.480		1.435	0.100	Egg masses/branch/tree within plots	
			E	64	8	0.632	−0.612		1.221	0.611	Egg masses/branch/plot within sites	
"Macrolepidoptera"												
52	*Platyprepia virginalis* (Ranchman's tiger moth)	Lupines	T	10–20	5		−1.316		2.820		Wet sites – #/10 bushes	√
			T	10–21	8		0.309		2.820		– Dry sites	
53	*Helicoverpa zea* and *Spodoptera frugiperda* (panicle caterpillars)	Sorghum	E	48–96	15	0.970	0.540	0.170	1.160	0.080	Texas coastal plain	√
			E	48–96	23	0.949	0.440	0.170	1.100	0.080	Texas high plan	
			E	48–96	40	0.944	0.470	0.110	1.060	0.060	Central Oklahoma	
			E	48–96	115	0.899	0.290	0.070	1.090	0.050	Combined + central Kansas	
54	*Lymantria dispar* (gypsy moth)	Forest	S	180–373	11	0.682	−1.077	0.759	4.341	1.554	Wisconsin	
			S	170–264	11	0.887	0.214	0.134	2.270	0.393	West Virginia	
			S	170–373	11	0.647	0.010	0.264	2.388	0.629	Combined	
Trichoptera (caddis flies)												
55	*Philopotamus montanus*	Freshwater	S	10	48	0.989	0.164	0.074	1.269	0.029	Total larvae	
			S	10	16	0.978	0.595	0.261	1.524	0.096	Prepupae	
			S	10	12	0.991	0.649	0.281	1.620	0.078	Pupae	
56	*Silo pallipes*	Freshwater	S		20	0.991	0.176	0.069	1.258	0.040	Total larvae	
			S		20	0.991	0.836	0.405	1.738	0.056	Prepupae and pupae	
57	*Silo pallipes*	Freshwater	S	30–40	13	0.987	0.814	0.210	1.851	0.102	Host pupae	
	Agriotypus armatus (parasitic wasp)	Silo pallipes	S	30–40	13	0.975	0.407	0.248	1.391	0.108	Parasitized pupae	
Diptera												
58	Assorted fruit flies	Fruit	M-S		34	0.943	0.719	0.065	1.777	0.110	Flies emerging from collections of fruit	
	Drosophila melanogaster (fruit fly)		E		28	0.900	0.607	0.074	1.522	0.144	Eggs – females with virgin females	
			E		28	0.876	0.521	0.072	1.354	0.146	– Females with males	

Continued

Ref.	Latin name (common name)	Habitat/host/prey	Type	NQ	NB	r	A	SE(A)	b	SE(b)		Comments (√=sampling plan)
59	Cricotopus bicinctus (chironomids)	Bulrush	S	14–69	55	0.959	6.270	0.108	1.470	0.056		Spring
	Cricotopus fuscus		S	14–69	43	0.943	2.000	0.096	1.560	0.087		Autumn
			S	14–69	54	0.889	8.220	0.094	1.270	0.092		Spring
	Orthocladius sp.		S	14–69	46	0.917	2.340	0.028	1.560	0.107		Autumn
			S	14–69	31	0.954	9.440	0.085	1.320	0.082		Spring
			S	14–69	46	0.843	3.150	0.034	1.330	0.133		Autumn
	Synorthocladius semivirens		S	14–69	73	0.927	10.920	0.108	1.240	0.061		Spring
			S	14–69	43	0.889	2.820	0.140	1.260	0.102		Autumn
	Parafanytarsus natvigi		S	14–69	66	0.943	8.300	0.122	1.370	0.061		Spring
			S	14–69	47	0.883	2.650	0.039	1.330	0.107		Autumn
60	Stomoxys calcitrans (stable fly)	Cattle	S	10	67	0.818			1.490	0.130		2004
			S	11	87	0.837			2.110	0.150		2005
			S	19	75	0.832			1.540	0.120		2010
			S	21	90	0.761			1.650	0.150		2011
61	Aedes aegypti (yellow fever mosquito)	Mammals	T	40	240	0.940	0.310		1.290	0.030	√	Adults/trap
			T	40	240	1.000	1.326		1.320	0.000		Eggs/ovitrap
62	Aedes albopictus (Asian tiger mosquito)	Mammals	S	63	35	0.946	1.184	0.206	1.622	0.097		Eggs/ovitrap/week – spatial
			T	35	63	0.997	0.881		1.560	0.014	√	– Temporal
63	Aedes albopictus (Asian tiger mosquito)	Mammals	T	25–145	40	0.868	1.296	0.231	1.376	0.128		Eggs/ovitrap/week
64	Culex tritaeniorhynchus (invasive mosquito)	Mammals	S	20	11	0.892	1.544	0.215	1.052	0.178	√	2009 – Sampled rice paddy in Greece
			S	20	11	0.983	0.693	0.114	1.522	0.096		2010
			S	20	10	0.973	0.863	0.122	1.433	0.121		2011
Hymenoptera												
65	Pseudacteon tricuspis (parasitoid)	Fire ant	S		15	0.932	−0.490	0.185	2.342	0.253		No zeros
			S		15	0.947	0.422	0.082	1.473	0.138		Zeros counted
66	Ooencyrtus kuvanae (parasitic wasp)	Gypsy moth eggs	T	28	65	0.970	0.350		1.270	0.040		Egg parasitoid

Dermaptera

67	*Forficula auricularia*	Apple	S		33	0.970			1.430	0.060	Orchard 1
			S		25	0.980			1.480	0.060	Orchard 2
			S		25	0.985			1.730	0.070	Orchard 3
			S		17	0.985			1.730	0.080	Orchard 4
	Forficula pubescens		S		16	0.894			1.240	0.170	Orchard 3
			S		7	0.656			0.920	0.480	Orchard 4
68	Various arthropods	Soil	M-S		16	0.982	0.049	0.063	1.446	0.074	Reserve – introduced
			M-S		28	0.975	0.081	0.045	1.507	0.068	– Endemic
			M-S		30	0.980	0.139	0.041	1.532	0.059	– Native
			M-S		50	0.983	0.376	0.026	1.666	0.044	Island – introduced
			M-S		37	0.974	0.269	0.040	1.606	0.064	– Endemic
			M-S		64	0.984	0.263	0.023	1.521	0.035	– Native
			M-S		95	0.986	0.479	0.015	1.712	0.030	Archipelago – introduced
			M-S		81	0.978	0.393	0.021	1.602	0.039	– Endemic
			M-S		118	0.985	0.379	0.015	1.557	0.025	– Native
	Various arthropods	Canopy	M-S		15	0.970	0.192	0.070	1.641	0.115	Reserve – introduced
			M-S		30	0.981	0.085	0.043	1.461	0.054	– Endemic
			M-S		28	0.987	0.242	0.032	1.698	0.054	– Native
			M-S		26	0.971	0.218	0.055	1.489	0.076	Island – introduced
			M-S		101	0.977	0.198	0.023	1.472	0.032	– Endemic
			M-S		38	0.988	0.287	0.025	1.602	0.042	– Native
			M-S		43	0.964	0.337	0.041	1.516	0.065	Archipelago – introduced
			M-S		56	0.983	0.354	0.021	1.595	0.041	– Endemic
			M-S		59	0.989	0.355	0.019	1.589	0.032	– Native
69	*Baetis rhodani*	Freshwater	S	20	20	1.000	0.146	0.025	1.100	0.005	Day
			S	20	20	1.000	-0.678	0.042	1.990	0.005	Night
	Ephemerella ignita		S	20	20	1.000	0.225	0.010	1.400	0.005	Day
			S	20	20	1.000	0.021	0.033	1.970	0.010	Night
	Rhithrogena semicolorata		S	20	20	0.999	0.127	0.026	1.110	0.010	Day
			S	20	20	0.994	0.127	0.168	1.990	0.051	Night

Continued

Latin name (common name) Ref. name	Habitat/host/prey	Type	NQ	NB	r	A	SE(A)	b	SE(b)	Comments (√=sampling plan)
Perlodes microcephalus		S	20	20	0.990	0.009	0.061	1.070	0.036	Day
		S	20	20	0.995	−0.009	0.045	1.080	0.026	Night
Isoperla grammatica		S	20	20	0.998	0.127	0.013	1.120	0.015	Day
		S	20	20	0.992	0.143	0.070	1.850	0.056	Night
Siphonoperla torrentium		S	20	20	1.000	0.143	0.012	1.300	0.005	Day
		S	20	20	1.000	0.146	0.018	1.310	0.005	Night
Leuctra fusca		S	20	20	1.000	0.146	0.012	1.300	0.005	Day
		S	20	20	0.999	0.152	0.037	1.280	0.015	Night
Protonemura meyeri		S	20	20	0.999	−0.009	0.026	1.300	0.010	Day
		S	20	20	1.000	−0.009	0.035	1.700	0.010	Night
Amphinemura sulcicollis		S	20	20	0.999	0.009	0.043	1.290	0.010	Day
		S	20	20	1.000	0.000	0.026	1.700	0.005	Night

Arachnida

Acari (ticks and mites)

Latin name (common name) Ref. name	Habitat/host/prey	Type	NQ	NB	r	A	SE(A)	b	SE(b)	Comments (√=sampling plan)
70 Ixodes ricinus (sheep ticks)	Sheep	E	20–86	10	0.977	0.240	0.138	1.457	0.114	Ticks/sheep
		S		33	0.946	0.150	0.087	1.594	0.098	Ticks/2 × 50yd2 – with Poisson points
		S		26	0.958	0.244	0.091	1.575	0.097	– No Poisson points
71 Dermacentor andersoni (Rocky Mountain wood tick)	Mammals	E	86–250	222	0.975	0.458	0.048	1.330	0.020	√ Ticks/10m2
72 Kampimodromus aberrans	Mites	E	50	38	0.980			1.310	0.045	Predatory mites/leaf
Amblyseius andersoni	Mites	E	50	70	0.975			1.270	0.035	
Typhlodromus pyri	Mites	E	50	34	0.959			1.230	0.064	
73 Varroa destructor (bee parasite)	Bees	E	24–57	31	0.977	0.345	0.029	1.239	0.050	√ #Mites/100 adult bees
74 Eutetranychus orientalis (oriental red mite)	Ornamentals	S	40	24	0.977	2.230	0.250	1.630	0.071	Eggs
		S	40	24	0.985	2.250	0.170	1.670	0.209	Immatures and males
		S	40	24	0.974	1.550	0.130	1.530	0.066	Females
		S	40	24	0.984	2.090	0.170	1.690	0.061	Mobile stages

No.	Organism	Host	Stage	n1	n2						✓	Description
	Phytoseiids	Eutetranychus orientalis	S	40	24	0.981	2.400		1.620	0.061		All stages
			S	40	24	0.932	0.150		1.030	0.077		Predatory mites
	Scolothrips longicornis	Eutetranychus orientalis	S	40	24	0.974	1.170		1.370	0.082		Predatory thrips
75	Eutetranychus frosti	Apple	S	~150	11	0.938	0.482		1.310	0.160	√	Prey and competitor
	Tydeus longisetosus		S	~150	11	0.889	0.493		1.090	0.170		
	Zetzellia mali (predatory mite)	Mites	S	~150	11	0.964	0.387		1.270	0.110		Predator
76	Halotydeus destructor (redlegged earth mite)	Canola	E		60	0.908	0.409		1.753	0.106	√	Operator 1
			E		30	0.958	0.415		1.732	0.098		Operator 2
			E		30	0.929	0.912		1.561	0.118		Operator 3
77	Mononychellus progresivus (cassava green mite)	Cassava	H-S		43	0.975	0.690	0.060	1.760	0.060		Competitors
	Oligonychus gossypii (cotton red mite)		H-S		23	0.975	0.730	0.070	1.570	0.080		
78	Tetranychus urticae (red spider mite)	Cucumber	S	30-36	67	0.994	1.303		1.943	0.027		Prey
79	Phytoseiulus persimilis	Red spider mite	S	30-36	63	0.992	1.193		1.900	0.030		Predator
	Tetranychus urticae (red spider mite)	Tomato	S	45	12	0.964	-0.229		1.670	0.145		T. urticae alone
			S	45	12	0.943	-0.469		1.770	0.197		T. urticae + P. macroplis
			S	45	12	0.964	-0.377		1.750	0.152		T. urticae + M. pygmaeus
			S	45	12	0.943	-0.481		1.840	0.205		T. urticae + P. macroplis + M. pygmaeus
	Phytoseiulus macroplis (predatory mites)	Red spider mite	S	45	12	0.954	0.410		1.610	0.160		T. urticae + M. pygmaeus
			S	45	12	0.954	-0.337		1.660	0.165		T. urticae + P. macroplis + M. pygmaeus
	Macrolophus pygmaeus (predatory mirid)	Red spider mite	S	45	12	0.860	0.030		1.070	0.201		T. urticae + M. pygmaeus
			S	45	12	0.877	-2.000		1.070	0.185		T. urticae + P. macroplis + M. pygmaeus

Continued

Ref.	Latin name (common name)	Habitat/host/prey	Type	NQ	NB	r	A	SE(A)	b	SE(b)	Comments (√=sampling plan)
80	Panonychus ulmi (European red mite)	Apple	E	6057					1.594	0.005	√ Active stage
	Amblyseius fallacis	European red mite	E	3646					1.336	0.007	Predator all stages
81	Panonychus citri (citrus red mite)	Lemon	E	30	125	0.964	0.100		1.690	0.042	√ Immature
			E	30	125	0.954	0.120		1.590	0.045	Adult male
			E	30	125	0.917	0.230		1.500	0.059	Adult female
			E	30	25	0.938	0.410		1.820	0.140	Total – Site 1
			E	30	29	0.970	0.560		1.560	0.076	– Site 2
			E	30	21	0.933	0.360		1.720	0.153	– Site 3
			E	30	30	0.995	0.410		1.450	0.028	– Site 4 – winter
			E	30	20	0.990	0.680		1.530	0.052	– summer
82	Panonychus citri and P. ulmi	Lemon	*						1.320	0.080	√ Average of 19 studies
	P. ulmi		*						1.340	0.090	Average of 16 studies – validation
	Tetranychus spp, mostly		*						1.490	0.100	Average of 23 studies
	T. urticae		*						1.480	0.060	Average of 22 studies – validation

Crustacea

Ref.	Latin name (common name)	Habitat/host/prey	Type	NQ	NB	r	A	SE(A)	b	SE(b)	Comments (√=sampling plan)
83	Ankylocythere sinuosa	Freshwater crayfish	E	4–20	24	0.959			1.670	0.100	#Commensals/host – rivers in Spain
84	Lepeophtheirus salmonis (salmon louse)	Sea trout	H-S		15	0.780	−0.798	0.312	1.823	0.406	Sessile copepodids/fish host
			H-S		15	0.557	−0.186	0.300	0.957	0.396	Mobile chalimus/fish host
85	Lepeophtheirus pectoralis (plaice louse)	Plaice	S	30–199	12	0.969	0.239	0.041	1.139	0.091	1972
			S	30–199	12	0.980	0.191	0.030	1.211	0.077	1973
			S	30–199	24	0.973	0.217	0.025	1.166	0.059	Combined
86	cirripeds	Littoral	S		23	0.972	1.246	0.322	1.498	0.079	Recovery from hypoxia
87	Percnon gibbesi (Sally Lightfoot)	Littoral	E	5–49	20	0.909	−0.548	0.162	2.133	0.230	Invasive to Malta
88	multiple species	Littoral	M-S	26	12	0.957	0.362	0.272	1.561	0.150	Each point a species-stage (#/m³)

Key to Appendix 8.M

Ref	Author	Ref	Author	Ref	Author
1	Taylor et al. (1998)	24	Darbemamieh et al. (2011)	47	Agboka et al. (2013)	70	Milne (1943)
2	Steiner (1990)	25	Sanchez and Ortin-Angulo (2011)	48	Park and Perring (2010)	71	Rochon et al. (2012)
3	Navarro-Campos et al. (2012)	26	Allsopp and Bull (1990)	49	Minarro and Jacas (2011)	72	Tixier et al. (2014)
4	Badieritakis et al. (2015)	27	Alyokhin et al. (2001)	50	Peña and Schaffer (1997)	73	Lee et al. (2010)
5	Reisig et al. (2011)	28	Andow (1992)	51	Serra et al. (2002)	74	Gonzalez-Zamora et al. (2011)
6	Sedaratian et al. (2010)	29	Bentz and Townsend (2004)	52	Karban et al. (2012)	75	Khodayari et al. (2010)
7	Aliakbarpour and Salmah (2011)	30	Munyaneza et al. (2008)	53	Elliott et al. (2014)	76	Arthur et al. (2014)
8	Elliott et al. (2003)	31	Moura et al. (2007)	54	Tobin and Blackburn (2008)	77	Bonato et al. (1995)
9	Badenhausser (1996)	32	Castle and Naranjo (2008)	55	Elliott (1981)	78	Nachman (1981)
10	Hutchison et al. (1988)	33	Naranjo and Castle (2010)	56	Elliott (1982)	79	Gigon et al. (2016)
11	Tesfaye et al. (2016)	34	Bisseleua et al. (2011)	57	Elliott (1983)	80	Croft et al. (1976)
12	Celini and Vaillant (2004)	35	Espino et al. (2008)	58	Atkinson and Shorrocks (1984)	81	Jones and Parrella (1984)
13	Meats and Wheeler (2011)	36	Seiter et al. (2013)	59	Drake (1983)	82	Jones (1990)
14	Urías-López et al. (2016)	37	Stubbins et al. (2014)	60	Taylor (2013)	83	Mestre et al. (2014)
15	Berlinger et al. (1999)	38	Nyrop and Wright (1985)	61	Silva-Lima et al. (2016)	84	Arechavala-Lopez et al. (2016)
16	Beltra et al. (2013)	39	Carvalho et al. (2013)	62	Žitko and Merdić (2014)	85	Boxshall (1974)
17	Schuster (1998)	40	Elmouttie et al. (2013)	63	Carrieri et al. (2011)	86	Hansen et al. (2002)
18	Chaubey et al. (2014)	41	Moradi-Vajargah et al. (2011)	64	Lytra and Emmanouel (2014)	87	Sciberras and Schembri (2008)
19	Xu (1985)	42	Pérez et al. (2015)	65	Henne et al. (2009)	88	Vogedes et al. (2014)
20	Monzo et al. (2015)	43	Hamid et al. (1999)	66	Brown and Cameron (1982)		
21	Rakhshani and Saeedifar (2013)	44	Abd-Elgawad (2014)	67	Lordan et al. (2015)		
22	Prager et al. (2013)	45	Overholt et al. (1990)	68	Elliott (2002)		
23	Crespo-Herrera et al. (2012)	46	Sétamou et al. (2000)	69	Gaston et al. (2006)		

References

Abd-Elgawad, M.M., 2014. Spatial patterns of *Tuta absoluta* and heterorhabditid nematodes. Russ. J. Nematol. 22, 89–100.

Agboka, K., Schulthess, F., Tounou, A.K., Tamo, M., Vidal, S., 2013. The effect of leguminous cover crops and cowpea planted as border rows on maize ear borers with special reference to *Mussidia nigrivenella* Ragonot (Lepidoptera: Pyralidae). Crop Prot. 43, 72–78.

Aliakbarpour, H., Salmah, M.R.C., 2011. Seasonal abundance and spatial distribution of larval and adult thrips (Thysanoptera) on weed host plants in mango orchards in Penang, Malaysia. Appl. Entomol. Zool. 46, 185–194.

Allsopp, P.G., Bull, R.M., 1990. Sampling distributions and sequential sampling plans for *Perkinsiella saccharicida* Kirkaldy (Hemiptera: Delphacidae) and *Tytthus* spp. (Hemiptera: Miridae) on sugarcane. J. Econ. Entomol. 83, 2284–2289.

Allsopp, P.G., Ladd, T.L., Klein, M.G., 1992. Sample sizes and distributions of Japanese beetles (Coleoptera: Scarabaeidae) captured in lure traps. J. Econ. Entomol. 85, 1797–1801.

Alm, S.R., Yeh, T., Campo, M.L., Dawson, C.G., Jenkins, E.B., Simeoni, A.E., 1994. Modified trap designs and heights for increased capture of Japanese beetle adults (Coleoptera: Scarabaeidae). J. Econ. Entomol. 87, 775–780.

Alyokhin, A.V., Yang, P., Messing, R.H., 2001. Distribution and parasitism of *Sophonia rufofascia* (Homoptera: Cicadellidae) eggs in Hawaii. Ann. Entomol. Soc. Am. 94, 664–669.

Andow, D.A., 1992. Population density of *Empoasca fabae* (Homoptera: Cicadellidae) in weedv beans. J. Econ. Entomol. 85, 379–383.

Arechavala-Lopez, P., Uglem, I., Berg, M., Bjørn, P.A., Finstad, B., 2016. Large-scale use of fish traps for monitoring sea trout (*Salmo trutta*) smolts and sea lice (*Lepeophtheirus salmonis*) infestations: efficiency and reliability. Mar. Biol. Res. 12, 76–84.

Arthur, A.L., Hoffmann, A.A., Umina, P.A., 2014. Estimating densities of the pest *Halotydeus destructor* (Acari: Penthaleidae) in canola. J. Econ. Entomol. 107, 2204–2212.

Atkinson, W.D., Shorrocks, B., 1984. Aggregation of larval Diptera over discrete and ephemeral breeding sites: the implications for coexistence. Am. Nat. 124, 336–351.

Badenhausser, I., 1996. Sequential sampling of *Brachycaudus helichrysi* (Homoptera: Aphididae) in sunflower fields. J. Econ. Entomol. 89, 1460–1467.

Badieritakis, E.G., Thanopoulos, R.C., Fantinou, A.A., Emmanouel, N.G., 2015. A qualitative and quantitative study of thrips (Thysanoptera) on alfalfa and records of thrips species on cultivated and wild *Medicago* species of Greece. Biologia 70, 504–515.

Beall, G., 1939. Methods of estimating the population of insects in a field. Biometrika 30, 422–439.

Bell, J.F., 2002. Log Scaling and Timber Cruising, Revised Edition Oregon State University Book Stores, Corvallis, OR.

Beltra, A., Garcia-Mari, F., Soto, A., 2013. Seasonal phenology, spatial distribution, and sampling plan for the invasive mealybug *Phenacoccus peruvianus* (Hemiptera: Pseudococcidae). J. Econ. Entomol. 106, 1486–1494.

Bentz, J., Townsend, A.M., 2004. Spatial and temporal patterns of abundance of the potato leafhopper among red maples. Ann. Appl. Biol. 145, 157–164.

Berlinger, M.J., Segre, L., Podoler, H., Taylor, R.A.J., 1999. Distribution and abundance of the oleander scale (Homoptera: Diaspididae) on jojoba. J. Econ. Entomol. 92, 1113–1119.

Bisseleua, D.H.B., Yede, Vidal, S., 2011. Dispersion models and sampling of cacao mirid bug *Sahlbergella singularis* (Hemiptera: Miridae) on *Theobroma cacao* in Southern Cameroon. Environ. Entomol. 40, 111–119.

Bliss, C.I., 1941. Statistical problems in estimating populations of Japanese beetle larvae. J. Econ. Entomol. 43, 221–232.

Bliss, C.I., Owen, A.R.G., 1958. Negative binomial distributions with a common k. Biometrika 45, 37–58.

Bonato, O., Baumgärtner, J., Gutierrez, J., 1995. Sampling plans for *Mononychellus progresivus* and *Oligonychus gossypii* (Acari: Tetranychidae) on cassava in Africa. J. Econ. Entomol. 88, 1295–1300.

Boxshall, G.A., 1974. The population dynamics of *Lepeophtheirus pectoralis* (Müller): dispersion pattern. Parasitology 69, 373–390.

Brown, D.W., Cameron, E.A., 1982. Spatial distribution of Ooencyrtus kuvanae (Hymenoptera: Encyrtidae), an egg parasite of Lymantria dispar (Lepidoptera: Lymantriidae). Can. Ent. 114, 1109–1120.

Campbell, R.W., 1963. Some factors that distort the sex ratio of the gypsy moth, *Porthetria dispar* (L.) (Lepidoptera: Lymantriidae). Can. Entomol. 96, 465–474.

Campbell, R.W., 1967. Studies on the sex ratio of the gypsy moth. For. Sci. 13, 1–22.

Carrieri, M., Albieri, A., Angelini, P., Baldacchini, F., Venturelli, C., Zeo, S.M., Bellini, R., 2011. Surveillance of the chikungunya vector *Aedes albopictus* (Skuse) in Emilia-Romagna (northern Italy): organizational and technical aspects of a large scale monitoring system. J. Vector Biol. 36, 108–116.

Carvalho, M.O., Faro, A., Subramanyam, B., 2013. Insect population distribution and density estimates in a large rice mill in Portugal—a pilot study. J. Stored Prod. Res. 52, 48–56.

Castle, S.J., Naranjo, S.E., 2008. Comparison of sampling methods for determining relative densities of *Homalodisca vitripennis* (Hemiptera: Cicadellidae) on citrus. J. Econ. Entomol. 101, 226–235.

Celini, L., Vaillant, J., 2004. A model of temporal distribution of *Aphis gossypii* (Glover) (Hem., Aphididae) on cotton. J. Appl. Entomol. 128, 133–139.

Chaubey, R., Naveen, N.C., Andrew, R.J., Ramamurthy, V.V., 2014. Bionomics and population dynamics of *Bemisia tabaci* on subabul (*Leucaena leucocephala*). Indian J. Agric. Sci. 84, 291–294.

Clark, S.J., Perry, J.N., 1994. Small sample estimation for Taylor's power law. Environ. Ecol. Stat. 1, 287–302.

Clark, S.J., Perry, J.N., Marshall, E.J.P., 1996. Estimating Taylor's power law parameters for weeds and the effect of spatial scale. Weed Res. 36, 405–417.

Crespo-Herrera, L.A., Vera-Graziano, J., Bravo-Mojica, H., Lopez-Collado, J., Reyna-Robles, R., Pena-Lomeli, A., Manuel-Pinto, V., Garza-Garcia, R., 2012. Spatial distribution of *Bactericera cockerelli* (Sulc) (Hemiptera: Triozidae) on green tomato (*Physalis ixocarpa* (Brot.)). Agrociencia 46, 289–298.

Croft, B.A., Welch, S.M., Dover, M.J., 1976. Dispersion statistics and sample size estimates for populations of the mite species *Panonychus ulmi* and *Amblyseius fallacis* on apple. Environ. Entomol. 5, 227–234.

Darbemamieh, M., Fathipour, Y., Kamali, K., 2011. Population abundance and seasonal activity of *Zetzellia pourmirzai* (Acari: Stigmaeidae) and its preys *Cenopalpus irani* and *Bryobia rubrioculus* (Acari: Tetranychidae) in sprayed apple orchards of Kermanshah, Iran. J. Agric. Sci. Technol. 13, 143–154.

Dewer, A.J., Dean, G.J., Cannon, R., 1982. Assessment of methods for estimating the number of aphids (Hemiptera: Aphididae) in cereals. Bull. Entomol. Res. 72, 675–685.

Dietrick, E.J., Schlinger, E.I., Bosch, R.V.D., 1959. A new method for sampling arthropods using a suction collecting machine and modified Berlese funnel separator. J. Econ. Entomol. 52, 1085–1091.

Drake, C.M., 1983. Spatial distribution of chironomid larvae (Diptera) on leaves of the bulrush in a chalk stream. J. Anim. Ecol. 52, 421–437.

Duffey, E., 1980. The efficiency of the Dietrick vacuum sampler (D-vac) for invertebrate population studies in different types of grassland. Bull. Ecol. 11, 421–431.

Eisler, Z., Bartos, I., Kertész, J., 2008. Fluctuation scaling in complex systems: Taylor's law and beyond. Adv. Phys. 57, 89–142.

Elkinton, J.S., 1987. Changes in efficiency of the pheromone-baited milk-carton trap as it fills with male gypsy moths (Lepidoptera, Lymantriidae). J. Econ. Entomol. 80, 754–757.

Elliott, J.M., 1981. A quantitative study of the life cycle of the net-spinning caddis *Philopotamus montanus* (Trichoptera: Philopotamidae) in a Lake District stream. J. Anim. Ecol. 50, 867–883.

Elliott, J.M., 1982. The life cycle and spatial distribution of the aquatic parasitoid *Agriotypus armatus* (Hymenoptera: Agriotypidae) and its caddis host *Silo pallipes* (Trichoptera: Goeridae). J. Anim. Ecol. 51, 923–941.

Elliott, J.M., 1983. The responses of the aquatic parasitoid *Agriotypus armatus* (Hymenoptera: agriotypidae) to the spatial distribution and density of its caddis host *Silo pallipes* (Trichoptera: Goeridae). J. Anim. Ecol. 52, 315–330.

Elliott, J.M., 2002. A quantitative study of day-night changes in the spatial distribution of insects in a stony stream. J. Anim. Ecol. 71, 112–122.

Elliott, N.C., Brewer, M.J., Giles, K.L., Backoulou, G.F., McCornack, B.P., Pendleton, B.B., Royer, T.A., 2014. Sequential sampling for panicle caterpillars (Lepidoptera: Noctuidae) in sorghum. J. Econ. Entomol. 107, 846–853.

Elliott, N.C., Giles, K.L., Royer, T.A., Kindler, S.D., Tao, F.L., Jones, D.B., Cuperus, G.W., 2003. Fixed precision sequential sampling plans for the greenbug and bird cherry-oat aphid (Homoptera: Aphididae) in winter wheat. J. Econ. Entomol. 96, 1585–1593.

Elmouttie, D., Flinn, P., Kiermeier, A., Subramanyam, B., Hagstrum, D., Hamilton. G., 2013. Sampling stored-product insect pests: a comparison of four statistical sampling models for probability of pest detection. Pest Manag. Sci. 69, 1073–1079.

Embody, D.A., Bachlor, C., 1980. Estimating gypsy moth egg mass densities using moth catches of 2-quart traps baited with racemic Disparlure in Hercon® wicks. Unpublished report of the U.S. Department of Agriculture, Animal and Plant Health Inspection Service.

Embree, D.G., 1961. Studies on the population dynamics of the winter moth *Operophtera brumata* (L.) (Lepidoptera: Geometridae) in Nova Scotia. (PhD Thesis), The Ohio State University.

Embree, D.G., 1965. The population dynamics of the winter moth in Nova Scotia, 1954–1962. Mem. Entomol. Soc. Can. 97 (S46), 5–57.

L. Espino, L., Way, M.O., Wilson, L.T. 2008. Determination of Oebalus pugnax (Hemiptera: Pentatomidae) spatial pattern in rice and development of visual sampling methods and population sampling plans. J. Econ. Entomol. 101, 216–225.

Finney, D.J., 1941. Wireworm populations and their effect on crops. Ann. Appl. Biol. 28, 282–295.

Finney, D.J., 1946. Field sampling for the estimation of wireworm populations. Biometrics 2, 1–7.

Fleming, W.E., Baker, F.E., 1936. A method for estimating populations of larvae of the Japanese beetle in the field. J. Agric. Res. 53, 319–331.

Fox, H., 1937. Seasonal trends in the relative abundance of Japanese beetle populations in the soil during the annual life cycle. J. NY Entomol. Soc. 45, 115–126.

Fukaya, K., Okuda, T., Hori, M., Yamamoto, T., Nakaoka, M., Noda, T., 2013. Variable processes that determine population growth and an invariant mean-variance relationship of intertidal barnacles. Ecosphere 4, 48.

Fukaya, K., Okuda, T., Nakaoka, M., Noda, T., 2014. Effects of spatial structure of population size on the population dynamics of barnacles across their elevational range. J. Anim. Ecol. 83, 1334–1343.

Gaston, K.J., Borges, P.A.V., He, F., Gaspar, C., 2006. Abundance, spatial variance and occupancy: arthropod species distribution in the Azores. J. Anim. Ecol. 75, 646–656.

Gigon, V., Camps, C., Le Corff, J., 2016. Biological control of *Tetranychus urticae* by *Phytoseiulus macropilis* and *Macrolophus pygmaeus* in tomato greenhouses. Exp. Appl. Acarol. 68, 55–70.

Gonzalez-Zamora, J.E., Lopez, C., Avilla, C., 2011. Population studies of arthropods on *Melia azedarach* in Seville (Spain), with special reference to *Eutetranychus orientalis* (Acari: Tetranychidae) and its natural enemies. Exp. Appl. Acarol. 55, 389–400.

Hamid, M.M., Perry, J.N., Powell, W., Rennolls, K., 1999. The effect of spatial scale on interactions between two weevils and their food plant. Acta Oecol. 20, 537–549.

Hamilton, W.D., 1971. Geometry for the selfish herd. J. Theor. Biol. 31, 295–311.

Hansen, B.W., Stenalt, E., Petersen, J.K., Ellegaard, C., 2002. Invertebrate re-colonisation in Mariager Fjord (Denmark) after severe hypoxia. I. Zooplankton and settlement. Ophelia 56 (3), 197–213.

He, F., Gaston, K.J., 2003. Occupancy, spatial variance, and the abundance of species. Am. Nat. 162, 366–375.

Healy, M.J.R., Taylor, L.R., 1962. Tables for power-law transformations. Biometrika 49, 557–559.

Henne, D.C., Hilbun, W.S., Johnson, S.J., 2009. Spatio-temporal population sampling of a fire ant parasitoid. Entomol. Exp. Appl. 129, 132–141.

Hutchison, W.D., Hogg, D.B., Poswal, M.A., Berberet, R.C., Cuperus, A.W., 1988. Implications of the stochastic nature of Kuno's and Green's fixed-precision stop lines: sampling plans for the pea aphid (Homoptera: Aphididae) in alfalfa as an example. J. Econ. Entomol. 81, 749–758.

Johnson, C.G., 1950. The comparison of suction trap, sticky trap and tow-net for the quantitative sampling of small airborne insects. Ann. Appl. Biol. 37, 268–285.

Jones, E.W., 1937. Practical field methods of sampling soil for wireworms. J. Agric. Res. 54, 123–134.

Jones, V.P., 1990. Developing sampling plans for -spider mites (Acari: Tetranychidae): those who don't remember the past may have to repeat it. J. Econ. Entomol. 83, 1656–1664.

Jones, V.P., Parrella, M.P., 1984. Dispersion indices and sequential sampling plans for the citrus red mite (Acari: Tetranychidae). J. Econ. Entomol. 77, 75–79.

Jordan, T.A., Youngman, R.R., Laub, C.L., Tiwari, S., Kuhar, T.P., Balderson, T.K., Moore, D.M., Saphir, M., 2012. Fall soil sampling method for predicting spring infestation of white grubs (Coleoptera: Scarabaeidae) in corn and the benefits of clothianidin seed treatment in Virginia. Crop Prot. 39, 57–62.

Karban, R., Grof-Tisza, P., Maron, J.L., Holyoak, M., 2012. The importance of host plant limitation for caterpillars of an arctiid moth (*Platyprepia virginalis*) varies spatially. Ecology 93, 2216–2226.

Kendal, W.S., 1995. A probabilistic model for the variance to mean power law in ecology. Ecol. Model. 80, 293–297.

Kendal, W.S., 2002. Spatial aggregation of the Colorado potato beetle described by an exponential dispersion model. Ecol. Model. 151, 261–269.

Khodayari, S., Fathipour, Y., Kamali, K., Naseri, B., 2010. Seasonal activity of *Zetzellia mali* (Stigmaeidae) and its preys *Eotetranychus frosti* (Tetranychidae) and *Tydeus longisetosus* (Tydeidae) in unsprayed apple orchards of Maragheh, Northwestern of Iran. J. Agric. Sci. Technol. 12, 549–558.

Knight-Jones, E.W., Stevenson, J.P., 1950. Gregariousness during settlement in the barnacle *Elminius modestus* Darwin. J. Mar. Biol. Assoc. UK 29, 281–297.

Ladd, T.L., Klein, M.G., Tumlinson, J.H., 1981. Phenethyl propionate + eugenol + geraniol (3:7:3) and Japonilure: a highly effective joint lure for Japanese beetles. J. Econ. Entomol. 74, 665–667.

Lamp, W.O., Nielsen, G.R., Danielson, S.D., 1994. Patterns among host plants of potato leafhopper, *Empoasca fabae* (Homoptera: Cicadellidae). J. Kansas Entomol. Soc. 67, 354–368.

Lee, K.V., Moon, R.D., Burkness, E.C., Hutchison, W.D., Spivak, M., 2010. Practical sampling plans for *Varroa destructor* (Acari: Varroidae) in *Apis mellifera* (Hymenoptera: Apidae) colonies and apiaries. J. Econ. Entomol. 103, 1039–1050.

Lordan, J., Alegre, S., Moerkens, R., Sarasua, M.J., Alins, G., 2015. Phenology and interspecific association of *Forficula auricularia* and *Forficula pubescens* in apple orchards. Span. J. Agric. Res. 13, 1–12.

Lytra, I., Emmanouel, N., 2014. Study of *Culex tritaeniorhynchus* and species composition of mosquitoes in a rice field in Greece. Acta Trop. 134, 66–71.

Matui, I., 1932. Statistical study of the distribution of scattered villages in two regions of the Tonami Plain, Toyama Prefecture. Jpn. J. Geol. Geogr. 9, 251–266.

McGuire, J.U., Brindley, T.A., Bancroft, T.A., 1957. The distribution of European corn borer larvae *Pyrausta nubilalis* (Hbn.) in field corn. Biometrics 13, 65–78.

Meats, A., Wheeler, S., 2011. Dispersion, contagion, and population stability of red scale, *Aonidiella aurantii*, in citrus orchards with low or zero insecticide use. Entomol. Exp. Appl. 138, 146–153.

Mestre, A., Monros, J.S., Mesquita-Joanes, F., 2014. The influence of environmental factors on abundance and prevalence of a commensal ostracod hosted by an invasive crayfish: are 'parasite rules' relevant to non-parasitic symbionts? Freshw. Biol. 59, 2107–2121.

Meyers, M.T., Patch, L.H., 1937. A Statistical study of sampling in field surveys of the fall population of the European corn borer. J. Agric. Res. 55, 849–871.

Milne, A., 1943. The comparison of sheep-tick populations (*Ixodes ricinus* L.). Ann. Appl. Biol. 30, 240–250.

Minarro, M., Jacas, J.A., 2011. Pest status of leafminers in cider-apples: the case of orchards in Asturias (NW Spain). Crop Prot. 30, 1485–1491.

Monzo, C., Arevalo, H.A., Jones, M.M., Vanaclocha, P., Croxton, S.D., Qureshi, J.A., Stansly, P.A., 2015. Sampling methods for detection and monitoring of the Asian citrus psyllid (Hemiptera: Psyllidae). Environ. Entomol. 44, 780–788.

Moradi-Vajargah, M., Golizadeh, A., Rafiee-Dastjerdi, H., Zalucki, M.P., Hassanpour, M., Naseri, B., 2011. Population density and spatial distribution pattern of *Hypera postica* (Coleoptera: Curculionidae) in Ardabil, Iran. Not. Bot. Horti Agrobot. Cluj Napoca 39, 42–48.

Moura, M.F., Picanço, M.C., Guedes, R.N.C., Barros, E.C., Chediak, M., Morais, E.G.F., 2007. Conventional sampling plan for the green leafhopper *Empoasca kraemeri* in common beans. J. Appl. Entomol. 131, 215–220.

Munyaneza, J.E., Jensen, A.S., Hamm, P.B., Upton, J.E., 2008. Seasonal occurrence and abundance of beet leafhopper in the potato growing region of Washington and Oregon Columbia basin and Yakima valley. Am. J. Potato Res. 85, 77–84.

Myers, J.H., Boettner, G., Elkinton, J., 1998. Maternal effects in gypsy moth: only sex ratio varies with population density. Ecology 79, 305–314.

Nachman, G., 1981. Temporal and spatial dynamics of an acarine predator-prey system. J. Anim. Ecol. 50, 435–451.

Naranjo, S.E., Castle, S.J., 2010. Sequential sampling plans for estimating density of glassy-winged sharpshooter, *Homalodisca vitripennis* (Hemiptera: Cicadellidae) on citrus. Crop Prot. 29, 1363–1370.

Navarro-Campos, C., Aguilar, A., Garcia-Mari, F., 2012. Aggregation pattern, sampling plan, and intervention threshold for *Pezothrips kellyanus* in citrus groves. Entomol. Exp. Appl. 142, 130–139.

Nyrop, J.P., Wright, R.J., 1985. Use of double sample plans in insect sampling with reference to the Colorado potato beetle, *Leptinotarsa decemlineata* (Coleoptera: Chrysomelidae). Environ. Entomol. 14, 644–649.

Overholt, W.A., Knutson, A.E., Smith, J.W., Gilstrap, F.E., 1990. Distribution and sampling of southwestern corn borer (Lepidoptera: Pyralidae) in preharvest corn. J. Econ. Entomol. 83, 1370–1375.

Pan, W., 2001. Akaike's information criterion in generalized estimating equations. Biometrics 57, 120–125.

Park, J.J., Perring, T.M., 2010. Development of a binomial sampling plan for the carob moth (Lepidoptera: Pyralidae), a pest of California dates. J. Econ. Entomol. 103, 1474–1482.

Park, S.-J., Taylor, R.A.J., Grewal, P.S., 2013. Spatial organization of soil nematode communities in urban landscapes: Taylor's power law reveals life strategy characteristics. Appl. Soil Ecol. 64, 214–222.

Peña, J.E., Schaffer, B., 1997. Intraplant distribution and sampling of the citrus leafminer (Lepidoptera: Gracillariidae) on lime. J. Econ. Entomol. 90, 45–64.

Pérez, M.L.P., Isas, M.G., Salvatore, A.R., Gastaminza, G., Trumper, E.V., 2015. Optimizing a fixed-precision sequential sampling plan for adult *Acrotomopus atropunctellus* (Boheman) (Coleoptera: Curculionidae), new pest on sugarcane. Crop Prot. 74, 9–12.

Prager, S.M., Butler, C.D., Trumble, J.T., 2013. A sequential binomial sampling plan for potato psyllid (Hemiptera: Triozidae) on bell pepper (*Capsicum annum*). Pest Manag. Sci. 69, 1131–1135.

Rakhshani, E., Saeedifar, A., 2013. Seasonal fluctuations, spatial distribution and natural enemies of Asian citrus psyllid *Diaphorina citri* Kuwayama (Hemiptera: Psyllidae) in Iran. Entomol. Sci. 16, 17–25.

Ramsayer, J., Fellous, S., Cohen, J.E., Hochberg, M.E., 2012. Taylor's Law holds in experimental bacterial populations but competition does not influence the slope. Biol. Lett. 8, 316–319.

Reisig, D.D., Godfrey, L.D., Marcum, D.B., 2011. Spatial dependence, dispersion, and sequential sampling of *Anaphothrips obscurus* (Thysanoptera: Thripidae) in timothy. Environ. Entomol. 40, 689–696.

Rochon, K., Scoles, G.A., Lysyk, T.J., 2012. Dispersion and sampling of adult *Dermacentor andersoni* in rangeland in western North America. J. Med. Entomol. 49, 253–261.

Routledge, R.D., Swartz, T.M., 1991. Taylor's power law re-examined. Oikos 60, 107–112.

Sanchez, J.A., Ortin-Angulo, M.C., 2011. Sampling of *Cacopsylla pyri* (Hemiptera: Psyllidae) and *Pilophorus gallicus* (Hemiptera: Miridae) in pear orchards. J. Econ. Entomol. 104, 1742–1751.

Schuster, D.J., 1998. Intraplant distribution of immature lifestages of *Bemisia argentifolii* (Homoptera: Aleyrodidae) on tomato. Environ. Entomol. 27, 1–9.

Sciberras, M., Schembri, P.J., 2008. Biology and interspecific interactions of the alien crab *Percnon gibbesi* in the Maltese Islands. Mar. Biol. Res. 4, 321–332.

Sedaratian, A., Fathipour, Y., Talebi, A.A., Farahani, S., 2010. Population density and spatial distribution pattern of *Thrips tabaci* (Thysanoptera: Thripidae) on different soybean varieties. J. Agric. Sci. Technol. 12, 275–288.

Seiter, N.J., Reay-Jones, F.P.F., Greene, J.K., 2013. Within-field spatial distribution of *Megacopta cribraria* (Hemiptera: Plataspidae) in soybean (Fabales: Fabaceae). Environ. Entomol. 42, 1363–1374.

Serra, G., Luciano, P., Lentini, A., Gilioli, G., 2002. Spatial distribution and sampling of *Tortrix viridana* L. egg-clusters. IOBC/WPRS Bull. 25, 155–158.

Sétamou, M., Schulthess, F., Poehling, H.-M., Borgrmeister, C., 2000. Spatial distribution of and sampling plans for *Mussidia nigrivenella* (Lepidoptera: Pyralidae) on cultivated and wild host plants in Benin. Environ. Entomol. 29, 1216–1225.

Silva-Lima, A.W., Honório, N.A., Codeço, C.T., 2016. Spatial clustering of *Aedes aegypti* (Diptera: Culicidae) and its impact on entomological surveillance indicators. J. Med. Entomol. 53, 343–348.

Southwood, T.R.E., 1966. Ecological Methods with Particular Reference to the Study of Insect Populations, first ed. Methuen, London.

Steiner, M.Y., 1990. Determining population characteristics and sampling procedures for the western flower thrips (Thysanoptera: Thripidae) and the predatory mite *Amblyseius cucumeris* (Acari: Phytoseiidae) on greenhouse cucumber. Environ. Entomol. 19, 1605–1613.

Stubbins, F.L., Seiter, N.J., Greene, J.K., Reay-Jones, F.P.F., 2014. Developing sampling plans for the invasive *Megacopta cribraria* (Hemiptera: Plataspidae) in soybean. J. Econ. Entomol. 107, 2213–2221.

Taylor, D.B., Friesen, K., Zhu, J.J., 2013. Spatial-temporal dynamics of stable fly (Diptera: Muscidae) trap catches in Eastern Nebraska. Environ. Entomol. 42, 524–531.

Taylor, L.R., 1962a. The absolute efficiency of insect suction traps. Ann. Appl. Biol. 50, 405–421.

Taylor, L.R., 1962b. The efficiency of cylindrical sticky insect traps and suspended nets. Ann. Appl. Biol. 50, 681–685.

Taylor, L.R., 1965. A natural law for the spatial disposition of insects. In: Freeman, P. (Ed.), Proceedings of the XIIth International Congress of Entomology. Royal Entomological Society, London, pp. 396–397.

Taylor, L.R., 1984. Assessing and interpreting the spatial distributions of insect populations. Annu. Rev. Entomol. 29, 321–357.

Taylor, R.A.J., 1987. On the accuracy of insecticide efficacy reports. Environ. Entomol. 16, 1–8.

Taylor, L.R., Taylor, R.A.J., 1977. Aggregation, migration and population mechanics. Nature 265, 415–421.

Taylor, L.R., Woiwod, I.P., Perry, J.N., 1978. The density-dependence of spatial behaviour and the rarity of randomness. J. Anim. Ecol. 47, 383–406.

Taylor, R.A.J., 1987. On the accuracy of insecticide efficacy reports. Environ. Entomol. 16, 1–8.

Taylor, R.A.J., 2018. Spatial distribution, sampling efficiency and Taylor's power law. Ecol. Entomol. 43, 215–225.

Taylor, R.A.J., McManus, M.L., Pitts, C.W., 1991. The absolute efficiency of gypsy moth, *Lymantria dispar* (Lepidoptera, Lymantriidae), milk-carton pheromone traps. Bull. Entomol. Res. 81, 111–118.

Taylor, R.A.J., Lindquist, R.K., Shipp, J.L., 1998. Variation and consistency in spatial distribution as measured by Taylor's power law. Environ. Entomol. 27, 191–201.

Taylor, R.A.J., Park, S.-J., Grewal, P.S., 2017. Nematode spatial distribution and the frequency of zeros in samples. Nematology 19, 263–270.

Tesfaye, A., Wale, M., Azerefegne, F., 2016. Dispersion patterns and sampling plans for the pea aphid, *Acyrthosiphon pisum* (Harris), on grass pea, *Lathyrus sativus* L. Int. J. Pest Manag. 62, 30–39.

Tixier, M.S., Lopes, T., Blanc, G., Dedieu, J.L., Kreiter, S., 2014. Phytoseiid mite diversity (Acari: Mesostigmata) and assessment of their spatial distribution in French apple orchards. Acarologia 54, 97–111.

Tobin, P.C., Blackburn, L.M., 2008. Long-distance dispersal of the gypsy moth (Lepidoptera: Lymantriidae) facilitated its initial invasion of Wisconsin. Environ. Entomol. 37, 87–93.

Trumble, J.T., 1985. Implications of changes in arthropod distribution following chemical application. Res. Popul. Ecol. 27, 277–285.

Tumlinson, J.H., Klein, M.G., Doolittle, R.E., Ladd, T.L., Proveaux, A.T., 1977. Identification of the female Japanese beetle sex pheromone: inhibition of male response by an enantiomer. Science 197, 789–792.

Urías-López, M.A., Nava-Camberos, U., González-Carrillo, J.A., Hernández-Fuentes, L.M., García-Álvarez, N.C., 2016. Development of a sampling program for the white mango scale. *Aulacaspis tubercularis* Newstead. Southwest Entomol. 41, 115–126.

Vogedes, D., Eiane, K., Båtnes, A.S., Berge, J., 2014. Variability in *Calanus* spp. abundance on fine- to mesoscales in an Arctic fjord: implications for little auk feeding. Mar. Biol. Res. 10, 437–448.

Williams, C.B., 1951. Comparing the efficiency of insect traps. Bull. Entomol. Res. 42, 513–517.

Xu, R.-M., 1985. Dynamics of within-leaf spatial distribution patterns of greenhouse whiteflies and the biological interpretations. J. Appl. Ecol. 22, 63–72.

Xu, X.-M., Madden, L.V., 2002. Incidence and density relationships powdery mildew on apple. Phytopathology 92, 1005–1014.

Yates, F., Finney, D.J., 1942. Statistical problems in field sampling for wireworms. Ann. Appl. Biol. 29, 156–167.

Žitko, T., Merdić, E., 2014. Seasonal and spatial oviposition activity of *Aedes albopictus* (Diptera: Culicidae) in Adriatic Croatia. J. Med. Entomol. 51, 760–768.

Chapter 9

Other invertebrates

In this chapter, we examine the distributions of metazoan animals not included in the preceding chapters. Although populous and ubiquitous, other invertebrates receive less attention than worms and arthropods because they are of generally lower direct economic importance.

Rotifers

The Rotifera are a phylum of multicelled microorganisms that range from 50 μm to ~2 mm, although most are 0.1–0.5 mm. They are predominantly planktonic in freshwater, although there are marine species and some are sessile. Rotifers are colloquially called "wheel animals" because of the ciliated corona surrounding the mouth that move in waves giving the appearance of circular movement. They are predominately detritivores, drawing dead bacteria, algae, and protozoans into their mouths with the coronal cilia. Rotifers are a major food source for larger aquatic invertebrates, such as arthropods, and also some vertebrates.

LAKE EUFAULA, OKLAHOMA

The North Canadian River and the deep Fork River join the Canadian River at Lake Eufaula, a 410-km^2 reservoir in east Central Oklahoma. The reservoir is held back by a flood control and hydroelectric dam and the lake provides fresh water and recreation. At approximately 4-week intervals from April to September 1968, Bowles (1972) took water samples at depths of 1, 5, 10, and 15 m and at the bottom at 5 stations in the middle of Lake Eufaula. The 6-L samples were filtered and rotifers identified and counted in 1-cc subsamples. Bowles' Tables 2–7 provide estimates of the abundance (#/L) of 8 species of rotifer at 5 stations and 5 depths on 6 sampling occasions. These tables provide data for both temporal and spatial community analyses of rotifer abundance segregated by depth (Fig. 9.1; Appendix 9.A).

In hybrid spatial and temporal TPL analyses of rotifer communities, neither the slopes nor the intercepts differ significantly (slope: $P > 0.30$, intercept: $P > 0.35$). However, a bigger difference between the plots is the residual

Taylor's Power Law. https://doi.org/10.1016/B978-0-12-810987-8.00009-4

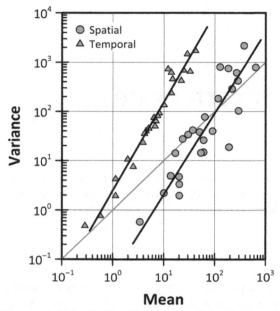

FIG. 9.1 Hybrid temporal ($NQ=25$, $NB=30$) and spatial ($NQ=30$, $NB=25$) TPLs of rotifer communities at 5 depths in Lake Eufaula, Oklahoma, are almost identical. Temporal TPL is offset by one cycle for clarity. *(Data from Tables 2–7 in Bowles (1972).)*

variance, which is much greater in the spatial case. The two plots have a common spatial component, the depth of sample, which could account for the similarity, but not the difference in correlation coefficients, unless the difference is due to greater differences in temporal heterogeneity between depths. In that case, more points per depth would resolve into individual plots at each depth.

UPPER PARAÑA RIVER BASIN, BRAZIL

The Baia River is a tributary of the Paraña River in the Brazilian state of Mato Grosso do Sul. It has a number of overflow lakes along its length. One of them, Guaraná Lake, is separated from the river by a permanent channel that maintains the level of the lake according to the river flow. Bini et al. (2001) sampled the center of the lake, the river, and the channel for rotifers for 12 months from March 1992. Samples were taken from 3 layers of the water column in each water body by filtering 1000 L of water at each of 9 stations. Rotifers were identified, counted, and the density estimated (#/m³). 138 species in 39 genera were recorded, of which 17 species accounted for 75% of individuals. Their Table 2 gives the logarithms of the monthly mean and variance of total density for each of the 3 habitats. Bini et al. conducted a TPL by ODR on the data from the 3

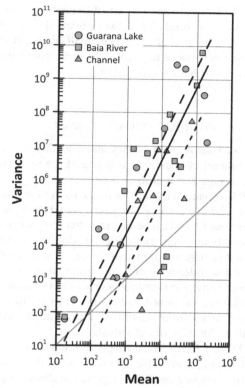

FIG. 9.2 Spatial TPL ($NQ = 3$, $NB = 12$) of total rotifers (#/cc) in Upper Parana River basin, Brazil. *Fitted lines* are parallel and in the order of the legend. *(Data from Table 2 in Bini et al. (2001).)*

habitats and concluded a common slope fit the data, but with different intercepts. Their analysis is repeated here using GMR and computed on #/L (Fig. 9.2; Appendix 9.B).

Rotifers in this study in the Upper Paraná basin were highly aggregated in all 3 habitats, including the river where vertical mixing might be expected. The channel had the lowest overall vertical variation in density and the lake had the highest variation. Temperature measurements at the 9 stations showed a strong temperature differential in the lake's water column but not in either the channel or river, which may have contributed to the higher variances. These results show significantly higher aggregation than Bowles (1972) results in Lake Eufala. In both studies, the number of data points in the combined spatial analyses ($NB = 30$ or 36) is similar, but the number of samples is small ($NQ = 3$ or 5), contributing to comparatively low correlation coefficients and additional uncertainty in slope. The major difference between the 2 studies is the enormous difference in densities between these sites in the Paraña River basin and the

PART | II

colder water of the Oklahoma reservoir. The difference in slopes between the 2 sites may also be due to a correlation between b and the overall mean density that can occur when the number of samples is small (Engen et al., 2008). In this instance, the correlation between the geometric mean population and the slope is $r = 0.5$ for $b_{GMR} = 2.19$ but $r = -0.4$ for $b_{ODR} = 1.64$.

RIVER ELBE ESTUARY, GERMANY

The Elbe Estuary is a coastal-plain estuary that extends 140 km from the North Sea to ~40 km southeast of Hamburg. It is one of the most polluted rivers in Europe with both sewage and industrial waste.

Holst et al. (1998) conducted a survey of the rotifer community in the tidal reaches of the Elbe west of Hamburg. They took weekly samples from March to July 1995 at low tide and 1 and 2 h before and after low tide in a shallow backwater ~15 km downstream from Hamburg. In addition, 4 samples were also taken at low tide at 8 stations ~10 km apart in the main channel between Hamburg and the sea. All samples were of 2.25 L water taken from the top 1 m and sieved to extract organisms. Rotifers in all samples were counted under a microscope. Subsamples were examined and all rotifers identified. More than 70, mostly freshwater species were identified and a single species, *Keratella cochlearis*, accounted for 32% of individuals.

Holst et al.'s Table 1 lists 75 taxa caught at the backwater site on 11 sampling occasions. Their Fig. 2 gives the average density (#/L) estimated from the 5 samples bracketing low tide and Table 1 gives the abundance of taxa on each sampling occasion, coded 1–5 as <1%, 1%–2%, 2%–5%, 5%–10%, and >10%, respectively. Given the abundance codes and the average density per sample date, approximate densities may be computed for each taxon-date combination. Approximate densities were validated against Holst et al.'s Fig. 4 with the mean densities of the 12 most abundant species. As the samples were taken from the same location, but at intervals before and after low tide, TPLs derived from these data are hybrid TPLs.

In Holst et al.'s Table 1, dates on which a taxon was not recorded were left blank. Failure to include zero counts in TPL analysis can have a profound effect on the results. Mixed-species means and variances with (Fig. 9.3A) and without (Fig. 9.3B) zero counts are based on different numbers of taxa: 76 with zeros and 44 without (Appendix 9.C). Zeros not included also distort the low-density end of the regression reducing the correlation coefficient from $r = 0.99$ to 0.94 and substantially increasing the TPL slope from $b = 1.70 \pm 0.03$ to $b = 2.57 \pm 0.13$. The same procedure applied to the community TPL also increases the slope but without loss of data points or distortion at the low end because most points are above the Poisson line.

The effect excluding zeros has on TPL prompts the question, when should the absence of an individual from a sample be counted as a zero or ignored.

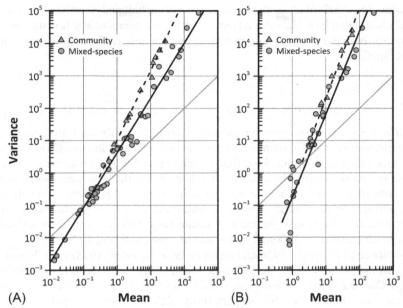

FIG. 9.3 Community ($NQ = 76$, $NB = 11$) and mixed-species ($NQ = 11$, $NB = 76$) TPLs of rotifers in the River Elbe estuary are different if zero counts are included in the analysis. (A) Species found in at least one site are listed as zero counts instead of missing entries (Zeros included). (B) As reported by the authors with zero counts listed as no count (Zeros excluded). *(Data from Table 1 and Figs. 2 and 4 in Holst et al. (1998).)*

A taxon's absence from a sample may be because it is really absent from the location or it is present at a density below the samplers' detection threshold. Increasing the sampler's efficiency may make it sensitive enough to catch the rare individual but will have no effect when sampling outside the target's range. Pragmatically, if a taxon is absent from all samples at a site, it seems reasonable to assume it is truly absent and should not be counted in any sample. But if it occurs in one or more samples, it should be counted as a zero in the others. Counting zeros, the mixed-species estimate is comparable to the temporal and spatial community TPL estimates of rotifers in Bowles' (1974) samples from Lake Eufaula.

Molluscs

Mollusca is one of the largest phyla with ~85,000 described extant species and as many fossil species. Molluscs are found on land and in freshwater as well as the ocean where they account for >20% of described marine animals. They are anatomically highly diverse with 10 classes recognized. These include the bivalves (clams, mussels, and oysters), gastropods (slugs and snails), which account for ~80% of all molluscan species, and cephalopods (cuttlefish, octopi, and squids) that have the most advanced nervous systems of any invertebrates. These three classes include species of gastronomic importance and are therefore well studied.

TELLINA TENUIS IN THE FIRTH OF CLYDE, SCOTLAND

Tellina tenuis is a marine bivalve mollusc endemic to Northwest Europe and the Mediterranean; it is also found in the eastern Indian Ocean. Its trocophore larvae are free-swimming zooplankton that develops into veliger larvae prior to settling on the sea floor. The adult bivalve lives in benthic zone sand and feeds on detritus and microplankton such as diatoms via a siphon that extends to the surface. In the littoral zone, it migrates to ~10 cm below the surface at low tide. It is frequently the most abundant large species in the benthos and can reach densities >4000/m^2 in the littoral zone (Stephen, 1928).

Between September 1926 and October 1927, Stephen (1928) took 39 samples in the intertidal zone at stations at Kames Bay, Garrison Bay and White Bay, Cumbrae Island, Castle Bay, Little Cumbrae Island, and Fairlie Sands on the south shore of the Clyde estuary (Firth of Clyde) opposite Cumbrae. Samples were taken with a 0.25-m^2 square quadrat. Kames Bay was selected for intensive study with 33 samples taken at 6 stations over a 12-month period. In Kames Bay, *T. tenuis* densities increased progressively from zero at the spring-tide high-water mark to ~4000/m^2 at the spring low-water mark. In offshore samples, none were found at depths >4 fathoms (7.3 m). Stephen measured the sizes of shells (equivalent to age classes) of all *T. tenuis* recovered and presented the frequency distributions in his Tables 7 and 8. Table 7 is of the intensively sampled Kames Bay site with shell sizes arranged in 16 1-mm bins from 3 to 18 mm at 33 station dates. Table 8 reports 6 additional site dates in the Clyde estuary arranged in 19 bins from 3 to 21 mm and a 24-mm bin. The size-frequency distributions are closely similar at the 5 sites with the exception of the largest size classes' absence at Kames Bay, indicating fewer age classes there than at the other sites.

Spatial and age-class ensemble TPLs of *T. tenuis* were computed from Stephen's Tables 7 and 8. The ensemble at Kames Bay is of $NQ = 16$ size/age classes by $NB = 33$ station-dates and $NQ = 33$ station-dates by $NB = 16$ size/age classes. The other sites comprise $NQ = 20$ size/age classes by $NB = 6$ site-dates and $NQ = 6$ site dates by $NB = 20$ size/age classes (Fig. 9.4 left; Appendix 9.D). The plots of station dates at Kames Bay and the other sites are coincident with slopes not different ($P > 0.24$) with a common slope of $b = 2.05 \pm 0.15$. The plots of size/age classes (Fig. 9.4 right) also have common slope ($P > 0.35$) but are separated with different intercepts ($P < 0.001$). As is usually the case with spatial versus temporal TPLs, the slopes of the spatial and size/age class TPLs differ ($P < 0.05$).

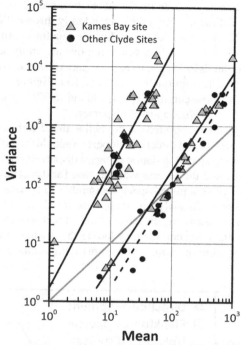

FIG. 9.4 Spatial ensemble TPLs of *Telina tenuis* by site averaged over size class *(left)* and by size/age class averaged over site and date *(right)*. A single ensemble TPL ($NQ = 16$) fits the Kames Bay and other sites averaged over size class of 33 samples at 6 stations at Kames Bay, Cumbrae, and of 6 samples taken at 4 other sites in the Clyde estuary ($NB = 39$). Two parallel TPLs fit the age-class points averaged over site and date at Kames bay ($NQ = 33$, $NB = 16$) and the other sites($NQ = 6$, $NB = 20$). *(Data from Tables 7 and 8 in Stephen (1928).)*

INTERTIDAL MOLLUSCS ON THE ISLE OF MAN

Many marine molluscs are sessile or have limited motility as adults (bivalves such as *Tellina tenuis*, for example) and must disperse as free-living larvae. Other species, such as snails, are mobile as adults and may not have a motile larval stage. Species not having larval stages develop directly and likely have a different spatial distribution as adults than species with a dispersal stage. Johnson et al. (2001) tested the proposition that colonization by direct developing species have relatively greater fine-scale patchiness than those species with a larval dispersal stage. They tested this by collecting molluscs at different spatial scales at shores on the Isle of Man, an island in the Irish Sea. Prior studies suggested densities of dispersing species were higher on the west coast than on the east coast, so direct developers on east and west coasts were expected to show smaller differences in abundance and diversity than dispersers. Samples were collected from 4 shores in July 1999: 2 each on the east and west coasts. To maximize the number of individuals and species, the samples were taken just above the spring-tide low-water mark. At each shore, 2 sites separated by ~100 m were selected for sampling with 6 randomly thrown 0.25-m^2 quadrats; 48 quadrats total. Molluscs and algae within the quadrats were removed. The algae were washed and then wash sieved to remove any molluscs on the plants. A total of 8580 individuals in 34 species and 5 genera were classified as dispersers or direct developers. The reproductive status of 10 individuals in 4 genera was unknown. One of the most common species found in 77% of quadrats was the edible mussel (*Mytilus edulis*), a larval disperser.

Johnson et al.'s results supported the prediction that differences in abundance and diversity of dispersers and direct developers would differ within and between shores. Specifically, the spatial variation of larval dispersers was greater between shores than within, while the reverse was the case for direct developers. Part of their analysis consisted of a mixed-species ensemble TPL based on all 48 quadrats from the 4 shores. Reanalyzing their data by GMR, the mixed-species TPLs do not differ between dispersers and direct developers ($P > 0.25$; Fig. 9.5; Appendix 9.E). Several taxa, including the 4 whose mode of development was unknown, were quite rare with points lying on the Poisson line.

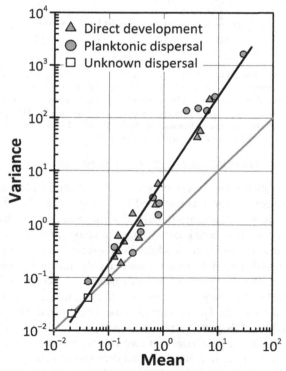

FIG. 9.5   Ensemble mixed-species TPL ($NQ = 48$, $NB = 39$) of intertidal molluscs sampled on 4 shores of the Isle of Man. (*Data from Table 1 and Fig. 2 in Johnson et al. (2001).*)

MARINE BIVALVES AND GASTROPODS IN DENMARK

Following a severe hypoxia event in summer 1997, a sampling program the following spring and summer was initiated in the affected area of Mariager Fjord on the east coast of Denmark. The 1997 anoxic episode had apparently extinguished the entire benthic community in the central part of the estuary. Hansen et al. (2002) documented the subsequent recolonization by zooplankton from the open water of the Kattegat. As part of the study, they monitored settlement rates of meroplankton (molluscs, polychaetes, and crustaceans), using strips of material used for cultivating mussels commercially. At each of 3 stations, 12 settlement strips were exposed for 14–28 days. Strips were retrieved and settlers identified and counted on 48 cm^2 in the middle of the strips. Four randomly picked strips per station were examined for bivalves and 3 strips for gastropods. Hansen et al's Table 1 gives the mean and SD of number of bivalves and gastropods per m^2 for 8 or 9 sample dates from 7 May to 26 August 1998. Their data show that recolonization of Mariager Fjord occurred very quickly from adjacent parts of the estuary and the open sea following the defaunation.

A spatial TPL of Hansen et al.'s Table 1 data of bivalve and gastropod molluscs (Fig. 9.6; Appendix 9.F) showed settling to be similarly distributed. The comparatively low correlation coefficients are probably due to the variation in SD resulting from the small number of samples. It is noteworthy that the residual variance of the fitted lines is greater for gastropods ($NQ = 3$) than bivalves ($NQ = 4$). Considering that dispersal of meroplankon is largely controlled by external forces (currents and tides) contributing to mixing, settling by mollusc larvae might be expected to be random. Hensen et al's data, however, show the distributions to be moderately aggregated at all densities observed. The TPL slopes of bivalves and gastropods are not significantly different ($P > 0.45$), suggesting that the same processes control the distribution of settling in both taxa.

TERRESTRIAL GASTROPODS IN ALBERTA

Concern over an outbreak of lungworm in a herd of bighorn sheep (*Ovis canadensis*) led Boag and Wishart (1982) to conduct a quantitative survey of the intermediate molluscan hosts of *Protostrongylus* spp., the nematode causative agents. The bighorns occupy different seasonal habitats in their range on the flanks of the Rocky Mountains in Alberta. The summer range consists of 9 habitat types in a complex mosaic of grasslands, sedge meadows, willow fens, and aspen, poplar, spruce, and pine woodlands in various combinations. Boag and Wishart established 160 30x30cm permanent plots in April 1979 in the 9 habitats. In 1979 the plots were visited every 7 days, and in 1980 and 1981 a slightly different set of 160 plots was visited every 4 days. All molluscs were removed every visit and returned to the laboratory for identification. The total number of collections ranged from 170 in an aspen copse (17 collections at 10 plots) in 1979 to 2040 in open grasslands in 1980 (51 collections at 40 plots).

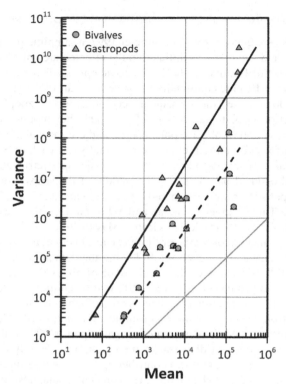

FIG. 9.6 Ensemble spatial TPLs of bivalve ($NQ = 4$, $NB = 25$) and gastropod mollusc ($NQ = 3$, $NB = 14$) settlement (#/m^2) during recolonization of a hypoxic section of Meriager Fjord, Denmark, in 1998. Gastropods are shifted 1 cycle to the right for clarity. *(Data from Hansen et al. (2002).)*

The number of specimens collected belonging to 13 taxa is recorded in Boag and Wishart's Table 1. They recorded 14 species of pulmonate snail and a single slug with the greatest abundance in the sedge meadow and willow fen. The fewest molluscs were collected from the dryer grassland and deciduous tree habitats.

The collections over 3 years and 9 habitats provide a total of 22 ensemble sets of molluscan communities. Boag and Wishart's Table 1 records the total number of each taxon collected, the number of plots per habitat-year, and the number of collections per year. Community ensemble TPLs of the total number collected and density estimated from the number per 100 collections are shown in Fig. 9.7 (Appendix 9.G). Curiously, the TPLs computed from total number ($A = 0.78 \pm 0.076$, $b = 1.81 \pm 0.083$) and density ($A = 0.64 \pm 0.050$, $b = 1.85 \pm 0.088$) are almost identical despite the fact the density estimates are obtained from totals divided by factors ranging from 1.7 to 20.4. Noticeably, the density TPL (#/100 collections) has a narrower range and slightly lower

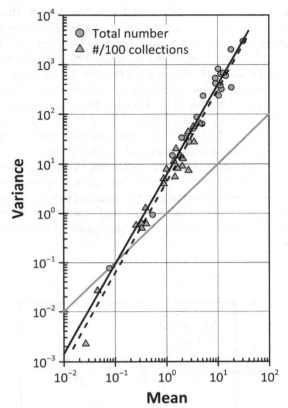

FIG. 9.7   Community ensemble TPLs ($NQ = 13$, $NB = 22$) of pulmonate molluscs in 9 Alberta big-horn sheep habitats over 3 years. TPLs of the total number of specimens collected from 10 to 70 plots on 15–51 collecting visits over 3 years *(solid line)* and the average #/100 collections *(dashed line)* are almost identical. *(Data from Table 1 in Boag and Wishart (1982).)*

intercept than the total TPL, but the slope and correlation coefficients differ only in the second and third decimal places. Careful examination shows the clusters of points to be closely similar, with minor changes to their relative positions. In this instance with $NQ = 13$ and $NB = 22$, community TPL is remarkably robust and minimally influenced by essentially random adjustments.

SLUGS IN NORTHUMBERLAND

Between January 1963 and March 1965, Hunter (1966) took soil samples every 4 weeks to study the spatial and temporal distribution and abundance of slugs at a market garden near Newcastle upon Tyne in Northumberland, England. The procedures changed several times during the study. On each sampling occasion, a cubic foot (28,317 cc) of soil was removed and divided into four 7.6 cm layers.

Nine 4.6×4.6 m plots were sampled in 1963 and 12 in 1964–5. Three species of slug (*Agriolimax reticulatus*, *Arion hortensis*, and *Milax budapestensis*) were extracted from each layer by either soil washing or flooding. The latter method replaced soil washing from March 1964 but was ~90% as efficient as soil washing. Fig. 9.8 (Appendix 9.H) shows the TPLs for the total number of slugs extracted from the samples and slug density (#/m³) after correcting for number of samples and extraction efficiency. Like the previous example, conversion from #/sample ($A = -0.48 \pm 0.250$, $b = 2.69 \pm 0.241$) to density ($A = -0.64 \pm 0.247$, $b = 2.83 \pm 0.233$) resulted in only small changes to TPL. A small decrease in range was accompanied by small increases in both the slope and correlation coefficient. While the difference in slopes is not significant ($P > 0.6$), both are significantly >2 ($P < 0.001$). As with the previous example, conversion to density preserved the general pattern of points, this time bringing them closer to the fitted line, resulting in a slight increase in correlation coefficient. Again, TPL robustly measured the aggregation pattern of both raw and adjusted data.

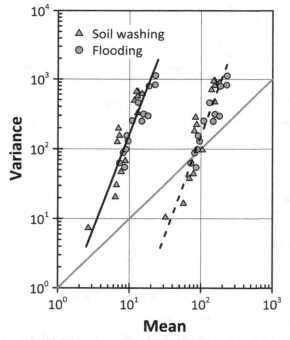

FIG. 9.8 Three species of slugs sampled at 4 depths combined in a hybrid community TPL over sample dates in an arable field ($NQ = 12$, $NB = 28$). Number of slugs in samples (left). Density estimates (#/m³) after correction for efficiency and number of samples (right, offset one cycle for clarity). *(Data from Table 1 in Hunter (1996).)*

Echinoderms

The phylum Echinodermata is an exclusively marine taxon—no freshwater or terrestrial species are known to have existed. The ~7000 species include starfish, sea urchins, and crinoids. Like rotifers and many molluscs, most echinoderms have a planktonic larval stage, although in some species females brood. Reproduction may be sexual or asexual and in some species both may occur. Adults often exhibit five-fold radial symmetry and are present at all depths from the littoral to the abyssal. Most are mobile, but some crinoids are sessile.

STARFISH IN NORTH WALES

The starfish *Astropecten irregularis* live on sandy, muddy, or gravel seabeds between 1 and 1000 m below the surface of the Atlantic Ocean and Mediterranean Sea. They generally remain buried in the sediment during the day, migrating to the sediment surface to hunt at night. Their preferred prey are bivalve molluscs such as clams, mussels, and oysters. Their prey are caught with the starfish's arms and taken to the mouth where the prey are trapped by long mobile spines.They will also eat snails and will scavenge if necessary. Spawning occurs several times during summer resulting in planktonic larvae that metamorphose into juvenile starfish distinguishable by their stubby arms.

In a study of the seasonal distribution and abundance of *A. irregularis*, Freeman et al. (2001) trawled Red Wharf Bay, Anglesey, North Wales, monthly at neap tide from October 1995 to May 1998. The study site was an area ~1 km^2 divided into sixteen plots of 6.3 ha. The length of each trawl was determined by GPS and the area sampled computed. All specimens of *A. irregularis* recovered were counted and measured. Catches showed a marked seasonal cycle, peaking in summer. Freeman et al. concluded that the spatiotemporal pattern of abundance indicated that starfish moved out of the study area to deeper water in early winter and returning in spring.

Freeman et al. used Lloyd's mean crowding index (Eq. 3.19) to characterize the spatial distribution. Their Table 1 gives Lloyd's index, the mean and SD of number/ha, and the number of trawls for 23 months. Values of Lloyd's index indicated high aggregation throughout the study period and oscillations reflected the seasonal cycle of catch. Spatial TPL computed from Freeman et al.'s Table 1 confirms Lloyd's index result of the high aggregation at all densities (Fig. 9.9; Appendix 9.I). Each point in the regression is the monthly average density (#/ha) based on $NQ = 16–36$ trawls. Over 23 months, the density cycled in response to mortality, emigration in winter, return to the summer range in spring, and summer recruitment. Over 2 population cycles, the variance remained proportional to the mean squared over a range of $M = 6–160$/ha.

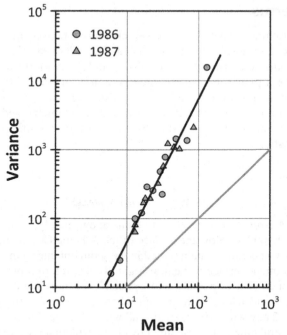

FIG. 9.9   Spatial TPL ($NQ = 16$–36, $NB = 23$) of starfish *Astropecten irregularis* (#/ha) in Red Wharf Bay, Anglesey, North Wales. Seasonal variation is likely due to migration into and out of the study area in addition to recruitment and mortality. *(Data from Table 1 in Freeman et al. (2001).)*

CRINOIDS IN SÃO PAULO STATE, BRAZIL

Crinoids are echinoderms found in both shallow water and at depths to 9000 m. They may be free living as adults or connected to the substratum by a stalk (sea lilies) or without a stalk (feather stars). Male and female crinoids release gametes into the water and fertilized eggs develop into free-swimming vitellaria larvae that settle within a few days.

The feather star *Tropiometra carinata* is very abundant along the Brazilian coast where it can occur in dense populations in the sublittoral zone on gravel or rocks. MacCord and Duarte (2002) surveyed *T. carinata* at 2 sites in the São Sebastião Channel (Ponta do Baleeiro and Cabelo Gordo Beach) on the coast of São Paulo State. Two rocky sites <1 km apart were each divided into 50 1-m² contiguous quadrats at an average depth of 2 m. The sites were visited at approximately 2-month intervals from December 1997 to October 1998 and the crinoids were collected from each quadrat for measuring. Thus, the population each month was a new one of recently settled vitellaria.

FIG. 9.10   Spatial TPL ($NQ = 50$, $NB = 14$) of crinoids *Tropiometra carinata* at 2 sites in the São Sebastião Channel, Brazil. *(Data from Table 1 in MacCord and Duarte (2002).)*

MacCord and Duarte computed Morisita's Index (Eq. 3.14) to measure the spatial distribution, which, with the total number of specimens and population density (#/m²), was recorded in their Table 1. Morita's index (Eq. 3.14) indicated strongly aggregated populations within a narrow range of means. Knowing the sample size, $NQ = 50$, and mean density, M, variance may be calculated from Morista's index. Fig. 9.10 (Appendix 9.J) shows the spatial TPL of *T. carinata*. At $NQ = 50$, these data meet Clark and Perry's (1994) sample size criterion, while $NB = 14$ points is marginal. With only 14 points, combined with a very narrow range of means ($1.82–2.80$/m²), it is likely b may be correlated with the mean and the slope overestimated. Although caution should be exercised in interpreting results with so narrow a range of means, using the result to transform the data for analysis is still appropriate.

Other Phyla

BRYOZOANS IN THE GREENLAND SEA

Bryozoa is a phylum of small aquatic invertebrates that filter feed with tentacles lined with cilia. Most species are marine and live in tropical seas, although many are in temperate or cold seas, and some live in brackish or freshwater. Of the ~4000 extant species, all but one genus is colonial.

The structure of bryozoan assemblages from two dissimilar Arctic regions was studied by Kuklinski and Bader (2007) at sites off the east coast of Greenland and the west coast of Spitsbergen. The east Greenland and west Spitsbergen sites are at the same latitude (79°N) on the Greenland Sea but differ in hydrology, distance from the land, and breadth of the shelf. They also differ markedly in temperature as the Greenland site is strongly influenced by the East Greenland Current, which originates in the Arctic Ocean, whereas the Spitsbergen site is warmer due to the West Spitsbergen Current, a branch of the Norwegian Current, which is itself a branch of the warm North Atlantic Current (the Gulf Stream).

Samples of bryozoans were taken by dredging the same range of depths of 75–260 m at both sites. 100 rocks with bryozoans were selected at random and the bryozoans identified to species or genus and counted at each of 9 stations off Greenland and 8 stations off Spitsbergen. The number of rocks per station ranged from 3 to 33 at the Greenland site and 3–21 at Spitsbergen. Rocks without bryozoans were not included in the survey; thus, there were no zero samples. The site in the cold East Greenland current yielded 86 taxa, while the warmer West Spitsbergen current yielded 59 taxa; 31 taxa were common to both sites. The abundance of brozoans taken ranged from 180 to $1521/m^2$ and 532–1464/ m^2 at Greenland and Spitsbergen sites, respectively. Altogether 14,922 specimens were recorded.

Kuklinski and Bader's Table 1 gives the density and SD of each taxon recovered at the 2 sites. Fig. 9.11 (Appendix 9.K) shows the ensemble mixed-species TPLs for the Greenland and Spitsbergen survey sites. Although the species compositions and abundances differed between the sites, the TPLs differed in intercept ($P < 0.05$) but not in slope ($P > 0.6$). It seems unlikely that the sampling efficiency differed between the sites, so the higher variance at the Greenland site is likely due to the differences in oceanography and hydrology, which also account for the differences in abundance and diversity. Because only rocks with bryozoans were examined, the absence of sample units (rocks) with no specimens likely inflated the estimates of slope: the common estimate of $b = 1.65 \pm 0.026$ may therefore be an overestimate.

HYDROIDS IN THE ARGENTINE SEA

The Hydrozoa are mostly marine predators that constitute a class in the Cnidaria, to which the corals and jellyfish also belong. Hydroids are a polyp life stage of most hydrozoans, which may be colonial. Polyps bud and branch, the exact form being specific. Colonies may be quite large and members may be specialized and unable to survive alone. Examples of hydrozoans include the freshwater *Hydra* species and the marine Portuguese man o' war (*Physalia physalis*), both of which have specialized stinging structures called nematocysts.

The coast near Mar del Plata, Argentina, is characterized by outcroppings of quartzitic rocky blocks extending from the intertidal several kilometers out to sea. The blocks provide settlement areas for many benthic invertebrates, including hydrozoans. Starting in November 2000, Genzano et al. (2002) collected 11 approximately monthly samples of hydroids along a 10-m line transect on one

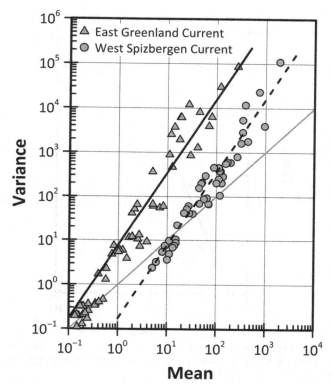

FIG. 9.11 Ensemble mixed-species TPLs of bryozoans at sites off Spitzbergen ($NQ = 3$–33, $NB = 59$) and Greenland ($NQ = 3$–29, $NB = 86$) on either side of the Greenland Sea. The West Spitsbergen Current TPL is shifted one cycle to the right for clarity. *(Data from Table 1 in Kuklinski and Bader (2007).)*

block, Banco del Medio, 18-20 m below the surface. They laid a 10-m chain with 2-cm links arranged in 1-m segments on the rock and identified and counted hydroid clumps in contact with the chain in each segment to estimate cover. Of the 13 species of hydroid recorded, the most common were *Amphisbetia operculata*, *Sertularella mediterranea*, and *Plumularia setacea*. The average coverage and SD of these species are given in their Table 3.

Spatial TPLs of the 3 species in Fig. 9.12 (Appendix 9.L) show *S. mediterranea* and *P. setacea* are very similar; not significantly different in either intercept ($P > 0.45$) or slope ($P > 0.70$). TPL for *A. operculata* is less steep with a pronounced curvature (Bartlett's test is significant at $P < 0.05$. *A. operculata* is the largest of the 3 species at ~10 cm, *S. mediterranea* is the smallest at ~2 cm, and *P. setacea* is ~7 cm. TPL curvature may be a consequence of size, as there is only limited space for individuals on a line transect; the bigger the individual, the fewer can touch the transect so the increase in variance becomes

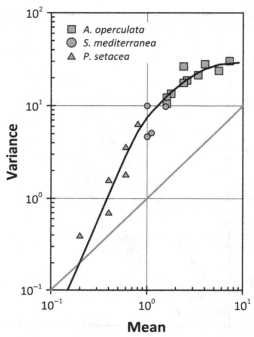

FIG. 9.12   TPLs of hydroids in Argentina is linear for the less abundant *P. setacea* but curved for the more abundant *A. operculata* possibly a consequence of the limited sampler size used in a line transect. *(Data from Table 3 in Genzano et al. (2002).)*

relatively smaller as the mean increases (Chapter 15), resulting in a left-hand curve with increasing density as seen with these hydroids. This highlights a difficulty with line transects, although the same problem can occur if quadrats are too small in relation to the size of the target (Figs. 5.12A and 15.7).

JELLYFISH IN OREGON-WASHINGTON COASTAL WATERS

Scyphozoan jellyfish are principally planktivorous. In seasons of high abundance, they may compete indirectly with commercial fish whose prey are predominantly planktivores. When sea nettle *Chrysaora fuscescens* populations are high, Ruzicka et al. (2016) found they adversely affect the growth of salmon off the coast of Oregon and Washington. Significantly, their study of the population dynamics of sea nettles and salmon found a negative association between salmon stomach contents and sea nettle abundance.

Trawl samples were taken in the upper 20 m of the water column at 5 stations on 8 transects in June and September from 1999 to 2013. Jellyfish were identified to species and counted: all were counted at low densities but large catches were subsampled. Sea nettle biomass was estimated from the diameter of the bell using an empirical relationship. Ruzicka et al.'s Fig. 4 gives the mean and SD of salmon abundance (#/km^2) and sea nettle biomass (T/km^2). These data were used for a spatial analysis of sea nettle biomass. (See Chapter 10 for the salmon TPL analysis.) The spatial regression of $NB=22$ months based on $NQ=40$ trawls is highly significant ($r=0.99$) with a very steep slope ($b=2.56\pm0.08$) over 2 orders of magnitude of mean (Fig. 9.13; Appendix 9. M). Nearly half the line is below the Poisson line where the residual variance is slightly greater than above. In addition, there is a hint of curvature, confirmed by Bartlett's test ($P<0.01$). The curvature is resolved by fitting separate lines above and below the Poisson line. The difference in slopes (significant at $P<0.005$) suggests a difference in sampling efficiencies (see Chapter 15), which may be due to the practice of subsampling the larger catches with efficiency declining with catch volume.

   In general, the species sampled from surfaces (bryozoans, crinoids, hydroids, and starfish) have lower TPL slopes than those sampled from

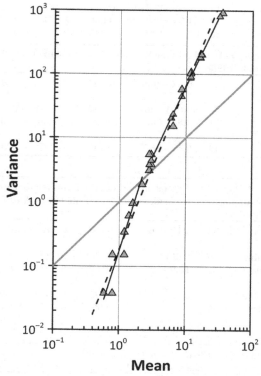

FIG. 9.13 Spatial TPL ($NQ=40$, $NB=20$) of biomass (T/km^2) of sea nettles off the Oregon-Washington coast. The two *solid lines* are TPLs fitted to biomass above and below the Poisson line, the *hatched line* is fitted to the full set. *(Data from Fig. 4 in Ruzicka et al. (2016).)*

volumes. The additional dimension may contribute to slopes significantly greater than $b = 2$ for jellyfish taken from the sea, slugs from the soil, and some rotifer samples from water.

Appendix: TPL estimates for other invertebrates

	NB	Range of means	r	A	SE[A]	B	SE[b]
Rotifera							
A Spatial and temporal TPLs ($NQ = 6$) of rotifers (#/L) at 5 depths in Eufaula Reservoir, Oklahoma. Data from Tables 2–7 in Bowles (1972)							
Temporal	25	0.283 - 41.8	0.982	0.396	0.042	1.797	0.068
Spatial	30	0.340 - 63.6	0.913	0.314	0.082	1.650	0.123
B Spatial TPLs of rotifers (#/L) at $NQ = 3$ depths in the Upper Paraña River basin, Brazil. Data from Table 2 in Bini et al. (2001)							
Guaraná Lake	12	17.4 - 199,526	0.908	−0.766	1.699	1.945	0.236
Channel	12	427 - 72,444	0.647	−6.187	1.196	2.959	0.651
Baía River	12	17.4 - 138,038	0.686	−2.174	2.212	2.185	0.459
Combined	36	17.4 - 199,526	0.746	−2.381	1.073	2.193	0.243
C Mixed-species ($NQ = 11$) and community ($NQ = 76$) TPLs of rotifers in the Elbe Estuary. Data in Holst et al. (1998)							
Zeros included – Community	11	0.82 - 26.9	0.995	1.014	0.038	2.156	0.065
Mixed species	76	0.014 - 267	0.990	0.621	0.027	1.695	0.027
Zeros excluded – Community	11	3.88 - 60.4	0.985	−0.486	0.070	2.788	0.144
Mixed species	44	0.69 - 267	0.946	−0.757	0.099	2.568	0.125
Mollusca							
D Ensemble TPLs ($NQ = 16$-30) of littoral bivalve mollusc *Telina tenuis* size class frequencies at sites in the Firth of Clyde, Scotland, in 1926–27. Data from Tables 7 and 8 in Stephen (1928)							
Kames Bay – Station-date ensemble	33	1.19 - 56.63	0.891	0.252	0.067	2.026	0.160
Size class ensemble	16	0.061 - 93.7	0.990	0.599	0.064	1.656	0.058
Other sites – Site-date ensemble	6	10.9 - 38.3	0.939	−0.307	0.079	2.476	0.347
Size class ensemble	20	0.67 - 69.0	0.962	0.230	0.069	1.771	0.108
Combined sites – Site-date ensemble	39	1.19 - 56.6	0.893	0.223	0.057	2.047	0.148
Size class ensemble	36	0.061 - 93.7	0.973	0.396	0.054	1.717	0.066
E Ensemble mixed-species TPL ($NQ = 48$) for marine molluscs sampled (#/quadrat) on 4 shores of the Isle of Man. Data from Table 1 and Fig. 2 in Johnson et al. (2001)							
Planktonic dispersal	21	0.021 - 6.89	0.985	0.769	0.053	1.517	0.057
Direct development	14	0.021 - 28.4	0.979	0.866	0.092	1.640	0.088
Combined	39	0.021 - 28.4	0.985	0.836	0.046	1.573	0.044

Continued

	NB	Range of means	r	A	SE[A]	B	SE[b]
F Ensemble spatial TPL (NQ=3-4) of bivalve and gastropod mollusc settlement (#/m²) during recolonization of Meriager Fjord, Denmark, following hypoxia-induced defaunation. Data from Hansen et al. (2002)							
Bivalves	25	68.0 - 194,325	0.946	0.497	0.465	1.713	0.111
Gastropods	14	33.0 - 14,828	0.897	1.044	0.474	1.556	0.183
G Community ensemble TPLs (NQ =13) of pulmonate molluscs in 9 bighorn sheep habitats in Alberta. The same data were used to compute TPLs of the total number of specimens collected regardless of the number of collections (solid line) and the average number per 100 collections (dashed line). Data from Table 1 in Boag and Wishart (1982)							
Total number	22	0.077 - 31.1	0.979	0.780	0.052	1.813	0.079
#/100 collections	22	0.026 - 4.57	0.977	0.639	0.050	1.853	0.084
H Hybrid spatial community TPL (NQ =12) of monthly samples of 3 slugs in samples at 4 depths from soil in an arable field in Northumberland, England. Data from Table 1 in Hunter (1966)							
Total number	28	2.67 - 23.1	0.889	−0.475	0.048	2.690	0.232
Density (#/m³)	28	3.19 - 23.0	0.908	−0.642	0.046	2.830	0.224

Echinodermata

	NB	Range of means	r	A	SE[A]	B	SE[b]
I Spatial TPL (NQ =16-36) of starfish *Astropecten irregularis* in Red Wharf Bay, Anglesey, North Wales. Data from Table 1 in Freeman et al. (2001)							
Astropecten irregularis	23	6 - 130	0.974	−0.412	0.036	2.076	0.098
J Spatial ensemble TPL (NQ =50) of crinoid *Tropiometra carinata* at two sites in São Paulo State, Brazil. Data from Table 1 in MacCord and Duarte (2002)							
Tropiometra carinata	14	1.82 - 2.80	0.794	−0.187	0.022	2.824	0.459

Other Phyla

	NB	Range of means	r	A	SE[A]	B	SE[b]
K Mixed-species ensemble TPLs (NQ=100) of bryozoans sampled on the east and west sides of the Greenland Sea. Data from Table 1 in Kuklinski and Bader (2007)							
East Greenland Current	86	0.10 - 85.4	0.985	0.926	0.025	1.648	0.030
West Spitsbergen Current	59	0.50 - 192	0.975	0.832	0.035	1.675	0.048
Combined	112	0.10 - 192	0.982	0.892	0.021	1.653	0.026
L Spatial TPLs (NQ =10) of hydroids on Banco del Medio, Argentina. Data from Table 3 in Genzano et al. (2002)							
Amphisbetia operculata	10	0.20 - 7.20	0.908	0.768	0.076	1.218	0.162
Sertularella mediterranea	6	0.20 - 2.40	0.980	0.714	0.050	1.507	0.123
Plumularia setacea	10	0.20 - 4.00	0.974	0.662	0.046	1.448	0.104
M Spatial TPLs (NQ =40) of sea nettles, *Chrysaora fuscescens*, off the Oregon-Washington coast. Data from Table 4 in Ruzicka et al. (2016)							
Sea nettles	24	0.60 - 34.8	0.990	−0.732	0.056	2.555	0.075
Above the Poisson line	13	2.78 - 34.8	0.993	−0.388	0.072	2.206	0.072
Below the Poisson line	11	0.60 - 2.19	0.933	−0.759	0.068	3.371	0.366

References

Bini, L.M., Bonecker, C.C., Lansac-Tôha, F.A., 2001. Vertical distribution of rotifers on the Upper Paraná River floodplain: the role of thermal stratification and chlorophyll-*a*. Stud. Neotropical Fauna Environ. 36, 241–246.

Boag, D.A., Wishart, W.D., 1982. Distribution and abundance of terrestrial gastropods on a winter range of bighorn sheep in southwestern Alberta. Can. J. Zool. 60, 2633–2640.

Bowles, L.G., 1972. A description of the spatial and temporal variations in species composition and distribution of pelagic net zooplankton in the central pool of Eufaula Reservoir, Oklahoma, with comment on forced aeration destratification experimentation. Trans. Kans. Acad. Sci. 75, 156–173.

Clark, S.J., Perry, J.N., 1994. Small sample estimation for Taylor's power law. Environ. Ecol. Stat. 1, 287–302.

Engen, S., Lande, R., Saether, B.-E., 2008. A general model for analyzing Taylor's spatial scaling laws. Ecology 89, 2612–2622.

Freeman, S.M., Richardson, C.A., Seed, R., 2001. Seasonal abundance, spatial distribution, spawning and growth of *Astropecten irregularis* (Echinodermata: Asteroidea). Estuar. Coast. Shelf Sci. 53, 39–49.

Genzano, G.N., Zamponi, M.O., Excoffon, A.C., Acuña, F.H., 2002. Hydroid populations from sublittoral outcrops off Mar del Plata, Argentina: abundance, seasonality and reproductive periodicity. Ophelia 56, 161–170.

Hansen, B.W., Stenalt, E., Petersen, J.K., Ellegaard, C., 2002. Invertebrate re-colonisation in Mariager Fjord (Denmark) after severe hypoxia. I. Zooplankton and settlement. Ophelia 56 (3), 197–213.

Holst, H., Zimmermann, H., Kauscha, H., Koste, W., 1998. Temporal and spatial dynamics of planktonic rotifers in the Elbe Estuary during spring. Estuar. Coast. Shelf Sci. 47, 261–273.

Hunter, P.J., 1966. The distribution and abundance of slugs on an arable plot in Northumberland. J. Anim. Ecol. 35, 543–557.

Johnson, M.P., Allcock, A.L., Pye, S.E., Chambers, S.J., Fitton, D.M., 2001. The effects of dispersal mode on the spatial distribution patterns of intertidal molluscs. J. Anim. Ecol. 70, 641–649.

Kuklinski, P., Bader, B., 2007. Comparison of bryozoan assemblages from two contrasting Arctic shelf regions. Estuar. Coast. Shelf Sci. 73, 835–843.

MacCord, F.S., Duarte, L.F.L., 2002. Dispersion in populations of *Tropiometra carinata* (Crinoidea: Comatulida) in the São Sebastião Channel, São Paulo State, Brazil. Estuar. Coast. Shelf Sci. 54, 219–225.

Ruzicka, J.J., Daly, E.A., Brodeur, R.D., 2016. Evidence that summer jellyfish blooms impact Pacific Northwest salmon production. Ecosphere 7 (4), e01324.

Stephen, A.G., 1928. Notes on the biology of *Tellina tenuis* da Costa. J. Mar. Biol. Assoc. UK 15, 683–702.

Chapter 10

Vertebrates

Compared to most invertebrates, vertebrates are long lived with long generation times and low abundances. This presents a challenge to obtaining data adequate for TPL analysis. Exceptions are fish because of their economic importance, and birds are well represented because of the popularity of birding, and some are game animals of economic importance. Several large-scale bird surveys are conducted annually in UK and USA that provide a trove of quantitative data on the distribution and abundance of hundreds of species, some common to North America and Western Europe. Surprisingly, there are some data of amphibians and reptiles adequate for TPL analysis. It is mammals, especially large mammals, which are most difficult to find data sufficient for analysis. An exception is cetaceans, which attract considerable interest and research on population dynamics, primarily for conservation efforts, now that whaling is strictly controlled by international agreement.

In previous chapters, samples were typically obtained at points in space with some kind of sampling mechanism that physically isolates and/or removes the organisms from the environment—quadrat, trap, or auger—and that frequently kills the specimens. For some vertebrates such as small mammals live-trapping methods exist, but for most members of the Phylum Chordata, trapping methods have limited utility. For these species, line transects offer a rigorous sampling method that interferes minimally with the target population.

In this chapter, we examine TPL analyses of fish, herptiles, birds, and mammals except *Homo sapiens*, which is covered in Chapter 11. Most of the data reported in this chapter were abstracted from the literature and much obtained by distance sampling, which we describe briefly first.

Distance sampling

Distance sampling is one of several methods generalized from quadrat sampling. In one approach, the sample area is a defined strip along which an observer moves recording all target species within a fixed distance of the path. Thus, strip sampling can be regarded as a long thin quadrat. Two of the herptile studies are examples of this approach. Density, D, is estimated from

$$\hat{D} = \frac{N}{2wL}$$

Taylor's Power Law. https://doi.org/10.1016/B978-0-12-810987-8.00010-0

where N is the number of specimens encountered, L is the length of the transect, and w is the strip's half width. Typically there will be k transects and N and L are the sum of n_i and l_i, $i = 1 \ldots k$, respectively.

Another method used in these case studies, line transect sampling, like the strip sample is a long thin quadrat, but the edges are not clearly defined. Instead, the distance to observed targets is estimated, often with range-finding binoculars, and used to determine an encounter probability from which a density estimate may be arrived at. Variance estimates, usually in the form of confidence intervals, are made by bootstrap techniques (see Good, 2010). A convenient feature of bootstrapped confidence intervals (CI) is the standard error of the estimate and the standard deviation of the sample are numerically the same. Thus, variance is simply $V = (CI/1.96)^2$. The willow ptarmigan, cetacean, and herding ungulate studies used this approach. Density is estimated from

$$\hat{D} = \frac{N}{2wL\hat{P}_a}$$

where P_a is the detection probability, which is estimated from

$$P_a = \frac{\int_0^w g(x)dx}{w}$$

and $g(x)$ is a function of x, the measured distance to the animal, $g(x) = 1 - 0.0052x^2$.

A third approach, frequently used for birds, is a point transect sampling system in which point distance samples are taken at intervals along a transect. This is the approach used by the US Department of Interior's Breeding Bird Survey and the Audubon Society's Christmas Bird Count, examples of which are included. The density estimate for point-transect samples is

$$\hat{D} = \frac{N}{2\pi k \int_0^w rg(r)dr}$$

where w is now the radial distance measurement and k is the number of sample points. The bird sampling system used by the British Trust for Ornithology is a line sample based on sight and sound, but of variable width.

Detailed accounts of distance sampling are in Buckland et al. (2001, 2007), and software called DISTANCE has been developed by Thomas et al. (2010) to estimate abundance with confidence intervals.

Fish

HADDOCK AND WHITING OFF MASSACHUSETTS

In the 1930s and 1940s, fisheries scientists were concerned about sources of variation in fish samples (e.g., Winsor and Walford, 1936). Two sources of variation were considered: variation in the sampler itself and variability in the spatial distribution of fish stocks. It was found that the former source was generally insignificant in a comparison of plankton samples taken by net and samples taken by a metered pump (Barnes, 1949). Clearly, accurate population estimates of fish stocks were then and are still important for setting fishing policy and conserving important stocks. From 1948 to 1950, CC Taylor (1953) examined the issue of fish catch variability by trawl net on the Georges Bank off Cape Cod, Massachusetts, an active area for New England fishermen. A US Fish and Wildlife Service research vessel made 382 30-minute passes or tows in the vicinity of Georges Bank, sampling in three depth zones (0–60, 60–120 and >120 m) in a stratified random selection of six stations. All species caught were recorded and representative samples taken for morphometric analysis

CC Taylor fit the Poisson and negative binomial distributions to his abundance data of whiting (*Merluccias bilinearis*) and haddock (*Melanogrammus aeglefinus*) and considered several other distributions, concluding that the best overall fit was the lognormal distribution. He even plotted variance against mean for both species, but not on log scales. The means and variances at the 6 stations and 3 depths for these species are given in his Tables 10 and 11. Data of haddock were used in LRT61. Fig. 10.1 (Appendix 10.A) shows the two TPL plots with the haddock plot on the left and the whiting plot displaced to the right by a factor of 10 for clarity.

Haddock were most abundant at the middle depth and least at the lowest depth, but all the means and variances of all three depths conform well to a single line. Like haddock, whiting's aggregation was generally similar across all three depths, except that whiting has a lower correlation coefficient with variances at high density at the lower depths slightly higher than expected from the fitted line. Whiting had the widest range of densities at the lowest depth.

The number of trawls at the six stations and three depths varied from $NQ = 4$–45. TPL analysis of these data is temporal/ensemble as each point averages time, but the $NB = 18$ sample points constitute an ensemble as the points are of structurally different samples. Despite the differences in abundance at different depths and times, the conformity to TPL is remarkably consistent for both species.

HERRING AND MACKEREL IN THE NORWEGIAN SEA

Herring (*Clupea harengus*) and mackerel (*Scomber scombrus*) are pelagic planktivorous fish both preying mostly on arthropods: crustaceans, euphausiids, amphipods, and copepods. Consequently, they might be expected to be in direct competition. A survey by Langøy et al. (2012) to investigate spatial distribution, potential diet overlap, and environmental preferences suggested that these schooling predators avoid competition by preferring waters of different temperatures, mackerel preferring warmer water than herring. Even where the populations overlapped in the Norwegian Sea, the food preferences were slightly different, again limiting niche overlap. Fish were sampled by trawl at 44 stations

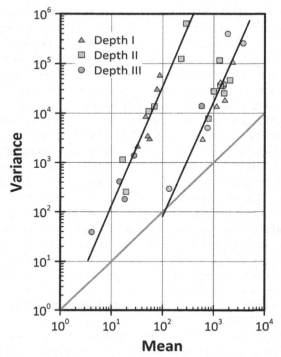

FIG. 10.1 Hybrid spatial-ensemble TPLs ($NQ = 4$–45, $NB = 18$) of haddock *(left)* and whiting *(right)* taken in samples from 1948 to 1950 at 3 depths at 6 stations on Georges Bank off Cape Cod, Massachusetts. Whiting is shifted right by one cycle for clarity. *(Data from Tables 10 and 11 in CC Taylor (1953).)*

in July 2004 and 68 stations in July and August 2006. The stations were arranged on 6 and 8 roughly parallel east-west transects between Norway and Iceland. At each station, 100 of each species were selected at random and body length, wet weight, sex, age, and stomach contents were recorded

The abundance data of herring and mackerel are not in a form suitable for analysis; however, the data for the mass of stomach contents for 5 stations in each year are given in Langøy et al.'s Table 5 (mg ± SD). The gut contents show a strong dependence of variance on mean (Appendix 10.B). These data are noisier than abundance data generally due largely to second-order effects. Gut content estimates depend on availability of food as well as the size of the predators in the sample. Thus, there are two sources of variation that may not be entirely independent, leading to increased variability and lower TPL correlation coefficients. A single outlier in both species influences both slope and intercept, which when removed the lines to be almost identical. This example reveals the utility of TPL for nonspatial data as it is frequently used to select an appropriate transformation for statistical analysis of experimental and observational data (Chapter 14). Thus, a single transformation based on $b = 1.56$ would be appropriate for statistical comparison of these data.

SALMON IN THE NORTHEAST PACIFIC OCEAN

Variations in salmon returns to the Columbia River in the US Pacific Northwest are thought to be driven by oceanographic variability and availability of zooplankton prey for juvenile salmon. Scyphozoan jellyfish also prey on zooplankton and in years of high jellyfish abundance, may be in competition for food with juvenile salmon. To test this, Ruzicka et al. (2016) compared the abundance of the sea nettle *Chrysaora fuscescens*, the dominant jellyfish off the Oregon and Washington coast, with juvenile coho (*Oncorhynchus kisutch*) and Chinook (*O. tshawytscha*) salmon abundance between 1999 and 2013. They found abundance of sea nettle biomass to be negatively correlated with adult coho and Chinook salmon returning to the Columbia River. Furthermore, salmon stomachs were less full at locations with high sea nettle biomass. Ruzicka et al. concluded that sea nettles and salmon may be indirect competitors for plankton, which can suppress salmon production in years of high jellyfish abundance.

Ruzicka et al.'s Fig. 4 shows the biomass of sea nettles $(T/km^2 \pm SD)$ and juvenile salmon abundance $(\#/km^2 \pm SD)$ obtained from 40 trawl samples taken in the upper 20 m of the water column in June and September from 1999 to 2013. The sea nettle data are analyzed in Chapter 9 and the salmon TPL slopes are steep, averaging $b = 2.01 \pm 0.089$, with June slopes steeper than September (Appendix 10.C). Although not significantly different $(P > 0.35)$, the slight decrease in slope is consistent in both Chinook age classes and the coho data. Recruitment of subyearling Chinook from the natal rivers (principally the Columbia River) occurs during the summer adding to the density and reducing aggregation slightly (Fig. 10.2C and D). The density and spatial variance of both yearling Coho and Chinook also declined over the summer (Fig. 10.2A and B, E and F, respectively). By June of the following year, the slope of Chinook yearling age class had increased to its prior level only to decrease again over the summer. Summer mortality and/or emigration reduced the average densities of yearlings with a slight decrease in aggregation.

The reader is reminded that lack of statistical significance does not necessarily denote lack of biological significance. These small nonsignificant differences in TPL may be random fluctuations, but as the same pattern appears in both Chinook age classes and in coho suggests a common biologically-based explanation. This reduction in aggregation during the summer may be purely numerical as the immigrants leaving their natal rivers "dilute" the resident populations in the ocean.

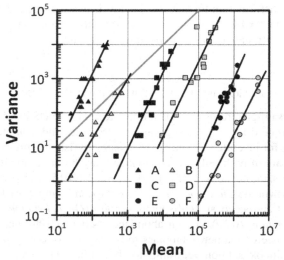

FIG. 10.2   Spatial TPLs ($NQ = 40$, $NB = 13$) of juvenile coho and Chinook salmon off the coast of Oregon and Washington. A, Yearling coho in June; B, yearling coho in September; C, sub-yearling Chinook in June; D, subyearling Chinook in September; E, yearling Chinook in June; F, yearling Chinook in September. Plots offset to the right of previous by one cycle for clarity. *(Data from Fig. 4 in Ruzicka et al. (2016).)*

DEMERSAL FISH IN A TROPICAL BAY IN BRAZIL

In a study designed to precisely partition sources of variation in fish communities between spatial, temporal, and environmental components, Costa de Azevedo et al. (2007) trawled three zones in Sepetiba Bay in southeastern Brazil's Rio de Janeiro State. Sepetiba is a large bay, ~450 km² in extent with a 40-km long sand bar and some small islands almost enclosing it. Costa de Azevedo et al defined three bay zones according to depth and salinity: inner, middle, and outer bay. The inner bay is shallow and muddy with lower salinity resulting from river discharge, while the outer bay, adjacent to the Atlantic, is sandy and cooler with a slightly higher salinity

Three replicate 1.5-km trawls were made monthly in each zone from October 1998 to September 1999. 20,483 specimens in 93 species, 73 genera and 37 families were caught in 108 trawls. Species composition varied between zones and the most abundant species differed between zones. Greater variation in abundance and composition was attributed to zone than to season. Costa de Azevedo et al.'s Table 4 lists the mean and SE of abundance of the 30 most abundant species found by zone. All but two of these species are common to all zones. Mixed-species analyses of their data show numerical variability decreasing with proximity to the opening to the ocean (Fig. 10.3; Appendix 10.D). Aggregation increased slightly from outer to inner bay as the average depth decreases from 30 m to about 8.6 m. Although the estimates are slightly different, neither intercept nor slope differ significantly (intercept, $P > 0.87$; slope, $P > 0.86$). In short, the community-level aggregation is the same in all three zones, except for three outliers (circled).

One outlier, the Atlantic moonfish *Selene setapinnis*, is a schooling species usually found near the bottom. Its schooling habit may account for the higher-than-expected variance in the plot of the outer zone. Two species are outliers in the inshore zone having lower-than-expected variance suggestive of a more regular distribution. Those two species are puffer fish *Sphoeroides tyleri* and *S. testudineus*, both bottom dwellers. They can be locally very abundant but aggressively defend nesting territories, a habit tending to reduce spatial variance as territories become compressed. Their variance/mean ratios conform closely to the fitted TPL line in the middle and outer zones where abundance was lower. These three outliers suggest different spatial behavior relative to the rest of the demersal community in different zones. All three species are present in all three zones, but note that the schooling species has higher-than-

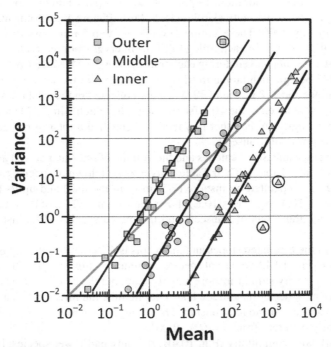

FIG. 10.3 Mixed-species TPLs ($NQ = 36$, $NB = 27$–30) of demersal fish caught in trawls in three zones of a tropical bay in Brazil. Each point represents abundance of one species (all but 2 species were caught in all three zones). Three outliers are circled: two species of territorial puffer fish and schooling Atlantic moonfish. *(Data from Table 4 in Costa de Azevedo et al. (2007).)*

expected variance where it is most common near the ocean while the territorial species have lower-than-expected variance in their preferred nesting habitat closer inshore.

PELAGIC FISH LARVAE IN PORTUGAL

Wildlife conservation is a significant role of National Parks. The Arrábida Marine Park west of Lisboa, Portugal, is a rocky shallow subtidal habitat extending ~ 50 m offshore and dominated by large calcareous boulders. It has a very high fish diversity

Borges et al. (2007) surveyed the composition and distribution patterns of larval fish assemblages and their temporal dynamics at Arrábida between May and October 2000. Their purpose was to obtain a good understanding of the spawning patterns, larval distribution and abundance as well as larval dispersal, which are all important for effective management of fish stocks. Monthly replicate (>11) 5-min samples were taken in a pair of transects parallel to the rocky shore with a tow net at about 1 m below the surface; one transect adjacent to a rocky reef <50 m off shore and the other ~ 3.70 km offshore (here called near-shore). In addition, 17 parallel transects perpendicular to the shoreline from the reef to 10 nautical miles (18.52 km) offshore and 0.6 km apart were sampled in July 2000. These samples comprised two 5-min tows (toward and away from the shore) in each mile (1.85 km) interval from the reef. The mean volume of sea water sieved during each transect was 29.44 ± 2.26 (m³ $\pm$ SD). A total of 843 pelagic fish larvae in 61 taxa were taken in the in-shore and near-shore samples and 178 larvae in 29 taxa were collected in the offshore samples. 229 larvae were taken in the 10 offshore transects, which with 397 larvae taken in July and August in the other two reef transects had a total of 626 larvae in 52 taxa caught in all transects.

The average and SD of catches and the volume of sea water sieved are given in Borges et al.'s Table 1. Their Table 2 gives a breakdown of species caught by mean and SD, separated by inshore (<50 m from the reef) and up to 18.5 km offshore. Fig. 10.4 (Appendix 10.E) shows the ensemble TPL of the total larvae caught in the three sets of transects and the mixed-species TPLs inshore and offshore.

None of the estimated slopes and intercepts is different whether considering total larvae (Fig. 10.4A) or larval species community (Fig. 10.4B). The range of abundances is very similar inshore and offshore, but the near-shore peak abundances are an order of magnitude less. Despite this, the three sets clearly have the same variance–mean relationship. The average value of $b = 1.323 \pm 0.083$ is significantly different from 1.0 ($P < 0.001$).

The offshore community (Fig. 10.4B) not only had fewer species, but their abundance was somewhat lower than inshore possibly due to dispersal by sea currents. Larvae off Arrábida are subject to strong tidal flows (tidal range is ~ 3 m) and other currents, including outflow from the Sado River at the eastern

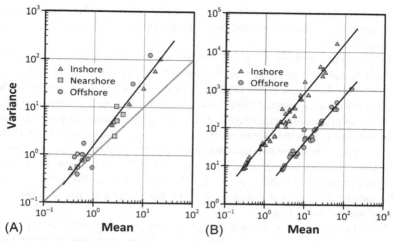

FIG. 10.4   (A) Ensemble TPL ($NQ = 11$–30, $NB = 22$) of total number of pelagic fish larvae caught in transects off Arrábida Marine Park, Portugal, between May and October 2000. Points are number of larvae per transect per month. (B) Mixed-species TPL ($NQ = 11$–30) of larvae caught in-shore and off-shore (includes the near-shore samples south of the reef). Each point is a single taxon (most to species, others to genus). The offshore plot is shifted one cycle to the right for clarity. *(Data from Tables 1 and 2 in Borges et al. (2007).)*

end of the Park. With a limited ability to control their movement, we might expect sea currents to increase mixing and reduce aggregation. These pelagic fish larvae out to 18.5 km offshore maintained nonrandom spatial distributions despite potent mixing forces. Furthermore, the larval totals and community displayed the same level of aggregation, suggesting that whatever forces regulate spatial distribution, they are approximately the same for all members of the community regardless of distance from shore. There are no obvious outliers in the larval community plot as in the previous example, suggesting that all members of this pelagic larval community are subject to the same forces influencing their distribution.

PELAGIC FISH LARVAE IN NEW JERSEY

Comparing the composition and abundance of larval fish to recently settled juvenile fish on the continental shelf and in an adjacent estuary in southern New Jersey, Able et al. (2006) sought to evaluate the relative contribution of the ocean and the estuary as settlement areas for benthic species. They conducted surveys from May to November 1992 in the Great Bay Little Egg Harbor estuary and on and near the adjacent Beach Haven Ridge on the inner continental shelf. Larvae and recently settled juveniles were sampled every 2 to 4 weeks at three fixed locations relative to the offshore ridge by plankton nets and beam trawls, respectively. Sixty benthic trawl samples were taken at three locations, landward, over, and seaward of the ridge at depths from 8 to 15 m. Ninety plankton samples were taken at two locations, landward and seaward of the ridge, at 8–15 m depth

In the estuary, 288 benthic trawl samples were taken at depths 1–8 m and 55 plankton samples were taken weekly from a bridge spanning a tidal creek for three separate half-hour hauls during night flood tides. Larvae and juveniles were identified and counted. Twenty-two dominant species are listed, and there was some overlap in species caught at sea and in the estuary. In general, more larvae were caught in the estuary than at sea while the total number of juveniles caught was similar in the two habitats, although they peaked earlier in the estuary perhaps because of higher water temperature. They found that the species sampled comprise two habits: species that develop and settle only in the estuary, and those that settle in both estuary and at sea. No larvae were found in the estuary that settled only at sea.

Able et al.'s Fig. 5 gives means and SEs of abundance of juveniles expressed as number per $1000\,m^2$ of sea floor trawled and larvae expressed as number per $1000\,m^3$ of water sieved by plankton net.

Unlike the previous example of pelagic larvae in the top 1 m of sea, TPL analyses of Able et al.'s data show these populations are highly aggregated (Appendix 10.F). The TPL of juveniles trawled in the estuary has the steepest slope, but as it has only 6 points, it is perhaps anomalous. The other examples, larvae in the estuary and larvae and juveniles at sea with $NB = 19$, 14 and 10, respectively, have slopes not different from 2.0. The principal difference between this and the previous example is the depth of sampling. Borges et al.'s (2007) sea samples were taken in the top 1 m, whereas Able et al.'s trawl samples were taken deeper, including from the bottom. Thus, the populations of larvae sampled by these two groups may be different. In addition, mixing by wave action is stronger in the surface waters, which likely reduces aggregation of plankton there.

FISH LARVAE ENTERING PAMLICO AND ALBEMARLE SOUNDS, NORTH CAROLINA

In a study to determine strategies used by winter-spawning fish to enter the protected waters of North Carolina's Pamlico-Albemarle Sound lagoon system, Joyeux (1998) sampled larvae entering from the Atlantic at four major inlets: Beaufort, Hatteras, Ocracoke, and Oregon. Sampling was conducted at intervals from January to April in 1992 and 1993 by netting from a research vessel moored or motoring in the inlets. Samples were taken at different depths, times of day, tides, and speeds relative to the tide. Three species were abundant enough for analysis to determine their mode of entry to the Sounds: Atlantic menhaden (*Brevoortia tyrannus*), Atlantic croaker (*Micropogonias undulatus*), and spot (*Leiostomus xanthurus*). The magnitude of Joyeux's catches at the tidal inlets depended on several factors: month, inlet, luminosity (day, dusk or night), tidal flow (speed and direction), and depth. The strategies for entering the lagoon system differed between species. Spot and croaker adjusted their depth according to the direction of tidal flow, while menhaden immigrated on flood tides at dusk or on weather-related flows into the lagoon.

Joyeux's Table 2 gives the means and SEs of transformed data aggregated according to classifications of interest: inlet, month, flow, etc. Thus, each category's statistics were computed from the same raw abundance data. The estimates in Table 2 have been back transformed to mean and variance in the original scales using the delta technique (Appendix 14). Joyeux's classifications include from 2 to 6 sets equivalent to NB. I have selected the two classifications with $NB = 6$ (flow and luminosity) and $NB = 4$ (inlet and month). Despite the comparatively low NB, the regressions (Appendix 10.G) are all significant with correlation coefficients $r > 0.930$. The range of slopes both within and between species is large: the steepest slope is menhaden and inlet ($b = 2.16 \pm 0.30$): the shallowest is menhaden and flow ($b = 1.26 \pm 0.25$). The ranges for croaker and spot are less, $1.49 \leq b \leq 1.94$ and $1.56 \leq b \leq 1.90$, respectively.

Joyeux's results showed major differences in spatial behavior by these species. The differences may be familial: menhaden belong to the family Clupeidae, to which herrings also belong, and both croaker and spot are members of the family Sciaenidae otherwise known as drums. The Sciaenidae are small-to-medium-sized inshore and estuarine fish that feed primarily on arthropods and smaller fish, whereas the Clupeidae are predominately oceanic fish. Menhaden's habit of migrating into inland seas is not characteristic of the family. Its habit of migrating into the Sound on flood tides at dusk would lead to aggregation in time, and therefore steeper TPL curves. The drums by contrast had more opportunities for immigration into the Sound by adjusting their depth to take advantage of the tidal flow, a habit that would lead to lower aggregation in time.

One of the most important features of this study is the fact that the four examples per species are all derived from the same data aggregated differently according to the factor of interest. Thus, the heterogeneity in b results from the particular statistical (as opposed to biological) population being analyzed. Statistically each of the four ensembles, flow, light, inlet, and month, is a different population.

SEA TROUT FRY IN ENGLAND'S LAKE DISTRICT

Dr. Malcolm Elliott of the Britain's Freshwater Biological Association studied Britain's freshwater fauna, primarily in the north of England. He monitored migratory brown trout, also called sea trout, *Salmo trutta* in Black Brows Beck in the Lake District from 1967 to 2000 and the earlier data are given in three papers (Elliott, 1984, 1986, 1989). Sea trout spend their adult lives at sea but spawn in upland streams. Adults migrate into freshwater and spawn in headwaters in late autumn, laying their eggs in gravel nests. The eggs hatch by February and the neonates, called alevins, remain in the nest, their yolk sacs providing food for about three months. By the end of April, the young fry emerge from the nest and start to forage, feeding mainly on aquatic arthropods. The fry, now called parr, establish territories and remain in the stream for two years and migrate down stream as smolts. Elliott established a 60-m^2 area of Black Brows Beck, divided into six 10-m^2 sections. Each April, he took 5 random samples of alevins in their nests in each stream section using 0.09-m^2 cylinders. In late May/early June and again in late August/early September, parr were surveyed in each of the 10-m^2 sections by electrofishing. Parr were classified as 1, 2, and 3 year-olds. Third-year parr were only taken in the May/June samples as they had migrated downstream by August

Elliott (1986) analyzed $NB = 15$ to 18 years' data of alevins ($NQ = 30$) and parr ($NQ = 6$) by TPL. Fig. 10.5A (Appendix 10.H) shows spatial TPLs of the four life stages rescaled for clarity. Alevin abundance is expressed as number per 0.1 m^2, 1st and 2nd year parr as number per 10 m^2, and 3rd year

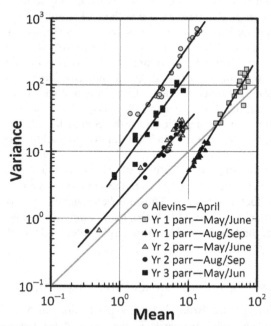

FIG. 10.5 Spatial TPLs ($NB = 15$–18) of juvenile sea trout in Black Brows Beck in England's Lake District. Alevins ($NQ = 30$) number/0.1 m^2; 1st an 2nd year parr ($NQ = 6$) number/10 m^2; 3rd year parr ($NQ = 6$) number/25 m^2. Each point is one year observation. *(Adapted from Fig. 2 in Elliott (1986).)*

parr as number per $25\,m^2$. As one would expect, each age class is smaller than its predecessor due to mortality. This is also evident in the regressions of 1st and 2nd year parr identified as May/June and August/September. Despite the large decline in number between June and August, intercept and slope of 1st year parr did not differ between the two time periods (intercept, $P >$ 0.93; slope, $P > 0.97$). Similarly, 2nd year parr did not differ between the two time periods (intercept, $P > 0.57$; slope, $P > 0.37$). Taken collectively, the four age classes differ in both intercept and slope (intercept, $P \ll 0.001$; slope, $P < 0.001$), indicating dramatic changes in spatial organization of different age groups. This heterogeneity is due to a small increase in aggregation from alevin to 1st year parr and a large decrease in b in year 2, which recovered in year 3. The stage with the steepest slope is 1st year parr, which is the stage that experiences the greatest mortality. After leaving the nest in May/June, parr are very vulnerable until they have established territories, as Elliott reported later.

Elliott (1989) reported on 8 years in which Black Brows Beck was sampled five times following fry emergence from the nest for up to 3 months. In this study, he identified a critical survival period. In the nest alevin mortality is very low, but after leaving the nest, 1st year parr suffered a decline in survival, leveling off 33–70 days after leaving the nest. During this critical survival time (CT), as many as 90% of the cohort was lost. Prior to CT, 1st year parr were aggregated, but afterwards the variances were closer to the means. But despite this quite dramatic change in density and apparent change in frequency distribution of abundance, the spatial distribution as registered by TPL remained the same. It was mostly parr that failed to establish a territory that perished during their first spring and early summer. Elliott concluded that the duration of the CT was an important density-dependent regulator of population. After CT, mortality was essentially density independent as parr migrated from the fast-moving shallow stream in summer and autumn to deeper slower-moving waters in their first winter.

As populations declined, TPL registered a small difference in A before and after CT ($P < 0.06$) no doubt due to the high mortality, but no change in slope ($P > 0.33$). Thus, this age class remained strongly aggregated even as it suffered approximately 90% mortality.

Elliott's investigations have shown that the spatial distribution in one trout population is essentially density dependent, even when the mortality is also density dependent. As is evident in Fig. 10.5, he also showed that TPL can vary between different life stages within the same population. Furthermore, he concluded that "changes in the parameters of the Power Law can be used to detect major spatial changes which can be related to the behavioral movements of the trout" (Elliott, 1986). These results were used to develop a new model of trout population dynamics based on Ricker curves with time-variable parameters.

ADULT SEA TROUT CATCHES IN ENGLAND AND WALES

In a large-scale study of adult sea trout, Elliott (1992) analyzed annual rod and commercial catches for 14 to 37 years in 67 rivers in 5 regions of England and Wales. In this study, he examined both spatial and temporal variation in catches for each river. Elliot used records of annual catches from the annual reports of the Regional Water Authorities for all rivers supporting large populations of sea trout. Although many factors affect catch size, Elliott argued that population density is likely the dominant one and used the annual catch records to estimate the dependence of population variation on mean density (Fig. 10.6A). Annual catches were used to estimate either variation in catches between rivers for each year (spatial variance) or variation in catches between years for each river (temporal variance). Data for temporal and spatial TPL analysis of rod catches were available for more rivers than commercial catches, but the commercial data were sufficient for analyses of several regions. One purpose of the study was to characterize the population variability of sea trout populations on a regional basis with a view to assisting sea trout population conservation and management. For this, he used the coefficient of variation (CV), which he expressed as a function of TPL:

$$\log(CV) = \{A + (b - 2) \cdot \log(M)\}/2 \qquad (10.1)$$

Examining Eq. (10.1), we see that when $b = 2$, CV is constant with respect to mean and when $b > 2$, CV increases with mean and conversely decreases when $b < 2$. Elliott reported both ODR and GMR estimates for spatial and temporal variation in Tables 2 and 3, respectively. Appendix 10.I reports the GMR results and as LRT and Woiwod (1982) found, the spatial and temporal TPL estimates differed. Elliott used the CV to classify the catch method and regions as increasing, decreasing, or flat. Commercial catch variability as measured by CV was negative temporally but either flat or positive spatially. Rod catches were mixed: spatially, CVs were either flat or positive while temporally they were either flat or negative: the number of data points for the negative regions was very low drawing into question their utility. Elliott included them in his analysis at regional scale, which resulted in a negative CV, in line with the commercial temporal analysis. He also combined rod and commercial catches in a single temporal analysis resulting in a negative CV also (Fig. 10.6B).

For those regions and catch method with b not significantly different from 2.0 and relative spatial variability remaining fairly constant log transformation of catches effectively removes the relationship between variance and mean catch, simplifying subsequent statistical analyses despite large fluctuations in mean catch between years and rivers. This is especially important for temporal analyses as these data are used for life table analysis important for conservation and management.

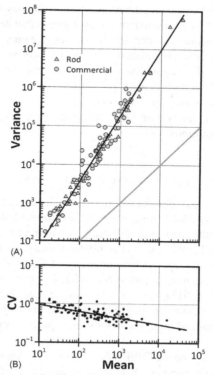

FIG. 10.6   (A) Temporal TPL ($NQ = 12$–39) of rod ($NB = 36$) and commercial ($NB = 67$) catches of sea trout in 67 rivers in England and Wales coincide exactly with a slope $b = 1.73 \pm 0.072$ leading to declining CV with density (B). *(Adapted from Fig. 4 in Elliott (1992).)*

The significantly higher spatial slopes obtained for Wessex ($b = 2.23 \pm 0.053$) and NE England ($b = 2.82 \pm 0.286$) resulting in CVs increasing with mean annual catch in these rivers are particularly interesting. In these two regions, differences between rivers were greater in high-catch years and relatively small in low-catch years indicating that the highest-catch rivers perform either very well or very badly. Elliott offered no explanation for why fluctuations in these regions were so extreme.

Temporal CV appears to be a useful means for classifying the major sea trout rivers. With temporal CV declining with mean catch, rivers with a low temporal CV generally have a high annual catch and catch sizes varying little between years. This implies a stable sea trout population regulated predominantly by density-dependent factors. In the rivers with a high CV and low annual catches, the more variable population is likely more susceptible to density-independent factors like fluctuations in climate. Despite the differences in catching mechanism by rod and commercial netting, the consistency of temporal results from different regions for both catch methods suggests that catch data are a good estimate of relative population density. It also suggests a common underlying dynamic process regulating populations. In his previous work on juvenile trout (above), Elliott found that regulation of low-density populations is chiefly density independent, while high-density population regulation is density dependent. These conclusions are repeated in this study with important implications for the conservation and management of sea trout and possibly other game fish populations.

CALIFORNIAN COMMERCIAL FISHERIES

Reducing the population density of fish by mechanized fishing might be expected to reduce spatial heterogeneity of exploited fish, just as pesticide application reduces the measured TPL slope of agricultural pest populations (Trumble, 1985; RAJT, 1987). Kuo et al. (2016) tested the proposition that commercial fishing alters the spatial heterogeneity of fish and that this would be reflected in TPL slope. In addition, if b is related to life history traits, as suggested by Park et al. (2013), species with smaller size, higher fecundity, and shorter generation time (high r) might be expected to exhibit larger b, and this effect might be enhanced in comparisons of exploited and unexploited fish species.

Using sample data from the California Cooperative Oceanic Fisheries Investigations long-term fish surveys,[1] Kuo et al. examined the effects of fishing and life-history traits on TPL slope by comparing spatial distributions of exploited and unexploited fish living in the same environment. They found that in general exploited species exhibited a slightly positive but not significant relationship between b and life history traits (Fig. 10.7A), although the significance of the correlation between b and log (length at maturity) of demersal fish is marginal at $r = 0.74$, $P < 0.06$. By contrast b, for unexploited species (Fig. 10.7B) exhibited significant negative relationships with the logarithms of length at maturity ($r = -0.69$, $P < 0.01$), maximum length ($r = -0.70$, $P < 0.01$), and marginally with trophic level ($r = -0.51, P < 0.08$). As size often increases with increasing carrying capacity, K, b is inversely proportional to K for unexploited fish but may increase slightly with K in exploited demersal fish. Kuo et al. suggested the weak relationship between b and life-history traits may be the result of fishing increasing spatial aggregation by destabilizing the population's size/age structure.

---

1. CalCOFI is a Long Term Ecological Research site. Its core area from San Diego to San Francisco consists of 11 transects perpendicular to the California coast comprising 66 sample sites to 400–800 km offshore. Survey cruises have been made most years since 1949 and are typically conducted in the spring. Other transects up to the Canadian border are surveyed periodically (http://www.calcofi.org/).

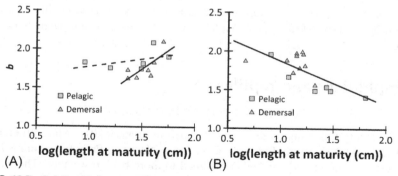

FIG. 10.7 Relationship between TPL b and size of (A), exploited and (B), unexploited fish off the California coast. As expected, b declined with the life-history traits (size and trophic level) in unexploited fish, but was either flat or increased with size in exploited fish. *(Data from Supplementary Table S1 in Kuo et al. (2016).)*

A more detailed analysis of 4 exploited species (Pacific hake (*Merluccius productus*), chub mackerel (*Scomber japonicas*), South American pilchard (*Sardinops sagax*) and bocaccio (*Sebastes paucispinis*)) computed cross-correlations between b and two indices of age structure: average age weighted by biomass and the evenness of age classes calculated using Simpson's index. Correlations between b and the age structure indices were computed for 5-year moving windows and in three species the moving window cross-correlation analysis supported the conclusion that b is negatively correlated with mean age and age evenness. The exception was bocaccio, a demersal species, whereas the others are pelagic and migratory. Chub mackerel and South American pilchard are schooling species feeding primarily on copepods. Pacific hake undergoes a diurnal migration from the seabed to the surface to feed at night, making it partially demersal. That the specific results and the mixed-species analysis seem to contradict each other suggests that at the community level, exploitation increases spatial variability, blurring any connection between life-history trait and spatial distribution as measured by TPL.

The shift in relationship between b and life-history trait seen in the comparison of exploited and unexploited fish, Kuo et al. suggested may be due to fishing-induced age/size-truncation effects, which can increase spatial aggregation without apparent reduction in population mean abundance (Hsieh et al., 2006). Age truncation may also lead to reduced larval survival and dispersal ability, and increased fluctuation in abundance and intraspecific competition (Hsieh et al., 2006). Thus, fishing may make otherwise K-selected species behave more like r-selective species. If so, exploited species may be more vulnerable to environmental changes rendering them more at risk of extinction. While density-dependent regulation in the most abundant species maintains populations, low abundance species although more variable, may replace core species should environmental conditions change (Henderson and Magurran, 2014). Kuo et al.

suggest that "sound fishery management strategy will be concerned with changes to the age and spatial structure of exploited fishes" as well as their purely numerical abundance.

Amphibians and reptiles

REPTILES IN THE FLORIDA EVERGLADES

Immediately following an extreme cold spell in January 2010, Mazzotti et al. (2016) surveyed South Florida's Everglades National Park to examine its impact on nine reptile species accustomed to more benign weather. The data all come from areas within the park that occupies 6110km^2 (about 20%) of the large saltwater and freshwater wetland of the Florida Everglades. South Florida is a semitropical ecotone with both temperate and tropical wildlife. This includes the tropical American crocodile (*Crocodylus acutus*) and the warm temperate American alligator (*Alligator mississippiensis*), both of which are found throughout the Everglades. Also in the Everglades is the invasive Burmese python (*Python bivittatus*) and a number of native snakes, including five featured in this study: the venomous cottonmouth (*Agkistrodon piscivorus*) and nonpoisonous southern water snake (*Nerodia fasciata*), green water snake (*N. floridana*), eastern ribbon snake (*Thamnophis sauritus*), and common garter snake (*T. sirtalis*).

Mazzotti et al. surveyed coastal and inland waterways from 20 January to 6 March 2010 to document crocodile and python mortality. Combining their data with reports from individuals and agencies working in the Everglades, Mazzotti et al. compared the population trends of the tropical crocodile with the warm-temperate alligator, and the introduced python with the native snakes. They estimated encounter rates of these eight species from the number of live sightings and road kills per kilometer of highway or river examined in the Park over a period of 6–13 years. Although a clear decline in python abundance was evident in the data, the cold snap apparently had little impact. Similarly, American alligators have also been in decline in this area of Florida, but the cold snap did not apparently contribute to its decline. The cold weather did appear to reduce the abundance of the five native snakes, which recovered the following year, except for the common garter snake, which had not recovered by 2013. The other species to respond negatively to the cold was the semitropical American crocodile, which showed a marked decline in abundance from 2009 to 2010, followed by a recovery in 2011 and 2012. In contrast, the alligator showed no decline in 2010, but an increase in 2011 after which the population continued its long-term decline. Mazotti et al. published four figures giving abundance data expressed as mean and SD of number per km surveyed for the eight reptiles.

Results of TPL analyses of data rescaled to number per 100 km are in Appendix 10.J. Six of the eight species' analyses resulted in TPL plots with correlation coefficients $0.77 \leq r \leq 0.99$. The correlation coefficients for two species, the American alligator and the common garter snake, were $r < 0.67$. Two regressions, alligator and ribbon snake, have TPL slopes less than but not significantly different from 1.0, indicative of a random distribution of encounters. The water snake by contrast has a slope greater than, but not significantly different from, 2.0. The crocodile, with a slope of 3.64, is significantly >2.0 ($P<0.01$). There is no distinction between live and dead observations, which are likely to have different sampling efficiencies. The observations taken along roads and waterways through the Everglades Park are similar to sampling by distance. The unusually steep crocodile plot may be the consequence of a very short range of means and highly biased sampling. The very poor TPL alligator plot is partly due to the high variance in 2011, the year following the freeze; another year, 2015, also has a high variance and 2006 an unusually low variance. Removing these outliers (circled) increased the correlation, but the slope was unchanged.

The possibility arises that the higher-than-normal mortality disrupted the spatial organization of the garter snake; eliminating 2010–1012 raises the correlation from $r=.61$ to 0.91. There are insufficient data following the freeze to know if the spatial distribution recovered. Although the number of data points is low, the cold spell clearly dramatically reduced the variance as the abundance declined, a phenomenon observed in some pest species following pesticide application. In contrast, despite increased mortality of crocodiles in 2010, the spatial distribution as measured by TPL was unaffected.

The American crocodile and American alligator are similar animals, with overlapping life-history strategies. The difference in TPL plots begs the question: why did the crocodile produced a very nice TPL when the alligator produced a poor one? Susceptibility to freezing likely differs, but it was the tropical crocodile that retained its spatial pattern through adversity when the subtropical alligator, which is near the center of its range did not. Probably 2011, with the highest variance, was influenced by the freeze, although we would expect the mean to decline as it did for the garter snake. The unexpectedly low and high variances for 2006 and 2015, respectively, are likely due to differences in sampling efficiency, although the data collection methods for the alligator were no different from those employed for the other reptile species in this study. Alternatively, undocumented events in those years may have affected American alligator's population distribution and abundance. Outliers like these can be useful indicators of ecologically important events.

HERPTILES IN ARIZONA'S RINCÓN MOUNTAINS

In 2001 and 2002, Flesch et al. (2010) surveyed the herpetofauna across an elevation range of 900m in the Rincón Mountains in southeastern Arizona's Saguaro National Park. They estimated the species richness, distribution, and relative abundance of 4171 individuals in 46 species. Using their observations, expert opinion, and a probabilistic model, they concluded there were likely 57 species present in the Rincón Mountains. They documented a decline in species richness with elevation, and both increases and decreases in relative abundance of some species with elevation in two different surveys. In 2001, Flesch et al. conducted intensive visual-encounter surveys for 1 h in 1-ha plots along 17 line transects at 3 elevation ranges (936–1218 m, 1219–1829 m, and >1829 m) in spring and again in summer. They also conducted 85 extensive visual-encounter line-transect surveys ($n = 85$) unconstrained by time or area in the same three elevation ranges in spring and summer 2001 and 2002.

Flesch et al.'s Tables 1 and 2 give abundance data with SEs for the herptiles in the intensive and extensive surveys: 23 species are given as number/ha/h in the intensive survey and the extensive surveys observed 34 species, which are presented as number/10 h. Two species, *Sceloporus clarkii* and *Urosaurus ornatus*, were recorded at all times and elevations in both surveys, providing 8 means and variances for analysis. They were also recorded in all three elevation zones in the intensive survey. Variances were calculated from the standard errors in Tables 1 and 2 and TPL analysis for the community of species at the 3 elevation ranges and total for all elevations are given in Appendix 10.K. All species combined and *S. clarkii* and *U. ornatus* in both surveys were also analyzed. All regressions have very high correlations ($r > 0.92$).

Fig. 10.8 shows the mixed-species TPL segregated by elevation. The wide ranges of means are due to differences in abundance in spring and summer in

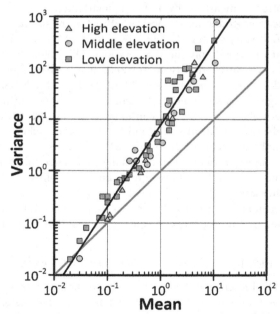

FIG. 10.8   Mixed-species TPL ($NQ = 12$–50, $NB = 67$) of number of individuals of 40 herptile species seen per 10 h observation during extensive visual-encounter surveys in the Rincón Mountains, Arizona, in 2001 and 2002. *(Data from Table 2 in Flesch et al. (2010).)*

addition to the elevation differences. Taken individually, the mixed-species TPLs at three elevations are not significantly different from each other (intercept, $P > 0.19$; slope, $P > 0.98$). Combining the data prior to analysis results in an insignificant change in b ($P > 0.67$), but a significant increase in A ($P \ll 0.001$), reflecting an increase in variance with little change in mean abundance as the numbers of each species was combined over the three elevation zones.

The moderate aggregation of $S.$ *clarkii* ($b = 1.60$) is in contrast to the high degree of aggregation of $U.$ *ornatus* ($b = 2.55$), which is more characteristic of all herptiles combined ($b = 2.51$). Both $S.$ *clarkii* and $U.$ *ornatus* are tree lizards; the latter is noted for the highly variable color of the male's throat while the former is substantially larger perhaps making it more visible and increasing its sampling efficiency relative to $U.$ *ornatus*.

FROGS IN AN ALPINE HABITAT IN CALIFORNIA

Pope and Matthews (2001) studied the movement ecology of mountain yellow-legged frogs (*Rana muscosa*) by tagging 500 frogs and monitoring their movements between ponds during the summers of 1997 and 1998. Their field study was conducted in one large (5.3 ha) lake and 10 smaller ($<3000\,m^2$) ponds in Kings Canyon National Park, California. The ponds at an elevation of 3470 m are frozen from October to July. The frogs overwinter in the lake and deeper ponds with well-oxygenated water or in crevices with air bubbles. Of the 500 tagged frogs, 97% were recorded at their overwintering lake in October. Although most did not stray far from their overwintering lake, movement up to 1 km was detected with 17% recorded $>66\,m$ from their overwintering lake. In common with other species and populations, the number distance distribution apparently declined sharply with a long tail (RAJT, 1978, 1980).

To estimate overall abundance, Pope and Matthews surveyed weekly the lake, the ponds, and the streams linking them. One person walked the perimeter of each pond and stream counting all adult and juvenile yellow-legged frogs. These data are presented in their Fig. 2 as mean and SEs of number observed at each pond or stream by date. Spatial TPLs of the time series of lake and stream counts are moderately aggregated ($b = 1.51 \pm 0.096$). The pond and stream points are not significantly different, but the stream data have a narrower range of means (Appendix 10.L). Despite the obvious habitat differences between lakes and streams, the distribution of observations clearly forms a continuum over three orders of magnitude, while habitat differences are reflected in abundance. A low number of observations per stream sample ($NQ = 2–7$) contributed to the comparatively large scatter of points ($r = 0.85$) while possibly contributing to bias.

AMPHIBIAN LARVAE IN TWO EPHEMERAL PONDS IN OHIO

In a project to relate amphibian larval abundance to environmental measurements using the standardized Morisita index (Eq. 3.14), Smith et al. (2003) found that the larvae of four species were significantly aggregated on all sampling occasions. In this study, they sampled amphibian larvae in two ephemeral ponds in Licking County, Ohio, by establishing a sampling grid of points 10 m apart along the long axis and 5 m apart along the short axis of the ponds. A single sweep of $0.117m^3$ with a 3-mm mesh net was made at each sample point on 3 occasions from late May to mid-July. As the ponds dried up, the number of sweeps at one pond declined from 33 to 31 to 20, and from 21 to 17 at the smaller pond, for a total of 5 sampling occasions. Tadpoles netted were identified and their abundances recorded. Two species were caught on all 5 nonzero sampling occasions: the American toad (*Bufo americanus*) and the spring peeper (*Pseudacris crucifer*). Spatial TPLs of these two species are parallel and moderately aggregated with common $b = 1.77$ (Appendix 10.M).

Smith et al. noted that larval abundances changed over time in the larger pond but not in the smaller pond. Spring peeper tadpole abundance was higher in the shallows than deeper water, while toad tadpole densities varied less across the ponds. Although spring peeper and toad tadpole abundances were strongly correlated on some sample dates, the range of abundance of the toad tadpoles is three times that of the spring peeper's while the peeper's variance is nearly twice as high, but the slopes are identical. Thus, the spatial distribution at the scale of these ephemeral ponds is almost identical, suggesting that environmental gradients influence the distribution and abundance of tadpoles across the ponds more than species-specific behavior.

EAGLE PREY IN NORTHERN GREECE

In northeastern Greece, the Dadia Forest is the breeding area for a number of raptors, including the short-toed eagle *Circaetus gallicus*. The 370-km^2 area is characterized by a mosaic of habitats dominated by woodlands; pine forests, mixed pine and oak forests, and old oak forests. Other areas are rocky outcroppings, shrubland, grassland, and agricultural land, some of it intensively cultivated. With human encroachment on the area, there are concerns for the conservation of the eagle and a number of other species. Short-toed eagles prey primarily on snakes. Examination of remains at nest sites suggests its most important prey species is the grass snake *Natrix natrix*, with the Montpellier snake *Malpolon monspessulanus* and large whip snake *Coluber jugularis* also commonly taken. Bakaloudis et al. (1998) assessed the distribution and abundance of prey reptiles in nine habitats to determine their importance for foraging by short-toed eagles. The grass snake was found predominately in agricultural areas, while the other two species were found in all habitats studied. The eagles concentrated their foraging efforts in cultivated areas where grass snakes are most abundant, but also in grasslands where they are less common. From April to September 1996, Bakaloudis et al. conducted line-transect surveys in 5 randomly selected $10 \times 100\,\text{m}$ (0.1 ha) plots in each of the habitats. Two observers made 25 weekly surveys by walking 5-m apart along the length of the plots recording the reptiles spotted. Eight lizard and 10 snake species were observed in 1703 sightings in all. Their Table 1 shows the mean numbers of sightings of the 18 species by habitat. The average number of sightings out of 125 surveys ranged from zero for 14 species in at least one habitat to 55.4 for one species, European green lizard (*Lacerta viridis*) in oak forest: four species were seen in all habitats.

TPL analyses of the reptiles, presumed prey of the short-toed eagle, observed in Dadia Forest, compare TPLs (Appendix 10.N) where each point is a species abundance averaged over habitat, and community TPL with habitat abundance averaged over species. The range of means in both approaches is approximately the same: 2 orders of magnitude for snakes and a factor of 10 for lizards. In both approaches, TPL is steeper for snakes than lizards. Where snakes were present, they were generally quite abundant but elsewhere they were absent or rare and this is true for both individual species and as a community. Lizard distribution, by contrast, both specifically and by habitat were present in a continuum from Poisson and rare to aggregated and abundant. It is not obvious why snakes were so gregarious unless it is because many species brumate communally in winter, increasing their aggregation year round.

Birds

WILLOW PTARMIGAN IN NORWAY

In a 15-year study of willow ptarmigan (*Lagopus lagopus*) population dynamics, Kvasnes et al. (2015) compared willow ptarmigan populations at 42 sites in five spatially separated populations in a mountainous region of south-central Norway. Their purpose was to examine spatial and temporal patterns of variation in densities of adult willow ptarmigan and test whether the ideal free distribution (IFD) described willow ptarmigan spatial distribution. The IFD theory predicts that the distribution of animals among patches will minimize resource competition and maximize fitness (Fretwell and Lucas, 1969). Under the IFD, the expected value of b is ~ 2 (Gillis et al., 1986). Part of the study was to examine TPL and test this IFD prediction.

Line-transect surveys using a well-established method with pointer dogs to flush birds were conducted in August from 1996 to 2011. Flushed birds were recorded as juvenile, adult male, adult female or uncertain, and their distance from the transect estimated. The method is described in detail by Pedersen et al. (2004). Between 5 and 179 observations were made per year at between 3 and 15 areas; the number of surveys ranged from 4 to 51/year and number of birds per transect ranged from 5 to 179. Density of birds (adults/km² and juveniles/pair) was estimated and these data were used to test IFD. Comparison of reproductive success, measured as juveniles/pair, with the number of adults/km^2 showed no relationship ($r = -0.039$) in agreement with the IFD.

Kvasnes et al. conducted a hybrid spatial TPL analysis of their data. To minimize small sample bias, only data of a minimum of 5 surveys/region and 5 years/region were included; one region failed to meet this requirement and was excluded. Means and variances were computed for 4 mountain regions for 5 years (Fig. 10.9) and analysis of parallelism conducted with mountain region as the factor. A single line was fit with gradient $b = 2.83 \pm 0.27$; parallel lines with region as the factor resulted in a slightly steeper slope, $b = 3.07 \pm 0.42$. Statistics recomputed from their published data are given in Appendix 10.O. Combining the site and year data in an ensemble analysis found willow ptarmigan in the mountains of south central Norway to be highly aggregated ($b = 3.09 \pm 0.285$). Year-to-year abundance of willow ptarmigan within a region varied less than abundance between regions.

The variation in density of adult birds was primarily attributable to variation between survey areas, which could arise from spatial heterogeneity in adult survival or as a consequence of spacing behavior of juveniles during the settlement stage. Yearlings may be attracted to areas of high densities during the settlement period in spring, implying that the presence of conspecifics might represent a

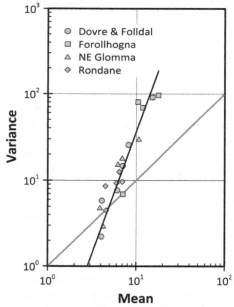

FIG. 10.9 Hybrid spatial TPLs ($NQ = 5$–10, $NB = 21$) of willow ptarmigan surveyed by line transects in 4 separate mountain regions in Norway over 5 years show ptarmigan to be highly aggregated in August. *(Data from the Appendix to Kvasnes et al. (2015).)*

cue signaling high adult survival and thus high habitat quality. Thus, one explanation for willow ptarmigan spatial distribution is the presence, activity, and number of conspecifics, especially of experienced birds.

Willow ptarmigan is the type species of a cluster of 19 subspecies that includes the red grouse of Scotland (*L. lagopus scoticus*). All are game birds with hunting seasons from late summer through autumn or winter. Males are highly philopatric, females much less so, and defend their territories of 2–10 ha during the breeding season. During August, when the surveys were conducted, ptarmigan are in the process of aggregating into flocks prior to migrating to lower elevations for the winter. Thus, a high degree of aggregation would be expected. With territorial behavior during the breeding season, aggregation was likely much lower.

JAYS IN WESTERN USA

Samaniego et al. (2012) investigated the spatial distributions of two closely related species of jay with different social systems and behavior, piñón jays (*Gymnorhinus cyanocephalus*) and western scrub jays (*Aphelocoma californica*), in a project to interpret *b*. Piñón jays are social and nonterritorial; they have a restricted diet consisting mainly of seeds of the piñón pine. In contrast, the scrub jay is comparatively polyphagous, is aggressively territorial and breeds in pairs without helpers. These striking differences in behavior of species occupying similar semiarid habitats, differing principally in elevation, provided Samaniego et al. an opportunity to compare modeled variance-mean slopes with field data derived from the Breeding Bird Survey (BBS; Robbins et al., 1986). BBS surveys are conducted by volunteers every spring on established routes in the contiguous 49 US states and southern Canada by observing and listening for birds for 3 min at 50 stations, 0.8 km apart. Because road density is closely associated with human population density, areas of sparse human population like the desert and mountain States are underrepresented in the surveys.

The piñón jay data were taken from 111 BBS routes and the scrub jay from 256 routes from 1968 to 2005 subject to the conditions that the routes were sampled in >10 years consecutive years, the birds were recorded in >5 years, and the piñón and scrub jay routes were >8 km apart. Mean and variance were calculated for each route and TPL slopes were calculated for each species separately using GMR. They modeled colonization-extinction process using a 2-state (colonized or not colonized BBS route) Markov chain (Norris, 1998) to obtain simulated TPL *b* for both species. The model generates transitions between the four states: persistence, continued absence, colonization, and extinction. These states can be derived from the colonizations and extinctions observed on the BBS routes. An additional 5 parameters distinguished between species: expected number of colonists; probability of zero abundance; probability of positive abundance; habitat suitability; and density dependence. By manipulating these parameters, the model was tuned to mimic the observed variance mean slopes (Fig. 10.10). Samaniego et al.'s purpose was to explore factors influencing the slope of TPL. The success of the model in mimicking the field data shows they captured many essential features of these birds' spatial dynamics. Their success supports the conclusions that the difference in slope between species was at least partly the result of behavioral differences. In particular, they found that flocking and population growth rate are important variables contributing to TPL. Although the model could mimic the TPL of the field data they could not recover the parameters from the field data, except by simulation. In short, life-system parameters can generate unique TPL relationships, but the ability of TPL to recover the life-system parameters is very limited. (A case study in Chapter 7 has b correlate roughly with life system.) Samaniego et al. emphasize that while birth and death must contribute to observed colonization and extinction rates, they must do so in a nontrivial fashion that requires mapping demographics with behavior onto TPL. The behavioral features of these models successfully mapped three behavioral traits, flocking, vagility, and colonization, onto TPL.

Temporal TPLs of Samaniego et al.'s field and model data are plotted together to compare the Markov chain model agreement with the field data (Appendix 10.P). This provides a useful means for testing model performance. Model estimates for both species overestimated the intercept and had somewhat higher scatter around the fitted lines. The slopes differed significantly between

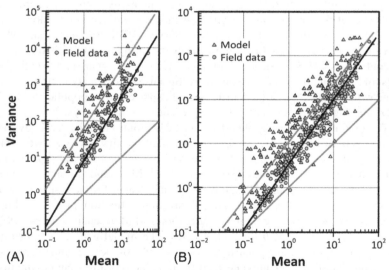

FIG. 10.10 Temporal TPLs of field data and model output of pinõn (A: $NQ > 10$, $NB = 111$) and western scrub jays (B: $NQ > 10$, $NB = 256$) in western USA are compared here as a test for model performance. The model for both species overestimated the intercept and had somewhat higher scatter (lower r) but closely similar slopes indicating the model's good variance structure. *(Adapted from Figs. 3 and 8 in Samaniego et al. (2012).)*

species but not between model and field data confirming the model's excellent variance structure. Because TPL intercepts are strongly influenced by sample size, the models could be tuned to have the intercepts coincide by incorporating a sampling constant. The higher residual variance (represented by lower r) of the model output is likely due to population and/or spatial dynamics factors that have an effect on distribution in the field but are not included in the model.

This study, while important in its own right and interesting in its approach and conclusions, also illustrates a use for TPL, not previously employed. Comparisons between model output and field data are typically made using single-valued statistics that compare a single aspect of the observed and modeled data: Moriasi et al. (2007) review the methods.

GRASSLAND SPARROWS IN CONTINENTAL USA

Curnutt et al. (1996) used TPL to investigate the distribution and abundance of nine species of grassland sparrows at continental scale. Using data from the Breeding Bird Survey (Robbins et al., 1986 and see previous study), they marked out the margins of the ranges of these species in America north of Mexico to examine their population dynamics at the edges versus the centers of population range. They chose sparrows for this study for the length (23 years) of continuous data available and because of their similarity in feeding and nesting behavior, and their ranges constitute a spectrum from restricted (McCown's longspur, *Calcarius mccownii*) to continental (Savannah sparrow, *Passerculus sandwichensis*) as reflected in the number of routes recording the species (Table 1 in Curnutt et al., 1996).

A major motivation of this study was to examine how population abundance and variation in spatial distribution as measured by TPL varied across the range and in relation to distance from the edge. Because of the integer effect at low densities (Fig. 15.2), Curnutt et al. restricted their analyses to sites with mean abundance >1.5 ($\log(M) = 0.176$) to avoid the bias at low mean abundances caused by sites recording zero due to failure to detect birds actually present. The issue of zeros in samples and their influence on TPL is examined in Chapter 15.

In their study, Curnutt et al. used ODR of standard deviation instead of variance against mean to estimate temporal TPL. Converting their estimates in Table 2 from SD on M to V on M and ODR to GMR (Eq. 4.13), revised estimates of r, b and SE(b) are reported in Appendix 10.Q. Using $\log(CV)$ to test for temporal variation over spans of 5–10, 10–20, and 5–20 years for the 6 most abundant species, the time spans averaged over sites and tested against zero variation increased with span ($P < 0.01$). Thus, population numbers as a whole became increasingly variable over time, but this increase depended on only part of the total population.

To examine populations at the center relative to the periphery of the range, Curnutt et al. marked out species ranges using the common sense approach of examining the occurrence of birds in "grid squares" 1 degrees of latitude by 1 degrees of longitude and linking contiguous and near-contiguous squares. The 6 most abundant species (those detected on >330 routes) were used to test for a gradient of abundance with distance from edge. They found that abundance increased with distance from the edge while variability measured as log(CV) declined with distance, implying that *b* also declined with distance from the center. Thus, these sparrow populations consist of centrally located areas of high abundance with relatively low numerical variability but high spatial aggregation, surrounded by sites of low abundance with relatively high numerical variability, but lower aggregation. From these results, Curnutt et al. proposed that the spatial and temporal populations exhibit a "source and sink" dynamic where the core of the range "feeds" the less productive peripheral areas. This is in contrast to data from Britain where Blackburn et al. (1999) analyzed British Trust for Ornithology data of distribution and abundance of 32 passerines and "found no convincing evidence that passerine bird densities are usually lower towards range edges in Britain." The difference is likely due to the fact that Britain, unlike continental USA, is an island with "hard" edges.

BIRDS IN URBAN AND NONURBAN ENVIRONMENTS

Using data from the Audubon Christmas Bird Count (CBC), Murthy et al. (2016) investigated the influence of urbanization on winter bird diversity in the contiguous United States by comparing species richness (alpha diversity) and turnover (beta diversity) of 378 species and their relative abundance at urban and adjacent nonurban sites. They used CBC data of 42 pairs of urban (defined as CBC circles >60% urban in the 2001 MODIS satellite dataset (Schneider et al., 2010) and nonurban (defined as <10% urban or agricultural) sites in their comparisons. The study supported the proposition that urban conditions could contribute to a homogenizing effect of bird communities in cities because urban resource subsidies (bird feeders, garbage, etc.) and shelter (landscaping, buildings, etc.) and the urban heat island could increase the occurrence of widespread species and reduce their variation in relative abundances in cities. Murthy et al. chose the CBC, which is conducted in winter in order to examine the effect of urban heat islands and anthropogenic foraging opportunities, which are more pronounced in winter.

Murthy et al. found that alpha diversity standardized for latitude did not differ between urban and nonurban sites. However, the analysis did reveal a more rapid decline in community similarity with distance from nonurban sites than from urban, suggesting lower beta diversity of winter bird communities in urban areas. In addition, using TPL they found the variation of relative abundance of bird communities in cities was lower than in neighboring nonurban sites at all densities. Murthy et al.'s Fig. 4 shows the computed means and variances over about six orders of magnitude of relative abundance and about 10 orders of magnitude of variance.

Adeline Murthy kindly made her original data available to me to examine Murthy et al.'s Fig. 4 in more detail. Fig. 10.11 (Appendix 10.R) shows the two TPL regressions have almost identical slopes ($P > 0.50$) but differ significantly in intercept ($P < 0.005$). They concluded that both increased occurrence of widespread species in cities and lower variation in relative abundances at urban sites contribute to a "homogenizing effect of urban areas on avian communities."

The lower variances at mean relative abundances in urban areas could be due to more abundant resources in the urban areas relative to the adjacent non-urban areas. Combined with the well-established heat island effect of cites, increased foraging opportunities in cities could reduce the spatial variation by buffering against environmental uncertainty. It could also be due to

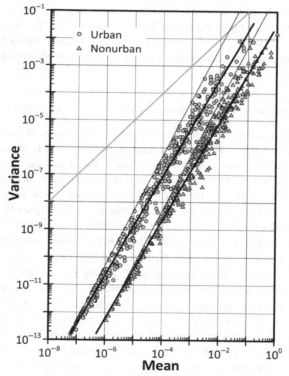

FIG. 10.11 Mixed-species TPLs of relative abundance of $NB = 378$ bird species recorded by CBC in $NQ = 42$ US cities and in adjacent nonurban areas. The significantly smaller intercept ($P < 0.005$) of the urban regression indicates reduced variability of city bird populations in winter; the slopes are not significantly different ($P > 0.50$). Both regressions are well below the Poisson (45 degrees) line. The *light lines* immediately above the scatter plots have Eq. $V = 50 \cdot M^2$ and represent an upper limit to variance equivalent to the upper limit visible in Fig. 15.2. *(Data generously provided by Adeline Murthy used in her Fig. 4 in Murthy et al. (2016).)*

differential reporting efficiency in cities versus the hinterland. Sampling efficiency can influence the value of TPL regression intercept and density-dependent sampling efficiency can modify slope (RAJT, 2018). The fact that the estimates of slope are so similar argues against density-dependent efficiency. However, the detection rates of birds in different habitats would also lead to parallel TPL regressions.

Two key points emerge from Murthy et al.'s analysis. Their elegant analysis to show how TPL can be used to compare community dynamics in two places or habitats, specifically that the lower urban variance at all (relative) densities, supports the proposition that beta diversity is indeed decreased by urban areas. The second point emerges from examination of the individual plots in Fig. 10.11. Both sets have a more or less random distribution of points relative to the fitted line below the fitted lines. But the points appear to be bounded above the line. This is clearest in the nonurban plot, but also apparent in the urban plot. LRT and Woiwod (1982) showed this phenomenon in their Fig. 1, which has an upper bound of $V = M^2$. In this figure, the upper bound is $V = 50 \cdot M^2$. Furthermore, all points are well below the Poisson line because of the transformation of data from abundance to relative abundance.

BRITISH TRUST FOR ORNITHOLOGY ANNUAL SURVEY

In a comparative study of 97 species of aphids, 263 species of moths, and 111 species of birds, LRT et al. (1980) examined the fits of TPL to samples taken simultaneously that covered most of the island of Great Britain, an area of about $243,000 \, km^2$. It was the largest longitudinal study of TPL to date. Bird populations were surveyed at 64–210 sites for up to 15 years by the British Trust for Ornithology (BTO), an independent charitable research institute that combines citizen and professional scientists to make wildlife, particularly bird population data available for public and governmental use. To this end, the BTO has sponsored an annual survey of breeding birds in field and forest habitats since 1961.

The spatial distribution of singing males during the breeding season is the basic measure for the bird census, which is taken from April to June. Each site receives at least eight intensive inspections each year, using a grid system for recording, with supplementary visits to check on nocturnal or crepuscular species (Preston, 1979). The farmland sites are approximately 75 ha and are mostly embedded in a larger agricultural landscape. Woodland sites are smaller (about 22 ha), more difficult to survey, and vary considerably in tree species composition.

The BTO made their raw data available to the Rothamsted Insect Survey for a comparative analysis of avian and insect spatial distribution that LRT et al. published in 1980. These data were collected by birders acting as citizen scientists (although that term was not used then) of 111 species of birds they sampled by direct observation at 19 to 98 woodland and 45 to 123 farmland sites between

1962 and 1976. LRT et al. standardized the bird counts to numbers/100 ha. Table 10.1 summarizes their results.

One objective of the study was to determine if variance plateaued at a high density, but no such case was observed. Another was to look for exceptions that might refute the idea that a species' behavior is a major contributor to population stability. If extrinsic factors in the habitat were more important than behavior in controlling spatial behavior, it was argued TPLs for the same species could be expected to differ between field and forest. The woodland and farmland reports were initially analyzed separately and only combined when found not to differ. Approximately half the species fit a single TPL, with estimates of the slope and intercept not differing between farmland and woodland, the remainder differing significantly in intercepts. Eight species also differed

TABLE 10.1 Summary of analysis of BTO data in the Appendices in LRT et al. (1980)

		Regressions	
		Single	Double
Number of species		56	55
Regressions significant (P<0.01) linear and parallel (if double)		55	42
Regressions significant (P<0.01) linear, not parallel		–	6
Regressions not significant (P>0.01)		0	0
Regressions with significant curvature (P<0.01)		1	7
Regressions not significantly nonrandom		1	–
Slopes for single and parallel regressions	Range		1.19–2.69
	Mean±SD		1.68±4.60
Intercepts for single and parallel regressions	Range		−0.73 to 0.96
	Mean±SD		0.96±3.17
Range of slopes for nonparallel regressions	Low ± SE		0.98±0.06
	High ± SE		3.38±0.35
Range of means			2.14–1820
% Variance accounted for by single and parallel regressions	< 50%		1.0%
	50%–80%		11.6%
	>80%		87.4%
Number of data points per plot (years)			5–29[a]

[a]Where the birds are treated separately in woods and farms, there are two data sets per year.

significantly in slope between field and forest. No species plot was nonsignificant, although 8 showed some curvature in one habitat. In many species, the range of means was similar in both habitats, but some species, illustrated in LRT et al.'s Figs. 3 and 4, were more abundant in one habitat so that the points for field and forest were separated with the forest means generally higher than the fields'. This was likely due to the fact that most species had evolved originally as forest species and were secondarily adapted to hedgerows, copses, and marginal areas around arable land. Thus not only differences in abundance might be expected, but also differences in behavior and life-history strategies.

Of the 8 of 111 species with slopes that differed between field and forest, estimates of slope ranged from $b = 0.98 \pm 0.059$ for kingfisher (*Alcedo atthis*) to $b = 3.38 \pm 0.349$ for yellow wagtail (*Motacilla flava*) both in field habitats. The distributions of slope and intercept are quite similar (Fig. 10.12), but not Gaussian. In all but 2 parallel cases, intercepts were higher in wood than field ($A_{wood} > A_{field}$): the exceptions were willow warbler (*Phylloscopus trochilus*) and nuthatch (*Sitta europaea*). In the light of Murthy et al.'s (2016) results, we might wonder if fields offer a resource subsidy, despite the higher plant diversity of the forest. Alternatively, a higher sampling efficiency in fields than forest would produce the same result. The fact that in 45 of 47 cases $A_{wood} > A_{field}$ supports this proposition. However, of the 8 species with nonparallel TPLs 4 had $b_{wood} > b_{field}$ and 4 species had $A_{wood} > A_{field}$, suggesting that sampling efficiency alone could not account for the differences in *TPLs*. If the 8 nonparallel TPLs were due to differences in sampling efficiency, efficiency must be density dependent for these species in either field or forest (Chapter 15). It is difficult to envisage a situation where sampling efficiency might be density dependent: it would require that the ability to discern and distinguish bird song differ systematically with density, such that the probability of detection in one habitat declined with density, but not with the other.

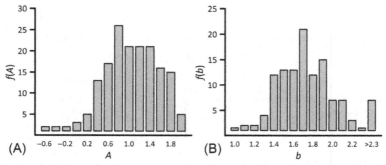

FIG. 10.12 Frequency distributions of A (A) and b (B) of TPL fitted to 111 species of bird surveyed in the BTO's breeding bird survey between 1962 and 1979. (*Data from the Appendix in LRT et al. (1980).*)

The consistency with which TPL fit across 111 species of birds in two habitats with slopes ranging from about 1 (Poisson) to over 3 (superaggregated), LRT et al. argued, cannot be explained by habitat differences alone as half the species' spatial distributions were the same in both habitats. Of those species with parallel regressions the spatial behavior organizing the distributions must be common to both habitats and that property must be intrinsic to the species. Of the 8 nonparallel regressions, they concluded that if the differences were due to behavioral differences in the two habitats, b must contain some information about the spatial behavior, a conclusion reiterated by Curnutt et al. (1996). But as Samaniego et al. (2012) observed, while b may be influenced by life-history traits, recovering life system information from b is generally not possible.

AUDUBON SOCIETY'S CHRISTMAS BIRD COUNT

LRT et al.'s (1980) study examined bird populations in field and forest, while Murthy et al. (2016) used Christmas Bird Counts (CBC) data to compare bird communities in urban and nonurban settings. The Audubon Society's CBC is conducted in a wide variety of habitat types and includes >1000 species providing an opportunity to replicate LRT et al.'s study at larger scale. Both Murthy et al.'s (2016) and LRT et al.'s (1980) studies suggested that habitat's influence on TPL was primarily in the position of the intercept; in other words, the variation in population distribution was influenced by the nature of the habitat, resulting in one population intrinsically more variable than the other.

With that in mind, Norelli et al. (unpublished) have examined the CBC data from 1960 to 2015 for variations in TPL across five habitat types and three scales. The CBC has been conducted annually since 1900. It is a standardized survey conducted at 557 locations in 1960 increasing to 1837 in 2015. Each CBC survey is conducted within a 12-km radius circle by volunteers who count all individuals seen or heard on line transects within the circle in one 24-h period between mid December and early January. Each circle records the number of sightings of each species and variables describing the efforts of the participants. The number of "effort hours" for a survey circle is the sum of the number of hours each volunteer participated in the survey. In a preliminary analysis, Norelli et al. found that TPL analyses of raw data and effort corrected data in most species differed little in intercept and not at all in slope. Their analyses therefore used raw data.

Norelli et al. examined bird distributions at three scales by computing mean and variance of bird numbers in circles within each state, within 6 regions (Northeast, South, Midwest, Great Plains, Mountain, and Pacific), and nationally. Several states were excluded from the state-scale analysis as there were fewer than 15 circles; all circles were included at the regional and national levels. Of the 48 contiguous states, 20 failed to reach the 15 circle threshold in the first decade and Delaware and Rhode Island were never included in the state-level analysis. Thus, the points in the analysis are means and variances per year with a maximum of 59 points at the national level, >300 at the regional level, and ~1200 at the state level. A limitation of 15 circles per year for inclusion in the analysis reduced the number of species qualifying at all scales. A second restriction was imposed, that the range of means should be at least one order of magnitude. Thus, to be included in the study, a species had to be both common and broadly distributed. This reduced the number of species included in the study to 371, 406, and 313 at the state, regional, and national scales, respectively.

Norelli et al. found that the frequency distribution of b was slightly skewed to the right with a small number of slopes $b > 4.0$ (Fig. 10.13B), a proportion that increased with change of scale from national to state as the average value of b increased with scale (Table 10.2). The correlation coefficients were very similar across changes of scale, but the left-hand tail (negative skewness) increased with scale, indicating the proportion of high r decreased with increasing scale.

The effect of landuse on TPL was examined using the US Department of Agriculture's Natural Resource Conservation Service (USDA-NRCS) landuse land cover classifications. NRCS's 13 classes were reduced to five: in addition to urban, water included open water and 3 classes of wetlands, rangeland included range and scrub, forest combined deciduous, evergreen and mixed, and field included cultivated fields and pastures. Of 1182 species (including subspecies and suspected hybrids) listed in the CBC database, 406 species at the regional scale were used for this analysis. The landuse class for each circle was determined using a Geographic Information System (GIS) to compute the proportion of each of the 5 classes in each CBC circle. Circles with ≥60% of area in a single class were assigned to that class. Circles with no class ≥60% in a single class were discarded. This reduced the number of circles by 50%.

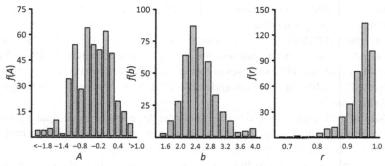

FIG. 10.13 Frequency distributions of A (A), b (B), and r (C) for bird TPLs at the regional scale computed from Audubon Society's Christmas Bird Surveys 1960–2015. Means and variances computed for 406 species with a minimum of 15 circles and a range of means of at least one order of magnitude. *(Unpublished analysis by Norelli et al.)*

TABLE 10.2 Frequency distribution statistics of *b*, *A*, and *r* at three spatial scales

	Nation	Region	State
Number of species	313	406	372
Slope, $\bar{b}$ - Mean	2.862	2.490	2.358
Standard deviation	0.645	0.462	0.408
Skewness	0.788	1.027	1.320
Intercept, $\bar{A}$ - Mean	−0.689	−0.233	−0.241
Standard deviation	1.155	0.596	0.552
Skewness	−2.646	−0.999	−1.302
Correlation, $\bar{r}$ - Mean	0.921	0.944	0.941
Standard deviation	0.057	0.044	0.041
Skewness	−1.958	−1.804	−1.432

Murthy et al. (2016) used a standard of >60% urban land cover for their urban class and their nonurban class contained <10% agricultural or urban areas. It should be noted that Murthy's classification was human modified versus minimally modified land, whereas Norelli et al. also considered the agricultural modification of land (arable, orchard, vineyard, and pasture land) separate from minimally managed or unmanaged land (forest, wetland, and rangeland). One uncontrolled factor is the shift of landuse from nonurban to urban that occurred in large parts of the country between 1960 and 2015. By using landuse classes established in 2001, the number of urban circles is overestimated in the decades preceding and underestimated in the years since 2001. Because urban land is rarely deurbanized, the other classifications are probably fairly stable over 56 years, although some low-lying agricultural land has reverted to wetland in some areas. The impact on TPL parameters of the systematic shift to urbanization that may have occurred in some circles is uncertain, but Murthy et al.'s results suggest that the intercept in those sites would be reduced over time. In combination with sites that did not transition to urban, this would tend to increase the residual variance and decrease the correlation coefficient.

Unlike the BTO analysis, Norelli et al.'s analysis revealed differences in 50% of habitat comparisons in *A* and 55% in *b* (Table 10.3). Thus, the degree of aggregation varies between habitats as well as the degree of heterogeneity. Species with high *b* tended to have low *A*.

LRT et al. (1980) found that the forest intercepts were generally greater than the fields', a result echoed in Norelli et al.'s analysis. However, unlike LRT

TABLE 10.3 Number of species with significantly different regression coefficients in the comparison of habitats

Habitat comparison	Comparisons	A	b
Rangeland × Urban	181	92	100
Forest × Urban	174	90	98
Field × Urban	178	89	98
Field × Forest	167	91	92

et al., Norelli et al. found about half the species differed in slope as well as intercept in all comparisons, including field and forest. The differences are likely due to the extreme difference in scale of the BTO and CBC analyses. The much larger-scale CBC analysis by habitat shows aggregation as measured by b to increase from rangeland to forest: Rangeland < Water < Urban < Field < Forest. The order of intercepts, Water < Forest < Rangeland < Fields < Urban, suggests that heterogeneity increases from the least disturbed habitats to the most in agreement with LRT et al.'s (1980) findings of field and forest but differing from Murthy et al.'s (2016) findings.

The average slopes of urban and rangeland (Table 10.4) are very similar, in agreement with Murthy et al.'s findings while the intercepts are reversed: urban distributions are more heterogeneous than rangeland in this analysis. The important difference between Murthy et al.'s and Norelli et al.'s analyses is Murthy et al.'s focus on the community of birds instead of the comparison of individual species in the two habitats. The larger differences in slopes between the habitats of Norelli et al.'s analysis are also the result of difference in precision and the longer time scale than Murthy et al.'s analysis.

It has sometimes been suggested that $b \leq 2$ is the norm and that $b > 2$ is in some sense aberrant (e.g., Kilpatrick and Ives, 2003; Ballantyne, 2005). The examples collected by LRT et al. (1978) show a range of $1 < b < 3$ with a mean of $\bar{b} \approx 1.8$ and ~20% of cases with $b > 2$. Norelli et al.'s results are startling in that $\bar{b}$ ranges from 2.36 to 2.90 (Table 10.2) over a range of scales and from 2.34 to 2.54 (Table 10.4) for habitats at the regional scale. These large slopes are undoubtedly due in part to the very large scale of Norelli's study. However, a more potent cause is due to the "magnification" of b that occurs when zero samples are excluded from the analysis. The CBC data have very few zeros because zeros are only recorded if a species known to be present because it was seen in the previous week is not observed on survey day. This means that zeros are underrepresented. The much steeper slopes observed in this study are partly a consequence of the absence of zeros. RAJT et al.'s (2017) study shows that for nematodes, removing zeros increases b by 20%–50%. Assuming a

TABLE 10.4 Frequency distribution statistics of b, A, and r in 5 landuse classes

	Water	Rangeland	Forest	Fields	Urban
Number of species	187	249	214	233	205
Slope, $\overline{b}$ - Mean	2.465	2.338	2.535	2.422	2.355
Standard deviation	0.432	0.457	0.457	0.522	0.469
Skewness	0.830	0.852	1.047	1.394	0.796
Intercept, $\overline{A}$ - Mean	−0.462	−0.327	−0.431	−0.209	−0.122
Standard deviation	0.683	0.814	0.586	0.694	0.593
Skewness	−0.864	−1.522	−1.007	−0.895	−0.157
Correlation, $\overline{r}$ - Mean	0.916	0.919	0.922	0.922	0.935
Standard deviation	0.066	0.074	0.069	0.071	0.055
Skewness	−1.311	−1.846	−2.088	−2.606	−1.718

similar effect with birds at this very large scale, yields an average $\overline{b} \approx 1.8$ in agreement with LRT et al. (1978), but retaining >35% of species with $b > 2$. The small increase in b Norelli et al. found with increasing scale is also likely due to the zero-censoring effect. Reporting should be the same at all scales, but clearly with zero censoring, this is not the case as samples outside the normal range are censored as the number of zero samples remains the same, but the proportion gets progressively smaller.

Mammals

CETACEANS NEAR THE AZORES

Silva et al. (2014) investigated the spatial and temporal distribution of cetaceans in waters near the Azores in mid-Atlantic using data collected from boats and land-based observations over a 10-year period from 1999. Boat surveys covered an area of 258,228 km^2 and land-based observations from a lookout on Pico totaled 4944 h.

The land-based surveys from the lookout on Pico were by trained observers who could observe about 800 km^2. Observations were made from May 1999 to March 2001 and January 2008 to September 2009 and were for 1–3 h. The boat surveys collected data year round from 1999 and 2009 from dedicated cetacean surveys as well as platforms of opportunity (Azores-based fishing boats) near the islands and nearby seamounts as well as areas of steep island slopes, canyons, and deep abyssal plains. Dedicated boat surveys were transects and observations from fishing boats were essentially random. Expressing their data as number per 100 h observation and encounter rates as number/100 km traveled, Silva et al. analyzed their data in relation to proximity to the Azores, depth of ocean, and time of year. The high diversity of marine habitats and high productivity of the oligotrophic ocean in and around the Azores made this area a center for whaling for centuries, and now a center for whale watching. Their analysis of the spatial and temporal distribution of whale sightings is a valuable asset for this leisure industry.

Of 2968 observations, 24 species in all were recorded. These include 7 species commonly observed in groups. Most numerous were short-finned pilot whale, *Globicephala macrorhynchus*, Risso's dolphin, *Grampus griseus*, false killer whale, *Pseudorca crassidens*, Atlantic bottlenose dolphin, *Tursiops truncatus*, short-beaked common dolphin, *Delphinus delphis*, Atlantic spotted dolphin, *Stenella frontalis* and striped dolphin, *Stenella coeruleoalba*. Observations of smaller groups included sperm whale *Physeter macrocephalus* and blue whale *Balaenoptera musculus*. Several species, including humpback whale *Megaptera novaeangliae*, were only recorded singly.

One of the products of this study is a table of numbers of individuals per school. Silva et al.'s Table 2 gives the mean and SD of school size of 14 species. Mixed-species TPL of school size varying from 1.7 to 43.3 individuals (Fig. 10.14A; Appendix 10.S) has a steep slope of $b = 2.66 \pm 0.158$. With each point representing the average school size of these 14 species, the dominant source of variation is the observer's ability to estimate the size of the school accurately. As the characteristic school size increases, the observer's ability to discriminate and count each individual must decline. Thus, the steep TPL is likely due to the product of two sources of variation: natural variability and observer error, which is likely to grow with school size. However, sampling efficiency tends to decline with density, not increase, resulting in underestimation of b (Chapter 15), so these sources of variation would tend to cancel each other.

CETACEANS AROUND THE BRITISH ISLES

A survey of cetaceans in the North Sea supported by the European Commission and 8 countries was conducted in summer 1994. A major purpose of the Small Cetacean Abundance in the North Sea (SCANS) survey was to estimate the absolute abundance of small cetaceans as a baseline for future monitoring of the environmental health of the North Sea and to provide information to policymakers for conservation and management. The North Sea extends from the Dover Straight north to Scotland and east to the northern tip of Denmark on the European mainland and covers about 411,000 km². In addition to the North Sea survey from Scotland to the English Channel, reports by Hammond et al.

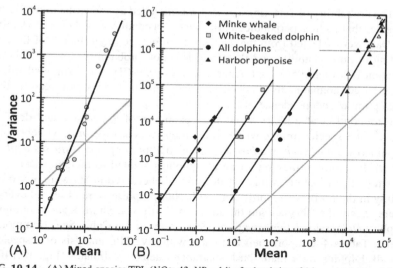

FIG. 10.14   (A) Mixed-species TPL ($NQ > 43$, $NB = 14$) of school size of 14 species in the vicinity of the Azores scales steeply with mean size. (B) Ensemble TPLs ($NB = 5$–14) of density (#/100km^2) of cetaceans observed in the 1993 SCANS survey. Minke whale white-beaked dolphin, all dolphins, which includes white-beaked, and harbor porpoise mean and variance of density estimates are boot-strapped estimates from surface and aerial surveys made in Sumer 1994. Harbor porpoise solid tri-angles are surface estimates, shaded triangles are aerial estimates. *Lines are displaced by one log cycle for clarity. (A: Data from Table 2 in Silva et al. (2014). B: Data from Tables 4–6 in Hammond et al. (2002).)*

(2002, 2013) also covered the Celtic Sea west of Great Britain and south of Eire to Brittany, the Orkney and Shetland Islands east to Sweden, and the western-most part of the Baltic Sea.

The 1994 survey consisted of 20,000 km surface cruises covering an area of 890,000 km^2 and 7000-km aerial surveys covering 150,000 km^2. The survey was highly intensive and analysis used robust methods to estimate the probabil-ity of detection of animals on the transects. For both the surface and aerial sur-veys, the study area was divided into 9 zones in which zigzag tracks were defined to provide a known, nonzero, coverage probability for each zone to pro-vide an unbiased estimate of abundance to be made using line-transect sampling methods (Buckland et al., 2001, 2007). The methods incorporate corrections for animals missed on the transect line and for movements triggered by the observers. On the surface cruises, observers made simultaneous searches from two platforms over different ranges. Abundance estimates from the aerial sur-veys were based on zigzag tracks defined in each zone with survey transects replicated on east-west and west-east flights. In most surveys, two aircraft were flown at 90 knots 600 ft. asl (111 kph at 183 m), the second 9 km behind the first to maximize the probability of detection.

The probability of detecting a school was modeled as a function of predictor variables that include perpendicular distance, sea condition, and overall visibility. Data were selected from the database to minimize the influence of factors leading to poor visibility and therefore detection probability. Coefficients of variation (CV) and 95% confidence intervals (95%CI) were estimated using a log-based nonparametric bootstrap procedure that does not require the assumption of independence of the transects sampled.

Hammond et al. (2002) report their estimates of abundance and density (#/km^2) of 3 species: minke whale (*Balaenoptera acutorostrata*), whitebeaked dolphin (*Lagenorhynchus albirostris*), and harbor porpoise (*Phocoena phocoena*). In addition, they report on all dolphins sighted to include species of uncertain identification. Minke whale and whitebeaked dolphin were observed mainly in the north west North Sea and harbor porpoise were observed everywhere except the Channel and the southern North Sea. Short-beaked common dolphin, *Delphinus delphis*, were observed only in the Celtic Sea. Hammond et al.'s Tables 4–6 report their estimates with CV. TPL analysis of 3 species and all dolphins were made on bootstrapped density estimates (Fig. 10.14B; Appendix 10.T). The lines are displaced by 1 log cycle for clarity; aerial and surface estimates for harbor porpoise are analyzed together but marked differently to show their consistency.

A second survey, SCANS II, was conducted in July 2005 (Hammond et al., 2013). This covered the entire European Atlantic continental shelf to establish estimates of abundance for more cetacean species. The survey used line-transect sampling methods similar to those described above. Surface transects of 19,725 km covered an area of $>10^6$ km^2; aerial transects covered an area of ~364,000 km^2 in 15,802 km of transect. Again data were selected for runs with good visibility and moderate waves. Of 13 species observed, abundance and density estimates with 95%CI were published for minke whale, white-beaked dolphin, bottlenose dolphin (*Tursiops truncatus*), short-beaked common dolphin, and harbor porpoise.

Estimates of abundance and density are given in Tables 4–8 in Hammond et al. (2013) for 8–16 zones and CV estimated using a bootstrap method. Variances were calculated from the mean and CV of number of schools, average school size, total abundance, and density data (#/km^2) analyzed (Appendix 10.U). Although overall abundance in 1994 and 2005 was similar for minke whale, white-beaked dolphin, and harbor porpoise, the spatial distributions differed between years. In this survey there are small differences in slope between counts and density estimates and between surface and aerial observations: the differences in *A* are expected.

HARBOR PORPOISE IN THE NORTH SEA

The harbor porpoise, *P. phocoena*, at <2 m in length is one of the smallest cetaceans. It is found in coastal waters of the North Atlantic and North Pacific and frequently ventures into fresh water. It is common on the European continental shelf, including the North Sea. Major prey items are the large group of bottom-dwelling fish known colloquially as sand eels (*Ammodytes* spp.). Concern over declines in predator populations, including sea birds, prompted European officials to place restrictions on human exploitation of sand eel. As a major top predator in the North Sea, harbor porpoise is considered to be an important environmental health indicator. Consequently, a number of surveys have been conducted of harbor porpoise populations in the central and southern North Sea by the countries bordering it: Belgium, Denmark, Germany, the Netherlands, and the United Kingdom.

The central and southern North Sea is a shallow continental shelf sea with depths typically 30-200 m except at its northern margin where it is 600 m. The shallowest part is central, the 18,000 km^2 Dogger Bank at 15–35 m deep.

Gilles et al. (2016) assembled 27 sets of harbor porpoise abundance data from surveys by North Sea countries as the basis for developing a seasonal habitat-based statistical model of harbor porpoise numbers in the North Sea to assist policy makers in regulating human activities in the North Sea. Their seasonal maps of harbor porpoise distribution and abundance have assisted conservation efforts. The data for the model were records of visual surveys by line transect from 2005 to 2013. These data consist of 14,356 sightings of porpoise schools made in cruises totaling 156,630 km on transects conducted by a large-scale international survey in 2005 (see the previous study; Hammond et al., 2002, 2013) and 22 small-scale national surveys from 2008 to 2013 (citations in Gilles et al., 2016).

In their paper, Gilles et al. developed and tested against the survey data a statistical model intended to predict harbor porpoise abundance in the North Sea from environmental data. Predictor variables used include latitude, longitude, depth, sea floor slope and aspect, mean and variance of sea surface temperature, day length, and distance to shore and to sandeel grounds. The standard deviation of surface temperature was used as a proxy for weather.

The means and 95% CIs for 27 estimates of North Sea abundance of harbor porpoise and the estimates from their statistical model are given in Gilles et al.'s Fig. 8. These data include averages from Hammond et al. (2013) described above. Abundance and density data were obtained from the original publications cited in Gilles et al. and model data from Gilles et al.'s Fig. 8. In the original publications, density was expressed as #/km; it has been rescaled to #/100 km^2, a more useful scale that changes only the position of the plot on V-M domain. Results of TPL analysis are shown in Fig. 10.15 (Appendix 10.V). The rescaled plot has all points above the Poisson line, where the original scale had all points below the Poisson line.

Like Samaniego et al.'s (2012) modeling, Gilles et al.'s model resulted in a slope not different from the observed data. Unlike Samaniego et al.'s model, Gilles et al.'s predicted variances less than observed, although the

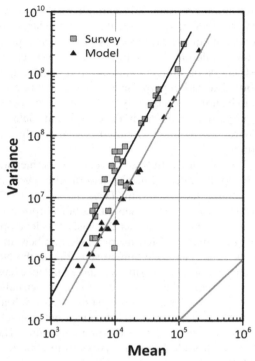

FIG. 10.15 Ensemble TPLs ($NB = 27$) of abundance of the North Sea harbor porpoise population from line transects and statistical model-based estimates are parallel with lower variance of the model-based estimates. *(Data from citations in Fig. 8 in Gilles et al. (2016).)*

estimates intercept are not significantly different ($P > 0.30$). The comparison indicates the statistical model's high fidelity in capturing the most important factors governing the spatial distribution of harbor porpoise in the North Sea.

The TPL analysis of the density estimates, derived from the same data as the abundance estimates, resulted in a substantially lower slope than the abundance estimates. Two outliers in the density plot are responsible for the lower slope: the graph has points, at ~1000 and ~10,000 with variances substantially greater and less than expected. Removing the outliers fixes the discrepancy, bringing the slope close to those computed for the abundance and model estimates (Appendix 10.V). However, it raises the question "why are these points off the line when the abundance data from which they were computed conform?" It is possible the areas these two points represented were misattributed. This comparison illustrates the use of TPL to identify possible data errors or hypothesis exceptions when data points are significantly different from expectation.

FIN WHALES IN THE LIGURIAN SEA

The fin whale or common rorqual (*Balaenoptera physalus*) is a baleen whale growing up to ~26 m: only the blue whale is larger. It is found in all oceans where several allopatric subspecies are recognized. Like the other large whales, the fin whale was a whaling target and is now listed as endangered with <11,000 individuals remaining worldwide. Although commercial whaling by most countries has been halted, two (Japan and Iceland) continue to hunt whales. In addition to the major oceans, fin whales are also found in the Mediterranean Sea where as many as 30% of the planetary population are thought to reside.

A monitoring study started in 2009 used ferries as platforms of opportunity to assess the presence and distribution of fin whales in the Pelagos Sanctuary, a pelagic protected area in the Ligurian Sea in the northwestern Mediterranean between Corsica and the European mainland. Positions of sightings were recorded by GPS and distance to the target by the method of Kinzey and Gerrodette (2003) that estimates distance using the angle between the target and horizon and the height of the observer above the sea which differed between ferry boats. In 306 crossings of almost 42,000 km, 662 sightings were made during the months of June to September 2009 to 2013.

Using these observations, Cominelli et al. (2016) analyzed the data as fixed line-transects by the methods in Buckland et al. (2001, 2007) using Thomas et al.'s (2010) DISTANCE software. In their report, Cominelli et al. compared two analytical methods (linear encounter rate and a density index), but also reported their raw data of mean and SD of density (#/100 km^2) by route, month, and year (potentially 40 points) in a supplemental figure (Fig. S2). Fin whale density along the transects varied between and within years, with peaks in 2012 and 2013 occurring during the first half of the season. Whale density was higher on the route crossing a bathyal plane of 2–2.5 km depth than the other route over more complex topography with sea mounts and canyons.

Mean and SD from 30 nonzero density estimates in Cominelli et al.'s Fig. S2 were rescaled to #/1000 km^2 for TPL analysis (Appendix 10.W). The 30 month-route observations indicate fin whale populations are highly aggregated, at least in the Ligurian sea. However, the mean is greater than the variance at the scale of 1000 km^2 suggesting regularity or Poisson, making the square root an appropriate transformation for analysis. Results of analysis based purely on the variance-mean ratio would likely be in error as TPL analysis shows the full extent of data are aggregated ($b = 2.28 \pm 0.234$), suitable for a log transformation. It is important to determine the dependence of mean on variance over the full range of means to select an appropriate transformation for analysis.

HERDING UNGULATES IN KENYA'S RIFT VALLEY

Research by Schuette et al. (2016) to provide insights in guiding conservation efforts in multiple-use rangelands found that population densities of herding ungulates in Kenya's southern Rift Valley had population densities comparable to those in protected game parks. In their study, Schuette et al. surveyed five species in a landscape used by humans and characterized by bush, woodland, and open grassland. Their objective was to assess the relative influences of bottom-up effects (e.g., vegetation) versus top-down effects from predators such as the African lion (*Panthera leo*) and spotted hyena (*Crocuta crocuta*), and human disturbance including livestock of the seminomadic Maasai herders of the region. The target ungulates were chosen for their range in body size and herd size: small, Grant's gazelle (*Nanger granti*) and impala (*Aepyceros melampus*); medium-sized, zebra (*Equus quagga*) and wildebeest (*Connochaetes taurinus*); and large, giraffe (*Giraffa camelopardalis tippelskirchi*). They concluded that these species coexisted with the Maasai herders and each other by occupying different niches made possible by the spatial and temporal variation in human land use as well as the natural spatial heterogeneity of the landscape.

Schuette et al. established 44 approximately 1.4-km line transects and surveyed them by day and night every 6 weeks during the dry season (13 surveys) and wet season (8 surveys) for a total of 2572 km over a 2 ½-year period. In each survey, a driver and two spotters drove the transect at <15 kph. Animal sightings were recorded by species, herd size and density, position, and distance from the transect. Population density was estimated by methods described in Buckland et al. (2001, 2007). Because a test for bias in herd size detection was not significant for any species, animal densities ($\#/km^{-2}$) were estimated from the product of herd density and herd size.

The density estimates with 95% CIs for the five species separated by wet and dry season and by day and night are given in Schuette et al.'s Tables 2 and 3. TPL analyses by species and herd computed assuming densities are based on 44 transects for impala and giraffe are plotted in Fig. 10.16 (Appendix 10.X).

TPL slopes of zebra do not differ between day and night observations ($P > 0.70$), nor between wet and dry seasons ($P > 0.95$); wet/dry intercepts are not different ($P > 0.20$), although day/night intercepts differ significantly ($P < 0.01$). Differences in A between day and night are probably due to differences in sample efficiency resulting from the difference in visibility between day and night observations. The similarity in slopes of herd and individual observations is due to the application of a single density estimate to all observations. Grant's gazelle herds' $b < 1.0$ while $b > 1.0$ for individuals; neither is significantly different from 1.0, indicating this species is Poisson at all densities. A similar result was obtained for impala, gazelle, and wildebeest. Only for giraffe (Fig. 10.16) was b significantly >1 (individuals, $P < 0.05$; herds, $P < 0.05$). If herds were all the same size, only A would be affected. If detectability depends on herd size, larger herds would have a higher probability of being seen and counted, although with very large herds the number might be underestimated, which would tend to depress b.

These results suggest that the Rift Valley ungulate herds are approximately randomly distributed in space, and the similarity of results for individuals

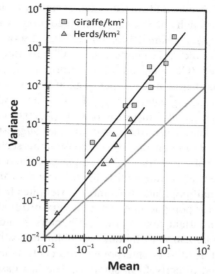

FIG. 10.16   Spatial TPLs ($NQ = 44$, $NB = 8$) of giraffe individuals and herds result in closely similar slopes but different intercepts reflecting their herd sizes. *(Data from Table 3 in Schuette et al. (2016).)*

suggests that herd size variation is small. Contrast this result with the ptarmigan example, which used a similar sampling method to estimate abundance during the period when the birds were actively aggregating into flocks.

KANGAROO RAT MOUNDS IN NEW MEXICO

Banner-tailed kangaroo rats (*Dipodomys spectabilis*) build distinctive mounds at the entrance to their burrows in the arid southwestern USA. They build extensive burrows in the sandy soils of the short-grass prairie of New Mexico, which they occupy by day and from which they forage by night. Active mounds are large (0.5 m high by 5 m in diameter) and mark the center of a clearing in the vegetation; unoccupied mounds erode and are consequently smaller and may be overgrown. Thus, it is easy to distinguish occupied from unoccupied mounds.

Schroder and Geluso (1975) surveyed a 47-ha area of Socorro County where kangaroo rats are locally abundant. They identified and mapped 79 active and 42 inactive mounds. Given that these communal animals maintain their burrows at the center of a foraging area, we might expect the distribution of mounds to be more regular than Poisson. A nearest-neighbor test confirmed the distribution of mounds to be significantly more regular than random ($P < 0.01$). At this scale and overall abundance, TPL's ability to detect departures from Poisson is limited.

TPL analysis of Schroder and Geluso's data is similar to the example of *Arenicola* (Chapter 7) in which quadrat size is changed progressively. Subdividing the area into progressively smaller quadrats from 2.93 ha (16 quadrats) to 521 m^2 (900 quadrats), the variance/mean ratio (CD) for three of the 14 sample sizes is overdispersed (CD > 1) and 11 underdispersed, three significantly. Ensemble TPL of number of mounds per quadrat for the 14 sizes has a slope of $b = 1.052$, significantly greater than $b = 1.0$ ($P < 0.01$) while the intercept, $A = -0.44$ is not significantly different from 0 ($P > 0.65$) (Appendix 10.Y). The conflicting TPL and nearest-neighbor analyses suggest that for practical purposes the distribution of kangaroo rat mounds (at this scale) is indistinguishable from Poisson. Substantially larger sample sizes and scale of sampling would be needed to reveal the likely fairly regular spatial distribution with $b < 1$.

The relatively small populations of most vertebrates have led to development of techniques for population estimation very different from those used for economically important invertebrates (Chapters 7 and 8). Line-transect data combined with bootstrap techniques for obtaining estimates of variation conform to TPL but need to be interpreted with caution, as the derived variances are not derived directly from samples. The last case study highlights another problem of small data sets: it is very difficult to detect nonaggregated distributions with TPL. Even though in principle it is possible to measure $b < 1$, in practice the Poisson line limits the visibility of low variance distributions.

Appendix: TPL estimates for vertebrates abstracted or calculated from the literature

	NB	Range of means	r	A	SE[A]	b	SE[b]
Fish							
A Hybrid ensemble TPLs ($NQ = 4$-45) of standardized trawl catches of haddock and whiting on Georges Bank off Cape Cod, Massachusetts, in 1948–1950. Data from Tables 10 and 11 in C.C. Taylor (1953)							
Haddock	16	4.10 - 288	0.962	−0.293	0.211	2.419	0.177
Whiting	18	13.4 - 376	0.899	−0.412	0.387	2.324	0.255
B Hybrid temporal TPLs ($NQ = 100$) of the stomach contents of herring and mackerel in the Norwegian Sea. Data from Table V in Langøy et al. (2012)							
Herring gut weight (mg)	10	20 - 833	0.899	0.295	0.464	1.798	0.279
Mackerel gut weight (mg)	10	61 - 729	0.865	1.509	0.459	1.403	0.249
Herring gut weight—no outlier	9	20 - 833	0.923	0.828	0.391	1.596	0.232
Mackerel gut weight—no outlier	9	61 - 729	0.952	0.803	0.354	1.650	0.192

Continued

	NB	Range of means	r	A	SE[A]	b	SE[b]
C Spatial TPLs ($NQ = 40$) of salmon off the Pacific Northwest coast before and after salmon enter the sea from their natal rivers. Data from Table 4 in Ruzicka et al. (2016)							
Yearling coho – June	15	24.5 - 254	0.918	−1.264	0.340	2.211	0.244
September	13	2.45 - 95.5	0.941	−0.491	0.161	1.747	0.179
Subyearling chinook – June	15	4.68 - 164	0.949	−1.326	0.241	2.264	0.198
September	14	9.35 - 295	0.902	−0.869	0.382	2.179	0.272
Yearling chinook – June	15	10.9 - 125	0.956	−1.812	0.242	2.401	0.195
September	13	1.22 - 48.6	0.970	−0.678	0.117	2.045	0.149
D Mixed-species TPLs ($NQ = 3$) of demersal fish in an enclosed bay in Brazil. Data from Table 4 in Costa de Azevedo et al. (2007)							
Outer zone	30	0.03 - 68.6	0.975	0.414	0.044	1.678	0.071
Middle zone	29	0.03 - 29.1	0.975	0.375	0.044	1.827	0.078
Inner zone	27	0.14 - 45.8	0.902	0.105	0.103	2.027	0.175
Combined	86	0.03 - 68.6	0.954	0.336	0.035	1.792	0.058
E Ensemble TPLs ($NQ = 11$-30) of pelagic fish larvae caught in transects off the coast of Portugal at the Arrábida Marine Park in May to September 2000. Means and standard deviations from Tables 1 and 2 in Borges et al. (2007)							
Total larvae	22	0.34 - 21.3	0.963	0.197	0.037	1.374	0.083
Larvae inshore (<50 m from reef)	6	0.34 - 21.3	0.995	0.230	0.039	1.257	0.060
Larvae near-shore (∼3.2 km from reef)	5	0.47 - 3.87	0.912	0.103	0.108	1.373	0.324
Larvae offshore (3–16 km from reef)	11	0.41 - 13.2	0.952	0.258	0.055	1.570	0.160
Species inshore (<50 from reef)	57	0.28 - 62.5	0.990	1.637	0.012	1.296	0.024
Species near-shore and offshore	31	0.28 - 14.9	0.984	1.628	0.013	1.240	0.041
F Spatial TPLs ($NQ = 3$ or 7) of pelagic fish larvae and juveniles caught in a New Jersey estuary and near an off-shore ridge in May to November 1992. Means and standard errors digitized from Fig. 5 in Able et al. (2006)							
Estuary – Larve/1,000m^3	19	13.8 - 676	0.876	−1.315	0.344	2.202	0.257
Juveniles/1000 m^2	6	66.5 - 453	0.677	−3.668	1.098	3.286	0.669
Offshore ridge – Larve/1,000m^3	14	10.3 - 103	0.913	−1.832	0.778	2.968	0.630
Juveniles/1000 m^2	10	24.6 - 421	0.914	−1.315	0.344	2.202	0.257
G Spatial TPLs ($NQ =$ variable) of pelagic fish larvae of 3 species caught in North Carolina's Pamlico and Albemarle Sounds on 10 occasions in winter 1992-93. The same data were used to compute the means and variances but combined in different ways according to when, where, and how the samples were taken. Data from Table 2 in Joyeux (1998)							
Menhaden – Flow	6	1.08 - 8.25	0.931	0.850	0.197	1.351	0.246
Light	6	0.040 - 3.99	0.999	0.833	0.377	1.560	0.032
Inlet	4	1.32 - 6.10	0.981	0.470	0.535	2.200	0.299
Month	4	0.592 - 4.64	0.987	0.528	0.521	2.047	0.231

Continued

	NB	Range of means	r	A	SE[A]	b	SE[b]
Croaker – Flow	6	2.24 - 9.00	0.996	0.717	0.438	1.913	0.090
Light	6	0.663 - 5.12	0.996	0.746	0.291	1.889	0.083
Inlet	4	0.028 - 12.2	0.997	0.902	0.649	1.494	0.083
Month	4	0.352 - 16.0	0.999	0.656	0.715	1.896	0.015
Spot – Flow	6	2.27 - 6.06	0.979	0.662	0.368	1.941	0.197
Light	6	0.294 - 4.78	0.992	0.818	0.225	1.569	0.100
Inlet	4	0.380 - 6.13	0.997	0.694	0.482	1.755	0.089
Month	4	0.622 - 6.67	0.996	0.641	0.539	1.901	0.114

H Spatial TPLs of juvenile sea trout fry in Black Brows Beck in England's Lake District monitored for 18 years: $NQ = 30$ samples of alevins in their gravel nests in April, $NQ = 6$ for 1st and 2nd year parr in May/June and August/September, and 3rd year parr sampled in May/June. Intensive sampling of 1st year parr following their exit from the nest. Data from Elliott (1986, 1989)

	NB	Range of means	r	A	SE[A]	b	SE[b]
Alevins – April	18	1.41 - 14.2	0.976	1.084	0.046	1.516	0.083
Yr 1 parr – May/June	18	24.8 - 72.7	0.895	−1.211	0.252	1.830	0.204
Yr 1 parr – August/September	17	10.2 - 21.2	0.954	−1.011	0.111	1.707	0.132
Yr 1 – combined	35	10.2 - 72.7	0.983	−1.050	0.059	1.737	0.056
Yr 2 parr – May/June	18	0.50 - 9.17	0.971	0.237	0.043	1.275	0.076
Yr 2 parr – August/September	17	0.33 - 8.17	0.974	0.289	0.035	1.175	0.069
Yr 2 – combined	35	0.33 - 9.17	0.972	0.264	0.027	1.227	0.050
Yr 3 parr – May/June	15	0.83 - 8.33	0.971	0.727	0.039	1.490	0.098
Before critical time (CT)	16	19.8 - 144	0.947	−0.905	0.100	1.603	0.110
After critical time (CT)	24	10.0 - 29.9	0.979	−0.451	0.102	1.416	0.077
CT – Combined	40	10.0 - 144	0.989	−1.000	0.045	1.694	0.041

I Spatial and temporal TPLs of rod and commercial catches of sea trout in English and Welsh rivers. TPL estimates of ODR from Tables 2 and 3 in Elliott (1992): Only the temporal regression of rod catches in NorthEast England was not significant (*). Data from Tables 2 and 3 in Elliott (1992)

	NB	Range of means	r	A	SE[A]	b	SE[b]
Spatial analysis							
Rod catches – Wales	16	-	0.945	−0.134	0.585	2.151	0.206
NE England	26	-	0.882	−2.365	0.674	2.824	0.286
NW England	14	-	0.912	0.144	0.670	2.008	0.265
SW England	37	-	0.978	−0.333	0.210	2.144	0.078
Wessex	30	-	0.993	−0.181	0.132	2.233	0.053
NW, W, and SW England	67	-	0.954	−0.653	0.235	2.293	0.087
Commercial – Wales	16	-	0.963	0.062	0.443	2.099	0.166
NW England	12	-	0.955	−0.399	0.743	2.309	0.245
SW England	37	-	0.916	−0.783	0.489	2.479	0.176
All regions	65	-	0.951	−0.809	0.272	2.455	0.097
Temporal analysis		-					
Rod catches – Wales	28	-	0.960	0.473	0.242	1.619	0.093
NE England*	5	-	0.733	−1.039	3.448	2.289	1.458
NW England	13	-	0.977	0.510	0.250	1.516	0.109
SW England	17	-	0.981	0.057	0.248	1.799	0.099
Wessex	4	-	0.988	−0.246	1.044	2.036	0.477
All regions	67	13.6 - 4704	0.964	0.269	0.142	1.697	0.057

Continued

	NB	Range of means	r	A	SE[A]	b	SE[b]
Commercial − Wales	16	-	0.975	−0.071	0.269	1.759	0.114
NW England	6	-	0.997	0.205	0.223	1.698	0.095
SW England	10	-	0.994	0.195	0.191	1.725	0.080
All regions	36	10.9 - 29,350	0.990	0.045	0.111	1.750	0.044
Rod + Commercial − All regions	103	10.9 - 29,350	0.977	0.158	0.091	1.729	0.037

Amphibians and reptiles

J Ensemble TPLs (NQ = variable) of reptiles in the Florida Everglades before and after a major freeze in January 2010. Data from Figs. 2–5 in Mazzotti et al. (2016)

	NB	Range of means	r	A	SE[A]	b	SE[b]
American crocodile (#/100 km)	9	24.9 - 41.0	0.919	−3.888	0.641	3.962	0.592
American alligator (#/100 km)	13	25.8 - 132	0.671	0.337	0.383	1.159	0.260
Minus outliers	10	28.6 - 132	0.938	0.158	0.260	1.228	0.135
Burmese python (#/100 km)	5	0.015 - 0.22	0.989	−0.783	0.104	1.442	0.122
Cottonmouth (#/100 km)	7	0.94 - 1.84	0.765	−1.048	0.056	1.813	0.522
Southern water snake (#100/km)	7	0.55 - 1.95	0.938	−0.846	0.049	2.553	0.397
Green water snake (#/100 km)	7	0.19 - 0.92	0.992	−0.674	0.039	1.847	0.107
Eastern ribbon snake (#/100 km)	7	0.27 - 1.29	0.831	−0.946	0.053	1.072	0.266
Common garter snake (#/100 km)	7	0.42 - 1.45	0.607	−0.764	0.174	2.134	0.759

K Mixed-species ensemble TPLs of herptiles at 3 elevations in the Rincón Mountains, Arizona, and ensemble TPLs of 2 lizards. Data from Tables 1 and 2 in Flesch et al. (2010)

	NB	Range of means	r	A	SE[A]	b	SE[b]
Low elevations ($NQ = 50$)	37	0.02 - 10.0	0.980	0.967	0.031	1.584	0.053
Middle elevations ($NQ = 23$)	18	0.03 - 11.2	0.971	0.843	0.042	1.561	0.094
High elevations ($NQ = 12$)	12	0.10 - 6.20	0.986	0.770	0.038	1.667	0.089
Common regression ($NQ = 85$)	67	0.02 - 11.2	0.976	0.894	0.023	1.581	0.043
All elevations combined	40	0.01 - 8.96	0.984	1.086	0.030	1.596	0.047
All individuals	8	2.00 - 53.7	0.969	−0.896	0.193	2.510	0.253
Sceloporus clarkii	8	0.07 - 4.59	0.925	0.272	0.108	1.729	0.269
Urosaurus ornatus	8	0.63 - 10.5	0.946	−0.303	0.144	2.696	0.357

L Spatial TPLs (NQ = 2-7) of yellow-legged frogs in alpine ponds and streams in Kings Canyon National Park, California. Data from Fig. 2 in Pope and Matthews (2001)

	NB	Range of means	r	A	SE[A]	b	SE[b]
Yellow-legged frogs − combined	44	0.33 - 172	0.911	0.171	0.066	1.512	0.096
Lake and ponds	33	0.33 - 172	0.919	0.067	0.115	1.553	0.110
Streams	11	0.33 - 6.23	0.849	0.370	0.136	1.653	0.291

Continued

	NB	Range of means	r	A	SE[A]	b	SE[b]
M Spatial TPLs ($NQ = 17$-33) of spring peepers and American toads in 2 ephemeral ponds in Ohio. Data from Table 2 in Smith et al. (2003)							
Spring peeper	5	26.0 - 176	0.922	0.651	0.539	1.845	0.413
American toad	5	0.38 - 58.4	0.991	1.173	0.136	1.726	0.137
N Mixed-species temporal TPLs ($NQ = 125$) of presumed reptile prey of the short-toed eagle in Dadia Forest in Greece. Data from Table 1 in Bakaloudis et al. (1998)							
Species – Lizards	8	0.60 - 26.9	0.938	0.0814	0.1555	1.9241	0.272
Snakes	7	0.58 - 6.76	0.951	−0.3651	0.1512	3.1698	0.439
Habitat – Lizards	9	1.61 - 19.2	0.961	0.4827	0.1077	1.7469	0.181
Snakes	9	0.24 - 11.0	0.967	0.1824	0.0629	2.2000	0.213

Birds

	NB	Range of means	r	A	SE[A]	b	SE[b]
O Hybrid temporal TPL ($NQ = 5$-10) of willow ptarmigan in the mountains of southern Norway. Data from the Supplement to Kvasnes et al. (2015)							
Willow ptarmigan	21	2.77 - 17.8	0.916	−1.483	0.174	3.090	0.285
P Temporal TPLs ($NQ > 10$) of field data and model results for Piñon and Western scrub jays. Data from Figs. 3 and 8 in Samaniego et al. (2012)							
Piñon jay – field data	111	0.29 - 51.0	0.948	0.888	0.029	1.771	0.054
Probability model	111	0.12 - 33.3	0.875	1.789	0.038	1.662	0.077
Western scrub-jay – field data	256	0.13 - 60.3	0.966	0.521	0.013	1.492	0.024
Probability model	256	0.05 - 71.3	0.878	1.031	0.024	1.330	0.040
Q TPL of Breeding Bird Survey data of 9 species of sparrow in North America observed or heard on survey routes with continuous data from 1980–1989. Slopes and SEs have been converted from ODR of standard deviation on mean to GMR of variance on mean regressions. Column 3 gives the number of routes the birds were detected as a proxy for abundance. Data from Table 2 in Curnutt et al. (1996)							
Chestnut-collared sparrow	11	150	0.953	-	-	1.554	0.156
Discissel	140	764	0.896	-	-	1.547	0.058
Grasshopper sparrow	163	1393	0.822	-	-	1.633	0.072
Lark bunting	20	335	0.966	-	-	1.582	0.096
Lark sparrow	84	878	0.895	-	-	1.678	0.082
LeConte's sparrow	9	147	0.948	-	-	1.526	0.182
McCown's longspur	5	63	0.954	-	-	1.498	0.224
Savannah sparrow	191	1461	0.863	-	-	1.491	0.054
Vesper sparrow	163	1451	0.882	-	-	1.492	0.054
R TPL of relative abundance of bird species recorded by CBC in $NQ = 42$ US cities and adjacent nonurban areas. The higher intercept of the nonurban regression indicates decreased variability in cities during winter. Data courtesy of Adeline Murthy were originally published as Fig. 4 in Murthy et al. (2016)							
Urban	378	6×10^{-8} - 0.147	0.991	0.126	0.039	1.798	0.012
Nonurban	378	1×10^{-7} - 0.103	0.990	0.355	0.039	1.805	0.013

Continued

	NB	Range of means	r	A	SE[A]	b	SE[b]
Mammals							
S Mixed-species ensemble TPL ($NQ > 43$) of cetaceans observed in the vicinity of the Azores. Data from Table 2 in Silva et al. (2014)							
Atlantic whales	14	1.7 - 43.3	0.979	−0.946	0.112	2.663	0.158
T Spatial TPLs (bootstrapped) of cetaceans in the North Sea and Celtic Sea. Data from Tables 4-6 in Hammond et al. (2002)							
Harbor porpoise (surface) — Abundance	8	4211 - 92,340	0.931	−2.239	1.088	2.280	0.339
Density (#/100 km²)	8	9.50 - 77.6	0.849	3.070	0.479	1.957	0.422
Harbor porpoise (aerial) — Abundance	6	588 - 24,335	0.973	−0.639	0.599	1.917	0.223
Density (#/100 km²)	6	10.1 - 81.2	0.933	3.552	0.375	1.700	0.305
White-beaked dolphin — Abundance	5	115 - 2443	0.991	1.092	0.253	1.488	0.116
Density (#/100 km²)	5	0.11 - 5.38	0.987	3.570	0.060	1.595	0.149
All dolphins — Abundance	6	116 - 4063	0.961	1.006	0.472	1.531	0.210
Density (#/100 km²)	6	0.11 - 9.29	0.975	3.584	0.077	1.597	0.176
Minke whale — Abundance	7	49.0 - 2920	0.978	0.436	0.320	1.636	0.153
Density (#/100 km²)	7	0.1 - 2.86	0.967	3.350	0.059	1.596	0.183
U Spatial TPLs (bootstrapped) of cetaceans observed around the British Isles. Data from Tables 4–8 in Hammond et al. (2013)							
Harbor porpoise — Abundance	16	1473 - 93,938	0.951	−0.085	0.452	1.846	0.153
Density (#/100 km²)	16	1.70 - 59.8	0.929	3.997	0.150	1.478	0.147
Minke whale — Abundance	9	833 - 4515	0.795	0.868	0.916	1.634	0.375
Density (#/100 km²)	9	0.90 - 5.70	0.915	3.498	0.106	2.641	0.403
White-beaked dolphin — Abundance	8	273 - 7557	0.991	0.672	0.208	1.695	0.095
Density (#/100 km²)	8	0.30 - 10.5	0.982	3.767	0.051	1.834	0.141
Common dolphin — Abundance	8	392 - 18,039	0.945	1.067	0.540	1.571	0.209
Density (#/100 km²)	8	1.20 - 30.2	0.895	3.751	0.185	1.697	0.309
Bottlenose dolphin — Abundance	11	151 - 7665	0.993	0.569	0.131	1.721	0.066
Density (#/100 km²)	11	0.10 - 3.88	0.988	3.763	0.033	1.727	0.089
V Ensemble TPLs (bootstrapped) of abundance and model predictions of harbor porpoise abundance in the North Sea. The slopes and correlations are in good agreement, but the model's predicted variance is half the variance of observed abundance, although the intercepts are not significantly different. Data from Gilles et al. (2016) and references therein							
Harbor porpoise density (#/100 km²)	27	12.5 - 299	0.897	−0.267	0.213	1.674	0.148
Abundance — Survey	27	971 - 116,500	0.924	−0.538	0.450	1.972	0.151
Abundance — Model	27	2616 - 200,000	0.989	−0.841	0.166	1.914	0.057

Continued

	NB	Range of means	r	A	SE[A]	b	SE[b]
W Spatial TPL (*NQ* variable) of fin whales observed from ferries crossing the Pelagos Sanctuary in the Ligurian Sea between Corsica and the European mainland in Summers 2009–2013. Data digitized from Fig. S2 in Cominelli et al. (2016)							
Finwhale (#/1000 km^2)	30	2.18 - 22.0	0.827	−1.287	0.178	2.277	0.234
X Spatial TPLs (*NQ* = 44) of large African ungulate herds observed on transects Kenya's southern Rift Valley. Variances estimated by bootsrapping. Data from Table 2 in Schuette et al. (2016)							
Giraffe herds/km^2	8	0.02 - 1.35	0.968	0.818	0.070	1.315	0.135
Giraffe/km^2	8	0.15 - 16.2	0.964	1.442	0.076	1.337	0.144
Grant's gazelle herds/km^2	8	0.06 - 0.71	0.867	0.068	0.114	1.108	0.225
Grant's Gazelle/km^2	8	0.30 - 5.56	0.961	0.684	0.045	1.237	0.139
Impala herds/km^2	8	0.06 - 1.40	0.948	0.108	0.083	1.376	0.179
Impala/km^2	8	0.38 - 13.9	0.967	0.783	0.066	1.236	0.128
Wildebeest herds/km^2	8	0.61 - 3.35	0.897	0.531	0.072	1.629	0.294
Wildebeest/km^2	8	2.00 - 12.2	0.920	0.783	0.157	1.583	0.253
Zebra herds/km^2 (day/night)	8	0.02 - 1.45	0.971	0.982	0.077	1.312	0.129
Zebra herds/km^2 (wet/dry)	8	0.12 - 0.50	0.828	0.404	0.175	1.559	0.357
Zebra/km^2 (day/night)	8	0.28 - 9.07	0.934	1.497	0.063	1.276	0.186
Zebra/km^2 (wet/dry)	8	0.28 - 1.38	0.880	0.597	0.062	1.478	0.286
Y Kangaroo rat mounds in a semiarid habitat in New Mexico. TPL conducted on a series of quadrats ranging in size from 2.93 ha to 521 m^2 (*NQ* = 16-900). Occupied, abandoned, and all mounds. Data from Table 1 in Schroder and Geluso (1975)							
Kangaroo rat mounds/ quadrat	14	0.134 - 7.56	0.999	-0.004	0.006	1.053	0.015
Active mounds only	14	0.088 - 4.94	0.999	0.006	0.007	1.053	0.015
Abandoned mounds only	14	0.047 - 2.63	0.999	0.020	0.010	1.053	0.015

References

Able, K.W., Fahay, M.P., Witting, D.A., McBride, R.S., Hagan, S.M., 2006. Fish settlement in the ocean vs. estuary: comparison of pelagic larval and settled juvenile composition and abundance from southern New Jersey, U.S.A. Estuar. Coast. Shelf Sci. 66, 280–290.

Bakaloudis, D.E., Vlachos, C.G., Holloway, G.J., 1998. Habitat use by short-toed eagles *Circaetus gallicus* and their reptilian prey during the breeding season in Dadia Forest (north-eastern Greece). J. Appl. Ecol. 35, 821–828.

Ballantyne, F., 2005. The upper limit for the exponent of Taylor's power law is a consequence of deterministic population growth. Evol. Ecol. Res. 7, 1213–1220.

Barnes, H., 1949. On the volume measurement of water filtered by a plankton pump, with some observations on the distribution of planktonic animals. J. Mar. Biol. Assoc. NS 28, 651.

Blackburn, T.M., Gaston, K.J., Quinn, R.M., Gregory, R.D., 1999. 3 Do local abundances of British birds change with proximity to range edge? J. Biogeogr. 26, 493–505.

Borges, R., Ben-Hamadou, R., Alexandra Chícharo, M., Rí, P., Gonçalves, G.J., 2007. Horizontal spatial and temporal distribution patterns of nearshore larval fish assemblages at a temperate rocky shore. Estuar. Coast. Shelf Sci. 71, 412–428.

Buckland, S.T., Anderson, D.R., Burnham, K.P., Laake, J.L., Borchers, D.L., Thomas, L., 2001. Advanced Distance Sampling: Estimating Abundance of Biological Populations. Oxford University Press, Oxford.

Buckland, S.T., Anderson, D.R., Burnham, K.P., Laake, J.L., Borchers, D.L., Thomas, L. (Eds.), 2007. Introduction to Distance Sampling: Estimating Abundance of Biological Populations. Oxford University Press, Oxford.

Cominelli, S., Moulins, A., Rosso, M., Tepsich, P., 2016. Fin whale seasonal trends in the Pelagos Sanctuary, Mediterranean Sea. J. Wildl. Manag. 80, 490–499.

Costa de Azevedo, M.C., Araújo, F.G., Gomes da Cruz-Filho, A., Pessanha, A.L.M., Silva, M.A., Guedes, A.P.P., 2007. Demersal fishes in a tropical bay in southeastern Brazil: partitioning the spatial, temporal and environmental components of ecological variation. Estuar. Coast. Shelf Sci. 75, 468–480.

Curnutt, J.L., Pimm, S.L., Maurer, B.A., 1996. Population variability of sparrows in space and time. Oikos 76, 131–140.

Elliott, J.M., 1984. Numerical changes and population regulation in young migratory trout *Salmo trutta* in a Lake District stream, 1966–83. J. Anim. Ecol. 53, 327–350.

Elliott, J.M., 1986. Spatial distribution and behavioural movements of migratory trout *Salmo trutta* in a Lake District stream. J. Anim. Ecol. 55, 907–922.

Elliott, J.M., 1989. Mechanisms responsible for population regulation in young migratory trout, *Salmo trutta*. I. The critical time for survival. J. Anim. Ecol. 58, 987–1001.

Elliott, J.M., 1992. Variation in the population density of adult sea-trout, *Salmo trutta*, in 67 rivers in England and Wales. Ecol. Freshw. Fish 1, 5–11.

Flesch, A.D., Swann, D.E., Turner, D.S., Powell, B.F., 2010. Herpetofauna of the Rincón Mountains, Arizona. Southwest. Nat. 55, 240–253.

Fretwell, S.D., Lucas, H.L., 1969. On territorial behavior and other factors influencing habitat distribution in birds. I. Theoretical development. Acta Biotheor. 19, 16–36.

Gilles, A., Viquerat, S., Becker, A., et al., 2016. Seasonal habitat-based density models for a marine top predator, the harbor porpoise, in a dynamic environment. Ecosphere 7(6).

Gillis, D.M., Kramer, D.L., Bell, G., 1986. Taylor power law as a consequence of Fretwell ideal free distribution. J. Theor. Biol. 123, 281–287.

Good, P.I., 2010. Permutation, Parametric, and Bootstrap Tests of Hypotheses. Springer-Verlag, New York.

Hammond, P.S., Berggren, P., Benke, H., Borchers, D.L., Collett, A., Heide-Jørgensen, M.P., Heimlich, S., Hiby, A.R., Leopold, M.F., Øien, N., 2002. Abundance of harbour porpoise and other cetaceans in the North Sea and adjacent waters. J. Appl. Ecol. 39, 361–376.

Hammond, P.S., Macleod, K., Berggren, P., et al., 2013. Cetacean abundance and distribution in European Atlantic shelf waters to inform conservation and management. Biol. Conserv. 164, 107–122.

Henderson, P.A., Magurran, A.E., 2014. Direct evidence that density-dependent regulation underpins the temporal stability of abundant species in a diverse animal community. Proc. R. Soc. B 281, 20141336.

Hsieh, C.-H., Reiss, C.S., Hunter, J.R., Beddington, J.R., May, R.M., Sugihara, G., 2006. Fishing elevates variability in the abundance of exploited species. Nature 443, 859–862.

Joyeux, J.-C., 1998. Spatial and temporal entry patterns of fish larvae into North Carolina estuaries: comparisons among one pelagic and two demersal species. Estuar. Coast. Shelf Sci. 47, 731–752.

Kilpatrick, A.M., Ives, A.R., 2003. Species interactions can explain Taylor's power law for ecological time series. Nature 422, 65–68.

Kinzey, D., Gerrodette, T., 2003. Distance measurements using binoculars from ships at sea: accuracy, precision and effects of refraction. J. Cetacean Res. Manag. 5, 159–171.

Kuo, T.-C., Mandal, S., Yamauchi, A., Hsieh, C.-H., 2016. Life history traits and exploitation affect the spatial mean-variance relationship in fish abundance. Ecology 97, 1251–1259.

Kvasnes, M.A.J., Pedersen, H.C., Solvang, H., Storaas, T., Nilsen, E.B., 2015. Spatial distribution and settlement strategies in willow ptarmigan. Popul. Ecol. 57, 151–161.

Langøy, H., Nøttestad, L., Skaret, G., Broms, C., Ferno, A., 2012. Overlap in distribution and diets of Atlantic mackerel (*Scomber scombrus*), Norwegian spring-spawning herring (*Clupea harengus*) and blue whiting (*Micromesistius poutassou*) in the Norwegian Sea during late summer. Mar. Biol. Res. 8, 442–460.

Mazzotti, F.J., Cherkiss, M.S., Parry, M., Beauchamp, J., Rochford, M., Smith, B., Hart, K., Brandt, L.A., 2016. Large reptiles and cold temperatures: Do extreme cold spells set distributional limits for tropical reptiles in Florida? Ecosphere 7 (8), e01439.

Moriasi, D.N., Arnold, J.G., van Liew, M.W., Bingner, R.L., Harmel, R.D., Veith, T.L., 2007. Model evaluation guidelines for systematic quantification of accuracy in watershed simulations. Trans. ASABE 50, 885–900.

Murthy, A.C., Fristoe, T.S., Burger, J.R., 2016. Homogenizing effects of cities on North American winter bird diversity. Ecosphere 7 (1), e01216.

Norris, J.R., 1998. Markov Chains. Cambridge University Press, Cambridge.

Park, S.-J., Taylor, R.A.J., Grewal, P.S., 2013. Spatial organization of soil nematode communities in urban landscapes: Taylor's power law reveals life strategy characteristics. Appl. Soil Ecol. 64, 214–222.

Pedersen, H.C., Steen, H., Kastdalen, L., Brøseth, H., Ims, R.A., Svendsen, W., Yoccoz, N.G., 2004. Weak compensation of harvest despite strong density-dependent growth in willow ptarmigan. Proc. R. Soc. B Biol. Sci. 271, 381–385.

Pope, K.L., Matthews, K.R., 2001. Movement ecology and seasonal distribution of mountain yellow-legged frogs, *Rana muscosa*, in a high-elevation Sierra Nevada basin. Copia 2001, 787–793.

Preston, F.W., 1979. The invisible birds. Ecology 60, 451–454.

Robbins, C.S., Bystrak, D., Geissler, P.H., 1986. The Breeding Bird Survey: Its First Fifteen Years, 1965–1979. U.S. Fish & Wildlife Service Resource Pub 157, USGS, Interior Department, Washington, DC.

Ruzicka, J.J., Daly, E.A., Brodeur, R.D., 2016. Evidence that summer jellyfish blooms impact Pacific Northwest salmon production. Ecosphere 7 (4), e01324.

Samaniego, H., Sérandour, G., Milne, B.T., 2012. Analyzing Taylor's scaling law: qualitative differences of social and territorial behavior on colonization/extinction dynamics. Popul. Ecol. 54, 213–223.

Schneider, A., Friedl, M.A., Potere, D., 2010. Monitoring urban areas globally using MODIS 500-m data: new methods and datasets based on 'urban ecoregions'. Remote Sens. Environ. 114, 1733–1746.

Schroder, G.D., Geluso, K.N., 1975. Spatial distribution of *Dipodomys spectabilis* mounds. J. Mammal. 56, 363–368.

Schuette, P., Creel, S., Christianson, D., 2016. Ungulate distributions in a rangeland with competitors, predators and pastoralists. J. Appl. Ecol. 53, 1066–1077.

Silva, M.A., Prieto, T., Cascéo, I., Seabra, M.I., Machete, M., Baumgartner, M.F., Santos, R.S., 2014. Spatial and temporal distribution of cetaceans in the mid-Atlantic waters around the Azores. Mar. Biol. Res. 10, 123–137.

Smith, G.R., Dingfelder, H.A., Vaala, D.A., 2003. Distribution and abundance of amphibian larvae within two temporary ponds in central Ohio, USA. J. Freshw. Ecol. 18, 491–496.

Taylor, C.C., 1953. Nature of variability in trawl catches. U.S. Fish Wildl. Serv. Fish. Bull. 29, 158–166.

Taylor, L.R., Woiwod, I.P., 1982. Comparative synoptic dynamics. I. Relationships between inter- and intra-specific spatial and temporal variance/mean population parameters. J. Anim. Ecol. 51, 879–906.

Taylor, L.R., Woiwod, I.P., Perry, J.N., 1978. The density-dependence of spatial behaviour and the rarity of randomness. J. Anim. Ecol. 47, 383–406.

Taylor, L.R., Woiwod, I.P., Perry, J.N., 1980. Variance and the large scale spatial stability of aphids, moths and birds. J. Anim. Ecol. 49, 831–854.

Taylor, R.A.J., 1978. The relationship between density and distance of dispersing insects. Ecol. Entomol. 3, 63–70.

Taylor, R.A.J., 1980. A family of regression equations describing the density distribution of dispersing organisms. Nature 286, 53–55.

Taylor, R.A.J., 1987. On the accuracy of insecticide efficacy reports. Environ. Entomol. 16, 1–8.

Taylor, R.A.J., 2018. Spatial distribution, sampling efficiency and Taylor's power law. Ecol. Entomol. 43, 215–225.

Taylor, R.A.J., Park, S.-J., Grewal, P.S., 2017. Nematode spatial distribution and the frequency of zeros in samples. Nematology 19, 263–270.

Thomas, L., Buckland, S.T., Rexstad, E.A., Laake, J.L., Strindberg, S., Hedley, S.L., Bishop, J.R.B., Marques, T.A., Burnham, K.P., 2010. Distance software: design and analysis of distance sampling surveys for estimating population size. J. Appl. Ecol. 47, 5–14.

Trumble, J.T., 1985. Implications of changes in arthropod distribution following chemical application. Res. Popul. Ecol. 27, 277–285.

Winsor, C.P., Walford, L.A., 1936. Sampling variations in the use of plankton nets. J. Cons. Int. Explor. Mer 11, 190–204.

## Chapter 11

Other biological examples

In this chapter, we examine TPL analyses of nontaxonomic biological examples and data of human demographics and social behavior. Most of the data reported were abstracted from the literature, although several sources published TPL analyses, and a new analysis of the United States' decennial censuses from 1790 to 2010 is presented.

General biology

RATE OF EVOLUTION

Recent studies of phenotypic change under natural as opposed to human selection have shown that contrary to expectation, evolution can occur quite quickly. The decades-long research by Rosemary and Peter Grant in the Galapagos, described with other examples by Weiner (1995), has documented sometimes profound phenotypic changes in Darwin's finches in response to changing environmental conditions driven by weather variations. Examples of heritable phenotypic changes in populations in conjunction with changes in population abundance were found by DeLong et al. (2016) to be well described by TPL. They developed a database of population and phenotypic rates of change in 6 microorganisms and 11 vertebrates. The original data obtained from both field and laboratory studies were of body size dimension such as volume, mass, and length. DeLong et al. refer to the rates of change in traits as rates of phenotypic change rather than evolutionary change because it is not always clear to what degree the changes were genetic rather than arising from phenotypic plasticity. Ecological rates were based on changes in the population abundance, density, or other indicator of population size such as number of nests.

DeLong et al. calculated means and variances of absolute rates of change over the period of each study. Several studies contributed more than one trait, resulting in a total of 21 points in their TPL analysis. They used ODR to fit TPL, but their Fig. 3 and supplementary Table S1 provide sufficient information to recompute the TPLs using GMR (Fig. 11.1, Appendix 11.A). Besides increasing the estimated slopes very slightly, the estimated intercepts are also included.

Taylor's Power Law. https://doi.org/10.1016/B978-0-12-810987-8.00011-2

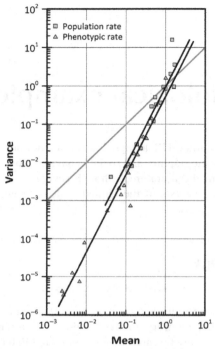

FIG. 11.1   Temporal TPLs of rates of change per unit of phenotypes ($NB = 20$) and populations ($NB = 19$) have slopes $b \sim 2$ and not significantly different ($P > 0.75$) with the phenotypic variances significantly less than the population variances ($P > 0.20$). Each point is a different species. *(Data from Fig. 3 and Supplemental Table S1 in DeLong et al. (2016).)*

Heritable changes are the stuff of evolution. DeLong et al's analysis draws a direct connection between evolution and TPL. They relate the mean and variance of rates of phenotypic change and population abundance via.

$$\frac{E^2\left[\dfrac{1}{z}\dfrac{dz}{dt}\right]}{E^2\left[\dfrac{1}{N}\dfrac{dN}{dt}\right]} = \frac{V\left[\dfrac{1}{z}\dfrac{dz}{dt}\right]}{V\left[\dfrac{1}{N}\dfrac{dN}{dt}\right]} \tag{11.1}$$

where $E[\cdot]$ and $V[\cdot]$ are the expectation and variance, respectively, z is the phenotypic metric, and N is population abundance. The examples found by DeLong et al. are all of fairly rapid phenotypic change, exemplified by the Grant's Galapagos finch studies (Grant and Grant, 2014), averaging about 25% the population rate of change. Thus, $dN/dt > dz/dt$, the ratio of variances is <1, and the phenotypic TPL's intercept is less than the population's as shown in Fig. 11.1.

The most important result in this study may be that at least some of the time ecological and evolutionary change can occur on similar time scales, although the results clearly show populations change to be a minimum of 3–4 times faster than traits. The fact that both phenotypic and population rates of change obey TPL is intriguing but more interesting is the phenotypic rate of change, while slower than the population change, is less variable at any given population density. This may be due to more constraints on phenotypic change than population change. Like Murthy et al.'s (2016) analysis of relative abundance of birds in two habitats (Chapter 10), most of the points are below the Poisson line. As these data are of rates of change and not density, the restrictions on the position of the plot in the variance-mean domain do not apply.

EXOENZYMES IN SOIL

Very often, it is not practical to assess populations or their properties directly. This is especially problematical with microorganisms that are hard to reliably enumerate because they are difficult to see and/or so numerous as to be difficult to count. The following examples used chemical traces to document population and spatial dynamics of microorganisms in the environment. They are included here because the variables subjected to TPL are essentially chemical and the taxonomy of the organisms generating the proxies is often uncertain.

Exoenzymes are enzymes secreted by cells to catalyze chemical reactions extracellularly to breakdown macromolecules into smaller molecules that can be absorbed by the cell. Many bacteria secrete exoenzymes, some of which are used in environmental bioremediation. Fungi also secrete exoenzymes useful for bioremediation. Of the many organisms used for bioremediation, those that produce hydrolases are particularly useful.

Kim et al. (2015)measured the activity of six exoenzymes in three soils from fields in France: clay from a conventionally managed field and loam from low input and conventionally managed field plots. Soil cores taken from the three fields were air-dried and sieved into 5 size classes: 0.25, 0.5, 1, 2, and 3 mm diameter. Replicated samples of each size class and soil type were incubated and the activity of the enzymes was monitored for 9 h. The mean and variance were computed from 48 replicates for each enzyme, soil size class, and soil type and an ensemble TPL analysis conducted for each enzyme. Because enzyme activity was not detected in some of the samples, Kim et al. conducted a sensitivity analysis to determine if enzyme activity below the detection threshold affected the calculation of b. Exclusion of zeros can raise the estimate of b (Chapter 15), but Kim et al. concluded there was no discernible effect. They found that enzyme activity increased with increasing soil size class ($\equiv$ quadrats of unequal size) and was generally highest in the clay soil and lowest in the low-input loam. They reported ensemble TPLs with $r > 0.93$, although 2 of 18 regressions were not significant at $P > 0.05$. The slopes ranged as widely between soil types as between enzymes ($0.95 < b < 1.67$; Fig. 11.2). The example of β-D-glucosidase in their Fig. 1 was reanalyzed by GMR (Appendix 11.B).

The wide range of TPL slopes of enzyme activity among the six exoenzymes suggests that different spatial strategies may be used by members of the microbial community, as Park et al. (2013) found with nematodes (Chapter 7). Enzymes associated with phosphorus, carbon, and nitrogen metabolism showed different degrees of aggregation. In particular, alkaline phosphatase ($1.09 < b < 1.27$) had comparatively low level of aggregation and a widespread distribution, suggesting the importance of phosphorus for microorganisms. Cellobiohydrolase, associated with degradation of cellulose, which is ubiquitous in soil, was only moderately aggregated in all three soils ($1.29 < b < 1.42$). By comparison, the more aggregated xylosidase used in the carbon cycle was comparatively highly aggregated ($1.47 < b < 1.67$), suggesting it is important for a subset of the community. Comparison of the overall enzyme activity in the three soils reveals activity more aggregated in the conventionally managed clay than the conventionally managed loam, but there was no difference between the conventionally managed and the low-input loams. Higher TPL slopes of enzymes

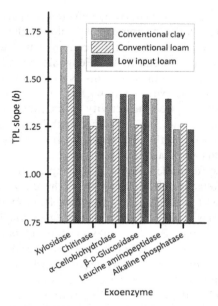

FIG. 11.2 Exoenzymes in low-input and conventionally managed soils: estimates of b for 6 exeoenzymes in the 3 soils range $0.95 < b < 1.67$. *(Data from Figs. 1 and 2 in Kim et al. (2015).)*

associated with the carbon and nitrogen cycles in the low-input soil than the conventionally managed soil were possibly due to differences in the distribution of inputs and/or to differences in organic matter quality.

METABOLISM IN A RIVER SYSTEM

The majority of primary producers in water bodies as elsewhere use chlorophyll to fix carbon dioxide. Chlorophyll-*b*, a yellow pigment, uses photons of blue light to store energy as ATP to drive the sugar production cycle catalyzed by hlorophyll-*a*, which uses red wavelengths. The concentration of chlorophyll-*a* in water (both freshwater and oceanic) is used as a proxy for the amount of phytoplankton and therefore primary production in water bodies. Because the movement of water in rivers and streams depends critically on their channel geomorphology and hydrology, observations are highly site specific making generalizations difficult.

Naiman (1983) described a study to establish if the activity of metabolic components occurred in a predictable manner so that combined with geomorphological data of stream number, size and length the productivity of an entire watershed may be estimated. The Moisie River watershed in Quebec occupies almost $20{,}000\,km^2$, consists of $\sim\!200\,km^2$ surface water, contains $>\!38{,}000$ headwater streams, typically a few tens of hectares in extent, and $>\!10{,}000$ rivers of intermediate size. Among other parameters, Naiman estimated the concentration of chlorophyll-*a* associated with mosses, periphyton (algae, cyanobacteria, and heterotrophic microbes) and detritus in five representative rivers. They are First Choice Creek (1st order), Beaver Creek (2nd order), Muskrat River (5th order), Matamek River (6th order), and the Moisie River (9th order).

Chlorophyll concentration was estimated from 15 samples taken and incubated in each river in rotation, one river per week, sampling each river 6 times between April and November 1979. Chlorophyll concentration in samples was measured in the laboratory following 24-h incubation. Naiman's Table 3 gives the mean and standard error of concentration (mg/m^2) of chlorophyll for mosses, periphyton, and detritus in each river for six sampling periods. The total chlorophyll increased from $<\!1\,mg/m^2$ early in the season to $>\!60\,mg/m^2$ at peak season in the Matamek and Moisie Rivers. The presence of mosses only in the higher order rivers contributed to their higher production. Ensemble TPL ($NB = 15$–29) of chlorophyll concentration shows it to be strongly aggregated over 2.5 orders of magnitude (Fig. 11.3; Appendix 11.C). The slopes for detritus ($b = 2.30 \pm 0.134$) and periphyton ($b = 1.74 \pm 0.186$) are significantly different ($P < 0.05$), but the slope for moss ($b = 2.08 \pm 0.363$) is not significantly different from either detritus or periphyton ($P > 0.40$). As the concentration of chlorophyll increased with size of river, the strong dependence of the variance on the mean implies increasing variability in chlorophyll and therefore primary production with descent down the watershed.

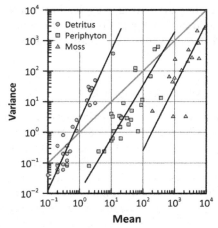

FIG. 11.3   Ensemble TPL of intraseasonal concentration of chlorophyll-*a* originating from detritus (*NB* = 29), periphyton (*NB* = 28), and moss (*NB* = 15) sampled in 5 Canadian creeks and rivers on 6 sampling occasions (*NQ* = 15–30). Plots shifted right by one cycle for clarity. *(Data from Table 3 in Naiman (1983).)*

PHOSPHORUS IN LAKES

Like chlorophyll-*a,* phosphorus, a nutrient essential for plant growth, is an indicator of primary production in water bodies. In pristine lakes, phosphorus is often growth limiting when sunlight, heat, water, and nitrogen are abundant. However, agricultural activities can deliver phosphorus in excess to lakes, such as Lake Erie, causing severe and sometimes dangerous algal blooms (Wang et al., 2018). Other sources of phosphorus include soil erosion, detergents, septic systems, and animal manure (Wang et al., 2019). The phosphorus in lakes includes the phosphorus in plant and animal fragments suspended in the water column and dissolved phosphate used by plants. Dissolved phosphorus fluctuates daily, while the other sources are more stable. Because it promotes algal growth and eutrophication, phosphorus is used as a major indicator of water quality. Sampling programs for phosphorus in lakes are routinely used by agencies charged with protecting the quality and quantity of surface waters, especially those used for human consumption.

Contradictory results in the literature led France and Peters (1992) to use TPL in a comparison of the temporal stability of total phosphorus and chlorophyll-*a*. Using published data, they computed mean and variance of total phosphorus and chlorophyll at *NQ* = 7 sample points between April and October in *NB* = 65 north temperate lake-years. France & Peters used ODR to fit an intra-annual temporal TPL to total phosphorus and chlorophyll-*a*. TPL analysis by GMR using information in the text, and their Fig. 1 shows total phosphorus to be aggregated (Appendix 11.D). Data of total phosphorus from another experiment (Marshall et al., 1988) fit into the cloud of points resulting in a slightly steeper TPL. The estimate of $b = 2.11 \pm 0.055$ for chlorophyll-a is very close to the estimates obtained for mosses and detritus in Naiman's study.

With its higher b, France and Peters results show greater seasonal variability of chlorophyll-a than total phosphorus, and the lower overall variability (higher r) suggests that factors determining chlorophyll-a concentration also control its seasonal variation. The divergence of slopes suggests that total phosphorus and chlorophyll-a may be coupled only weakly under eutrophic conditions. The main point, however, is the greater precision of the chlorophyll-a estimates suggests it may be the preferred statistic for heterotrophic abundance, but the steep TPL also suggests the need for a large number of samples; they recommend $NQ \geq 7$ compared with Clark and Perry's (1994) recommendation of $NQ \geq 15$ (Chapter 4).

Physiology

METASTATIC CANCERS

Noting that the numbers of metastatic cancer tumors per individual appear to be aggregated, Kendal and Frost (1987) examined the frequency distribution of induced metastases in laboratory mice with the particular objective of testing a model of Liotta et al. (1976). This model depends on the rate of entry of metastatic cells into the circulatory system, the life span of tumor cells, and the accretion of metastatic cells. Assuming these processes are all random, the Liotta model predicts that metastases per animal should be Poisson. Kendall and Frost tested this prediction with 10 groups of 7–8 laboratory mice injected with 2 clones of murine melanoma cells. 21 days post injection, the number of metastatic melanoma tumors in the lungs was counted.

Kendal & Frost analyzed their data of tumors per animal using ODR. Reanalyzed by GMR, the slope is slightly higher of $b = 1.45 \pm 0.106$ and 1.71 ± 0.216 for the two clones (Fig. 11.4A; Appendix 11.E). One use of TPL discussed in Chapter 14 is to find potential missing or errant data points as these can create outliers. The outlier in Fig. 11.4A on the Poisson line (circled) was a group with 7 mice; it is not stated whether the original set included 7 or 8 mice. If 8, the rejected mouse was not considered in the calculation of mean and variance, possibly biasing the variance (see Chapter 15). Adding a value between 20 and 30 (in line with the other B16-F10 groups) to the data set moves the point (square) in line with the main body of points. The same result is achieved by including a zero in the set.

In addition to their own experiment, Kendal and Frost analyzed data of 8 similar melanoma and sarcoma experiments from the literature. These data were reanalyzed individually and combined by GMR (Appendix 11.E): the combined data are plotted in Fig. 11.4B with a slope of 1.606 ± 0.029. All

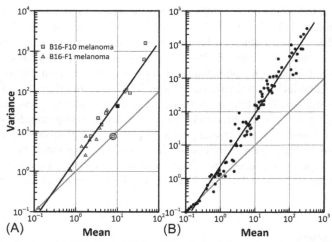

FIG. 11.4   (A) Ensemble TPL ($NQ = 8$, $NB = 20$) analysis of the number of metastatic melanomas per individual developed in mice injected with tumor cells. The outlier *(circled)* may represent a biased variance estimate resulting from the omission of one mouse; the *black square* is the same point with an additional high tumor count mouse. (B) Combined data of 10 studies ($NB = 95$) analyzed by Kendall and Frost lie on the line $V = 2.25 M^{1.61}$. Note the points below the intersection of the fitted and Poisson lines. *(Data from Tables 1 and 2 and references cited by Kendal and Frost (1987).)*

10 experiments reported in their Table 1 had slopes significantly different from 1, averaging $b = 1.61$. This result led Kendal and Frost to reject the Liotta hypothesis. In its place, they suggested that TPL scaling in metastases may be fractal behavior resulting from diffusion-limited aggregation (DLA; Chapter 16). DLA results in a skewed distribution of aggregate sizes; the successful formation of metastatic cancers is thus determined by the probability of the larger aggregates forming. In subsequent articles, Kendal has suggested this probability is Tweedie distributed (Chapter 2). Kendal et al. (2000) and Kendal (2000) present evidence of Tweedie distributed clustering in human hematogenous metastases.

PHYSIOLOGICAL RESPONSES TO STIMULI

Human emotional and psychological status can be gauged by micromovements of facial muscles and sweating. The former are almost certainly adaptations for nonverbal communication that emerged early in hominid evolution. They are now used by skilled interrogators as "tells," indicating the truthfulness of responses to interrogation. They can also be used by psychologists to investigate attitudes to psychosocial interactions. It is possible to quantitatively measure micromovements of the so-called smiling muscle (*zygomaticus major* or ZYG) and the frowning muscle (*corrugator supercilii* or COR) using electromyography (EMG). Sweating or phasic skin conductance (PHSC) is another tell measured by electrodermal activity (EDA). Chołoniewski et al. (2016) enlisted 65 volunteers to test whether biological subsystems that are sensitive to human emotions follow temporal TPL and whether they could distinguish between the human emotions elicited by the stimuli using TPL slopes.

The volunteers were offered a range of visual stimuli on computer monitors and asked to score their emotional experience positively or negatively (valence) and the intensity of the emotion (arousal). The volunteers' valence responses to the emotional stimuli were scored in 7 classes from very negative to very positive and arousal responses were scored in 7 classes from very calm to very aroused. While the volunteers were viewing the stimuli, their facial muscles and PHSC were monitored by EMG and EDA at high frequency. The data therefore comprise time series of ZYG, COR, and PHSC for each volunteer.

To obtain a TPL from a time series of observations, a window of length Δ is moved progressively along the series, and mean and variance are calculated from

$$V_\Delta = E(f_\Delta^2) - M_\Delta^2 \quad \text{where}$$

$$M_\Delta = E(f_\Delta) = \sum_{j=1}^{d} \left(\sum_{i=1}^{n} x_{ij} \right) \quad \text{and} \quad E(f_\Delta^2) = \sum_{j=1}^{d} \left(\sum_{i=1}^{n} x_{ij} \right)^2 \quad (11.2)$$

where $n =$ the number of observations in the jth window and $d =$ the number of windows of length Δ. Ideally all windows contain the same number of observations, n, and the total number of observations is $N = nd$; otherwise, $N = \Sigma n_j$ summed over d windows.

The EMG and EDA signals were split into time windows of $\Delta = 0.5$ ms (the smallest practical window) and mean and variance of COR, ZYG, and PHSC computed for each volunteer. TPL of the three signals was fitted by ODR with $NB = 65$ (Appendix 11.F). Chołoniewski et al. give the estimates of b and r in their Fig. 3. The GMR estimates are COR, $b = 1.95 \pm 0.10$; ZYG, $b = 2.50 \pm 0.18$; and PHSC, $b = 1.87 \pm 0.12$, all significantly different from Poisson. It had been thought that facial EMG responses were the result of quasirandom (variance proportional to the mean) firing of muscles. Chołoniewski et al's TPL results clearly show muscle firing is aggregated in time, suggesting a degree of synchronization of muscle firings.

The three physiological responses corresponding to the seven valence and arousal states also obey TPL in time windows ranging from 5 ms to 1.5 s. TPL slopes of EMG signals increased with the length of the sample window for ZYG, but not for COR. TPL exponents of valence and arousal of ZYG signal increased with increasing window size, while COR exponents remained unaffected. The PHSC TPL slopes proved unsuitable as proxies for emotional response. Chołoniewski et al. concluded that facial tells could be quantified using b to identify high valence and arousal levels in ZYG and COR signals.

392 PART | II

Human demography

UNITED STATES DECENNIAL CENSUS

Article I, Section 2, of the United States Constitution requires a census be conducted every 10 years. The first such census was conducted in 1790 one year after the Constitution came into force in the original 13 States. Each decade since then, the total population and the number of people belonging to several demographic groups have been determined. The first census was conducted in 250 counties and in the 2010 census there were 3142 counties. Since the first census, the population has grown from 3.86 million to 309 million. Data for the 22 decennial censuses are available at the county level on the Census Bureau's website (Census Bureau, 2018).

This was one of the examples in LRT et al. (1978) in which the 1790 and 1900–1980 censuses were used. Here we use all censuses since 1790 and in addition use GMR instead of ODR. The result is a steeper estimate of slope and lower intercept: LRT et al. obtained $A = 0.26 \pm 0.02$, $b = 2.04 \pm 0.01$, here, $A = -1.473 \pm 0.155$, $b = 2.230 \pm 0.025$. LRT et al. considered only 10 censuses compared to this analysis with 23. The inclusion of the 10 censuses from 1800 to 1890 and the 3 from 1990 improved the fit while increasing the correlation close to unity so that the ODR and GMR estimates differ only in the third decimal place ($b_{ODR} = 2.227$). Taking the state as the sample unit (quadrat), the number of units has grown from 15^1 in the census of 1790 to 50 since the 1960 census and the average number of people per state has increased from 242 thousand to 6.2 million. The agreement between observed population and fitted line is excellent with only two points (1800 & 1810) conspicuously off the line (Fig. 11.5A). Normalizing the population per state by converting to density ($\#/km^2$) instead of using the state as the quadrat results in a significantly curved plot (Fig. 11.5B; $P < 0.001$ by Bartlett's method). Apparently, TPL is less consistent when applied to ratios than it is when applied to quadrats of different physical size.

Instead of using the states as quadrats for a temporal TPL, the county populations may be used to compute ensemble TPLs for each census. Ensemble b increased for the first century and since 1900 has decreased a result of both the increase in population and increasing difference in state populations over the period when population grew from 3.77 million in 1790 to 74.6 million in 1900. Territorial expansion from 1790 to 1900 far outstripped the increase in population, resulting in larger differences in number per county and therefore increased variance.

---

1. In addition to the original 13 states, the first census included Kentucky and Vermont which were in the process of entering the Union.

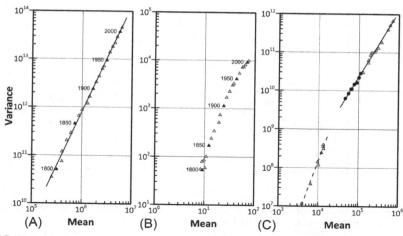

FIG. 11.5   (A) Spatial analysis of US Decennial Census, 1790–2010; $NB = 23$ means and variances of the total population per state based on $NQ = 15$–50 states produces a straight line clearly showing the increase in population from one census to the next. (B) The decennial census with the state populations recast as population density ($\#/km^2$) is distinctly curved where the TPL based on number per state is not. Clusters of points at top and bottom show that population density is changed little in the early decades and is changing little now. (C) African American population separated into slave and free Antebellum and Postbellum. Free Antebellum African Americans are the points in the lower left of the graph. *(Data from U.S. Census Bureau (2018).)*

Other demographics are considered in the US census include gender, continent of origin, (Americas, Europe, Africa, and Asia). African Americans, including slaves up to 1860, were included in the first 8 censuses. Several classifications are included in Appendix 11.G. Fig. 11.5C shows the spatial TPLs for three African American demographic groups: Antebellum slaves (which probably includes some non-African slaves), free Antebellum, and Postbellum African Americans. Where TPLs of people of African and European origins are significantly different ($P \ll 0.001$), the TPLs for slave and Postbellum African Americans are not significantly different ($P > 0.11$) while the difference in slopes of slave and free Antebellum African populations is significant ($P \ll 0.001$).

Census data at the county level enable analysis of the stability of TPL for these demographics through time. Ensemble analysis of each census at the resolution of the county ($= NQ$) has regressions with NB = the number of state mean-variance pairs. Fig. 11.6 shows the trajectories of b for total population and 4 demographic groups over the course of the 23 censuses. In this analysis, the TPL slope increased fairly consistently from 1790 to 1910 as the predominantly rural population grew, increasing the variance of states with major cities. A decline in slope coincides with the start of the migration to the cities that accelerated through the 20th century as agriculture

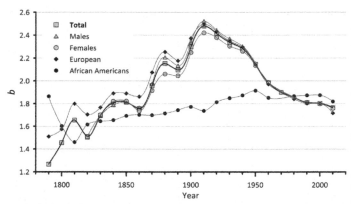

FIG. 11.6   Trajectory of *b* from 1790 to 2010 for different demographics groups based on the county level populations. *(Data from the Census Bureau (2018).)*

became more mechanized. The different trajectories of African and European origin populations in Fig. 11.6 reflect the difference in spatial results seen in Fig. 11.5.

The biggest effect is the difference between African American and European (and other ethnicities) ensemble TPLs. The slow growth in *b*, which has only recently reversed, in line with the total TPL suggests that the African American population was always less heterogeneous at the county level than the general population. Only in Antebellum years when most African Americans were slaves concentrated in the South was the distribution of black people equally or more aggregated than the rest of the population.

The pattern in Fig. 11.6 raises the question of how TPL reflects demographics, especially migration. For a given state population total, the mean number per county remains the same as individuals move from low-density counties to high-density counties (within a state), but the variance increases. Since 1900, the population increased from 74.6 million to 309 million in 2010. During this time, only five states have been added to the Union and none in the last 60 years, so the decline in *b* since 1900 is due to the gradual homogenization of number per county that has occurred with population growth and the spread of metropolitan areas into surrounding counties counteracting the trend for migration from rural to urban communities that has paralleled the industrialization of agriculture. Metropolitan populations expanding into neighboring counties, reduce the overall variance by increasing the number in suburban counties. The distribution is still long tailed, but the middle range increased at the expense of the lower-density range. Converting absolute numbers to density does not resolve this as it seems to distort TPL.

POPULATION OF NORWAY

Data of population and land area for Norway's 18 counties plus the capital Oslo were obtained for the years 1978–2010 from the Central Population Register of Statistics Norway by Cohen et al. (2013) who examined some properties of TPL. They used these data to compute population density (#/km^2) for each of the 19 political units and performed a spatial TPL analysis of the mean and variance of county population density instead of number per county as in the previous example. TPLs were analyzed with and without Oslo, by far the largest absolute population and highest population density, whose population declined during the first 7 years of the study.

Cohen et al. investigated a number of properties by performing several statistical experiments. They tested three weighting policies for calculating mean and variance of population density; examined the influence of Oslo on TPL, especially the change in demographics that took place about 1985; compared results with TPL predicted by an exponential growth model; examined how TPL changed from year to year by computing the "local slope" from the logs of mean and variance of adjacent years; and examined TPL slopes at several spatial scales.

The means and variances of population density of counties (with and without Oslo) in each year were calculated with each county weighted by a weight w: equal weights where $w = 1/n$, areal weights $w = a/A$ where $A =$ the sum of county areas, Σa, and population weights, $w = p/P$ where $P = \Sigma p$ with the summations over n counties (18 or 19 depending on whether Oslo was included). The equal weight and areal weights remained the same for each year, while the population weight for each county varied as the county populations changed from year to year.

Oslo is densely populated, accounting for 10%–12% of Norway's population and 10 times the population density of the next county. Its presence in analyses produces two distinct slopes for the periods 1978–1984 and 1985–2010 that disappear when Oslo is excluded. The discontinuity at 1984 was apparently caused by radically different growth rates of Oslo's population in the two time periods. Excluding Oslo resulted in highly significant TPLs using all three weighting policies (Appendix 11.H). However, they also exhibit significant curvature at high means. With Oslo included, TPLs for 1978–1984 and 1985–2010 differ with the equal and areal weights. However, area-weighting produces a negative TPL for the 1978–1984, indicating a reduction in variance of density with increasing mean that coincided with the Oslo's decline in population but increased in the national population. The TPL of population weighted means and variances with Oslo included showed no decline in variance with increasing mean for 1978–1984.

Cohen et al. developed an exponential growth model in which each county's population density grew or declined by a unique rate r_i from an initial density $D_i(0)$:

$$D_i(t) = D_i(0) \exp(r_i t) \tag{11.3}$$

where $D_i(t)$ is the density of the ith county at time t. The weighted density mean and variance of population density at time t are:

$$M(t) = \sum_{i=1}^{n} w_i(t) D_i(0) \exp(r_i t)$$

$$V(t) = \sum_{i=1}^{n} w_i(t) D_i^2(0) \exp(2 r_i t) - M^2(t) \tag{11.4}$$

where $w_i(t)$ is a county-year specific weight: 1, $1/n$, $a/\Sigma a$, or $p/\Sigma p$. Simulating county population densities, based on measured growth rates, Cohen et al. reproduced the population TPLs with remarkable fidelity. Excluding two outliers, the correlation between simulated and observed TPL slopes for all years, all weights, and with and without Oslo is $r = 0.922$ ($b_{sim} = 0.99 b_{obs} + 0.02$); including the outliers, correlation falls to 0.226. The two outliers are noteworthy: the slopes estimated from data are well outside the normal range of b, but the model predictions were in line with experience. One outlier, already mentioned, is the area weighted 1978–1984 period with $b = -2.10 \pm 0.233$ and the other is the unweighted 1978–1984 analysis with a value of $b = 4.94 \pm 0.372$. The population weighted analysis of 1978–1984 period and all the model predictions conformed to the longer 1985–2010 TPLs. The unweighted and area weighted analyses identified the 1978–1984 period as special. Cohen et al. attributed this to the large influence the capital city has on population dynamics of the country. However, it's not clear why the population weighted TPL did not also expose this special behavior.

The average TPL slope using the data with and without Oslo and including all three weighting policies but excluding the outliers is $b = 2.53$, slightly steeper than the spatial TPL of $b = 2.23$ for the total population of U.S.A. and coincidentally almost identical to the curved spatial TPL of U.S. population density, $b = 2.54$.

To examine spatial TPL's temporal stability, Cohen et al. computed what they called "local" slopes defined as the ratio of the difference of log variance of population density between adjacent years divided by the difference of log mean of population density. For the 33 years in the data set, there are 32 local slopes whose average should intuitively converge on the temporal estimate of b obtained from regression of the full set. With the exception of Oslo included and equal weights, the 95% CI of the slope of the growth model fell within the 95% CI of the slope of the data, indicating good agreement between model and data. Only the case of Oslo excluded with equal weights, the average of local slopes did not converge on the long-term spatial estimates of b: the comparatively large 95%CIs indicate very high heterogeneity in the local slopes, suggesting TPL is highly unstable in the short term, as suggested by Clark and Perry's (1994) analysis.

In most cases when analysis of TPL is planned, the spatial scale is dictated by the sampling method. In the case of human population census data, several scales may be available for study. Using the Norway data, Cohen et al. were able to study TPL at the scales of the municipality and region in addition to the county. The spatial and local slope TPLs of population density fit data disaggregated to the municipality and aggregated to the regional scales fit satisfactorily for all weightings with and without Oslo. Output from the exponential growth model also fits well at both scales. Spatial and local TPL slopes are unaffected by the spatial scale for any weighting policy. Exclusion of Oslo, however, always resulted in lower slopes for municipalities than for counties regardless of weighting. Demographic differences between the scales may account for increased variance at the county scale. As we saw with Clark et al.'s (1996) study of weeds in cereal crops (Chapter 6), the impact of scale on TPL may be unpredictable, even when the data are not sampled.

To my knowledge, this is the only paper to examine the influence of weighting the calculation of mean and variance of population density on the structure of TPL. Perry (1981) assessed weighting the estimated means and variances in his comparison of fitting methods. Cohen et al. suggest that what they called equal weighting may be appropriate only for political or administrative purposes as it ignores both land area and number of people affected by the local population density. In contrast, area weighting may be most appropriate for studies of land-use, agriculture, forestry, and conservation, while population weighting might be more appropriate for studies of urbanization, urban-rural dynamics, and migration.

As explained by Cohen et al., Norway's demographics were quite complicated over the study period. The discovery of North Sea oil and joining the European Economic Area resulted in both internal migration and immigration from outside suggesting that nondemographic factors may influence TPLs of population density. However, the results beg the question: why do these weighting systems applied to population density, including when applied to a simple model output, result in significantly curved TPLs? The answer is unlikely to be due to bias as Clark and Perry's (1994) requirement that both NB and $NQ \geq 15$ when using ODR of log mean and log variance was met with these data.

MORTALITY IN ENGLAND AND WALES

As a control for their investigation of variation in crime records (see later), Hanley et al. (2014) fit TPL to 2013 mortality data obtained from the U.K. Office of National Statistics. Mean and variance of the monthly mortality statistics were calculated for 348 local authorities in England and Wales. Using ODR of log transformed mean and variance, they obtained estimates for

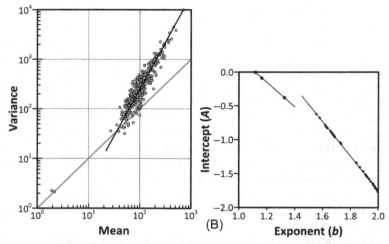

FIG. 11.7 (A) Spatial TPL ($NQ = 12$, $NB = 348$) of deaths per month is steeper if the two points on the Poisson line are excluded *(bold line)*. Including the very low mean points on the left reduces the slope by about 8%. (B) The relationship between intercept and slope for segments of 30 ordered means produces two distinct straight lines with a discontinuity at $1.35 < b < 1.55$. *(Based on Fig. 1 in Hanley et al. (2014) using data from UK Office of National Statistics.)*

$A = -1.187 \pm 0.041$ and $b = 1.799 \pm 0.019$. The computed means and variances provided in their supplementary table (Data File S1) were reanalyzed by GMR (Fig. 11.7A; Appendix 11.I) and obtained estimates of $A = -1.385 \pm 0.084$ and $b = 1.891 \pm 0.041$. TPL transitioned from $b \approx 1.8$ to Poisson ($b \approx 1$) at a mean of $\sim$25 deaths/month. The two very low mean points clearly on the Poisson line reduce the slope to $b = 1.75$ when included in estimating TPL. Also computed are TPLs for the mortality data aggregated by county and region. The county TPL is steeper than either the reporting authority or the region, which are almost the same, suggesting that as these data are increasingly aggregated the slope steepens to a maximum and then declines, the reverse of the pattern seen with plant diseases (Appendix 5.I) and nematodes (Appendix 7.R).

Hanley et al. also investigated the relationship between intercept and slope for different data segments obtained using a 30-point moving window of the data sorted by mean (Eq. 11.2). This analysis showed the exponent approached a lower limit of $b = 1.1$ and an upper limit of $b = 2.0$ and the intercept ranged from $1 < A < 0.018$. Reanalysis by GMR is shown in Fig. 11.7B. There appear to be two distinct lines, the majority of points ($n = 32$) are fit by the line $A = 3.346 - 2.550b$, $r = -0.999$ and 3 points are fit by $A = 2.014 - 1.806b$, $r = 0.999$, creating a discontinuity in the relationship at $b \approx 1.4$. The discontinuity coincides with the exclusion of the three smallest means, which lie close to the Poisson line. The longer line appears to have a slight curvature at the lower end, but it is not significant by Bartlett's method.

MOVEMENT IN CHINA

Zhao et al. (2014) analyzed data of human movements in both cyberspace (through browsing of websites) and physical space (through mobile towers) obtained from a database of millions of mobile phone users in China. The mobile phone database was randomly sampled for 20,000 users who appeared in the database >100 time. These users were then screened again and only those recorded by >50 mobile towers and who visited >50 different websites during a one-month period were retained. This resulted in a database of 3174 users with a significant level of activity in both cyberspace and physical space.

Analyzing the locations and website visits by TPL, Zhao et al. found a super-aggregated scaling relation between the mean frequency of visits and variance with temporal TPL slope $b = 2.4$. Appendix 11.J has approximate values for minimum and maximum means, the intercept and slope from Zhao et al's Fig. 1 and the text. This super-aggregated TPL suggested to Zhao et al. that heterogeneity increased by the visits of individuals with each successive visit. What is remarkable about Zhao et al's results is that the TPLs for physical movement as recorded by cell towers and website visits are virtually identical, suggesting a common behavioral basis to both activities.

To account for the high TPL slope and similarity in result, Zhao et al. developed a simple stochastic behavioral model incorporating preferential return to a previously visited site and exploration for a novel site. These behaviors, to return to a known place (congregate) and to go out and explore a new place (emigrate), are very ancient counterbalancing behaviors that contribute to a population's spatial distribution (LRT and RAJT, 1977; RAJT and LRT, 1979). In Zhao et al's model, an individual can either explore a new site with probability $p_e = \rho n^{-\lambda}$ or return to a previously explored site with the probability $p_r = 1 - \rho n^{-\lambda}$, where n is the number of prior visits. The parameters ρ and λ are not identified behaviorally but act as scale and frequency-dependent parameters, respectively. Simulations with $0.2 < \lambda < 0.6$ predicted $b \approx 2.4$, with the ranges of mean depending on the value of λ. Substituting frequency-dependent return with random return, the model predicted $b \approx 2$, again with the range of mean determined by the value of λ. Presumably intermediate values of b may be obtained with frequency-dependent return probability combined with a random component. How a random component added to the exploration term affects b is not discussed, but as it must increase randomness, it likely would reduce b. What is significant about the model is its ability to predict values of $b > 2.0$, something few other models have achieved (Chapter 17).

Human health

HUMAN IMMUNODEFICIENCY VIRUS

In a study of the epidemiology of HIV/AIDS, Anderson and May (1988) examined the rate of sexual partner change, an important factor in the transmission of all sexually transmitted diseases. They found that the rate of partner change obtained from surveys of different populations employing different sampling methods and over different time spans obeyed TPL. Surveys included hetero-sexual men and women in the general population of England and Wales; het-erosexual men and women attending sexually transmitted disease clinics (STD) in London; and gay men in England and Wales. Segregation of Anderson and May's data by gender suggested that although the TPL was common the rates of partner change differed between the sexes in some groups. In particular, gay men and clients at STD clinics, which included prostitutes, generally claimed higher numbers of partners than university students or the general public.

The rates of partner change per year among the various groups varied widely from ~0.5 to ~14, providing a range of means resulting in a TPL of $V = 0.555M^{3.231}$. Estimates by GMR analysis of the data cited by Anderson and May are slightly different: $V = 0.758M^{3.116}$ (Fig. 11.8A, Appendix 11. K). The very steep TPL and high variability of points at the high end reinforces

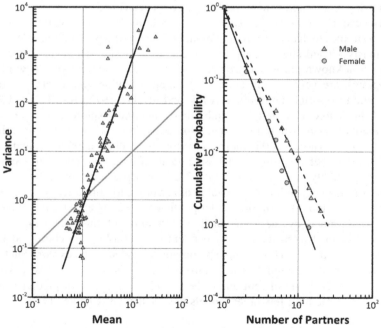

FIG. 11.8 (A) Ensemble TPL (*NB* = 96) of sex partners of gay men, attenders at STD clinics, university students, and the general public. (B) Cumulative probability of number of sexual partners per year among male (*n* = 20) and female (*n* = 14) Swedes surveyed in 1996 shows that a small proportion of the population account for a disproportionate number of partners and that males have more sex partners than females. *(A: Data from cited works in Anderson and May (1988). B: Adapted from Fig. 2 in Liljeros et al. (2001).)*

the conclusions reached by Liljeros et al. (2001) who analyzed a 1996 Swedish survey of the sexual behavior of 2810 adults aged 18–74 years.

Liljeros et al. found that the cumulative distribution of number of sex partners per year against number of partners is a negative power law, indicating the rate of partner exchange to be scale free (Fig. 11.8B). Similar to other scale-free networks, success in obtaining partners leads to increased success in a positive feedback. The cumulative distribution of partners of males was higher than females and the two curves diverge slightly suggesting a small difference in strength of feedback. The power laws of lifetime partner exchange rates also showed some divergence between males and females.

Both Anderson and May's and Liljeros et al's results have implications for epidemiology because epidemics spread faster in scale-free networks (Chapter 16). Because stable scale-free networks are maintained by the nodes with the most connections, preventing STD transmission by the most sexually active members of a community can radically change the epidemiology of STDs, including HIV/AIDS. In fact, some of the data used by Anderson & May showed a reduction in the tail of the distribution of partners of gay men following an anti-AIDS advertising campaign in the UK (Anderson and Johnson, 1988).

TYPHOID IN CAMBODIA

Typhoid is a bacterial infection caused by *Salmonella enterica* serotype Typhi. It is contracted by consuming food or water contaminated with the faeces of an infected person. Symptoms include high fever, weakness, abdominal pain, constipation, and headaches. Some people also develop a rash. Symptoms can last weeks or months without treatment. About 1% of cases are fatal. Given how it is spread, its occurrence is likely to be highly aggregated in areas with poor sanitation. It is endemic in many rural areas of southeast Asia. In a study to better understand the epidemiology of pediatric typhoid fever in rural Cambodia, Thanh et al. (2016) analyzed data of admissions and discharges of children <15 years of age at a children's hospital in Siem Reap province between 2007 and 2014. They examined the relationship between typhoid incidence and socioeconomic descriptors such as population density, literacy, schooling, availability of latrines, and quality of drinking water in districts of the province.

In those districts where typhoid occurred, it was most common in the monsoon season, April–July, and found to be spatially clustered in areas subject to flooding. Their Fig. 1 gives the mean and standard deviation of confirmed monthly typhoid cases at the hospital in 2007–2014. Temporal TPL analysis demonstrates the high degree of spatial clustering (Fig. 11.9; Appendix 11.L). The spatial TPL is highly aggregated ($b=2.1$) with points segregating into wet and dry seasons. Fig. 11.9 shows the variance increasing and decreasing as the average number of cases increased from March to a maximum in May and declined back to the base level in August. Despite low sample numbers

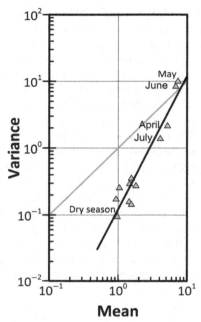

FIG. 11.9 Spatial TPL ($NQ = 8$, $NB = 12$) of monthly pediatric typhoid cases confirmed at a children's hospital in Siem Reap, Cambodia 2007–2014. Cases are strongly aggregated in time with most occurring during the monsoon season when the variance increased then decreased with the changing mean. *(Data from Fig. 1 in Thanh et al. (2016).)*

($NQ = 8$, $NB = 12$) and a slight curvature, the regression is highly significant ($r = 0.96$) and confirms the expectation that typhoid outbreaks are highly aggregated with the variance following a predictable temporal trajectory.

MEASLES AND WHOOPING COUGH

TPL has been found to describe the temporal incidence of other communicable diseases. Using ODR Keeling and Grenfell (1999) showed that the weekly case reports of whooping cough (*Bordetella pertussis*) in 60 cities in England and Wales from 1946 to 1974 exhibit a clear power law relationship (Appendix 11.M) with slope of $b \approx 1.56$. A GMR reanalysis of the case reports has a slope of $b = 1.57 \pm 0.033$. Whooping cough has a complex natural history, has had an inconsistent vaccination program and low case reporting in the UK. Measles or rubeola (*Measles morbillivirus*) has a less demanding natural history, better vaccination history, and higher reporting efficiency. It was analyzed and modeled by Keeling and Grenfell who showed that introduction of measles vaccination in 1967 not only reduced the overall incidence of measles but also changed the variance-mean relationship of measles incidence (Appendix 11.M).

Means and variances of weekly measles case reports in 366 communities in England and Wales and TPLs with $400 < NQ < 1200$ and $NB = 366$ were computed for periods corresponding to pre-vaccination and 3 time spans following the introduction of vaccination in 1967. Vaccination reduced the TPL line from $b = 1.7$ in the prevaccination period (1944–1966) to $b = 1.2$ when 90% of children were vaccinated (1990–1997). During the period of transition following the introduction of vaccination for children starting in 1967, the TPL slope was unchanged at 60% vaccination (1968–1980) but intermediate at 80% vaccination (1980–1990) when $b = 1.5$. Keeling and Grenfell note that the variance about the TPL has increased with increasing vaccination and decreasing slope, possible as a result of variations in implementing vaccination policy.

Like the examples of transitions shown in the TPLs for maize and other commodity crops in the U.S. (Figs. 6.15–6.18), the measles case shows transition with the implementation of child vaccination policy. The introduction of near-universal vaccination reduced clustering of measles outbreaks from highly aggregated to near Poisson. The implication of this is that at 100% vaccination, measles would not be transmissible and incidence should be rare. Under such conditions, we might expect the TPL to lie on the Poisson line as noncommunicable diseases such as childhood leukemia do (Schmiedel et al., 2010).

Measles is a communicable disease that doctors in the UK are required to report. But if a doctor is not called, the case is unlikely to be reported. Keeling and Grenfell considered the case of underreporting and concluded that even if only 60% of cases are reported there would be little change in b. Clearly the level of reporting is the same as sampling efficiency, the impact of which is considered in Chapter 15.

Keeling and Grenfell developed epidemiological models to investigate the processes underlying the TPL measles patterns observed. They found that there is a critical point at which the frequency of no cases of measles in the observations goes to zero and the TPL slope declines to $b \approx 1$. In practice, there is never a condition of no-measles (even with 100% vaccination), so TPL b approaches Poisson, but does not reach it. Historically, measles outbreaks have displayed a pattern of peaking every 2 years. This pattern has been disrupted by large-scale vaccination. The model results produced a steeper power law gradient for communities reporting >70 cases. This, combined with the transition seen in the models from $b \approx 1.7$ to $b \rightarrow 1$, suggests a critical community size above and below which the transmission probabilities differ widely, indicative of a percolation threshold (Chapter 16). Keeling and Grenfell conclude that this transition in b provides a method for identifying the critical community size.

DISEASE MONITORING

Detection of disease outbreaks relies on both reporting and modeling the incidence of diseases. A proper understanding of the variability to be expected in reporting is essential for reliable prediction. TPL has long been used for disease, nematode and insect sampling to assist prediction of the need for management of crop pests (Chapters 5, 7, and 8). Enki et al. (2017) applied TPL to disease surveillance data of infectious disease organisms to determine if they conform to the Tweedie exponential dispersion models with a view to informing statistical models of disease outbreaks.

Public Health England maintains a database (LabBase) of weekly reports of infectious disease organisms detected by laboratories in England, Wales, and Northern Ireland. The disease agents are identified from samples of blood, faeces, or urine taken for diagnostic purposes, primarily in hospitals. Although many of the disease organisms were identified as belonging to more than one biotype, for convenience, the disease agents are here referred to as species.

Enki et al. used 1066 weekly records of 1737 species from 1996 to mid-2001. The means and variances were calculated for $NQ = 26$ weekly reports for a maximum of $NB = 41$ 6-month periods. TPL regressions were performed using Perry's (1981) gamma model (Eq. 4.14). All analyses had slopes significantly greater than zero. Enki et al. provide plots of the 1737 species (722 species had insufficient points for analysis) in a supplement to their paper, but do not provide regression statistics. Instead the TPL slopes and 95% CI of all 1737 species and the smaller set of 1015 organisms are presented in "caterpillar plots," which showed that all had slopes of $0.75 < b < 2.5$, the majority having slopes $1.0 < b < 1.5$. The large number of species with $b \approx 1.0$ suggests that the distributions of many disease agents are close to Poisson and may therefore be quite rare.

Means and variances of the six species in Enki et al's Fig. 1 (*Acinetobacter baumanii*, *Chlamydia trachomatis*, *Giardia lamblia*, *Salmonella enteritidis*, *S. typhimurium* and norovirus) plus measles (*Measles morbillivirus*) and mumps (*Mumps rubulavirus*) and two biotypes of whooping cough (*Bordetella pertussis*) were reanalyzed by GMR. Results are presented in Appendix 11.N, three of which are reproduced in Fig. 11.10. The distribution of sexually transmitted bacterium *C. trachomatis* reports is highly aggregated with $b \approx 2$ and the two whooping cough biotypes have identical TPLs and are slightly more aggregated than Poisson but have very different abundances in laboratory reports. The two species of *Salmonella* are also identical with different but overlapping abundances. Like whooping cough, the measles virus is only slightly more aggregated than Poisson, both reflecting the effect of immunization to reduce aggregation and contrasting with *C. trachomatis*. The GMR estimates of b are 7%–12% greater than Enki et al's Gamma regression estimate. This contrasts the comparison of GMR against weighted least squares (WLS) where the GMR underestimated the WLS estimates (Fig. 6.4).

Enki et al. also tested the skewness-mean relationships in addition to the variance-mean relationship confirming that these surveillance data conformed to a Tweedie model suitable as a basis for disease outbreak modeling.

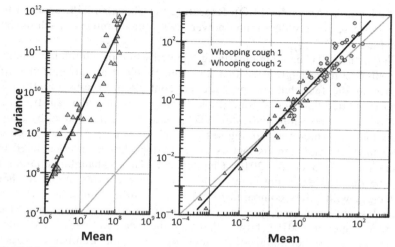

FIG. 11.10 Examples from Enki et al. (2017) illustrating TPL ($NQ=26$, $NB=41$) applied to disease agents identified by diagnostic laboratories in England. (A) Distribution of sexually transmitted bacterium *Chlamydia trachomatis* reports is highly aggregated with $b \approx 2$; (B) two biotypes of whooping cough virus with identical TPLs are slightly more aggregated than Poisson. The range of means of the whooping cough viruses have very different but overlapping abundances. *(Data from Public Health England's LabBase database cited by Enki et al. (2017).)*

Human behavior

STRESS IN AIR TRAFFIC CONTROLLERS

Air traffic control (ATC) is a demanding occupation, requiring intense concentration, precise communication, and quick responses in order to route aircraft safely through sometimes crowded skies. It is one of the most stressful occupations, determining, as it does the safety of hundreds of aircraft per hour and their thousands of passengers. Wang et al. (2016) used TPL to compare 2 data sets with 2 variables: the number of flights and the number of communication events in 14 Paris ATC simulator training sessions on 10 sectors in 2010 and 8 active Shanghai ATC sectors over a 2-week period in 2013. The number of flights and communication events ranged between 38.7 and 76.8 events/h and 169.3–323.0 planes/h at Paris and 23.3–55.9 events/h and 70.4–105.4 planes/h at Shanghai.

The basic data used by Wang et al. are the number of communications sent by each controller to each plane in each sector. They computed the means and variances of communications made by all controllers in the Paris and Shanghai sectors and found the variation in number of communications increases with the number of communications made to planes in the sector. The ODR slopes of TPLs of the Paris simulator and Shanghai actual data differed significantly ($P \ll 0.001$): simulator, $b = 1.08 \pm 0.02$ and actual

$b = 1.54 \pm 0.02$ (Appendix 11.O). Repeating the Shanghai TPL analysis for each sector separately they found power laws "quite similar to each other," with $1.2 < b < 1.8$.

The actual ATC data are clearly aggregated while the simulator data are only slightly more aggregated than Poisson. Wang et al. attribute the difference to the fact that the pressure on controllers is substantially less under simulated than actual conditions. They argue that under real-life conditions, the steeper slope indicates ATC processes are dominated by exogenous factors, which the training session lacked. The relative influence of exogenous and endogenous factors on TPL slope has been considered by others (e.g., Sato et al., 2010, Chapter 16). Thus, Shanghai's larger b may be due the extra pressure of the real safety considerations in their exchanges with pilots compared to controllers during simulation who are under less stress and may pay less attention to safety considerations. Elsewhere Wang et al. (2010) showed that both the mean and variance of number of controller communications increases with the number of flights entering a sector, suggesting appropriately that it is the number of flights being managed that is the exogenous factor increasing the slope, and the probable cause of stress.

Wang et al. include a table with overall means and variances of communication events/h and number of flights/h for the 18 sectors in the study. A Procrustean analysis of these data produces a TPL with $b = 2.28 \pm 0.209$. When all the data are identified, it is clear there are 2 plausible TPLs and 2 clusters. It's not obvious why the simulator flights and the actual communication events are aggregated, while the simulator communication events/h and Shanghai flights/h are not. Presumably it results from the real difference between simulation and actual ATC conditions. It does demonstrate, however, that different but related data sets if they cover a wide enough range of means can produce a plausible TPL.

CRIME IN BRITAIN

It has been observed that certain types of crime appear to be clustered in certain locations or times. For example, in China, burglary correlates with past history in an area (Chen et al., 2013), and certain cities are notorious for violence (Economist, 2017). Hanley et al. (2014) asked if clustering of crimes is unusual and whether clusterings were outliers. They used TPL to examine the variability in crime statistics in England and Wales to establish if a relatively easily calculated and robust statistical method could be applied to data collected by different methods and locations and used to evaluate and perhaps guide public policy. They found variations in crime reports indeed scale with the average incidence using data obtained from the UK Home Office, the Office of National Statistics, and the Economic Policy Centre. The data are of monthly reports for the period November

2012 to October 2013 at two scales. The smaller scale was of neighborhood policing in 95 Derbyshire districts and 56 Nottinghamshire districts. The larger scale was of 42 regional constabularies in England and Wales, including the London Metropolitan Police but excluding the City of London Police, which is really a neighborhood scale force. Crimes analyzed at both scales were antisocial behavior (ASB), burglary, violent crime, and total crimes. In addition, criminal damage & arson, drug offenses, and other crimes were examined at the regional scale.

At the local neighborhood scale, the data include reports of 84,165 ASB incidents, 16,369 burglaries, 24,759 violent crimes, and 205,857 total crimes. Hanley et al. fitted TPL by ODR ($NQ = 12$, $NB = 151$, their Fig. 3) and showed that different crime categories exhibit specific TPL slopes and intercepts and that these differed little between the two counties. Their supplementary data were reanalyzed by GMR (Appendix 11.P), which results agree broadly with the ODR estimates. Fig. 11.11A shows the plot for total crimes with Derbyshire and Nottinghamshire identified, showing they are completely coincident. Parallel-line analysis by Hanley et al. showed no differences in county slopes in any class. GMR analysis confirmed this. They reported that the ODR estimate of total crimes' slope differed from Poisson

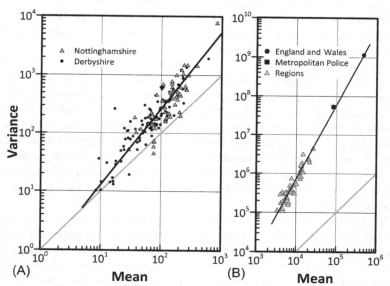

FIG. 11.11   Ensemble TPLs of crime statistics in England and Wales at two scales. (A) Total crime reports for $NB = 95$ Derbyshire and $NB = 56$ Nottinghamshire local police stations completely coincide. (B) Total crime reports for $NB = 42$ regional constabularies plus the countrywide total. London Metropolitan Police is included, but City of London Police is excluded. Both figures are for the reporting period November 2012 to October 2013. *(Data from supplementary data in Hanley et al. (2014).)*

at $P < 0.05$; the GMR slope differs from Poisson at $P < 0.0001$. No intercepts were significantly different from 0 ($P < 0.05$). Both ODR and GMR intercepts differed between the two counties in burglary ($P < 0.01$) and ODR intercepts for ASB also differed at $P < 0.05$.

Means and variances of 42 regional constabulary reports for 7 categories of crime plus the countrywide totals ($NQ = 12$, $NB = 43$) were computed and TPL analyzed by ODR (Hanley et al's Fig. 5). Their supplementary data were reanalyzed using GMR. Results are in Appendix 11.P and the TPL for total crimes is shown in Fig. 11.11B. The GMR estimates agree broadly with Hanley et al's ODR results. At this scale, slopes were steeper and displayed a wider range ($1.61 < b < 2.16$) than those at the neighborhood scale. Intercepts also were more variable and all were significantly < 0 ($P < 0.01$).

Hanley et al. point out that reporting standards vary between and probably within some constabularies. This manifests itself as differing thresholds for reporting, below which crimes may not be entered in central databases. They simulated this and found that TPL slopes increased and intercepts decreased with higher reporting thresholds. Closer examination of the plots showed the introduction of reporting thresholds created subtle discontinuities in the TPL, obviously distorting the signal. This is a sampling efficiency issue examined in Chapter 15. In addition to question of efficiency, we see changes in slope at different scales of neighborhood and regional police reports. A (usually unpredictable) difference in TPL slopes between scales is seen in almost all data where multiple scales have been analyzed (see U.S. census data, this chapter and Chapters 5, 7, 8 and 10).

In an examination of the relationship between TPL slope, b, and intercept, A, Hanley et al's Fig. 6 shows that A tends to decline with increasing b, which agrees with other results (see Mortality earlier). They also show how the introduction of reporting thresholds and rescaling affects the exact relationship between A and b but not the overall pattern. They also suggest that the "other crimes" category may be fundamentally different from the principle crime classes. However, GMR analysis shows other crimes to be parallel to ASB, differing only in intercept. What is different about other crimes is that A and b appear not to be correlated.

It's a truism that serious crime, violence, and homicides are more prevalent in urban areas than rural. This being the case the expectation might be that data from city centers are fundamentally different from those from more rural communities. The fact that the mean-variance relationship for Nottingham Town Centre and Derby City neighborhoods is the same as for the villages of Hulland and Brailsford and the Buxton Rural West neighborhoods provides a common metric for these crimes in these very different locales. Thus, despite differences between scales and possible differences in sampling efficiency, a thorough understanding of the variance structure of crime reports, Hanley et al. conclude, would be useful aids to public safety policy.

EUROPEAN WARS

Quincy Wright, Professor of International Law at the University of Chicago, published in 1942 his two volume opus *A Study of War*. Wright (1942) collected together the work of the students and faculty of the interdisciplinary program he established at Chicago to study war and its historical, legal, and cultural institutions. The *Study of War* analyzes hundreds of accounts of wars and battles around the world from circa 500 BCE up to the 1940s. Although almost all of the data are quantitative and are reproduced in scores of tables and charts, one table stands out as amenable to TPL analysis. Table 22 lists the number of battles engaged in by 10 European powers between 1480 and 1940. Battles are catalogued by decade and by country: Austria, Britain (England, initially), Denmark, France, Prussia (which includes modern Germany), Russia, Spain, Sweden, The Netherlands, and Turkey. Table 22 lists 2659 battles of which many were fought between belligerents in the list of countries. It does not list participants that were not then recognized in "the modern family of nations." Thus, Italy, which was mostly a collection of small provinces and city states until unification was completed in 1871, is not included in Wright's list.

Not all combinations of 10 European powers and 46 decades are populated in Wright's table. Only five countries are included in the first two decades (1480–1499) and Prussia's first entry is the decade 1740–1749. Thus, there are 9 powers listed from 1500 to 1739 and 10 from 1740 to 1940. Despite the gaps, these data are sufficient for both temporal and spatial TPL analyses. They provide means and variances of the number of battles per decade calculated for $NQ = 5$–10 powers and $NB = 46$ decades and by $NQ = 20$–46 decades and $NB = 10$ powers (Fig. 11.12; Appendix 11.Q).

The fact that some battles were fought between two or more belligerents in the table is a potential source of bias. The multiple inclusion of the same battle would constitute an example of $>100\%$ sampling efficiency, in which case the slope would be overestimated (see Chapter 15) only if the battles common to two or more belligerents were systematic rather than random. If random, the effect would be to increase the residual variance around the fitted line. Also, the effect is more likely to be apparent when the decadal mean and variance per country are computed (temporal heterogeneity) than when computed for each country per decade (spatial heterogeneity). Fig. 11.12 Shows the spatial TPL to be slightly steeper than temporal plot (significantly different at $P = 0.08$) but with very little variation about the fitted line. The temporal TPL with more points is also more variable, especially at the lower means. Two points appear as outliers: 1490–1499 and 1590–1599, both decades with only 8 total battles of which 6 were unique. Removing the common battles from one country reduces both mean and variance with little effect on the outliers.

Richardson (1944) also analyzed these data. He tested the distributions of outbreak and duration of the wars in which the battles occurred (Tables 31–41 in Wright) and concluded that, depending on bin size, the distribution

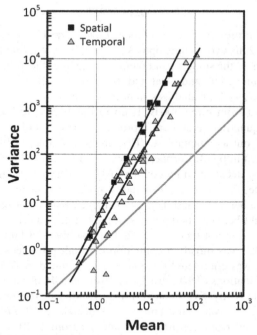

FIG. 11.12 TPL of the spatial ($NB = 10$) and temporal ($NB = 46$) distributions of battles fought by the major European powers between 1480 and 1940. *(Data from Table 22 in Wright (1942).)*

may be Poisson for small bins (number of years <3), but for larger bins the Poisson law did not hold. We see from the TPL analysis that the number of battles per decade is quite strongly aggregated both spatially and temporally with variance-mean gradients not significantly different from 2 ($P > 0.18$). Only at low mean number of battles per decade does the temporal distribution approach random.

As we have seen elsewhere, temporal and spatial analyses derived from the same data do not necessarily reveal the same TPL. In this case, their close similarity is surprising, suggesting that battles cluster very similarly in space and time.

Socioeconomic

CONVENIENCE STORE SALES IN JAPAN

Some TPL data show a transition from aggregated to Poisson at an observable mean value (Figs. 4.2 and 5.11). This transition has implications for both fitting and interpretation of TPL-type data. Failure to take into account points on or parallel to the Poisson line will result in underestimation of the TPL slope of points above the transition. Conversely, TPL will overestimate the slope of points below the transition and analysis of curvature may indicate curvature where none exists.

Fukunaga et al. (2016) applied TPL to point-of-sale data for 158 convenience stores in Kawasaki City, Kanagawa Prefecture, Japan. The stores were open for business 24 h a day, 7 days a week, during the period from June 1 to October 31, 2010. Two types of data were analyzed: receipt data and daily data. Receipt data were recorded at each store every time a customer completed a purchase comprising sales date and time, the sequential customer number, the product bought, and quantity. Daily data included the number and quantity of each product sold. The complete record includes 60 categories of product of which 8 were analyzed by Fukunaga et al. These data were used to create time series for the 8 product categories (soft drinks, rice wine (sake), eggs, cheese, hair-styling products, tissues, magazines, and stamps), which were analyzed by TPL. The time series were structured several ways, by location and time frame. The published data are of number and quantity of sales recorded as means and variances in 15–20 ensemble bins.

To account for scaling in number and quantity of sales, they used TPL to combine the number of sales and quantity of sales by defining a mean A_M below which the variance-mean relationship is Poisson and above which it is not. A_M is defined as:

$$A_M = \frac{V_X}{M_X CV(n)^2} \tag{11.5}$$

where the subscript X refers to the quantity sold so that M_X and V_X are the mean and variance of quantity sold per purchase, and $CV(n) = SD(n)/M(n)$ and n is the number of sales of a product category. Using this definition for A_M, Fukunaga et al. created a compound TPL of quantity and sales:

$$\begin{aligned} V_S &= M_S(1 + M_X), & M_S \ll A_M = CV(n)^{-2} \\ V_S &\approx M_S^2 - CV^2(n), & M_S \gg A_M = CV(n)^{-2} \end{aligned} \tag{11.6}$$

where subscript S refers to the total quantity of a product category sold. The variance of total sold is proportional to the mean squared above a mean of A_M and proportional to the mean below A_M. Below A_M the constant of proportionality therefore depends on the distribution of the quantity sold per purchase, which must be >1.0. The variance-mean relationship for quantity of hair-styling products sold has $V \approx 1.16 M$ for the lowest 7 points; below $A_M \approx 20$ and $V = 1.24 M^{1.83}$ above (Fig. 11.13, Appendix 11.R), closely similar to the sake example (Fig. 4.2) $V = 0.14 M^{1.59}$.

The 8 product categories analyzed all fit TPL very well ($r > 0.99$) above the transition $M > A_M$. At $M < A_M$ estimates of $b \approx 1.0$ with proportionality constant $0.66 < a < 3.4$ ($-0.18 < A < 0.52$). Of the three products for which both number of sales and quantity sold are given, the TPL slopes differ with the slope for

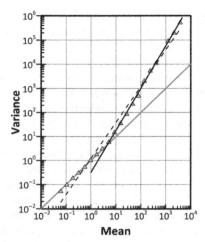

FIG. 11.13 Quantity of hair-styling products sold at convenience stores in Kawasaki City, Japan. Below a mean of ~5 the number of sales is indistinguishable from Poisson ($NB = 7$); above $M \approx 5$ the quantity sold fit a TPL ($NB = 13$) with equation $V = 0.32 M^{1.75}$. *(Adapted from Fig. 4 in Fukunaga et al. (2016).)*

quantity of soft drinks sold greater than the slope for number of sales. The pattern is reversed for cheese and hair-styling products. Apparently demand for soft drinks grows faster with number of soft drink customers than demand for either cheese or hair-styling products with their customers. This seems to suggest that the amount of soft drink purchased accelerates with increasing number of customers.

The model's formal structure (Eq. 11.6) effectively constrains the $M > A_M$ regime to TPL slope of $b = 2.0$, although the data show $1.28 < b < 2.0$. Fukunaga et al. acknowledged their model broke down for the number of stamps purchased because $b = 1.28 \pm 0.013$. In fact, the model was probably only valid for the number of hair-styling products sold ($b = 1.92 \pm 0.036$). From the perspective of understanding TPL, the important contribution of this report is the modeling of data with a crossover from Poisson behavior to aggregated, although the formal definition of A_M in terms of number of sales and quantity sold, may not readily translate to other systems.

SIZE OF CORPORATIONS

The Pareto distribution has long been identified with the size of quantities such as wealth and city sizes, but is robust only when the data are aggregated above some minimum resolution and so holds only above a certain size threshold. For entities that do not reach this threshold some other distribution, generally the lognormal distribution may be used to describe the size of entities such as corporations belonging to a particular commercial sector or political unit. The

Pareto implies a scale-free distribution of sizes above the threshold. Gaffeo et al. (2012) used TPL to investigate the distribution of sizes of corporations in commercial and political divisions, smaller than the Pareto minimum. They also developed models to account for their results.

Gaffeo et al. extracted company performance data from the AMADEUS[2] database for 18 commercial sectors in three Mediterranean countries (France, Italy, and Spain) for 1996–2001. These Mediterranean countries have much in common, but their commercial histories and regulatory regimes differ substantially. Differences in institutional and regulatory regimes are likely to create different opportunities for company growth and consequently lead to differences in the variance-mean relationships of measures of company size. They used three proxies to measure company size: total assets, the number of employees, and value added (compensation of employees plus earnings before taxes). The data set thus comprised an ensemble TPL derived from the large number of companies in each of the 108 sector years. The results of fitting TPL by ODR to the three measures of size of company in three countries were presented in their Fig. 1 and Table 1: they are reproduced in Appendix 11.S.

To account for their results, Gaffeo et al. tested four models of firm growth: a demographic model, a Markov chain, a stochastic difference equation, and a biased random walk model. The demographic model failed to produce a TPL, only a band of variance at a single mean. The others produced TPLs, although the Markov chain produced a TPL with very high scatter around the line and only the latter two models produced convincing TPLs, but with evidence of curvature in the stochastic difference model. They concluded that the Markov chain model may be most appropriate for the Spanish commercial system where the principal cause of increases in value-added had been due to increased use of temporary workers rather than productivity growth; the Italian production system, which favors industrial districts comprised of clusters of small- and medium-sized firms specializing in production of similar goods or services fits the assumptions of the stochastic difference model; and French companies may fit the biased random walk model as there are clusters of innovation in certain technological areas reflecting a high level of innovation combined with recent institutional and regulatory developments.

Gaffeo et al. tested the fit of their data of company size data to the Pareto and rejected it at $P < 0.05$ in all 108 cases. Where Pareto does fit size data, it denotes data with very long tails. The application of TPL to the range of sizes where Pareto apparently fails suggested to Gaffeo et al. the general applicability of the TPL to the analysis of industrial dynamics.

---

2. AMADEUS is a database of financial and business information on Europe's largest 522,000 public and private companies in 43 countries. It is published by Bureau van Dijk *at* https://amadeus.bvdinfo.com/.

SIZE OF CITIES

The distribution of city size, X, has often been found to follow the scale-invariant power law known as Zipf's law: $N = A \cdot X^{-\lambda}$ where N is the number of cities $\geq X$ in size, and A and λ are parameters. Zipf's law is a discrete form of the Pareto distribution, which has cumulative distribution $F(x) = 1 - \left(\dfrac{k}{x}\right)^{\lambda}$ and density function

$$p(x) = \frac{\lambda k^{\lambda}}{x^{\lambda+1}} \quad (\lambda > 0, x \geq k > 0) \tag{11.7}$$

The rth central moments of Eq. (11.7) exist only for $r < \lambda$ (Johnson et al., 1994). Thus, the Zipf law with $\lambda < 2.0$ does not have finite variance. Applied to city size (also stock price fluctuations, income, and company size), Pareto/Zipf law generally fits the available data at the extremities of city size, income, etc. but deviates at more modest levels. On the other hand, the lognormal distribution often fits at the lower end of these variables' ranges, but does not have a long enough tail to fit the extremes. For this reason, Fluschnik et al. (2016) considered TPL as model for city size.

As Cohen et al. (2013) point out, the political boundaries may not be appropriate for some questions so that defining the city unambiguously may be difficult. Fluschnik et al. used an objective measure of city size based on percolation theory. By dividing a map into pixels of size l, pixels of urban land cover belong to the same urban cluster only if they have a common edge. When applied to a region or country, this definition determines the areas and boundaries of its cities. Varying the parameter l changes the urban map. As l increases in size, clusters containing urban land-use undergo a transition from many small separated clusters to a country spanning cluster. The value of l at this transition is called the critical threshold, l_c.

Fluschnik et al. used remote-sensing classification of urban land cover data obtained from the GlobCover 2009 land cover map (ESA, 2010) at a grid resolution of about 0.308 km at the equator. Special software (City Clustering Algorithm; Kriewald et al., 2011) compensated for the changes in size and shape of the pixels as they approached the poles. Using the City Clustering Algorithm, Fluschnik et al. found the probability density of cluster area (city size) for all urban clusters on the globe followed Zipf's law when applied to city areas $>0.1 \, \text{km}^2$ with $\lambda = 1.93$ for $l = 0.4 \, \text{km}$ (249,512 clusters) and $\lambda = 1.75$ for $l = 4 \, \text{km}$ (46,754 clusters). Zipf's law did not apply for areas $<0.1 \, \text{km}^2$. In a similar study of populations $>10^4$ per pixel, they estimated $\lambda = 1.85$ and 1.75 for $l = 0.4 \, \text{km}$ and $4 \, \text{km}$, respectively. In both studies, Zipf's law approximately holds, but with exponents $\lambda < 2$ at pixel sizes $l > 0.4 \, \text{km}$ implying distributions with no finite moments except the mean.

To elucidate the cluster size scaling at the country scale, Fluschnik et al. examined the changes in sample mean and SD that occur in urban cluster sizes when the resolution is progressively coarsened by increasing l. To do this, they applied TPL to a range of cluster sizes in two countries by varying l in the ranges 0.4–30 km for Austria and 0.4–80 km for Spain. The ensemble TPLs used the mean and SD of cluster size $SD(A_l) \propto M(A_l)^d$, where A_l km^2 is the cluster area corresponding to pixel size l and $d \equiv b/2$. Here changing the value of l to obtain means and variances is equivalent to changing the quadrat size in the *Arenicola* example in Chapter 7.

Fig. 11.14 shows how changing pixel size (l) creates TPLs over a limited range, but with changes in slope, transitions, and discontinuities at certain values of l. In Austria, the slope seems to hold at $b \approx 1.58$ up to $l \approx 15$ km but separated by a transition in the range $10 < l < 13$ with steeper b, reverting to $b \approx 1.58$ from $13 < l < 15$ km. The ensemble TPL for Spain shows two different power laws. For the range $0.4 < l < 4$ km $b \approx 3.26$ and for $4 < l < 16$ km $b \approx 1.52$. Above $l \approx 15$ km and 16 km for Austria and Spain, respectively, TPL breaks down completely. The values of l at which TPL breaks down match the percolation threshold $l_c = 15$ km, suggesting a relationship between the percolation threshold l_c and TPL, where the presence of a maximum standard deviation (or variance) could identify the threshold. Within limits, TPL apparently holds for city size. Although there appears not to be a unique scaling exponent, the evidence suggests the scaling relationship could be country specific just as TPL for measures of company size proved to vary between countries.

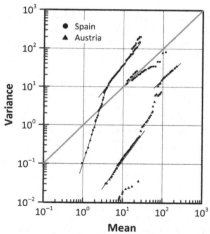

FIG. 11.14 Each point in the plot is the mean and variance of city size determined with a different pixel size l ($\equiv$ quadrat). The lines fitted by eye show TPL is stable for wide ranges of l but with changes in slope (Spain), transitions (Austria), and discontinuities (both countries) occurring at certain values of 1. Austria is shifted one cycle to the right for clarity. *(Adapted from Fig. 6 in Fluschnik et al. (2016).)*

Appendix: TPL estimates for other biological examples

	NB	Range of means	r	A	SE(A)	b	SE(b)
General biology and physiology							

A Ensemble TPLs (*NQ* variable) of rate of phenotypic and population change. Data from Fig. 3 and Table S1 in DeLong et al. (2016)

	NB	Range of means	r	A	SE(A)	b	SE(b)
Rate of population change	20	0.0024 - 1.0103	0.979	−0.203	0.133	2.091	0.094
Rate of Phenotypic change	19	0.0413 - 1.6879	0.945	0.003	0.087	2.148	0.162

B Ensemble TPLs (*NQ* = 48) of exoenzyme β-D-glucosidase in 3 soils. Data from Figs. 1 and 2 Kim et al. (2015)

	NB	Range of means	r	A	SE(A)	b	SE(b)
Conventional clay	5	1×10^{-6} - 404	0.996	−0.273	0.199	1.418	0.056
Conventional loam	5	2×10^{-7} - 73.2	0.975	−0.653	0.460	1.259	0.125
Low-input loam	5	9×10^{-6} - 559	0.996	−0.829	0.193	1.167	0.048

C Hybrid ensemble TPL (*NQ* = 15) of production of chlorophyll-*a* by 3 substrates in 5 rivers in Quebec Naiman (1983)

	NB	Range of means	r	A	SE(A)	b	SE(b)
Periphyton	28	0.20 - 37.0	0.824	−0.198	0.139	1.740	0.186
Mosses	15	2.00 - 87.8	0.737	−0.588	0.495	2.083	0.363
Detritus	29	0.10 - 11.4	0.949	0.397	0.074	2.298	0.134

D Ensemble TPL (*NQ* = 7) of phosphorus in north temperate lakes. Data from Fig. 1 in France and Peters (1992)

	NB	Range of means	r	A	SE(A)	b	SE(b)
Total phosphorus − France and Peters	65	3.05 - 253	0.823	−1.130	0.238	2.002	0.141
Marshall et al	12	9.88 - 125	0.931	−1.817	0.384	2.447	0.257
Combined	77	3.05 - 253	0.838	−1.224	0.212	2.057	0.128
Chlorophyll-*a*	65	-	0.949	-	-	2.107	0.055

E Ensemble TPLs (*NQ* = 8) of metastatic cancer in clones of laboratory mice injected with cancer cells obey TPL. The analyses are of different clones of cancer cells. The B16-F1 and B16-F10 melanoma data are from Table 1 in Kendal and Frost (1987). The remainder is of data from the original literature referenced in Table 2 in Kendal and Frost. All were reanalyzed by GMR

	NB	Range of means	r	A	SE(A)	b	SE(b)
B16-F1 murine melanoma	10	0.125 - 5.57	0.973	0.305	0.053	1.446	0.106
B16-F10 murine melanoma	10	2.29 - 46.9	0.917	0.037	0.240	1.714	0.216
B16-F1 and B16-F10 combined	20	0.125 - 46.9	0.965	0.260	0.077	1.523	0.089
B16 melanoma	15	5.70 - 248	0.966	0.176	0.192	1.664	0.111
UV-2237 fibrosarcoma	21	2.80 - 311	0.946	0.492	0.187	1.563	0.110
KHT-35 sarcoma	8	0.714 - 7.64	0.904	0.226	0.157	1.899	0.287
KHT-13 sarcoma	8	0.100 - 7.00	0.996	0.386	0.031	1.420	0.044
KHT-3 sarcoma	12	0.111 - 12.6	0.965	0.426	0.096	1.551	0.118
KHT-24 sarcoma	5	0.071 - 11.0	0.986	·0.185	0.098	1.363	0.102
Murine sarcoma virus 1	11	0.400 - 17.8	0.947	0.094	0.095	1.820	0.176
Murine sarcoma virus 2	15	1.40 - 36.2	0.959	0.251	0.140	1.778	0.131
All studies combined	95	0.071 - 311	0.984	0.353	0.036	1.606	0.029

F Ensemble TPLs (samples windowed) of physiological responses to stimuli and valence and arousal states. Estimates are GMR adjusted from ODR in the text and Fig. 3, valence and arousal states from Figs. 4 and 5 in Chołoniewski et al. (2016)

	NB	Range of means	r	A	SE(A)	b	SE(b)
XYG − smiling muscle	65	0.59 - 5.75	0.825	-		2.50	0.18
COR − frowning muscle	65	1.50 - 37.2	0.911	-		1.95	0.10
PHSC − sweating	65	0.005 - 0.093	0.927	-		1.87	0.14

Continued

	NB	Range of means	r	A	SE(A)	b	SE(b)
XYG – smiling muscle				Valence b and SE(b)		Arousal b and SE(b)	
Score 1 – Negative valence/ calm	25	-	-	2.37	0.13	2.58	0.10
Score 2	25	-	-	2.73	0.12	2.37	0.16
Score 3	25	-	-	2.66	0.09	2.49	0.09
Score 4 – Median valance/ arousal	25	-	-	2.78	0.09	2.76	0.09
Score 5	25	-	-	2.65	0.10	2.42	0.09
Score 6	25	-	-	2.35	0.07	2.42	0.09
Score 7 – Positive valence/ aroused	25	-	-	2.25	0.07	2.16	0.15
COR – frowning muscle				Valence b and SE(b)		Arousal b and SE(b)	
Score 1 – Negative valence/ calm	25	-	-	1.95	0.11	2.10	0.17
Score 2	25	-	-	1.80	0.12	1.86	0.11
Score 3	25	-	-	1.87	0.12	2.04	0.10
Score 4 – Median valance/ arousal	25	-	-	1.85	0.18	1.97	0.15
Score 5	25	-	-	2.03	0.13	1.83	0.08
Score 6	25	-	-	1.98	0.12	1.64	0.11
Score 7 – Positive valence/ aroused	25	-	-	1.45	0.15	1.77	0.12

Human demography

G Spatial TPLs of the decennial censuses of the USA, 1790–2010. Mean and variance of population per state and population density per state ($NQ = 15$–50 states with 4 demographic categories and 3 categories of the African American population). Population density TPL has significant concave curvature. Data from US Census Bureau (2017)

	NB	Range of means	r	A	SE(A)	b	SE(b)
Total population	23	241,545 - 6.18 $\times 10^6$	0.999	−1.473	0.155	2.230	0.025
Population density (#/km²)	23	8.38 - 74.8	0.991	−0.556	0.098	2.535	0.069
Males	18	249,370 - 3.04 $\times 10^6$	0.999	−1.132	0.161	2.185	0.027
Females	18	242,161 - 3.14 $\times 10^6$	0.999	−0.896	0.121	2.148	0.020
European origin	22	195,044 - 4.94 $\times 10^6$	0.995	−1.290	0.275	2.212	0.045
African origin	22	46,501 - 8.05 $\times 10^5$	0.992	1.639	0.325	1.744	0.059
Postbellum African Americans	14	127,252 - 8.05 $\times 10^5$	0.993	2.155	0.328	1.638	0.067
Antebellum free African Americans	8	3685 - 13,492	0.992	−3.148	0.499	2.831	0.125
Antebellum slaves	8	43,050 - 110,980	0.980	1.232	0.411	1.811	0.077

H Spatial TPLs of population density of Norway with ($NQ = 19$) and without ($NQ = 18$) Oslo using three weighting policies to compute mean and variance of county densities all show curvature. * TPLs with significant concave curvature; † TPLs with significantly convex curvature. Approximate range of means read from Figs. 3 and 4 and statistics from Tables 1 and 2 in Cohen et al. (2013)

	NB	Range of means	r	A	SE(A)	b	SE(b)
Excluding Oslo – Equal weight*	33	22.9 - 29.0	1.000	−0.977	0.021	2.767	0.015
Area weight*	33	11.8 - 14.0	0.998	−1.955	0.046	3.952	0.042
Population weight*	33	30.9 - 40.2	1.000	−0.332	0.016	2.192	0.010
Including Oslo – Equal weight*	33	78 - 100	0.998	0.736	0.052	2.111	0.027

Continued

	NB	Range of means	r	A	SE(A)	b	SE(b)
Area weight, 1978–2010-	33	13.3 - 16.0	0.985	−0.269	0.118	3.099	0.102
1978–1984*	7	13.3 - 13.6	0.983	5.611	0.263	−2.099	0.233
1985–2010	26	13.6 - 16.0	0.999	−0.686	0.046	3.454	0.040
Population weight*	33	140 - 190	1.000	1.317	0.020	1.718	0.009
Model − Equal weight†	33	75 - 95	1.000	0.822	0.002	2.066	0.003
Area weight†	33	13.2 - 16.0	1.000	−0.168	0.009	3.011	0.021
Population weight†	33	130 - 185	1.000	1.369	0.002	1.695	0.002

I Temporal TPL (NQ=12) of monthly mortality in England and Wales. Data from UK Office of National Statistics and cited by Hanley et al. (2014)

Mortality − Reporting District	348	2.30 - 701	0.915	−1.385	0.084	1.891	0.041
County	35	329 - 2713	0.979	−2.038	0.209	2.089	0.073
Region	10	2205 - 6482	0.994	−1.186	0.227	1.832	0.063

J Ensemble TPLs (NQ>50) of the frequency of visits by humans to geographic locations and internet sites as recorded by cell phone towers in China. Their similarity suggests a fundamental feature of human spatial behavior is common to both activities. Approximate estimates from the text and Fig. 1 in Zhao et al. (2014)

Internet sites	3174	∼33 - ∼1,075	-	∼−0.24	-	∼2.4	-
Geographic locations	3174	∼21 - ∼1,500	-	∼−0.07	-	∼2.4	-

Human health

K Ensemble TPL (NQ variable) of HIV/AIDS. Data from cited works in Anderson and May (1988); cumulative probability of number of sex partners per year. Data from Fig. 2 in Liljeros et al. (2001)

Anderson and May − Partners	96	0.474 - 28.8	0.930	−0.171	0.053	3.116	0.117
Liljeros et al − Partners of males	12	1 - 20	−0.997	−0.008	0.475	−2.155	0.050
Partners of females	9	1 - 14	−0.996	−0.023	0.629	−2.702	0.081

L Ensemble TPL (NQ=8) of monthly pediatric typhoid cases in Siem Reap, Cambodia, 2007–2014. Data from Fig. 1 in Thanh et al. (2016)

Typhoid cases	12	0.932 - 7.37	0.958	−0.943	0.081	2.081	0.172

M Case reports of childhood diseases in England and Wales obey TPL. ODR estimates for measles reports for pre- and postvaccination periods. GMR estimate of whooping cough reports in NB = 60 cities 1946–1974 (NQ ≈ 1500). Measles intercepts estimated by eye from Fig. 1 and whooping cough data digitized from Fig. 6 in Keeling and Grenfell (1999)

Measles − prevaccination, 1944–1966	-	0.020 - ∼25	-	∼0.10	-	1.7	-
80% vaccination, 1980–1990	-	0.375 - ∼150	-	∼0.15	-	1.5	-
90% vaccination, 1990–1997	-	1.05 - ∼1060	-	∼0.25	-	1.2	-
Whooping cough − 1946–1974	60	0.252 - 83.1	0.986	0.69	0.024	1.57	0.033

N Temporal TPLs (NQ = 26) of weekly disease reports from diagnostic laboratories in England. Data from Public Health England's LabBase database cited by Enki et al. (2017)

Acinetobacter baumanii	41	3.2×10^{-3} - 327	0.981	0.017	0.059	1.064	0.032
Chlamydia trachomatis	41	1.5×10^6 - 1.27×10^8	0.935	−4.353	0.782	1.976	0.110
Giardia lamblia	41	10,092 - 79,433	0.855	−3.497	0.700	1.956	0.158
Salmonella enteritidis	41	6.7×10^{-4} - 250	0.918	0.217	0.137	1.339	0.083
S. typhimurium	41	7.44 - 28,303	0.921	−0.357	0.251	1.466	0.089
Salmonella combined	82	6.7×10^{-4} - 28,303	0.940	0.118	0.115	1.326	0.050
Norovirus (gastroenteritis)	41	153 - 2.51×10^5	0.927	−0.805	0.409	1.770	0.104

Continued

	NB	Range of means	r	A	SE(A)	b	SE(b)
Measles virus	41	0.040 - 205	0.926	0.206	0.072	1.198	0.070
Mumps virus	41	0.130 - 65,323	0.945	−0.036	0.157	1.422	0.073
Whooping cough biotype #1	41	0.551 - 113	0.893	0.012	0.103	1.228	0.086
Whooping cough biotype #2	40	4.9×10^{-4} - 3.82	0.959	0.201	0.064	1.203	0.054
Whooping cough combined	81	4.9×10^{-4} - 113	0.971	0.125	0.037	1.155	0.031

Human behavior

O Hybrid-ensemble TPLs (windowed samples) of communication events per hour by air traffic controllers in simulated and actual ATC situations in Paris and Shanghai. Estimates of b are ODR; number of data points (NB) and range of means estimated from Figs. 2 and 3 in Wang et al. (2016)

	NB	Range of means	r	A	SE(A)	b	SE(b)
Paris simulator ATC record	>100	~1.15 - ~216	-	-	-	1.08	0.02
Shanghai actual ATC record	>1000	~2.35 - ~221	-	-	-	1.54	0.02
Procrustean analysis – Paris flights/h	10	38.7 - 76.8	0.801	−2.00	0.987	2.18	0.577
Shanghai flights/h	8	23.3 - 55.9	0.017	1.66	0.619	0.02	0.397
Paris communications/h	10	102 - 323	0.152	2.72	1.291	0.24	0.554
Shanghai communications/h	8	70.4 - 221	0.885	−2.67	1.285	2.91	0.626
Combined	36	23.3 - 323	0.882	−1.88	0.407	2.28	0.209

P Temporal TPLs ($NQ = 12$) of 4 classes of crime reported by 56 policing neighborhoods in Nottinghamshire and 95 in Derbyshire and 7 classes at regional and national scale. Slopes of classes of crime did not differ between counties .Means and statistics for England and Wales are for 42 policing regions (excluding the City of London) also include the country-wide means and variances. Data reanalyzed with GMR from Supplementary data in Hanley et al. (2014)

	NB	Range of means	r	A	SE(A)	b	SE(b)
Policing neighborhoods							
Derbyshire – Antisocial behavior	95	2.50 - 261	0.934	−0.27	0.079	1.41	0.052
Burglary	95	0.25 - 27.8	0.888	−0.07	0.052	1.40	0.066
Violent crime	95	0.25 - 77.5	0.932	−0.06	0.037	1.13	0.042
Nottinghamshire – Antisocial behavior	56	7.33 - 1978	0.882	−0.45	0.165	1.57	0.099
Burglary	56	2.42 - 34.7	0.866	−0.43	0.116	1.60	0.107
Violent crime	56	3.75 - 106	0.779	−0.30	0.141	1.32	0.111
Both counties – All crimes	151	8.33 - 835	0.890	−0.20	0.095	1.31	0.049
Regional and Country Scale							
Antisocial behavior	43	1642 - 177,429	0.967	−1.95	0.295	2.10	0.082
Burglary	43	116 - 37,616	0.962	−1.64	0.220	1.84	0.077
Criminal damage and arson	43	279 - 42,698	0.964	−1.77	0.220	1.80	0.074
Drug offenses	43	67.6 - 15,824	0.891	−0.85	0.278	1.61	0.111
Violent crime	43	241 - 53,952	0.910	−1.68	0.363	1.89	0.120
Other crimes	43	22.2 - 8144	0.970	−0.85	0.173	2.16	0.080
Total crimes	43	3390 - 465,114	0.975	−1.82	0.261	1.92	0.065

Q TPL of the spatial ($NQ = 20$–46) and temporal ($NQ = 5$–10) distributions of battles fought by the 10 major European powers between 1480 and 1940. Data from Table 22 in Wright (1942)

	NB	Range of means	r	A	SE(A)	b	SE(b)
Number of battles – Temporal	44	0.44 - 108	0.950	0.29	0.081	1.87	0.095
Spatial	10	0.75 - 30.8	0.993	0.62	0.089	2.10	0.088

Socioeconomic

R Ensemble TPLs sales at convenience stores in Kawasaki City, Japan. Values for NB are the number of points above or on the Poisson line and the total number of points. Poisson and aggregated points diverge at $M = A_M$. Data from Figs. 3–7 in Fukunaga et al. (2016)

	NB	Range of means	r	A	SE(A)	b	SE(b)
Quantity of soft drink sold – $M < A_M \approx 12$	5/19	0.785 - 8.95	0.994	0.241	0.028	0.965	0.047
$M > A_M \approx 12$	14/19	16.4 - 151,000	0.996	−0.834	0.137	1.812	0.042

Continued

	NB	Range of means	r	A	SE(A)	b	SE(b)
Quantity of cheese sold ($A_M \approx 15$)	9/19	20.9 - 1014	0.998	−0.446	0.066	1.395	0.026
Quantity of hair product sold ($A_M \approx 5$)	12/20	5.17 - 3748	0.999	−0.495	0.042	1.752	0.019
Number of soft drink sales ($A_M \approx 10$)	9/16	50.0 - 75,405	0.999	−1.319	0.074	1.671	0.021
Number of cheese sales ($A_M \approx 125$)	5/15	410 - 7852	0.985	−1.372	0.414	1.653	0.128
Number of hair product sales ($A_M \approx 20$)	13/18	33.5 - 14,125	0.998	−1.126	0.101	1.915	0.036
Number of rice wine sales ($A_M \approx 20$)	9/16	26.2 - 18,276	0.997	−0.856	0.127	1.593	0.042
Number of egg sales ($A_M \approx 10$)	9/16	21.0 - 11,391	0.999	−0.475	0.070	1.497	0.024
Number of tissue sales ($A_M \approx 30$)	8/16	36.9 - 8957	0.998	−0.474	0.090	1.307	0.031
Number of magazine sales ($A_M \approx 5$)	11/16	6.28 - 37,847	0.992	−0.501	0.200	1.759	0.068
Number of stamp sales ($A_M \approx 0.5$)	13/16	1.14 - 46,017	0.999	0.709	0.036	1.284	0.013

S ODR estimates for TPL (*NQ* variable) of three measure of company size in France, Italy, and Spain for the years 1996–2001. Range of means is estimated approximate. Data from Fig. 1 and Table 1 in Gaffeo et al. (2012)

France − Total assets	108	30 - 1000	0.903	1.713	0.442	2.056	0.108
Value added	108	950 - 1.6×10^5	0.815	1.200	0.342	2.161	0.170
# Employees	108	2000 - 3.6×10^5	0.891	1.324	0.301	2.228	0.169
Italy − Total assets	108	20 - 3600	0.929	−3.177	0.562	3.089	0.135
Value added	108	400 - 6.7×10^5	0.941	−3.427	0.475	3.326	0.132
# Employees	108	900 - 2.0×10^6	0.936	−2.095	0.287	3.822	0.159
Spain − Total assets	108	35 - 200	0.952	1.191	0.261	1.905	0.068
Value added	108	1600 - 22,000	0.953	1.820	0.27	1.905	0.067
# Employees	108	3000 - 65,000	0.931	1.327	0.151	1.940	0.084

References

Anderson, R.M., Johnson, A.M., 1988. Rates of sexual partner change in homosexual and heterosexual populations in the United Kingdom. In: Voeller, B., Reinisch, J.M., Gottlieb, M. (Eds.), AIDS and Sex: An Integrated Biomedical and Biobehavioral Approach. Oxford University Press, Oxford, UK, pp. 121–154.

Anderson, R.M., May, R.M., 1988. Epidemiological parameters of HIV transmission. Nature 333, 514–519.

Census Bureau, 2018. United States Department of Commerce, Census Bureau. https://factfinder.census.gov/faces/nav/jsf/pages/guided_search.xhtml. Verified January 2019.

Chen, P., Yuan, H., Li, D., 2013. Space-time analysis of burglary in Beijing. Secur. J. 26, 1–15.

Chołoniewski, J., Chmiel, A., Sienkiewicz, J., Hołyst, J.A., Küster, D., Kappas, A., 2016. Temporal Taylor's scaling of facial electromyography and electrodermal activity in the course of emotional stimulation. Chaos, Solitons Fractals 90, 91–100.

Clark, S.J., Perry, J.N., 1994. Small sample estimation for Taylor's power law. Environ. Ecol. Stat. 1, 287–302.

Clark, S.J., Perry, J.N., Marshall, E.J.P., 1996. Estimating Taylor's power law parameters for weeds and the effect of spatial scale. Weed Res. 36, 405–417.

Cohen, J.E., Xu, M., Brunenborg, H., 2013. Taylor's law applies to spatial variation in a human population. Genus 69, 25–60.

DeLong, J.P., Forbes, V.E., Galic, N., Gibert, J.P., Laport, R.G., Phillips, J.S., Vavra, J.M., 2016. How fast is fast? Eco-evolutionary dynamics and rates of change in populations and phenotypes. Ecol. Evol. 6, 573–581.

Economist, 2017. The world's most dangerous cities. The Economist. 31 March 2017.

Enki, D.G., Noufaily, A., Farrington, P., Garthwaite, P., Andrews, N., Charlett, A., 2017. Taylor's power law and the statistical modelling of infectious disease surveillance data. J. R. Stat. Soc. A. Stat. Soc. 180, 45–72.

ESA, 2010. The Ionia GlobCover Project. GlobCover Land Cover 2009 v2.3. European Space Agency website. http://due.esrin.esa.int. Verified January 2019.

Fluschnik, T., Kriewald, S., Cantú Ros, A.G., Zhou, B., Reusser, D.E., Kropp, J.P., Rybski, D., 2016. The size distribution, scaling properties and spatial organization of urban clusters: a global and regional percolation perspective. ISPRS Int. J. Geo-Inf. 5, 1–14.

France, R.L., Peters, R.H., 1992. Temporal variance function for total phosphorus concentration. Can. J. Fish. Aquat. Sci. 49, 975–977.

Fukunaga, G., Takayasu, H., Takayasu, M., 2016. Property of fluctuations of sales quantities by product category in convenience stores. PLoS ONE 11 (6), e0157653.

Gaffeo, E., Di Guilmi, C., Galleti, M., Russo, A., 2012. On the mean/variance relationship of the firm size distribution: evidence and some theory. Ecol. Complex. 11, 109–117.

Grant, P.R., Grant, B.R., 2014. 40 Years of Evolution: Darwin s Finches on Daphne Major Island. Princeton University Press, Princeton, NJ.

Hanley, Q.S., Khatun, S., Yosef, A., Dyer, R.-M., 2014. Fluctuation scaling, Taylor's law, and crime. PLoS ONE 9 (10), e109004.

Johnson, N.L., Kotz, S., Balakrishnan, N., 1994. Distributions in Statistics. Continuous Univeriate Distributions—1, second ed. Wiley, New York.

Keeling, M.J., Grenfell, B.T., 1999. Stochastic dynamics and a power law for measles variability. Philos. Trans. R. Soc. Lond. 354, 769–776.

Kendal, W.S., 2000. A frequency distribution for the number of hematogenous organ metastases. J. Theor. Biol. 217, 203–218.

Kendal, W.S., Frost, P., 1987. Experimental metastasis: a novel application of the variance-to-mean power function. J. Natl. Cancer Inst. 79, 1113–1115.

Kendal, W.S., Lagerwaard, F.J., Agboola, O., 2000. Characterization of the frequency distribution for human hematogenous metastases: evidence for clustering and a power variance function. Clin. Exp. Metastasis 18, 219–229.

Kim, H., Nunan, N., Dechesne, A., Juarez, S., Grundmann, G., 2015. The spatial distribution of exo-enzyme activities across the soil micro-landscape, as measured in micro- and macro-aggregates, and ecosystem processes. Soil Biol. Biochem. 91, 258–267.

Kriewald, S., Fluschnik, T., Reusser, D., Rybski, D., 2011. OSC: Orthodromic Spatial Clustering, R Package Version 1.0.0. The Comprehensive R Archive Network at.https://CRAN. R-project.org/.

Liljeros, F., Edling, C.R., Amaral, L.A.N., Stanley, H.E., Yvonne, A., 2001. The web of human contacts. Nature 411, 907–908.

Liotta, L.A., Saidel, G.M., Kleinerman, J., 1976. Stochastic model of metastases formation. Biometrics 32, 535–550.

Marshall, C.T., Morin, A., Peters, R.H., 1988. Estimates of mean chlorophyll-*a* concentration: precision, accuracy, and sampling design. Water Resour. Bull. 24, 1027–1034.

Murthy, A.C., Fristoe, T.S., Burger, J.R., 2016. Homogenizing effects of cities on North American winter bird diversity. Ecosphere 7 (1), e01216.

Naiman, R.J., 1983. The annual pattern and spatial distribution of aquatic oxygen metabolism in boreal forest watersheds. Ecol. Monogr. 53, 73–94.

Park, S.-J., Taylor, R.A.J., Grewal, P.S., 2013. Spatial organization of soil nematode communities in urban landscapes: Taylor's power law reveals life strategy characteristics. Appl. Soil Ecol. 64, 214–222.

Perry, J.N., 1981. Taylor's power law for dependence of variance on mean in animal populations. Appl. Stat. 30, 254–263.

Richardson, L.F., 1944. The distribution of wars in time. J. Roy. Statist. Soc. 107, 242–250.

Sato, A.-H., Nishimura, M., Hołyst, J.A., 2010. Fluctuation scaling of quotation activities in the foreign exchange market. Phys. A 389, 2793–2804.

Schmiedel, S., Blettner, M., Kaatsch, P., Schiiz, J., 2010. Spatial clustering and space-time clusters of leukemia among children in Germany, 1987–2007. Eur. J. Epidemiol. 25, 627–633.

Taylor, L.R., Taylor, R.A.J., 1977. Aggregation, migration and population mechanics. Nature 265, 415–421.

Taylor, L.R., Woiwod, I.P., Perry, J.N., 1978. The density-dependence of spatial behaviour and the rarity of randomness. J. Anim. Ecol. 47, 383–406.

Taylor, R.A.J., Taylor, L.R., 1979. A behavioural model for the evolution of spatial dynamics. In: Anderson, R.M., Turner, B.D., Taylor, L.R. (Eds.), Population Dynamics. Blackwell Scientific Publications, Oxford, UK, pp. 1–27.

Thanh, D.P., Thompson, C.N., Rabaa, M.A., et al., 2016. The molecular and spatial epidemiology of typhoid fever in rural Cambodia. PLoS Negl. Trop. Dis. 10 (6), e0004785.

Wang, Y., Hu, M., Duong, V., 2010. Fluctuation scaling in the air traffic controller communication activities. In: Proceedings of the ENRI International Workshop on ATM/CNS, Tokyo, Japan.

Wang, Y., Zhang, Q., Zhu, C., Duong, V., 2016. Human activity under high pressure: a case study on fluctuation scaling of air traffic controller's communication behaviors. Physica A 441, 151–157.

Wang, Z., Zhang, T.Q., Tan, C., Taylor, R.A.J., Wang, X., Welacky, T., 2018. Simulating crop yield, surface runoff, tile drainage and phosphorus loss in a clay loam soil of the Lake Erie region using EPIC. Agric. Water Manag. 204, 212–221.

Wang, Z., Zhang, T.Q., Tan, C.S., Wang, X., Taylor, R.A.J., Qi, Z.M., Yang, J.W., 2019. Modeling the impacts of manure on P loss in surface runoff and subsurface drainage. J. Environ. Qual. 48, 39–46.

Weiner, J., 1995. The Beak of the Finch. Jonathan Cape, London.

Wright, Q., 1942. A Study of War. The University of Chicago Press, Chicago, IL.

Zhao, Z.-D., Huang, Z.-G., Huang, L., Liu, H., Lai, Y.-C., 2014. Scaling and correlation of human movements in cyberspace and physical space. Phys. Rev. E90, 050802.

Chapter 12

Nonbiological examples

Here we look at examples of TPL from the nonbiological domain. Included are examples from astronomy, geology, meteorology, networks, and number theory.

Astronomical

CYANOGEN IN A COMET HALO

In July 1995, Comet Hale-Bopp was first observed at about 1.2×10^9 km from the sun. Three months later, when the comet was 10^8 km closer, Wagner and Schleicher (1997) took measurements of the gas coma with a spectrograph at the Multiple Mirror Telescope in Arizona. In their Fig. 2, Wagner and Schleicher show their measurements of the density of the highly toxic gas cyanogen (CN) at intervals up to $\sim$17,000 km from the comet nucleus. The distribution of CN around Hale-Bopp is much broader than the distribution of dust, reflecting the comparatively rapid diffusion of lighter molecules.

The measurement of CN spatial distribution is presented as 13 means and SDs suitable for TPL analysis (Fig. 12.1; Appendix 12.A). The slope of $b = 3.84 \pm 0.38$ is steep compared to most ecological results, possible due to the three-dimensional nature of the samples. It should be noted that the SDs in this case measure the variation in signal from Hale-Bopp integrated over the 20-min observation window. Thus, TPL is measuring the temporal mean and variance of CN density at 13 distances from the comet nucleus which combine the measurement error of the instrument and the spatiotemporal variation in density at each sample point.

Wagner and Schleicher used their data to estimate the rate of diffusion of CN from this comet and found them to be a good fit to the Haser model of gas diffusion from small comets. Thus, the TPL is the result of diffusion of material from a source under both attractive and repulsive forces: gravity and heating, respectively. At 1.1 billion km from the sun, Hale-Bopp's emission of CN was considered by Wagner and Schleicher to be in a steady state. Thus, in this instance, the TPL analysis is a direct measurement of a diffusion process.

Taylor's Power Law. https://doi.org/10.1016/B978-0-12-810987-8.00012-4

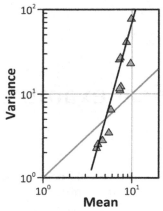

FIG. 12.1   Ensemble TPL ($NB = 13$) analysis if cyanogen density at intervals up to 17,000 km from the nucleus of Comet Hale-Bopp. *(Data from Fig. 2 in Wagner and Schleicher (1997).)*

The distribution of heavenly bodies

In the following case studies, we consider the distribution of stars and galaxies visible from Earth obtained from star and galaxy catalogs. The catalogs include names, celestial coordinates, galactic coordinates, and various physical properties. Coordinates of stars in the Milky Way galaxy are recorded in degrees north and south of the galactic equator and along the galactic equator east and west of a line from Earth to the galactic center. Because our solar system is close to the galactic equator, galactic and celestial latitudes are similar. Both systems provide a framework for a sampling frame suitable for TPL in which Right Ascension (RA) and Declination (DC) are equivalent to terrestrial longitude and latitude projected onto the celestial sphere. DC is measured in degrees, minutes, and seconds with points on the celestial sphere north of the celestial equator having positive DC, while those south are negative. RA is the object's position (longitude) at the vernal equinox recorded as hours minutes and seconds from GMT. A star or galaxy's position relative to the Earth changes with time, so coordinates are recorded relative to a standard datum; the current one is 1 January 2000. The units of hours of RA and degrees of DC comprise a system of 24×180 apertures within which celestial objects may be considered to lie. Of course, this ignores their distance from Earth. We can consider a celestial object, a star or galaxy, to be sampled by this 3-dimensional quadrat arrangement.

Whether a celestial object is included in a catalog depends on its brightness. The brightness of stars and galaxies is also recorded in the catalogs. The brightness of a star or galaxy is its magnitude, a logarithmic measure of the luminosity at a specified range of wavelengths, including the visible spectrum, which is denoted by M_v. The more luminous an object, the smaller the numerical value of its magnitude. For example, a magnitude 1 star is exactly a 100 times brighter than a magnitude 6 star. Astronomers use two different definitions of magnitude: apparent magnitude and absolute magnitude. The apparent magnitude is the brightness of an object as it appears in the night sky from Earth, while the absolute magnitude describes the intrinsic brightness an object would have if it were placed at 32.6 light years (or 10 parsecs) from us. Very luminous bodies may have negative magnitudes. The sun, for example, has an apparent magnitude of −27, but an absolute magnitude of 4.83. By comparison, the Milky Way galaxy has an absolute magnitude of about −21.

YALE BRIGHT STAR CATALOG

The Yale Bright Star Catalog (YBSC; Hoffleit and Warren, 1991) contains data on 9096 stars in the Milky Way Galaxy visible with the naked eye (apparent magnitude ≤ 6.5). It is a widely used source of basic astronomical and astrophysical data, including names, magnitudes, colors, and dimensions for single, double-, and multiple-star identifications. Star positions (as of 1 January 2000) are given in galactic coordinates. It is maintained and available on line from Harvard University (YBSC, 1991).

Using 1-h RA ($\equiv 15°$ of longitude) by $10°$ DC (latitude) quadrats the YBSC stars were analyzed by TPL along 18 RA transects of 24 quadrats and 24DC transects of 18 quadrats. The 18 RA transects parallel the equator while the 24 DC transects cross the galactic equator from south pole to north pole (Fig. 12.2A; Appendix 12.B). It is immediately clear that stars in quadrats on

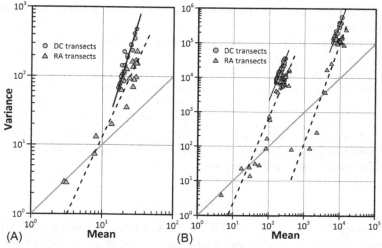

FIG. 12.2 (A) Ensemble TPLs of magnitude 6.5 stars listed in the Yale Bright Star Catalog counted in $NQ = 18$, $NB = 24$ for the DC transects and $NQ = 24$, $NB = 18$ for the RA transects. (B) Smithsonian Astrophysical Observatory Catalog of magnitude 10 stars counted in $NB = 18$ blocks of $NQ = 18$ quadrats (right) and $NB = 36 \times NQ = 36$ (left) on RA and DC transects. *(Data from http://tdc-www.harvard.edu/catalogs/bsc5.html and http://tdc-www.harvard.edu/catalogs/sao.html.)*

the DC transects covering a narrower range of means are more highly aggregated ($b = 3.54 \pm 0.190$) than those in the RA transects ($b = 1.90 \pm 0.109$). This could be anticipated as each DC transect samples stars across a very wide range of means that includes the periphery and the galactic center, but variation between transects is limited. By contrast, the RA transects sample the low-density peripheries and high-density equator with limited within transect range but greater range between transects. We see here, as with the barnacle *Balanus balanoides* on rocky shores at Cumbrae Island (Chapter 8), how TPL can identify variations in aggregation in orthogonal directions.

SAO STAR CATALOG

The Smithsonian Astrophysical Observatory Catalog (SAOC) contains 258,944 stars to about 10th magnitude compiled from various previous catalogs. The online version of the catalog, created by NASA's High Energy Astrophysics Science Archive Research Center from an earlier version of the Smithsonian catalog (Smithsonian, 1971), revised in 1989, is available from Harvard University (SAOC, 2001). These data with the much finer resolution of the Milky Way's stars were analyzed with 18 RA by 18 DC ($20° \times 10°$) and 36 RA by 36 DC ($10° \times 5°$) quadrats. The higher magnification of the SAOC included more stars than YBSC and produced a lower slope in the DC ($b = 2.79 \pm 0.149$ versus $b = 3.54 \pm 0.190$) and higher in the RA directions ($b = 2.94 \pm 0.206$ versus $b = 1.90 \pm 0.109$). The higher resolution of the SAOC permitted a finer resolution analysis with four times as many quadrats as the YBSC analysis. At this finer resolution, the range of means in the DC transects is about the same as the range in the YBSC analysis, although numerically greater while the range of means in the RA transects is an order of magnitude greater. Changing the resolution of SAOC analysis made little difference to the DC slopes ($b = 2.68 \pm 0.105$ versus $b = 2.79 \pm 0.149$), but the RA slopes differ markedly ($b = 2.50 \pm 0.124$ versus $b = 2.94 \pm 0.206$). It also generated 8 mean-variance points on or near the Poisson line (Fig. 12.2B left) compared to only 3 points with the larger quadrat size (Fig. 12.2B right). It is noteworthy that slopes of all 4 SAOC transects lie between the extremes of the YBSC. As the magnification and resolution increased, the difference between the orthogonal RA and DC transects diminished, even though the difference in ranges remained unchanged or increased.

PRINCIPAL GALAXY CATALOG

The Principal Galaxy Catalog (PGC, 2003) is a catalog of about one million galaxies, brighter than absolute magnitude 18. It constitutes the framework of the HYPERLEDA database (Paturel et al., 2003) and like the star catalogs gives properties, including the presumed morphological type, as well as accurate positions of the recorded galaxies. As with Milky Way stars in the SAO catalog, TPL was applied to the galaxies in the PGC in RA and DC transects

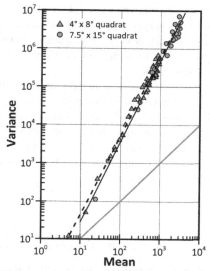

FIG. 12.3 Ensemble TPLs of galaxies listed in the HYPERLEDA Galaxy Catalog counted in windows of $4° \times 8°$ ($NB=NQ=45$) and $7.5° \times 15°$ ($NB=NQ=24$) are closely similar with the smaller window slightly more variable (higher A). *(Data from https://heasarc.nasa.gov/W3Browse/all/ pgc2003.html.)*

of two sizes: 45×45 ($8°$ RA $\times 4°$ DC) and 24×24 ($15°$ RA $\times 7.5°$ DC) quadrats. Fig. 12.3 (Appendix 12.C) shows the DC transects for the two quadrat sizes. Despite the larger quadrats having an order of magnitude higher means the two TPLs are remarkably close with a common line, $A = -0.38 \pm 0.085$ and $b = 2.04 \pm 0.031$ ($r=0.992$).

These orthogonal TPLs, almost identical at both scales, indicate no preferred direction but provide clear indication of structure in the universe as the slopes of $b=2.0$ indicate aggregation at all scales. In other words, the distribution of galaxies is not random, but highly structured. Although the variation in abundance of galaxies grows as the square of abundance at both scales in both directions, there is a small difference in the variation of RA transects at the two scales: the intercept ($A=-0.10 \pm 0.231$) of RA transects with $15°$ RA $\times 7.5°$ DC quadrat-sized samples is greater than the intercept of the transect with $8°$ RA $\times 4°$ DC samples ($A=-0.41 \pm 0.426$), although not significantly different from each other or zero ($P < 0.35$). Thus, although the overall abundance of galaxies does not appear to differ with direction (at these scales), there is a difference in the variance structure in orthogonal directions indicating a nonisotropy in the structure of the visible universe. The TPL analyses of galaxies recorded in PGC show the variation to be proportional to slightly more than the square of the mean, but the variance is not the same in all directions. It is known that at very large scales, galaxies are assembled in clusters and clusters of clusters. The universe is apparently organized into networks of large-scale filaments of matter where ordinary (baryonic) matter is aggregated at a range of scales (Carroll and Ostlie, 2017). Observation shows there is very little ordinary matter between the networks, so quadrats that happen to sample more of these gigantic voids will on average register higher variances.

An important feature of these astronomical catalogs is that the positions of stars and galaxies are known with considerable accuracy; typically, RA is known to <0.1 s and declination to <2 arcsec. Furthermore, unlike plants, they are discrete entities and unlike animals they move very slowly across the celestial sphere. Consequently, these catalogs offer a unique tool for investigating TPL properties by increasing the population (YBSC to SAOC) and changing the quadrat size (SAOC and PGC).

Geophysical

EARTHQUAKES OFF THE EAST COAST OF JAPAN

The instability of the Earth's crust that leads to earthquakes is difficult to monitor and predict because the dynamics governing the crust's stability comprise a set of "mutually dependent mechanisms acting at different spatial and temporal scales" (de Arcangelis et al., 2016). Many attempts have been made to model and predict earthquake occurrence, but these forces interacting over a range of spatial and temporal scales present a challenge. To find order in the disorder of earthquake events, Ogata (1988) applied several biostatistical models to data of timing and magnitude of earthquakes. His motivation in applying biostatistics to earthquake frequency and magnitude was to investigate the utility of using the period between earthquakes for predicting major earthquake events. He found long-range correlation in the data with an inverse power decay of autocovariance at large time lags, necessitating a change of time scale to make the data fit his models. In particular, he found that the actual curve of variance against time lag is much steeper than expected from a stationary Poisson process. In other words, earthquakes are aggregated in time. In addition, the cumulative number of earthquakes against magnitude on the Richter scale is a negative power law like the Pareto distribution.

Ogata included in his paper a table of 483 earthquakes by date and magnitude recorded in Japan from 1885 to 1980. In addition, the events were annotated as main shock, foreshock, or aftershock. The data are of earthquakes of <100 km depth with Richter magnitude ≥6.0 in one of the most active seismic areas in Japan where the Pacific plate is subducting beneath northeastern Japan.

Ogata's data of mainshock, fore-, and after-shock arranged by decade from 1880 to 1970 were analyzed by TPL individually and combined (Fig. 12.4; Appendix 12.D). The strong dependence of variance on mean of aftershocks supports the hypothesis of aggregation in time. However, the mainshock and foreshock TPLs with rather narrow ranges of means are poor. Combining all

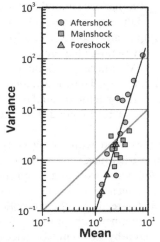

FIG. 12.4 Spatial TPL ($NQ = 2$–10, $NB = 24$) of earthquakes of magnitude >6.0 in the subduction zone off the northeast coast of Japan. A common line fits data of mainshocks, foreshocks, and aftershocks recorded from 1885 to 1980. *(Data from Table 1 in Ogata (1988).)*

three categories, however, shows that all categories of earthquake are aggregated, supporting the hypothesis that the incidence of earthquakes is correlated in the long term.

TORNADOES IN THE CONTINENTAL USA

Tippett and Cohen (2016) analyzed the frequency of "outbreaks" of multiple tornadoes occurring within a short period of time. They found that although the number of tornadoes reported per year from 1954 to 2014 had declined very slightly, the number of tornadoes in outbreaks had increased since 1954 by about 47%. The impact of tornadoes is profound, but the human and economic impact of tornadoes is greatly increased when several occur more or less simultaneously. An EF2 tornado in September 2010 hit The Ohio State University's agricultural experiment station (OARDC) at Wooster, Ohio. This event, one of 10 in northeast Ohio that afternoon, resulted in substantial damage to OARDC and neighboring residential areas, but was without serious injury. Another tornado a few miles southeast claimed a life.

The power of tornadoes is registered on a scale based on the amount of damage. The scale, called the Enhanced Fujita scale, ranks the weakest tornadoes EF0 and the strongest EF5. Tippett and Cohen selected clusters of tornadoes of EF1 and above separated by no more than 6 h regardless of where they occurred in the continental USA. They found the mean number of tornadoes per year to have been relatively constant in the years 1954–2014, but the number of tornadoes per outbreak for the same period had increased 6.6% per decade.

Furthermore, the variance of tornadoes per outbreak increased four times faster than the mean in line with a TPL slope of $b = 4.33 \pm 0.44$, which they compared to predictions from Lewontin and Cohen's (1969) model of multiplicative growth (Eq. 17.9). In the Lewontin-Cohen model, a quantity N_{t+1} at time $t+1$ is related to N_t by $N_{t+1} = A_t N_t$, where A_t is random multiplicative factor that grows or declines at each time step. Thus, it is anticipated that A_t is a measure of the growth in variance with respect to mean. Using ODR to estimate b, Tippett and Cohen found TPL to be in good agreement with the Lewontin-Cohen model.

Tornadoes are often associated with strong convection and atmospheric rotation. Tippett and Cohen used measures of these properties to compute an index for the number of tornadoes expected given the potential for supercell thunderstorms, which are frequently associated with tornado formation. Both indices also fit TPL well and agreed with the multiplicative growth prediction. The variable combining EF number with kilometers on the ground was fit well by TPL but not well predicted by Lewontin-Cohen, suggesting that this TPL relationship may be better explained by sampling variation.

Concern that the criteria for assessing tornadoes having changed in 1977 might lead to differences in the periods 1954–1976 and 1977–2014, Tippett and Cohen recomputed TPLs for the latter period, but found no grounds for separating the two periods: they presented results for 1954–2014.

As of this writing, The National Climatic Data Center's tornado database includes 1950–2016. Tippett and Cohen's tornado cluster analysis was repeated using the longer database and GMR instead of ODR, resulting in somewhat steeper TPL than they reported (Fig. 12.5; Appendix 12.E). In addition, analyses were run with different thresholds for cluster size: minimum of 3, 4, and 5 tornadoes as well as 6+ tornadoes within 6h. Lengthening the time period

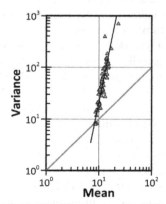

FIG. 12.5   Spatial TPL ($NQ = 3$–43, $NB = 67$) of clusters of ≥ 6 EF1–5 tornadoes in an outbreak are extremely highly aggregated ($b = 5.06 \pm 0.25$) for the period 1950–2016. *(Data of tornado occurrence downloaded from the National Climatic Data Center (http://www.nws.noaa.gov/om/hazstats.shtml).)*

increased b very slightly, but lowering the threshold defining the outbreak cluster size systematically reduced the slope from $b = 5.06 \pm 0.25$ at the 6+ threshold to $b = 4.02 \pm 0.22$ at 3+. Simultaneously, the intercept increased while the correlation coefficient also declined. The very high values of b obtained in this analysis remain to be explained, but strongly suggest that if the number of tornadoes per outbreak continues to increase the extreme growth in variation will lead to ever larger outbreaks.

PRECIPITATION ACTUAL AND SIMULATED

In a study of the possible effect of climate change on insect pest pressure on crops, RAJT et al. (2018) used linked simulation models to examine changes in insect population dynamics and crop production as the climate warms. The plant model EPIC is a well-known tool for studying the environmental footprint of agriculture by simulating crop growth and production methods. It has been used to answer both pure research and policy questions (Gassman et al., 2005; Wang et al., 2012). The insect population dynamics model GILSYM simulates cohorts of insects through their life cycle and from generation to generation (Rasche and Taylor, 2019). Both models are driven by weather data; wind, solar radiation, maximum and minimum temperatures, relative humidity, and precipitation. In their climate change study, RAJT et al. simulated populations of nine insect species, pests of corn (*Zea mays*) and soybeans (*Glycine max*), using actual data of temperature and precipitation from 1901 to 2000 and simulated data for 2001 to 2100 to drive their joint model.

The simulated data were generated by Geophysical Fluid Dynamics Laboratory (GFDL) using the CM2.0 General Circulation Model (GCM; Stouffer et al., 2006) simulating the comparatively benign CM2-SRES-B1 carbon scenario developed for the Intergovernmental Panel on Climate Change Fourth Assessment (IPCC-4A). This scenario simulates observed climate conditions from 1861 to 2000 very well and projects CO_2 stabilization at 550 ppm by 2100. Daily (simulated) records of maximum and minimum temperatures (°C) and precipitation (mm) from 1 January 2001 to 31 December 2100 were downloaded from the GFDL website at 6 nodes 2° of latitude apart by 12 nodes 2.5° of longitude apart covering the continental US. The actual data were taken from the NOAA database (NOAA, 2010) at sites as close as possible to the simulated nodes. As they were of varying lengths and with gaps, some interpolation and extrapolation was necessary to create continuous daily records to drive the models. The built-in weather generation function in EPIC was used for this purpose. As a test of the recreated and simulated data, TPL was conducted on the daily temperature and precipitation inputs at decade intervals. Much of the first decade (1901–1910) had to be estimated by EPIC's weather generator, very few site-days needed to be estimated for 1991–2000, and the 2091–2100 projection was entirely simulated using the CM2.0 GCM.

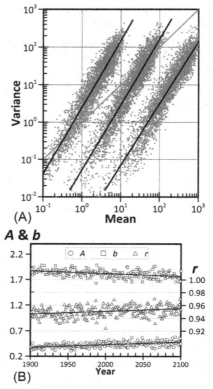

FIG. 12.6 (A) Spatial TPLs ($NQ = 72$, $NB = 3652$) of measured and simulated daily precipitation averaged over sites for three decades: left, measured daily rainfall with missing values simulated, 1901–1910; center, measured daily rainfall with missing values simulated, 1901–1910; right, daily rainfall simulated using the GFDL CM2.0 GCM, 2091–2100. Plots are staggered by one cycle for clarity. (B) Estimates of intercept and correlation coefficient trend upward and slope trends down for actual and simulated precipitation computed from daily records at 72 sites in each year from 1901 to 2100. *(Data used by RAJT et al. (2018).)*

Means and variances of the precipitation records at the 72 nodes for 3652 days were analyzed and are shown in Fig. 12.6A (Appendix 12.F) for the decades 1901–1910, 1991–2000, and 2091–2100. Clearly spatial variance scales with mean in all three decades. The three plots look very similar, but there are small differences that probably reflect the genesis of the three sets. All three are good fits although the first decade has a slightly lower correlation than the latter two; the completely simulated last decade and the partly simulated first decade have similar slopes, somewhat greater than the near complete 1991–2000 record, which also had the highest intercept indicating a greater overall variability. This latter point may be significant for climate simulations as real world precipitation is more variable than the simulated world but has weaker dependence on the mean. The range of precipitation is almost the same for all three decades, although there are more points in the middle decade below an average of 0.1 mm/day than the other two. A warmer atmosphere can hold more water vapor, which translates to increased rainfall and fewer dry days. The absence of very low means in the first decade may be due to lower precision in recording instruments in the early 20th century. There are clear trends in A, b, and r over the two centuries, with no obvious break at the transition from actual to simulated precipitation (Fig. 12.7B), reflecting the ongoing changes in spatio-temporal patterns of precipitation occurring with oceanic evaporation. The steady decline in b is interesting as it suggests reduced spatial variation with increasing rain/day as the number of consecutive rain days increases (Roque-Malo and Kumar, 2017).

A companion analysis of the measured and simulated maximum and minimum temperature is examined in the next chapter.

Numerical

TRAFFIC THROUGH NETWORKS

The Internet is a network of linked routers directing packets of information to and from client and host computers. de Menezes and Barabási (2004a) analyzed the relationship between mean and variance of Internet traffic and accesses of World Wide Web (WWW) pages. They analyzed daily traffic through 374 routers of different sizes with average numbers of packets over

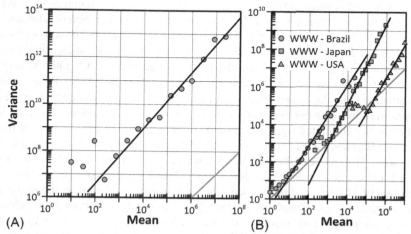

FIG. 12.7   (A) Ensemble TPL ($NQ = 374$, $NB = 15$) of activity of 374 Internet routers of various sizes over a 2-day period at 5-min resolution. (B) Ensemble TPLs ($NQ = 1000$, $NB = 14$–21) of daily visits to websites over a 30-day period for 1000 sites in Brazil, Japan, and USA. Plots are staggered by one cycle for clarity. *(Adapted from Figs. 2 and 3 in de Menezes and Barabási (2004a).)*

6 orders of magnitude. The slope estimated by ODR is $b \approx 1$. Analyzing accesses to 3000 WWW pages by people in three countries over a 30-day period, they obtained an average slope of $b \approx 2$. These analyses suggested to de Menezes and Barabási that characteristic values of $b = 1$ and $b = 2$ result from two different processes, a conclusion bolstered by a model of transport over a scale-free network. Additionally, they analyzed data of activity of a microprocessor and streamflow through a network of 3945 US rivers (see Hurst scaling in Chapter 16) and daily automobile traffic over 127 highways in Colorado and Vermont and found all three sets to obey TPL. The activity in a microprocessor resulted in a $b \approx 1$ and the other networks produced $b \approx 2$. They concluded that the observed fluctuation scaling was the net result of competing internal and external dynamics, a concept they explored in a second paper (de Menezes and Barabási, 2004b).

The results of a reanalysis of the Internet and WWW cases are reproduced in Fig. 12.7 (Appendix 12.G). Data from de Menezes and Barabási's Figs. 3 and 4 and analyzed by GMR obtained slightly higher slopes than in the original report. More significantly, the new analyses omit the outliers at low means clearly visible in the figure. In the case of the WWW accesses in Brazil and Japan, the lowest three points appear to parallel to the Poisson line, but the low mean values of other WWW example (USA) appears to form a line perpendicular to the fitted line. Similarly, the lowest three points in the Internet router plot clearly do not fit TPL. de Menezes and Barabási fitted TPL by ODR and included these low mean outliers and considered the WWW as a single set, whereas it is clear that the three TPLs differ both in slope and intercept, suggesting differences in behavior of Internet users in the three countries.

Duch and Arenas (2006) also analyzed Internet traffic through routers and found b to vary from 1.42 to 1.72 for the Abeline backbone routers (also analyzed by de Menezes and Barabási) over time spans of 2 days to 2 months using a window of 5 min. They found the value of b decreased with increasing time span: TPLs of both 1- and 2-month spans resulted in $b = 1.42$, whereas time spans of 1 week and 2 days gave $b = 1.54$ and 1.72, respectively. To account for these results, Duch and Arenas developed a network model in which agents were introduced at random and traversed the network at random. The length of time the simulated agents persisted in the network and the length of the observation window (Δt) determined the TPL parameters. In short persistence models, TPL slope followed a sigmoid curve from $b = 1.0$ (Poisson) to $b \approx 1.45$ as the window was lengthened. For long persistence models, b approached higher asymptotes with increasing window length; at the highest persistence $b \approx 2.0$. Clearly in their model b was sensitive to both the window and the population of agents in the simulation network. Other authors have analyzed automobile traffic by TPL: Fronczak et al. (2010) analyzed traffic in Minnesota and found $b = 1.43$ in aggregate but varied temporally as traffic flow varied diurnally.

FOREIGN EXCHANGE MARKETS

The mathematician Benoit Mandelbrot (1963) showed that time series of cotton prices were correlated over long time spans so that images of the time series looked similar regardless of the time scale. This implies (and he demonstrated) that large price changes (both positive and negative) are more likely to be followed by large changes than small; likewise, small price changes are more likely to be followed by small changes. Thus, the trace of commodity prices is not a Gaussian random walk but one he called Paretian from the Pareto distribution, which has no finite moments above the mean. Of course, any finite time series of such events has a finite sample variance, but the distribution has long tails and is highly leptokurtic. In the case of commodity, equity and foreign exchange (forex) prices, changes can be either positive or negative, while the equivalent data for the distribution of plants, animals, stars, and galaxies can only be positive and highly skewed to the right.

Since Mandelbrot's (1963) paper, the time series of equity and commodity prices have been explored by a variety of means and Mandelbrot's suggestion has been confirmed many times over. Fluctuation scaling properties have been found in time series of river discharge (see Chapter 16) and computer traffic through routers among other systems (Eisler et al., 2008). A study by Sato et al. (2010) of forex activities is a good illustration of Mandelbrot's theory.

The forex market, the largest financial market in the world, is a network of traders (brokers, bankers, and speculators) linked electronically. Their trades are recorded continuously by computers that also match bid and ask prices of the traders. The time series of number of trades resembles what Mandelbrot called fractional Brownian motion (Fig. 16.1).

A major difference in these analyses compared to ecological applications of TPL is that the data are typically not sampled. While forex buy and sell orders are discrete events in time, their recording is continuous and complete. Thus, data for investigating scaling with forex pair prices are not discrete entities like animals and galaxies but a near-continuous stream of real numbers. The question of how to examine scaling is answered by accumulating the signal over a specified time interval or window. Sato et al. demonstrated a relationship between TPL slope and the size of the window in their study of scaling behavior of time series of forex currency pairs.

Sato et al. analyzed data of transactions of 45 currency pairs recorded at 1-min resolution provided by CQG Inc.[1] For each currency pair, quotation activities were extracted from the CQG database. They calculated the mean and SD of number of transactions within a set of windows of length Δt over a one-week period in June 2007 and performed ODR of log transformed mean and SD and estimated the slope α $(=b/2)$ and intercept. To establish the

---

1. CQG Inc. is a company headquartered in Colorado that provides real-time and historical forex data as well as sophisticated software to analyze and execute forex transactions.

relationship between slope and window length, Sato et al. examined several consecutive periods with window lengths of $\Delta t = 1$ min, 10 min, and 120 min (Appendix 12.H). At window lengths of $\Delta t > 100$ min, the market is strongly influenced by information flowing to traders (by phone, email, or other electronic means), but at very short windows (Δt of a few minutes) fluctuations in market dynamics are random. However, transaction activities are not Poisson but aggregated in time even for windows ~ 1 min with $b = 1.6$. They suggested that temporally or mutually correlated activities occur even at this short time scale.

TPL slope at 12 window lengths ranging from $1 < \Delta t < 120$ min, ranged from $0.8 < \alpha < 0.9$ and varied from week to week (Fig. 12.8A). As a function of window length, $\alpha = 0.8 + 0.04\Delta t$ up to about $\Delta t = 100$ min, above which α changed little: Sato et al. called this "saturation." Examination of TPL derived from daily ($\Delta t = 1440$ min) windows confirmed that slope varied between these limits over the longer time frame (Fig. 12.8B). They noted the similarity in forex market quotation fluctuations and stock market value fluctuations (Eisler et al., 2008) in their TPL behavior, including the tendency to increase with window length (Fig. 12.8A). A big difference between forex and equity markets is the increased noise in b over time (Fig. 12.8B), no doubt due to the extreme volatility of Forex prices over quite short time spans. The values of $1.6 < b < 1.8$ suggest that forex market transactions are influenced by both endogenous and exogenous factors and because they are strongly correlated in time they are related to Hurst scaling (Chapter 16). Exponents of individual currency pairs

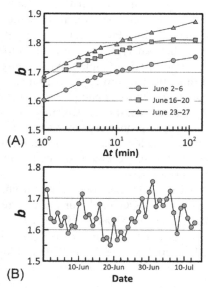

FIG. 12.8 (A) Stability of forex temporal TPLs in relation to window length (Δt) for three dates in June 2007. (B) Variation in temporal b over 2-month period. *(Adapted from Figs. 3 and 4 in Sato et al. (2010).)*

increase with the mean activity of the pair in a pattern very similar to the variation in b with window length.

The observed temporal fluctuations are probably due to the interaction of a stochastic external driving force (forex transactions initiated after information received) and complex dynamics internal to the forex network. The scaling exponent is very dependent on the length of the observation window providing information on the correlations on different time scales.

PRIME NUMBERS

One of the most intriguing discoveries about TPL was made by Kendal and Jørgensen (2015) who showed a connection between TPL, the distribution of prime numbers, and the zeros of the Riemann zeta function. To explain their contribution, it is necessary to review some prime number theory, which I draw from the very readable Inaugural Lecture by Prof. Don Zagier of Bonn University and translated by R. Perlis (Zagier, 1977).

The earliest formal proof that there are an infinite number of prime numbers, that is numbers divisible only by themselves and 1, is due to Euclid (3rd century BCE). Contrary to intuition, 1 is not included in the list of primes that starts 2, 3, 5, 7, 11, 13, 17 … The intervals between prime numbers obey no discernable pattern, making them hard to predict. It has long been a goal to discover a formula for determining the positions of primes. The graph of the accumulated distribution of primes is a step function called $\pi(x)$, the number of primes less than or equal to x. It was suggested by Carl Friedrich Gauss in about 1792 that $\pi(x)$ is approximated by the logarithmic integral, $\text{Li}(x)$:

$$\pi(x) \approx \text{Li}(x) = \int_2^x \frac{du}{\ln(u)}$$

Although $\text{Li}(x) > \pi(x)$ for all known primes, it has been proved that $\text{Li}(x)$ and $\pi(x)$ do cross at some point. An improvement to Gauss's prime number equation was proposed 67 years later by Bernhard Riemann:

$$\pi(x) \approx R(x) = \sum_{n=1}^{\infty} \frac{\mu(n)}{n} \text{Li}\left(x^{1/n}\right) \tag{12.1}$$

where $\mu(n)$ are the Möbius numbers, which take values of 1 if $n = 1$, 0 if n has multiple instances of the same prime factor(s), or $(-1)^k$ if n is the product of k different primes. Eq. (12.1) is known as the prime number theorem and $R(x)$ is called the Riemann function. It was proved independently by de la Vallée Poussin and Hadamard in 1896. Zagier remarked that "the approximation is amazingly good": the absolute difference between $\pi(x)$ and $R(x)$, $|D(x)| = 1946$ in the nearly 51 million primes below $x = 10^9$. An alternative form for $R(x)$ is.

$$R(x) = 1 + \sum_{k=1}^{\infty} \frac{(\ln(x))^k}{kk!\zeta(k+1)},$$

where $\xi(k+1)$ is the Riemann zeta function whose arguments are complex numbers involving the imaginary constant $i = \sqrt{(-1)}$. Riemann showed that an exact formula for $\pi(x)$ exists:

$$\sum_n^\infty \frac{1}{n} \pi\left(x^{1/n}\right) = \text{Li}(x) - \sum_\rho \text{Li}(x^\rho),$$

where the sum on the right runs over the roots of the Riemann zeta function (ξ). The roots are the points on the graph where $\xi = 0$. There are two sets of zeros: the so-called trivial zeros are the negative even integers and the nontrivial zeros all lie in the complex plane between 0 and 1 on the real axis: they are conjectured to lie exactly on the line whose real part $= \frac{1}{2}$. This is the Riemann hypothesis and is one of the oldest and most intractable mathematical mysteries, a proof of which "would have far-reaching consequences for number theory" (Zagier, 1977).

The fluctuations between the prime counting function and Riemann's equation can be expressed in terms of the entire set of zeros of the complex zeta function:

$$D(x) = R(x) - \pi(x) = \sum_\rho R(x^\rho). \tag{12.2}$$

The sum of zeros is given by

$$\sum_\rho R(x^\rho) = T_0(x) + \sum_{k=1}^n T_k(x),$$

where $T_0(x)$ is the infinite sum of the trivial zeros, which is a constant and makes only a minor contribution to the equation and

$$T_k(x) = -R(x^{\rho_k}) - R(x^{\rho_{-k}})$$

the kth pair of roots of the Riemann zeta function. These are the nontrivial zeros, whose imaginary parts above and below the real axis cancel so the $T_k(x)$ are all real. The approximation to $\pi(x)$, which this function yields, is

$$R_n(x) = R(x) - \left\{ T_0(x) + \sum_{k=1}^n T_k(x) \right\} = \pi(x) \tag{12.3}$$

As $n \to \infty$, this sequence of functions approaches the $\pi(x)$ step function. As low as $n = 100$, Zagier called the fit "astonishingly good."

Kendal and Jørgensen developed a computational analysis of the deviations between the prime number step function $\pi(x)$ and the positions predicted by Riemann's counting formula $R(x)$ (Eq. 12.2). They computed the absolute value of the difference between $\pi(x)$ and $R(x)$, $|D(x)|$, for integers up to 10^{12} and assigned $|D(x)|$ to sequential equal-sized nonoverlapping bins. The contents of each bin were summed and the means and variances calculated for the set of bins. This procedure was repeated for progressively larger bin sizes and the set of means and variances subjected to TPL analysis. Thus, the TPL analysis was an ensemble of different quadrat sizes (NQ). Their first experiment placed $|D(x)|$ for the first 50,000 integers in bins initially 10 integers long and increasing in steps of 10. They do not specify the number of bin sizes (equivalent to NB), but clearly the number of discrete samples (equivalent to NQ) gets smaller as the size of the bins increases. The first two points on the graph have a mean of about 1.5 and variance 5. The next bin size had mean and variance of approximately 25 and 400. The largest mean-variance pair is approximately 3000 and 2×10^6. The experiment was repeated for three other ranges of integers, 10^6–10^7, 10^8–10^9, and 10^{11}–10^{12}, as a check that the first result was not a local phenomenon. Their estimates of a and b are given without standard errors or correlation coefficients in Table 12.1.

Kendal and Jørgensen show empirically that the deviations $|D(x)|$ can be described by the Tweedie Poisson-gamma exponential distribution model (see Chapter 3) and has the appearance of $1/f$ noise (Fig. 16.1). $1/f$ noise appears to be random but possesses long-range correlations that can generate TPLs with a limited range of b (Mandelbrot and van Ness, 1968). Kendal and Jørgensen suggest that $1/f$ noise is a consequence of the statistical convergence behavior associated with Tweedie's probability generating function.

An interesting aspect of the study is the similarity in estimates of b in this study and those of forex activity: Table 12.1 shows b increasing with bin (window) size over a similar range (Appendix 12.G). At least part of this variation is probably attributable to variations in long-range correlation of prime numbers. The difference in b at different scales they suggest is evidence the distribution of primes is multifractal. Kendel and Jørgensen emphasize that although the TPL nature of $|D(x)|$ is entirely empirical, the Riemann deviations $D(x)$ are intimately related to the positions of the prime numbers and therefore the TPL analysis is saying something about the distribution of primes. If there is

TABLE 12.1 Estimates of TPL for ensemble analysis of the deviations between the prime counting function $\pi(x)$ and Riemann's approximation $R(x)$

Experiment	Interval	Bin steps	a (=10^A)	b
1	0–50,000	10	1.37	1.83
2	10^6–10^7	10^3	3.53	1.66
3	10_8–10^9	10^5	9.28	1.67
4	10^{11}–10^{12}	10^8	8.57	1.74

The number of samples (NQ) and bin sizes (NB) used in the regressions are not given. TPL estimates from Kendal and Jørgensen (2015).

a connection between the distribution of primes and TPL, consideration of Eqs. (12.2) and (12.3) suggest TPL may also describe the distribution of zeros of the Riemann zeta function,

$$|D(x)| = |R(x) - \pi(x)| = \left| \left\{ T_0(x) + \sum_{k=1}^{n} T_k(x) \right\} \right|.$$

The examples in this chapter show TPL fits data streams unconnected to biological phenomena. Whatever the rules governing the generation of TPL for these nonbiological examples, they must surely apply to biological phenomena also. It is not obvious what the deviations between observed and predicted prime numbers and the price fluctuations of currency pairs have in common with the incidence of tornadoes and earthquakes, but they are all the result of complex processes, something they have in common with the spacing behavior of organisms.

Appendix: TPL estimates for nonbiological phenomena

	NB	Range of Means	r	A	SE[A]	b	SE[b]
Astronomical							
A Ensemble TPL of cyanogen in the coma of comet Hale-Bopp. Data digitized from Fig. 2 in Wagner and Schleicher (1997)							
CN density (10^{10} cm^{-2})	13	4.01 - 9.82	0.934	−2.095	0.315	3.840	0.380
B Ensemble TPL of the stars in the Milky Way Galaxy. Yale Bright Star Catalog of 9096 stars of magnitude 6 and brighter (Hoffleit and Warren, 1991); Smithsonian Astrophysical Observatory Star Catalog of 258,997 stars (Smithsonian, 1971).							
YBS catalog − 15° × 10° DC transects	24	22.4 - 39.1	0.974	−2.713	0.274	3.540	0.190
RA transects	18	3.83 - 40.9	0.970	−0.667	0.154	1.901	0.109
SAO catalog − 20° × 10° DC transects	18	646 - 1,090	0.974	−2.767	0.430	2.789	0.149
RA transects	18	43.9 - 1,461	0.955	−3.777	0.580	2.936	0.206
10° × 5° DC transects	36	159 - 279	0.972	−2.027	0.240	2.684	0.105
RA transects	36	4.72 - 366	0.955	−2.157	0.276	2.504	0.124
C Ensemble TPL of 983,261 galaxies brighter than absolute magnitude 18 from the Principal Galaxy Catalog (PGC, 2003), of the HYPERLEDA database (Paturel et al., 2003)							
PGCcatalog − 8° × 4° DC transect	45	5.50 - 965	0.991	−0.452	0.105	2.074	0.041
RA transect	45	180 - 786	0.957	−0.101	0.231	2.018	0.087
15° × 7.5° DC transect	24	49.5 - 3106	0.990	−0.496	0.184	2.059	0.059
RA transect	24	762 - 2734	0.950	−0.405	0.426	2.094	0.133
8° × 4° and 15° × 7.5° DC transect	69	5.50 - 3106	0.992	−0.379	0.085	2.036	0.031

Continued

	NB	Range of Means	r	A	SE[A]	b	SE[b]
Geophysical							

D Temporal TPLs of earthquakes per decade off the northeast coast of Japan between 1885 and 1980. Data from Table 1 in Ogato (1988)

	NB	Range of Means	r	A	SE[A]	b	SE[b]
Mainshock	10	2.00 - 4.44	0.390	−0.552	0.246	1.837	0.535
Foreshock	5	1.33 - 2.50	0.263	−0.801	0.307	2.746	1.185
Aftershock	9	1.20 - 8.00	0.935	−0.872	0.232	3.432	0.407
Combined	24	1.20 - 8.00	0.842	−1.026	0.175	3.382	0.373

E Spatial TPLs of tornado outbreaks follows TPL. Updated NOAA data similar to Fig. 2d in Tippett and Cohen (2016)

	NB	Range of Means	r	A	SE[A]	b	SE[b]
1954–2014 – Outbreaks of 6+ tornadoes	61	8.35 - 23.2	0.913	−3.624	0.278	4.968	0.259
1950–2016 – Outbreaks of 6+ tornadoes	66	8.35 - 23.2	0.916	−3.728	0.267	5.062	0.250
5+ tornadoes	66	6.22 - 19.6	0.910	−3.028	0.240	4.627	0.237
4+ tornadoes	66	5.18 - 18.2	0.898	−2.515	0.224	4.377	0.237
3+ tornadoes	66	4.48 - 15.8	0.895	−1.890	0.190	4.022	0.221

F Spatial TPLs of precipitation at 72 points in central North America for 3 decades: 1901–1910 and 1991–2000 actual data and 1991–2000 simulated data using the GFDL CM2.0 GCM. Data used by Taylor et al. (2018)

	NB	Range of Means	r	A	SE[A]	b	SE[b]
1901–1910	3652	0.0047 - 16.2	0.949	0.370	0.005	1.864	0.010
1991–2000	3652	0.0025 - 16.7	0.949	0.439	0.005	1.837	0.010
2090–2100	3652	0.0001 - 18.9	0.958	0.484	0.005	1.716	0.008

Numerical

G Frequency of foreign exchange transactions of 45 currency pairs over a one-week period in June 2007 using window lengths Δt = 1, 10, and 120 min. Estimation of TPL parameters was by ODR; correlation coefficients were obtained from the standard errors and the range of means was obtained by digitizing the extreme values in Fig. 2 in Sato et al. (2010) which plotted standard deviation against mean

	NB	Range of Means	r	A	SE[A]	b	SE[b]
Window length, Δt = 1 min	45	0.8 - 120	0.973	1.95	0.007	1.60	0.04
Δt = 10 min	45	0.9 - 1250	0.976	1.82	0.015	1.70	0.04
Δt = 120 min	45	12 - 15,850	0.989	1.78	0.027	1.76	0.04

H Temporal TPL of the number of node connections in networks and web page accesses in 3 countries. Data from Figs. 3 and 4 in de Menezes and Barabasi (2004a)

	NB	Range of Means	r	A	SE[A]	b	SE[b]
Internet traffic through routers	12	250 - 2×10^7	0.992	4.219	0.169	1.189	0.046
WWW accesses – Brazil	21	6.95 - 22,214	0.993	−0.413	0.084	1.630	0.045
Japan	20	88.0 - 89,072	0.996	−1.363	0.106	2.138	0.043
USA	14	988 - 86,964	0.991	−0.789	0.201	1.816	0.071

References

Carroll, B.W., Ostlie, D.A., 2017. An Introduction to Modern Astrophysics, second ed. Cambridge University Press, Cambridge, UK.

de Arcangelis, L., Godano, C., Grasso, J.R., Lippiello, E., 2016. Statistical physics approach to earthquake occurrence and forecasting. Phys. Rep. 628, 1–91.

de Menezes, M.A., Barabási, A.-L., 2004a. Fluctuations in network dynamics. Phys. Rev. Lett. 92, 028701.

de Menezes, M.A., Barabási, A.-L., 2004b. Separating internal and external dynamics of complex systems. Phys. Rev. Lett. 93, 068701.

Duch, J., Arenas, A., 2006. Scaling of fluctuations in traffic on complex networks. Phys. Rev. Lett. 96, 218702.

Eisler, Z., Bartos, I., Kertész, J., 2008. Fluctuation scaling in complex systems: Taylor's law and beyond. Adv. Phys. 57, 89–142.

Fronczak, A., Fronczak, P., Bujok, M., 2010. Taylor's power law for fluctuation scaling in traffic. Acta Phys. Polon. 3, 327–333.

Gassman, P.W., Williams, J.R., Benson, V.W., Izaurralde, R.C., Hauck, L., Jones, C.A., Atwood, J.D., Kiniry, J., Flowers, J.D., 2005. Historical Development and Applications of the EPIC and APEX Models. Working Paper 05-WP 397. Iowa State University, Center for Agricultural and Rural Development, Ames, IA. http://www.card.iastate.edu/publications/synopsis.aspx?id=763. Verified January 2019.

Hoffleit, D., Warren, C., 1991. The Bright Star Catalogue, Fifth Revised Edition Yale University Observatory, New Haven, CT.

Kendal, W.S., Jørgensen, B., 2015. A scale invariant distribution of the prime numbers. Computation 3, 528–540.

Lewontin, R.C., Cohen, D., 1969. On population growth in a randomly varying environment. Proc. Natl. Acad. Sci. U. S. A. 62, 1056–1090.

Mandelbrot, B.B., 1963. On the variation of certain speculative prices. J. Bus. 36, 394–419.

Mandelbrot, B.B., van Ness, J.W., 1968. Fractional Brownian motions, fractional noises and applications. SIAM Rev. 10, 422–437.

NOAA, 2010. https://www.ncdc.noaa.gov/cdo-web/datatools/findstation. Verified Jaunary 2019.

Ogata, Y., 1988. Statistical models for earthquake occurrences and residual analysis for point processes. J. Am. Stat. Assoc. 83, 9–27.

Paturel, G., Petit, C., Prugniel, P., Theureau, G., Rousseau, J., Brouty, M., Dubois, P., Cambresy, L., 2003. HYPERLEDA—I. Identification and designation of galaxies. Astron. Astrophys. 412, 45–55.

PGC, 2003. https://heasarc.nasa.gov/W3Browse/all/pgc2003.html. Verified January 2019.

Rasche, L., Taylor, R.A.J., 2019. EPIC-GILSYM: modelling bitrophic processes in agriculture. J. Appl. Ecol. 56. in press.

Roque-Malo, S., Kumar, P., 2017. Patterns of change in high frequency precipitation variability over North America. Sci. Rep. UK 7, 10853.

SAOC, 2000. http://tdc-www.harvard.edu/catalogs/sao.html, 2001, Verified January 2019.

Sato, A.-H., Nishimura, M., Hołyst, J.A., 2010. Fluctuation scaling of quotation activities in the foreign exchange market. Physica. A 389, 2793–2804.

Smithsonian, 1971. Star Catalog: Positions and Proper Motions of 258,997 Stars for the Epoch and Equinox of 1950.0. Smithsonian Astrophysical Observatory Publication 4652. Smithsonian Institution of Washington, DC.

Stouffer, R.J., Broccoli, A.J., Delworth, T.L., et al., 2006. GFDL's CM2 Global coupled climate models. Part IV: idealized climate response. J. Clim. 19, 723–740.

Taylor, R.A.J., Herms, D.A., Cardina, J., Moore, R.H., 2018. Climate warming and pest management: unanticipated consequences of trophic dislocation. Agronomy 8, 1–23.

Tippett, M.K., Cohen, J.E., 2016. Tornado outbreak variability follows Taylor's power law of fluctuation scaling and increases dramatically with severity. Nat. Commun. 7, 10668.

Wagner, R.M., Schleicher, D.G., 1997. The spectrum and spatial distribution of cyanogen in comet Hale-Bopp (C/1995 01) at large heliocentric distance. Science 275, 1918–1920.

Wang, X., Williams, J.R., Gassman, P.W., Baffaut, C., Izaurralde, R.C., Jeong, J., Kiniry, J.R., 2012. EPIC and APEX: model use, calibration, and validation. Trans. ASABE 55, 1447–1462.

YBSC, 1991. http://tdc-www.harvard.edu/catalogs/bsc5.html. Verified January 2019.

Zagier, D., 1977. The first 50 million prime numbers. Math. Intell. 1, 7–19.

Chapter 13

Counter examples

Taylor's power law has such wide application that a chapter on exceptions is necessarily short. Many examples in the literature can be traced to inconsistent sampling or poorly defined statistical population being sampled. As Clark and Perry (1994) showed, the number of quadrats and points in the regression can bias the estimation of parameters. But even when NQ and NB are inadequate for unbiased estimation, the resulting regression may be convincing (e.g., Figs. 7.29 and 10.9A). However, there is no shortage of examples where poor fit can be traced to inadequate NQ and/or NB: Clark & Perry give the example of meadow brome (*Bromus commutatus*) with $NQ = 2$. But there are other classes of failure and paradoxically perhaps, these failures of TPL are more informative than the successes in understanding TPL. This chapter is devoted to counter examples that document the limitations of TPL and help in discerning how and why it exists. Although means and variances can be computed from lists containing negatives, clearly TPL cannot work with data that contain negative means as the logarithms do not exist. Therefore, only data bounded by zero can be considered candidates for TPL.

Sampling

The most common cause of failure of TPL to fit data is a failure to collect the data in a manner amenable to TPL analysis. There are three main causes: too few quadrats per data point (NQ), too few data points for the regression (NB) and poorly defined population being sampled. The following case studies illustrate these issues. It should be noted that none of the data were collected with fitting TPL in mind. They are data in the literature that should fit but don't. We look at inadequate NQ and NB first and then an example of a biological population that occupies two different environments that are statistically distinct.

Inadequate NQ *or* NB

LRT et al. (1988) and Clark and Perry (1994) have recommended that $NQ \geq 15$. Failure to reach this ideal may lead to biased estimates even when the power law relationship is clear and the fit looks good. However, Clark and Perry caution that $NQ \geq 15$ is no guarantee of an unbiased estimate. The reason estimates may be biased is that systematic bias in sampling is more likely to occur with small

Taylor's Power Law. https://doi.org/10.1016/B978-0-12-810987-8.00013-6

samples than large ones. With small NQ, TPL is frequently poorly defined with low correlation coefficient. Examples in which NQ is very small and NB is barely adequate but producing a plausible result, include a number of nematode analyses in which $NQ < 7$ and $NB < 10$ (e.g., Appendices 7.H, 7.J, 7.M, 7.Q, and 7.W). Parameter estimates, even if biased, may still be useful for transformation and for developing sampling plans, provided the same protocols are used in practice. However, the bias will render comparisons between data sets, populations, or species problematical.

MEIOBENTHOS IN THE BALEARIC ISLANDS

A survey of the meiobenthic communities of the Balearic Islands in the Mediterranean by Deudero and Vincx (2000) found little variation in communities in seven locations varying in anthropogenic influence. The least influenced were four stations in Cabrera National Park and the most closely associated with human activity was off a tourist destination on Mallorca. Five sediment samples were taken by scuba diver at each site and the fauna extracted. Nematodes dominated the samples, accounting for $>55\%$ of specimens. These benthic communities show little variation between stations and the nematode distribution was little affected by environmental conditions at the sites. This is not surprising given the high variation and narrow range of means between sites (Fig. 13.1A). One national park station stands out; without it, there would be no TPL. Given the very high V/M ratios, $NQ = 5$ is quite inadequate to define

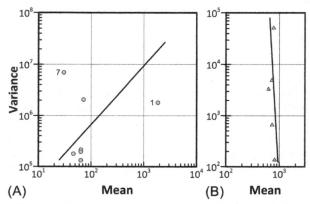

(A)   Mean   (B)   Mean

FIG. 13.1 (A) Ensemble TPL ($NQ = 5$, $NB = 7$) of benthic nematode communities at 7 sites in the Balearic Islands. Station 7 is the most anthropogenically influenced site and station 1 is one of the national park sites: 6 sites span a narrow range of means but a large range of variances result in an unconvincing TPL. (B) Variance-mean plot of density (#/10 cm^2) of nematodes recovered from topographically different locations in the Darwin Mounds in the Atlantic. The sites are characterized less by nematode density than the extreme variability. *(A: Data from Table 2 in Deudero and Vincx (2000); B: Data from Table 2 in Van Gaever et al. (2004).)*

TPL efficiently, and by implication, separate the stations along the presumed anthropogenic gradient.

NEMATODES ON THE DARWIN MOUNDS

The Darwin Mounds are coral-topped sandy mounds 900–1060 m below the surface of the Atlantic between Scotland and Iceland. Van Gaever et al. (2004) took boxcore sediment samples 900–960 m below the surface at five topographically different stations (top, edge, tail or off mound, or "pockmarks," depressions of 1–2 m deep and approximately the same area as the mounds). Two box core samples were taken per station type, except 4 were taken in mound tail areas: thus, $NQ = 2$ or 4 stations per physiographic feature. Meiofauna were extracted from the samples and density expressed as $\#/10 \, cm^2$. Nematodes comprised >93% of the meiofauna extracted. Van Gaever et al. report no significant differences in the relative abundances of the meiobenthos between stations. This is not surprising given the extremely large range of variances compared to means. Fig. 13.1B shows the variance-mean relationship for the nematodes: the total meiofauna is virtually identical. Such a steep variance-mean slope ($b = -21.78 \pm 12.281$, $r = -0.22$) is probably beyond the ability of any transformation to stabilize (Healy and Taylor's (1962) power transformation is approximately $Z = X^{12}$). The density of nematodes at these sites is similar but the variation is extreme, at least in part due to the small number of samples per topographical feature, but may also reflect real differences in community structure from littoral and land communities.

Cohen (2014) developed a model for TPL that predicts a singularity in TPL at which the slope becomes very steep, both positive and negative (Chapter 17). This case is possibly an example of b close to the singularity.

Inconsistent sampling

THRIPS IN A CUCUMBER CROP

The spatial distribution of the western flower thrips (*Frankliniella occidentalis*) is discussed in Chapter 8 (Appendix 8.M1). RAJT et al. (1998) compared their own data collected in six commercial greenhouses with data from the literature and obtained closely similar TPLs with one exception.

Samples taken with yellow sticky cards stationed within the canopy of a cucumber crop did not produce a convincing TPL, while samples taken above the canopy did (Fig. 13.2). These regressions compare the spatial pattern of western flower thrips in two distinctly different habitats: in free air and in a tightly enclosed space. Differences in TPL cannot be due to demographic differences, because the populations within and above the canopy were reproductively the same; the thrips differed only in where they were flying. Both different pattern(s) of flight and variable catchability of traps within the canopy

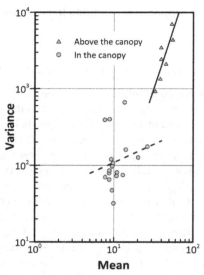

FIG. 13.2 Spatial TPLs of western flower thrips caught within and above a cucumber crop. Counts of thrips caught above the crop form a significant TPL $(A = -2.70 \pm 1.443, b = 3.74 \pm 0.883, r = 0.850, P < 0.01)$, but the TPL of thrips caught within the canopy is not significant $(A = -0.38 \pm 0.602, b = 0.47 \pm 0.576, r = 0.198, P > 0.2)$. *(Unpublished data of Bogaers and Lindquist in RAJT et al. (1998).)*

would produce unconvincing TPLs. If the sticky trap samplers inside the canopy were variably visible, either spatially or temporally, each sampler would have a unique sampling efficiency and therefore could not produce consistent abundance estimates (RAJT, 2018). Only data of the same behavioral population, collected systematically, result in the high level of significance and goodness of fit typical of TPL. In this instance, statistically speaking, the population is defined by where the thrips fly.

Trap saturation

POWDERY MILDEW ON APPLES

Xu and Madden's (2002) study of powdery mildew on apple trees examined the relationships between disease incidence and colony density. They recorded the progression of the powdery mildew infestation by counting the number of disease colonies on leaves of 10 apple trees approximately twice a week. In 3 of 4 years, they obtained good spatial TPLs of colonies/leaf through the season. Those 3 years' TPLs were sufficiently similar to justify a single TPL. But in the fourth year, a year of unusually high and early incidence of powdery mildew, they obtained a very narrow but high count of colonies/leaf with a narrow range of means and variances. Consequently, TPL did not fit the pattern seen in the previous years (Fig. 5.12). When average colony number is >2/leaf, the

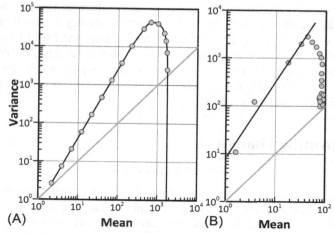

FIG. 13.3   (A) TPL of simulated trap saturation is similar to Fig. 5.12A. (B) TPL fitted to first five generations of a Monte Carlo simulation of populations subject to birth, death, and migration is similar is similar to the simulation. *(B: Adapted from Fig. 1 in Anderson et al. (1982).)*

incidence of infected leaves is ∼75%, a level well above the action threshold, so this level is rarely seen. At this level of incidence, Xu and Madden concluded that a point had been reached beyond which the number of colonies/leaf could not increase as all leaf area had become occupied by powdery mildew colonies. Therefore, at colony densities much >2/leaf, the variance must drop to zero.

A simple simulation serves to illustrate this. Simulate the growth of a population whose members occupy finite area in an arena of finite area. Allow the population to reproduce and grow in number and/or size. Even allowing for death and movement, a point will come when all available space is taken and no further growth is possible. Computing the mean and variance of 500 realizations of such a model produces a TPL with variance declining at the highest means. Fig. 13.3A shows the simulated trap saturation: its similarity to Fig. 5.12A is clear.

A MODELING EXAMPLE

Fig. 13.3A is also very similar to Figs. 1d and 2d (Fig. 13.3B) in Anderson et al. (1982) in which they made the case that TPL could be accounted for by simple stochastic models of numerical change in birth, death and migration without reference to space. Their models produced approximately sigmoidal trajectories of mean density with time with variances declining when the mean reached the asymptote. Which leads to the question, do populations approaching their carrying capacity become spatially less variable to the point the TPL breaks down? This was a primary question asked by LRT et al. (1980) and for which they could find no evidence. In the powdery mildew case where the number of colonies/apple leaf is defined by the size of the leaf, the answer would seem to be yes. Data cited by Anderson et al. of populations of the vole *Microtus pennsylvanicus* show a decline in spatial variation with increasing density (Grant and Morris, 1971) in apparent agreement with their results and Fig. 13.3B. Grant and Morris's data are of captures of the vole in several 0.5 acre (0.2 ha) enclosed grasslands caught by 25 Longworth mammal traps. Longworth traps are sprung when entered by a single animal. With so few traps, even quite small populations would exhibit trap saturation.

Catastrophic change

EFFECT OF PESTICIDES

In a study of the impact of pesticide applications on the spatial distribution of arthropods, Trumble (1985) found that sampling plans based on data of high populations in untreated fields did not necessarily work well in populations already knocked down in treated fields. One reason for this was the spatial distributions of pre- and post-treatment populations generally differed. He sampled strawberry plants for red spidermite (*Tetranychus urticae*) weekly for 15 weeks. A miticide applied after the fifth sample obtained a density-independent knockdown of >99%. Every one of Trumble's 6 post-treatment spatial TPLs had lower slopes than pretreatment spatial TPLs and with little change in correlation coefficient. Following knockdown, population recovery resulted from immigrating females producing short-lived highly aggregated populations on a few plants. Subsequent dispersal and population redistribution caused a change in aggregation as new plants were colonized. Although the spatial TPLs declined following pesticide application, none approached the Poisson distribution: the distributions were still slightly aggregated.

As the Poisson is the distribution of rare events, and surviving a pesticide attack is (unless pesticide resistance has evolved) a comparatively rare event favoring those individuals in refugia, the spatial distribution might be expected to approach the Poisson state at least in the short term. As many sampling plans were, and still are, based on TPLs of high-density pretreatment population estimates, it seems likely that such sampling plans are probably not optimal in assessing populations of treated fields.

RAJT (1987) considered the situation where the post-application spatial distribution changed from aggregated to Poisson. He found that one practical consequence of failing to consider the different spatial distributions in pesticide efficacy trials (as well as actual pest management) is that many pests are likely to be inconsistently sampled, confirming Trumble's findings. Transforming the data and use of measures of mortality such as Abbott's method did little to overcome the sampling bias, and consequently some pesticide efficacy reports were probably misleading. Taylor recommended using pretreatment counts as a covariate with different transformations for pre- and post-treatment counts as necessary. The practical consequence for efficacy trials is the likelihood of incorrect conclusions being made with higher probability than the nominal significance level. In the field, this amounts to making the wrong decision to spray or not to spray.

The basic problem with sampling populations whose distributions change radically following catastrophic mortality is that even quite large samples ($NQ > 30$, say) may be unreliable. For example, when individuals can survive in refugia, the degree of aggregation may increase dramatically and, when this happens, samples that include a number of refugia show an increase in density when the population has in fact declined dramatically.

Ratios

Bounded ratios (%)

SEX RATIO

The example in Chapter 8 of gypsy moth (*Lymantria dispar*) showed ensemble TPLs of all stages including male and female pupae found under burlap bands (Fig. 8.3B). Their TPLs were not different in either slope or intercept. Overall, the sex ratio differed slightly with 46% males and 54% females. At individual sampling stations, however, the sex ratio ranged from 30% to 80% male. Fig. 13.4A shows the ensemble TPL of average number of male pupae per tree at $NB = 168$ stations and the mean and variance of the proportion of males, Π(Male pupae). The TPL regression of Π(Male pupae) has slope, $b = -0.86 \pm 0.191$ ($r = -0.115$). Clearly, sex ratio does not conform to TPL.

PARASITE PREVALENCE

One of the statistics used in epidemiology and parasitology is prevalence. This is the proportion of a population afflicted by a parasite or disease and one minus prevalence is the proportion of potential hosts not parasitized or diseased. The density of a parasite is often recorded as incidence (#/host), which excludes potential hosts not parasitized or diseased. Abundance (#/unit area or habitat) includes the unparasitized hosts. Incidence and abundance data are described by different TPLs (Fig. 7.22). Prevalence is an example of a bounded ratio that is often not well fit by TPL.

Deter et al. (2007) provide data of the prevalence of the gastrointestinal nematodes *Trichuris arvicolae* and *Syphacia nigeriana* in the fossorial water vole *Arvicola terrestris,* which exhibits regular outbreaks in eastern France. The data come from a study of the vole in 13 locations in which they found heterogeneity in prevalences was best explained by the phase of the vole population fluctuations. Prevalence of the two nematodes ranged 2%–13% and within this fairly narrow range, TPLs were negative: $A = 0.99 \pm 0.075$, $b = -0.40 \pm 0.105$, $r = -0.62$ for *T. arvicolae* and $A = 0.98 \pm 0.085$, $b = -0.39 \pm 0.113$, $r = -0.25$ for *S. nigeriana* (Fig. 13.4B). The estimates are so close that clearly heterogeneity of prevalence is virtually identical for both nematodes even though their incidences were uncorrelated.

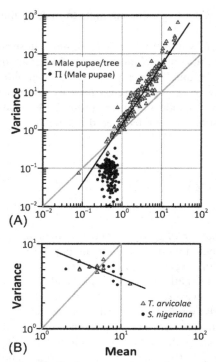

FIG. 13.4   (A) The proportion, $\prod$, of male gypsy moth pupae in a population does not form a TPL where the density (#/tree) of male pupae does ($b = 1.60 \pm 0.038$). (B) Parasite prevalence is also a proportion that generally fails to obey TPL. Examples of nematode parasites of voles. *(A: Data from a 1985 survey by the author; B: Data from Table 2 in Deter et al. (2007).)*

In the gypsy moth example, sex ratio ranged from 30% to 80%, a fairly broad range, but the range of variance is about 33-fold. In the second example, the range of prevalence is much narrower and the range of variance is also narrower. The negative regressions are significant ($P < 0.003$). It seems that numbers bounded by zero and one (or 0 and 100) do not typically obey TPL and may produce negative slopes. The cumulants of proportions are described by quantal response models such as the logit, probit, and complementary log-log functions, all of which are similar in appearance to the logistic model of population growth. Such functions necessarily have mean-variance functions that approach zero as the mean approaches 100%, similar to those seen with trap saturation.

Unbounded ratios

COMMODITY CROPS

The raw data for TPL are counts that are by definition bounded at zero. Although there may be good reasons to suppose there is an upper limit to the counts, provided the actual observations do not approach the theoretical upper limit, TPL usually fits. However, ratios that are not bounded above may also perform poorly. For example, crop yield per unit area is a ratio of production quantity and area harvested. Both variables are bounded below and only theoretically bounded above; they both obey spatial TPL for a wide range of commodity crops (Chapter 6), sometimes highlighting changes in agricultural methods or policy that influence area cultivated and/or production (Figs. 6.15–6.18). The Green Revolution and changes in policy and surveying methods caused changes to the spatial TPLs, but the TPLs over long periods retained their essential character and in many cases the same slope. By contrast, the TPLs of yield are mixed. The two most extreme examples are tobacco and peanuts.

Fig. 13.5A and B shows yield of tobacco and peanuts. The number of states growing tobacco and peanuts has declined from 16 to 9 by 2000, so in recent years NQ is a little low at $NQ \geq 9$ but $NB \geq 56$ for tobacco and $NB \geq 38$ for peanuts. Both commodities show two main periods separated by a transition

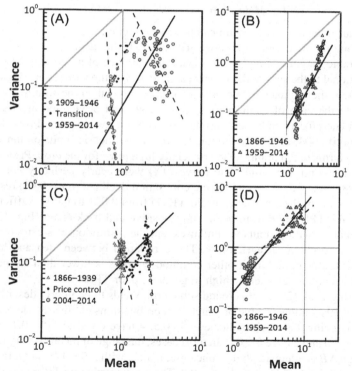

FIG. 13.5 Examples of poorly fitting TPLs: variance-mean relationships for yields (metric tons/ha) of commodity crops for two periods with different sampling or agricultural policies. (A) Peanuts. (B) Tobacco. (C) Wheat. (D) Maize.

corresponding to periods of disruption. In this respect, they reflect TPLs for area harvested and production, but the slopes are negative. They do not obey Taylor's power law in the main sequences, but curiously, the transition periods do because. Like the decline in variance at high density in Figs. 5.12A and 13.3, these examples have a very narrow range of means but a large range of variances.

Yield TPLs of wheat and maize (Fig. 13.5C and D) clearly conform better to our expectations, given the TPLs of area harvested and production (Figs. 6.15 and 6.17), but the variation around the regression is much higher than their parents'. In common with the previous examples, the early periods show greater range in the variance than the mean.

Greater range in variance is to be expected under certain circumstances. The approximate mean and variance of a quotient are given by

$$M\left(\frac{X}{Y}\right) \approx \frac{M(X)}{M(Y)}$$

$$V\left(\frac{X}{Y}\right) \approx \frac{M^2(X)}{M^2(Y)} \left\{ \frac{V(X)}{M^2(X)} + \frac{V(Y)}{M^2(Y)} - 2\frac{Cov(X,Y)}{M(X)M(Y)} \right\}$$

(13.1)

Clearly, the covariance term can have an effect on the variance of the ratio depending on the correlation between X and Y. If the means of X and Y are of comparable magnitude and uncorrelated the variance of a quotient, $V(X/Y)$, is proportional to the sum of the variances. If, however, the variates are negatively correlated, the variance of the ratio will grow faster than the sum of the variances. In the case of commodities, the acreage and production are likely to be correlated for a number of reasons. Tobacco is a case in point. From the late 1930s until 2004, the US government imposed production limits on individual tobacco farms but guaranteed an artificially high price for the crop. Prior to and after price control, area (Y) and harvest (X) were nearly perfectly correlated ($r > 0.98$), but with price control the correlation declined to $r = 0.61$, resulting in a decrease in the covariance in Eq. (13.1) from 0.031 to 0.004, sufficient to change the TPL slope from negative to positive and back again (Fig. 13.5A). The situation with peanuts is similar except the disruption was caused by the transition to chemical agriculture. The correlation between area and harvest for wheat, maize, and most other commodity crops is lower than for peanuts and tobacco, but sufficiently high to generate poor TPLs.

Döring et al. (2015) also found some crop yields to be poorly described by TPL. Of 19 examples (some of which are combinations of other entries), only 5 are convincing TPLs with adjusted $r^2 > 0.5$: 4 have $0.0 < $ adj. $r^2 < 0.1$, 5 have adj. $r^2 < 0.0$ and 2 have $b < 0.0$. In most cases, the range of means is very narrow although NB is adequate, but in one case, maize, with $NB = 128$ and a range of means of $0.5 \leq M \leq 16$, the adj. $r^2 = 0.0$. The most convincing TPLs are conventionally and organically grown wheat (data from Jones et al. (2010)) and conventional lentils (data from Vlachostergios et al. (2011)). These examples are of yields from different cultivars grown at different times and places and under different management strategies and are thus ensemble TPLs. They have slopes of $b = 2.51 \pm 0.407$ and $b = 1.30 \pm 0.311$, respectively. Several examples are from a large meta-analysis of conventional and organic crop yields by Seufert et al. (2012). Mixed-species ensemble analyses of Seufert et al's data in which each TPL point represents a crop managed organically or conventionally extend over nearly 3 orders of magnitude of mean. The slopes ($b \approx 1.8$) are not different ($P > 0.44$) but the organic intercept is slightly higher than the conventional TPL's intercept, indicating greater variability of the organic yield at a given mean. The difference between the observed and fitted $\text{Log}(V)$, the residual, Döring et al. used to interpret yield stability, a factor useful in plant breeding programs. It is functionally the same as Shiyomi et al.'s (2001) spatial heterogeneity index that measures heterogeneity of species in relation to the community (Chapter 6).

Uncountable numbers

Number theory recognizes several different kinds of number of which rational numbers are numbers that can be expressed as the ratio of two integers and these includes all integers. They have the property that numerator and denominator are discrete, and although there an infinite number of them they can be enumerated: they are said to be countable. Irrational numbers are real numbers that are not a ratio of integers. They include $\sqrt{2}$, π, and e and unlike the rational numbers they cannot be enumerated and are called uncountable.

TEMPERATURE

In Chapter 12, TPL fit both actual and simulated rainfall extremely well, but it does very poorly with another meteorological variable, temperature. The climate databases used to drive insect and crop growth simulations under climate change included daily maximum, minimum and average temperatures for 1901–2000 actual and 2101–2100 simulated data. Fitting power laws to the daily average temperature at 72 weather stations and simulation grid points for 200 years produced no recognizable spatial TPLs: slopes were highly variable with very low correlations (Fig. 13.6A). Fig. 13.6B is typical of plots for all 200 years. Slopes range from $b = -12$ to $+4.8$ with only one correlation coefficient greater than $r = 0.2$ and more than half negative. This compares with $1.57 < b < 1.91$ and $r > 0.91$ for precipitation in all years (Fig. 12.5).

Why then is there a qualitative difference between these two basic meteorological variables? One important difference between precipitation and temperature is that fundamentally rainfall is a countable quantity as it could in principle be resolved to an integer at the molecular level. By contrast, temperature is a real number not derived from a countable quantity.

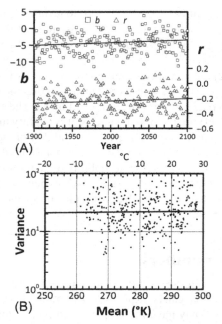

(A)

(B)

FIG. 13.6   (A) Estimates of spatial TPL *b* and *r* for the years 1901–2000 (actual temperatures) and 2001–2100 (simulated) computed from average daily temperatures at 72 points across the American Midwest. TPL slopes range $-10 < b < 5$ and correlation coefficient averages $r = -0.2$. (B) Spatial TPL of daily temperatures for 1984 is typical. The *solid line* is the fitted TPL ($A = 0.46 \pm 2.28$, $b = 0.36 \pm 0.94$, $r = 0.02$). Note the mean axis is linear for clarity. *(Actual data from NOAA and simulated from GFDL.)*

Aside from the known issues of small samples, the nature of the numbers subjected to TPL analysis plays a part in whether variance scales. Altogether it looks like TPL does not work well with real uncountable numbers or ratios, bounded or unbounded. Although rational numbers like yield per unit area are countable, their upper bounds, like those imposed by trap saturation, impose some limits on TPL's applicability. In this case, however, the distribution of the statistics describing both numerator and denominator plays a larger part in the ratio's variance-mean relationship, and can create a variance-mean relationship not described by TPL, or none at all.

References

Anderson, R.M., Gordon, D.M., Crawley, M.J., Hassell, M.P., 1982. Variability in the abundance of animal and plant species. Nature 296, 245–248.

Clark, S.J., Perry, J.N., 1994. Small sample estimation for Taylor's power law. Environ. Ecol. Stat. 1, 287–302.

Cohen, J.E., 2014. Taylor's law and abrupt biotic change in a smoothly changing environment. Theor. Ecol. 7, 77–86.

Deter, J., Chaval, Y., Galan, M., Berthier, K., Salvador, A.R., Garcia, J.C.C., Morand, S., Cosson, J.-F., Charbonnel, N., 2007. Linking demography and host dispersal to *Trichuris arvicolae* distribution in a cyclic vole species. Int. J. Parasitol. 37, 813–824.

Deudero, S., Vincx, M., 2000. Sublittoral meiobenthic assemblages from disturbed and non-disturbed sediments in the Balearics. Sci. Mar. 64, 285–293.

Döring, T.F., Knapp, S., Cohen, J.E., 2015. Taylor's power law and the stability of crop yields. Field Crop Res. 183, 294–302.

Grant, P., Morris, R.D., 1971. The distribution of *Microtus pennsylvanicus* within grassland habitat. Can. J. Zool. 49, 1043–1052.

Healy, M.J.R., Taylor, L.R., 1962. Tables for power-law transformations. Biometrika 49, 557–559.

Jones, H., Clarke, S., Haigh, Z., Pearce, H., Wolfe, M., 2010. The effect of the year of wheat variety release on productivity and stability of performance on two organic and two non-organic farms. J. Agric. Sci. 148, 303–317.

Seufert, V., Ramankutty, N., Foley, J.A., 2012. Comparing the yields of organic and conventional agriculture. Nature 485, 229–232.

Shiyomi, M., Takahashi, S., Yoshimura, J., Yasuda, T., Tsutsumi, M., Tsuiki, M., Hori, Y., 2001. Spatial heterogeneity in a grassland community: use of power law. Ecol. Res. 16, 487–495.

Taylor, L.R., Woiwod, I.P., Perry, J.N., 1980. Variance and the large scale spatial stability of aphids, moths and birds. J. Anim. Ecol. 49, 831–854.

Taylor, L.R., Perry, J.N., Woiwod, I.P., Taylor, R.A.J., 1988. Specificity of the spatial power-law exponent in ecology and agriculture. Nature 332, 721–722.

Taylor, R.A.J., 1987. On the accuracy of insecticide efficacy reports. Environ. Entomol. 16, 1–8.

Taylor, R.A.J., 2018. Spatial distribution, sampling efficiency and Taylor's power law. Ecol. Entomol. 43, 215–225.

Taylor, R.A.J., Lindquist, R.K., Shipp, J.L., 1998. Variation and consistency in spatial distribution as measured by Taylor's power law. Environ. Entomol. 27, 191–201.

Trumble, J.T., 1985. Implications of changes in arthropod distribution following chemical application. Res. Popul. Ecol. 27, 277–285.

Van Gaever, S., Vanreusel, A., Hughes, J.A., Bett, B.J., Kiriakoulakis, K., 2004. The macro- and micro-scale patchiness of meiobenthos associated with the Darwin Mounds (north-east Atlantic). J. Mar. Biol. Assoc. UK 84, 547–556.

Vlachostergios, D., Lithourgidis, A., Roupakias, D., 2011. Adaptability to organic farming of lentil (*Lens culinaris* Medik.) varieties developed from conventiona lbreeding programmes. J. Agric. Sci. 149, 85–93.

Xu, X.-M., Madden, L.V., 2002. Incidence and density relationships powdery mildew on apple. Phytopathology 92, 1005–1014.

Part III

Chapter 14

Applications of TPL

The distribution and abundance of organisms are intimately related, and together define a population's size and structure in space. TPL states that the shape of the frequency distribution of abundance at a point in space changes systematically with abundance; that is to say, it is density dependent. Fig. 1.1 of European corn borer and parasitism of gypsy moth eggs shows this clearly. The transition from one frequency distribution to another with density has important implications for the interpretation of sample data because so much ecological theory has been derived from or assumes population distributions conform to particular statistical frequency distributions. Factors affecting the efficiency of sampling and the number of samples required for precise population estimation of natural and experimental populations depend intimately on the details of TPL, especially if the population is highly aggregated.

Population densities change in two ways: by change in population number, and by change in population distribution. If a population expands its range as it reproduces, the average population density may grow more slowly than the population number. Thus, for the sessile stages of species such as corals, an increase in total population must translate to an increase in range, with little change in density. However, for most animal species, the density increases as the population grows, but not necessarily at the same rate. For highly dispersive populations, the range grows rapidly with respect to population number, and a large proportion of the available suitable habitat becomes colonized. Conversely, populations that grow numerically faster than they can spread their range (disperse), tend to become highly aggregated with large areas of (nearly) empty space between. This is the situation with entomopathogenic nematodes, for example (Chapter 7).

In this chapter, we look at the application of TPL to data transformations, development of sampling programs, and their role in environmental assessment. We start with numerical methods using TPL to obtain and analyze sample data. Sampling programs are developed by assuming a frequency distribution, which may be appropriate only within a particular range of densities. Fig. 1.1 exemplifies this: at low densities the Poisson is a good model, but at higher densities

Taylor's Power Law. https://doi.org/10.1016/B978-0-12-810987-8.00014-8
461

the negative binomial or lognormal would be appropriate models for a sampling program. All three distributions may apply simultaneously within a comparatively small area.

Statistical analysis of survey or experimental data makes assumptions about the structure of those data. Real data rarely meet the requirements of normality and uniformity of variance (homoscedasticity) required by parametric statistics. TPL says that variance is a function of mean, so any data set that obeys TPL is by definition heteroscedastic over some or all of its range. A common and often effective solution is to bring data into conformity with assumptions by transformation. The data of Brown and Cameron suggest a square root transformation at low densities where the Poisson distribution fits well and a logarithmic transformation at higher densities where the negative binomial and lognormal fit better. Thus, there may not be a single *distribution-based* transformation for a dataset. However, the slope of TPL suggests an exact transformation, which is frequently effective in stabilizing variance and normalizing the data.

Transformations

With the need for data to conform to certain standards for analysis of variance, transformations were developed to satisfy those requirements. Typically, they were based on reasoning from the presumed underlying statistical distribution, until Bartlett (1947) summarized the art and developed an argument for a general approach. Very often, the variance is a function of the mean in experimental data. By defining that function, the experimental variance can be stabilized by a suitable change of scale. He proposed that variance is a function of the mean, $v_x = f(m)$, where v_x is on the scale of measurements of x and m is the mean of x. For another function $g(x)$, the variance v_g is approximately

$$v_g = f(m) \cdot \left\{ \frac{dg}{dm} \right\}^2 \quad \text{so that } if \ v_g = C^2 \text{a constant, then}$$

$$g(m) = \int \frac{C}{\sqrt{f(m)}} dm \qquad (14.1)$$

Bartlett considered the variance proportional to the mean, $f(m) \propto m$, and the standard deviation proportional to the mean, $f(m) \propto m^2$ for which the square root and log transformations are appropriate. He did not consider other powers of the mean, $f(m) \propto m^3$, for example, which has the reciprocal of the square root as appropriate transformation. A frequency distribution with this relationship is the inverse Gaussian distribution (Chapter 2). Nor did Bartlett consider fractional exponents, perhaps because the conventional wisdom was that all data should conform to a definite frequency distribution and none were known with fractional power relationships between mean and variance.

As we saw in Chapter 2, there are known quadratic relationships between mean and variance based on the negative binomial distribution with parameter k. Beall (1942) developed transformations for experimental data of the lady beetle *Leptinotarsa decemlineata* and the crambid moth *Ostrinia* (=*Pyrausta*) *nubilalis* based on $f(m) = m + km^2$. A somewhat more effective transformation was also proposed by Beall (1942) for negative binomial distributed variates

$$z = k^{\frac{1}{2}} \sinh^{-1} (x/k)^{\frac{1}{2}}. \tag{14.2}$$

A simpler one proposed by Anscombe (1948), $z = \log(x + k/2)$, was used by Bliss and Owen (1958) in their treatise on the negative binomial with common k. Both transformations require a common k to be estimated.

TPL61 proposed a transformation based on TPL b: substituting aM^b for $f(m)$ in Eq. (14.1) and integrating suggests the general transformation of $z = x^p$ where

$$p = (1 - b/2) \tag{14.3}$$

except for the case $b = 2$, which has $p = 0$. Integration of Eq. (14.1) for $f(m) = aM^2$ has $z = \ln(x)$. The other special case of $b = 1$ results in $p = \frac{1}{2}$, the square root transformation appropriate for Poisson distributed variates. Healy and Taylor (1962) published a table of transformations for noninteger values of b in the range $0.4 \leq b \leq 3.6$ or $(-0.8 \leq p \leq 0.8)$. Today, such tables are redundant as statistical programs can calculate an exact value of p for any data set.

STABILIZING VARIANCE

Simultaneously and independent of LRT61, Hayman and Lowe (1961) in New Zealand estimated the slope of variance on mean for 182 colonies of the cabbage aphid *Brevicoryne brassicae* on kale and derived a variance stabilizing transformation using the same argument. Their estimated variance-mean relationship of $V = 3.42M^{1.65}$ gave a power transformation of $p = 0.175$ resulting in a stabilized variance of 0.105 (Fig. 14.1A). They make the point that the power relation they observed "is not evidence against a negative binomial distribution for cabbage aphids in individual samples." They go on to suggest that over the range of sample means the negative binomial "parameter k may vary [with the mean] in such a way as to imply [the observed] relation between V and M." Apparently they expected the negative binomial distribution to describe the spatial distribution of the aphid colonies. They found the hyperbolic sine transformation (Eq. 14.2), which required four different values for negative binomial k to be slightly poorer at stabilizing variance than the power transformation. Fig. 14.1B shows the effectiveness of the power transformation. The vast majority of points have no dependence on the mean, but there remain some residual Poisson dependence on about two dozen points.

LRT (1970) analyzing three sets of counts of black bean aphid, *Aphis fabae*, colonies on bean stems found the TPL slope to be the same ($b = 1.72$) for all sets, but with different intercepts. The data were all taken at Rothamsted Experimental Station in the 1950s: an unpublished set of C.G. Johnson's in which approximately weekly samples of stems were cut and aphids washed off and

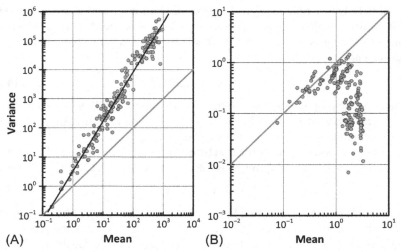

FIG. 14.1 Ensemble TPL of cabbage aphid on kale (A) and (B) variance-mean plot after tansformation based on the TPL slope of $b = 1.65$. *(Adapted from Figs. 1 and 2 in Hayman and Lowe (1961).)*

counted separately as apterae and alatae ($NQ = 48$–51 stems, $NB = 16$); a published set of C.J. Banks (1954) of weekly in situ stem counts of plants in 3 plots made by a subsample and estimation procedure ($NQ = 50$ stems, $NB = 17$); and hitherto unpublished data of LRT in which all stems in 67 small plots were excised and all aphids counted on all stems ($NQ = 7$–100, $NB = 67$). He used the tabulated (Healy and Taylor, 1962) power transformation $z = x^{0.2}$, instead of the exact transformation $z = x^{0.14}$, and found this transformation slightly more effective than the logarithmic transformation $z = \log(x + 1)$. As the log transformation is equivalent to a power transformation of $z = x^0$, LRT's result suggests that the power transformation may be sensitive to differences of ~ 0.2 in the power used with the sensitivity depending on the range of M.

Fig. 14.2A shows TPLs of the three datasets, two spatial and one ensemble, with a common gradient and separate intercepts. Fig. 14.2B shows the results of the transformation. Like Hayman and Lowe's post-transform data, the range was greatly reduced but the majority of data points retained a Poisson distribution. LRT suggested that the residual variance-mean relation at small means remaining after transformation may be associated with the frequency of zero counts and advocates relocating counts by adding 1 before taking logs to improve the overall effectiveness of the transformation or alternatively using more than one transformation. With TPLs with two slopes, as in Fig. 11.13, clearly a single transformation is not appropriate. However, there is no sign of a bend in either Figs. 14.1A or 14.2A as all but a couple of points are above the Poisson line. The failure to remove all dependence of variance on the mean may also indicate more than one source of variation in the sample data: in addition to the intrinsic spatial and temporal variation in the populations being counted, there is also the variation introduced by the sampling process.

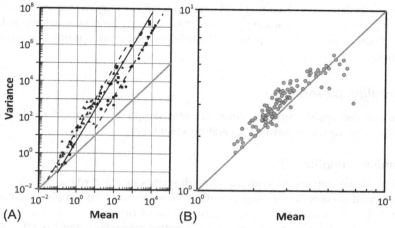

FIG. 14.2 Ensemble and spatial TPLs of three datasets of black bean aphid on beans (A) and (B) variance-mean plot after tansformation based on the common TPL slope of $b = 1.72$. (*Adapted from Figs. 1 and 2 in LRT (1970).*)

Sampling

Ecologists have long recognized this duality of distribution and abundance of natural populations: two populations of the same absolute size may occupy the same range at one spatial scale, but be distributed so differently that their densities at smaller scales differ by orders of magnitude: one population is aggregated in tight clusters, while the other is more randomly distributed. Paoletti et al. (1991) found this to be so with the microfauna of bromeliads and Park et al. (2013) found evidence of this with nematodes (Chapter 7). Population density is a measure of how close together individuals are packed (Fig. 2.1), an attribute generally determined by sampling. We can imagine the sampling process as a set of photographs of the population taken simultaneously at different places, and count the individuals portrayed. This instantaneous record of the population provides information on the number of animals per photograph (relative density) and their relative positions. Knowing the area of each photograph, we can also determine the absolute density. The number of organisms per photograph, $\{X_I\}$, forms a frequency distribution $\Phi(x)$ characterized by its moments (Eqs. 2.1, 2.2).

Sampling populations of organisms is time consuming and expensive. Knowing how best to estimate abundance while incurring minimal cost is therefore important. In designing a sampling program, we need to consider several factors: with what device will the samples be taken and how representative of the population will those samples be; where and in what pattern will the samples be taken; what kind of sampling and how many samples do we need to take. Much of the research on animal and plant sampling has been in pursuit of methods for assessing pest populations. A review of pest management decision making by Binns and Nyrop (1992) enumerated three required pieces of information: an assessment of the pest population density; an economic threshold; and knowledge of the phenological state of the crop to determine when to sample.

Sampling patterns

We can distinguish between three different sampling patterns. Each has its unique properties: its advantages and disadvantages.

Random sampling

As the name implies, random sampling describes the case where samples are positioned at random with respect to the area or habitat being sampled and each other. The classic means for doing this is to throw the sampler into the air and sample where it lands. A safer and more objective method is to mark out the area to be sampled as a rectangle or square with the sides marked at intervals from 0 to S_1 and 0 to S_2 Then select pairs of uniform random numbers in the range 0 to 1 either from tables or with a computer program, and rescale them to S_1 and S_2 use the pairs of random numbers to define the position of the sample point and take the sample using an appropriate sampler.

Systematic sampling

With systematic sampling, the sample area is laid out in a grid and samples are taken at the intersection of the grid lines using an appropriate sampler. This method is frequently more convenient and sometimes less expensive than the random method, and is suitable for analysis by a wider range of analytical techniques, including geostatistical methods. Where random samples suffer from the possibility that samples may chance to be aggregated so that an important part of the sample range remains unsampled, systematic samples suffer from the possibility that the sample spacing may coincide with a periodic variation in the population under investigation, and thereby go undetected.

Stratified random sampling

This method combines the advantages of both random and systematic sampling plans. The area is marked out into a grid, and the position of the sample(s) is selected at random within each grid square. In this way, the likelihood of the sample pattern coinciding with an underlying pattern in the population is eliminated while the entire area is sampled efficiently.

Binomial sampling

Sampling can often be simplified by scoring populations as having a density more or less than T individuals per sample. This is called binomial sampling, and in its simplest form has $T = 0$, and is then called presence-absence sampling. Analysis of these data is based on the relationship between mean density, M_x per sample, and the proportion of samples with T or less per sample, $P(T)$. Binomial sampling is attractive because it is usually much faster, and therefore less expensive, than total enumeration.

Binomial sampling models developed and tested by Wilson and Room (1983) incorporated the clumping patterns of different species and stages of cotton arthropods. Inserting TPL into an expression for the proportion of plants infested given a clumped distribution described by the negative binomial distribution produced a model using TPL estimates and average density. This model reliably estimated abundance purely on the basis of whether a cotton plant was infested or not. Sample sizes required to estimate the mean with a given level of precision (D) using binomial sampling were small for low densities but increased rapidly for highly clumped populations at higher densities as the proportion of infested plants approached 100%. Although binomial sampling requires large sample sizes at high densities, they may be more cost effective as the time taken for each sample is low. The rapid increase in sample size required of binomial sampling is due to the increased uncertainty entailed by essentially throwing away $100 \cdot (1 - P(T))\%$ of the information. Thus, its applicability may be limited to pest management decision making. A binomial analog of TPL, the Hughes-Madden power law (Chapter 16), is used widely in plant pathology as the presence or absence of fungal colonies or virus lesions is easily ascertained.

Table 14.1 gives the probability statements for binomial sampling assuming $T = 0$ (presence-absence), the negative binomial and Poisson distributions, and TPL. An inappropriate choice of distribution or threshold (T) can have unpredictable consequences (Binns and Bostanian, 1988, 1990).

Using the estimated values of A and b for entomopathogenic nematodes *Steinernema feltiae* and *S. glaseri* (Chapter 7), binomial probabilities for threshold

TABLE 14.1 Probability statements for presence-absence sampling assuming the negative binomial and Poisson distributions and TPL

Assuming the sampling distribution is negative binomial, the proportion of nonempty samples is:	$P(x > 0) = 1 - P(0) = 1 - \exp\left\{\frac{-\mu^2}{\sigma^2 - \mu} \cdot \ln\left(\frac{\sigma^2}{\mu}\right)\right\}$	(14.4)
Assuming the Poisson distribution:	$P(x > 0) = 1 - \exp(-\mu)$	(14.5)
Assuming TPL as the variance-mean relationship:	$P(x > 0) = 1 - \exp\left\{-\mu \cdot \frac{\ln\left(a \cdot \mu^{b-1}\right)}{a \cdot \mu^{b-1} - 1}\right\}$	(14.6)

TABLE 14.2 Binomial sampling probabilities for entomopathogenic nematodes

			Mean density (number/sample)					
Species	A	b	*1*	*3*	*10*	*30*	*100*	*300*
Steinernema feltiae	−0.260	2.900	0.735	0.728	0.587	0.394	0.210	0.104
S. feltiae	−2.400	3.140	0.996	1.000	1.000	1.000	0.997	0.920
S. glaseri	0.170	2.230	0.558	0.670	0.737	0.761	0.759	0.741

of $T = 0$ are given in Table 14.2. The values in the body of the table represent the probability of obtaining a nonempty sample when the actual population densities are 1, 3, 10, etc. per sample for the power law parameters given. Notice that the sampling probabilities depend quite sensitively on the values of both A and b. The estimated slopes of *S. feltiae* are similar but the intercepts are very different, resulting in very different binomial probabilities both in value and pattern with increasing density. *S. glaseri* with lower b and higher A has a maximum sampling probability at ~30/sample and is intermediate between the two *S. feltiae* probabilities.

Sequential sampling

A generalization of binomial sampling, sequential sampling has no predetermined sample size: sampling is continued until a predefined level of precision is reached (e.g., Kuno, 1969; Green, 1970). Data $\{X_I\}$ are obtained from samples, a statistic (usually the mean, M_x) is continually recalculated and plotted against the current sample size, n. Sampling continues until the trajectory of M_x crosses a predetermined threshold, or stop line. One or two stop lines can be generated to select between alternative actions, for example, to spray or not to spray (Fig. 14.3). These two critical thresholds are defined by the probabilities of making erroneous decisions (failure to spray when necessary or spray when unnecessary) assuming an underlying distribution (e.g., negative binomial) or variance mean relationship. No decision is made until the trajectory of M_x at sample i crosses one of the decision lines. In general, this method like binomial sampling is better suited to decision making than parameter estimation. However, there are two types of fixed precision decision: Wald's (1947) and Iwao's (1975). The former (classical) approach is specifically intended for decision making, while Iwao's method uses a difference test to classify population density levels and so may be appropriate for parameter estimation in certain cases.

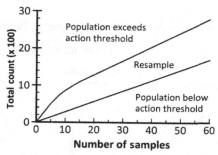

FIG. 14.3 Acceptance and rejection lines for sequential samples.

TABLE 14.3 Definitions of reliability

Coefficient of variation of the sample mean, $M[X]$, CV:	$CV = \sigma \cdot \frac{\sqrt{n}}{\mu}$	(14.7)
For constant $CV = C$, solve for n:	$n = \left(\frac{\sigma}{\mu \cdot C}\right)^2$	(14.8)
Confidence interval as a proportion of the mean, m, estimating μ:	$\Phi\left(m - z_{\alpha/2} \cdot \frac{\sigma}{\sqrt{n}} < \mu < m + z_{\alpha/2} \cdot \frac{\sigma}{\sqrt{n}}\right) \approx (1 - \alpha)$	(14.9)
Confidence interval is a fixed positive number, H:	$H = z_{\alpha/2} \cdot \frac{\sigma}{\sqrt{n}}$	(14.10)

Sequential sampling can proceed indefinitely, leading to the requirement of a stopping rule. Many of the examples of TPL in Chapters 7 and 8 developed sequential sampling schemes for pests include a stopping rule. From prior knowledge of the variation in animal abundance, it is possible to determine the optimum number of samples required to detect populations of a chosen size with a given level of precision. This requires adoption of a definition of reliability, that is, a formal statement of how often we will incorrectly reject the hypothesis (Type I error) and accept an unsupported decision. The basic definitions of reliability are given in Table 14.3, where μ and σ are population mean and standard deviation, respectively, and α is the defined probability of error.

Optimum sample size

The simplest definition of reliability is given by the coefficient of variation (CV) of the sample mean m (Eq. 14.7). CV is estimated by the ratio of the sample standard deviation to the mean m. By specifying a fixed value of CV, we are

TABLE 14.4 Optimum sample size (n) for different definitions of reliability

Parent distribution	Coefficient of variation	Confidence interval based on probability	
		Fixed positive number	*Proportion of the mean*
General	$n=\left(\dfrac{\sigma}{\mu \cdot C}\right)^2$	$n=z_{\alpha/2}^2 \cdot \left(\dfrac{\sigma}{H}\right)^2$	$n=z_{\alpha/2}^2 \cdot \left(\dfrac{\sigma}{\mu \cdot D}\right)^2$
Poisson	$n=\dfrac{1}{\mu \cdot C^2}$	$n=z_{\alpha/2}^2 \cdot \dfrac{\mu}{H^2}$	$n=z_{\alpha/2}^2 \cdot \dfrac{1}{\mu \cdot D^2}$
Negative binomial	$n=\dfrac{1}{C^2} \cdot \left(\dfrac{1}{k}+\dfrac{1}{\mu}\right)$	$n=z_{\alpha/2}^2 \cdot \dfrac{\sigma}{H^2} \left(\dfrac{k \cdot \mu + \mu^2}{k \cdot H^2}\right)$	$n=z_{\alpha/2}^2 \cdot \left(\dfrac{1}{k}+\dfrac{1}{\mu}\right) \cdot \dfrac{1}{D^2}$
TPL	$n=\dfrac{a \cdot \mu^{b-2}}{C^2}$	$n=z_{\alpha/2}^2 \cdot \dfrac{\sqrt{a \cdot \mu^{b-2}}}{H^2}$	$n=\left(\dfrac{z_{\alpha/2}}{D \cdot \mu}\right)^2 \cdot a \cdot \mu^b$

defining the amount of variation in our estimate of m that we will accept. Solving for n then gives us the number of samples required for that level of precision.

Another way to define reliability is to assume we are taking a random sample of size n from a distribution having mean μ and standard deviation σ, then for sufficiently large sample (>50) the probability of the true population mean, μ, lying in an interval defined by the estimate m and its $100 \cdot \alpha\%$ confidence interval is approximately equal to $1—\alpha$ is given by Eq. (14.9), where $z_{\alpha/2}$ is the upper $\alpha/2$ point of the standard normal distribution. Eq. (14.9) states that the confidence interval for μ will include the true mean μ with probability approximately $\alpha - 1$ regardless of the form of the parent distribution. The larger the sample size used to obtain m, the better this approximation. Eq. (14.10) replaces the proportionate confidence interval with a number, H. To apply these definitions, we also require a further assumption regarding the underlying frequency distribution of counts. Several special cases have been defined, and are given in Table 14.4 with the most general case based on TPL.

Number of samples

Using the TPL results of western flower thrips (WFT) in greenhouses (Appendix 8.A), the number of sticky traps, n, required for $(1 - \alpha)$ confidence that the true mean density (μ) lies in the interval $m \pm D\mu$, where m is the estimated mean number per sticky trap is

$$n = \left(\frac{z_{\alpha/2}}{D \cdot \mu}\right)^2 \cdot a \cdot \mu^b \qquad (14.11)$$

where $a = 1.20$, $b = 1.84$, and D is the precision expressed as a proportion of the true mean μ, which is estimated by m, and $z_{\alpha/2}$ is the upper $\alpha/2$ point of the

TABLE 14.5 Number of weekly samples per 1000 m² of greenhouse required for fixed-precision sampling using different levels of precision and TPL estimated for WFT caught on yellow sticky cards

Powerlaw parameters	Precision (D)	Number of WFT per sticky card per week					
		1	*3*	*10*	*30*	*100*	*300*
a = 1.20	10%	66	55	46	38	33	26
(A = 0.08),	15%	29	25	20	16	14	12
b = 1.84	20%	16	14	11	10	9	7

standard normal distribution. Table 14.5 gives the sample sizes required for 95% confidence that the estimated number of WFT per sticky trap lie within $\pm D = 10$, 15 and 20% of the true means for 1–300 WFT/card/week.

The comparative stability of TPL slope across multiple data sets for a species confers practical utility, as witnessed by the enormous number of sample programs that use TPL as their basis (see LRT, 1984; Jones, 1990; Binns and Nyrop, 1992; and Chapters 5, 7, and 8). But, no matter how stable b proves to be, it is the intercept A that has the larger impact on required sample size. Small variations in gradient have only a modest effect on n, and RAJT et al.'s (1998) data of b for WFT were closely similar over a range of greenhouse crops, but the intercepts differ significantly (Appendix 8.M1). The slopes are not significantly different ($P > 0.40$), while the intercepts differ at $P < 0.055$. Sensitivity analysis shows how much more sensitive n is to variation in A than b (Fig. 14.4). The smaller standard error of b (± 0.04 versus ± 0.48 for A) contributes greatly to the lower sensitivity of n to b, but in addition, the exponential

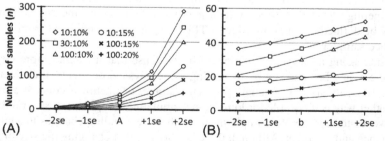

FIG. 14.4 Sensitivity analysis of TPL parameters A and b on the required sample size, n, using the estimates of combined sample data of western flower thrips (Appendix 8.M1) and Eq. (14.11). (A) $A = \log(a) = 0.08 \pm 2$ *SE* for constant $b = 1.84$. (B) $b = 1.84 \pm 2$ *SE* at constant $A = 0.08$. The key refers to the population densities and precision levels used in the analysis.

dependence of n on A is clearly evident in Fig. 14.4A. The advantage of the comparative stability of b (Fig. 14.4B) over multiple data sets can be completely offset by variations in A.

Variation in A can be caused either by a shift in the regression line corresponding to increased variance, or by a change in b reflecting a shift in the dependence of variance on mean. In the WFT example, the number of samples, n, declines with density, but for extremely aggregated species, the reverse is true. Table 14.6 shows the number of samples required for precision, $D = 10$, 15, and 20% of the estimated means for (relative) population densities of 1, 3, 10, 30, 100, and 300 individuals per sample, given the results obtained in the TPL analyses of *Steinernema feltiae* and *S. glaseri* (Appendix 7.F1). As in the WFT example, the optimum number of samples declines as D increases, but in these examples with very highly aggregated populations ($b \approx 3$) n increases sharply with density.

Sampling efficiency

Although there is some evidence that the intercept A is an index of aggregation in its own right (Clark et al., 1996), there is no doubt that it is also a scale factor that can depend strongly on the sampling details (LRT et al., 1983, LRT, 1984). The sensitivity of n to A shown in Fig. 14.4A can be used to create a more effective sampling program.

Developers of sampling programs for agriculture or ecology usually have a good knowledge of the biology and habits of their target, and often have a choice of suitable samplers. Although the process whereby they arrive at the number of samples is often more intuitive than scientific, it is frequently based on a knowledge of the likely numbers obtained by a particular sampler, and an educated guess of how many to use to obtain adequate population estimates. The use of fixed-precision sampling techniques to determine n makes this more objective, but the choice of sampler remains somewhat subjective. With knowledge of a sampler's efficiency, this subjectivity can be removed by comparing competing samplers. Furthermore, a sampler of requisite characteristics might now be defined using knowledge of TPL using estimates obtained from a reference sampler.

Increasing a sampler's efficiency (in the sense that it catches more target species) may be achieved by increasing some physical or chemical characteristic (e.g., its size, wavelength (color), wattage, or chemical concentration). Another way is to sample for longer—accumulating samples over several days or weeks to increase the sample mean, with corresponding changes to sample variance and intercept. Although there is evidence that changing the size of the sample can modify b (e.g., Appendices 7.Y and A2), the many cases of consistency of b over several data sets (e.g., RAJT et al., 1998) suggest that this approach may be viable in many situations.

TABLE 14.6 Sample sizes for power law sampling of entomopathogenic nematodes

Data set	TPL parameters	Precision (D)	Mean density (number/sample)						
			1	3	10	30	100	300
Steinernema feltiae (Spridonov and Voronov, 1995)	a = 0.55	10%	211	567	1677	4507	13,320	35,803
	(A = − 0.26),	15%	94	252	745	2003	5920	15,913
	b = 2.90	20%	53	142	419	1127	3330	8951
S. feltiae (Peters, 1994)	a = 0.004	10%	2	5	21	74	291	1020
	(A = − 2.40),	15%	1	2	9	33	130	453
	b = 3.14	20%	0	1	5	18	73	255
S. glaseri (Smits, 1996)	a = 1.48	10%	568	732	965	1242	1639	2110
	(A = 0.17),	15%	253	325	429	552	728	938
	b = 2.23	20%	142	183	241	311	410	527

Assuming the intercept, A, is a scale factor largely dependent on the sampling details, it offers a means of manipulating n by changing the sampling efficiency. Two sampling methods such as pheromone traps that differ only in their relative efficiencies normally catch insects in the proportion.

$$\phi = y/x \tag{14.12}$$

If TPL is known for sampler **X**, then the power law for **Y** may be calculated from the relationships between means and variances:

$$m_y = \phi \cdot m_x \tag{14.13}$$

and

$$v_y = \phi^2 \cdot v_x. \tag{14.14}$$

Substituting Eqs. (14.12) and (14.13) into the TPL for **X** and rearranging, we obtain the corresponding power law for **Y**:

$$v_y = a \cdot \phi^{(2-b)} \cdot m_x^b \tag{14.15}$$

where $a\phi^{2-b}$ is the power law intercept for sampler **Y**. Thus, changing the efficiency of the sampler by a factor of ϕ moves the intercept by a factor of ϕ^{2-b}. For values of $b \neq 2$, n is highly dependent on the magnitude of the intercept, and it is important to choose a sampler that optimizes the efficiency. Replacing a with 10^A in Eq. (14.11),

$$n = 10^A \cdot \left(\frac{z_{\alpha/2}}{D}\right)^2 \cdot \left(\frac{\phi}{m_x}\right)^{2-b} \tag{14.16}$$

we see that n grows rapidly with A and, unless $b \geq 2$, with ϕ also. When $b > 2$, n and ϕ are inversely proportional and the efficiency of the sampler must go up in order to decrease the value of n without increasing A. In the case of attractant traps like pheromone and light-mediated traps, this can be achieved by changing the light intensity or spectrum or the concentration or release rate of pheromone to obtain a larger sample. Simultaneously, increasing the power of an attractant trap is likely to move the TPL line up and to the right. Appropriate choice of sample power therefore can decrease A and therefore n. In practice, situations for manipulating ϕ in this way may be limited to when an action threshold is known in terms of sampler **X**, and a sample program can be defined to detect the threshold using a second sampler **Y** specially tuned for that density. Furthermore, it suggests that different samplers may be required to estimate populations of the same species at different times and places (McArdle et al., 1990; Southwood and Henderson, 2000), complicating population estimation.

The foregoing discussion implicitly assumes that ϕ is density independent. RAJT et al. (1991) have shown that in at least one instance, this assumption is not valid and RAJT (2018) has explored the consequences for interpreting TPL (Chapter 15).

With the large number of samples required to estimate reliably the magnitude of highly aggregated populations, the standardization and quantification of sampling methods is vitally important. Attempts to trace known populations of nematodes applied more or less uniformly to soil only a short time after application show how quickly they become highly aggregated (Appendices 7.F1, 7.H1, and 7.K1). The rapid increase in aggregation may be due to a rapid reduction in recovery in samples following application. If we assume the application process is not flawed, then there are two possibilities: they do not survive long following application, a numerical response; or they redistribute themselves very rapidly, a behavioral response. While it seems implausible that entomopathogenic nematodes could actively aggregate themselves within a few hours of being deposited uniformly onto the surface of the soil, it is even less likely that nematodes sprayed onto their natural habitat would have lower survival than those sprayed onto collectors. Redistribution detected by changes in b does occur following a randomizing process such as pesticide application (RAJT, 1987; Trumble, 1985). It is to be expected that infective juveniles sprayed onto bare earth are at most Poisson distributed ($b \leq 1$), confirmed by collectors on the surface (RAJT, 1999). If they remained in the same place until sampled, then the level of aggregation would be low and sampling would be efficient. As this seems not to be the case, and high values of b are obtained in both natural and induced populations, the inescapable conclusion is that they redistribute themselves very quickly.

Postscript

It is obvious that a TPL line with gradient $b \neq 1$ must cross the line $V = M$ (the Poisson line) somewhere. Furthermore, a line with gradient $b \neq 2$ must intersect the line with gradient $b = 2$ ($V = M^2$). This observation has implications for both transformations and sampling plans. A square root transformation of experimental data at the lower end of the power law line would be appropriate, but at means close to the square root of variance a more powerful transformation would be needed. Thus, we might conclude that no single transformation is adequate. Generally, the logarithmic transformation seems to work very well as shown by Williams (1937). Knowledge of the variance-mean relationship offers an exact transformation, although this does not always work perfectly as Fig. 14.2 shows. LRT's (1970) data still contain some dependence on the mean after transformation possibly due to a preponderance of zero counts. Hayman and Lowe's (1961) exact transformation (Fig. 14.1) worked almost perfectly.

Similarly, the efficiency of sampling assuming a frequency distribution may not be constant as the mean density changes. A sampling plan based on negative binomial distribution in which the variance is approximately proportional to the square of the mean is tuned to sample an aggregated population. But that same population at a different time and density may be indistinguishable from Poisson, with the result that a greater sampling intensity is demanded than is necessary. Conversely, a sampling plan based on the Poisson binomial distribution will be efficient and reliable at low densities but inefficient and unreliable at higher densities where the population is more aggregated. As Trumble (1985) pointed out, care must be taken when a sampling plan designed at one density is applied at another.

Environmental assessment and monitoring

As we saw in Chapter 7, micro- and meiofauna are widely used as bioindicators of environmental health. Metrics used include the absolute and relative abundances of species, genera or families of nematodes, arthropods, or amphibians, as well as measures of community diversity (e.g., Shannon or Simpson indices), or the mix of feeding guilds: specialist or generalist, herbivores, predators or parasites. Early studies considered only the counts, regardless of the distribution of abundance. Thus, the power of the data to distinguish between environments is limited by the sampling variance, which varies seasonally and from site to site. Legislation requiring risk assessments to include a measure of confidence has led to a surge in statistical assessments of bioindicators, especially with respect to aquatic habitats (Monaghan, 2015). TPL has been an important component of these assessments.

Nematodes occur in all habitats with available organic carbon and occupy all niches but primary producer. Their lifestyles span the range from highly r-selected cp-1 opportunists to K-selected cp-5 persisters (see Chapter 7 for Colonizer-Persister scale). They vary in sensitivity to pollutants and environmental disturbance along the same continuum with cp-1 species far more likely to be found in polluted habitats, than species cp-3 and above. The cumulative effects of pollution on high cp species, their generally lower genetic variability, and slow recolonization rates relative to cp-1 species make them much more sensitive to environmental perturbations, including pollutants like metals and agrichemicals. The species composition of the nematode community is a useful bioindicator as pollution tends to shift community structure toward dominance by opportunistic cp-1 species, a shift that can occur rapidly. An average of nematode abundance weighted by their cp status, called the maturity index (MI; Bongers, 1990), varies from -4 to $+4$ with $MI = 2$, indicating nutrient-enriched disturbed systems. Nematodes respond rapidly to disturbance and enrichment and increased microbial activity can lead to changes in the proportion of bacterial feeders in a community.

Detecting environmental perturbations

Case studies of nematodes in freshwater habitats (Chapter 7: Beier and Traunspurger, 2003; Lazarova et al., 2004; Heininger et al., 2007; Ristau and Traunspurger, 2011) found differences in community assemblages in polluted and non- or less-polluted habitats, but the sample sizes were generally too small to identify differences in TPL. A study in India by Mukhopadhyay and Sarkar (1990) monitored plant-parasitic nematodes *Tylenchorhynchus zeae* and *Hoplolaimus indicus* in a field receiving effluent from a cotton mill and a nearby untreated control field. During the 2½-year period ($NB = 30$) of the study, $NQ = 10$ soil samples were taken in the fields in fallow and sown to jute, mustard or rice. One objective was to determine if TPL was sensitive to pollution. Only *T. zeae* was sufficiently abundant for analysis and the crop-specific TPLs are inconclusive as the number of points and range of means is limited. Analysis of parallelism of the spatial TPLs of all the *T. zeae* data (ignoring the crop) in the two fields shows the slopes to be significantly different ($P < 0.05$) with the nematodes in the polluted field more aggregated than the control field; $b = 1.73 \pm 0.051$ and $b = 1.54 \pm 0.082$, respectively (Fig. 14.5). In this example, the introduction of a pollutant modified the spatial organization of *T. zeae* sufficiently to be detectable by TPL.

The ability of surveys to detect the impact of disturbance, including pollution, on communities surely depends on the magnitude of the disturbance. To examine the effect of habitat and disturbance on nematode community composition, Neher et al. (2005) took 70–75 soil samples for nematodes over a two-year period in each of 6 habitat types: relatively undisturbed and disturbed wetland, forest and agricultural ecosystems at 18 sites in three ecoregions of North Carolina (coastal plain, piedmont, and mountain). Neher et al. present the mean and SE of number of nematodes by family recovered from samples in the six habitat types and the number by genus in the wetland habitats.

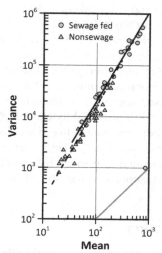

FIG. 14.5 Spatial TPLs ($NQ = 10, NB = 30$) of plant-parasitic nematode *Tylenchorhynchus zeae* in a sewage-treated and a control field differ in slope ($P < 0.05$): $\log(V) = 0.84 + 1.73\log(M)$ ($r = 0.988$) and $\log(V) = 0.99 + 1.54\log(M)$ ($r = 0.963$), respectively. *(Data from Tables 4–7 in Mukhopadhyay and Sarkar (1990).)*

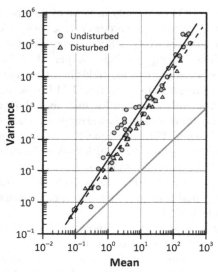

FIG. 14.6 Mixed-species spatial TPLs of nematodes in relatively undisturbed and disturbed wetlands differ in slope $(P=0.05)$: $\log(V)=1.37+1.59\log(M)$ $(r=0.975)$ and $\log(V)=1.20+1.51\log(M)$ $(r=0.987)$, respectively. Each point is a genus. *(Data from Table 3 in Neher et al. (2005).)*

At the family level, nematode composition differed between the three ecosystems and disturbance levels, the differences varying seasonally with much larger differences between habitats than between disturbed and undisturbed habitats. Mixed-species spatial TPLs did not differ in slope or intercept between disturbed and undisturbed for any habitat at the family level $(0.10 < P < 0.55)$, but disturbed and undisturbed wetland habitats differed at the genus level $(P < 0.05;$ Fig. 14.6). The wetlands mixed-species TPLs were significantly more aggregated than either agriculture or forest habitats $(P < 0.005)$, but there was no difference between the agriculture and forest communities $(P > 0.65)$. In this instance with mixed species, it is the relatively undisturbed wetland community that is steeper; it is also slightly curved $(P < 0.10)$.

The ability of TPL to discriminate between the disturbed and undisturbed wetlands at the genus level but not the family level supports the view that TPL has a role in environmental assessment but sample sizes need to be much larger, and the lower the taxonomic status, the better.

The cost of conservation

Looking for cost-effective conservation decision tools, Bach et al. (2012) used TPL to relate the population density to the rate of loss of genetic diversity in order to develop a metric for return on investment (ROI) for conservation efforts. ROI has been applied to species diversity, but Bach et al. applied it to individual species populations using population genetics arguments. Motivating this is the difficulty in prioritizing conservation management efforts. The amount of genetic variability in a population is related to the effective population size, N_e, estimated by the harmonic mean,[1] H, of the population size (Frankham, 1996), and to which the rate of genetic drift is related. Bach et al. related the harmonic mean to TPL by $H = M - aM^{(b-1)}$ and the inbreeding rate by $\Delta F = 0.5/(M - aM^{(b-1)})$. Because b is a measure of the population variation in relation to density, these two statistics provide information on the risk of genetic diversity loss in relation to population density fluctuations. The two special cases $b = 1$ and $b = 2$ emphasize the importance of the intercept $a\,(= 10^4)$ to both effective population size and inbreeding rate, just as A has far greater impact on the number of samples needed for pest management decision making (Fig. 14.4A).

Differentiating H gives the turning points of H as a function of M from which the domains of ROI can be inferred as a function of population size, which, in turn, can be related to cost of conservation, assuming cost is related to the population size. The shape of dH/dM with respect to M depends on the value of b with the result that ROI can increase or decrease with cost. For the common range of b near 2, the increase in genetic diversity has a diminishing return: the marginal cost of conservation increases as the increase in genetic diversity diminishes. As a result, Bach et al. suggest that ranking actions that increase mean population size according to their associated marginal increases in genetic diversity (cost) provides a metric for prioritizing conservation efforts. What is required for this is time series of surveys providing spatial mean and variances, data that are available for many endangered species.

This case study shows the value TPL's aggregation property, combined with abundance, in marking out risk spatially. It has the advantage of combining data in a parsimonious and easily visualized fashion, while providing feedback to planners of the consequences of potentially disruptive human activities.

Stream water quality

The earliest use of TPL in aquatic environmental bioassessment appears to be Clarke et al.'s (2002), who used TPL to quantify the sampling variation in ecological indices used to evaluate water quality. The indices are number of taxa or species richness, BMWP a river quality score, and the ratio of score per taxon (RSPT). The BMWP scoring system, developed by the Biological Monitoring Working Party (Hawkes, 1998), uses the presence of a set of macroinvertebrate taxa scored by their assumed tolerance to organic pollution. Clarke et al. selected 16 sites differing in 4 levels of ecological quality in a study of the assessments taken each of spring, summer, and autumn in 1994. Using TPL to characterize the variance structure of replicated samples were roughly

---

1. The reciprocal of the average of reciprocal values of the data set.

Poisson distributed ($b \approx 1$) for richness and BMWP, but mean and variance of RSPT were uncorrelated ($b < 0.2$). Square root transformation of the number of taxa and score result in roughly constant variance at each site regardless of its classification or health. This enabled Clarke et al. to estimate confidence intervals for the values of the number of taxa, score and ratio based on one, two, or three seasons. They extended their study to examine the sampling distributions of other bioindicators after transformation in comparisons of 17 stream types.

Extreme values of one or two components of a metric can badly bias the metric, but transformation of bioindicators usually removes the skewness of raw data and improves the statistical properties of the metric. Improving the metrics is an important goal but doing so without developing new sampling protocols is desirable. Monaghan (2015) found the stability and fidelity of metrics is greatly improved by applying TPL to the raw data before computing metrics. This has important relevance to the design and interpretation of bioassessment protocols because it is backwards compatible, requiring no change to existing protocols.

Monaghan et al. tested the effectiveness of transforming raw abundance data on the fidelity of bioassessment metrics by comparing the metrics using transformed and untransformed data of abundance, Simpson's diversity index H, MBWP, ASPT, and proportions of several key freshwater arthropod taxa (Ephemeroptera, Plecoptera, Trichoptera, and Chironomidae) and the proportions of four feeding guilds (shredders, scrapers, filter feeders, and predators). They used the UK Environmental Change Network (ECN; http: //www.ecn.ac.uk/) database of 15–30 years data for 29 rivers. Slopes of TPLs for taxa at species ($1.73 \leq b \leq 2.72$), order ($1.75 \leq b \leq 2.90$), and class ($1.78 \leq b \leq 2.21$) were used to choose transformations: square root for $b \leq 1.5$ and logarithmic for $b > 1.5$. The power transformation (Eq. 14.3) was not used.

Metric fidelity measured by correlation with a reference dataset of 635 sites in the United Kingdom (Wright et al., 2000) found fidelity to be consistently higher for metrics based on transformed data than those based on raw data. Metric precision was improved two to fourfold by transforming the raw data. For example, Simpson's diversity index, H, computed from transformed abundances reduced metric variation by a factor of 4.2 and was significantly correlated with all environmental variables while untransformed data produced no significant correlations with environmental variables. Monaghan et al. further tested their approach by applying the log-transformation to a species abundance database of indicator species in US streams (Hilsenhoff, 1988), resulting in a 17%–60% increase in metric fidelity in relation to a range of variables for water quality and a 15%–55% improvement in fidelity for watershed characteristics. Monaghan et al. emphasized the value of TPL in guiding transformations in river bioassessment and conclude that their application of TPL "reduces the risk of misclassifying biological quality for a suite of environmentally contrasting rivers from 80% to 4%."

In the context of assessing possible environmental and health risks of genetically modified crops (GM crops), Perry et al. (2009) emphasized the importance of appropriate experimental design and analysis, including the necessity for transformation to conform to statistical assumptions. The error type relevant in environmental risk assessments is not the typical statistical test's Type I error rate, the chance of rejecting a correct null hypothesis. Instead it is the Type II error that occurs when the null hypothesis is not rejected when it is actually false. In the context of environmental protection, the Type II is the more serious as decisions based on it can expose the environment and/or the population to hazard. Poorly designed experiments with insufficient replication and analysis with inappropriate transformations may lack the ability to discriminate in a difference test between the target GMO or environmental metric and a standard. One minus the Type II error rate is the test's statistical power, which unlike the Type I error rate cannot be manipulated. Thus, Monaghan et al.'s reduction in variability of environmental metrics is an important contribution to environmental assessment.

Environmental assessment of wind farms

Christel et al. (2013) mapped the slope of TPL of seabird abundance to identify areas of high risk of collision with a proposed windfarm offshore of Catalonia, Spain. The area of the planned wind farm was divided into 16 blocks in which 45 transects surveyed seabird abundance monthly from March to September. Ensemble TPLs were computed from the mean and variance of seabird abundance in each block and month. The monthly ensemble slope oscillated about an annual mean of $b \approx 1.93$ with a minimum in June ($b \approx 1.71$) maximum in July ($b \approx 2.05$). As a result, aggregation was significantly lower in the breeding season ($b = 1.89 \pm 0.02$) than the postbreeding season ($b = 1.97 \pm 0.01$). Values of b for each block in breeding and postbreeding season quantified aggregation in the areas of the proposed wind farm. Having identified blocks with high seabird aggregation, Cristel et al. developed a risk map showing the main traveling and feeding areas in the breeding and post-breeding months. The traveling areas are areas of high collision risk and in the feeding areas the presence of turbines would additionally impose a habitat loss to foraging seabirds. Obviously, the methodology also has application to terrestrial wind farms.

Other uses

Testing vaccines

Evaluating the safety and efficacy of vaccines and pharmaceuticals is an important and expensive part of the approval process. For both safety and efficacy tests, control of Type II errors and the power of statistical tests is paramount. Three important tests for vaccine efficacy involve the comparison of treated and control sets: infection intensity at the end of the trial; incidence of infection at the end of the trial; and the incidence rate of first infection during the trial. Alexander et al. (2011) evaluated the statistical power of these measures of efficacy for vaccination against helminth parasite infections. Statistical power was evaluated by modeling the impact of trial interventions on the intensity of nematode infections. To do this, the negative binomial distribution with mean m and parameter k was used to simulate the distribution as incidence of parasitism declines from its pretrial value. Shaw and Dobson's (1995) estimate of TPL slope, $b \approx 1.5$, was used to predict the change in variance with changing k in relation to the mean. This permitted evaluation of different degrees of aggregation of parasite incidence in the vaccine-treated and control sets. Eliminating variance from the equation for k (Eq. 3.11) by substituting Shaw and Dobson's TPL estimate,

$$k = \frac{m}{am^{b-1} - 1} \approx \frac{m^{2-b}}{a} = \frac{\sqrt{m}}{a}$$

provides a means for controlling for the degree of aggregation in treated and control sets:

$$\frac{k_t}{k_c} = \sqrt{\frac{m_t}{m_c}} \tag{14.17}$$

Simulating nematode infections over a range of situations and using Eq. (14.17) to test for differences between treated and control sets Alexander et al. were able to determine the power of several alternative statistical tests for efficacy. They found that under most circumstances the ratio of means of treated and control is more powerful than the alternative tests.

Model calibration and validation

Comparison of simulation results with observed data, such as the case studies of Samaniego et al.'s (2012) jays and Gilles et al.'s (2016) harbor porpoise, illustrates TPLs use as a tool to calibrate models and validate results. In both case studies, the TPLs of observed and modeled had identical slopes and differed only slightly in intercept. Samaniego et al. were the first to my knowledge to publish such a comparison. Standard techniques for comparing model fit to observations include regression of observed against predicted, correlation coefficient, Nash-Sutcliffe efficiency, and several tests based on mean absolute error, mean square error, and root mean square error (see Moriasi et al., 2007 for a critical comparison). They each examine one source of discrepancy between observed with predicted using a statistic that attempts to compare the data over a range of values and sometimes multiple scales. The use of TPL can simultaneously assess correlation, accuracy, precision and confidence by comparing the intercepts, slopes and residual variances, while also providing information useful in recalibrating the model. In addition, the use of TPL takes the variance structures into account, not just the mean values and can account for scale differences. It is also possible to adjust the model, either by varying a parameter or introducing a scale factor to bring the intercepts into alignment, provided the slopes agree. Also, if the slopes agree as they do in the examples in Chapter 10 (Figs. 10.10 and 10.15), the correlation coefficients measure the relative variation about the best fit lines. Finally, comparison by TPL has the added advantage of being visual in a way that percent bias and Nash-Sutcliffe are not.

Quality control

As alluded to in Chapter 11, TPL can sometimes be used to identify errors in data sets. If experimental or survey data produce a TPL with outliers, the cause is usually due to a real difference in behavior of data elements as seen in the mixed-species TPL of fish in Brazil's Sepetiba Bay (Fig. 10.4). But inconsistencies in the data set can also produce outliers. The following example illustrates TPL's use in quality control of research data. Fig. 14.7 shows the results of an experiment in which a decimal point of one reading was misplaced (bold entry in the X column) leading to anomalous mean and variance and an outlier, the square instead of the triangle. This trick works only when the scatter around the fitted line is small, and works best when the slope is not $b \approx 2$.

I have used this method to error check data many times, but to my knowledge it has not been published before. However, TPL has been used for error checking in another context: to identify defects in raw silk (Niu et al., 2011). Data from detectors checking for defects recording the thickness of silk strands obey TPL and using this information Niu et al. developed spot tests for variation

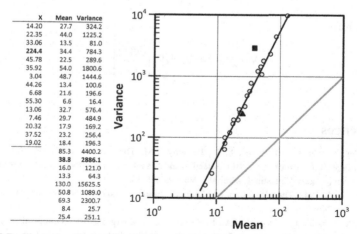

X	Mean	Variance
14.20	27.7	324.2
22.35	44.0	1225.2
33.06	13.5	81.0
224.4	34.4	784.3
45.78	22.5	289.6
35.92	54.0	1800.6
3.04	48.7	1444.6
44.26	13.4	100.6
6.68	21.6	196.6
55.30	6.6	16.4
13.06	32.7	576.4
7.46	29.7	484.9
20.32	17.9	169.2
37.52	23.2	256.4
19.02	18.4	196.3
	85.3	4400.2
	38.8	**2886.1**
	16.0	121.0
	13.3	64.3
	130.0	15625.5
	50.8	1089.0
	69.3	2300.7
	8.4	25.7
	25.4	251.1

FIG. 14.7 Illustration of using TPL to find an error in a dataset. The *bold entry* in column 1 should read 22.4 but contributes to an unusually high variance (*bold* in column 3) and produces an outlier (*square*) in the TPL plot. The true mean and variance for this point is represented by the *triangle*.

in silk strand thickness. Combined with simulations, their TPL analysis provided a basis for establishing grading standards for electronically tested raw silk.

As variable as the topics of the preceding case studies are, they have one thing in common. All use TPL to identify variations in data streams with a view to identifying environmental, experimental, or property differences for decision making, a use they have in common with pest management sampling.

Appendix: The delta technique for variance of a function of a random variable

Assume a random variable x has mean θ. A function of x, $y = g(x)$ has mean

$$E[g(x)] = g(\theta) + O(n^{-r}), \tag{A.1}$$

where n is the number of observations in the sample and r is typically equal to one (Stuart and Ord, 2006). Expanding $g(x)$ around θ by the Taylor series, we have:

$$g(x) = g(\theta) + \frac{\partial g(x)}{\partial x}(x - \theta) + O(n^{-r}) \tag{A.2}$$

$$g(x - \theta) = \frac{\partial g(x)}{\partial x}(x - \theta) + O(n^{-r}) \tag{A.3}$$

Squaring (A.3) and taking expectations, we have:

$$E[g(x - \theta)]^2 = E\left[\frac{\partial g(x)}{\partial x}(x - \theta)\right]^2 + o(n^{-r}) \tag{A.4}$$

and therefore

$$Var[g(x)] \cong Var[x] \cdot \left[\frac{\partial g(x)}{\partial x}\right]_\theta^2 \tag{A.5}$$

References

Alexander, N., Cundill, B., Sabatelli, L., Bethony, J.M., Diemert, D., Hotez, P., Smith, P.G., Rodrigues, L.C., Brooker, S., 2011. Selection and quantification of infection endpoints for trials of vaccines against intestinal helminths. Vaccine 29, 3686–3694.

Anscombe, F.J., 1948. The transformation of poisson, binomial and negative-binomial data. Biometrika 35, 246–254.

Bach, L.A., Pertoldi, C., Vucetich, J.A., Koeschke, V., Lundberg, P., 2012. Diminishing return of investment in genetic diversity. Evol. Ecol. Res. 14, 793–801.

Banks, C.J., 1954. A method for estimating populations and counting large numbers of *Aphis fabae* Scop. Bull. Entomol. Res. 45, 751–756.

Bartlett, M.S., 1947. The use of transformations. Biometrics 3, 39–52.

Beall, G., 1942. The transformation of data from entomological field experiments so that the analysis of variance becomes applicable. Biometrika 32, 243–262.

Beier, S., Traunspurger, W., 2003. Temporal dynamics of meiofauna communities in two small submountain carbonate streams with different grain size. Hydrobiologia 498, 107–131.

Binns, M.R., Bostanian, N.J., 1988. Binomial and censored sampling in estimation and decision making for the binomial distribution. Biometrics 44, 473–483.

Binns, M.R., Bostanian, N.J., 1990. Robustness in empirically based binomial decision rules for integrated pest management. J. Econ. Entomol. 83, 420–427.

Binns, M.R., Nyrop, J.P., 1992. Sampling insect populations for the purpose of IPM decision making. Annu. Rev. Entomol. 37, 427–453.

Bliss, C.I., Owen, A.R.G., 1958. Negative binomial distributions with a common *k*. Biometrika 45, 37–58.

Bongers, T., 1990. The maturity index: an ecological measure of environmental disturbance based on nematode species composition. Oecologia 83, 14–19.

Christel, I., Certain, G., Cama, A., Vieites, D.R., Ferrer, X., 2013. Seabird aggregative patterns: a new tool for offshore wind energy risk assessment. Mar. Pollut. Bull. 66, 84–91.

Clark, S.J., Perry, J.N., Marshall, E.J.P., 1996. Estimating Taylor's power law parameters for weeds and the effect of spatial scale. Weed Res. 36, 405–417.

Clarke, R.T., Furse, M.T., Gunn, R.J.M., Winder, J.M., Wright, F.J., 2002. Sampling variation in macroinvertebrate data and implications for river quality indices. Freshw. Biol. 47, 1735–1751.

Frankham, R., 1996. Relationship of genetic variation to population size in wild-life. Conserv. Biol. 10, 1500–1508.

Gilles, A., Viquerat, S., Becker, A., et al., 2016. Seasonal habitat-based density models for a marine top predator, the harbor porpoise, in a dynamic environment. Ecosphere. 7(6).

Green, R.H., 1970. On fixed precision level sequential sampling. Res. Popul. Ecol. 12, 249–251.

Hawkes, H.A., 1998. Origin and development of the biological monitoring working party score system. Water Res. 32, 964–968.

Hayman, B.I., Lowe, A.D., 1961. The transformation of counts of the cabbage aphid (*Brevicoryne brassicae* (L.)). NZ J. Sci. 4, 271–278.

Healy, M.J.R., Taylor, L.R., 1962. Tables for power-law transformations. Biometrika 49, 557–559.

Heininger, P., Höss, S., Claus, E., Pelzer, J., Traunspurger, W., 2007. Nematode communities in contaminated river sediments. Environ. Pollut. 146, 64–76.

Hilsenhoff, W.L., 1988. Rapid field assessment of organic pollution with a family-level biotic index. J. N. Am. Benthol. Soc. 7, 65–68.

Iwao, S., 1975. A new method of sequential sampling to classify populations relative to a critical density. Res. Popul. Ecol. 16, 281–288.

Jones, V.P., 1990. Developing sampling plans for -spider mites (Acari: Tetranychidae): those who don't remember the past may have to repeat it. J. Econ. Entomol. 83, 1656–1664.

Kuno, E., 1969. A new method of sequential sampling to obtain population estimates with a fixed level of precision. Res. Popul. Ecol. 11, 127–136.

Lazarova, S.S., de Goede, R.G.M., Peneva, V.K., Bongers, T., 2004. Spatial patterns of variation in the composition and structure of nematode communities in relation to different microhabitats: a case study of *Quercus dalechampii* Ten. forest. Soil Biol. Biochem. 36, 701–712.

McArdle, B.H., Gaston, K.J., Lawton, J.H., 1990. Variation in the size of animal populations: patterns, problems and artifacts. J. Anim. Ecol. 59, 439–454.

Monaghan, K.A., 2015. Taylor's Law improves the accuracy of bioassessment; an example for freshwater macroinvertebrates. Hydrobiologia 760, 91–103.

Moriasi, D.N., Arnold, J.G., van Liew, M.W., Bingner, R.L., Harmel, R.D., Veith, T.L., 2007. Model evaluation guidelines for systematic quantification of accuracy in watershed simulations. Trans. ASABE 50, 885–900.

Mukhopadhyay, M.C., Sarkar, P.K., 1990. Population behavior and distribution pattern of *Tylenchorhynchus zeae* and *Hoplolaimus indicus* in cultivated soil receiving industrial pollutants. Indian J. Nematol. 20, 152–160.

Neher, D.A., Wu, J., Barbercheck, M.E., Anas, O., 2005. Ecosystem type affects interpretation of soil nematode community measures. Appl. Soil Ecol. 30, 47–64.

Niu, J.T., Hu, Q., Xu, J.M., Dong, S.Z., Bai, L., 2011. Research on the grading theory of thick and thin defects in the electronic testing for raw silk. Silk: inheritance and innovation—modern silk road, 7th China International Silk Conference on Inheritance and Innovation—Modern Silk Road. Adv. Mater. Res. 175–176, 439–444.

Paoletti, M.G., Taylor, R.A.J., Stinner, D.H., Stinner, B.R., Benzing, D.H., 1991. Diversity of soil fauna in the canopy and forest floor of a cloud forest in Venezuela. J. Trop. Ecol. 7, 373–383.

Park, S.-J., Taylor, R.A.J., Grewal, P.S., 2013. Spatial organization of soil nematode communities in urban landscapes: Taylor's power law reveals life strategy characteristics. Appl. Soil Ecol. 64, 214–222.

Perry, J.N., ter Braak, C.J.F., Dixon, P.M., 2009. Statistical aspects of environmental risk assessment of GM plants for effects on non-target organisms. Environ. Biosaf. Res. 8, 65–78.

Peters, A., 1994. Interaktionen zwischen den Pathogenitätsmechanismen entonopathogener Nematoden und den Abwehrmechanismen von Schnakenlarven (*Tipula* spp.) sowie Möglichkeiten zur Virulenzsteigerung der Nematoden durch Selektion [Interactions between the pathogenicity mechanisms of entomopathogenic nematodes and the defense mechanisms of leatherjackets (*Tipula* spp.) as well as opportunities for increasing the virulence of nematodes by selection] (PhD Thesis). University of Kiel, Germany.

Ristau, K., Traunspurger, W., 2011. Relation between nematode communities and trophic state in southern Swedish lakes. Hydrobiologia 663, 121–133.

Samaniego, H., Sérandour, G., Milne, B.T., 2012. Analyzing Taylor's scaling law: qualitative differences of social and territorial behavior on colonization/extinction dynamics. Popul. Ecol. 54, 213–223.

Shaw, D.J., Dobson, A.P., 1995. Patterns of macroparasite abundance and aggregation in wildlife populations: a quantitative review. Parasitology 111, S111–S133.

Smits, P.H., 1996. Post-application persistence of entomopathogenic nematodes. Biocontrol Sci. Tech. 6, 379–387.

Southwood, T.R.E., Henderson, P.A., 2000. Ecological Methods, third ed. Blackwell Science, Oxford, UK.

Spridonov, S.E., Voronov, D.A., 1995. Small scale distribution of *Steinernema feltiae* juveniles in cultivated soil. In: Griffin, C.T., Gwynn, R.L., Masson, J.P. (Eds.), Ecology and Transmission Strategies of Entomopathogenic Nematodes. COST 819. European Commission, Luxembourg, pp. 36–41.

Stuart, A., Ord, J.K, 2006. Kendall's advanced theory of statistics. In: Distribution Theory, sixth ed., Vol. 1. Hodder Arnold, London.

Taylor, L.R., 1970. Aggregation and the transformation of counts of *Aphis fabae* Scop. on beans. Ann. Appl. Biol. 65, 181–189.

Taylor, L.R., Taylor, R.A.J., Woiwod, I.P., Perry, J.N., 1983. Behavioural dynamics. Nature 303, 801–804.

Taylor, L.R., 1984. Assessing and interpreting the spatial distributions of insect populations. Annu. Rev. Entomol. 29, 321–357.

Taylor, R.A.J., 1987. On the accuracy of insecticide efficacy reports. Environ. Entomol. 16, 1–8.

Taylor, R.A.J., 1999. Sampling entomopathogenic nematodes and measuring their spatial distribution. In: Gwynn, R.L., Smits, P.H., Griffin, C., Ehlers, R.-U., Boemare, N., Masson, J.-P. (Eds.), Application and Persistence of Entomopathogenic Nematodes (EUR 18873 EN). European Commission, Brussels, pp. 43–60.

Taylor, R.A.J., 2018. Spatial distribution, sampling efficiency and Taylor's power law. Ecol. Entomol. 43, 215–225.

Taylor, R.A.J., McManus, M.L., Pitts, C.W., 1991. The absolute efficiency of gypsy moth, *Lymantria dispar* (Lepidoptera, Lymantriidae), milk-carton pheromone traps. Bull. Entomol. Res. 81, 111–118.

Taylor, R.A.J., Lindquist, R.K., Shipp, J.L., 1998. Variation and consistency in spatial distribution as measured by Taylor's power law. Environ. Entomol. 27, 191–201.

Trumble, J.T., 1985. Implications of changes in arthropod distribution following chemical application. Res. Popul. Ecol. 27, 277–285.

Wald, A., 1947. Sequential Analysis. John Wiley, New York.

Williams, C.B., 1937. The use of logarithms in the interpretation of certain entomological problems. Ann. Appl. Biol. 24, 404–414.

Wilson, L.T., Room, P.M., 1983. Clumping patterns of fruit and arthropods in cotton, with implications for binomial sampling. Environ. Entomol. 12, 50–54.

Wright, J.F., Sutcliffe, D.W., Furse, M.T. (Eds.), 2000. Assessing the Biological Quality of Freshwaters: RIVPACS and Other Techniques. Freshwater Biological Association, Ambleside, UK.

Chapter 15

Properties of TPL

The ubiquity of TPL in realms as far removed as nematodes in a sandy beach, cetaceans in the English Channel, and packets of information traveling through the Internet, suggests a common origin. However, it is clear from consideration of these case studies that the results are not independent of the manner in which they were collected. In this chapter we examine some properties of TPL that are intimately associated with the methods of data collection, the sampler or measuring stick used and how its properties may influence the resulting TPL plot. In any survey conducted by sampling there are likely to be samples with no target organisms. The number of empty samples, which is not independent of the sampling method or device, can have a profound effect on the TPL estimated. This in turn has implications for all disciplines using TPL, especially field ecology and ecological theory as the variability of populations are important components of population dynamics and the practical implementation of ecological theory in conservation biology (see Lepš, 1993; Gaston and McArdle, 1993 and references therein). We start with a consideration of a possible connection between TPL and fractal geometry via self-similarity.

Self-similarity

Some natural features look much the same regardless of the scale at which they are viewed or measured. Cumulus clouds viewed from an airplane, for example, comprise bumps and hollows on a range of scales that without a reference object such as another plane it is impossible to determine the size of the features. A now-familiar example, made famous by Mandelbrot (1967, 1982), is the length of the coastline of Great Britain. The measured length depends on the length of the ruler used to measure it. As the ruler used is reduced in length, the measured length of the coastline increases (Fig. 15.1). This is a characteristic of many natural features that Mandelbrot called fractal and that is explored more fully in Chapter 16.

Mandelbrot developed the idea of fractal geometry by considering geometries that repeat indefinitely. One such is the triadic Koch curve, a line of four segments of equal length: two segments separated by two sides of an equilateral triangle, __/__. Superimposing this image or constructor on each edge increases the number of edges to 16. Continuing this process indefinitely creates an area filling structure whose parts do not touch—at least until the width of the

Taylor's Power Law. https://doi.org/10.1016/B978-0-12-810987-8.00015-X

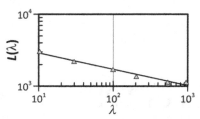

FIG. 15.1 The measured length, $L(\lambda)$, of the west coast of Great Britain increases as the length (λ) of the ruler used to measure the coastline decreases in length. The slope of -0.227 indicates a fractal dimension of $D_F = 1.227$. The intercept 4898 is the measured length of coastline when the ruler is $\lambda = 1$ km. (*Adapted from Fig. 33 in Mandelbrot (1982) based on data of Richardson (1961).*)

constructed line becomes larger than the space between the vertices. Mandelbrot showed that this structure has a dimension, $D_F > 1$, even though it is a line and therefore by Euclidean definition has $D_E = 1$. The fractal dimension of the Koch curve is obtained by considering the ratio of the length of the line (4 units) and the distance between the ends (3 units). As the number of constructions, n, is increased, the ratio also increases: $R_\lambda = (4/3)^n$. The length of each edge at the nth generation is $\lambda_n = 3^{-n}$ so $n = -\ln\lambda_n/\ln3$ and

$$R_\lambda = \left(\frac{4}{3}\right)^n = \exp\left\{\frac{\ln\lambda(\ln4 - \ln3)}{\ln3}\right\} = \lambda^X$$

where $X = 1 - \ln4/\ln3 = 1.263$, which is the fractal dimension, D_F, of the Koch curve. Examining Koch curves and coastlines at a range of scales results in images that are so similar it is impossible to determine the scale. Such features are said to be self-similar, a characteristic of which, as evidenced by Fig. 15.1, is

$$X_\lambda = X_0 L^\lambda,$$

which says that the value of the variable X is self-similar if it grows relative to a baseline (X_0) as a power of a measure L. If the similarity is seen as a statistical property as, for example, by sampling, it is said to be statistically self-similar. TPL is statistically self-similar as is easily established. Given $V = aM^b$ and a baseline mean, M_0, the variance at M_0 is V_0 and the variance at a mean kM_0 ($k > 0$) is:

$$V_k = a(kM_0)^b = ak^b M_0^b = ak^b\left(\frac{V_0}{a}\right) = k^b V_0 \qquad (15.1)$$

The implication of this is that populations that obey TPL are statistically self-similar and they appear similar at a range of scales above a baseline value. The baseline, M_0, is the density at which TPL intersects the Poisson ($V = M$) line. At densities below M_0, sample data return variances equal to the mean, except in a small realm near where $V = M \approx 1$. In this realm, where samples are all small, a pattern of variance-mean points resembling a Moiré pattern emerges (Fig. 15.2).

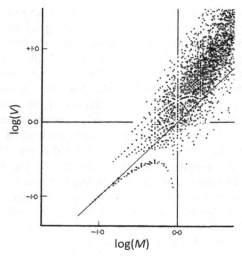

FIG. 15.2 Samples with few specimens and many with no specimens are subject to an integer distortion as only some combinations of mean and variance exist. The curve of points in the third quadrant of the mean-variance domain is bounded by the Poisson line and a quasiparabola curving down to the ordinate at $M = 1$. Fig. 1b in LRT and Woiwod (1982). *(Copyright 1982 British Ecological Society, Courtesy of John Wiley and Sons.)*

Another way to express the population variance is in relation to the variance of the Poisson distribution. Because the mean and variance of the Poisson distribution are equal, TPL is also the relationship between the observed variance and the expected variance assuming the observations had been Poisson distributed:

$$V_{observed} = cV_{Poisson}^{b} \tag{15.2}$$

In tests of the slope of TPL, the null hypothesis is frequently a comparison with slope to the Poisson line, $b = 1.0$. Hughes and Madden (1992) were the first to make this connection and used it as the basis for their development of the binomial power law used to describe the distribution of disease incidence (Chapter 16).

Effect of small samples

In their comparison of spatial and temporal TPLs, LRT and Woiwod (1982) showed how when a survey of n quadrats comprises $(n - 1)$ empty quadrats and one quadrat contains a single individual the sample mean and variance are both $1/n$. With only one nonzero quadrat, a series of means each with a corresponding identical variance appears in the third quadrant of the mean-variance domain as is clearly seen in Fig. 15.2. A survey comprising samples of different sizes $\{n_i\}$ with some samples containing only a single specimen and others a few, areas in the mean-variance domain that contain no points becomes apparent. A quasiparabola $\log(V) = \log[n/(n-1)] + \log(1-M) + M$ and the Poisson line in the third quadrant ($\log M < 1$, $\log V < 1$) define the lower limit of permissible variance-mean points. If the singleton quadrat has two individuals, the mean is $2/n$ and the variance is $4/n$. As with the singleton case, there is a quasiparabola above and to the right. A series of samples with few specimens and a large number of empty samples will have variance-mean points constrained to the patterning clearly seen in Fig. 15.2. The gaps are, to use LRT and Woiwod's words, "forbidden regions" resulting from the integer nature of the data. This integer constraint will distort regressions down in this region, tending to reduce b and elevate a. Low integer counts in samples with few specimens describe similar quasiparabola patterns intersecting with vertical patterns at M that are powers of 2. These patterns, evident at intervals as M increases, disappear as M approaches 10. Caution should be exercised interpreting TPLs in this region.

Effect of zeros

Direct effect

In the first large-scale longitudinal study of a soil nematode community to produce data sufficient for a comparative analysis of TPL (Appendices 7.D and 7.E), Park et al. (2013) examined the idea that TPL is sensitive to life-history strategy such as nematode cp-classes and other ecological groupings. Their data suggested that the aggregation patterns adopted by taxa with otherwise similar demographic and feeding styles can reduce their niche overlap and avoid direct competition by exploiting space differently and these patterns are captured by TPL.

In a subsequent analysis of the same data, RAJT et al. (2017) deduced a prediction of the behavior of TPL. Their data of nematode abundance in a set of $NQ = 10$ samples taken at $NB = 36$ site-times that included species on a continuum from ubiquitous to rare. They pointed out that if all quadrats in a block are empty, resulting in both mean and variance being zero, the block is not included in the analysis. The rationale for "throwing away" a data point is that zero mean and variance contain no useable information, their logarithms being negative infinity. Somewhat whimsically, they also make the point that we do not search for morpho butterflies in Antarctica because we know they are not to be found there, and ask if we are deliberately biasing our results by limiting the number of zero samples.

A block in which species **A** is absent from all samples is also ignored, but other species **B**, **C**, **D**, etc. of the same ecological grouping may be present in one or more of the quadrats comprising that block. Thus, in an ensemble of species, all samples may contain members of some but not all of the taxa in the ensemble. Similarly, some blocks may contain members of some but not all of the taxa in the ensemble. If we aggregate the members of the ecological grouping, we will find that there are fewer or no empty blocks, perhaps no empty quadrats. This was the case with Park et al.'s (2013) data. Considering the community as a whole, there were no empty quadrats and aggregating the genera at the trophic group level, none of the 36 blocks were empty. By comparison, even the most abundant genus (*Aphelenchoides*) was not found in 14 of 360 quadrats and only 10 of the 28 genera studied were found in all 36 blocks.

They observed that combining the genera into an ensemble of ecological groupings resulted in generally higher *b* than the average of the gradients obtained for the individual genera comprising the groupings, and the difference in slopes of a grouping is positively correlated with the number of genera in the grouping ($r = 0.51$, $P < 0.02$). These observations suggested to RAJT et al. that "filling in" the empty quadrats by using an ensemble of taxa resulted in a steeper power law gradient. Conversely, for any ensemble (taxon or ecological grouping), removing empty quadrats from the sample set should also produce a steeper power law gradient.

RAJT et al. tested the prediction that eliminating zero samples from a spatial survey increases the TPL slope by calculating the means and variances of the 28 genera in Park et al.'s (2013) data with (Zero set) and without empty quadrats (Nonzero set). All 28 genera showed strong increases in *b*. The smallest increases occurred with the most abundant genera with the fewest empty quadrats. In the 14 most abundant genera, *b* ranged from 1.51 to 2.35 in the Zero set and from to 1.71 to 2.55 in the Nonzero set. Removing empty quadrats increased the estimate of *b* by an average of 0.92, from below 2 to above 2.

While removing empty quadrats increased *b*, the intercepts decreased by amounts ranging from 92% to 3500%. With the increase in *b* and decrease in *a*, the regression lines appear to swivel around a common point. Fig. 15.3 illustrates it for *Helicotylenchus* and *Rhabditis*. As a result of removing empty quadrats, there are 82 and 67 fewer quadrats (*NQ*) and 2 and 1 fewer points (*NB*) on the graph, respectively. The coefficients of determinations declined from 0.93 to 0.90 and increased from 0.88 to 0.89, respectively. Overall, r^2 declined by 0.063, not significantly different from zero. The range of means increased for 18 of 28 genera when empty quadrats are ignored and decreased only slightly for most other genera.

Examination of Fig. 15.3 shows that removing zeros increases the mean by a larger proportion at the low end of the graph than the high end. Removing zeros can decrease the range of means; comparatively more for the lower than the higher values and the variance is reduced, generally by a small amount. As the variance decreases it creates the appearance of a counterclockwise rotation. It is easily seen why the variance declines when the mean increases. Decreasing the number of samples (*NQ*) at any point on the graph leaves $\sum x$ and $\sum x^2$ unchanged while increasing $E[x] = \frac{1}{NQ}\sum x$ and $E[x^2] = \frac{1}{NQ}\sum x^2$ by the same

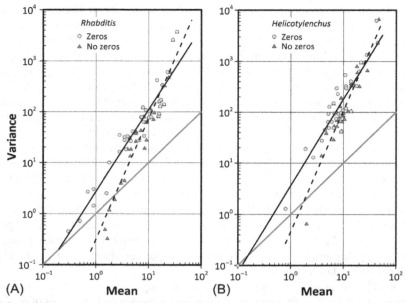

FIG. 15.3 Plots of nematodes *Rhabditis* (A) and *Helicotylenchus* (B) illustrate how TPL slope increases when empty samples are excluded from calculation of means and variances. Regression estimates changed from $b = 1.61$ to 2.37 and $A = 0.43$ to -0.50 (*Rhabditis*) and $b = 1.70$ to 2.37 and $A = 0.56$ to -0.34 (*Helicotylenchus*). The regression lines swivel around the points $M \approx 17$, $V \approx 260$ (*Rhabditis*) and $M \approx 20$, $V \approx 600$ (*Helicotylenchus*) as the regressions shift rightward and the variances decline.

proportion. Provided $\sum x > 1.0$, $E^2[x]$ appreciates faster than $E[x^2]$ and $V[x] = E[x^2] - E^2[x]$ declines as $E[x]$ increases.

By removing empty quadrats, we may be increasing bias of the regression estimates. Of the several potential sources of bias estimation in small-sample TPLs Clark and Perry (1994) found the most important to be the exclusion of samples for which $M = V = 0$ and exclusion of samples for which $V = 0$, but $M > 0$. In their simulations, Clark and Perry (1994) found that both a and b were underestimated when there were a large number of zero means and variances. In RAJT et al.'s analyses, by contrast, a decreased and b increased by amounts proportional to the number of zeros removed (Fig. 15.4). Clark and Perry's expectation of changes to a are also not reflected in these results.

We use samplers appropriate to the target in habitats we might reasonably expect the organism to be found. Considering the shifts in the regression shown in Fig. 15.3 and TPL parameters' dependence on the number of zeros (Fig. 15.4), RAJT et al. (2017) proposed that TPL measures the space between clusters or somehow "counts" zeros. This is not an alternative to the conventional interpretation, but an extension. The strong influence of the proportion of zeros in an ensemble of samples on b, and the systematic way in which this

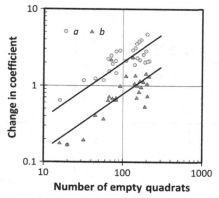

FIG. 15.4   Changes in TPL regression coefficients are correlated with the number of empty quadrats. On log scales, the changes in coefficients have nearly identical slopes. Triangles = the change in intercept (*a*), Zero set minus Nonzero set; circles = change in slopes (*b*), Nonzero set minus Zero set.

proportion changes with abundance to preserve a regression pattern while changing its gradient, suggested to RAJT et al. that it may be the space between individuals of a population that is subject to some control by that population's members and that this is intimately associated with their life-history strategy.

Anatomy

The number of zero samples in the nematode survey varied substantially between genera, resulting in a spectrum of responses as Zero samples were transformed to Nonzero. The pattern of change in TPL plots is illustrated in a TPL analysis of European corn borer (*Pyrausta nubilalis*) larvae in corn (*Zea mays*). Nine fields in Iowa were surveyed for the pest insect with number of samples per field from 324 to 3205 and totaled 11,924 (McGuire et al., 1957). The frequency distribution of number of larvae per sample differed between fields from Poisson at low density to Neyman's A at the highest density (Fig. 1.1A). The total number of zero samples was 3677 (31.7%). Progressively removing 3.17% of zeros at random and recalculating TPL produces 11 graphs of progressively steeper slopes. The estimated TPL parameters are plotted against $P(0)$, the proportion of zeros in the survey, which ranges from 3677 to 0 in Fig. 15.5A. As we have seen in the nematode study, *a* and *b* are inversely proportional (Fig. 15.5B). This analysis shows *a* and *b* related by the power function: $b = 1.313a^{-0.297}$ ($r = 0.999$). The locus of the point where the TPL lines cross, M^*, V^* (see Fig. 15.3) is a straight line ($V^* = 3.756 - 15.5M^*$, $r = 0.999$) that moves down the domain as $P(0)$ declines (Fig. 15.5C), but the average means, $x = \log(M)$ and variances $y = \log(V)$ form a maximum curve, well approximated by a cubic function (Fig. 15.5D; $y = -25.06x^3 + 20.03x^2 - 4.96x + 0.78$, $r = 0.993$).

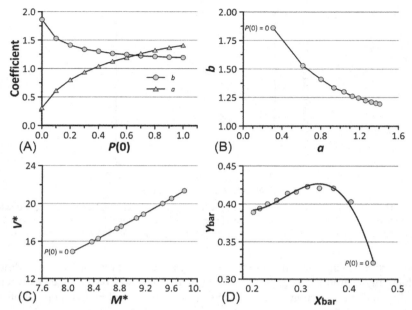

FIG. 15.5 TPL structure and parameters change as the proportion of zeros, $P(0)$, in a sample set is reduced; European corn borer in $NB = 9$ fields in Iowa. (A) The slope increases and the intercept decreases as $P(0)$ declines to zero. (B) a and b are related exponentially with increasing $P(0)$. (C) The crossover point (M^*, V^*) increases linearly with increasing $P(0)$. (D) The average values of $\log(M) = X$bar and $\log(V) = Y$ bar are related approximately as a cubic curve with increasing $P(0)$.

We see from this example that the proportion of zeros in a survey has a profound effect on both the slope and the position of TPL in the V-M domain. Furthermore, the relationship can be predicted with a high degree of precision. The number of zero quadrats in a survey is determined by the ability of the sampler to detect low density populations: the higher the sampling efficiency, the smaller the detectable population.

Sampling efficiency

Using samplers that rely on the behavior of the target organism can result in underestimation of b because the sampling efficiency can change with density (RAJT, 2018). The efficiency of insect traps sampling the atmosphere was defined by LRT (1962) as the proportion of insects extracted from the air. By knowing the volume of air being filtered by a suction trap or aircraft tow net and its extraction efficiency, the absolute aerial density of a population can be estimated. If the extraction efficiency is 50% and the number of insects extracted per cubic meter of air passing through a suction trap or tow net is N, the absolute aerial density is $2N$ m^{-3}. Such an estimate LRT called an absolute estimate and the efficiency of the sampler absolute efficiency. The samples of nematodes taken by core (Chapter 7) or the samples of plants taken by quadrat (Chapter 6) are therefore absolute estimates. But insect samples taken by light traps or pheromone traps (Chapter 8) that rely for their operation on the response of insects to light or pheromone can only provide relative estimates because their efficiency is unknown.

The efficiency of one sampler relative to another in terms of logarithms of the counts x and y obtained by traps **X** and **Y** is given by

$$E_r = \log(\varepsilon_r) = \log(y) - \log(x) \tag{15.3}$$

If the absolute efficiency of one trap, **X** say, is known then the absolute efficiency of **Y** may also be determined from $E_a = \log(\varepsilon_a) = \log(y) - \log(\rho)$, where ρ is the absolute density estimated by trap **X**.

Density-dependence

Using this simple relationship between samples taken by two traps (a suction trap and a pheromone trap), RAJT et al. (1991) found that the efficiency, ε_a, of Gyplure® baited USDA "milk carton" traps sampling flying male gypsy moths declined as gypsy moth density increased. They found that at an absolute density of one gypsy moth in 10^6m^3 of air ($\rho = 1/\text{mcm}$), the absolute efficiency of gypsy moth pheromone traps is $\varepsilon_a = 36\%$ ($E_a = -0.44$), and at $\rho = 10/\text{mcm}$, $\varepsilon_a = 16\%$ ($E_a = -0.79$), declining to $\varepsilon_a = 3.2\%$ ($E_a = -1.51$) at $\rho = 1000/\text{mcm}$. This discovery has implications for estimates of b obtained from attractant traps, as well as practical importance in forest pest management.

Defining the density-dependent efficiency of trap **Y** in terms of trap **X** as $M_Y = \varepsilon_M M_X^{\kappa_M}$ with a corresponding relationship between the sample variances, $V_Y = \varepsilon_V V_X^{\kappa_Y}$, where ε is the efficiency from Eq. (15.3) and κ is the degree of density dependence, RAJT (2018) proved a relationship between TPLs for the target organism sampled by both samplers:

$$b_x = b_y \frac{\kappa_M}{\kappa_V} \text{ and } a_x = \left(a_y \frac{\varepsilon_M^{b_y}}{\varepsilon_V}\right)^{1/\kappa_V}, \tag{15.4}$$

where a_x, b_x, a_y, and b_y are the TPL parameters for samplers **X** and **Y**. For situations where one variance is unknown, as for example when a sampler is not replicated, an alternative approach based on the variance of a function may be used (Appendix 14A). Knowing the power law obtained using one sampler, and the relative density-dependent efficiency of the samplers, the power law for the other sampler can be predicted:

$$b_y = (b_x + 2\kappa - 2)/\kappa \text{ and } a_y = a_x \cdot \kappa^2 \cdot \varepsilon^{(2-b)/\kappa}$$

where the subscripts for ε and κ have been omitted. Alternatively, κ and ε can be estimated from two power laws:

$$\kappa = \frac{b_x - 2}{b_y - 2} \text{ and } \varepsilon = \left\{ \frac{a_x \cdot \kappa^2}{a_y} \right\}^{1/(2-b_y)} \tag{15.5}$$

The value of this approach may be limited as it is possible that sampler **X** may be more efficient (captures more insects) than sampler **Y** and is simultaneously less variable than sampler **Y**, thereby lowering the measured variance. Alternatively, the responsiveness of insects to an attractant trap may not be fixed, but distributed nonsymmetrically in the population. Both possibilities would impose a separate trap-specific influence on variance, situations that are captured by Eq. (15.4).

It is easily seen that if sampling is density independent, that is, if $\kappa_M = \kappa_V$ or $\kappa = 1$, then $b_x = b_y$ and the power law lines are parallel, separated by $a_x(\varepsilon^{(2-b)} - 1)$. Notice that the power laws must diverge if $\kappa_M \neq \kappa_V$ or $\kappa \neq 1$. Furthermore, if sampling efficiency is density dependent ($\kappa \neq 1$), a and b must be correlated. Consideration of Eq. (15.5) shows that if a sampler's efficiency declines systematically with density, the estimate of b will be biased, and by how much. For example, a measured value of $b = 2.5$ when $\kappa = 0.8$, then the true value, $b^* = 2.4$, and when $\kappa = 0.5$, $b^* = 2.25$.

RAJT (2018) illustrated and explored the relationship between density-dependent sampling efficiency and TPL with sample data of adult gypsy moth counts in pheromone traps, two light traps, and a suction trap in a Pennsylvania forest during a gypsy moth (*Lymantria dispar*) outbreak in 1987.

RAJT et al. (1991) conducted their comparison experiment with a suction trap of known absolute efficiency and 15 spatially separated USDA gypsy moth pheromone traps over a 27-day period in the early summer of 1987. The primary TPL data used are the daily mean and variance of gypsy moths caught in the 15 pheromone traps. In addition to the suction trap, a Rothamsted 200-W tungsten trap and a 15-W Pennsylvania UV trap were used in the experiment. All three attractant traps when standardized against the suction trap were found to have density-dependent sampling efficiency. The two light traps' density dependence as measured by κ was similar but ε differed, indicating the Pennsylvania trap was a more efficient sampler than the Rothamsted trap. RAJT (2018) computed TPL parameters for the light traps and converted their counts to density.

Rescaling average number per sample by a constant (number per day to number per 10 days, for example) only affects TPL's intercept. Rescaling by converting counts to density given a density-dependent sampling efficiency reveals how both a and b change and how the correlation between variance and mean can increase as the TPL plot is "stretched out" with increasing range of means, and little change in the variance about the fitted line (Fig. 15.6); the slope increased from $b_C = 1.685$ for the actual counts to $b_D = 1.771$ for counts converted to density (Table 15.1).

RAJT (2018) argued that the density-dependent sampling efficiency of attractant traps shows how TPL is sensitive to spatial aggregation behavior.

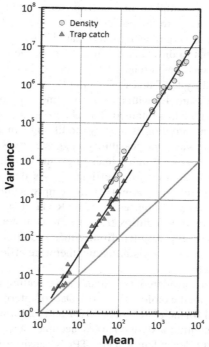

FIG. 15.6 Converting trap catches to density increases the range and shifts the plot up and to the right while preserving the overall pattern of points. The slope increased from $b = 1.68$ to 1.77.

TABLE 15.1 TPL estimates for adult male gypsy moths caught in pheromone traps and light traps, actual counts, and catch converted to aerial density

	N	r	b	A
Pheromone trap counts	27	0.991	1.685 ± 0.046	-0.111 ± 0.105
Pheromone trap counts converted to density	27	0.995	1.771 ± 0.037	0.315 ± 0.217
Calculated TPL parameters for light trap counts and density			Count	Density	Count	Density
Pennsylvania trap	19		1.727	1.765	0.537	0.290
Rothamsted trap	21		1.734	1.786	0.630	0.378

If behaviorally neutral samplers like suction traps produce unbiased power laws, the adjustment made to the power law by a targets' behavioral response to the trap is measuring a component of the spatial behavior. In the case of traps that use sex attractants to draw insects into the trap, the effect of aggregative behavior on the variance-mean relationship is clear. TPL is sensitive to spatial behavior as proposed by LRT and RAJT (1977).

As described here, samples with their zeros removed produce steeper slopes than the same data with zeros retained. The phenomena of both zero censoring and density-dependent sampling efficiency modifying b in a systematic fashion may be linked in some way. One possibility, suggested by RAJT et al. (2017), is the space and/or time between individuals or events to which TPL is sensitive gets distorted when a sampler loses efficiency. As density or frequency of events increases, the relative weight of the empty samples must increase, thereby depressing b. The differences found by RAJT et al. (1998) in TPL slope in samples of western flower thrips (Fig. 13.2) taken above and within in a greenhouse cucumber crop resulted from either the effect of the canopy on the pattern of flight or variable visibility and therefore efficiency of sticky card traps within the canopy.

RAJT (2018) also considered how quantitative scaling in the nonbiological world might differ from the biological world. Data of internet traffic through the network's nodes and financial data (Figs. 12.7 and 12.8), which typically have $b \leq 2$, are collected continuously with 100% accuracy and so are unlikely to be influenced by density-dependent effects. TPL's sensitivity to density dependence may be what makes it important in ecology. Whatever constraints physical phenomena place on quantitative scaling, evolutionary processes may relax those constraints and extend the range of permissible slopes. Furthermore, since density-dependent behavior apparently influences the value of b, it may be that it is density dependence, which is absent from physical phenomena that forces the higher values of b often encountered with biological data.

Trap saturation

In the foregoing discussion, we have seen that the size and structure of the sampling program interacting with the density of the target population can markedly influence the slope of TPL. In the case of density-dependent sampling efficiency of attractant traps, slope depression may be a consequence of the behavior of the targets. Another influence on TPL is limitations of sampler design. It is easily shown by simulation that if the sampler has small capacity and the average population exceeds a critical density close to the capacity, the variance recorded by samplers must decline as the mean increases above samplers' ability to record the population. The simulation of trap saturation (Fig. 13.3A) shows how variance increases as mean population density increases up to a maximum and then declines as the traps approach saturation. In Fig. 15.7, data of male gypsy moths collected in USDA milk carton traps at locations

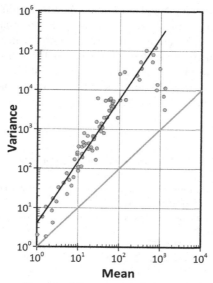

FIG. 15.7   Data from several surveys of Gypsy moth in South-Central Pennsylvania in the 1980's show the variance decline as the moth's population density increases above the ability of the 1.9 L capacity USDA gypsy moth traps to collect more than 2000 moths.

in South-central Pennsylvania over a period of years as the expanding wavefront of the invasive insect approached its peak. The USDA traps have a capacity of 2 quarts ($\sim$1.9 L) and can hold $\sim$2000 moths. At about 1500 moths/trap, the variance begins to decline, approaching 2000/trap the variance declines rapidly. At population densities significantly above 2000/trap when all traps have reached capacity, the variance must inevitably decline to near zero. This effect is in addition to the density-dependent efficiency described here. In the calibration experiment, the traps were emptied daily, and so never approached capacity; the survey data were from traps left out for the entire 5-week flight season and so at high densities had ample time to fill.

It is interesting to note that this pattern is the converse of a pattern seen in some network TPLs where, at the low mean end of the plot, a spike upward is evident (Fig. 12.7). It seems likely that some of the counter examples in Chapter 13 and some simulation studies in Chapter 17 with nonlinear TPLs result from a process equivalent to trap saturation.

In Chapters 5–8, we saw how changing the spacing of samples could influence the value of TPL parameters. However, the actual relationship between b and sample separation did not necessarily change consistently (Fig. 7.1B). The size and/or duration of sampling (sampling effort) also could influence the parameters (e.g., Appendices 8.M35 and 8.M42), although the intercept was usually more sensitive than the slope. The inconsistent influence of scale on slope (e.g., Appendices 6.D and 6.E) undoubtedly results from interactions of sample spacing and sample size or efficiency. In this chapter, two powerful processes are seen as influencing b: the presence or absence of empty quadrats and the possible density-dependent efficiency of the sampler used. It seems likely that the variation in slope seen in some case studies was due in part to these causes of variation in slope, in addition to inadequate sample number and sampler saturation.

References

Clark, S.J., Perry, J.N., 1994. Small sample estimation for Taylor's power law. Environ. Ecol. Stat. 1, 287–302.

Gaston, K.J., McArdle, B.H., 1993. Measurement of variation in the size of populations in space and time: some points of clarification. Oikos 68, 357–360.

Hughes, G., Madden, L.V., 1992. Aggregation and incidence of disease. Plant Pathol. 41, 657–660.

Lepš, J., 1993. Taylor's power law and the measurement of variation in the size of populations in space and time. Oikos 68, 349–356.

Mandelbrot, B.B., 1967. How long is the coast of Britain? Statistical self-similarity and fractional dimension. Science 156, 636–638.

Mandelbrot, B.B., 1982. The Fractal Geometry of Nature. W. H. Freeman & Co, San Francisco, CA.

McGuire, J.U., Brindley, T.A., Bancroft, T.A., 1957. The distribution of European corn borer larvae Pyrausta nubilalis (Hbn.) in field corn. Biometrics 13, 65–78.

Park, S.-J., Taylor, R.A.J., Grewal, P.S., 2013. Spatial organization of soil nematode communities in urban landscapes: Taylor's power law reveals life strategy characteristics. Appl. Soil Ecol. 64, 214–222.

Richardson, L.F., 1961. The problem of contiguity: an appendix to Statistics of Deadly Quarrels. General System Yearbook 6, 139–187.

Taylor, L.R., 1962. The absolute efficiency of insect suction traps. Ann. Appl. Biol. 50, 405–421.

Taylor, L.R., Taylor, R.A.J., 1977. Aggregation, migration and population mechanics. Nature 265, 415–421.

Taylor, L.R., Woiwod, I.P., 1982. Comparative synoptic dynamics. I. Relationships between inter- and intra-specific spatial and temporal variance/mean population parameters. J. Anim. Ecol. 51, 879–906.

Taylor, R.A.J., 2018. Spatial distribution, sampling efficiency and Taylor's power law. Ecol. Entomol. 43, 215–225.

Taylor, R.A.J., McManus, M.L., Pitts, C.W., 1991. The absolute efficiency of gypsy moth, Lymantria dispar (Lepidoptera, Lymantriidae), milk-carton pheromone traps. Bull. Entomol. Res. 81, 111–118.

Taylor, R.A.J., Lindquist, R.K., Shipp, J.L., 1998. Variation and consistency in spatial distribution as measured by Taylor's power law. Environ. Entomol. 27, 191–201.

Taylor, R.A.J., Park, S.-J., Grewal, P.S., 2017. Nematode spatial distribution and the frequency of zeros in samples. Nematology 19, 263–270.

Chapter 16

Allometry and other power laws

That animals' organs grew at different rates was known by Galileo in the 17th century and probably by da Vinci a century earlier. In the 20th century, Thompson (1917), followed by Huxley (1932), subjected the growth and form of organs to detailed quantitative study, leading Huxley and Tessier (1936) to coin the terms allometric growth and allometry to describe the power law relating the relative growth of two organs. Although the term allometry was coined to describe an organic phenomenon, its use is not restricted to anatomy. TPL and other power laws are also allometric relationships if the logarithms of two measurements are linearly related. Most power laws are necessarily statistical in nature and are characterized by the properties of scale invariance (Eq. 15.1), some have ill-defined means with infinite variance. In some branches of physics, powers laws are also characterized by universality in which the behavior of a system is independent of the dynamical details of the system.

Science recognizes more than a hundred power-law distributions in physics, biology, and the social sciences (Andriani and McKelvey, 2007). This chapter reviews some in each of those disciplines, the majority of which have a plausible connection to TPL. The chapter is split into four sections: mathematical and statistical relationships; power laws in physics; power laws in biology; and a power law in the social sciences. While some relationships can be derived theoretically, most results are purely empirical and some are common to several disciplines.

Mathematical

Pareto distribution

The Pareto distribution originally formulated as the Pareto law of income distribution $N = Ax^{-\alpha}$ has cumulative distribution function

$$P(x) = \Pr(X \geq x) = \left(\frac{k}{x}\right)^{\alpha}, \ \alpha > 0; x \geq k.$$

where $k > 0$ is the smallest value of x. The mean and variance of Pareto distributed variates are:

Taylor's Power Law. https://doi.org/10.1016/B978-0-12-810987-8.00016-1
503

$$M = \frac{\alpha k}{\alpha - 1}, \; \alpha > 1 \text{ and } V = \left(\frac{k}{\alpha - 1}\right)^2 \frac{\alpha}{\alpha - 2}, \; \alpha > 2$$

(Johnson et al., 1994). Although the Pareto is explicitly a power law, the mean and variance (when they both exist) scale as TPL with $b = 2.0$.

Zipf's law

Zipf's law is the discrete form of the Pareto distribution. As originally described, it is an empirical law that states the frequency, F, of any word (in any language) is inversely proportional to its rank, R, in a table of the frequency of occurrence of all words, $F \propto R^{-\alpha}$ where $\alpha \approx 1$. The most frequent word occurs about twice as frequently as the next most frequent word and three times as often as the third most frequent word, and so on. It has been used to describe the distribution of city sizes (Chapter 11) with exponent $1.0 < \alpha < 2.0$. The populations of cities are often lognormally distributed as predicted by Gibrat's (1931) rule of proportionate effect that states the proportional rate of growth of an organization like a company is independent of its absolute size. This rule gives rise to lognormal distributions of size and has been applied to city size and growth rate. While city size distribution is often associated with Zipf's law, this holds only in the upper tail as noted in Chapter 12.

Not limited to words and cities, Zipf's law also fits the frequency of genes expressed in a variety of organisms and tissues, such as yeast, nematodes, human normal and cancer tissues, and embryonic stem cells (Furusawa and Kaneko, 2003). Simulations suggest that abundance of chemicals in intracellular reaction networks also follows Zipf's law, suggesting a universal feature of cells.

Spectra

Many signals, such as the price of commodities, equities, and foreign exchange (Chapter 12), when plotted form a spectrum with frequency f:

$$S(f) \propto \frac{1}{f^{\alpha}} \tag{16.1}$$

where $0 \leq \alpha \leq 3$ is a parameter that relates to the degree of temporal correlation. The value of the exponent governs the qualitative as well as quantitative behavior of the spectra (Fig. 16.1). While the exact value of α specifies the form, four integer values are recognized and named:

White noise $= 1/f^0$ is a signal in which each successive observation is independent of frequency and previous observations resulting in a spectrum with no pattern (Fig. 16.1A).

Pink noise $= 1/f$ is a signal with a frequency spectrum with spectral density is inversely proportional to the frequency of the signal. It is frequently observed in

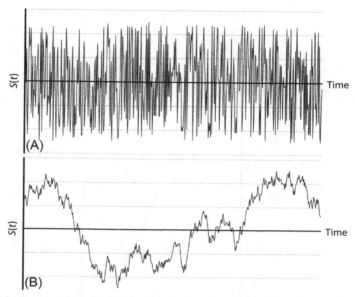

FIG. 16.1   Traces of spectra. (A) White noise: $S(t)$ is independent of f. (B) Fractional red or Brown noise: $S(t) \propto f^{-2.8}$.

systems in which halving or doubling frequency (octave) carries an equal amount of energy.

Red noise $= 1/f^2$, also known as Brown noise named after Robert Brown who first described the random walk called Brownian motion, is the kind of signal noise produced by Brownian motion. It decreases in power by 6 dB per octave and sounds like heavy rain.

Black noise $= 1/f^3$ has a frequency spectrum of predominantly zero power level over all frequencies except for a few narrow bands or spikes. Signals with $1/f^\alpha$ where $2 < \alpha < 3$ are used to model the frequency of natural disasters, such as earthquakes, floods, and droughts, because once started they tend to persist for a while (Fig. 16.1B).

The minute-to-minute fluctuations in stock prices and the example of foreign exchange price fluctuations (Sato et al., 2010; Chapter 12) are examples of pink noise and are called self-affine sequences.

Scale-free networks

A scale-free network is a connected graph or network in which the fraction $P(x)$ of nodes with x connections to other nodes is

$$P(x) \approx x^{-\gamma}$$

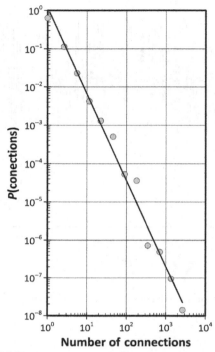

FIG. 16.2 The World Wide Web is a scale-free network with the probability of links between pages following a negative power law with slope $\gamma = -1.54 \pm 0.088$ ($r = -0.994$). *(Adapted from Fig.1 in Barabási and Albert (1999).)*

where the exponent is typically in the range $2 \le \gamma \le 3$ and x is the degree of the node. If $\gamma < 3$, the variance of x diverges and if $\gamma < 2$ the mean also diverges. Network nodes having a large number of links are called hubs and networks with large hubs are typically scale-free. A scale-free network can be constructed by progressively adding nodes to an existing network and introducing links to existing nodes in proportion to the number of existing links. Examples of scale-free networks include the citations in peer-reviewed scientific publications and the hyperlinks in web pages (Fig. 16.2), and the number of partners (Fig. 11.8).

Diffusion-limited aggregation

Diffusion-limited aggregation (DLA) is a process whereby clusters build up by random walkers stopping when they encounter another walker. Clusters build up with filaments spreading from the center and lesser filaments adjoining the main lines. DLA structures on two-dimensional lattices resemble extremely spidery leaf veins spreading from a single point that eventually become area filling. This type of aggregation process results in a structure with fractal dimension $1 > D > 2$. The decrease in density ρ, with distance from the center r, is described by a power law, $\rho(r) \propto r^{D-E}$, where E is the Euclidean dimension. Physical examples of DLA typically have $1.5 < D < 1.8$. A biological example is Kendal's (1995) application of DLA to Colorado potato beetle (Fig. 8.6). His demonstration that TPL follows naturally from a DLA-like distribution process is described in Chapter 17.

Fractals

The preceding three examples are characterized by having fractional or fractal dimension. Mandelbrot (2004) defined a fractal as "a set for which the Hausdorff-Besicovitch dimension strictly exceeds the topological dimension," although an informal definition attributed to him is "a fractal is a shape made of parts similar to the whole in some way" (Feder, 1988). The magnitude of fractal shapes and structures, when measured by rulers or boxes of decreasing size, appear to get larger. The classic example is the length of the coastline of Great Britain shown in Fig. 15.1. Its significance for TPL is that Kendal (1992, 1995) has shown that if populations were fractal aggregates, TPL would follow naturally.

Repetition as a source of self-similarity

A number of simple geometric rules can create structures with fractional dimensions. The Koch snowflake is an area-filling line structure that has a Hausdorff dimension of $D_H = 1.26$ (Chapter 15). The Cantor dust is an infinite number of unconnected points generated by removing the middle third of a unit line, and repetitively removing the middle third of each remaining line segment. As the number of erasures approaches infinity, the number of points with Euclidean dimension ($D_E = 0$) approaches infinity but fills the unit line. The Cantor dust has Hausdorff dimension $D_H = \log(2)/\log(3) = 0.63$. In three dimensions, the Sierpinski pyramid has a tetrahedron as initiator and a generator that removes a half-sized inverted tetrahedron from the pyramid leaving 4 tetrahedra. In successive cuts, half-sized tetrahedral are excised from each solid part, creating a spidery lattice that is mostly space. It has dimension $D_H = \log(4)/\log(2) = 2 < D_E = 3$. These fractal structures are constructed from a simple starting condition or Initiator and a simple process rule or Generator iterated a very large number of times. Biological examples of structures obeying scaling laws are the bifurcations of blood vessels with a space-filling dimension of $D_H = 2.7$ (Gabryś et al., 2005) and the bifurcations of the bronchi of lungs that are closer to $D_H = 3$. Mandelbrot (1977) proposed a model of bronchi and blood vessel bifurcation: an initiator of a bud and a generator of bud elongation forming a tube and production of two new buds. This scenario, simpler than Thompson's (1917) explanation, iterated 15–20 times requires only two parameters, the width/length ratio of branches, and the tube diameter, to produce the space filling surface area of lungs and the nearly space-filling vascularization of capillaries.

Physical

A number of familiar physical relationships are power laws. Unlike allometry and most of the other power laws, they are not statistical but deterministic, derived from established mathematical and physical principles.

Inverse square law

Both gravity and electromagnetic radiation decline in intensity as $E = kd^{-2}$, where k is a constant appropriate to the energy form. Both gravitational attraction and light energy radiating from a point are spread over the surface of an expanding sphere whose area is proportional to the radius squared. Thus, the energy, E, at a point on the expanding sphere must decline as the square of the distance from the center. In general, the inverse-square law applies whenever a conserved quantity such as energy radiates outward from a point source.

Stefan–Boltzmann law

The total energy/unit area radiated on all wavelengths by a black body radiator is proportional to the fourth power of the body's temperature, $J = \sigma T^4$, where σ is the Stefan–Boltzmann constant, which can be derived from other known physical constants.

Self-organized criticality

Self-organized criticality relates to natural complexity that emerges from comparatively simple rules. Importantly, for self-organization, the complexity emerging does not depend on the specific values of parameters as when a phase transition occurs as a material changes its properties. An example is the transition of liquid water to vapor at a critical temperature. As the temperature (θ) approaches the critical value (θ_c), the density (ρ) of water is well described by a power law of the form $\rho = a|\theta - \theta_c|^\alpha$, where α is a critical exponent. As the transition is approached, scale-dependent parameters like density become less important and scale-invariance dominates. The behavior of such systems are simplified and may be well approximated by an exactly solvable, power law model. Very different systems exhibit this behavior in the vicinity of a critical point.

Mandelbrot (1977) demonstrated linkages between processes producing self-organization noting that these processes frequently produce power laws with fractional exponents. Wolfram's (2002) cellular automata produce similar results.

Percolation

In a seminal paper, Broadbent and Hammersley (1957) developed the concept of percolation as an alternative form of diffusion in a heterogeneous medium: the spread of disease through an orchard. Their thesis has since evolved into an important experimental and theoretical branch of physics (Stauffer and Aharony, 1992).

There are many phenomena in which a fluid or particles spread or migrate through a medium. The simplest case is the diffusion of particles along a line from a point of origin. In one dimension, a flea jumping to right or left with equal probability at each time step has probability of it being at a given distance from its starting point approximately normally distributed after a large number of jumps. When the number of jumps is infinite, the flea visits every point of the medium infinitely often with probability 1. In contrast, percolation in one dimension analogous to the flea example, the particle and medium are as before, but the stochastic mechanism resides in the medium rather than the particle. Specifically, each point of the medium has, independent of the other points, a probability, p, of being a "right-sense" or "left-sense" point. In this case, movement will end when the particle encounters successive points of opposite sense, whereupon it will oscillate indefinitely. The distribution of terminal position is nothing like normal: the probability is 1 that the particle will visit only a finite number of sites.

In a two-dimensional lattice, the probability of transit to an adjacent point, the percolation probability, is p. Below a critical value $p = p_c$, a finite number of particles placed in the lattice at random explore disconnected clusters of sites. At p_c, the clusters rapidly coalesce into a single "spanning cluster." The number of sites in the largest cluster grows with the size of the lattice in a predictable fashion:

$$N(L) \propto \begin{cases} \ln(L), & \text{for } p < p_c \\ L^D, & \text{for } p = p_c \\ L^E, & \text{for } p > p_c \end{cases}$$

for a square lattice of side L, as $L \to \infty$ (Feder, 1988). Exponents D and E are the cluster fractal dimension and Euclidian dimension, respectively.

Meteorology

The number of rain clouds in a cluster declines with radius of the cluster to the power of -2 over a range of $\sim$20 to $\sim$200 km (Machado et al., 1992; Machado and Rossow, 1993) and the energy dissipation in cyclones also follows a power law with negative sign (Corral et al., 2010). The energy in tropical cyclones is measured by the power dissipation index (PDI), defined as the sum of the maximum one-minute sustained wind speed cubed (v_t^3), at intervals (Δt) over the lifespan of the storm, $\text{PDI} = \Sigma \, v_t^3 \Delta t$. A plot of the normalized PDI probability density, Φ(PDI), against PDI is similar to the Pareto distribution. PDI power laws have negative slopes over ~ 2 orders of magnitude for cyclones in four oceans: the North Atlantic, Northeast Pacific, Northwest Pacific, and the Southern Oceans with exponents of $-1.19\pm 0{:}06$, -1.175 ± 0.05, -0.96 ± 0.02, and -1.11 ± 0.04, respectively.

Hydrology

Harold Hurst and colleagues (Hurst et al., 1965) analyzed the time series of 1080 observations of the water level of the River Nile at Roda, north of Cairo. Rainfall in the headwaters of the Nile is sporadic but extreme leading to periods of low water and flooding in the lower reaches. Hurst et al. devised a method for characterizing Nile flow in order to determine optimum flood control dam sizes. The method relates the rate at which autocorrelations in a time series, $\{x_t\}$ of length T, decrease as the lag between pairs of values increases:

$$\mathrm{E}\left[\frac{R(n)}{S(n)}\right] \propto n^H \text{ as } n \to \infty,$$

where $\mathrm{E}[\cdot]$ denotes the expected value or mean of the rescaled range $R(n)/S(n)$, n is the number of observations in the time series, and H is the Hurst exponent. The ranges, $R(n)$, are calculated from the difference between the maxima and minima of the accumulated sums, $Z_t = \sum_{i=1}^{t} y_i$ of the mean adjusted series $y_t = (x_t - m)$, where $m = \frac{1}{n}\sum_{t=1}^{n} x_t$. The standard deviations of the partial time series are computed from $S(n) = \sqrt{\frac{1}{n}\sum_{i=1}^{n}(x_i - m)^2}$. $R(n)$ and $S(n)$ are computed for all values of the time series of length $n = 2 \dots T/2$.

The Hurst exponent H is an index of long-range correlation. A value of $0.5 < H < 1.0$ indicates a time series with long-term positive autocorrelation. Fig. 16.1B shows a spectrum with long-term positive autocorrelation: highs and lows tend to be sustained for longish periods. Although $H = 0.5$ indicates a completely uncorrelated series, it is true only in the long run. In the short run, positive or negative autocorrelations with short time lags can occur but decay exponentially quickly to zero. Decay times for series with $0 < H < 0.5$ and $0.5 < H < 1$ is much longer.

Analysis of the level of the Nile at Roda from AD 622–1469 by Mandelbrot and Wallis (1969) estimated $H = 0.91$. This high value based on river height contrasts with the more orderly flow of the River Rhine at Basel (1808–1966), where $H = 0.55$ or the Loire (1863–1966) with $H = 0.69$, both based on monthly river flow. Hurst's H is related to the Hausdorff fractal dimension D_H by

$$D_H = 2 - H \tag{16.2a}$$

and to the spectral frequency exponent by

$$\alpha = 1 + 2H \tag{16.2b}$$

(Mandelbrot, 1977). The power spectrum of the River Nile is $S(t) \propto f^{-2.8}$ implying very long-term persistence: drought years most likely follow drought years and flood years most likely follow flood years.

Geophysics

The occurrence of earthquakes is notoriously difficult to predict. Even so, there are statistical regularities evident in earthquake incidence. de Arcangelis et al. (2016) review the power law properties of earthquakes that emerge in size, time, and space. The Gutenberg–Richter law is an exponential function relating the number of earthquakes with magnitude $M \geq m$ to magnitude, $N(M \geq m) = 10^{(a - bm)}$, where a and b are empirical parameters: a is a measure of the local seismicity and $0.5 < b < 2.5$ measures of the relative frequency of small earthquakes to large ones and is typically $b \approx 1.0$. Plots of $N(M \geq m)$ against m for California, Italy, and Japan are roughly parallel over 2–3 orders of magnitude with common slopes of $b \approx 1.0$ and different intercepts.

For fault-line earthquakes, the slippage measured by scalar seismic moment, S, is also an exponential function of m, $S = 10^{(1.5m+16.1)}$, which is proportional to the energy released. The energy released by a fault-line earthquake of moment S is a power law distribution with exponent $\beta = 1 + 2b/3$. Another power law emerges as the distribution of the interval Δt between seismic events of magnitude M, which is approximated by the Weibul distribution. The Weibul is a probability function with a power law with exponential cutoff. Its parametric form is $f(x) = Ax^{\beta} \exp(-x/\alpha)$, where A is a normalization constant, and for values of x less than $\approx \alpha$, the power law dominates and for $x < \alpha$ the exponential decay modifies the power law. The fact $f(x)$ is not a pure exponential function suggests correlations over the short time period. Another power law, known as Omori's law, has the frequency of aftershocks decreasing roughly with the reciprocal of time after the main shock: $f(t) = k/(c + t)^p$, where k and c are empirical constants and $p = 1.0$ (pink noise) in the original formulation. It is now recognized that p is variable in the range $0.7 < p < 1.5$. The magnitude of earthquakes of Richter magnitude ≥ 6.0 obeys TPL (Fig. 12.4; Ogata, 1988).

Biological

Allometric growth

Allometric growth is defined as differential growth of organs leading to a change in shape with size (Lincoln et al., 1982). The mathematical relationship defined by Huxley and Tessier (1936) expresses the size (length, area, or volume) of organ Y in terms of the size of organ X: as $y = ax^b$, where a and b are empirical constants. When $b = 1.0$, growth is isometric and when $a = 1.0$, the two organs not only grow at the same rate but are identically sized. In a few organisms, for example frogs after the tadpole stage, growth occurs isometrically, but in most species growth is allometric with some organs growing faster than others. For example, the growth of both left and right human femurs is isometric and identical, but allometric with respect to cranial width. In isometric growth, an organism's surface area increases fourfold and its volume and mass increase eightfold, for a doubling of linear dimension (a square-cube law). The scaling of body mass with respect to body length changes to the ⅓ power, and to surface area to the ⅔ power. There are clear limits to isometric growth as energetic constraints will limit size as physical demands outgrow metabolism. To compensate for the disproportionate growth of mass with length, mammal skeletons grow stronger and more robust as the linear dimension of body size increases. Allometric growth of some organs is faster than others and may express itself in the growth of individuals, in the comparison of members of a population, and of species changing over evolutionary time.

A classic example of allometry is the biometrical study of 338 specimens of *Micraster coranguinum* taken from chalk beds in Kent, England (Kermack, 1954). *Micraster* is an extinct genus of echinoids whose fossils display a continuous gradual evolution over 10^7 years in chalk beds of the Late Cretaceous. Kermack's measurements of the total length and height to the apical system were used as an example in Kermack and Haldane's (1950) paper on fitting allometric equations (Eq. 4.11–4.13).

Dimensional relationships for flying animals

In a Smithsonian Institute monograph, Crawford Greenewalt (1962) analyzed data of the dimensional relationships of flying animals gleaned from 16 sources, including a paper of his own. His intention was to bring together all the available data on insect, bird, and bat flight dynamics into a single source. The monograph features 17 figures of the allometric relations between size and power of animal flight and includes the original data in 15 tables. Greenewalt's results are summarized in Table 16.1.

Total arm length of birds (humerus+ulna+manus) scales with body mass to the power ~0.38. The individual wing-bones scale with similar exponents but isometrically with total length. Larger birds tend to have longer wings, wing bones, and total wing length with total length contributing most to the scaling power (Nudds, 2007). As total bone length grows, there is a general trend for the primary feathers to contribute proportionally less to overall wingspan as wingspan increases. Wingspan in birds is constrained close to $M^{1/3}$ (Nudds et al., 2011). The extreme wing area allometry of hummingbirds is likely a consequence of selection for relatively large wings to maximize performance while minimizing energy cost (Skandalis et al., 2017). In hummingbirds, the oxygen cost per wingbeat scales with mass to the 1.4 power and with wing area to the 0.6 power, so that increasing wing size rather than frequency is likely to be selected (Groom et al., 2018).

TABLE 16.1 Allometric relations in the dimensional relations in flying animals

Allometric variables	Taxon	Equation	Comment
Wing length × body weight	Hummingbirds	$L = 19.5B^{2/3}$	
	Other birds and insects	$L = aB^{1/3}$	$a_{birds} > a_{insects}$
Wing area × body weight	Hummingbirds	$A = aB^{3/4}$	
	Other birds and insects	$A = aB^{3/2}$	$a_{birds} > a_{insects}$
	Other birds	$A = aB^{3/2}$	3 values of a for families
Wing length × wing area	Birds	$L^2 = 1.93A$	
	Insects	$L^2 = cA$	5 values of c ranging from silkworms and swallowtails $c = 0.66$ to small dipterans $c = 3.39$; most $c = 1.88$
	Bats	$L^2 = 1.76A$	
Wing length × wing span	Birds	$L = 0.31S$	$L = $ length of "hand" the first articulated joint
Wing weight × wing area	Insects and birds	$2570\,W = A^{1.67}$	Hummingbirds below the line—high wing beat frequency—soaring birds, vultures and albatross above the line—low wing beat frequency
Wing beat frequency × wing length	Insects and birds	$f = 3540\,L^{-1.15}$	Maximum line: spread of points below the maximum diminishes with increasing length but with outliers
	Hummingbirds	$f = 5830\,L^{-1.25}$	
	Insects	$f = cL^{-1}$	4 values of c; groups not taxonomic
Flight muscle weight × body weight	Birds	$P = 0.115B$	

Continued

TABLE 16.1 Allometric relations in the dimensional relations in flying animals—cont'd

Allometric variables	Taxon	Equation	Comment
Flight muscle weight × wing weight		$P \propto W^{1.1}$	
Flight muscle weight × body weight	Insects	$M \propto B^b$	$b > 1$ with variation between insect orders

Data from figures in Greenewalt, C.S. 1962. Dimensional Relationships for Flying Animals. Smithsonian Miscellaneous Collections 144#2, Smithsonian Institution, Washington, DC.

The maximum body mass of flying animals is limited by the flight muscle mass and wingbeat frequency. In bats, the maximum wingbeat frequency attainable and the minimum frequency required for flight together define the maximum body mass for flight. New wingbeat frequency data for 65 bat species ranging from 2.0 to 870 g show wingbeat frequency decreases with increasing body mass to the power -0.26, close to that of birds' -0.27 but with a lower intercept (Norberg and Norberg, 2012).

Wing allometries are closely associated with flight patterns and behavior in the odonate suborders, Anisoptera (dragonflies) and Zygoptera (damselflies). The exponents of wing length on body length are $b = 1.20 \pm 0.031$ for damselflies and $b = 0.85 \pm 0.020$ for dragonflies. Migrants have generally longer wings than nonmigrants in the same family (Saachi and Hardersen, 2013).

One of Greenewalt's sources was a review of the range and speed of insect flight by Hocking (1953) who considered the physiological parameters involved in the flight of several biting flies (blackflies, mosquitoes, and horseflies). Using flight mills and wind tunnels, Hocking examined their flight energetics. He appears to have been the first to note a power law relationship between power developed by the different species and their mass, $P \propto M^{0.73}$, with intercepts differing by family. In general, the mass-specific power required for flight scales with body mass to a power of 0 to ⅙, while the mass-specific power available from the flight muscles scales approximately as the ⅓ power. Combining these two relationships, suggests the mass-specific power required for flight varies with speed according in a U-shaped curve. However, metabolic measurements suggest a J-shaped curve, with little change in power from hovering to intermediate flight speeds (Ellington, 1991).

Resting metabolic rate of flying insects scales with body mass as $M^{1.10}$ while flying but as $M^{0.66}$ while at rest. However, separating insects into those greater and less than 10 mg reveals two parallel lines with flying metabolic rates scaling as $M^{0.86}$. Insects weighing >10 mg have metabolic rates four times those <10 mg. The scaling exponents of resting and flying metabolic rates in insects are closely similar to birds and bats, suggesting that they might be determined by similar factors (Niven and Scharlemann, 2005).

Fractal movement

Considering the movement of the slug-parasitic nematode, *Phasmarhabditis hermaphrodita*, Hapca et al. (2007) found that the presence of sand particles on the substrate changed the temporal correlation in step length and turning angles, affecting the rate of dispersal. Using a box counting algorithm to quantify the length and turning of nematode trails, Hapca et al. plotted the number of boxes of size l needed to completely cover a trail against l in logs to estimate the trail's fractal dimension, D: $\log(N_l) \propto D\log(l)$. A value of $D=1$ measures a straight trail, with D increasing with increasing sinuosity, and $D=2$ corresponds to Brownian motion. The nematode's changed dispersal behavior with the presence of sand changed the estimate of D. Additionally, the mean square displacement is a power law of time: MSD $\propto t^{\alpha}$ with the exponent that changed also with surface roughness.

Frequency of species size

The frequency of size of organisms is a lognormal-like distribution with a nearly linear decline on log scales from the maximum (May, 1978). Ignoring the small number of vertebrates smaller than ~2 cm long, the log number of species in each length class, S_L scales with, L, for terrestrial vertebrates to the power $z=-1.612$ ($r=0.963$), which is somewhat lower than the value predicted on theoretical grounds by Hutchinson and MacArthur (1959), but not much different from the power law

$$S_L \propto L^{-3/2}, \tag{16.3}$$

that Southwood et al. (2006) suggested arises from a fractal measure that maps body length into the numbers of species in an environment. In addition, Southwood et al. pointed out that this power law combined with the species-area law implies a third power law relating animal size to the home range area, $M_{max} \propto A^{1/2}$, where M_{max} refers to the size of the top vertebrate, a relationship investigated by Burness et al. (2001).

Species-area

A power law relationship between the number of species (S) in a defined area and the area (A) was first noted by Arrhenius (1921) in relation to plants, and expressed in power law form,

$$S \propto A^Z, \qquad (16.4)$$

by Preston (1962a, b). In general, the analyses have focused on particular types of organisms, clades, or guilds. The exponent is typically $z > \frac{1}{4}$ on continents $z < \frac{1}{4}$ and islands (Drakare et al., 2006). Reasons advanced for the relationship include the balance of immigration and emigration (MacArthur and Wilson, 1967) and predator–prey interactions (Brose et al., 2004), neither of which are mutually exclusive. Habitat heterogeneity and sampling may also play a part, but neither alone explain the species-area relationship as populations are often more strongly aggregated than expected by habitat heterogeneity (Storch et al., 2003).

A particularly impressive example spanning 15 orders of magnitude of area is the species-area relationship between freshwater and marine phytoplankton algae and the area of 142 natural ponds, lakes, and oceans and 239 experimental mesocosms. Smith et al. (2005) estimated slopes for natural systems of $Z = 0.114 \pm 0.009$ $(r = 0.714)$ and for mesocosms $Z = 0.139 \pm 0.013$ $(r = 0.566)$. The natural and experimental systems were not different in either slope or intercept $(P > 0.13)$. Analyzed together, they extended the range to 15 orders, resulting in a combined plot with $Z = 0.134 \pm 0.004$ $(r = 0.860)$. The small difference in Z was likely due to the fact that the mesocosms were closed systems with little or no migration, compared to the highly permeable natural systems.

Related to the species-area relationship is species occupancy, which refers to the areas occupied by a species within its normal range. This intraspecific relationship is a function of the commonness and rarity of species. Abundant species tend to occupy more territory within their range than rare species, which frequently occupy only small isolated patches. Kevin Gaston and colleagues have explored this relationship for both intra- and interspecific abundance-occupancy relationships (Gaston and Blackburn, 2000; He and Gaston, 2003).

An example of the interspecific abundance-occupancy relationship is the power function relating the breeding population sizes (N) of British birds to their geographical breeding range measured as the number of 10×10 km National Grid squares (A) in which a species was recorded, $N \propto A^c$ (Blackburn et al., 1999; Gaston et al., 1997). The slope of $\mathrm{Log}(N) = -1.04 + 1.94 \mathrm{Log}(A)$ $(r = 0.911)$ is significantly greater than 1.0 $(P < 10^{-10})$, indicating the abundance of widespread species is higher than expected on the basis of their higher spatial distribution.

In reviewing the distribution patterns of the flora and fauna of the British Isles, Gaston et al. (2000) reported that density varies with occupancy primarily because maximum density does. Most species are rare at some sites, but only the most widespread species attain high densities at some sites. Like LRT et al.'s (1980) TPL study of birds in field and forest (Chapter 10), the interspecific abundance-occupancy slopes for British birds do not differ between farmland and woodland sites, despite consistently lower densities in farmland than woodland sites. There appears to be little seasonal or annual difference in abundance occupancy: neither the slope nor the intercept differ between winter and summer although abundance varies substantially between seasons for both resident and migrant species. Apparently as abundance changes between seasons and years, each species' occupancy moves up and down the power law with changing abundance. There are differences in the slopes of the interspecific curves between summer and winter but residents and migrants lie on the same curve within each season. That migrant and resident populations are subject to different environmental pressures suggest that whatever the forces setting the abundance-occupancy curve, they are the same for both groups. Additionally, Gaston et al. note that despite differences in abundance of an order of magnitude, the relationships between population size and range size for birds and mammals in Britain are not different.

Studying populations of sparrows in the United States, Curnutt et al. (1996) found that peripheral populations were probably fed by higher central populations in a "source and sink" scenario (Chapter 10). In contrast, Blackburn et al. (1997) found no evidence of 32 passerine bird densities being lower near the edges of their range in Great Britain. As Curnutt et al.'s study was conducted in the middle of the North American continent and Blackburn et al.'s on a comparatively small island, it is possible that natural boundaries limit falloff of abundance with distance from a presumed population center. Another limitation may be energy availability, which influence species richness in an area (Storch et al., 2005).

The dynamics of island and continental populations clearly differ as the different exponents for island and continental species-area curves suggest. An examination of the spectra of ordinations of habitat and bird species compositions at 128 sample points in mixed forests, wetlands, villages, and fields of the Czech Republic showed a declining frequency with slopes close to -1.0, indicating $1/f$ or pink noise (Storch et al., 2002). The slopes of power spectra for individual species averaged -0.35 with 78% of species exhibiting slopes of $0.0 > \gamma > -0.94$, indicating pinkish spectra. As pink noise indicates scale invariance, species composition of assemblages varies at all spatial scales, but very large populations become increasingly rare: only a small proportion of sites contain dense populations (Gaston and Blackburn, 2000). Controlling for habitat composition, Storch et al. found that the bird species spectra lost most $1/f$ dependence, suggesting that much of the species-area variation is environmentally imposed.

A similar species abundance power law relates the numbers of microbes inhabiting animals. The number per host ranges with host mass over 12 orders of magnitude from nematodes to whales. An allometric analysis of animal-microbe relationships by Kieft and Simmons (2015) using estimates of microbial abundances per host in the literature found that prokaryotes/host abundance scales as a power function of host mass, $N \propto M^{1.07}$ ($r = 0.97$). Combining this power function with species area equations for land and sea revealed animal-associated microbes totaling $\sim 2.2 \times 10^{25}$ prokaryotes. The authors speculated that this allometric power function may reflect underlying mechanisms involving the transfer of energy and materials between microorganisms and their animal hosts.

Kleiber's law of metabolism

Based only on consideration of heat, Kleiber's (1932) law relating animal metabolic rate, Q, to mass, M, was originally thought to scale as the ⅔ power, but a power of ¾ may be more appropriate: $Q = aM^{\frac{3}{4}}$. However, neither the ⅔ nor the ¾ power fit all the data.

For example, small mammal (<10 kg) data fit the ⅔ power model better while the ¾ power model fits better for larger mammals. Over the whole size range, mammal data are not a pure power law as they show a clear curvature. Kolokotrones et al. (2010) rejected the fractal-based model of West, Brown and Enquist (see later) and proposed a quadratic that fit better: $\log(Q) = a + b\log(M) + c\log(M)$. Alternative models that preserve the definite power law structure are the "curved power law" model.

$$f(x) \propto x^{(\alpha + \beta x)}$$

and the "broken power law" model that consists of two power laws like the examples in Figs. 4.2, 5.11 and 11.13:

$$f(x) \propto x^{\alpha} \text{ for } x \leq x_T$$
$$f(x) \propto x^{\beta} \cdot x_T^{\alpha - \beta} \text{ for } x > x_T,$$

where x_T is the threshold on the abscissa where the curve switches from one power law to the other.

Respiration

McNeill and Lawton (1970) compared the annual population respiration, R, and annual population production (biomass produced, including eggs), B, of warm- (endothermic) and cold- (ectothermic) blooded animals. Measuring both R and B in kcals/m^2/year, they found that R scaled with B as power laws with $b = 1.0$ for endotherms, but a more complex picture for ectotherms. They computed ODR of log transformed R on B and B on R, but did not compute the GMR. For ectotherms, $\log(R) = 0.376 + 1 \cdot 073 \log(B)$ and $\log(B) = -0 \cdot 237 + 0 \cdot 823 \log(R)$ ($r = 0.940$), and for endotherms, $\log(R) = 1.742 + 0 \cdot 981 \log(B)$ and $\log(B) = -1.776 + 1 \cdot 014 \log(R)$ ($r = 0.997$). The GMR estimates of slope, b, are 1.142 ± 0.062 and 0.984 ± 0.024 for ectotherms and endotherms, respectively. The ectothermic slope is significantly different from $b = 1.0$ ($P < 0.05$), but the endothermic slope is not ($P > 0.3$).

Analyzing only the short-lived ectotherms, comprising mostly arthropods, McNeill and Lawton found a substantial difference. Both ODRs were significantly different from $b = 1.0$ ($P < 0.001$); $\log(R) = 0.135 + 1\cdot174\log(B)$ and $\log(B) = -0\cdot095 + 0\cdot826\log(R)$ ($r = 0.985$). The estimated GMR slope of $b = 1.192 \pm 0.040$ is significantly different from 1.0 at $p < 10^{-4}$. Relative to endotherms and long-lived ectotherms, short-lived ectotherms' net production efficiencies evidently increase at low production levels. McNeill and Lawton explained this difference in long- and short-lived ectotherms by noting that most short-lived ectotherms with $P < 1$, overwinter as eggs when R and B are virtually zero while those with $B > 1$ overwinter as juveniles with a higher respiration rate while B is virtually zero.

Self-thinning and space-filling

Crowding of smaller and younger plants by larger older plants in an even age stand decreases their survival probability. The self-thinning rule describes this suppression of smaller and younger plants; its expression is seen in the power law with exponent $c = -\frac{3}{2}$ relating biomass/unit area to the number of plants/unit area. In field experiments, Yoda et al. (1963) found the $-\frac{3}{2}$ power thinning rule to apply to dense stands of horseweed (*Erigeron canadensis*) starting from different initial conditions and growing in soils of different fertility levels. Plants in all plots followed the same power law. Weller (1987) found support in a combined mixed-species analysis of 63 datasets, but individual analyses supported the $-\frac{3}{2}$ rule in only 20 single species datasets. Since then, it has been challenged on both theoretical and empirical grounds.

The presumption of $c = -\frac{3}{2}$ has been challenged by West, Enquist, and colleagues (West et al., 1997, 1999a, b; Enquist et al., 1998, 1999) who developed a fractal-based model of vascular systems. West et al.'s model of a space-filling network of branching tubes was developed to explain the structure of mammalian circulatory systems. In addition, it predicts properties of other natural branching systems such as vertebrate cardiovascular and respiratory systems, plant vascular systems, and insect tracheal systems. Their model has features in common with Mandelbrot's (1977) branching network model. However, where Mandelbrot's model relies solely on an Initiator and Generator, West et al.'s model derives from the specifics of fluid flow through the network of N branchings from aorta to capillaries. They assume Mandelbrot's space-filling fractal-like branching network that capillary size is independent of body size (i.e., the final branch of the network is size invariant), and the energy required to distribute resources is minimized.

For a self-similar fractal the number of branches, N_k, increases geometrically as their size decreases from level 0 (aorta) to the Nth level ($N_k = n^k$). As the size of capillaries is independent of size of the animal, the number of capillaries must scale with size, $N_c \propto M^\alpha$. For the energy requirement to be minimized, the total volume of blood should be proportional to mass, $V_T \propto M$, which implies that the volume of capillaries also scales with size, $V_c \propto M^\alpha$. Because the blood transports oxygen and nutrients for metabolism, the rate of flow must be proportional to the volume of capillaries, which implies that metabolism is proportional to rate of flow and therefore must also scale with size: $B \propto M^{\alpha}$.

The system's total volume (and the volume of blood), V_T, is a function of the volume of capillaries, V_c, the number of branchings, N, and a scale factor, γ, relating the relative diameters and lengths of blood vessels: $V_T = F(V_c, N, \gamma)$. West et al. show that

$$V_T = \frac{(n\eta)^{-(N+1)} - 1}{(n\eta)^{-1} - 1} n^N V_c, \qquad (16.5)$$

which implies a fractal nature of the system. The scale factor is a combination of the relative radii, $R_k = \frac{r_{k+1}}{r_k}$ and lengths $L_k = \frac{l_{k+1}}{l_k}$, $\gamma = L_k R_k^2$. For scale invariance, R_k and L_k must be independent of k, so $\gamma = LR^2$ and $\gamma^N \propto 1/M$. From Eq. (16.5), West et al. go on to show that $\alpha = \frac{3}{4}$. From $B \propto M^{3/4}$, they demonstrate that other scaling laws follow: the aorta radius scales as $M^{3/8}$ and its length scales as $M^{1/4}$.

Using their model of vascular systems as a basis for further investigations, West, Enquist and colleagues developed other allometric theories. The self-thinning law they explained in terms of xylem transport as a function of stem diameter and predicted average plant size should scale as the $-\frac{4}{3}$ power of maximum population density, significantly less than the $-\frac{3}{2}$ power self-thinning rule (Enquist et al., 1998). Average plant mass and maximum population density ranging over eight orders of magnitude in 251 populations has slope, $c = -1.34 \pm 0.017$ ($r = 0.981$), which is not significantly different from $c = -\frac{4}{3}$ ($P > 0.50$). As a consequence, the rates of resource use in individual plants scale as approximately the $\frac{3}{4}$ power of body mass, which is the same as metabolic rates of animals.

West et al. (1999a) further developed their model for the hydrodynamics and biomechanics of plants. They predict scaling laws between and within individual plants that are multiples of $\frac{1}{4}$. The length of the trunk scales as $M^{1/4}$, the radius as $M^{3/8}$, and the number of leaves as $M^{3/4}$. In addition, they deduce that conducting tubes taper so that the fluid flow per tube is independent of plant size. Their model predicts the maximum height of trees and explains why the energy use of plants in ecosystems is independent of size. In Enquist et al. (1999), they developed the framework further by relating plant life-history variables to rates of production (dM/dt). They argued that metabolic rate limits production as a plant grows forcing $dM/dt \propto M^{3/4}$ and relative growth rate $(dM/df)M^{-1} \propto M^{-1/4}$. Their "universal growth law" fits data for a large sample of tropical tree species with diverse life histories. Combined with evolutionary life-history theory, the growth law also predicts several qualitative features of tree demography and reproduction.

West et al. (1999b) suggest that the $\frac{1}{4}$ power scaling laws are emblematic of fractal processes and that fractal-like networks "effectively endow life with an additional fourth spatial dimension." Recall that fractal structures have noninteger dimensions, so the fourth dimension need not be taken literally but applies to the evolution of structures that effectively have fractional dimension. They propose that natural selection favors adaptations that maximize rate at which energy and nutrients are acquired and utilized because this will enhance survival and reproductive fitness. This is equivalent to maximizing metabolism, which is limited by the geometry of exchange interfaces and transport systems. Thus, natural selection maximizes metabolic capacity by employing fractal geometry to pack as much surface area into limited volumes for gas and nutrient exchange while minimizing the time and distance for transporting gas and nutrients. This leads to fractal dimensions of vascular and circulatory systems of $2 < D < 3$. These principles are independent of the details of the dynamics and apply to virtually all organisms.

West et al. (2001) derive a general model based on energy allocation between production and maintenance of biomass. Using the principle of conservation of energy, allometric scaling, and the energetic cost of producing and maintaining biomass, they derived a general growth equation. The model relates the dimensionless mass ratio, $R = (m/M)^{1/4}$, to the dimensionless time variable, $T = (at/4M^{1/4})$-ln$(1 - (m_0/M)^{1/4})$. R represents the lifetime metabolic energy expended, m is the total body mass at time t, and M is the maximum body size. T is a dimensionless measure of time derived from the maximum mass, the mass at birth (m_0) and a constant (a) relating the metabolic energy required to create a cell, to the mass of the cell and the rate of energy flow into the cell. The relationship between R and T is well described by the parameterless curve $R = 1 - \exp(-T)$. The fit is astonishingly good for 13 animals with determinate and indeterminate growth, oviparous and viviparous species, ectotherms and endotherms, vertebrates and invertebrates. The model attributes the slowing of growth as body size increases to limitations on the capacity of networks to supply sufficient resources to support further increases in body mass. It provides the basis for deriving allometric relationships for growth rates and the timing of life-history events and can be extended to plants: Enquist et al.'s (1999) model for trees is a simpler version.

A simpler, alternative model proposed by Banavar et al. (1999), based on general features of networks also, accounts for the 1/4 power allometric scaling without assuming a specific geometry. Their approach applies to both biological systems such as cardiovascular and respiratory networks, plant vascular and root systems, and to river drainage networks. They considered the scaling of total nutrient and nutrient delivery at sites with respect to the mean distance between sites, L. The scaling of the metabolic rate B on the total amount of nutrients C is obtained from $B \propto L^\alpha$ and $C \propto L^{\alpha+1}$, where α is the number of dimensions, 3. Thus, B scales as $C^{\alpha/(\alpha+1)}$ and assuming total nutrients is proportional to total mass, the metabolic rate, B, scales with mass as the 3/4 power.

Soil fertility and crop yields

In crop protection, fertilization, or cultivar evaluation trials, it is critically important that the soil fertility over the area of the trials be uniform. In fact, soil fertility can exhibit extreme variation over quite small distances, making plot trials sometimes difficult to evaluate reliably. In order to find an optimal plot size, Smith (1938) evaluated the yields of control or "blank" plots in trials in relation to plot size. He found that the size and sometimes the shape of replicated plots influenced not only the average yield but also the variance of yield. Using his own data and data from the literature, Smith investigated the form of the dependence of variability on plot size. Graphs of 46 crop trials showed that both *SD* and *CV* were power laws of plot size, *S*:

$$SD = aS^b \text{ and } CV = cS^d.$$

where both b and d are negative (Fig. 16.3). A little algebra to eliminate S leads to the conclusion that the variance of yield for fixed S obeys TPL,

$$V_s = kM_s^{2b/(b-d)} \tag{16.6}$$

provided $b > d$. As $d \rightarrow b$, variance grows rapidly in relation to mean and TPL blows up. To examine the effect of scale, Smith derived an exponent for infinitely large plots, $b' \leq b$. The difference between b' and b increases as the S/n ratio increases. Some of the counter cases in Chapter 13 in which the range of V greatly exceeds the range of M may be due to this condition.

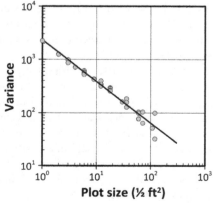

FIG. 16.3 Power law relationship between variance of yield and size of trial plot. *(Adapted from Fig. 3 in Smith (1938).)*

Binomial power law

The binomial power law (BPL) is a variant of TPL that works for bounded data. In Chapter 13, we saw that TPL often breaks down when the range of data is bounded above. Hughes and Madden (1992) developed the BPL to describe the distribution of disease incidence in plants using the proportions of individual diseased leaves, fruit or plants. They had noticed that the abscissa of TPL could be regarded not only as the mean density but also the variance of density assuming a Poisson distribution (Eq. 15.2). From this, they deduced that the variance of the binomial distribution could be used in the same way to compare the observed variance of binomially distributed data. As the Poisson is a limiting condition of the binomial when the probability is small and the sample size large, for suitably rare events TPL should work on binomial data.

BPL was conceived as a model describing the heterogeneity of disease incidence measured as the proportion diseased in each sample or the number of diseased plants of n plants sampled. Formally, BPL is

$$\log(V_{\text{observed}}) = \kappa_x + b \log(V_{\text{binomial}}) = \kappa_x + b \log(np(1-p))$$

for counts, or

$$\log(V_{\text{observed}}) = \kappa_p + b \log(p(1-p)/n)$$

for proportions where the intercepts (κ) differ but b is unchanged with model.

Since 1992, Hughes and Madden's BPL has been used over 200 times to describe the spatial heterogeneity of plant disease epidemics. Madden et al. (2018) review the development and use of BPL in empirical and theoretical studies. They report that BPL provided a very good fit to heterogeneity data, in most studies with $1.0 > b > 1.5$ in $>80\%$ of 230 studies reviewed.

Madden et al.'s review covers only applications to diseases, although it has also been used in other contexts. Similar to Ma's (2015) mixed-species TPL, a mixed-species BPL has been defined by Shiyomi et al. (2001) and used by Tsuiki et al. (2005) to analyze the changes in plant community under three levels of livestock grazing in a BPL study similar to Zhang et al.'s (2016) TPL study (Appendix 6.Q).

Tsuiki et al. surveyed grasslands with light, intermediate, and heavy grazing intensities in the steppe of Heilongjiang Province, China. Villagers use the steppe to raise cattle, horses, sheep, and goats. Each of the three grasslands was surveyed with 100 quadrats on a 50-m line transect. Each quadrat was divided into four 25 cm × 25 cm subquadrats and all plant species occurring in each of the four subquadrats were identified and recorded. Applying the BPL to the observed and binomial variances, they evaluated the spatial heterogeneity and frequency of occurrence for each species in the grassland community. The lightly grazed grassland exhibited high spatial heterogeneity and the

524 PART | III

highest species diversity, while the heavily grazed grassland exhibited high spatial heterogeneity and low species diversity. Although the BPL slopes of the three grassland communities were close to $b = 1.0$ (lightly grazed, $b = 1.03x + 0.27$, $r = 0.975$; medium $y = 1.16x + 0.51$, $r = 0.988$; heavy, $y = 1.16x + 0.57$, $r = 0.983$), all were significantly steeper ($P < 0.001$). Tsuiki et al. used the deviations of the individual species from the equality and fitted lines to infer the impact of grazing on the plant species in the three grassland communities.

Density-size and variance-size laws

Similar to the self-thinning law, the abundance of organisms scales negatively with their size: $M \propto S^{\nu}$. The density-size law applies to single species and mixed-species collections. Cohen et al. (2012a) combined the density-size law with TPL to create a relationship they call "variance-mass allometry":

$$V = aM^b S^{b\nu},$$

where S is the mean body mass per individual. Obviously, because b is >0 and if $\nu < 0$, $b\nu$ is also negative. Cohen et al. used abundance and size data of oak trees (*Quercus* spp.) in the Black Rock Forest preserve in New York State to test the validity of variance-mass allometry. TPL of $NB = 12$ plots of $NQ = 9$ subplots had slope of $b = 2.26 \pm 0.370$ ($r = 0.910$); a second survey in which some of the trees had been girdled (and were dead or dying) produced similar results ($b = 2.04 \pm 0.419$, $NB = 9$, $r = 0.912$). The equivalent density-size analysis of mean density against biomass obtained slope estimates of $\nu = -0.95 \pm 0.054$ ($r = 0.988$) for ungirdled trees and $\nu = -1.22 \pm 0.283$ ($r = 0.891$) for the girdled trees. Predicted slopes for variance-mass allometry were therefore $k = b\nu = -2.18$ and -2.49 for ungirdled and girdled tress, respectively. The corresponding measured values were $b\nu = -2.10 \pm 0.405$ ($r = 0.881$) and -2.61 ± 0.467 ($r = 0.851$), which are well within statistical error. Cohen et al. used both ODR and nonlinear regression to test their theory obtaining very little difference in estimates. Xu et al. (2015) used the same data to test the robustness of TPL under different policies of grouping data into blocks for analysis. They found that all policies obeyed TPL with slopes not significantly different from one another. Randomizations of the data did not change slopes either, although intercepts varied. Organizing data of weeds and insects in different ways affecting the scale of analysis, Perry and colleagues (Clark et al., 1996; Hamid et al., 1999) obtained mixed results with the insect study producing consistent slopes and the weed study both consistent and variable slopes. Clark et al. (1996) concluded that the effect of scale is unpredictable. Xu et al.'s finding with forest trees, if confirmed elsewhere, would provide foresters with a powerful tool for predicting the variations of basal area.

In a further test of the variance-mass law, Cohen et al. (2016) used data of censuses of single species stands of mountain beech (*Fuscospora cliffortioides*) trees in New Zealand forests. The stands are unmanaged and unexploited, but had been censused 10 times between 1974 and 2004. The structure of the censuses permitted TPLs to be fitted at two scales with slope at the larger spatial scale much steeper ($b \approx 3.57$; $NB = 13$) than the fine scale ($b \approx 1.41$; $NB = 246$–250). Significant curvature was detected in the large scale plots in the later years, but not as a consequence of collision with the Poisson line. The density-mass relationships' slopes averaged $\nu \approx -0.91$ and 0.85 and the variance-mass slopes averaged $k = -3.26$ and -1.20 for large and small scales, respectively. The products $b\nu \approx -3.25$ and -1.20 are indistinguishable from the measured values of k. However, a bootstrap estimate of confidence intervals resulted in $k < b\nu$ for fine-scale analysis. Because sample size and spatial scale are confounded, it is unclear which might have caused the discrepancy. Overall, the three power laws were remarkably stable year to year, so that although the variation between computed and measured variance-mass slopes was somewhat greater than the averages, they were remarkably close. The fitted lines were apparently uninfluenced by environmental conditions although variation about the fitted lines probably was.

Applying this trio of laws to metazoan communities in lakes in New Zealand, Lagrue et al. (2015) found that mixed-species power law analyses for parasites, free-living parasitized species, and free-living unparasitized species differed. TPL for parasites was steeper than for either free-living group. Similarly, the parasites' slopes for density-mass and variance-mass were also larger (more negative) than for the free-living species. As before, the composite exponent $b\nu$ was in good agreement with the estimated exponent, k, for variance-mass. Lagrue et al. suggest that the power-law parameters may be specific to each lifestyle. Interestingly, the parameters for parasitized free-living species were closer to the parasite parameters than to those of the unparasitized organisms.

The variance-mass law has practical application. Applying the variance-size analysis to the known spectrum of fish size and output from a fisheries model, Cohen et al. (2012b) found that the variance-mass relationship of modeled fish populations is more stable with balanced harvesting across a wide range of body sizes than fishing only for fish above a certain size, in agreement with a different approach of Law et al. (2012). The implications of this for fish conservation and sustainability are profound.

Xu (2016) has formalized the relationship between TPL and size variation in what he calls density mass variance allometry (DMVA) by invoking the connections between population variance and mean, individual body mass variance and mean, and individual body mass mean and population mean to deduce DMVA. In effect, the four connections form a square with TPL variance (V_p) and body size variance (V_b) related by.

$$V_p = \frac{a\alpha^b}{c^{b\beta/d}} V_b^{b\beta/d},$$

where a and b are the TPL parameters, c and d are the mass power law parameters, and α and β are the population density mass power law parameters. Xu tested the theory with Black Rock Forest data, fish data from Lake Kariba in East Africa, and freshwater fish and forest data from 20 US National Science

Foundation–funded long-term experiments (NSF-LTER). Fitting the four laws at least one, usually one of the mass laws, was significantly curved in ~27% of 108 test cases. The nonlinearity of some of the mass allometries suggests another factor may be involved, warranting further study. This original concept emphasizes the importance of integrating existing ecological laws to create new scaling patterns, thereby adding to our understanding of existing laws.

Genetics and physiology

The distribution of number of proteins coded by duplicate genes or paralogs, scales with the number (k) of paralogs: $f(k) \propto k^{-\alpha}$ in both unicellular and multicellular organisms (Padawer et al., 2012). The estimates for yeast *Saccharomyces cerevisiae* and the nematode *Caenorhabditis elegans* are $\alpha = 2.34$ and 1.74, respectively. Furthermore, paralog abundance in specific *C. elegans* cells scales with exponents similar to those of the yeast. Together with the positive correlation of gene expression and paralog count in different *C. elegans*, these results inform the evolutionary basis of multicellularity.

With few exceptions, such as *C. elegans*, the number of cells in organs of multicellular organisms varies between individuals. Azevedo and Leroi (2001) estimated the variability in organ cell number for a variety of animals and plants and found that the variance of cell number scaled with mean with exponents ranging from $b = 1.68 \pm 0.394$ for slime mold fruiting bodies to $b = 3.65 \pm 0.872$ for terrestrial nematode epidermis. A procrustean analysis of 2216 estimates of mean and variance of cell numbers in multiple species and tissues taken from the literature had an ODR estimate of $b = 2.03 \pm 0.006$ $(r = 0.991)$. A stochastic branching process model of cell division successfully simulated the procrustean analysis and provided insight into the limitations of cell size variation. The variation of individual species' organs may show deviation from $b = 2$ over short ranges, but it is the coefficient of variation that limits selection to a narrow range because very steep or shallow slopes quickly run up against geometrical limitations on cell size. Thus, they argued, the scale invariance of cell number is subject to natural selection.

Taxonomy

Burlando (1990, 1993) suggested that both biological diversity and the systems we use to classify organisms are fractal. By plotting the frequency of number of families within orders against the number of families using data from 44 checklists and catalogs of protists, fungi, plants, and animals, Burlando found evidence of a negative power law with slopes ranging $1.10 \leq D \leq 2.14$, averaging 1.57. The fossil record also shows the number of families within orders scales as a negative power law but with slightly higher exponents $(1.84 \leq D \leq 2.23)$ for 9 periods from the Ordovician to the Tertiary. Burlando suggests the exponents (D) are the fractal dimensions of taxonomic and evolutionary trees. The consistency of the results and the nonrandomness of D values among groups suggest a relationship with true biological diversity patterns and not artifacts of taxonomic criteria. Burlando argues that a fractal geometry of taxonomic systems reflects self-similar evolutionary patterns and his study of the frequency of families in the fossil record supports that view. Burlando's interpretation of fractality of cladograms and phylogenies was rejected by Ibanez et al. (2004), who compared the structures of the taxonomy of some plant parasitic nematodes (Tylenchina) to the USDA Soil Taxonomy. While rejecting the fractal dimension interpretation, they concluded that multifractal structure of biological and soil taxonomies appears to be due to bias resulting from the rules used to create classifications.

Sociological

Richardson's law of conflict

Mathematician and meteorologist Lewis Fry Richardson was a Quaker and pacifist who collected and analyzed data of wars and other conflicts including murders. Motivated by his pacifist principles, Richardson developed mathematical models of conflict (Richardson, 1960a) and developed a power law over nearly 7 orders of magnitude describing the frequency of deadly quarrels to the number of dead in conflicts as a proxy for its magnitude (Richardson, 1960b). Richardson's law of conflict fatalities is a negative power law resembling Zipf's law with slope $\alpha = -1.52 \pm 0.59$ (Fig. 16.4). Some of Richardson's data were also used by Quincy Wright (1942) in *A Study of War* (Chapter 11).

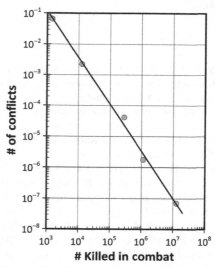

FIG. 16.4 Richardson's law of conflict severity has slope, $\alpha = -1.52 \pm 0.059$, and intercept $f_0 = 3.64 \pm 0.317$ ($r = -0.998$).

Cederman (2003) related $P(S > s)$, the probability of wars of severity $S > s$, the severity and found the cumulative probability scales as a power law: $P(S > s) = Cs^{\alpha}$ also similar to the Zipf law. Using data of 97 interstate wars between 1820 and 1997, Cederman estimated the slope $\alpha = -0.41 \pm 0.005$ ($r = -0.992$). With a slope of -0.41, a 10-fold increase in severity implies a decrease in the probability of war by a factor of 2.6. An agent-based simulation model based on principles of self-organized criticality produced a power law with slope $\alpha = -0.64 \pm 0.005$ ($r = 0.995$), suggesting that Richardson's law may be fractal. Studies of US casualties in the occupation of Iraq, the Vietnam War, and the Korean War, and the presumed casualties on both sides of the wars between Native Americans and the American army also produced Zipf-like graphs (Friedman, 2014). Friedman's objective was to refine the mortality estimates of the American Indian Wars and estimate the losses in small skirmishes that likely were underreported. The points in his analyses were of casualties in individual skirmishes or battles resulting in slope estimates of $\alpha = 2.2$ and 2.0 for Native Americans and American army, respectively. Richardson's and Cederman's analyses describe the global and long-term effects of war, while differences in slope for individual wars may be specific to those wars and reflect the technology of the time.

With over a 100 documented power laws to choose from, this short survey is necessarily eclectic. The major consideration in the choice was whether there is a discernible connection with TPL. The emerging connections between ecological and physiological measurements suggest most if not all biological power laws have a common foundation. TPL applied to sociological phenomena such as wars and criminality have counterparts in other power laws such as the Zipf law used by Richardson.

References

Andriani, P., McKelvey, B., 2007. Beyond Gaussian averages: redirecting international business and management research toward extreme events and power laws. J. Int. Bus. Stud. 38, 1212–1230.

Arrhenius, O., 1921. Species and area. J. Ecol. 9, 95–99.

Azevedo, R.B.R., Leroi, A.M., 2001. A power law for cells. Proc. Natl. Acad. Sci. U. S. A. 98, 5699–5704.

Banavar, J.R., Maritan, A., Rinaldo, A., 1999. Size and form in efficient transportation networks. Nature 399, 130–132.

Barabási, A.-L., Albert, R., 1999. Emergence of scaling in random networks. Science 286, 509–512.

Blackburn, T.M., Gaston, K.J., Quinn, R.M., Arnold, H., Gregory, R.D., 1997. Of mice and wrens: the relation between abundance and geographic range size in British mammals and birds. Philos. Trans. R. Soc. Lond. B 352, 419–427.

Blackburn, T.M., Gaston, K.J., Quinn, R.M., Gregory, R.D., 1999. Do local abundances of British birds change with proximity to range edge? J. Biogeogr. 26, 493–505.

Broadbent, S.R., Hammersley, J.M., 1957. Percolation processes. I. Crystals and mazes. Proc. Camb. Philos. Soc. 53, 629–641.

Brose, U., Ostling, A., Harrison, K., Martinez, N.D., 2004. Unified spatial scaling of species and their trophic interactions. Nature 428, 167–171.

Burlando, B., 1990. The fractal dimension of taxonomic systems. J. Theor. Biol. 146, 99–114.

Burlando, B., 1993. The fractal geometry of evolution. J. Theor. Biol. 163, 161–172.

Burness, G.P., Diamond, J., Flannery, T., 2001. Dinasaurs, dragons and dwarfs: the evolution of maximal body size. Proc. Natl. Acad. Sci. U.S.A. 98, 14518–14523.

Cederman, L.E., 2003. Modeling the size of wars: from billiard balls to sandpiles. Am. Polit. Sci. Rev. 97, 135–150.

Clark, S.J., Perry, J.N., Marshall, E.J.P., 1996. Estimating Taylor's power law parameters for weeds and the effect of spatial scale. Weed Res. 36, 405–417.

Cohen, J.E., Xu, M., Schuster, W.S.F., 2012a. Allometric scaling of population variance with mean body size is predicted from Taylor's law and density-mass allometry. Proc. Natl. Acad. Sci. U. S. A. 109, 15829–15834.

Cohen, J.E., Plank, M.J., Law, R., 2012b. Taylor's law and body size in exploited marine ecosystems. Ecol. Evol. 2, 3168–3178.

Cohen, J.E., Lai, J., Coomes, D.A., Allen, R.B., 2016. Taylor's law and related allometric power laws in New Zealand mountain beech forests: the roles of space, time and environment. Oikos 125, 1342–1357.

Corral, A., Ossó, A., Llebot, J.E., 2010. Scaling of tropical-cyclone dissipation. Nat. Phys. 6, 693–696.

Curnutt, J.L., Pimm, S.L., Maurer, B.A., 1996. Population variability of sparrows in space and time. Oikos 76, 131–140.

de Arcangelis, L., Godano, C., Grasso, J.R., Lippiello, E., 2016. Statistical physics approach to earthquake occurrence and forecasting. Phys. Rep. 628, 1–91.

Drakare, S., Lennon, J.L., Hillebrand, H., 2006. The imprint of the geographical, evolutionary and ecological context on species–area relationships. Ecol. Lett. 9, 215–227.

Ellington, C.P., 1991. Limitations on animal flight performance. J. Exp. Biol. 160, 71–91.

Enquist, B.J., Brown, J.H., West, G.B., 1998. Allometric scaling of plant energetics and population density. Nature 395, 163–165.

Enquist, B.J., West, G.B., Charnov, E.L., Brown, J.H., 1999. Allometric scaling of production and life-history variation in vascular plants. Nature 401, 907–911.

Feder, J., 1988. Fractals. Plenum Press, New York.

Friedman, J.A., 2014. Using power laws to estimate conflict size. J. Confl. Resolut. 59, 1–26.

Furusawa, C., Kaneko, K., 2003. Zipf's law in gene expression. Phys. Rev. Lett. 90.

Gabryś, E., Rybaczuk, M., Kędzia, A., 2005. Fractal models of circulatory system. Symmetrical and asymmetrical approach comparison. Chaos, Solitons Fractals 24, 707–715.

Gaston, K.J., Blackburn, T.M., 2000. Pattern and Process in Macroecology. Blackwells, Oxford, UK.

Gaston, K.J., Blackburn, T.M., Gregory, R.D., 1997. Abundance-range size relationships of breeding and wintering birds in Britain: a comparative analysis. Ecography 20, 569–579.

Gaston, K.J., Blackburn, T.M., Greenwood, J.J.D., Gregory, R.D., Quinn, R.M., Lawton, J.H., 2000. Abundance-occupancy relationships. J. Appl. Ecol. 37, 39–59.

Gibrat, R., 1931. Les Inégalités Économiques. Librairie du Recueil Sirey, Paris, France.

Greenewalt, C.S., 1962. Dimensional Relationships for Flying Animals. Smithsonian Miscellaneous Collections 144#2. Smithsonian Institution, Washington, DC.

Groom, D.J.E., Toledo, M.C.B., Powers, D.R., Tobalske, B.W., Welch, K.C., 2018. Integrating morphology and kinematics in the scaling of hummingbird hovering metabolic rate and efficiency. Proc. R. Soc. B 285, 20172011.

Hamid, M.M., Perry, J.N., Powell, W., Rennolls, K., 1999. The effect of spatial scale on interactions between two weevils and their food plant. Acta Oecol. 20, 537–549.

Hapca, S.M., Budha, P., Crawford, J.W., Young, I.M., 2007. Movement of the nematode, *Phasmarhabditis hermaphrodita*, in a structurally heterogeneous environment. Nematology 9, 731–738.

He, F., Gaston, K.J., 2003. Occupancy, spatial variance, and the abundance of species. Am. Nat. 162, 366–375.

Hocking, B., 1953. The intrinsic range and speed of flight of insects. Trans. Roy. Ent. Soc. 104, 223–345.

Hughes, G., Madden, L.V., 1992. Aggregation and incidence of disease. Plant Pathol. 41, 657–660.

Hurst, H.E., Black, R.P., Simaika, Y.M., 1965. Long-Term Storage. An Experimental Study. Constable & Co, London.

Hutchinson, G.E., MacArthur, R.H., 1959. A theoretical ecological model of size distributions among species of animals. Am. Nat. 93, 117–125.

Huxley, J.S., 1932. Problems of Relative Growth. Methuen, London.

Huxley, J.S., Tessier, G., 1936. Terminology of relative growth. Nature 137, 780–781.

Ibanez, J.J., Ruiz-Ramos, M., Tarquis, A.M., 2004. Mathematical structures of biological and pedological taxonomies. Geoderma 134, 360–372.

Johnson, N.L., Kotz, S., Balakrishnan, N., 1994. Distributions in Statistics. Continuous Univeriate Distributions—1, second ed. Wiley, New York.

Kendal, W.S., 1992. Fractal scaling in the geographic distribution of populations. Ecol. Model. 64, 65–69.

Kendal, W.S., 1995. A probabilistic model for the variance to mean power law in ecology. Ecol. Model. 80, 293–297.

Kermack, K.A., 1954. A biometrical study of *Micraster coranguinum* and *M. (Isomicraster) senonensis*. Philos. Trans. R. Soc. Lond. B 237, 375–428.

Kermack, K.A., Haldane, J.B.S., 1950. Organic correlation and allometry. Biometrika 37, 30–41.

Kieft, T.L., Simmons, K.A., 2015. Allometry of animal-microbe interactions and global census of animal-associated microbes. Proc. R. Soc. B 282, 20150702.

Kleiber, M., 1932. Body size and metabolism. Hilgardia 6, 315–353.

Kolokotrones, T., Savage, V., Deeds, E.J., Fontana, W., 2010. Curvature in metabolic scaling. Nature 464, 753–756.

Lagrue, C., Poulin, R., Cohen, J.E., 2015. Parasitism alters three power laws of scaling in a metazoan community: Taylor's law, density-mass allometry, and variance-mass allometry. Proc. Natl. Acad. Sci. U. S. A. 112, 1791–1796.

Law, R., Plank, M.J., Kolding, J., 2012. On balanced exploitation of marine ecosystems: results from dynamic size spectra. J. Mar. Sci. 69, 602–614.

Lincoln, R.J., Boxshall, G.A., Clark, P.F., 1982. A Dictionary of Ecology, Evolution and Systematics. Cambridge University Press, Cambridge, UK.

Ma, Z., 2015. Power law analysis of the human microbiome. Mol. Ecol. 24, 5428–5445.

MacArthur, R.H., Wilson, E.O., 1967. The Theory of Island Biogeography. Princeton University Press, Princeton, NJ.

Machado, L.A.T., Rossow, W.B., 1993. Structural characteristics and radiative properties of tropical cloud clusters. Mon. Weather Rev. 121, 3234–3260.

Machado, L.A.T., Desbois, M., Duvel, J.-P., 1992. Structural characteristics of deep convective systems over tropical Africa and Atlantic Ocean. Mon. Weather Rev. 120, 392–406.

Madden, L.V., Hughes, G., Moraes, W.B., Xu, X.-M., Turechek, W.W., 2018. Twenty-five years of the binary power law for characterizing heterogeneity of disease incidence. Phytopathology 108, 656–680.

Mandelbrot, B.B., 1977. The Fractal Geometry of Nature. Freeman, San Francisco, CA.

Mandelbrot, B.B., 2004. Fractals and Chaos: The Mandelbrot Set and Beyond. Springer-Verlag, New York.

Mandelbrot, B.B., Wallis, J.R., 1969. Some long-run properties of geophysical records. Water Resour. Res. 5, 321–340.

May, R.M., 1978. The dynamics and diversity of insect faunas. In: Mound, L.A., Waloff, N. (Eds.), Diversity of Insect Faunas. Blackwells, Oxford, UK, pp. 188–204.

McNeill, S., Lawton, J.H., 1970. Annual production and respiration in animal populations. Nature 225, 472–474.

Niven, J.E., Scharlemann, J.P.W., 2005. Do insect metabolic rates at rest and during flight scale with body mass? Biol. Lett. 1, 346–349.

Norberg, U.M.L., Norberg, R.A., 2012. Scaling of wingbeat frequency with body mass in bats and limits to maximum bat size. J. Exp. Biol. 215, 711–722.

Nudds, R.L., 2007. Wing-bone length allometry in birds. J. Avian Biol. 38, 515–519.

Nudds, R.L., Kaiser, G.W., Dyke, G.J., 2011. Scaling of avian primary feather length. PLoS ONE 6 (2), e15665.

Ogata, Y., 1988. Statistical models for earthquake occurrences and residual analysis for point processes. J. Am. Stat. Assoc. 83, 9–27.

Padawer, T., Leighty, R.E., Wang, D., 2012. Duplicate gene enrichment and expression pattern diversification in multicellularity. Nucleic Acids Res. 40, 7597–7605.

Preston, F.W., 1962a. The canonical distribution of commonness and rarity: part I. Ecology 43, 185–215.

Preston, F.W., 1962b. The canonical distribution of commonness and rarity: part II. Ecology 43, 410–432.

Richardson, L.F., 1960a. Arms and Insecurity. Boxwood Press, Pittsburgh, PA.

Richardson, L.F., 1960b. Statistics of Deadly Quarrels. Boxwood Press, Pittsburgh, PA.

Saachi, R., Hardersen, S., 2013. Wing length allometry in Odonata: differences between families in relation to migratory behaviour. Zoomorphology 132, 23–32.

Sato, A.-H., Nishimura, M., Hołyst, J.A., 2010. Fluctuation scaling of quotation activities in the foreign exchange market. Phys. A 389, 2793–2804.

Shiyomi, M., Takahashi, S., Yoshimura, J., Yasuda, T., Tsutsumi, M., Tsuiki, M., Hori, Y., 2001. Spatial heterogeneity in a grassland community: use of power law. Ecol. Res. 16, 487–495.

Skandalis, D.A., Segre, P.S., Bahlman, J.W., Groom, D.J.E., Welch, K.C., Witt, C.C., McGuire, J.A., Dudley, R., Lentink, D., Altshuler, D.L., 2017. The biomechanical origin of extreme wing allometry in hummingbirds. Nat. Commun. 8, 1047.

Smith, H.F., 1938. An empirical law describing heterogeneity in the yields of agricultural crops. J. Agric. Sci. 28, 1–23.

Smith, V.H., Foster, B.L., Grover, J.P., Holt, R.D., Leibold, M.A., de Noyelles, F., 2005. Phytoplankton species richness scales consistently from laboratory microcosms to the world's oceans. Proc. Natl. Acad. Sci. U. S. A. 102, 4393–4396.

Southwood, T.R.E., May, R.M., Sugihara, G., 2006. Observations on related ecological exponents. Proc. Natl. Acad. Sci. U. S. A. 103, 6931–6933.

Stauffer, D., Aharony, A., 1992. Introduction to Percolation Theory. Taylor & Francis, London.

Storch, D., Gaston, K.J., Cepák, J., 2002. Pink landscapes: 1/f spectra of spatial environmental variability and bird community composition. Proc. R. Soc. Lond. B 269, 1791–1796.

Storch, D., Šizling, A.L., Gaston, K.J., 2003. Geometry of the species–area relationship in central European birds: testing the mechanism. J. Anim. Ecol. 72, 509–519.

Storch, D., Evans, K.L., Gaston, K.J., 2005. The species–area–energy relationship. Ecol. Lett. 8, 487–492.

Taylor, L.R., Woiwod, I.P., Perry, J.N., 1980. Variance and the large scale spatial stability of aphids, moths and birds. J. Anim. Ecol. 49, 831–854.

Thompson, D.W., 1917. On Growth and Form. Cambridge University Press, Cambridge, UK.

Tsuiki, M., Wang, Y.-S., Yiruhan, Y., Tsutsumi, M., Shiyomi, M., 2005. Analysis of grassland vegetation of the southwest Heilongjiang steppe (China) using the power law. J. Integr. Plant Biol. 47, 917–926.

Weller, D.E., 1987. A reevaluation of the -3/2 power rule of plant self-thinning. Ecol. Monogr. 57, 23–43.

West, G.B., Brown, J.H., Enquist, B.J., 1997. A general model for the origin of allometric scaling laws in biology. Science 276, 122–126.

West, G.B., Brown, J.H., Enquist, B.J., 1999a. A general model for the structure and allometry of plant vascular systems. Nature 400, 664–667.

West, G.B., Brown, J.H., Enquist, B.J., 1999b. The fourth dimension of life; fractal geometry and allometric scaling of organisms. Science 284, 1677–1679.

West, G.B., Brown, J.H., Enquist, B.J., 2001. A general model for ontogenetic growth. Nature 413, 628–631.

Wolfram, S., 2002. A New Kind of Science. Wolfram Media Inc., Champaign, IL.

Wright, Q., 1942. A Study of War. The University of Chicago Press, Chicago, IL.

Xu, M., 2016. Ecological scaling laws link individual body size variation to population abundance fluctuation. Oikos 125, 288–299.

Xu, M., Schuster, W.S.F., Cohen, J.E., 2015. Robustness of Taylor's law under spatial hierarchical groupings of forest tree samples. Popul. Ecol. 57, 93–103.

Yoda, K., Kira, T., Ogawa, H., Kazuo, H., 1963. Self-thinning in overcrowded pure stands under cultivated and natural conditions. J. Biol. Osaka City Univ. 14, 107–129.

Zhang, J., Huang, Y., Chen, H., Gong, J., Qi, Y., Yang, F., Li, E., 2016. Effects of grassland management on the community structure, above ground biomass and stability of a temperate steppe in Inner Mongolia. China. J. Arid. Land. 8, 422–433.

Chapter 17

Modeling TPL

As with power laws in general, there are three broad categories of TPL model: statistical, physical, and biological with the models inevitably merging into each other and to a certain extent unifying them. In this chapter, we examine some of the many models generating linear regressions between $\log(V)$ and $\log(M)$ for some range of slope. Most have a restricted range of slope, but one class of model can generate slopes of $-\infty \leq b \leq \infty$. The majority of ecologically based models generate spatial b, while many physical models are temporal. We start with a consideration of the formal structure for TPL erected by Eisler et al. (2008) that applies to temporal, spatial, and ensemble data.

Physical models

In their review of TPL, Eisler et al. (2008) examined data from a range of disciplines and developed a formal mathematical structure to account for TPLs. The formal framework covers both temporal and ensemble scalings averaged over a collection of simultaneous samples (forming an ensemble set) or through time (a spatial set). In their framework, Eisler et al. refer to sample sites as nodes as though considering networks. In the following explanation, their use of node has been adopted.

The activity f_i at node i is the sum of events (animals, plants, stock transactions, packets of information, etc.) in the time interval $\delta t = [t, t + \Delta t)$. The number of events at node i in the interval is $N_i(\delta t)$ and the jth event contributes $C_{i,j}(\delta t)$ to the total activity f_i at time t:

$$f_i(t) = \sum_{j=1}^{N_i(\delta t)} C_{i,j}(\delta t) \qquad (17.1)$$

The contributions $\{C_{i,j}\}$ need not be independent, but their distribution must not depend on j. Assuming the variance of f_i at node i, $V(f) = E[\{f_i(t) - E[f_i(t)]\}^2]$, exists, Eisler et al. present a proof that

$$V(f) = V[C_{i,j}] E[N_i^{2H}] + V[N_i] E[C_i]^2, \qquad (17.2)$$

Taylor's Power Law. https://doi.org/10.1016/B978-0-12-810987-8.00017-3

where $E[\bullet]$ and $V[\bullet]$ denote mean and variance and H is the Hurst exponent (Chapter 16) that relates the variance of the contributions to the number:

$$N^{2H} \propto V[C_{i,j}] = E\left[\sum_{j=1}^{N} C_{i,j} - E\left[\sum_{n=1}^{N} C_{i,j}\right]\right]^2$$

Eq. (17.2) says the variance of all activity is the product of the variance of the contributions and the expected number of events to the $2H$ power, plus the product of the variance of the number of events and the expected contributions. If the $C_{i,j}(t)$ are uncorrelated, then $H = \frac{1}{2}$, and if they are long range correlated $H > \frac{1}{2}$. Thus, the variance of the contributions to the activity at the ith node (sample site) in the interval δt is proportional to the number of events in the interval to the $2H$ power.

In Eisler et al.'s formalism, each event may have a different contribution or size, C, to allow for different-sized packets arriving at the node in the interval δt. In ecological sampling, we may not distinguish between the number of packets and their size although in the case of insect stages we do, because experience has shown they may have different distributions (Chapter 8). If the number of events, N, is temporally or spatially correlated via the Hurst exponent and the size of events, C, is random with mean $= 0$ and variance $= \sigma^2$, then total activity is determined by the number of events not the size, and Eq. (17.1) becomes

$$V(f) = \sigma^2 E[N^{2H}] + 0 \cdot V(N)$$

and $b = 2H$ from which two important classes of systems are identifiable, $b = 1$ and $b = 2$, as highlighted by de Menezes and Barabási (2004).

Similarly, dependence of b on the size of the window Δt can be derived. If the activity time series are long term correlated with Hurst exponents $H(i)$ that are allowed to depend on the node i, the window-dependent variance at node i is

$$V_i(\Delta t) = E\left[f_i(\delta t) - E[f_i(\delta t)]^2\right] \propto \Delta t^{H(i)}$$

In this form, the time window size Δt is the scaling variable instead of the number of constituents, N, and V increases with $E[f]^b$. For a fixed signal, in the presence of long-range temporal correlations, the variance can also grow by changing the time window. $E[f_i]$ and Δt are analogous and consequently

$$\Delta t^{2H_i} \propto E[f_i]^{b(\Delta t)} \tag{17.3}$$

Differentiating (17.3)

$$\frac{dH(i)}{d\log(E[f_i])} \approx \frac{db(\Delta t)}{d\log(\Delta t)} \approx \gamma$$

The value of γ identifies three behaviors (Eisler and Kertész, 2006): $\gamma = 0$, b is independent of Δt, and the degree of temporal correlations measured by the Hurst exponent H^* is the same at all nodes; $\gamma > 0$, $b(\Delta t) = b^* + \log(\Delta t)$ and the Hurst exponent of the ith node $H_{(i)} = H^* + \log(E[f_i])$; γ takes two values in different ranges of Δt with nodes having separate Hurst exponents. The last is recognizable as the situation where TPL intersects the Poisson line (Figs. 4.2, 5.11 and 11.13).

A fractal model

Kendal (1992) considered a model for TPL in which an area divided into quadrats of side X and a population with density $\xi(x,y)$ at the point x,y has the number within a quadrat given by:

$$N_X = \int_0^X \int_0^X \xi(x,y)dydx$$

The mean and variance of number per quadrat are

$$M_X = E[N_X] = \alpha X \text{ and } V_X = E\left[(N_X - M_X)^2\right] = \int_0^X dx \int_0^X dy\phi(\lambda),$$

where $\phi(\lambda)$ is a covariance function relating the linear densities $\rho(x)$ and $\rho(x+\lambda)$, where λ is a lag:

$$\phi(\lambda) = E[\rho(x)\rho(x+\lambda)] - E[\rho(x)]E[\rho(x+\lambda)] = E[\rho(0)\rho(\lambda)] - E[\rho]^2$$

A characteristic of fractals is the covariance function is itself a power law related to the correlations in $1/f^{\alpha}$ spectra and Hurst statistics, $\phi(\lambda) = \beta\lambda^{D-E}$, where β is a constant, D is the Hausdorff dimension, and E is the Euclidean spatial dimension. Integrating V_X and substituting aX for M_X, the variance of number per quadrat is

$$V = \left\{\frac{2\beta}{(D-1)D\alpha^D}\right\}M^D = aM^b$$

The essence of Kendal's argument is that if the geographic distribution of populations has fractal geometry, then TPL emerges naturally from the spatial correlation of abundance declining with lag distance as a power law. A number of biological models examined in this chapter have been advanced that assume correlations in either environment or intrinsic characteristics of the model population.

Diffusion-limited aggregation

Another fractal based model proposed by Kendal (1995) assumes that the individuals in a population form randomly (Poisson) scattered clusters throughout quadrats. The clusters are assumed to form by diffusion-limited aggregation (DLA; Chapter 16), a process in which individuals migrate randomly until they encounter another individual. Clusters form by accretion of individuals: a positive feedback forms as the larger a cluster, the higher the probability the next migrant will join it. Assuming a compound Poisson negative binomial distribution for the number outside each annulus of a cluster, the probability generating function (PGF) of number per quadrat is

$$P(x) = \exp\left[\mu(x-1)\left\{\frac{1}{\rho} - \frac{(1-\rho)}{\rho}\exp(\alpha x - \alpha)\right\}^{-\beta}\right],$$

where α, β, and ρ are parameters and ρ is a function of μ, the product of the mean number of individuals per cluster, and the mean number of clusters per quadrat. This PGF has mean and variance that satisfy TPL with $b = 1 + 1/\beta$ and $a = \mu^{1-b}/\rho$. Note the intercept depends on μ and ρ, which, in turn, depend on the size of the quadrat in relation to the average size of clusters. Kendal tested this DLA model with data of the number of homes in a Japanese prefecture (Matui, 1932) and Beall's (1939) Colorado potato beetle data (Fig. 8.6), obtaining excellent agreement between the predicted and observed probabilities. He also noted that the case for TPL describing a fractal nature of population distribution is bolstered by its property of self-affinity (Eq. 15.1).

A network model

de Menezes and Barabási (2004) assumed a scale-free network, such as the Internet, in which there are N nodes (routers), each with L_i ($i = 1 \dots N$) links to other nodes and described by an adjacency matrix M_{ij}. To simulate people surfing the internet, they considered W random walkers on the network stepping from one node to another in each timestep, $t = 1 \dots T$. Each walker is placed on a randomly chosen node and makes T steps with the probability $1/L_i$ of choosing any particular link. Counters at each node record the number of visits by walkers. Repeating the simulations D times on the same network captures the temporal fluctuations on individual nodes. The number of visits to node i by walkers on day t is $f_i(t)$.

In this model, when the number of walkers $W < 100$, both the mean and variance of visits to each node grew linearly with time, $M[f] \propto t$ and $V[f] \propto t$ leading to the Poisson case, $b = 1$. A similar model in which packets are sent through the nodes at low rate also produced the Poisson case. In both models, as the expected number of walkers or packets per unit time (E[W]) increased from ~ 100 to ~ 1000, the temporal TPL slope increased from $b = \sim 1$ to ~ 2 (Fig. 12.8). The transition from $b = 1$ to $b = 2$ occurred gradually in this model, but as the size of the network increased the transition became steeper. At low E[W], the dynamics of the system are dominated by the detailed structure of the network. But, at a critical point, the volume of walkers or packets dominates triggering the transition to $b = 2$. In the particular network created by de Menezes and Barabási, that critical volume was E[W] ≈ 100. Their Fig. 4 shows TPLs getting steeper for a range of E[W] from 100 to 10,000.

de Menezes and Barabási observed there are two sources of fluctuations in their model. For E[W] = 0, visits can only occur as a result of the random deposition of walkers at $t = 0$. These they called internal fluctuations. For E[W] > 0, the fluctuations also result from the changing number of walkers (messages) from one day to the next. Consequently, the visits to individual nodes also change. If in a given interval the number of walkers increases, the visits at each node will grow proportionally, and at some point the growth in visits will exceed the internal fluctuations caused by the number of walkers placed on the network. Fluctuations caused by relative increase in node visits de Menezes and Barabási termed environmental.

If a series of pictures in time or space were taken of the system, we might expect to see $V = M$ for some and $V \propto M^b$ ($1 \leq b \leq 2$) for others. The observed scaling with $1 \leq b \leq 2$ observed in equity and forex transactions (Chapter 12) may be explained by competition between internal dynamics and external environmental forces.

Statistical physics

Fronczak and Fronczak (2010) developed an explanation for ensemble and spatial TPLs, which they see as equivalent, based on statistical physics. Their explanation derives from the number of states of a system obeying the laws of thermodynamics given specified values of macroscopic parameters. The macroscopic parameters they consider are M, V, and b and the number of states of the system is a function, $f(N)$, of the population size N. The model applies to entire populations and subpopulations as the main constraint is that M, V, and b be linked by TPL regardless of the scale of observation.

A fundamental equation of statistical physics is the Boltzmann distribution, $\mathcal{P}(\Omega; \mu)$ in which μ couples N to M, and Ω refers to the "microstate" of the system. Ordinarily, physicists are more interested in the macrostate of the system, but Fronczak and Fronczak's model concerns the relationship between Ω and TPL. In this model, Ω describes the local-scale spatial organization of the particles or organisms constituting the system, and the problem is to find the frequency distribution that obeys TPL at the macroscale by defining properties of the microstate represented by Ω. The Boltzmann distribution is related to the frequency distribution of N, $P(N; \mu)$ by the number of microstates, $g(N)$, by $P(N; \mu) = g(N) \, \mathcal{P}(\Omega; \mu)$. Fronczak and Fronczak derive an expression for $g(N)$:

$$g(N) \propto \frac{\exp(-N)}{N!} B_N(f_1, f_2, \ldots f_N), \qquad (17.4)$$

where $B_N(\cdot)$ is the Bell complete polynomial that describes the number of non-overlapping partitions of a set of size N (see Fig. 3.1C for illustration) into an arbitrary number of subsets and the N parameters f_i apply to subsets of size i. If all the f_i have the same value, then there is no preferential size and partitions correspond to a random distribution of elements, and the frequency distribution $P(N)$ is Poissonian.

The frequency distribution, $P(N)$, describing a system obeying TPL has the following form:

$$P(N) = g(N) \frac{\exp(-\mu N)}{\exp(F(\mu))} \qquad (17.5)$$

Fitting observed data to the formula, requires an estimate of μ, appropriate expressions for $F(\mu)$, and the number of states, $g(N)$. μ is estimated by M and $F(\mu)$, and is given by

$$F(\mu) = \frac{M^{2-b}}{a(2-b)} + Y, \text{ for } 1 \leq b \leq 2, \text{ and}$$

$$F(\mu) = \frac{\ln(M)}{a} + Y, \text{ for } b = 2,$$

where Y is a constant of integration. The number of states $g(N)$ underlying the considered systems has closed form for few values of b, which presents a problem. For very large N, Eq. (17.4) applies, but its solution is computationally intensive and unsuitable for most data. Fronczak and Fronczak proposed a workaround that uses TPL's scale invariance property by rescaling N to zN (which happens when we take samples anyway) and $V = z^2 a^*(M/z)^{b-2}$ to produce a new intercept, $a^* = az^{b-2}$ and new frequency distribution for $P(N/z)$ (Eq. 17.5).

The estimated frequency distributions of data for European corn borer (*Ostrinia nubilalis*), blue jay (*Cyanocitta cristata*), and traffic in Minnesota fit the observed data remarkably well. These three examples are spatial: temporal data of the daily activity of the New York Stock Exchange were not as close a fit. A case is made for an equivalence of the ensemble and temporal cases with this model.

A test compares two frequency distributions with different means, $\Delta\mu = (M_2 - M_1)$ and $h(M_1, M_2) = (M_2^{1-b} - M_1^{1-b})/(a - ab)$. Applied to estimates from two dozen sources, the identity $h(M_1, M_2) = \Delta\mu$ extends over 9 orders of magnitude with great accuracy. For the identity to hold, it is clear that a and b must be correlated: it is equally clear that the identity does not apply to the Poisson case as the denominator $(a - ab)$ disappears.

Fronczak and Fronczak's model is obviously based on the physical quantities and properties of energy. Kendal and Jørgensen (2011) have pointed out that their correspondence to biological systems is unproven, and question the applicability of statistical physics models to the range of spatial and temporal scales observed with biological data. They suggest that enquiries directed at Tweedie exponential dispersion models (see later) would be more productive as Tweedie models are characterized by TPL functions and fit ecological, physiological, genetic, and sociological data in addition to physical data.

The data requirements for Fronczak and Fronczak's model and tests are quite stringent, making many of the examples in Chapters 5–11 unsuitable. Although TPL operates on the macroscopic level of M and V, the basic unit of this model is the distribution, $P(N)$, which requires large data sets homogeneous in b to generate a smooth $P(N)$. Just because observed data obey TPL, does not mean that the underlying signal does not come from subsets each obeying a different b or a common b with different a. Such heterogeneity could be hidden by variability about the fitted line. Consequently, empirical data require careful processing before they can be used in this analysis. Even for large datasets, data standardization can reduce the data beyond what is usable, with a consequent loss in power of statistical tests. However, in addition to the US Interior Department's Breeding Bird Survey (Robbins et al., 1986) from which the blue jay example was drawn, there are other large, rigorously acquired datasets suitable for analysis by this method.[1]

A biophysical model

In Chapters 11 and 12, the case studies of physiological response to stress, traffic, and forex trading used the concept of a moving window to analyze the fluctuation scaling of signals in time. Koyama and Kobayashi (2016) examined a model of nerve transmission in which the window is fixed and the signal passes through it. They found that TPL can emerge from the excitation and inhibition of neurons during nerve activity creating fluctuations scaling with values of b determined by the ratio of incoming signals of excitation to inhibition. Their nerve transmission model is based on a stochastic differential equation (Ornstein-Uhlenbeck process) in which events occur only if the signal $S(t) > S_T$, a threshold value:

$$\frac{dS(t)}{dt} = -S(t) + \alpha + \beta\xi(t), \ S(0) = 0,$$

where α and β are parameters and $\xi(t)$ is Gaussian white noise with mean, $\mu = 0$ and temporal correlation proportional to separation in time, $E[\xi(t), \xi(t^*)] \propto (t - t^*)$. Integrating yields a solution for $S(t)$,

$$S(t) = \alpha(1 - e^{-t}) + \beta \int_0^t \xi(x)e^{x-t}dx,$$

from which mean and variance of $S(t)$ can be obtained:

$$M[S(t)] = \alpha(1 - e^{-t}) \text{ and } V[S(t)] = \frac{\beta^2}{2}(1 - e^{-2t})$$

Mean and variance of $S(t)$ scale as temporal TPL $V[S(t)] = cM[S(t)]^d$. Considering the sequence of nerve impulses or spike train into a neuron, Koyama and Kobayashi relate the parameters α and β to inputs from all connected neurons. There are two kinds of signals, excitatory impulses (A_{AMPA}) and

---

1. The Audubon Society's Christmas Bird Count, the British Trust for Ornithology's Breeding Bird Survey, the Marine Biological Association's Continuous Plankton Recorder Survey, various national Bottom Trawl Surveys, and the Rothamsted Insect Survey have data of hundreds of species collected over decades by well-established and verified methods.

inhibitory impulses (A_{GABA}) constituting the spike train into a neuron. Together with S_T and v_r, the cell's normal voltage (>0), the impulses define α and β:

$$\alpha = f(A_{GABA}, A_{AMPA}, S_T, v_r), \text{ and}$$

$$\beta = \frac{g(A_{GABA}, A_{AMPA}, S_T, v_r)}{h(A_{GABA}, A_{AMPA}, S_T, v_r)}$$

where $f(\cdot)$, $g(\cdot)$, and $h(\cdot)$ are functions defining the excitation and inhibition of a neuron. Koyama and Kobayashi's argument is specific to neural spike trains, but its significance lies in the fact that a and b can be defined in terms of measurable quantities ($A_{GABA}, A_{AMPA}, S_T, v_r$) and that a temporal TPL of intervals has slope $1 \leq b \leq 3$ that maps to a density TPL with slope $0 < d \leq 2$. The TPL of interspike interval predicted by the model and supported by simulations is $V_I = cM_I^d$. The number of spikes in time interval Δ has TPL $V[N_\Delta] = aT$ $[N_\Delta]^d$, where $a = \phi\Delta^{1-d}$ and $d = 3 - b$. In these simulations rates of inhibitory and excitatory signals to the neuron were λ_I and λ_E, respectively. Simulating constant spike input rate, that is, $\lambda_I = r\lambda_E$, b declined approximately exponentially from 3 to 2 as r increased from 0 (excitation dominant) to 1 (inhibition dominant).

Koyama and Kobayashi identified four parameter spaces (Fig. 17.1) that depend on the values of α and β relative to S_T that predict temporal TPL with slope b:

1. transmission when the threshold is exceeded ($S(t) > S_T$) and $\alpha \approx 1$ and $\beta \gg 1$, a condition they called threshold regime, where $b = 1$ and $a = 2\ln(2)$;
2. occasional transmission because of small fluctuations in $S(t)$ occurring when $1 - \alpha \gg \beta$ and $\beta \ll 1$, called subthreshold where $b = 2$ and $a = 1$;
3. a signal is transmitted when $\alpha \gg 1$ and $\beta \ll 1$, a condition they called suprathreshold, where $b = 3$ and $a = \beta^2$;

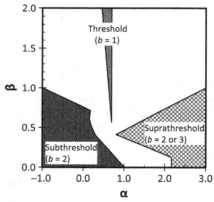

FIG. 17.1 Three regions of $\alpha\beta$ parameter space predicting interspike interval (temporal) TPL slopes of $b = 1$, 2 or 3. *(Adapted from Fig. 1 in Koyama and Kobayashi (2016).)*

4. and a second suprathreshold condition occurring when $\alpha \gg 1$ and $\beta = \kappa \sqrt{\alpha}$, where κ is a constant and $b = 2$ and $a = \kappa^2$.

With the correspondence between b and d, equivalent parameter spaces exist for d.

Statistical models

Reformulating TPL

A novel statistical argument by Soberón and Loevinsohn (1987) explored the statistical reasons leading to TPL. They argued, as did Hanski (1982) and Perry and Taylor (1985), that the proper null hypothesis for TPL slope is $b = 2.0$. Their argument is based on a reformulation of TPL. The sample variance of the ith sample set is $v_i = n_i^2 \, v_i^2 \, V_p$, where m_i and n_i are the mean and number of observations ($\equiv NQ$) and V_p is the variance of the proportions in the n_i sample units. Taking logs, they plot $2 \cdot \log(m_i) + n_i^2 \, V_p$ against $\log(m_i)$ for $p = NB$ sample points, obtaining $\log(V) = A + b \cdot \log(M)$, where the parameters are nontrivial functions of p, n_i, m_i, and the variance of $\log(M)$, V_m: $A = F(p, n_i, m_i, V_m^{-1})$ and $b = 2 + G(p, n_i, m_i, V_m^{-1})$. The decomposition of the slope into 2 plus a term containing information on the p regression points emphasizes that the proper null for slope is $b = 2.0$. In addition, the fact that n_i, m_i, and V_m appear in both functions suggests a correlation between the TPL parameters. Soberón and Loevinsohn note that no biological assumptions need be made to obtain straight lines with slope $b = 2 \pm s$ ($s < 2$) with any set of $\log(V)$ and $\log(M)$ as a consequence of their definition of variance. This does not mean that b is uninfluenced by biological processes via population size and/or variance of the proportions.

Considering the spatial form of TPL with a time series of number $x_i(t)$ for the ith of p sites at time t, a slope of $b = 2$ is expected if V_p is either a constant or a random variable independent of population size X. If V_p is positively correlated with X, then a slope $b > 2$ will be obtained and $b < 2$ if V_p decreases with X. A curved TPL implies a nonlinear relationship between V_p and X. V_p will remain constant if the numbers at each site increase and decrease proportionally resulting in $b = 2$. Even if there is random migration between sites, V_p and X will remain uncorrelated and $b = 2$ is expected. But if migration and/or mortality are density dependent, reducing numbers in overpopulated sites, $b < 2$ is expected. Density-dependent (social) aggregation on the other hand will lead to $b > 2$ in Soberón and Levinsohn's model. Both migration and aggregation occur in the mobile stage(s) of all animals, and if density dependent, their balance contributes to the observed value of b as proposed by LRT and RAJT (1977).

For temporal TPL, a slope of $b = 2$ is expected. If the sample sites share similar conditions, for example weather, their populations are likely to be correlated, tending to keep V_p constant. Distant sites not experiencing similar conditions with no correlation between good and bad years among sites also produces $b = 2$. However, if some sites experience good conditions every year resulting in high numbers and low V_p, $b < 2$, whereas a slope $b > 2$ would be expected if high numbers occur only in a few very good years. For the ensemble case, $b < 2$ is expected if V_p is smaller due to habitat in the higher density sites being more homogeneous, or if density-dependent movement leads to more even distribution of organisms, and $b > 2$ will occur if a few very high-quality sites are responsible for high means in those sites, whereas in others the numbers are lower but more even. As with the spatial case, social aggregation will also result in $b > 2$.

Higher moments

Even though there is no known *single* frequency distribution that fits TPL over a wide range of means (the Adès and Tweedie distributions are *families*), it has been found that TPL can be characterized by the first four central moments representing a general hypothetical distribution. Cohen and Xu (2015) showed that observations randomly sampled in blocks from any skewed frequency distribution with four finite moments obey TPL. If NQ_j ($j = 1 \dots NB$) quadrat samples are taken from NB blocks in which all samples, x_{ij} ($i = 1 \dots NQ_j$), are independently and identically distributed, the mean and variance of x_{ij} in the jth block are m_j and v_j, respectively. With $E[m_j] = M$ and $V[m_j] = V/NQ_j$, the mean and variance of the sample variance of the jth block are

$$E[v_j] = V \text{ and } V[v_j] = \frac{1}{NQ_j}\left\{\mu_4 - V^2\frac{NQ_j - 3}{NQ_j - 1}\right\},$$

where μ_4 is the fourth central moment from which kurtosis (Eq. 2.2) is computed, $K = \mu_4/V^2$. Cohen and Xu show that least squares estimates of A and b are obtained from

$$\hat{b} \approx \mu_3 \frac{M}{V^2} \text{ and } \hat{A} \approx \log(V) - \mu_3 \frac{M}{V^2}$$

where μ_3 is the third central moment, related to skewness (Eq. 2.2) by $S = \mu_3/V^{3/2}$. The standard error of b is

$$SE[\hat{b}] \approx \sqrt{\frac{M^2(K - 1 - S^2)}{(NB - 2)V}}.$$

Counts $\{x_{ij}\}$ were simulated for $NB = 100$ and $NQ = 100$ samples from five distributions for which the first four central moments are defined: the Poisson, negative binomial, exponential, gamma, and lognormal distributions. A sixth distribution, the Gaussian, shifted to the right to ensure no negative observations, was included as a control.

Estimates of $\hat{A}$, $\hat{b}$, and $SE[\hat{b}]$ were calculated directly from the six distributions' parameters, by estimating $\hat{A}$ and $\hat{b}$ from the four moments calculated from the sample data and by computing an ensemble TPL by ODR from the NB blocks of NQ samples. 10,000 random realizations of the sampling program were run to estimate the median and 95% CI of $\hat{A}$ and $\hat{b}$. The two estimates were compared to the TPL parameters computed directly from the distributions' parameters. In all cases, the calculated value of $\hat{A}$, $\hat{b}$, and $SE[\hat{b}]$ fell within the 95% confidence interval of the estimates. As expected, estimates were not different from zero in the Gaussian case.

The estimated values of $\hat{b}$ for the Poisson, negative binomial, exponential, and gamma distributions were all within 0.35% of the computed values of $b = 1.0$, 1.6, 2.0, and 2.0, respectively. However, the estimated values of $\hat{b}$ for the data generated by the lognormal distribution underestimated the calculated value of $b = 4.72$ by $>15\%$, a consequence of the lognormal's right-hand tail being too short to generate the really extreme samples needed to generate realistic distributions much over $b = 2.0$. Cohen and Xu's results show unequivocally that random sampling of skewed distributions having the first four central moments are sufficient to generate TPLs for $b \leq 2$, without any biological or behavioral assumptions. Generating TPLs with values of b much greater than 2.0 requires a more extreme distribution.

Simulated sampling

A number of authors have noted variation in spatial TPL slope with the physical size of sampler (e.g., Wheeler et al. (1987, 2000) in Chapter 7). Similar observations have been made with distance between samples as separation of samples and size of sampler are linked at least at very small scale (e.g., Fig. 7.2). An extensive series of simulations by Sawyer (1989) quantified this dependence for individuals distributed in a 1024×1024 cell arena. Five policies for distributing up to 10,000 individuals were used: Poisson distribution; randomly distributed clusters with comparatively high within-cluster aggregation, equivalent to a negative binomial distribution; and aggregated hierarchical distributions with clusters of individuals collected in groups of clusters in arrangements designed to give slight, moderate, and high aggregation. Populations of different size were created by successive removal of randomly chosen individuals. In the negative binomial simulation, the number of clusters was manipulated while the cluster size remained the same, equivalent to varying the number of founding mothers with constant mean egg number.

Randomly placed samples were taken with quadrats ranging from 3×3 cells to 100×100 cells. Means and variances per cell were computed from the individuals falling within a quadrat. Sample sizes, NQ, were chosen to ensure a precision level of $\sqrt{(V/NQ)} < 0.1M$. The number of quadrats required $(40 < NQ < 50,000)$ depended on the spacing policy and the quadrat size, with NQ increasing with decreasing quadrat size.

For the Poisson distributed simulations, the measured values of $b \approx 1$ for all quadrat sizes: slight variations in b were possibly due to imperfections in the random number generator. The measured b for the hierarchically aggregated populations increased with quadrat size. For populations with the highest aggregation, b increased quickly to $b \approx 2$ with increasing quadrat size, but for populations with lesser degrees of aggregation, the increases were slower. For the approximately negative binomial distributed populations with randomly distributed clusters, b increased slowly to a maximum at quadrat size 50×50 and then declined, presumably related to the spatial scale of aggregation. The hierarchical model, with individuals in clusters and clusters within groups, is presumably a more realistic way to look at population spatial distribution.

Sawyer comments that estimated values of b do not provide a complete picture of a species' spatial distribution without details of the sampler being specified. Instead the relationship between sample variance and sample mean over a range of densities depends on the spatial scale in relation to the size of the sampler. As is well known, the details of the sampling program are often folded into the intercept A as illustrated by Ferris et al. (1990) (Fig. 7.9, Table 7.3).

Sampling and feasible sets

Xiao et al. (2015) have proposed that TPL arises as a purely numerical, as opposed to statistical, consequence of taking NQ samples NB times. The mean-variance pairs (M_i, V_i, $i = 1 \dots NB$) obtained by sampling NB subpopulations results from sampling the "feasible sets" of numbers that constitute the NB subpopulation being sampled. If there are a total of $N_i = \Sigma x_i$ individuals in NQ_i samples the mean, $M_i = N_i/NQ_i$, in the ith subpopulation. Dropping the subscript and considering only one quadrat, the feasible set is the set of all possible configurations of individuals given the constraints of NQ numbers (samples ≥ 0) that sum to N. There are two different ways to make the arrangements, equivalent to combinations and permutations: partitions and compositions. A partition is an unordered list of integers (≥ 0) and a composition is an ordered list. The partitions approach describes how a total can be divided among groups without information on either the individuals or the groups (quadrats). In practice, it means assigning individuals at random to quadrats and focusing only on the possible patterns. In contrast, the compositions approach retains some information about the quadrats and individuals contained in them.

Given a list of numbers $\{x\}$ (whether ordered or not), the computed variance V is a consequence of the particular arrangement of the feasible set obtained by arranging a total of N individuals in NQ samples. At one extreme $V = N^2$ when all individuals are in one quadrat; at the other extreme $V = 0$ when all quadrats contain the same number of individuals (assuming NQ is a factor of N). To obtain a variance estimate all possible values of V must be calculated from the sample space of feasible set, but the number of possible values of V grows rapidly with N for a given NQ. For good TPL data sets over several orders of magnitude, with NQ and NB both ≥ 15, the computational power required to obtain meaningful estimates of V is daunting. To relieve the computational requirement Xiao et al. restricted sampling for V to 1000 samples.

Xiao et al.'s objective was to show that such a scheme could account for observed b, not necessarily that it does. They set stringent data requirements to ensure reliable results and computed feasible set slopes for 45 spatial and 70 temporal data sets from the literature for which means and variances were given. TPLs computed from the feasible sets were generally in good agreement with the empirical values of b and r, although the TPLs computed from the partitions feasible sets were somewhat better than those from the compositions sets. The actual estimates of variances did not match so well with 25%–75% of observed variances outside the 95% CI of the variances computed from feasible sets.

The TPL fits of the feasible sets averaged $r = 0.894$ for the partition method with 85% of slope estimates in the range $1 \leq b \leq 2$; for the composition method $r = 0.964$ and 75% of estimates were $1 \leq b \leq 2$, but about ¼ of cases had significant curvature. Both methods gave good range of slopes ($0 < b < \sim 3.5$) comparable to experience. The computed distribution of b agrees well with the observed distribution, producing the full range of b observed in field data. However, it is not clear how several replicates of a TPL experiment conducted at different times and places by different people (e.g., western flower thrips in Chapter 8) would produce closely similar realizations of such a scheme. This approach, which removes all necessity of, but without eliminating, biological assumptions and processes, is intriguing. There is a similarity of approach with Perry and Hewitt's (1991) moves index (Chapter 3).

A lattice model

Hanski (1980) created a simple 20×20 lattice model to explore how the movement of the scarab beetles he was studying might affect their spatial distribution. In his model, the probability of survival and the success of reproduction were independent of density and uncorrelated across the lattice, ensuring no spatial or temporal autocorrelation in the quality of the population sites. Movement between cells was either short distance to the eight adjacent cells or medium distance to the eight cells three steps away, the choice of direction being a uniformly distributed random variable. The number exiting a cell was also independent of density. The simple movement rule combined with spatially and temporally independent reproduction and survival was sufficient to generate convincing spatial TPLs over several orders of magnitude with slopes clustered around $b = 2$, but differing significantly in their intercepts. The simulations were similar to, but somewhat steeper than, TPLs at two spatial scales of mixed-species scarabs separated into *Aphodius* spp. and Hydrophilidae, which differed little from each other except in intercepts at the within field scale. Hanski succeeded in his aim "to show that a linear relationship between log spatial variance and log mean abundance ... follows from very simple alternative assumptions."

In a later purely statistical model, Hanski (1987) correlated local populations by generating random vectors from a multivariate normal distribution with a covariance matrix with net correlations ranging from $0 \le r \le 0.8$. The population dynamics within and between cells was subsumed into 5 randomly generated populations with different approximately lognormal random abundances for 30 generations. The means and variances used to generate the populations were selected to give realistic temporal TPLs in the range $1.75 < b < 3.0$. In this purely statistical model, spatial TPL slope increased with mean, which reflected changes in temporal TPL slope. This model generated slopes significantly smaller than $b = 2$ for rare species. In 50 simulations with crosscorrelation of $r = 0.98$ and five populations generated with $\log(M) = -1.0$, -0.75, -0.5, -0.25, and 0.0, $\log(V) = 0.7$, 0.6, 0.5, 0.4, and 0.3, the slopes of the resulting spatial TPLs were $b = 1.67 \pm 0.04$ ($r = 0.98$). Slopes of $b > 2.0$ were generated with populations with crosscorrelations of $r \le 0.40$. Fifty simulations of five populations generated with $\log(M) = 0.0$, 0.5, 1.0, 1.5, and 2.0 and zero crosscorrelation obtained $b = 2.48 \pm 0.11$ ($r = 0.96$). The steepest slope was generated with some populations negatively correlated to simulate populations living in different habitats: $b = 2.76 \pm 0.15$ ($r = 0.938$).

Hanski's results confirmed that species with more closely correlated dynamics would have relatively low variance across sites compared to species in more restricted or patchy distribution with high relative variance. Like Anderson et al. (1982), Hanski suggests that spatial $b > 2$ is unlikely and concludes that large-scale spatial variation can be explained by standard population dynamic arguments, not requiring "new, complex mechanisms to explain the observed patterns." Crosscorrelations in population dynamics exist, he argued, by simple mechanisms such as high rates of within-population migration, which would tend to decrease spatial variance. As migration is a complex population process (Johnson, 1969; Baker, 1978) conducted by individuals for their own selfish purposes (Hamilton and May, 1977; LRT and RAJT, 1977), the simple population dynamics leading to population crosscorrelations likely arise from complex mechanisms.

Biological models

Several classes of model have been proposed to account for spatial TPL in ecological populations: spatial heterogeneity in the mean abundance or carrying capacity (Anderson et al., 1982; Perry, 1988, 1994; Ballantyne, 2005); autocorrelated populations (Kilpatrick and Ives, 2003; Ballantyne and Kerkhoff, 2005, 2007); agent-based behavioral models (RAJT, 1981a, b, 1992; Perry, 1988).

Nonlinear maps

Models in time

A continuous time stochastic model of Anderson et al.'s (1982) of model populations subject to constant per capita rates of birth, death, emigration, and immigration (λ, μ, γ, and δ) predicts a negative binomial distributed population size at any point in time with mean and variance at time t given by

$$M(t) = \frac{\delta}{r}\{\exp(rt) - 1\} \text{ and } V(t) = M(t)\left\{1 + \frac{\lambda}{\delta}M(t)\right\},$$

where $r = (\lambda - \mu - \gamma)$. In this model, the ratio $V(t)/M(t)$ increases with slope λ/δ. At low population densities, when $\delta > \lambda$ TPL is approximately Poisson; as density increases TPL slope increases from $b = 1$ to $b = 2$. The exact trajectory of $\log(V(t))$, $\log(M(t))$ depends on the relative values of r, μ, and γ, which also determine the estimated value of A. With this model, slope can go no higher than $b = 2$.

An elaboration has multiple population patches with random immigration (δ) into patches dependent on the net rate of emigration from all other patches. When all other rates are constant and independent of population density, the mean abundance per patch, $M(t)$, and the variance of patch density, $V(t)$, are

$$M(t) = N_0 \exp\{(A - B)t\} \text{ and } V(t) = M(t)\left\{\frac{A+B}{A-B}\right\}\left\{\frac{M(t)}{N_0} - 1\right\},$$

where N_0 is the starting population, $A = \lambda + \gamma\delta P/(\eta + \delta P)$, $B = \mu + \gamma$, $\eta \gg \mu$ is the per capita death rate of migrants, and P is the number of patches. Population growth in all patches results from birth and immigration rates (A) and death and emigration rates (B) in which immigration from the other patches is $\gamma\delta P/(\nu + \delta P)$. This model produces a probability distribution of abundance per patch with a quadratic relationship between mean and variance, similar to the negative binomial distribution. The spatial TPL generated by this population model is not convincing because of the dramatic decline in variance at high mean (Fig. 13.3), which is not seen in sample data except when the sampler becomes saturated (Figs. 5.12 and 15.7). The drop off in Fig. 13.3 is due to all populations in the simulation reaching their carrying capacity resulting in maximal density with minimal variance, a rare condition in nature. LRT et al. (1980) found no evidence for a turning point in TPL corresponding to a maximum population density of aphids, moths, and birds.

Introducing habitat heterogeneity into the model with a density-dependent coefficient normally distributed with fixed mean and SD to assign different constant carrying capacities to each patch, the drop off seen in Fig. 13.3 disappeared except when a very homogeneous habitat (small SD) was simulated.

Like Hanski (1987) later, Anderson et al. concluded that the observed spatial distribution patterns depend purely on the dynamic balance between demographic forces of birth, death, immigration, and emigration and not requiring spatial behavior. And like Hanski, the success in producing convincing TPLs depended on the introduction of spatial variability in the habitat to which the modeled populations adopted a range of means.

Perry (1988) developed three models: two with abstract space and a stepping-stone model to simulate real space. A simple growth model with limits and a logistic-like model were run for 50 ($=NB$) generations in 150 ($=NQ$) population cells. The simple growth model was regulated only by density-dependent processes acting at lower and upper limits. Perry called this a "floor-and-ceiling" model in which emigration occurred when $N_i(t) > K_i$, the ith cell's ceiling, a population level somewhat less than carrying capacity. Each cell's population ceiling was assigned by random number drawn from a lognormal distribution; each generation, the reproductive rate, $r(t)$ was also drawn from a lognormal distribution. Populations were recorded as integers, which permitted extinction. Repopulation was by a low Poisson immigration rate (θ). The only difference between realized TPLs was the parameters used to generate the 150 values of K_i for different species at the beginning of each run. To estimate TPL standard errors, each notional species was replicated five times. This simple model produced convincing TPLs with slopes $1.54 \leq b \leq 2.54$ over several orders of magnitude of mean, agreeing very well with field data.

The second model was based on the discrete time difference logistic:

$$N_i(t+1) = \lambda N_i(t) \left\{ 1 + N_i(t) \left(\frac{\lambda - 1}{\alpha(t)K_i} \right) \right\},$$

where λ is the finite rate of increase, $\alpha(t)K_i$ is a density-dependent parameter-limiting $N_i(t)$ with $\alpha(t)$ allowed to vary randomly between years in some simulations and kept constant in others. In this model population, regulation operates with weak density dependence at all densities. The addition of weak density dependence and environmental heterogeneity resulted in poorer TPLs than the simpler floor-and-ceiling model. The range of means was apparently controlled by environmental heterogeneity but resulted in curved and autocorrelated TPLs. Desynchronizing the growth in cells by allowing density dependence to vary removed some curvature and serial correlation but drastically reduced the range of means.

Strong density dependence

Perry (1994) used Hassell et al.'s (1976) discrete time difference logistic with a density-dependent parameter β in addition to the weak density-dependence based on K to investigate the effect of strong density dependence on simulated TPL:

$$N_i(t+1) = \lambda_i N_i(t)(1 + \alpha_i N_i(t))^{-\beta} \tag{17.6}$$

where i and t are as before and λ_i is the finite rate of increase in the ith population cell. Parameters α_i and β define the degree of density dependence. This model with $\alpha_i = \alpha(t)K_i$ and $\beta = 1$ is equivalent to the previous model with the strength of density dependence controlled by β. Difference logistic models can be forced into a realm of chaotic dynamics by strong density dependence and/or high reproductive rate (May, 1976). Perry forced Eq. (17.6) by increasing $\beta \geq 6$.

A stochastic model based on Eq. (17.6) has equilibrium density, $E \approx (\lambda^{1/\beta} - 1)/\alpha$, a variable that measures the average density over time. Values of E across 25 cells, E^*, and $SE(E)$ could be defined by suitable choice of lognormal values of α_i. Setting α_i constant for all cells ($SE(E) = 0$) and a range of values of λ, β, E^*, and θ for 30–60 generations, the model generated chaotic dynamics and TPL slopes of $0.86 \leq b \leq 2.64$ for $1 < \beta < 6$. An additional simulation with 250 cells held α constant but the model forced with $\beta = 8$ and uniform random values of $50 \leq \lambda \leq 150$ also produced chaotic dynamics with cell extinctions and reinvasions over 250 generations. The frequency distributions of numbers/cell were highly skewed in common with real population counts and were well described by the Adès distribution (Chapter 3).

The relationship between the input variables and the output TPL parameters is not obvious. However, the highest values of b were obtained with the most complex models and in general TPL slopes increased with increasing E^* when $SE(E) = 0$. None of Perry's models, even the flawed second model, suffered the artifactual drop-off in Anderson et al.'s (1982) models by the populations reaching their carrying capacity. The combination of migration and strong density dependence contributed to the production of convincing TPL plots (Fig. 17.2). The feedback on numbers through mortality or emigration and the minimal linkage of cells by migration ensured that chaotic subpopulations going extinct could be revived ensuring global population persistence.

In the preceding models, space was abstract with the cells linked only to the extent that immigration was presumed to have occurred from emigration elsewhere. In his stepping-stone model, Perry (1988) introduced a spatial component by organizing the cells in a kite-shaped lattice of hexagons.

Perry developed an algorithm for distributing the individuals in a unit of seven hexagons according to a prescribed TPL with parameters A^* and b^*. Applying the algorithm repeatedly produced a spatially fluid population in which local groups of hexagons conformed to the desired TPL. The question Perry asked is: what is the global TPL if the local cells obey TPLs with the prescribed values of A^* and b^*? Under such circumstances, there might not be a TPL at the global scale or, if there is, scale issues alone might change the TPL estimates.

Simulations did yield realistic TPLs at the global scale with the global values of A differing from A^*, but the global estimates of $1.47 \leq b \leq 2.42$ did not differ significantly from the prescribed values of $1.5 \leq b^* \leq 2.5$. Perry noted that if individuals within a defined area always conform to a TPL locally, the results suggest they will also conform to it globally with the same value of b. Furthermore, simulations started with individuals concentrated in one hexagon

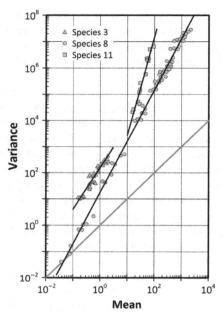

FIG. 17.2 Simulated TPLs generated by a discrete time difference logistic forced into chaos by strong density dependence with the following parameters from Table 1 in Perry (1994):

Species	λ	B	θ	E*	SE(E)	A	b
3	50	6	0.025	22.5	0	1.49	1.32
8	500	6	0.02	897	0	1.31	1.95
11	500	6	0.2	144	297	−0.33	2.56

Symbols are defined in the text. Notice that the largest range of means occurred with the TPL slope closest to $b = 2$, and the higher and lower slopes occurred at higher and lower densities, respectively. *(Figure adapted from Fig. 1 in Perry (1994).)*

generated movements toward realistic patterns of aggregation and dispersal respectively, restoring the programed TPL parameters at local and global scales.

Behaviorally induced emigration in all three models found b-values similar to those obtained with real data supporting the suggestion that under certain circumstances b could be species-stage specific (LRT, 1971; LRT et al., 1988). Perry's (1988) pseudospatial models produced moderately aggregated populations ($1.35 < b < 2.54$) by demographic rate processes alone, but they failed to account for extremely low levels of aggregation while his (Perry, 1994) chaotic model extended the range ($0.85 < b < 2.65$).

Effect of competition

Considering temporal TPL, Kilpatrick and Ives (2003) point out that for a population experiencing constant per-capita environmental variability, the expected value of temporal $b = 2$, which should be the null hypothesis in statistical tests. They developed an argument for competition between species in a community producing TPL slopes of $b < 2$. In a simple two species competitive interaction, one species may be less abundant because its carrying capacity K_1 is lower and/or it is the poorer competitor. The rarer species inevitably experiences greater environmental variation because the more abundant species constitutes a large part of its environment. Thus, the variability of the rare species resulting from competition is large relative to the more abundant species producing a slope of $b < 2$.

They considered a competitive community of n species each growing according to a discrete logistic equation in which the density of species i at time $t + 1$ is

$$N_i(t+1) = N_i(t) \exp \left\{ \frac{r_i}{K_i} \left(K_i - N_i(t) + \sum_{j \neq i} \alpha_{ij} N_j(t) \right) + \varepsilon_i(t) \right\}, \qquad (17.7)$$

where r_i is the intrinsic rate of increase, K_i is the carrying capacity, and ε_i is a normal random variable incorporating environmental stochasticity on r_i. The coefficient α_{ij} is the per-capita influence of species i on species j.

In communities of $n = 10$ species, simulations of Eq. (17.7) were run in which each species had a rate of increase, $0.5 < r_i < 1.0$, carrying capacity $50 < K_i < 1550$, and competition coefficient $0 < \alpha_{ij} < 0.5$. In each of five simulations, the r_i were fixed with one, the "focal species" with $r_i = 0.57$ against which the others were compared. Different values of K_i and α_{ij} produced trajectories of abundance of the focal species that differed relative to the others. Different values of α_{ij} produced different TPLs with slope decreasing from $b \approx 2$ when $\alpha_{ij} = 0$ (no competition) to $b \approx 1.25$ at $\alpha_{ij} = 0.15$. For the special case in which $r_i = r$ and $\alpha_{ij} = \alpha_j$ for all species and for the more general case of differing r_i and α_{ij}, b decreased with increasing strength of interspecific competition (α) but was relatively insensitive to the number of interacting species if $n \geq 5$. The extent to which competition decreased b also depended on how the stochastic variation affecting species densities was correlated. If environmental stochasticity acting on r affected all species the same (correlation, $\rho = 1$), then $b = 2$. Decreasing environmental correlation ρ reduced b from 2 by an amount dependent on r. If r was slightly greater than 1, the effect of environmental correlation was minimal ($b \approx 2$) but if $r = 1$, b increased from ~ 1 to ~ 2 as environmental correlation ρ increased from -1 to $+1$. For values of $r < 1$, the increase in b with ρ was correspondingly steeper. The foregoing suggests that b may be sensitive to how species respond to environmental stochasticity relative to each other as well as the direct effect of their interactions.

Temporal TPL and stability

A trio of papers (Ballantyne, 2005; Ballantyne and Kerkhoff, 2005, 2007) investigated temporal TPL. In the first, Ballantyne investigated temporal b predicted
by three deterministic single-species population dynamics models, the logistic
($N_{t+1} = \lambda N_t \{ 1 - N_t/\kappa \}$), the Ricker ($N_{t+1} = N_t \exp \{ \lambda (1 - N_t/\kappa) \}$), and the continuous time-lagged logistic growth model ($dN/dt = \lambda N \{ 1 - N(t - \tau)/\kappa \}$), in which
λ is intrinsic rate of increase, κ is carrying capacity, and τ is the lag in time t. The
first is essentially the same model Perry (1988, 1994) used to investigate spatial
TPL without the strong density-dependent parameter β. The three models,
which are related (May, 1974), predicted an upper limit for temporal TPL slope
of $b = 2$ with the implication that slopes $b < 2$ reflect a departure from
determinism.

Ballantyne and Kerkhoff (2005) examined the population consequences of
correlation between individual reproductive output R_i ($i = 1 \dots N$) for a population of $N = 5000$ trees in a model based on Satake and Iwasa's (2000) forest
model. The population mean and variance of reproductive output are

$$M_{\text{pop}} = N\mu \text{ and } V_{\text{pop}} = N \{ V_{\text{ind}} - \text{Cov}(R_i, R_j) \} + N^2 \text{Cov}(R_i, R_j) \qquad (17.8)$$

where $\text{Cov}(R_i, R_j)$ is the average reproductive covariance between individuals
and the R_i are identically distributed with mean μ and variance V_{ind}. For
the Satake and Iwasa forest model, Eq. (17.8) predicts a temporal TPL slope
of $b = 1$ because V_{pop} is a linear function of the number of trees, N if the R_i are
uncorrelated ($\text{Cov}(R_i, R_j) = 0$). If reproductive output is perfectly correlated,
$\text{Cov}(R_i, R_j) = V_{\text{ind}}$ and $V_{\text{pop}} \propto N^b$, $b = 2$, in agreement with Ballantyne (2005).
A parameter, k, in the forest model proportional to the energetic costs of fruiting
and flowering is an important driver of the model's prediction of TPL slope. For
$N = 5000$ and $k < 1.62$ the model predicts $b \approx 2$, but at $k = 1.62$ the model predicts
an abrupt transition to $b \approx 1$ for $k > 1.62$. For small values of k, reproduction is
synchronized across the model forest, equivalent to the phenomenon of masting.

The abrupt shift in b at $k \approx 1.62$ accompanies a shift from masting to unco-
ordinated or chaotic individual reproduction. Ballantyne and Kerkhoff suggest a
connection between TPL b and the Lyapunov exponent that measures the sensitivity of a dynamical system to small changes in initial conditions, and is used
to identify chaos. The rapid transition from a synchronized state to an unsyn-
chronized state is similar to phase transitions in physics and the sudden transition to a spanning cluster in percolation models (Chapter 16).

Ballantyne and Kerkhoff (2007) further developed their theory relating temporal patterns of TPL scaling to individual-level reproductive behavior. Having
established the boundaries for b in the previous paper, in this one they show how
$1 < b < 2$ may be generated through the decay of $\text{Cov}(R_i, R_j)$ as N increases. As
density-independent net reproduction leads to $b \approx 1$ and density dependence
leads to $b \approx 2$, this amounts to making the reproductive correlation dependent
on density to varying degrees. Introducing density dependence to Eq. (17.8)
leads to

$$M_{pop} = Nf_M(N) \text{ and } V_{pop} = AN\{V_{ind} - f_C(N)\} + N^2 f_C(N)$$

where A is a constant of proportionality. The functions f_M and f_C relate the mean and covariance of reproductive output to population size, and the variance of reproductive output is assumed independent of density. Assuming a steady-state $f_M(N)$ is constant equivalent to μ, then $f_C(N) \propto N^{b-2}$. For a given value of b, reproductive covariance decreases with increasing population size. For values of b close to 2, the reproductive correlation declines only slightly with increasing N, but for smaller values, $b \approx 1.5$, the correlation declines sharply at small N and then asymptotically approaches zero at a decreasing rate. Specifically, this model suggests that temporal fluctuations measured by TPL at the scale of a population result from variability of the individual. Thus, populations persist because individuals must balance their reproductive behavior between capitalizing on constant properties of their environment while ensuring the capacity for response to environmental change—Southwood's (1962) Principle.

The Lewontin-Cohen model

The Lewontin and Cohen (1969) model of stochastic population dynamics proposes that population size at time t is

$$N(t) = N(0) \prod_{i=0}^{t-1} A(i), \ t = 1, 2, \ldots, \tag{17.9}$$

where the $A(i)$ are random factors by which population density grows or declines in the interval t to $t+1$. The $A(i)$ are assumed to be independently and identically distributed with finite positive mean, μ, standard deviation, σ, and coefficient of variation, $CV[A(0)] = \sigma/\mu$. Cohen et al. (2013) use the Lewontin–Cohen model to derive a spatial TPL at large t. They define

$$\log(\alpha) := \lim_{t \to \infty} \frac{1}{t} \log E[N(t)] \text{ and } \log(\beta) := \lim_{t \to \infty} \frac{1}{t} \log V[N(t)]$$

and show that provided α and $\beta > 0$ in the limit as $t \to \infty$, the spatial mean and variance are

$$\log(\alpha) = \log(E[A(0)]) = \log(\mu) \text{ and}$$
$$\log(\beta) = \log\left(E[A(0)]^2\right) = \log(\sigma^2 + \mu^2)$$

If TPL applies to $N(t)$ for all t as $t \to \infty$, then $\log(\beta) = b\log(\mu)$ and $b = \log(\beta)/\log(\mu)$ then TPL parameters can be equated to the terms in the Lewontin-Cohen model:

$$a = \frac{E\left[(N(0))^2\right]}{(E[N(0)])^b} \text{ and } b = \frac{\log(\sigma^2 + \mu^2)}{\log(\mu)} = 2 + \frac{\log\left((CV[A(0)])^2 + 1\right)}{\log(\mu)}$$

The Lewontin-Cohen model assumes that the multiplicative factor $A(t)$ accounts for all demographic and environmental variation. Unlike other ecological models leading to TPL, there is no density dependence or autocorrelation of $A(t)$. Also, unlike many other models, Lewontin-Cohen permits values of b over a wide range. The model predicts (in the limit) $b > 2$ if and only if the population is growing, $b = 2$ if it is stable and deterministic, and $b < 2$ if it is declining. In particular, Cohen et al. show, with plausible values for $A(0)$, that the Lewontin-Cohen model can produce values of $0 < b < 1$, something that exponential dispersion models and Tweedie distributions (Chapter 2) cannot do (Jørgensen, 1987) and for which there is evidence. However, the stationarity conditions on b suggest that deviations from $b = 2$ indicate expansion or contraction of a population and that presumably oscillating populations will have TPL slopes oscillating in time.

Cohen et al. tested the Lewontin-Cohen model's prediction of spatial TPL using data of 15 tree censuses over a 75-year period in six plots at Black Rock Forest in New York State. They started by establishing that population density was independently and identically distributed over time and plots, then estimating the mean and variance of the changes in density between censuses, $A(p,t) = N(p,t+1)/N(p,t)$ for plots, $p = 1 \ldots 6$. Tests of the time-series of $A(p,t)$ revealed no significant autocorrelation between the successive censuses ~ 5 years apart. Next, they established that spatial TPL held, using only the last five censuses ($NQ = 6, NB = 5$). In the final step, the fitted TPL ($\log(V) = -0.33 + 2.62\log(M)$) was compared to the estimates of M and V predicted by the Lewontin-Cohen model (Eq. 17.9) for the last five censuses. The TPL 95% confidence zone fell entirely within the model's 95% confidence interval, indicating reasonable agreement. The importance of this model is less the goodness of fit than the fact that unlike other attempts to understand TPL, this approach is not arrived at by simulation and provides an explicit, exact interpretation of TPL's parameters.

In the Lewontin-Cohen model, autocorrelation is 0 and successive values of $A(t)$ are independent, equivalent to white noise. Cohen (2014a) makes the case that this is the situation in which transitions from good to bad weather and bad to good weather are equally likely and have probability $\lambda = \frac{1}{2}$. In good weather, the population density is predicted to increase by a factor of $A(t) = d_1 > 1$ and in bad weather the population density falls by a factor of $A(t) = d_2 < 1$. For values of $d_1 = 2$ and $d_2 = \frac{1}{4}$, this leads to $E[A(t)] = 1.125$ and $E[\log(A(t))] = -\ln(2)/2$. For all values of d_1 and d_2 controlled by transition probability λ, the autocorrelation of the time series of $A(t)$ is $\Phi(\lambda,t) = 1 - 2\lambda$. For $\lambda = \frac{1}{2}$ we have $\Phi(\lambda,t) = 0$, equivalent to white noise, but for $\lambda \to 0$ autocorrelation approaches 1 and the spectrum of $A(t)$ is reddened.

A singularity in TPL

Cohen (2014a) calculated an exact formula for TPL slope as a function of λ, and computed $b = b(\lambda)$, for a range of different climates, $0 < \lambda < 1$. He found that as λ increases, the slope $b(\lambda)$ increases from $b(\lambda) = 2$ increasingly rapidly until at $\lambda = 0.6$ $b(\lambda) = \infty$. Above $\lambda > 0.6$, $b(\lambda)$ increases rapidly from $-\infty$ and progressively more slowly back to $b(\lambda) = 2$. As the modeling methodology assumes λ is permanently fixed at different values in different climates and the singularity occurs at a precise value of λ, species in habitats with climates on either side of $\lambda = 0.6$ might be expected to have different TPL slopes. The model illustrates what might be expected to happen to b if λ changed slowly. It also suggests that a singularity in b might occur in other models if several conditions are satisfied. The long run growth of mean and variance of population density must be functions of a common parameter θ that can take two values such that a positive and a negative population growth rate are defined in terms of θ. Also, there must be at least one value of θ for which population growth is zero. This implies that θ be good for population growth at some value and bad for population growth at another. Cohen points out that in addition to temperature and precipitation, other environmental variables such as salinity and pH meet these conditions.

Using a Markov chain to choose values for of A_i in the Lewontin-Cohen model, Cohen (2014b) showed that TPL can be obtained with a and b expressed in terms of $A(t)$ in large time. In this model, an $s \times s$ matrix $\mathbf{D}$ with values of $d_i = A(t) > 0$ in the diagonal (zero everywhere else) has s environmental states with at least two states different. The chain's $s \times s$ transition probability matrix $\mathbf{P}$ whose elements $p_{ij} \geq 0$ sum to 1 by column are the probabilities that $A(t) = d_i$, given that $A(t-1) = d_j$ is p_{ij} for all i, j, and t.

Cohen simulated the model with $s = 2$ and d_i drawn from either an exponential or a lognormal distribution and the nonzero elements of $\mathbf{P}$ uniformly distributed. For each pair of randomly generated $\mathbf{D}$ and $\mathbf{P}$ matrices, b was calculated from an exact formula derived from the largest absolute value of the eigenvalues of the products $\mathbf{D}^2\mathbf{P}$ and $\mathbf{DP}$, represented by $r[\cdot]$ in

$$b = \frac{\log\left(r\left[\mathbf{D}^2\mathbf{P}\right]\right)}{\log\left(r[\mathbf{DP}]\right)}$$

Of 10^6 simulation runs with different values for $\mathbf{D}$ and $\mathbf{P}$, the proportion of b–values falling into the categories $b < 0$, $0 < b < 1$, $1 < b < 2$, and $b > 2$ was determined (Table 17.1).

TABLE 17.1 The proportion of 106 simulated TPL slopes falling into four categories with a Markovian version of the Lewontin-Cohen population model (Eq. 17.9)

Condition	Exponential	Lognormal
$-\infty < b < 0$	5.4%	3.7%
$0 < b < 1$	5.6%	3.2%
$1 < b < 2$	49.1%	34.9%
$b > 2$	39.9%	58.1%

Data from Table 1 in Cohen (2014b).

A small proportion of simulations produced estimates of $b<0$ in line with predictions from Cohen's formulation of the Lewontin-Cohen model; also a small proportion had $0<b<1$ in line with experience. The most striking result is that, unlike a majority of models, this one predicted $b>2$ almost ½ the time: a slight overestimate in the case of the lognormal distribution. With the exception of negative b, Cohen's Markovian version of the Lewontin-Cohen model is in qualitative and quantitative agreement with experience: Chapter 13 presents several cases with negative slope.

Singularities in other models

The Lewontin–Cohen model is a multiplicative random walk that assumes population density changes by multiplication of successive positive random variables that are independently and identically distributed. Discrete time models make different assumptions to the Lewontin-Cohen stochastic model, but Cohen (2013) showed that both lead to an asymptotic spatial TPL. Discrete time models with exponential growth, $N_i(t)=N_i(0)\cdot\exp.(r_i t)$, for $i=1\ldots n$ have both mean and variance of population density as exponential functions of time:

$$M(t)=A\cdot\exp(Bt),\ A>0,\ B\neq0,\ \text{and}$$
$$V(t)=C\cdot\exp(Dt),\ C>0,\ -\infty<D<\infty.$$

Eliminating t by substitution, $V(t)=C\cdot\exp((D/B)\cdot\ln(M(t)/A))=C\cdot(M(t)/A)^{D/B}\equiv aM(t)^b$. The mean and the variance of population density in the Lewontin-Cohen model also converge on TPL in long time and Cohen shows how the two models are equivalent in this respect. Both models are multiplicative with multipliers whose distribution does not change with time. In the Lewontin-Cohen model, the multipliers are independent and identically distributed random variables. In the exponential model, the multipliers are time invariant constants. Provided the multipliers converge sufficiently rapidly to a fixed distribution, the repeated multiplications give exponential change in the mean and the variance.

Cohen points out that convergence of exponential models on TPL suggests that although behavioral interactions may be sufficient to produce TPL they are not necessary. He goes on to suggest that the quantity $N_i(t)$ need not refer to population density. The number of shares times market price of equities, for example, should also produce a TPL, which is the case (Chapter 12; Eisler et al., 2008).

Jiang et al. (2014) investigated other factors that could create singularities in b in Pielou's (1977) simple birth and death process with $\lambda=$ per capita birth rate and $\mu=$ per capita death rate that predicts population at time t:

$$N(t)=N(0)\exp((\lambda-\mu)t).$$

The difference $(\lambda - \mu)$ is the rate of population increase or decrease and the mean and variance of population density $N(t)$ at time t are:

$$E[N(t)] = N_0 \exp((\lambda - \mu)t) \text{ and } V[N(t)] = 2N_0\mu t, \text{ for } \lambda = \mu$$
$$V[N(t)] = N_0 \frac{\lambda + \mu}{\lambda - \mu} \exp((\lambda - \mu)t)\{\exp((\lambda - \mu)t) - 1\}, \text{ for } \lambda \neq \mu$$

For finite t the instantaneous or "transient" slope of $\log(V)$ on $\log(M)$, $b(t)$ is given by

$$b(t) = \frac{2 - \exp((\mu - \lambda)t)}{1 - \exp((\mu - \lambda)t)}$$

in which $b(t) \gg 0$ in small time and $b(t) = 2$ at long time for λ slightly more than μ while for λ slightly less than μ, $b(t) \ll 0$ at small time and $b(t) = 1$ at long time. Transient b, $1 < b(t) < 2$ does not occur in this model.

The Pielou model was run simulating 100 patches, each with an independent birth and death process to determine the behavior of b in the vicinity of $\lambda = \mu$ with $\mu = 0.01$ and λ increased from 0.0085 to 0.0115 in small increments. 1000 samples taken in space and/or time resulted in different TPL slopes depending on sampling plan. At all values of λ, $b = 1$ was obtained when sampling was synchronized across all patches and plots, but assumed a U-shape with increasing λ when sampling was no longer synchronized. In the latter case, the slope ranged over $1.55 > b > 1.0 < b < 1.5$ as λ increased from 0.008 to 0.012. A similar response was obtained when there were no plots, and the patch populations were sampled multiple times at different ages. This time the range of slopes was slightly different $(1.35 > b > 1.0 < b < 1.75)$ as λ was varied. In both cases the minimum, $b \approx 1.0$, occurred at $\lambda = \mu = 0.01$. In a scenario in which all patches and plots started with the same density and samples were taken at different times in different patches, the singularity in b at $\lambda = \mu$ was observed.

Fig. S1 in Jiang et al.'s Appendix shows TPLs at intervals of λ changing from strongly linear with $b = 1.10$ ($r = 0.987$) at $\lambda = 0.0085$ to no regression ($r = 0.062$) at $\lambda = 0.01$, and increasing in both b and r to $b = 4.26$ ($r = 0.953$) at $\lambda = 0.0085$ and settling at $b = 2.14$ ($r = 0.998$) at $\lambda = 0.0115$. The range of means declined from about 1 order of magnitude of M to a factor of 0.1 at $\lambda = \mu = 0.01$ and back to nearly a factor of 10 at $\lambda = 0.0115$. At the singularity, the range of variances was roughly equal to the range means but in the approach the range of variances exceeded the range of means leading to steeply positive and negative slopes as seen in Fig. 13.5 of crop yield. The increase in variance of density as population growth $(\lambda \approx \mu)$ approaches zero is counterintuitive but may be a feature of patch models with different aged populations in which changes in mean density decline but the variance increases due to random births and deaths. The model examines populations close to long-term stability $(\lambda \approx \mu)$ but does not directly address short-term fluctuations typical of most invertebrate populations.

The surprising thing is how sensitive b is to very small changes in λ in the vicinity of the singularity. As the transition is very abrupt in terms of birth and death processes, detecting a singularity may prove difficult. However, as b approaches the singularity with $\lambda \to 0.01$, the plots exhibited slight curvature, which, when observed in nature, might be interpreted as indicative of an approach to a singularity. As noted by Cohen (2014a) if a real-world singularity in b were to occur, this could have serious repercussions for all aspects of applied ecology: fisheries, forestry, agriculture, conservation, and public health.

The Lewontin-Cohen model has also been used to show a scaling of the kth versus the jth moment. Giometto et al. (2015) applied large deviations theory and finite-sample arguments to suggest that $b \approx 2$ is a statistical artifact because the sample exponent $b(jk)$ depends predictably (using the Lewontin-Cohen model) on the number of samples and for finite samples $b(jk) \approx k/j$ in long time. The model and data show that power laws exist for combinations of moments (not the central moments) up to the fourth. In particular

$$\mathrm{E}\left[N_t^k\right] \propto \mathrm{E}\left[N_t^j\right]^{b(jk)}$$

where $b(jk) \to k/j$ as $t \to \infty$. They prove this empirically with a Markovian model based on the Lewontin-Cohen model where the multiplicative coefficients $A(t)$ are determined by a Markov chain with two states and transition matrix $\mathbf{P}$ with elements p_{ij} that give the probability that $A_{t+1} = j$ given that $A_t = i$. This prediction is supported by two empirical examples: the Black Rock Forest data previously used and data of carabid beetle abundance at sites in the Netherlands (den Boer, 1977).

Giametto et al. obtain remarkable agreement between the analytically predicted ratio k/j and $b(jk)$ values computed from the two data sets. For the Black Rock Forest tree censuses all combinations of the first four moments were computed, plus fractional moments (e.g., $j = 1$, $k = \frac{1}{4}$). The results for den Boer's carabid data were even better with differences between predicted k/j and calculated $b(jk)$ averaging $< 0.2\%$. The beetle analysis, equivalent to a hybrid ensemble mixed-species analysis, eliminated site-year species samples in which one or more sites reported zero. It is not clear what effect this might have on the moment power laws, but inclusion and exclusion of zeros in regular TPL analyses can have a profound effect (Chapter 15).

The use of moments rather than the central moments normally used in TPL analysis has little effect for data with a wide range of mean and $b > 1$ as the difference between $V = \mathrm{E}[X - M]^2$ and $\mathrm{E}[X^2]$ is usually small and its effect is difficult to detect in b. However, at low slopes approaching the Poisson case, $\mathrm{E}[X^2] = \mathrm{E}^2[X]$ and $V = M$ are not equivalent.

Giametto et al. suggest that the scaling of the first four moments supports the view that sample estimates of $b \approx 2$ may be statistical artifacts. More significantly, it draws into question the validity of predictions of b made by a broad class of population dynamics models.

Dispersal distance

Point processes, describing the spatial arrangement of points in space, are used by plant ecologists to analyze plant distributions in 2-dimensional space

(Diggle, 2014). A key variable in point processes is the population intensity $\lambda(x)$ at a point x defined as.

$$\lambda(x) = \lim_{A \to 0} \left\{ \frac{\mathrm{E}[N(\Delta x)]}{A} \right\},$$

where $\mathrm{E}[N(\Delta x)]$ is the expected number of individuals in a small area, A, in the vicinity of x. Given $\lambda(x)$, if N is the number of individuals in a bounded area, X, the mean and variance of $N(X)$ are

$$\mathrm{E}[N(X)] = \int_X \lambda(x)dx \text{ and}$$

$$V[N(X)] = \int_X \lambda(x)dx + \int_X \int_X \lambda(x)\lambda(y)[g(xy) - 1]dxdy,$$

where $g(x,y)$ is a function correlating the intensities at points x and y in the area A. For a spatially random point process, the intensity function $\lambda(x)$ is a constant and $g(x,y) = 1$ and the second integral collapses leaving $V[\cdot] = M[\cdot]$, the Poisson case. When $g(x,y) > 1$, the correlation function denotes attraction between individuals, and $g(x,y) < 1$ denotes repulsion, which can lead to $V[\cdot] < M[\cdot]$ or regularity.

Shi et al. (2016) used the Matérn and Thomas processes, special cases of the Neyman-Scott process introduced to describe the distribution of galaxies (Neyman and Scott, 1958), to model plant spatial distributions and computed TPL slopes in terms of their parameters. The results for both are similar. In the Matérn cluster process, the locations of offspring are independently and uniformly distributed inside a circle of radius R centered on each parent. The theoretical pair correlation function of a Matérn process is

$$g(r) = 1 + \frac{1}{4\pi R \kappa r} h(z) \text{ where } z = \frac{r}{2R}, \text{ and}$$

$$h(z) = \frac{16}{\pi} \left(z \cdot \arccos(z) - z^2 \sqrt{1 - z^2} \right) \text{ if } z \leq 1 \text{ and } h(z) = 0 \text{ otherwise,}$$

where κ is the intensity of parent points and r is the distance between offspring and parent. The intensity of offspring is $\lambda = \mu\kappa$ where μ is the number of offspring/parent, assumed Poisson distributed. Shi et al. applied the Matérn process to the positions of 434 pygmy bamboos (*Sasa pygmaea*) in an area of $1\,\mathrm{m}^2$, which they divided into 50 random tiles containing from 0 to 36 individual plants. The theoretical Matérn means and variances fit an ensemble TPL with estimates $b = 1.50 \pm 0.016$. By fixing κ and varying the scale parameter R, different arrangements of the 434 plants were simulated and corresponding values for ensemble b computed. As R increased, A declined approximately exponentially with b similar in shape to a gamma distribution, increasing rapidly to a maximum from which it decayed exponentially (Fig. 17.3). Varying the reproductive variable μ changes the shape of the $b(R)$ curve, but the maximum value attainable remains just under $b = 2$.

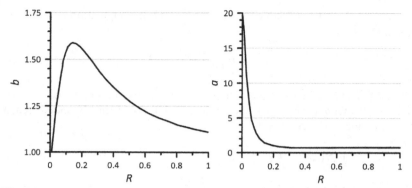

FIG. 17.3 TPL parameters A and b are related to the range of the cluster point process parameter R.

One advantage of the Matérn and Thomas cluster point processes is their parameters have biological interpretations. Specifically, the range parameter R is related to migration distance. The larger the migration distance the lower the competition, but the higher the reproductive and energetic costs. Conversely, if the dispersal distance is extremely short, clustering and competition for resources will be extreme. Thus, aggregation and migration need to be balanced (LRT and RAJT, 1977). An interesting point about this model is that b is not single valued with respect to the migration variable driving the spatial clustering.

Ideal free distribution

The ideal free distribution theory (IFD; Fretwell and Lucas, 1969) states that the number of animals in various patches is proportional to the patches' available resources. It predicts that animal populations will tend to minimize resource competition and maximize fitness. The theory assumes that all individuals are competitively equal and can assess the quality of each patch, that individuals are free to move between patches and will choose the best, or ideal, patch. A consequence of all individuals choosing the ideal patch is rapid patch deterioration via exploitation and increased competition, resulting in movement to other, previously inferior, patches. Ultimately, this game of musical chairs leads to a Nash equilibrium (Nash, 1965) in which all individuals end up with the same share of resources. Evidence for IFD is equivocal (Kennedy and Gray, 1993), although some species do distribute their foraging activities in proportion to resource availability, for example, bumblebees (*Bombus* spp.) (Dreisig (1995).

The IFD theory was invoked to account for TPL by Gillis et al. (1986) who showed that when animals form an IFD, $V \propto M^2$ corresponding to the average value of b documented for spatial TPL (LRT et al., 1983). They proposed that TPL results from resource tracking and that deviation from $b=2$ result from density-dependent effects acting on the population. Pointing out that the formula for sample variance (Eq. 2.1) can also be expressed in terms of the fraction X/M and taking expectations,

$$E[V] = \frac{1}{(n-1)} E[M^2] \cdot E\left[\sum \left(\frac{X}{M} - 1\right)^2\right], \qquad (17.10)$$

which is true only if X/M is constant for all samples, a condition for IFD. Accepting Eq. (17.10) as true, $V = aM^b$ with $b = 2$ and

$$a = \frac{1}{(n-1)} \sum \left(\frac{X}{M} - 1\right)^2$$

Assuming IFD, that is, distribution is controlled locally by behavior, Gillis et al. argued that behavior is likely to differ less between environments than habitat heterogeneity and demographics. Therefore, b should be expected to be 2 and a differ between samples of the same species taken at different times and places, in accordance with observation. The fact that TPL slope typically ranges from $1 < b < 3$ and is most commonly $b \approx 2$ can be explained by several mechanisms acting on the populations. Variations in a correlated with changes in M would result in exponents $b \neq 2$, a potential cause of which would be density-dependent changes in interactions between individuals. Other deviations from IFD, including resource defense by "despotic behavior," would cause variation in both a and b. At low densities, territorial defense would tend to increase a and decrease b as samples become more regular. Avoidance of dense aggregations or interference IFD leading to avoidance of the better patches would also tend to reduce b. Allee effects at low density would cause increased variation in the values of X/M among samples at low densities, resulting in higher a and as a increased, X/M among the samples would become more regular and lower b due to the greater variance at low densities. Slow or inhibited migration from over-populated preferred patches would tend to raise empirical estimates of b. Gillis et al. conclude by asserting that b provides a "poor comparative measure of aggregation" because according to IFD, distributions of organisms result from environmental tracking.

Agent-based models

Agent-based models, also called individual-based models, comprise a class of computational model in which the actions and interactions of autonomous agents are simulated in order to examine the system behavior. Computational techniques may include game theory, neural net or evolutionary programming or cellular automata, and frequently involve Monte Carlo methods to examine effects of stochasticity on the system. Applied to natural systems, the goal is usually to understand the collective behavior of agents obeying simple rules and to expose any emergent properties the system may possess. Agent-based simulations have been applied to computer universes of virtual organisms to find rules from which TPL might emerge.

Cellular automaton

A cellular automaton is a discrete model on a regular grid of cells, each with a set of states, such as Red, Amber, Green or On and Off, and subject to rules for changing state. The evolution of each cell is determined by the state of other, not necessarily contiguous cells. After initiation, the rules are applied to all cells simultaneously and their states changed or not as appropriate. Generally, the rules for updating cell states are the same for all cells and do not change with time, though this is not required. Cellular automata are usually deterministic, although stochastic cellular automata also exist. In *A New Kind of Science*, Stephen Wolfram (2002) described and classified deterministic cellular automata with very complex behavior according to the initial conditions and rules.

RAJT (1981b) developed a stochastic cellular automaton to test his ideas of the balance of migration with active aggregation developed in RAJT (1981a). The Δ-model (LRT and RAJT, 1977; RAJT and LRT, 1979) assumes that the tendencies for individuals to approach and separate from each other are stimulated by local population density: $\Delta = g\rho^p - h\rho^q$, where g, h, p, and q are parameters, ρ is density, and Δ is displacement that could be positive or negative, away from or toward a population center. The Δ-model has a stable point at $\rho = \rho_0 = (h/g)^{1/(p-q)}$ at which no net movement would occur. Movement was simulated in an array of cells within which the distance of movement Δ was determined by the number of agents in the cell. Movement could be internal or result in emigration to another cell. The arena of 130×120 cells comprised three types of cell: Benign in which reproduction was possible; Tolerant, sufficient for life but not reproduction; and Hostile, which was fatal. The agents were parthenogenic with variable reproductive rates, moved only once before reproduction and were not permitted to leave the arena as it was closed. Various proportions and patterns of Benign, Tolerant, and Hostile cells were simulated and in some simulations their status was static, in others status changed randomly or in response to the agent's population. Different starting populations and a range of Δ-model parameters were simulated. Simulations were iterated over many generations and ensemble TPLs were computed at intervals and temporal TPL at the end of a run.

The simulations produced ensemble TPLs with slopes from $1.04 \leq b \leq 4.21$ and temporal TPLs ranged from $1.3 \leq b \leq 2.1$. Once a population had stabilized at a spatial distribution characterized by ensemble TPL, it tended to oscillate $\pm 10\%–15\%$. Dramatic changes to the environment could reduce b toward Poisson, but recovery was generally rapid.

Changing the environmental matrix introduced nonselective mortality and mortality introduced the risk of population extinction when the environment was exceptionally hostile and the degree of mobility low. Highly mobile individuals could survive even in environments changing rapidly and drastically. In some simulations, the total population number fluctuated widely although density varied less and return time to a lower stable density after an epidemic was rapid. Comparison of direct and delayed density-dependent spatial behavior showed that movement in response to parental density resulted in more rapid buildup and higher peaks of population. All simulations, whether analyzed at an instant or continuously, produced population distributions obeying TPL.

Disappointingly, there was no discernible connection between the Δ-model parameters and either the ensemble or temporal TPL parameters. Thus, Δ-model had no predictive capacity. However, it appeared that the resulting TPLs were sensitively dependent on the population trajectories to a stable distribution in the first few generations, which were determined by the starting population and its environmental matrix. The Δ-model simulated as a rule for a cellular automaton was essentially a conservative process, tending to maintain a spatial structure once established, and not a preprogrammed density (ρ_0) as anticipated.

Arrangement in space

Perry (1996) presents an algorithm based on his SADIE analysis (Perry, 1995) for distributing agents in space. While it produces realistic spatial collections of individuals that obey TPL, it does not produce TPL populations for specified TPL parameters.

Arranging individuals on a plane so as to represent a population obeying specified TPL parameters is surprisingly difficult except for the case $b=2$. Simply assigning uniform random Cartesian coordinates between 0 and 1 places each individual on the unit square independent of every other. Taking random quadrat samples from the arena to estimate the mean and variance, and repeating the process with different total numbers, produces a set of mean-variance pairs. No matter what rule of sampling or total population size, the variance always grows as the square of the mean, as one would expect of a lognormal distribution. Generating a Poisson or pseudo-Poisson distribution with $V \propto M$ is more difficult. The arena must be divided into small cells (say 100×100) and into each cell assign a Poisson population, and then assign each member uniform coordinates within its cell. Provided the quadrat sampling the arena is larger than the cells (say 5×5), the calculated $b \approx 1$, but as the sampler approaches the cell size $b \rightarrow 2$. This emphasizes the fact made plain by others (e.g., Sawyer, 1989), that the measured value of b is not independent of the size of the sampler.

An indirect approach to assigning coordinates, based on box counting (Feder, 1988), uses the fact that changing the size of the box or quadrat, s, so that the number of the ith sample is sX_i changes TPL's intercept, $V_s = as^{2-b}M_s^b$. This can be simulated by recursively quartering the quadrats until only one individual remains in a quadrat, the last one is then assigned coordinates of the center of the quadrat. Repeating this sequence for a number of sites with different population densities chosen from a skewed distribution such as the lognormal distribution with parameters

$$\mu = \ln\left(M^2\right) - \frac{1}{2}\ln\left(aM^b + M^2\right) \text{ and}$$
$$\sigma^2 = \ln\left(aM^b + M^2\right) - \ln\left(M^2\right)$$

produces a population that obeys the ensemble TPL for a defined value of $1 \le b \le 2$, but not for $b > 2$ (Table 17.2).

To generate populations obeying TPL with $b > 2$ requires a further step. Moving individuals from one quadrat to another leaves the mean unchanged but changes the variance in a process like Perry and Hewitt's (1991) algorithm. At this point, the quadrats are small compared to the arena and contain 1 or 0 individuals. By reversing the quadrat quartering process and moving singletons to the highest density quadrats, the mean, $M = E[X]$, remains unchanged, but $E[X^2]$ and $V = E[X^2] - M^2$ are increased. A second round in which one individual in quadrats with two is moved to a higher density quadrat further increases V. This computationally intensive iterative process is continued until the desired b is reached.

The multiplicity of models generating TPL from different assumptions emphasizes how easy it is to generate TPLs in general. But obtaining predetermined values is more difficult, especially for extremely aggregated populations or processes ($b > 2$). Few models achieve this, but particularly significant is Perry's (1988) demonstrating that TPL slope defined locally could manifest at larger scale, as it demonstrates the scale invariance of spatial distribution.

TABLE 17.2 Comparison of observed and expected TPL slopes obtained with different simulation strategies

B_{input}	B_{output}	r
Uniform Cartesian coordinates		
	1.935	0.999
Lognormal random number per cell		
0.5	0.488	0.996
1.0	1.031	1.00
1.5	1.526	1.00
2.0	2.009	1.00
2.5	2.125	0.925
3.0	2.223	0.790
3.5	2.361	0.643

Adapted from Table 14.1 in RAJT (1992).

The ease with which TPLs can be generated from data streams generated by radically different models is also significant and suggests a common origin.

References

Anderson, R.M., Gordon, D.M., Crawley, M.J., Hassell, M.P., 1982. Variability in the abundance of animal and plant species. Nature 296, 245–248.

Baker, R.R., 1978. The Evolutionary Ecology of Animal Migration. Hodder & Stoughton, London.

Ballantyne, F., 2005. The upper limit for the exponent of Taylor's power law is a consequence of deterministic population growth. Evol. Ecol. Res. 7, 1213–1220.

Ballantyne, F., Kerkhoff, A.J., 2005. Reproductive correlation and mean-variance scaling of reproductive output for a forest model. J. Theor. Biol. 235, 373–380.

Ballantyne, F., Kerkhoff, A.J., 2007. The observed range for temporal mean-variance scaling exponents can be explained by reproductive correlation. Oikos 116, 174–180.

Beal, G., 1939. Methods of estimating the population of insects in the field. Biometrika 32, 243–262.

Cohen, J.E., 2013. Taylor's power law of fluctuation scaling and the growth-rate theorem. Theor. Popul. Biol. 88, 94–100.

Cohen, J.E., 2014a. Taylor's law and abrupt biotic change in a smoothly changing environment. Theor. Ecol. 7, 77–86.

Cohen, J.E., 2014b. Stochastic population dynamics in a Markovian environment implies Taylor's power law of fluctuation scaling. Theor. Popul. Biol. 93, 30–37.

Cohen, J.E., Xu, M., 2015. Random sampling of skewed distributions implies Taylor's power law of fluctuation scaling. Proc. Natl. Acad. Sci. U. S. A. 112, 7749–7754.

Cohen, J.E., Xu, M., Schuster, W.S.F., 2013. Stochastic multiplicative population growth predicts and interprets Taylor's power law of fluctuation scaling. Proc. R. Soc. B 280, 20122955.

de Menezes, M.A., Barabási, A.-L., 2004. Fluctuations in network dynamics. Phys. Rev. Lett. 92, 028701.

den Boer, P., 1977. Dispersal Power and Survival. Miscellaneous Papers 14. Landbouwhogeschool Wageningen, Wageningen, Netherlands.

Diggle, P.J., 2014. Statistical Analysis of Spatial Point Patterns, third ed. Chapman & Hall, London.

Dreisig, H., 1995. Ideal free distributions of nectar foraging bumblebees. Oikos 72, 161–172.

Eisler, Z., Kertész, J., 2006. Scaling theory of temporal correlations and size-dependent fluctuations in the traded value of stocks. Phys. Rev. E 73 (046109), 1–7.

Eisler, Z., Bartos, I., Kertész, J., 2008. Fluctuation scaling in complex systems: Taylor's law and beyond. Adv. Phys. 57, 89–142.

Feder, J., 1988. Fractals. Plenum Press, New York.

Ferris, H., Mullens, T.A., Foord, K.E., 1990. Stability and characteristics of spatial description parameters for nematode populations. J. Nematol. 22, 427–439.

Fretwell, S.D., Lucas, H.L., 1969. On territorial behavior and other factors influencing habitat distribution in birds. I. Theoretical development. Acta Biotheor. 19, 16–36.

Fronczak, A., Fronczak, P., 2010. Origins of Taylor's power law for fluctuation scaling in complex systems. Phys. Rev. E. 81, 066112.

Gillis, D.M., Kramer, D.L., Bell, G., 1986. Taylor power law as a consequence of Fretwell ideal free distribution. J. Theor. Biol. 123, 281–287.

Giometto, A., Formentin, M., Rinaldo, A., Cohen, J.E., Maritan, A., 2015. Sample and population exponents of generalized Taylor's law. Proc. Natl. Acad. Sci. U. S. A. 112, 7755–7760.

Hamilton, W.D., May, R.M., 1977. Dispersal in stable habitats. Nature 269, 578–581.

Hanski, I., 1980. Spatial patterns and movements in coprophagous beetles. Oikos 34, 293–310.

Hanski, I., 1982. On patterns of temporal and spatial variation in animal populations. Ann. Zool. Fenn. 19, 21–37.

Hanski, I., 1987. Cross-correlation in population dynamics and the slope of spatial, variance-mean regressions. Oikos 50, 148–151.

Hassell, M.P., Lawton, J.H., May, R.M., 1976. Patterns of dynamical behaviour in single-species populations. J. Anim. Ecol. 45, 471–486.

Jiang, J., DeAngelis, D.L., Zhang, B., Cohen, J.E., 2014. Population age and initial density in a patchy environment affect the occurrence of abrupt transitions in a birth-and-death model of Taylor's law. Ecol. Model. 289, 59–65.

Johnson, C.G., 1969. Migration and Dispersal of Insects by Flight. Methuen, London.

Jørgensen, B., 1987. Exponential dispersion models. J. R. Stat. Soc. B. 49, 127–162.

Kendal, W.S., 1992. Fractal scaling in the geographic distribution of populations. Ecol. Model. 64, 65–69.

Kendal, W.S., 1995. A probabilistic model for the variance to mean power law in ecology. Ecol. Model. 80, 293–297.

Kendal, W.S., Jørgensen, B., 2011. Taylor's power law and fluctuation scaling explained by a central-limit-like convergence. Phys. Rev. E. 83, 066115.

Kennedy, M., Gray, R.D., 1993. Can ecological theory predict the distribution of foraging animals? A critical analysis of experiments on the ideal free distribution. Oikos 68, 158–166.

Kilpatrick, A.M., Ives, A.R., 2003. Species interactions can explain Taylor's power law for ecological time series. Nature 422, 65–68.

Koyama, S., Kobayashi, R., 2016. Fluctuation scaling in neural spike trains. Math. Biosci. Eng. 13, 537–550.

Lewontin, R.C., Cohen, D., 1969. On population growth in a randomly varying environment. Proc. Natl. Acad. Sci. U. S. A. 62, 1056–1090.

Matui, I., 1932. Statistical study of the distribution of scattered villages in two regions of the Tonami Plain, Toyama Prefecture. Jpn. J. Geol. Geogr. 9, 251–266.

May, R.M., 1974. Stability and Complexity in Model Ecosystems. Princeton University Press, Princeton, NJ.

May, R.M., 1976. Simple models with very complicated dynamics. Nature 261, 459–467.

Nash, J.F., 1965. Essays on Game Theory. Edward Elgar, Cheltenham, UK.

Neyman, J., Scott, E.L., 1958. Statistical approach to problems of cosmology. J. Roy. Stat. Soc. B 20, 1–43.

Perry, J.N., 1988. Some models for spatial variability of animal species. Oikos 51, 124–130.

Perry, J.N., 1994. Chaotic dynamics can generate Taylor's power law. Proc. R. Soc. B 257, 221–226.

Perry, J.N., 1995. Spatial analysis by distance indices. J. Anim. Ecol. 64, 303–314.

Perry, J.N., 1996. Simulating spatial patterns of counts in agriculture and ecology. Comput. Electron. Agric. 15, 93–109.

Perry, J.N., Hewitt, M., 1991. A new index of aggregation for animal counts. Biometrics 47, 1505–1518.

Perry, J.N., Taylor, L.R., 1985. Adès: new ecological families of species-specific frequency distributions that describe repeated spatial samples with an intrinsic power-law variance-mean property. J. Anim. Ecol. 54, 931–953.

Pielou, E.C., 1977. Mathematical Ecology. John Wiley & Sons, New York.

Robbins, C.S., Bystrak, D., Geissler, P.H., 1986. The Breeding Bird Survey: Its First Fifteen Years, 1965–1979. U.S. Fish & Wildlife Service Resource Pub 157, USGS, Interior Department, Washington, DC.

Satake, A., Iwasa, Y., 2000. Pollen coupling of forest trees: forming synchronized and periodic reproduction out of chaos. J. Theor. Biol. 203, 63–84.

Sawyer, A.J., 1989. Inconstancy of Taylor's b: simulated sampling with different quadrat sizes and spatial distributions. Res. Popul. Ecol. 31, 11–24.

Shi, P.-J., Sandhu, H.S., Reddy, G.V.P., 2016. Dispersal distance determines the exponent of the spatial Taylor's power law. Ecol. Model. 335, 48–53.

Soberón, M.J., Loevinsohn, M., 1987. Patterns of variations in the numbers of animal populations and the biological foundations of Taylor's law of the mean. Oikos 48, 249–252.

Southwood, T.R.E., 1962. Migration of terrestrial arthropods in relation to habitat. Biol. Rev. 37, 171–214.

Taylor, L.R., 1971. Aggregation as a species characteristic. In: Patil, G.P., Pielou, E.C., Waters, W.E. (Eds.), Statistical Ecology. Spatial Patterns and Statistical Distributions Vol. 1. Penn State Press, University Park, PA, pp. 357–377.

Taylor, L.R., Taylor, R.A.J., 1977. Aggregation, migration and population mechanics. Nature 265, 415–421.

Taylor, L.R., Woiwod, I.P., Perry, J.N., 1980. Variance and the large scale spatial stability of aphids, moths and birds. J. Anim. Ecol. 49, 831–854.

Taylor, L.R., Taylor, R.A.J., Woiwod, I.P., Perry, J.N., 1983. Behavioural dynamics. Nature 303, 801–804.

Taylor, L.R., Perry, J.N., Woiwod, I.P., Taylor, R.A.J., 1988. Specificity of the spatial power-law exponent in ecology and agriculture. Nature 332, 721–722.

Taylor, R.A.J., 1981a. The behavioural basis of redistribution. I. The Δ-model concept. Anim. Ecol. 50, 573–586.

Taylor, R.A.J., 1981b. The behavioural basis of redistribution. II. Simulations of the Δ-model. J. Anim. Ecol. 50, 587–604.

Taylor, R.A.J., 1992. Simulating populations obeying Taylor's power law. In: DeAngelis, D.L., Gross, L.J. (Eds.), Individual-Based Approaches in Ecology. Routledge, Chapman & Hall, New York, pp. 295–311.

Taylor, R.A.J., Taylor, L.R., 1979. A behavioural model for the evolution of spatial dynamics. In: Anderson, R.M., Turner, B.D., Taylor, L.R. (Eds.), Population Dynamics. Blackwell Scientific Publications, Oxford, UK, pp. 1–27.

Wheeler, T.A., Kenerley, C.M., Jeger, M.J., Starr, J.L., 1987. Effect of quadrat and core sizes on determining the spatial pattern of *Criconemella sphaerocephalus*. J. Nematol. 19, 413–419.

Wheeler, T.A., Madden, L.V., Rowe, R.C., Riedel, R.M., 2000. Effects of quadrat size and time of year for sampling of *Verticillium dahliae* and lesion nematodes in potato fields. Plant Dis. 84, 961–966.

Wolfram, S., 2002. A New Kind of Science. Wolfram Media Inc., Champaign, IL

Xiao, X., Locey, K.J., White, E.P., 2015. A process-independent explanation for the general form of Taylor's law. Am. Nat. 186, E51–E60.

Chapter 18

Summary and synthesis

This book is primarily about the phenomenon of Taylor's power law in the biological realm, but the experience of researchers in the physical sciences, especially simulation experiments, has much to contribute to an understanding of the biological phenomena. In this summary chapter, the focus will be on the biological, especially ecological evidence, but will draw on the physical observations and experiments as well as mathematical models, where they help to understand the biological observations.

Williams' (1964) study of the distribution of species in genera, genera in families, etc. showed them to follow a common pattern well described by a single function, the log series. His title *Patterns in the Balance of Nature* underscored his belief that there is a strong pattern to nature and that this pattern is responsible for the balance we see. Williams saw the natural world as a "complex and endless interplay [in which] each species establishes temporarily an uneasy balance of numbers among all the others." The physical manifestation of this uneasy balance between the individuals in a species population and the environment, including competitors and natural enemies, is their spatial juxtaposition. Their survival and reproductive success depends, to a large extent, on where they are in relation to their resources, peers, and enemies. While the behavioral interplay creating a population's spatial pattern is most easily visualized in terms of mobile animals, it is just as crucial, but on a different time scale, for sessile animals and plants.

Environmental differences, demographic factors, and behavior all contribute toward the actual spatial pattern adopted by a population. All three factors are profoundly variable geographically. Thus, TPL could be a purely local phenomenon, but LRT et al. (1988) have shown that b varies between species far more than it does within species. Furthermore, it varies much less than expected by chance between different populations of the same species. The high degree of invariance in b is what makes it an important tool in both basic and applied ecology, in conservation and agriculture (LRT et al., 1988).

The different TPLs for western flower thrips within and above the canopy of cucumber plants (Fig. 13.2; Appendix 8.M1) were not due to demographic or geographic differences as the sampled populations differed only in where they were flying. RAJT et al. (1998) regard this as important evidence of what the power law actually measures, the nature of its specificity, and the limits of its applicability. Statistically speaking, the population was defined by where the

Taylor's Power Law. https://doi.org/10.1016/B978-0-12-810987-8.00018-5

samples were taken. A distinction between biological and statistical populations does not arise in the physical or mathematical experiments eliciting TPL. In the biological realm, where there are more degrees of freedom than individuals, the distinction between statistical and biological population is paramount. The power law's specificity over a wide range of conditions suggests that when the statistical and biological populations coincide, the power law measures an intrinsic property of the biological population's spatial structure, which the thrips example shows is largely under the control of its members. When they do not coincide, variance and mean may be totally uncorrelated.

The way in which individuals interact with each other, their enemies, and their competitors is what they have in common. The juxtaposition of individual members of a population at the local level is therefore mediated by their behavior with respect to their environment, including each other and their enemies. It seems probable that the specific distributional patterns that species adopt, and are measured by the power law, are the consequence of those behavior patterns, which contribute best to the selfish interests of the individual. Because the frequency distributions of abundance are themselves, and therefore the spaces between individuals, density dependent (Figs. 1.1 and 8.7), it is probable that spatial behavior dictating the spacing is also density dependent as proposed by LRT and RAJT (1977). Such behavior is entirely absent from the physical realm, except perhaps near the critical points of phase transition (see later).

Some models suggest TPL may be an inevitable result of sampling a certain kind of data (Cohen and Xu, 2015). While this is certainly true at one level, the deeper question is why certain natural systems adopt the skewed distributions necessary for TPL to emerge. Also, it's not clear that every realization of sampling from skewed models will always produce the consistency of results obtained with samples of natural populations taken at different times and places. Nor that the full range of slopes observed can be reproduced for either biological or physical phenomena. The great difficulty in generating very steep TPLs from statistical distributions suggests that those distributions do not capture the full behavior of TPL. That TPL consists of a suite of distributions that merge into each other as density increases (Figs. 1.1 and 8.7) confirms that there is no single frequency distribution capable of describing the spatial distributions of any organism – or tornadoes or earthquakes. To cover the full range of TPL experience, the Adès distribution uses the gamma distribution with TPL-tuned parameters and several Tweedie distributions are needed to cover the range of $b > 1$. While both approaches have practical utility, neither improves our understanding of TPL.

Various mechanisms have been advanced to explain TPL from different assumptions (Chapters 16 and 17): nonlinear maps of birth and death processes with migration (Anderson et al., 1982; Perry, 1988, 1994); spatial and temporal correlation of intrinsic and extrinsic (environmental) factors (Hanski, 1987; Ballantyne and Kerkhoff, 2005, 2007); models with random reproductive matrices (Cohen et al., 2013); and cellular automata with complex movement rules

(RAJT, 1981a,b). The convergence theorem of the Tweedie family of distributions (Kendal, 2004; Kendal and Jørgensen, 2011a,b) suggests variance-mean relationships should converge on a power law and sampling from feasible sets (Xiao et al., 2015) and analysis of primes (Kendal and Jørgensen, 2015) suggest number theoretic approaches to understanding TPL. Some nonlinear maps predicting the negative binomial distribution have $b \leq 2$ baked into them. Others either predict unobserved values, fail to predict observed values, or lack predictive capability altogether. With so many theories accounting for TPL, they cannot all be right; it's possible therefore that they are all wrong. I do not mean that they fail to capture some features of TPL's genesis, but that they are not both necessary and sufficient to reproduce the phenomenon of self-similar variance in all its postures.

The biological evidence

The spacing between individuals, represented as a density surface, has valleys and peaks with characteristic gradients that differ between species, frequently between stages of a species and sometimes between populations in different environments. Although individuals occupy points in space, density surface with its property of continuity is a convenient metaphor. Populations can be tightly clustered or loosely distributed, and the surface reflects this. The pattern adopted by a population's members leads to different surface slopes. The same species at different densities exhibits different slopes at different points in space or time. It is the consistent way the surface slope, or rate of change of density changes with density that Taylor's power law measures because the variation in numbers per unit area increases with increasing rate of change of density. Applied to nonecological populations, such as nerve impulses and stock market prices, it is the rate of change of the frequencies of those metrics that TPL measures.

For some domains, data are generated in a continuous stream and must be sampled with a window in order to compute means and variances. In the ecological domain, sampling is the only way to generate the data in the first place. Because order exists in those samples, it is reasonable to infer that it also exists in the populations sampled, because as Williams (1964) noted, "there is no known process of sampling that can turn chaos into order." This basic premise underpins this book's thesis; that if sample data show a consistent mathematical pattern or relationship it is because that relationship is a basic pattern of nature. Exactly what caused that pattern and what are its implications for the population and community is open to interpretation. It is my hope that this book, especially the collection of case studies, sheds some light on this.

At this point, it is worth pointing out that although Roy Taylor and colleagues, including me, have remained convinced that $V = aM^b$ describes an intrinsic property of the population or community, our interpretation has continued to evolve with the addition of new information and data. There have been many misstatements of our interpretations, some of which seem to have become irrevocably embedded in the lore of TPL. The title of a conference paper "Aggregation as a species characteristic" (LRT, 1971) has suggested to some that we believe TPL is a species-specific trait analogous to the absolute taxonomic distinctions between insect orders based upon wing structure, or between vertebrate orders on limb construction. We have never regarded the slope of TPL to be specific in that sense ("b ... is not a unique taxonomic character for separating species" [LRT et al., 1988]). In fact, the assertion made repeatedly by us is that b may be stage specific for species and influenced by environmental conditions, exactly as host specificity in parasitoids or phytophages is specific but variable (LRT et al., 1980; Elliott, 1981).

The 1971 conference paper was referring to the utility of b for the practical purpose of sampling for obtaining agricultural and ecological data. It is my hope that these misinterpretations and misquotations can be finally laid to rest with this volume. If b were specific in the taxonomic sense, then a single sampling program for a pest species would be adequate for that pest in all habitats, which we know not to be the case. TPL differs regionally (Trumble et al., 1989), in different habitats (LRT et al., 1979), at different phenological stages of the same plant (Drake, 1983), in different parts of the plant (Steiner, 1990), differentially within and above the crop canopy (RAJT et al., 1998), as a result of pesticide application (Trumble, 1985, RAJT, 1987), and at different times of the day (Elliott, 2002) or season (Drake, 1983).

TPL and the pattern of sampling

The kind of TPL analysis—temporal, spatial, or ensemble—depends on the pattern of sampling. Although Eisler et al.'s (2008) formalism (Eq. 17.1) is a general statement that covers all three data structures and unifies them, the appropriate TPL and the information gained from it depend on the pattern of sampling:

i. Sequential samples taken at a single site or node and averaged for NB time periods of length NQ samples. Many of the physical examples in Chapter 12 conform to this temporal TPL.

ii. Simultaneous samples taken at NB points in space at NQ times averaged over time. Temporal TPLs describe the stability of populations or communities such as the grassland examples in Chapter 6.

iii. Simultaneous samples taken at NB different times at NQ sites averaged over space usually through several population cycles. This yields spatial TPLs common with pest management data collection (Chapters 5–8).

iv. Multiple sets of NQ samples taken simultaneously once at NB sites with a fixed-sized sampler. An example is the ensemble TPL of *Balanus balanoides* on rocky shores at Millport, Scotland (Fig. 8.10).

v. Multiple sets of *NQ* samples taken simultaneously once at *NB* sites with variable sized samplers. An example is the ensemble TPL of *Arenicola* at Millport (Fig. 7.36).
vi. Some data are amenable to both spatial and temporal TPL analysis, such as the examples of rotifers in an Oklahoma lake (Fig. 9.1) and European wars (Fig. 11.12).

Variations in *b* are strongly connected to the sampling frame. In the physical realm this is a function of the size of window ($\Delta t \equiv NQ$) used as sampler, but in the biological realm it may also be a function of the target's response to the sampler. In Chapter 12, the use of a window for sampling the linear series of primes or foreign exchange quotations produced highly linear TPLs. Increasing the window size increases the length of sample, but reduces the number of windows from which mean and variance can be computed. These factors, combined with the changing temporal correlation between observations, alter the TPL obtained (Fig. 12.8A). Temporal variation in the signal captured taken on different occasions or places and forming a sequence of TPLs can also alter the slope (Fig. 12.8B). The differences observed in the behavior of physical systems' TPLs must also apply to biological systems, which being intrinsically more variable, may obscure differences.

Physical versus biological

There are some similarities, but also real differences between the physical examples in Chapter 12 and the biological systems of Chapters 5–11 that distinguish the systems and may account for differences in fluctuation scaling in biological and physical systems:

i. Physical systems permit much more rigorous data collection equivalent to censusing, rarely possible with biological systems, except national censuses (Chapter 11);
ii. Physical systems with different sample sizes (windows) produce different TPL slope with the same census data (Fig. 12.8);
iii. Biological sampling at different scales or separation of sample points (examples in Chapters 7 and 8) are equivalent to physical systems with different windows;
iv. The efficiency of sampling in biological systems may not be independent of the target's behavior, a factor not present in physical systems (Fig. 13.2);
v. Zero responses are more common in biological systems than physical systems and their exclusion can be a source of bias (Fig. 15.3);
vi. Some biological samplers have an upper limit of detection, resulting in the illusion of a decline in variance at high mean, an issue not present in the physical systems in Chapter 12.[1] Saturation of diseased leaves (Fig. 5.12A) and

---

1. Saturation may occur in some physical detectors.

gypsy moth adults in traps (Fig. 15.7) result in decreasing variance with increasing mean above a certain mean;

vii. Biological systems' data recorded as ratios often produce poor TPLs, especially if the ratio is bounded above (Fig. 13.4);

viii. Data based on irrational numbers, more common in physical systems, can also produce poor TPLs (Fig. 13.6).

Eisler et al.'s (2008) review of quantitative scaling showed that TPL slope for a variety of physical phenomena was limited to $b \leq 2$. TPLs of the physical systems are mostly temporal and there are good reasons for supposing $b = 2$ is a maximum for temporal TPL. It is expected that temporal $b = 2$ for fully deterministic systems, with $b < 2$ resulting from stochastic variation. This expectation is supported by the models of Kilpatrick and Ives (2003) and those of Ballantyne and Kerkhoff (2005, 2007). Another possibility is the dimensionality of the system. Time is one dimensional and space, two or three. The results of simulations modeling percolation and diffusion limited aggregation (Chapter 16) suggest that the fluctuation scaling of processes depends on the Euclidean dimension E in which they are embedded. If so, TPL should be restricted to dimension $\leq 2E$, which translates to $b \leq 2$ for temporal processes and $b \leq 4$ for spatial and ensemble processes. For processes in which $E = 3$, we might expect points to be volume filling and therefore produce TPL slopes up to a maximum of $b = 6$. Some empirical data support this (Figs. 5.1, 9.8, and 12.1).

Sources of range of means

A basic requirement for a TPL analysis is a wide range of means for efficient fitting and unbiased estimation. For physical systems, this can usually be achieved by suitable selection of the sample window size. For biological systems the size of sampler also plays a role, but many rare species are found only within a narrow range of densities. For them, aggregation may be random (Poisson) and, when it is not, TPL may be incapable of detecting aggregation as all samples have $V = M$. Some of the African ungulates in Chapter 10 have a narrow range of means and several are close to the Poisson line. Although these animals are not rare, their herds are widely scattered and the scale of sampling—44 transects of only 1.4 km—was insufficient to record a wide range of means. A box-counting approach at a larger scale would have recorded a wider range of means, if present.

Many species occur in a tight cluster at birth with a proportion dispersing from the natal site. Typically the dispersal distance declines quite rapidly, with only a few individuals traveling a large distance. The combination of dilution with distance and the rapid decline in dispersal from the concentration can produce a wide range of means. These distributions with long right-hand tails do not fit the Gaussian model of diffusion. The commonly observed long tails are the result of dispersal curves with either exponential or power law rate of decay (RAJT, 1980; Petrovskii et al., 2008). Weeds dispersing into a field from the hedgerow (Fig. 6.1; Appendix 6.A) are an example of range of means provided by passive dispersal. The example of barnacle larvae settling on glass plates (Fig. 8.11; Appendix 8.L) is an active process resulting in a similar pattern of exponential decline with distance.

The ability of populations to grow numerically with or without spatial expansion is frequently governed by the availability of resources. Variations in time and space of environmental factors, including weather, are the most common causes of numerical variation in abundance. Variations in weather, spatially and temporally, can provide the range of means required for analysis. Both year-to-year and site-to-site variations in abundance depend on annual reproductive performance. In temperate and sub-arctic zones, seasonality frequently imposes limits to population growth, and survival from year to year. High or low winter survival, the onset of the growing season, and the availability of resources all contribute to the within-year population growth of many invertebrates, particularly insects. Although populations can be "reset," temporal correlation is often an important component of the variation in means. Most of the examples in Chapter 8 are of samples taken at multiple sites over a period of time, many with a range of means representing abundance oscillating over time.

As intra-annual variation in abundance can provide a range of means, a high frequency of sampling can also contribute to a suitable range. This can include sampling a population over several months (e.g., gypsy moths, Appendix 8.C and potato leafhopper on maple trees, Appendix 8.M29), whether or not repeated in several years, can provide a range of means. The abundance of leeches sampled twice annually over 25 years (Chapter 7) varied widely, providing Elliott (2004) with means over almost two orders of magnitude.

Accumulation of the target at the sample site by use of an attractant trap can increase both the number caught and the range of means by effectively sampling very low densities that might be invisible to other samplers. However, a major source of bias is the interaction between the sampler and the sampled. Insects attracted to physiological traps (pheromone or wavelength of light) are not collected with the same efficiency as samplers not requiring a behavioral response: often the behavioral response is density dependent, making TPL density dependent (Chapter 15).

One surprising result is the TPL of *Longidorus elongatus* and trichodorid nematodes collected with different samplers (Fig. 7.1). Apparently, the different volumes of soil taken by the samples provided the range of means necessary to create the plot. The surprising thing is the consistency in the variance: all, but one, sampler lie close to the fitted TPL for both taxa individually and combined (Appendix 7.A).

The role of the sampler

The shape, size, and contiguity of soil voids contributes to differences in nematode spatial distribution in different soils (Appendix 7.B2). As soil structure is thought to be fractal (Young and Crawford, 1991), this suggests that the distribution of soil organisms that live in the voids or on the surface of soil particles may also be fractal. If so, then some TPLs may be measuring fractal dimension when derived from appropriate sampling schemes.

Sampling effort and efficiency

The size or duration of sampling affects its efficiency defined as the number caught (LRT, 1962). TPL and sampling efficiency are not independent (RAJT, 2018). For example, the time spent sampling the sugar cane weevil borer not only increased the efficiency, it also affected the slope (Appendix 8.M42). It has long been known that the intercept, A, is sensitive to the size or volume of sample taken as well as the type of sampler. Examples of comparisons between samplers for insects in Chapter 8 produced TPLs with different intercepts include the glassy-winged sharpshooter (Appendix 8.M32) and the kudzu bug on soybeans (Appendix 8.M37). Some also have different slopes as with the Asian citrus psyllid (Appendix 8.M20).

If the sampler is too small relative to the characteristic spacing of a population, it will always register a Poisson distribution. The sampler needs to be large enough to accumulate enough targets to get a meaningful estimate of abundance. Moens et al.'s (1999) data obtained by two samplers differing in volume were adequate for analysis but showed a difference in their TPLs. Comparison of their data from a 10-cm^2 cylindrical corer and a 1.25-cm^2 microcorer resulted in the expected difference in A ($P < 0.0001$), and the marginal difference in b ($P < 0.055$) suggests a possible density-dependent effect on sample size (Appendix 7.A2). As the extraction method was common to both sample sets, it is more likely that the difference in b was due to the effect seen when samples are taken at different separations (Fig. 7.2) with the samplers detecting spatial separations at different scales.

Changing the sample size can have a major effect on TPL. Examples include the ensemble TPLs of wireworms from Washington State, which show the effect of changing quadrat number and size (Fig. 8.4; Appendix 8.D). Variance and mean computed from groups of different number of quadrats conform to a single line, while changing the size of the sampler changes the slope (Fig. 8.4B). By contrast, changing the unit of area in density estimates has no effect on slope, only the intercept. Estimates of r, b SE(b) of wireworms in fields in England and Wales taken with a 4-in. (10 cm)-diameter core samples do not change with conversion of data from #/sample to absolute density estimates expressed as #/ft^2 or #/m^2, but A and SE(A) change with rescaling (Fig. 8.5; Appendix 8.D). Differences in sample size and sampling protocol do not necessarily change the slope or intercept: such effects may be subtle as with sugarcane weevil borer (Appendix 8.M35) or more obvious as with the rice stinkbug (Appendix 8. M42). Samples of *Aphis fabae* on beans taken at different times by different people using three protocols produced the same slope but different intercepts (Fig. 14.2A). In contrast, three methods for sampling gypsy moth egg masses produce similar but statistically different TPLs (Appendix 8.C), reflecting their relative efficiencies and possible density-dependent effects.

Converting #/sample to density can change the slope as well as the position as in Figs. 8.4B and 15.6. In the former case it results from the fact that the samples were of different size, and in the second it resulted from density-dependent efficiency. Comparing the two cases suggests that density-dependent samplers are equivalent to using samplers of different size in a survey.

The effort and time spent spatial sampling is equivalent to the window Δt in temporal sampling of physical systems. The impact on TPL can be seen comparing Figs. 7.1 and 12.8 in which b varies with separation and window size.

The sampling site

Where the sample is taken also influences TPL. Obvious examples are different host plants, for example, maize ear borer on maize and baobab trees (Appendix 8.M46). Even the variety of a host plant can influence TPL; for example, onion thrips on different varieties of soybean (Appendix 8.M6) and potato leafhopper on different red maple clones (Appendix 8.M29). The differences seen of leafhopper distribution on the maples are likely due to differences in plant chemistry as architectural differences are not discernible. Distributions differ between parts of the same plant, for example, mealy bugs on bougainvillea (Appendix 8.M16): again, chemistry may play a role in influencing the spatial distribution of these insects living on their host plant.

One important difference was seen with western flower thrips sampled above and within the canopy of cucumber plants (Fig. 13.2; Appendix 8.M1). TPL failed for the poorly defined statistical population inside the cucumber canopy that was part of the same biological population as those flying above the canopy. Variable sampling efficiency was the most likely cause. No difference in slope, and only slight differences in intercept were seen when western flower thrips were sampled above different plants using the same sampling method (yellow sticky cards, Appendix 8.M1). However, western flower thrips' distribution is not the same on different parts of the same plant (Appendix 8.M2 and 3) or on crops harvested at different intervals (Appendix 8.M4). Major population reduction due to cutting and removing alfalfa, or pesticide application changes the spatial distribution of insects on the crop, which may be reorganized as new recruits are added and survivors rearrange themselves (Trumble, 1985; RAJT, 1987).

The time of sampling

Most freshwater insects are active at night. Elliott (2002) has shown that TPLs differ with time of day in some freshwater insects: some species are more aggregated by day than night and vice versa (Appendix 8.M69). The distributions of *Verticillium dahliae* in potato fields (Appendix 5.I), potato leafhoppers on red maple (Appendix 8.M29), and chironomids on bulrushes (Appendix 8.M59) changed with season.

Transect sampling

Sampling for invertebrates usually results in their death (the earthworm example from Colombia (Fig. 7.33) is an exception), a practice not acceptable for most vertebrates. Alternatives to extraction sampling are the various transect sampling methods: strip sampling equivalent to a long thin quadrat, distance sampling estimating the distance to the target, and point transect samples taken at regular points on a transect. Because these samples often cannot be replicated ($NQ = 1$), variation is usually expressed as confidence intervals determined by bootstrapping. Examples of distance sampling with bootstrapping include the cetaceans in the seas around the British Isles (Fig. 10.14B; Appendix 8.T and U) and the African ungulates (Fig. 10.16; Appendix 8.X). The transect surveys of herptiles in Arizona (Fig. 10.8) and ptarmigan in Sweden (Fig. 10.9) were replicated.

Box counting

The box-counting approach to sampling employs samplers (quadrats) of increasing size. The TPL derived from a range of quadrat sizes directly mimics the use of boxes to estimate the fractal dimension of the coastline of Great Britain (Fig. 15.1) or the dimension of clusters obtained by diffusion-limited aggregation (DLA; Chapter 17). Kendal's (1995) example of Colorado potato beetle in Ontario and the example of *Arenicola* in Scotland (Fig. 7.36) suggest self-affine population structures. The *Arenicola* case study highlights two curious properties of the box-counting approach: it does not matter whether the quadrats overlap with the same observations counted at each box size or are nonoverlapping. Unlike the TPLs of number per fixed size quadrat, conversion of sample number/quadrat to density (number per unit area) can produce TPLs with significant curvature. An example is apparent in the conversion of the United States' population per state to number per unit area (Fig. 11.5). The states with their different areas correspond to a set of irregularly sized quadrats. Converting from number per sample to a density estimate changed TPL slope in Fig. 8.4B. However, numerous examples in Chapters 5–10 show convincing TPLs using a fixed quadrat size, and examples of a change of scale by conversion from one spatial unit to another simply shifts the plot's position in the variance-mean domain (Fig. 8.5).

Another example of changing the quadrat size changing TPL's slope is the Japanese beetle larvae taken from four plots in New Jersey (Fig. 8.7; Appendix 8.F). Box counting with a range of box sizes produces a single TPL. In this case, the plot of mean on box length and the box counting TPL are in excellent agreement (Fig. 8.7B). Subdividing the largest size (50 ft × 50 ft) into progressively smaller quadrats produces a range of TPL slopes (Appendix 8.F): it is possible there is one combination of quadrat size and number will produce the same slope as the box counting method.

Analysis in two directions

How data are arranged to compute means and variances can result in different TPLs. The abundance of the nematode parasite *Howardula aoronymphium* on drosophilids sampled on three species of mushroom at two sites is an example of hybrid ensemble TPLs in two different dimensions. Averaging the number of parasites per host over mushroom species, the TPL points are composite of site and fly species; averaging parasites per host by site, the points are composed of mushroom and fly species. The two arrangements produce TPL slopes of $b = 1.24 \pm 0.089$ by site and $b = 1.58 \pm 0.166$ by mushroom species, which differ at $P < 10\%$. Mixed-species and community ensembles computed from the same data also present TPLs in two different planes. They may be similar as with the nematodes in different forest types (Fig. 7.9) or dissimilar as with some of Ma's (2015) examples (Appendix 5.T).

Temporal versus spatial

In general, temporal and spatial TPLs derived from the same data do not have the same slopes. LRT and Woiwod (1982) found no correlation between spatial and temporal TPLs from the same data of 244 species of aphids, moths, and birds recorded at 9–84 sites over 10–15 years. The example of spatial and temporal TPLs of Japanese beetle caught on Terceira at 97–323 sites over 23 years (Fig. 8.9) is very weakly correlated in slope ($r = 0.30$) but more strongly correlated in intercept ($r = 0.60$), reflecting the changes in overall abundance year to year. However, quite by chance, some comparisons do produce the same slopes. Examples include rotifers in an Oklahoma reservoir (Fig. 9.1) and European battles between 1480 and 1940 (Fig. 11.12).

Orthogonal directions

Barnacles on two rocky shores had subtly different TPL slopes when analyzed parallel and perpendicular to the sea. The perpendicular TPL slopes were also influenced by the slopes of the shore: barnacles were more aggregated in the perpendicular than the parallel direction and the perpendicular slope on the steeper shore was significantly steeper than on the shallow shore. The slopes of analyses parallel to the sea were almost the same, as were the analyses of square instead of linear sample sets. Clearly, environmental gradients can also impact the distribution as well as abundance of organisms.

The Milky Way galaxy has a complex structure clearly recognizable in its barred-spiral shape. Its equatorial diameter is 80–100 times as long as its north-south axis, and the density declines (not uniformly because of the spiral arms) in the equatorial plane with distance from the center. This structure, like the barnacles on rocky substrates, affects TPL's slope when examined in the two orthogonal directions.

The ensemble TPLs of stars in the Yale Bright Star Catalog showed substantial difference in the distribution of naked eye visible stars in the polar and equatorial directions. That difference is less, but still evident with the higher magnification of the Smithsonian Astrophysical Observatory Catalog. In both catalogs, the stars recorded are those visible in quadrats—the shape of a truncated rectangular pyramid. Thus, the numbers are those within a volume projected onto a plane. With the exception of the Yale Right Ascension transect, all have slopes significantly greater than $b = 2$, in line with the higher dimensionality of the sample frame.

The ensemble analysis of the galaxy catalog resulted in TPL slopes not different from $b = 2$, confirming that galaxies are strongly clustered, but the lack of difference in Right Ascension and Declination indicates the degree of clustering is the same in all directions and distances into the past. TPL failed to detect any difference in the distribution of galaxies at two different resolutions. At both scales, the range of means was greater in the Declination transects than the Right Ascension transects, but neither resolution detected a difference in aggregation in the distribution of galaxies in either direction. Unlike the Milky Way galaxy, the degree of aggregation or clustering is the same in all directions even though the amount of visible matter is not isotropic.

In three dimensions

There are three ways to analyze the contents of a volume: count all the individuals in an entire block as with the earthworm study in Colombia (Fig. 7.33); project the contents of the profile onto the surface (#/unit area) as with most nematode studies in Chapter 7; or project all individuals in the profile onto a vertical plane (Fig. 7.2B). The abundance of most soil organisms declines with depth, potentially providing a range of means for analysis. While projection of the nematodes onto the surface did not distinguish between four species ($b \approx 2.0$; Fig. 7.2A), the samples projected onto the vertical plane showed differences in abundance and intercept with similar slopes, all between $2.0 < b < 2.5$. The difference between the vertical and horizontal projections suggests differences in population structure in the two directions similar to those seen in the star study. Somerfield et al.'s (2007) Scilly Isles study of mixed-species nematodes separated the samples into three depth ranges. Although the species composition varied by depth, the overall abundance was similar at the three depths and the three TPLs were virtually identical expressed as #/sample (Fig. 7.3). The presence of Poisson points suggests that some species differed in their distribution as well as abundance with depth.

Intersection of TPL and Poisson line

For a given sampler, there is a certain density below which the sampler cannot distinguish Poisson from aggregated because all sample returns are rare and therefore random. Thus, to borrow a term of Preston's (1948), there is a veil line beyond which we cannot see. Even perfectly recorded data like the sales at convenience stores (Figs. 4.2 and 11.13) exhibit this when the time between events is larger than the observation window.

Although often more difficult to see because of random noise, the transition at the Poisson also occurs with ecological data. Some reports of curvature in TPL may result from including points on the Poisson line. The discontinuity at the Poisson line of diseased blueberries (Fig. 5.11) shows a clear change in slope. Failure to allow for the change in slope at the Poisson line will underestimate the degree of aggregation. The example of Rocky Mountain wood tick (Appendix 8.M71) has only two Poisson points; their inclusion underestimates b by ~4%, enough to reduce the efficiency of sampling plans based on TPL. Milne's data of sheep tick (Fig. 8.12) also show samples with Poisson distributions overlapping the aggregated distributions obtained earlier in the season when populations were increasing.

The effect of scale

In general, samples are taken at scales larger than the behavioral scale, so much of the spatial detail resulting from behavior is lost. To capture spatial behavior and fully describe the distribution of counts in quadrats, the position is also needed. As Engen et al. (2008) point out, the size of the sampler in relation to the scale of local point processes is fundamental to interpreting spatial scaling. Perry's (1995) SADIE analysis attempts this, but at a fairly coarse resolution.

There is a clear effect of scale on TPL results, but as Clark et al. (1996) found, the effects may not be consistent. One possible reason for this is that both the size of sampler and the physical closeness of samples influence TPL by interacting with the population's natural spacing in different ways. However, differences associated with the sampler are real: both size of sampler and scale of sampling affect the TPL observed as seen with the wireworm and Japanese beetle results (Figs. 8.4 and 8.6; Appendices 8.D and 8.F). This is seen most clearly in a comparison of these studies with the nematode study in Fig. 7.2 in which all samples were taken with the same-sized quadrats at different separations. The TPL slopes of Japanese beetle larvae increased from $1.16 < b < 1.44$ as the quadrat size is increased (Appendix 8.F) just as the slope increased with separation of samples for *Longidorus elongatus* and *Heterodera avenae* in Fig. 7.2. The different pattern obtained for trichodorids may be a consequence of sampling multiple species.

The comparison of scale with the birds of the Audubon Christmas Bird Count (CBC) also showed a change in TPL slope with scale: b increased with scale from state to region to nation (Table 10.2). The TPL slopes from the British Trust for Ornithology (BTO; Fig. 10.12) data averaged less than the US state analysis, consistent with the observation that b increases with geographic scale.

It was at very small scale that Clark et al.'s (1996) results were inconsistent. Samples were all taken with the same quadrat size, but the scale was investigated by combining samples to change the quadrat size at each geographic scale, effectively confounding the effects of scale and separation.

The structure of populations often changes with life stage. In insects, b often declines with stage in concert with reductions in abundance. Egg masses cause initial aggregation to be high, but the mobile stages tend to disperse as they search for food, so that pupation sites are usually less aggregated than larvae or eggs. The higher aggregation of emerging winter moth adults than prepupae suggests congregation of prepupae at limited pupation sites (Fig. 8.2; Appendix 8.B). Aggregation of adults may be high as they come together for mating. The changing distribution with life stage can be equivalent to change in scale as the mean separation changes through time. This will not normally be obvious as the sampling method and scale used will be appropriate to the stage. Changes in b with life stage reflect the real changes in distribution as populations develop, but they also are affected by the scale of sampling, which differs with sampler. The measured spatial distribution of gypsy moth egg masses, pupae, and adults (Fig. 8.3; Appendix 8.C) reflect both the differences in life style and differences in sampling procedure. Egg mass density is expressed per unit effort or per tree; density of pupae as number per tree and adults as number per trap, both stages were aggregated by the sampler and hence operating at different spatial scales from each other and the egg mass surveys.

The differences in TPL between life stages of leeches were much larger than those of the same stage in the different habitats of the shallow stony and deep slow-moving reaches of a Cumbrian stream (Appendix 7.N2). Similarly, Elliott's (1986) calibrated sampling methods for trout larvae in streams, TPLs changed with age, stage, and habitat (Fig. 10.5; Appendix 10.H). By contrast, TPLs of adult sea trout caught by rod and commercial netting differed regionally but were indistinguishable within a region (Fig. 10.6; Appendix 10.I). The physical differences of rivers in different parts of England and Wales imposed differences on the spatial distribution of the fish and/or the distribution and scale of fishing.

Other causes of change of scale include the host plant architecture. Difference in TPL of maize ear borer on baobab trees and maize may be due to scale and spacing differences in host architecture as well as plant chemistry (Appendix 8.M46). Only differences in intercept of maize ear borer TPLs distinguished maize and three other tree hosts. Catastrophic population reduction by cutting crops not only imposes mortality on its insect community but also changes the geometry of the habitat (Appendix 8.M4).

Repeatability is also related to scale. If samples are always taken the same way, we generally get the same slope and often the same intercept. Examples include the western flower thrips over different greenhouse crops (Appendix 8. M1), moth larvae in sorghum fields in Texas' coastal plain and central Oklahoma (Appendix 8.M53) and the three life stages of a leech in an English stream

(Appendix 7.N2). Habitat differences played no part in the distribution, only in the leeches' abundance and regional differences in sorghum management and cultivar, as well as the larger climatic differences had no effect on either TPL parameter.

Super aggregation

For ensemble and spatial distributions, the evidence suggests that $b > 2$ is neither uncommon nor aberrant. In the collection of examples in Chapters 6–11, ~20% of biologically based data TPLs have slopes $b > 2.5$ and ~5% $b > 3$. More than half the TPLs of birds at very large scale based on the CBC data have $b > 2$ and 15% of data of the BTO surveys also have $b > 2$. While some certainly fail data quantity requirements of NQ and $NB \geq 15$, not all do, and some that do fail are probably not badly biased. In addition, by far the largest number of TPL analyses, both physical and biological, were conducted using ODR, which underestimates the functional slope unless the correlation between $\log(V)$ and $\log(M)$ is near perfect. Thus, many examples in the literature probably underestimate both b and the number of cases of $b > 2$.

The Rothamsted Insect Survey estimates of b for aphids ($1.29 \leq b \leq 2.95$) and moths ($0.95 \leq b \leq 3.32$) sampled over Great Britain (LRT et al., 1980) satisfied the requirements for $NQ > 15$ and the number of points, $NB = 11$ and 13, respectively, was adequate, if not ideal. The slope of TPLs of BTO bird data ranged over $0.98 \leq b \leq 2.75$ with NQ and $NB > 15$ (Fig. 10.12), and the analyses of the CBC bird counts range from $1.5 \leq b \leq 4.6$ were obtained from $15 \leq NQ \leq 212$ and $15 \leq NB \leq 56$ (Fig. 10.13). Both large-scale bird surveys counted only the birds actually observed on the sample dates. Birds present but not seen were not included as zeros, so the estimates are certainly inflated (Fig. 15.3). The aphid and moth data include zeros, so their slopes are not inflated; the birds' slopes may be inflated by as much as 50%, suggesting the maximum slope for birds is probably $b \approx 3$.

High values of b are often associated with low NQ or NB. For example, TPLs of gymnamoebae (Fig. 5.1) with $NQ = 4$ and three species of foraminifera (Fig. 5.4) with $NQ = 5$ have slopes $2.1 \leq b \leq 2.8$. *Phytophthora* in soil with $NQ = 8$ also have high estimates $2.23 \leq b \leq 3.12$, and so do *Phytophthora* spores in the air with $b = 2.23$ and $NQ = 16$ (Appendix 5.O). Other examples with high b with adequate NB and NQ include the observations of cetaceans off the Azores ($b = 2.66$; Fig. 10.14A), ciliates in high salinity seawater in the China Sea ($b = 2.6$; Fig. 5.6B) and several of Ma's (2015) studies of the human microbiome (Appendix 5.T). The number of sexual partners is also extremely aggregated ($b = 3.12$; Fig. 11.8A; Appendix 11.K). A Zipf/Pareto analysis of similar data (Fig. 11.8B) relates the number of partners to a scale-free network, showing evidence of increasing returns in obtaining partners.

Physical examples with high b are the data of dispersal of cyanogen from the nucleus of Comet Hale-Bopp ($b = 3.84$; Fig. 12.1), earthquakes ($b = 3.38$;

Fig. 12.4), and tornadoes ($b=5.06$; Fig. 12.5), all of which have NQ and $NB > 15$. The stars in the Milky Way galaxy also exhibit TPLs with slopes significantly >2.0.

The dimensionality of the environment is significant: all four physical examples exist in three-dimensional environments. The birds also exist in three dimensions, at least some of the time, and the aphids and moths were caught in flight. Whales, foraminifera, and gymnamoebae live in water, and the *Phytophthora* in soil and air were also obtained in three-dimensional environments. The high values of b estimated in these cases are likely a reflection of the three-dimensional world that they inhabit.

Trophic interactions

The interspecific interactions of competition, predation, and parasitism change the numbers in a population, but their presence or absence appears to have little influence on TPL. Because the strength of trophic interactions can change abundance and may be a major cause of variation from place to place or time to time, they can provide a range of means for TPL analysis.

Predation

The different predation pressures imposed by anglers and commercial fishermen of sea trout had no influence on TPLs of trout caught in English and Welsh rivers. However, regional differences in TPL may have been due to environmental differences in the rivers or differences in the level of predation in morphologically different rivers (Elliott, 1992). Gallucci et al.'s (2005) study of predatory and prey nematodes on a mud flat (Appendix 7.C2) showed the predator *Enoploides longispiculosus* to be only slightly more aggregated than Poisson, but a collection of several potential prey species was significantly ($P < 0.07$) more aggregated than the predator.

Including parasitoids as predators, the evidence in Chapter 8 finds almost all examples of predator and prey have the predator with lower TPL slope. Only one case is (not significantly) less steep (Appendix 8.M24). Examples of steeper prey TPLs include predatory mirid bugs on psyllids, leafhoppers and mites (Appendices 8.M25, 8.M26, and 8M79), predatory thrips on mites (Appendix 8.M74), mites on leafhoppers and mites (Appendices 8.M2, 8.M74, 8.M75, and 8.M78-80), and parasitoids and beetle predators on psyllid hosts (Appendix 8.M21). The examples of parasitoids *Ooencyrtus kuvanae* on gypsy moth egg masses and *Agriotypus armatus* on the larvae of caddisfly *Silo pallipes* are highly significant ($P < 0.001$). The evidence for prey and host species being generally more aggregated than their predators and parasitoids, including parasitoids, supports Hamilton's (1971) model of frogs trying to avoid a snake predator by aggregating.

Parasitism

While the evidence for higher aggregation in arthropod prey than predators is strong, the evidence with parasites (excluding parasitoids) is not so clear cut because the sampling unit is the host. The range of slopes for mixed-species TPLs of parasites $1.5 < b < 2.0$, averaging ~ 1.7 (Shaw and Dobson 1995; Morand and Guégan, 2000) and TPLs for individual species depend strongly on whether unparasitized hosts are included in the analysis (Figs. 7.22 and 7.31). A comparison of mixed-species parasite TPLs found parasites, free-living parasitized species, and free-living unparasitized species differed (Lagrue et al., 2015). TPL for parasites was steeper than for either free-living group and steeper than Shaw and Dobson's result. The parasitized group was also steeper than the unparasitized group suggesting the possibility that parasite status can change spatial behavior. Such a phenomenon is not unprecedented: for example, male *Macrosteles fascifrons* infected with the aster yellows phytoplasma are much more mobile than uninfected males (Hoy et al., 1999).

Competition

The act of competition does not seem to change TPLs slope, although competitors are likely to have different TPLs reflecting differences in the way they partition their space. Park et al. (2013) found differences in b for members of a trophic group indicating different degrees of aggregation of competing nematodes. For example, the range of slopes for genera in the two largest trophic groups, bacteria and plant feeders, was much greater than the range of slopes between trophic groups (Appendix 7.E). While their work showed differences between competitors, Park et al. could not tell whether competition changed TPL.

Sea nettle jellyfish off the Oregon coast compete with salmon for zooplankton. Ruzicka et al. (2016) reported salmon stomach contents were markedly lower in years of high abundance of sea nettles and a negative relationship between abundance of sea nettles and salmon. Both sea nettles and salmon were highly aggregated: sea nettles $b = 2.56$ (Fig. 9.13; Appendix 9.M), salmon species and age classes ranged $1.75 < b < 2.40$ (Fig. 10.2; Appendix 10.C). The small difference in slopes of the sea nettle TPL above and below the Poisson line could correspond to high and low competition years, but is more likely the result of the different sampling methods used in high and low years. There is no indication of a change in slope in years of low and high abundance of salmon. Seasonal differences are associated with the migration from natal rivers.

Park et al. (2013) suggested that competitors might distribute themselves differently to avoid competition. The example of spring peeper and American toad tadpoles competing for resources in vernal and semipermanent ponds produced parallel lines (Appendix 10.M). The spring peeper's intercept is significantly higher than the toad's indicating toad abundance was much less variable than the peeper's at any given density. The indistinguishable rates of change of spatial variance likely reflect the closely similar environmental influences on both species. That the peeper's abundance is more variable than the toad's suggests that the frequency distributions of abundance differ at the same density does not contradict Park et al., but does suggest that TPL is not very sensitive to competition. The studies of beetles in stored products (Elmouttie et al., 2013; Carvalho et al., 2013) also offer no evidence of sensitivity to competition, and Ramsayer et al.'s (2012) experiments confirmed the absence of differences between competing and noncompeting populations with bacteria cultures.

Mixed-species and community TPLs

Ma (2015) suggests that because intra- and interspecies interactions occur simultaneously at the population and community levels in both space and time, TPL and his four extensions are different manifestations of the same mechanisms and processes occurring in the community of interacting species. Furthermore, he suggests, and I agree, the several TPLs are emergent properties of the inter- and intraspecific interactions. Mixed-species plots with each point a taxon show spatial or temporal stability in relation to the rest of the community (Ma's Type III and IV TPL extensions). Community TPLs (Ma's Type I and II extensions) in which the mean is expressed as the average number of individuals per species are sensitive to changes in community dynamics in relation to total abundance. Ma found the human microbiome communities to be highly aggregated with Type I to II extensions having community averages of $b \approx 2.2$ and 2.5 for spatial and temporal, respectively. The spatial and temporal species TPL slopes and Type III & IV mixed-species TPL slopes were lower, averaging $b \approx 1.7$ similar to Shaw and Dobson's (1995) findings for worm parasites on vertebrates.

Mixed-species TPLs quantify the spatial or temporal characteristics of species in relation to their community (Shiyomi et al., 2001). Deviation from best fit line of a species highly variable in space or time will be above the line representing the community norm. Their distance from the Poisson line denotes the absolute degree of spatial heterogeneity or temporal stability. Guan et al. (2016) used mixed-species TPLs to assess rangeland species stability in dryland grass biomes. Deviations of demersal fish in a Brazilian bay (Fig. 10.3) identified territorial and schooling fish as different from the majority of the community. In community analyses, the average number per species is related to community diversity. Points above the best fit line indicate communities with a skewed distribution of species abundance in relation to the number of species. Such a pattern is exemplified by Fisher et al.'s (1943) log series, investigated by Williams (1964). The more even the distribution of individuals per species, the lower the variance in relation to the mean resulting in points below the best fit line.

A difficulty with both mixed-species and community TPLs is the data are generally based only on the individuals actually observed. Such samples generally excluded zeros as species not in the list are not included in the estimation of mean and variance, possibly inflating the slope.

Mixed-species analyses by Polley et al. (2007) excluded zeros; reanalysis with and without zeros showed b to be inflated by ~20%. Excluding zeros artificially inflated b, bringing the TPL plot into contact with the $V = nM^2$ upper limit (Fig. 6.12A). A similar inflation probably occurred with Murthy et al.'s (2016) mixed-species analysis of birds in urban and nonurban habitats as only birds actually seen on the survey day are included in the Audubon CBC surveys. Murthy et al's plots based on relative abundance show the limit $V = nM^2$ clearly. The points are below the Poisson line because they are relative abundance. In both studies, the estimate of n ($= NQ$) obtained by extrapolating the maximum variances to the $M = 1$ ordinate agreed with the number of samples actually used. In surveys in which NQ is variable, the cutoff is not visible (Fig. 6.12B).

The upper limit in variance is only obvious in mixed-species plots possibly because individual species rarely reach densities high enough for variance to reach the cutoff. LRT et al. (1980) failed to locate a variance limit at high mean, which was later identified by LRT and Woiwod (1982) only in a mixed-species analysis (Fig. 15.2). That populations rarely reach the forbidden zone suggests that a feedback prevents a species reaching so high a density. If so, the feedback must operate long before the carrying capacity is reached, either by mortality or movement away from a habitat being rapidly degraded.

Human demographics and sociology

Human ecology and behavior are obviously biological in nature, so finding TPL applies to national census data is not surprising. The US has conducted 23 censuses since the first in 1790. Since the first census, citizens and residents have been asked their ethnicity, and gender since 1830, providing a trove of information suitable for TPL analysis.

The spatial TPL of population per state (Fig. 11.5A) is highly linear, with minimal serial correlation, and super-aggregated ($b = 2.23$). It is similar to the box-counting approach of *Arenicola* (Fig. 7.36) or Japanese beetle larvae (Fig. 8.7) but with irregularly shaped and unevenly sized quadrats. Ignoring the fact that the number of states (quadrats) increased from 15 to 50 over 160 years, the growth of population within the quadrats is equivalent to increasing the size of quadrats in an ensemble study with a static population. Converting number per quadrat to density produces a curved TPL indicating that even as the population grew, the variation between the states declined.

Using county populations to compute ensemble TPLs (Fig. 11.6) for each census year ($NB =$ the number of states) is also equivalent to box counting with irregular, uneven quadrats. Ensemble b increased for the first century as a result of the increasing difference in state populations as territory expanded from 1790 to 1900 creating larger differences in number per county and therefore increased variance. Since 1900, b has decreased as the population increased from 74.6 million to 309 million in 2010 with small territorial increase. As only 5 states have been added to the Union since 1900 and none in the last 60 years, the decline in b since 1900 is due to the gradual homogenization of number per county that has occurred with population growth and the spread of metropolitan areas into surrounding counties counteracting the trend of migration from rural to urban communities.

The most obvious effect in the trajectory of ensemble TPLs is the difference between African American and European (and other ethnicities) ensemble TPLs. The slow growth in African American's ensemble b has only recently reversed, in line with the total TPL suggesting that the African American population was always less heterogeneous at the county level than the general population except in Antebellum years when African Americans were concentrated in the South.

Cohen et al.'s (2013) study of Norway's population shows clearly how a single sample unit can bias TPLs: the dominance of Oslo, the largest political unit, upset the TPL analysis, which different weighting policies failed to eliminate completely. Their "local" TPLs intended to present instantaneous ensemble TPL were less well defined than the county-level ensembles of the US censuses largely because of the greater scale and resolution of population data in the US. Cohen et al's use of different weighting policies emphasizes the importance of recognizing different constituencies in census data.

As human populations are aggregated so are crimes, and the types of crime differ in their TPL slopes. The aggregation constants of different classes of crime not differing between policing neighborhoods showed criminal behavior to be very uniform geographically while the national comparison showed classes of crimes to be aggregated differently. That most crimes are more common in large metropolitan areas than small villages reinforces the view that per-capita crime rates are density dependent.

The spatial and temporal analysis of battles (Fig. 11.12) shows human conflict to be highly aggregated in space and time. The small difference in spatial and temporal slopes is not significant, suggesting the two dimensions are equivalent. Richardson's law of conflict deaths (Fig. 16.4) produces a result similar to those described by $1/f^\alpha$ noise and the Zipf/Pareto law, which includes the frequency of network connections (Fig. 16.2) and sexual contacts which are known to be described by TPL (Fig. 11.8). Collectively, these results are strong indicators of a functional connection between TPL, $1/f^\alpha$ noise and the Zipf/Pareto law.

Other uses of TPL

Most studies of the same organism in different environments have shown little or no difference in their TPLs, which may explain why most efforts to detect the effects of pollution by changes in distribution have been unsuccessful. Most of the studies in Chapter 7 of nematodes in unimpaired and impaired waterways showed some difference in abundance of key trophic or cp-classes but failed to show differences in spatial distribution. The only difference evident in mixed-species analyses of nematodes in disturbed and undisturbed habitats was of nematodes in wetlands (Fig. 14.6); no difference was detectable in forests and agricultural land. An important exception is the variation seen in the plant-parasitic nematode *Tylenchorhynchus zeae* in a field receiving effluent from a cotton mill and a control field (Fig. 14.5). For most species minor stresses do not appear to affect their spatial distribution, but this example may be evidence for a stress level above which populations cannot respond fast enough to maintain their preferred spatial distribution.

Most of the environmental assessment examples in Chapter 7 were not intended to be analyzed by TPL—the examples in Chapter 14 were—so the sampling plan was not designed for TPL analysis. Very often, sufficient samples were taken but were composited to reduce variability in the hope of detecting differences between impaired and unimpaired habitats. Failure to properly account for variance in hypothesis testing usually reduces the power of the test. Monaghan (2015) use of TPL to transform the raw data of the diversity and abundance of indicator species in UK rivers and streams (Chapter 14) emphasizes the importance of including and understanding the natural variation in sample data, which can only be obtained with an adequate number of samples.

Not only do insufficient number of samples result in poor TPLs, but a narrow range of densities reduces confidence in the estimates. The examples of reptiles in Florida and ungulate herds in Africa demonstrated differences between species but have narrow ranges of means drawing into question those differences. Samples of opportunity like those taken of reptiles are likely to be poorly conditioned, and only long runs of data like those of cetaceans off the Azores (Appendix 10.S) and fin whales in the Mediterranean (Appendix 10.W) can produce convincing TPLs. The highly organized surveys of cetaceans around the British Isles (Appendix 10.T–V) produced excellent data, but the absence of replication required variance estimates to be made by bootstrapping.

Gilles et al.'s (2016) harbor porpoise model based on 27 studies produced an ensemble TPL in close agreement with the original abundance data, but with estimated variance half the observed variances of porpoise abundance (Fig. 10.15). The agreement in TPLs of a population model and sample data of the abundance of two species of jay (Fig. 10.10) was also very good, although in this case the model overestimated the variance slightly. The purpose of Samaniego et al.'s (2012) study was to compare the effect of social behavior on the jays' population dynamics and spatial distribution. It was the first comparison of model and data by TPL. Standard techniques for comparing model fit to observations include regression of observed against predicted, correlation coefficient, Nash-Sutcliffe efficiency, and several tests based on mean absolute error, mean square error, and root mean square error. Each has properties comparing observed with predicted, but are all single valued and examine only one source of discrepancy. The use of TPL can assess correlation, accuracy, precision, and confidence by comparing the intercepts, slopes, and residual variances, providing information useful in recalibrating models if necessary.

When TPL doesn't work

The failure of western flower thrips in a plant canopy to obey TPL (Fig. 13.2) emphasizes the need for the target population to be sampled consistently—in RAJT et al.'s (1998) words "the same statistical population." To this restriction, we should add "with an appropriate sampler at an appropriate scale."

When authors suggest that TPL is inappropriate or does not work, we should ask "were the data collected using the right tool, at the right scale and from a single, stable statistical population?" In ecology and especially in entomology, nematology, and plant pathology, no one deliberately samples their target organism with an inappropriate tool or at a scale not likely to give useful estimates of population distribution and abundance, but as we have seen changes in scale can and frequently do make a difference to the observed TPL.

Some of the failures of TPL to fit sample data in Chapter 13 resulted from inadequacies of the sampling plan. The studies of marine nematodes around Mallorca ($NQ = 5$, $NB = 7$; Fig. 13.1A) and the Darwin Mounds ($NQ = 2$ or 4, $NB = 5$; Fig. 13.1B) were not made with TPL in mind, but illustrate the problems with small sample sizes. Many of the case studies in Chapter 7 were not intended to be analyzed by TPL, but fit well nonetheless. Some actually produced convincing TPLs despite sample sizes as small as $NQ = 2$ (Appendix 7.Y). That they produced TPLs reflects the general robustness of the analysis even though estimates were possibly biased. Their use for selecting a transformation is unimpaired by the small sample size, but quite inadequate for comparative purposes and questionable for developing sampling plans, especially if they result in a preponderance of Poisson points.

Two situations in which TPL fails to fit or fits poorly are of means and variances derived from ratios and numbers that are not countable, i.e., numbers that are not reducible to integers or ratios of integers. Proportions are bounded by 0 and 1 and the shape is generally sigmoidal with the approach to the upper asymptote inevitably having variance approaching zero. At the lower end, variance approaches zero as the abscissa approaches zero, so that in the exponential part of the sigmoid or logistic curve, TPL may fit quite acceptably (Fig. 13.3). This condition is indistinguishable from trap saturation (Figs. 5.12 and 15.7), and the two situations are linked by the fact both possess an upper boundary. The failure of LRT et al.'s (1980) study of aphids, moths, and birds to find evidence in natural populations of a saturation effect strongly supports the conclusion that natural populations rarely reach their carrying capacity. Sex ratio is also bounded and, with few exceptions, does not vary much from 50%. However, at particular points in time and/or space, male and female abundance may vary over a wide range. As seen in Fig. 13.4 that variation translates to enormous variation in sex ratio per sample but over a very narrow range of sample means.

Cohen's (2014) discovery that under certain conditions the Lewontin-Cohen model (Eq. 17.9) predicts a singularity in b is intriguing. The fact that two data sets of marine nematodes produce very steep slopes, one positive (Fig. 7.29) and one negative (Fig. 13.1B), would support this surprising discovery if bolstered by more evidence. A simpler explanation exists for some other examples of very steep, positive or negative slopes of unbounded ratios of TPL compliant data: the degree and direction of correlation between numerator and denominator (Eq. 13.1). If the denominator is a constant, generally the TPL plot will simply move in the V-M domain (Fig. 8.5). The correlation between numerator and denominator can have a profound effect on the variance of a ratio (Eq. 13.1) with the result that variance can grow much faster than the mean, resulting in a TPL with very high slope, or no slope (Fig. 13.2). Both crop harvested and area planted obey TPL very well (Figs. 6.15–6.18; Appendix 6.V and W), although changes in policy and/or technology can change the TPLs with or without a period of transition. The yield or harvest per unit area may completely fail to fit TPL (Fig. 13.5A and C) or fit unconvincingly (Fig. 13.5B and D). In all four cases, TPLs fit the yields but failure to examine the plots or residuals could lead to erroneous conclusions. Other factors such as price support manipulating either acreage or production can further distort the relationship between variance and mean. Policy decisions can distort distributions and interfere with natural patterns.

Countable and noncountable number series

The other major class of number series for which TPL seems not to apply is with noncountable numbers. Fig. 12.6 shows examples of TPLs for daily precipitation averaged over 72 sites in North America. Annual estimates of TPL slope of actual and simulated daily rainfall decline slightly over 200 years, while the intercepts increase. The correlation coefficients indicate very good fits. In contrast, the variance mean plots for temperature over the same time period are essentially uncorrelated ($-0.6 < r < 0.2$) with $-10 < b < 5$ (Fig. 13.6). As remarked in Chapter 13, the major difference between rainfall and temperature is that temperature is an uncountable number where rainfall is countable. Thus, TPL may work only for countable numbers or numbers reducible to countable numbers. The unbounded quotients of crop yield are rational numbers derived from integers and are therefore rational numbers by definition. The average of rational numbers is also rational. The variances of ratios depend on the correlation coefficient, which is bounded, and can take any value between, $-1 \leq r \leq 1$.

Models

Models may be split into two broad groups analogous to the distinction between inductive and deductive reasoning. While deductive reasoning argues from the general to the specific, inductive reasoning argues from specific cases to general rules. Similar to induction, analytical or strategic models investigate general principles by sacrificing detail and creating a general framework. By contrast, pragmatic or tactical models are more like deductive reasoning by describing systems in detail in order to answer specific, usually applied, questions. Tactical models are almost always simulation models, while strategic models may use simulation to expose the general principles. The models in Chapter 17 all required some simulation, but their objectives were largely inductive in nature.

No matter the objective, an important purpose of models is to organize and interpret phenomena and if possible make predictions testable by experiment or new observations. When the predictions, whether strategic or tactical, disagree with empirical data, both model and data should be examined carefully. But rigorously collected and analyzed data should be impugned with care and the model assumptions questioned. The case of $b > 2$ not being supported by some models should be a salutary lesson in not taking model predictions too seriously. In many models, the maximum $b = 2$ is baked into the modeling paradigm as when the negative binomial distribution is assumed explicitly or is predicted by the model.

All the models producing power laws have in common the iteration of an equation or a set of rules to generate a distribution in space and/or time:

i. Iterated rules for spatial organization—cellular automata (Wolfram, 2002) and spatial simulations (RAJT, 1981b);
ii. Iterated rules in demographics—Markov chains (Anderson et al., 1982; Cohen, 2014);
iii. Fractal generators (Mandelbrot, 1982)—geometric fractal structures have fixed Hausdorff dimension and b;
iv. Chaos generators (Perry, 1994) can produce high b—the result of a high rate of change of density;
v. Networks can have extremely high b (Anderson and May, 1988).

Many models predict a maximum value of $b = 2$ for both temporal and spatial TPLs. Exceptions to models predicting a maximum value of $b = 2$ are Perry's (1988) lattice model and chaotic (1994) models, the Δ-model (RAJT, 1981a, b), and various manifestations of the Lewontin-Cohen model (Cohen et al., 2013; Cohen, 2014; Jiang et al., 2014; Giometto et al., 2015). Many models act in a virtual space in which parallel, interacting, temporal streams are iterated and interact through time. Hanski's (1980) and Perry's lattice models and the Δ-model are embedded in coordinate space permitting directed movement.

The problem with TPL is not how to account for or model $b = 2$, but to model and explain $b \neq 2$, especially $b > 2$, as many model systems work only for $1 \leq b \leq 2$. Strict application of statistical distributions to ensemble and spatial data would have variance scale with the mean to an integer power, 1, 2, or 3, depending on distribution. Statisticians accepted that variance of field data might not be proportional to the mean or its square and developed transformations for this situation. However, the idea that data should really conform to a theoretical distribution may have inhibited statisticians from recognizing the reality of variance-mean relationships with fractional powers. The existence of powers >2 describing spatial variation has not been universally accepted. The evidence for spatial distributions so acutely clumped that $b > 2$ is incontrovertible. Why the reluctance to accept this as fact? One factor is the extreme difficulty in modeling systems with $b > 2$. Another is the relative scarcity of long runs of variance-mean data with $b > 2$ because the range of means between the Poisson line and $V = nM^2$ is necessarily comparatively narrow. There are few super-aggregated populations with a convincingly long run of means. The previous section enumerated several examples in which high values of b occurred in three-dimensional environments. The dimensionality of networks like sex partners is not obvious.

All the models, whether simulating in time only or including space, have one thing in common, they all iterate an equation or set of rules in time and space. They do not all predict a range of TPL slopes in agreement with empirical data. Some, such as the geometric models creating fractal structures, the Koch snowflake, Cantor dust, and Sierpinski pyramid, have fixed fractional Hausdorff dimensions and a single variance mean relationship. Simple demographic models in time only (Anderson et al., 1982) generate negative binomial distributed populations having a maximum value of $b = 2.0$. Anderson et al. (1982), Hanski (1987), and Sawyer (1989) all asserted that $b = 2.0$ is the theoretical maximum value for b despite empirical data to the contrary.

Perry's lattice model in coordinate space and his chaotic model in virtual space succeeded in simulating $b > 2.0$. So also do the series of models based on the Lewontin-Cohen equation (Eq. 17.9). An intriguing property of these models is the singularity with b going to infinity. The Lewontin-Cohen models' random factors multiplication matrix and the various ways it can be parameterized (Cohen et al., 2013; Cohen, 2014; Giometto et al., 2015) provide insight into ways TPL can be generated, but lack clear biological meaning.

The fact that skewness increases with b has long been recognized as evidenced by Fig. 1.1A published in 1965. Its formal demonstration by simulation (Cohen and Xu, 2015) generate $b > 2.0$, but does not identify a biological meaning for the factor(s) determining b. By contrast, the parameter of the Matérn process controlling migration distance directly influences b by adding to or subtracting from the Poisson condition (Shi et al., 2016). This model too seems to be limited to $b < 2.0$.

Autocorrelation matrices define the correlation of elements separated by lags in time or space. Models using correlation matrices to relate features or process in time or space can also produce TPLs: reproductive correlation in time (Ballantyne and Kerkhoff, 2005, 2007) and environmental correlation in space

(Hanski, 1980). The competition matrix in Kilpatrick and Ives (2003) model is functionally similar to an autocorrelation matrix in that it relates subsets (species) to each other across an abstract space. All three generate slopes with $1 \leq b \leq 2$. For temporal TPL measuring stability, a maximum value of $b = 2$ is expected and confirmed by Ballantyne and Kerkhoff's (2005, 2007) models.

Fronczak and Fronczak's (2010) statistical physics explanation for TPL is rigorously defined within an established and tested physical system based on the second law of thermodynamics, but like so many other models it is stuck in the $1 \leq b \leq 2$ realm, and unlike most biologically based models it is hard to identify a correspondence of parameters or variables with biological systems. However, the model explicitly separates macro and microeffects in such a way that the macroeffects emerge from consideration of the micro in much the same way that TPL emerges from the spatial behavior of interacting organisms.

The Tweedie exponential dispersion models characterized by a TPL function are subject to a central limit theorem analog in which data are expected to converge on a Tweedie distribution (Kendal and Jørgensen, 2011a). One of the models in the Tweedie system, the Poisson-gamma model, predicts TPL with $1 \leq b \leq 2$ and the positive stable distribution has $b > 2$. The Tweedie models provide a probabilistic description of TPL for $b \geq 1$. The lack of a Tweedie distribution with $b < 1$ is a disadvantage to an otherwise powerful description of TPL data.

A correlation between b and fractal Hausdorff dimension, D_H, which approaches $E = 2$ when area filling, combined with Hurst exponent, H, suggests that b may have a maximum value of 4 in a two-dimensional world. In three dimensions, we might expect volume filling with $D_H \to E = 3$, so the maximum $b = 6$. The evidence for fractal interpretation, thanks to Wayne Kendal and Zoltán Eisler, is strong. If TPL is expressing a fractal property—at least with some sample sizes—and is a linear function of Hurst's H and Hausdorff's D_H, which have a definite relationship with the Euclidean dimension in which they act, it is difficult to see how the singularity predicted by Lewontin-Cohen models can coexist with a fractal interpretation that involves a finite number of dimensions.

However, the fractal interpretation does not explain the few instances of marine nematodes with $b > 6$ where the Lewontin-Cohen model can account for them. The steep marine nematode results are probably biased as a consequence of small NQ and NB, although the one (Fig. 7.29) has $r = 0.94$, so the bias is acting very systematically. If they prove to be evidence for a singularity in b generated, as suggested by Lewontin-Cohen model, the case for a fractal interpretation is considerably weakened.

Kendal's (1992) case for a fractal nature of TPL and Eisler et al's (2008) derivation of the connection between b and Hurst's exponent H explicitly invoke correlation over time or space. A key feature of signals displaying $1/f^{\alpha}$ ($\alpha > \frac{1}{2}$) noise (Fig. 16.1) is the signal's temporal correlation. Correlations across time (and space) generating signals that can be described by TPL suggest that TPL may be responsive to long-term temporal and spatial correlations. In the case of spatially explicit models incorporating directed migration between locations (e.g., RAJT, 1981b; Perry, 1988), this is literally true. It is also true for other models with migration between virtual spaces (e.g., Anderson et al., 1982; Perry, 1988, 1994), although the migratory flow is undirected and comparatively diffuse. While both explicit and virtual spatial models described in Chapter 17 produce convincing TPLs, only models that allow for reinvasion after extinction display TPLs with $b > 2$.

The physical systems of river discharge, foreign exchange prices, or internet connections have an equivalent to migration in terms of information flow. The difference is the signals are not repopulated in the same way that populations of organisms repopulate an environment. This is an important distinction between physical and ecological systems. When a physical system stops, restarts do not obtain information from the previous round. In ecological systems, when a population at a site goes extinct, it can be restarted by immigration from another place, or in the case of diapause, a previous time.

Are the power laws related?

A number of authors have noted that TPL is one of the few generalizations in ecology (e.g., Kilpatrick and Ives, 2003; Engen et al., 2008; Kalyuzhny et al., 2014; Ma, 2015). Other power laws relate frequency to size, allometry to population, and population to area. He and Gaston (2000, 2003) unified TPL and the relationship relating the proportion of area occupied by a species and its abundance (Eq. 8.1). Their unification suggests they are different expressions of species distribution. Southwood et al. (2006) combined the size frequency distribution with species area (Eq. 16.3). The connections between allometry and TPL exposed by Cohen and colleagues (Cohen et al., 2012a,b) encourage the conclusion that many naturally occurring (statistical) power laws are functionally connected. Connections between physiology and anatomy suggest a possible unification with West and colleagues' condensation of physiological power laws to multiples of ¼ (West et al., 2001). Fairfield Smith's (1938) soil fertility power law, resembling the Zipf/Pareto law, relates SD and CV to linear dimensions, which can be related to spatial TPL by a simple substitution (Eq. 16.6).

Excluding the deterministic physics laws, it seems likely that the statistical power laws are related mathematically, and some may describe fractal properties. Analyses of networks suggest connections (Chapter 12) and Sato et al. (2010) showed how TPL described temporal fluctuations in currency markets. Autocorrelations in time generate data described by the Hurst exponent, H, which is related to the exponent in $1/f^{\alpha}$ noise. For many natural processes Hurst's exponent, $0.5 < H \leq 1$ indicates long-range dependence or persistence and curves with $H < 0.5$ denote anticorrelation. Eisler et al.'s (2008) formalism relating TPL to Hurst also creates a connection with currency and equity markets and the

Brown noise spectra. Hurst's H is related to the Hausdorff fractal dimension, D_H by $D_H = 2 - H$, and the spectral frequency exponent α by $\alpha = 1 + 2H$ (Eq. 16.2). Eisler et al's model unifying temporal and ensemble/spatial TPLs connects spatial TPL with fractal dimension.

The density of structures created by diffusion-limited aggregation declines as negative powers of distance with exponent $D - E$, where D is a fractal dimension and the Euclidean dimension, E, of the cluster. If migration and dispersal of organisms produce clusters by aggregation in a similar manner, which seems plausible, then population clusters also have a fractal nature as proposed by Kendal (1992, 1995). The fact that some of the phenomena described by these power laws are inherently fractal with connections to TPL does not necessarily mean that all properties described by TPL and its derivatives are fractal in nature. Box-counting populations suggest the populations may be self-affine, but conversion of box-counted TPLs to density can produce a curved TPL and different quadrat areas produce different TPL slopes. The example of Japanese beetle larvae (Fig. 8.7) shows different quadrat sizes are statistically self-similar, but not necessarily self-affine. Depending on how the data stream is partitioned and analyzed, there is evidence for self-affinity in some TPLs describing the spatial distribution of organisms. It is tempting therefore to regard the known fractal sets of Cantor dust as metaphor for the spatial distribution of animals and plants on a surface and the Sierpinski pyramid a metaphor in three dimensions.

Assuming the identity of TPL b with the Hurst exponent H derived by Eisler et al. (2008), the following connections with the spectral exponent α and the Hausdorff dimension D_H can be made from Eq. (16.2):

$$b = 2H = \alpha - 1 = 4 - 2D_H. \tag{18.1}$$

As the value of b can depend on the quadrat size, it is likely that only one quadrat size corresponds to box counting and therefore qualifies as fractal. Thus, one estimate of b, call it b^*, corresponds to Eq. (18.1). The connections between TPL and the other biological power laws (Eqs. 8.2, 16.3, 16.4, and 16.6) suggest that many natural processes are scale invariant and may possibly have fractal components. The models of West and colleagues are explicit in this regard, and their connection via allometry to both community and population ecology supports the hypothesis that fractal behavior is ubiquitous in biology, and under certain circumstances TPL describes this.

Is TPL universal?

It has been suggested by Ma (2015), among others, that TPL may be universal because it is generated by the multitude of different systems described in Chapters 5–12 and models in Chapter 17. Feigenbaum's (1983) discovery of a universality in iterated maps supports the idea of universality in TPL as many of the models generating TPL in Chapter 17 are nonlinear iterated maps.

However, universality has a very specific definition: it applies to systems whose properties are independent of the dynamical details of the system. The term originated in the study of phase transitions in statistical physics. A key feature of universality is the concept of criticality. Many systems exist in more than one state and the transition between them occurs suddenly. A common feature is one state is highly ordered and the other more variable, exemplified by the transition of ice to water and in percolation experiments from a number of disconnected chaotic clusters to a single, ordered lattice spanning cluster as walkers join the separated clusters and suddenly link them (Chapter 17). As the transition is approached, scale-dependent parameters like density become less important and scale-invariance dominates. The behavior of the system is simplified and may be well approximated by an exactly solvable power law model. In particular, the probability of an event is related to its size by a power law, as for example the size of avalanches in a pile of sand.

Wolfram (2002) has shown that fractal geometric structures, $1/f^\alpha$ noise, and power laws can emerge spontaneously from simple local rules. If complex structures can emerge from simple models like cellular automata, natural complexity does not require fine tuning of parameters to exhibit complexity. The emergence of complex behavior independent of the value of model parameters (self-organized criticality) is a property of dynamical systems that converge on an attractor without requiring precise control by a variable. They appear to control themselves and close to the attractor they display spatial and/or temporal scale-invariance.

Many systems are maintained by feedback at the point between ordered and disordered states, a point called the edge of chaos where the complexity is maximal. Complex adaptive systems apparently evolve toward this boundary region between order and chaos.

Adaptation is important in so many natural systems that the edge of chaos concept has been adopted in many disciplines. For example, Kauffman (1993, 1995) has suggested that the rate of evolution is maximized near the edge of chaos with the balance maintained by the feedbacks inherent in natural selection. The currency of evolution is the survival and reproduction of the individual, but the concept of adaptation also applies, but with lesser strength, to population and community adaptation. Reproductive fitness can be visualized as a landscape of ranges of peaks of high fitness separated by valleys of lower fitness for the set of all possible genotypes. In the language of edge of chaos, the peaks represent order and the valleys disorder. This complex (multidimensional) landscape is explored by evolution with selection naturally pulling fitness to the nearest peak. The constant disturbances caused by changes to the other members of a community or extrinsic factors, such as climate, continuously reorganize the landscape's geography with natural selection navigating to a new peak as an old one subsides. In Kauffman's dynamical world, self-organization in which power laws could emerge is inevitable. His worldview has much in common with C.B. Williams' conviction that the natural pattern in the balance of nature was somehow inevitable.

Fronczak and Fronczak's (2010) statistical physics model for TPL suggests that under certain stringent conditions TPL may be universal in the strict dynamical sense. If indeed TPL is universal, in this sense it suggests that populations that obey TPL maintain their spatial distribution by behavioral feedback for only behavior is fast enough to maintain order within a generation, and it must be responsive to the others in its community. If the TPL pattern emerges naturally and inevitably from the kind of order visualized by Williams and Kauffman, the dynamical definition of universality may well apply. Even if TPL is not universal in the strict dynamical systems sense, there is certainly evidence of some features of universality to justify adoption of a "soft" universality.

Financial data and data of internet traffic through the network's nodes are collected continuously with 100% accuracy. We would not expect b for these nonbiological data to be influenced by density-dependent effects. That quantitative scaling occurs outside of ecology reinforces the suggestion that dependence of the variance on the mean is a near-universal phenomenon. Its sensitivity to density dependence is what makes it important in ecology, and this fact suggests that whatever the constraints physical phenomena place on quantitative scaling, evolutionary processes may relax those constraints and extend the range of permissible slopes. Furthermore, as density-dependent behavior can diminish the value of b, it may be that it is density dependence, which is absent from physical phenomena, that can force higher values of b encountered with biological data. The fact that TPL can be generated in both physical and biological systems suggests a deeper meaning to TPL than originally proposed by ecologists.

To summarize: TPL is ubiquitous, its expression depending on how you collect and organize data as well as the properties of the system being studied. There are strong indications of connections between TPL and all or most of the biologically based allometric laws, some of which have been united by connections to TPL. In ecological data, both the distance between sample sites and the scale of sampling influence the observed relationship between mean and variance. Because separation and scale have different effects on TPL and are linked, their combined effects may be unpredictable. In all studies, it is important to clearly identify the statistical population being studied and to sample it consistently and with an appropriate sampler. The distinction between biological and statistical population may not always be obvious, a problem absent from physical and mathematical experiments in which the choice of sampler is also limited. Inconsistent results or no result may be obtained when the population is poorly defined. Poor results may also be obtained when the target cannot be resolved to an integer. For this reason, ratios and irrational numbers sometimes fail to produce convincing TPLs. The slope of TPL is likely related to the fractional exponents of Hurst and Hausdorff, which have as upper limits the Euclidean dimension of the space in which the fractal property resides. Thus, in time the maximum value b can attain is 2, in space $b = 4$ is a maximum, and in a volume $b = 6$ would be maximum. TPL slopes tend to be higher in three- than two-dimensional systems and some natural systems involving networks may have higher dimensionality. The conclusion or assumption by some that the slope of spatial TPL is limited to $b = 2$ is supported neither by the empirical evidence nor the circumstantial evidence that b is related to fractal dimension. At least one data collection method, box counting, results in self-affine TPLs. Most do not, although variance is still statistically self-similar, and at least one sampling protocol with fixed sample size likely emulates box counting. The evidence for a connection with fractals is strong; the evidence for universality is weaker. Feigenbaum's discovery of universality in iterated maps and the emergence of TPL from iterated maps suggest TPL's slope may be universal. However, if TPL is universal, it is not strongly so as specific outcomes do depend on the details of the system under study, especially the sampling protocol. Whether or not Taylor's power law is universal, its appearance throughout science makes clear it describes patterns of order emerging in many natural systems exhibiting scaleless variation.

References

Anderson, R.M., May, R.M., 1988. Epidemiological parameters of HIV transmission. Nature 333, 514–519.
Anderson, R.M., Gordon, D.M., Crawley, M.J., Hassell, M.P., 1982. Variability in the abundance of animal and plant species. Nature 296, 245–248.
Ballantyne, F., Kerkhoff, A.J., 2005. Reproductive correlation and mean-variance scaling of reproductive output for a forest model. J. Theor. Biol. 235, 373–380.
Ballantyne, F., Kerkhoff, A.J., 2007. The observed range for temporal mean-variance scaling exponents can be explained by reproductive correlation. Oikos 116, 174–180.
Carvalho, M.O., Faro, A., Subramanyam, B., 2013. Insect population distribution and density estimates in a large rice mill in Portugal—a pilot study. J. Stored Prod. Res. 52, 48–56.
Clark, S.J., Perry, J.N., Marshall, E.J.P., 1996. Estimating Taylor's power law parameters for weeds and the effect of spatial scale. Weed Res. 36, 405–417.
Cohen, J.E., 2014. Stochastic population dynamics in a Markovian environment implies Taylor's power law of fluctuation scaling. Theor. Popul. Biol. 93, 30–37.
Cohen, J.E., Xu, M., 2015. Random sampling of skewed distributions implies Taylor's power law of fluctuation scaling. Proc. Natl. Acad. Sci. U. S. A. 112, 7749–7754.
Cohen, J.E., Xu, M., Schuster, W.S.F., 2012a. Allometric scaling of population variance with mean body size is predicted from Taylor's law and density-mass allometry. Proc. Natl. Acad. Sci. U. S. A. 109, 15829–15834.
Cohen, J.E., Plank, M.J., Law, R., 2012b. Taylor's law and body size in exploited marine ecosystems. Ecol. Evol. 2, 3168–3178.
Cohen, J.E., Xu, M., Brunenborg, H., 2013. Taylor's law applies to spatial variation in a human population. Genus 69, 25–60.
Drake, C.M., 1983. Spatial distribution of chironomid larvae (Diptera) on leaves of the bulrush in a chalk stream. J. Anim. Ecol. 52, 421–437.

Eisler, Z., Bartos, I., Kertész, J., 2008. Fluctuation scaling in complex systems: Taylor's law and beyond. Adv. Phys. 57, 89–142.

Elliott, J.M., 1981. A quantitative study of the life cycle of the net-spinning caddis *Philopotamus montanus* (Trichoptera: Philopotamidae) in a Lake District stream. J. Anim. Ecol. 50, 867–883.

Elliott, J.M., 1986. Spatial distribution and behavioural movements of migratory trout *Salmo trutta* in a Lake District stream. J. Anim. Ecol. 55, 907–922.

Elliott, J.M., 1992. Variation in the population density of adult sea-trout, *Salmo truta*, in 67 rivers in England and Wales. Ecol. Freshw. Fish 1, 5–11.

Elliott, J.M., 2002. A quantitative study of day-night changes in the spatial distribution of insects in a stony stream. J. Anim. Ecol. 71, 112–122.

Elliott, J.M., 2004. Contrasting dynamics in two subpopulations of a leech metapopulation over 25 year-classes in a small stream. J. Anim. Ecol. 73, 272–282.

Elmouttie, D., Flinn, P., Kiermeier, A., Subramanyam, B., Hagstrum, D., Hamilton, G., 2013. Sampling stored-product insect pests: a comparison of four statistical sampling models for probability of pest detection. Pest Manag. Sci. 69, 1073–1079.

Engen, S., Lande, R., Saether, B.-E., 2008. A general model for analyzing Taylor's spatial scaling laws. Ecology 89, 2612–2622.

Feigenbaum, M.J., 1983. Universal behavior in nonlinear systems. Physica D 7, 16–39.

Fisher, R.A., Corbett, A.S., Williams, C.B., 1943. The relation between the number of species and the number of individuals in a random sample of an animal population. J. Anim. Ecol. 12, 42–58.

Fronczak, A., Fronczak, P., 2010. Origins of Taylor's power law for fluctuation scaling in complex systems. Phys. Rev. E 81, 066112.

Gallucci, F., Steyaert, M., Moens, T., 2005. Can field distributions of marine predacious nematodes be explained by sediment constraints on their foraging success? Mar. Ecol. Prog. Ser. 304, 167–178.

Gilles, A., Viquerat, S., Becker, A., et al., 2016. Seasonal habitat-based density models for a marine top predator, the harbor porpoise, in a dynamic environment. Ecosphere. 7(6).

Giometto, A., Formentin, M., Rinaldo, A., Cohen, J.E., Maritan, A., 2015. Sample and population exponents of generalized Taylor's law. Proc. Natl. Acad. Sci. U. S. A. 112, 7755–7760.

Guan, Q., Chen, J., Wei, Z., Wang, Y., Shiyom, M., Yang, Y., 2016. Analyzing the spatial heterogeneity of number of plant individuals in grassland community by using power law model. Ecol. Model. 320, 316–321.

Hamilton, W.D., 1971. Geometry for the selfish herd. J. Theor. Biol. 31, 295–311.

Hanski, I., 1980. Spatial patterns and movements in coprophagous beetles. Oikos 34, 293–310.

Hanski, I., 1987. Cross-correlation in population dynamics and the slope of spatial, variance-mean regressions. Oikos 50, 148–151.

He, F., Gaston, K.J., 2000. Estimating species abundance from occurrence. Am. Nat. 156, 553–559.

He, F., Gaston, K.J., 2003. Occupancy, spatial variance, and the abundance of species. Am. Nat. 162, 366–375.

Hoy, C.W., Zhou, X.L., Nault, L.R., Miller, S.A., Styer, J., 1999. Host plant, phytoplasma, and reproductive status effects an flight behavior of aster leafhopper (Homoptera: Cicadellidae). Ann. Entomol. Soc. Am. 92, 523–528.

Jiang, J., DeAngelis, D.L., Zhang, B., Cohen, J.E., 2014. Population age and initial density in a patchy environment affect the occurrence of abrupt transitions in a birth-and-death model of Taylor's law. Ecol. Model. 289, 59–65.

Kalyuzhny, M., Schreiber, Y., Chocron, R., Flather, C.H., Kadmon, R., Kessler, D.A., Shnerb, N.M., 2014. Temporal fluctuation scaling in populations and communities. Ecology 95, 1701–1709.

Kauffman, S., 1993. The Origins of Order: Self Organization and Selection in Evolution. Oxford University Press, Oxford, UK.

Kauffman, S., 1995. At Home in the Universe: The Search for Laws of Self-Organization and Complexity. Oxford University Press, Oxford, UK.

Kendal, W.S., 1992. Fractal scaling in the geographic distribution of populations. Ecol. Model. 64, 65–69.

Kendal, W.S., 1995. A probabilistic model for the variance to mean power law in ecology. Ecol. Model. 80, 293–297.

Kendal, W.S., 2004. Taylor's ecological power law as a consequence of scale invariant exponential dispersion models. Ecol. Complex. 1, 193–209.

Kendal, W.S., Jørgensen, B., 2011a. Taylor's power law and fluctuation scaling explained by a central-limit-like convergence. Phys. Rev. E 83, 066115.

Kendal, W.S., Jørgensen, B., 2011b. Tweedie convergence: a mathematical basis for Taylor's power law, 1/f noise, and multifractality. Phys. Rev. E. 84.

Kendal, W.S., Jørgensen, B., 2015. A scale invariant distribution of the prime numbers. Computation 3, 528–540.

Kilpatrick, A.M., Ives, A.R., 2003. Species interactions can explain Taylor's power law for ecological time series. Nature 422, 65–68.

Lagrue, C., Poulin, R., Cohen, J.E., 2015. Parasitism alters three power laws of scaling in a metazoan community: Taylor's law, density-mass allometry, and variance-mass allometry. Proc. Natl. Acad. Sci. U. S. A. 112, 1791–1796.

Ma, Z., 2015. Power law analysis of the human microbiome. Mol. Ecol. 24, 5428–5445.

Mandelbrot, B.B., 1982. The Fractal Geometry of Nature. W. H. Freeman & Co, San Francisco, CA.

Moens, T., Van Gansbeke, D., Vincx, M., 1999. Linking estuarine nematodes to their suspected food. A case study from the Westerschelde Estuary (south-west Netherlands). J. Mar. Biol. Assoc. U. K. 79, 1017–1027.

Monaghan, K.A., 2015. Taylor's Law improves the accuracy of bioassessment; an example for freshwater macroinvertebrates. Hydrobiologia 760, 91–103.

Morand, S., Guégan, J.-F., 2000. Distribution and abundance of parasite nematodes: ecological specialisation, phylogenetic constraint or simply epidemiology? Oikos 88, 563–573.

Murthy, A.C., Fristoe, T.S., Burger, J.R., 2016. Homogenizing effects of cities on North American winter bird diversity. Ecosphere 7 (1), e01216.

Park, S.-J., Taylor, R.A.J., Grewal, P.S., 2013. Spatial organization of soil nematode communities in urban landscapes: Taylor's power law reveals life strategy characteristics. Appl. Soil Ecol. 64, 214–222.

Perry, J.N., 1988. Some models for spatial variability of animal species. Oikos 51, 124–130.

Perry, J.N., 1994. Chaotic dynamics can generate Taylor's power law. Proc. R. Soc. B 257, 221–226.

Perry, J.N., 1995. Spatial analysis by distance indices. J. Anim. Ecol. 64, 303–314.

Petrovskii, S., Morozov, A., Li, B.-L., 2008. On a possible origin of the fat-tailed dispersal in population dynamics. Ecol. Complex. 5, 146–150.

Polley, H.W., Wilsey, B.J., Derner, J.D., 2007. Dominant species constrain effects of species diversity on temporal variability in biomass production of tallgrass prairie. Oikos 116, 2044–2052.

Preston, F.W., 1948. The commonness and rarity of species. Ecology 29, 254–283.

Ramsayer, J., Fellous, S., Cohen, J.E., Hochberg, M.E., 2012. Taylor's Law holds in experimental bacterial populations but competition does not influence the slope. Biol. Lett. 8, 316–319.

Ruzicka, J.J., Daly, E.A., Brodeur, R.D., 2016. Evidence that summer jellyfish blooms impact Pacific Northwest salmon production. Ecosphere 7 (4), e01324.

Samaniego, H., Sérandour, G., Milne, B.T., 2012. Analyzing Taylor's Scaling Law: qualitative differences of social and territorial behavior on colonization/extinction dynamics. Popul. Ecol. 54, 213–223.

Sato, A.-H., Nishimura, M., Hołyst, J.A., 2010. Fluctuation scaling of quotation activities in the foreign exchange market. Physica A 389, 2793–2804.

Sawyer, A.J., 1989. Inconstancy of Taylor's b: simulated sampling with different quadrat sizes and spatial distributions. Res. Popul. Ecol. 31, 11–24.

Shaw, D.J., Dobson, A.P., 1995. Patterns of macroparasite abundance and aggregation in wildlife populations: a quantitative review. Parasitology 111, S111–S133.

Shi, P.-J., Sandhu, H.S., Reddy, G.V.P., 2016. Dispersal distance determines the exponent of the spatial Taylor's power law. Ecol. Model. 335, 48–53.

Shiyomi, M., Takahashi, S., Yoshimura, J., Yasuda, T., Tsutsumi, M., Tsuiki, M., Hori, Y., 2001. Spatial heterogeneity in a grassland community: use of power law. Ecol. Res. 16, 487–495.

Smith, H.F., 1938. An empirical law describing heterogeneity in the yields of agricultural crops. J. Agric. Sci. 28, 1–23.

Somerfield, P.J., Dashfield, S.L., Warwick, R.M., 2007. Three-dimensional spatial structure: nematodes in a sandy tidal flat. Mar. Ecol. Prog. Ser. 336, 177–186.

Southwood, T.R.E., May, R.M., Sugihara, G., 2006. Observations on related ecological exponents. Proc. Natl. Acad. Sci. U. S. A. 103, 6931–6933.

Steiner, M.Y., 1990. Determining population characteristics and sampling procedures for the western flower thrips (Thysanoptera: Thripidae) and the predatory mite *Amblyseius cucumeris* (Acari: Phytoseiidae) on greenhouse cucumber. Environ. Entomol. 19, 1605–1613.

Taylor, L.R., 1962. The absolute efficiency of insect suction traps. Ann. Appl. Biol. 50, 405–421.

Taylor, L.R., 1970. Aggregation and the transformation of counts of *Aphis fabae* Scop. on beans. Ann. Appl. Biol. 65, 181–189.

Taylor, L.R., 1971. Aggregation as a species characteristic. In: Patil, G.P., Pielou, E.C., Waters, W.E. (Eds.), Statistical Ecology. In: Spatial Patterns and Statistical DistributionsVol. 1. Penn State Press, University Park, PA, pp. 357–377.

Taylor, L.R., Taylor, R.A.J., 1977. Aggregation, migration and population mechanics. Nature 265, 415–421.

Taylor, L.R., Woiwod, I.P., 1982. Comparative synoptic dynamics. I. Relationships between inter- and intra-specific spatial and temporal variance/mean population parameters. J. Anim. Ecol. 51, 879–906.

Taylor, L.R., Woiwod, I.P., Taylor, R.A.J., 1979. The migratory ambit of the hop aphid and its significance in aphid population dynamics. J. Anim. Ecol. 48, 955–972.

Taylor, L.R., Woiwod, I.P., Perry, J.N., 1980. Variance and the large scale spatial stability of aphids, moths and birds. J. Anim. Ecol. 49, 831–854.

Taylor, L.R., Perry, J.N., Woiwod, I.P., Taylor, R.A.J., 1988. Specificity of the spatial power-law exponent in ecology and agriculture. Nature 332, 721–722.

Taylor, R.A.J., 1980. A family of regression equations describing the density distribution of dispersing organisms. Nature 286, 53–55.

Taylor, R.A.J., 1981a. The behavioural basis of redistribution. I. The Δ-model concept. Anim. Ecol. 50, 573–586.

Taylor, R.A.J., 1981b. The behavioural basis of redistribution. II. Simulations of the Δ-model. J. Anim. Ecol. 50, 587–604.

Taylor, R.A.J., 1987. On the accuracy of insecticide efficacy reports. Environ. Entomol. 16, 1–8.

Taylor, R.A.J., 2018. Spatial distribution, sampling efficiency and Taylor's power law. Ecol. Entomol. 43, 215–225.

Taylor, R.A.J., Lindquist, R.K., Shipp, J.L., 1998. Variation and consistency in spatial distribution as measured by Taylor's power law. Environ. Entomol. 27, 191–201.

Trumble, J.T., 1985. Implications of changes in arthropod distribution following chemical application. Res. Popul. Ecol. 27, 277–285.

Trumble, J.T., Brewer, M.J., Shelton, A.M., Nyrop, J.P., 1989. Transportability of fixed precision level sampling plans. Res. Popul. Ecol. 31, 325–342.

West, G.B., Brown, J.H., Enquist, B.J., 2001. A general model for ontogenetic growth. Nature 413, 628–631.

Williams, C.B., 1964. Patterns in the Balance of Nature. Academic Press, London.

Wolfram, S., 2002. A New Kind of Science. Wolfram Media Inc., Champaign, IL

Xiao, X., Locey, K.J., White, E.P., 2015. A process-independent explanation for the general form of Taylor's law. Am. Nat. 186, E51–E60.

Young, I.M., Crawford, J.W., 1991. The fractal structure of soil aggregates: its measurement and interpretation. J. Soil Sci. 42, 187–192.

Chapter 19

Epilogue

In the preceding chapters, I have attempted to present data, models, and theories of TPL without bias, criticism, or opinion. I have, where it seemed appropriate, pointed to connections between data and theories and described conflicting data without comment. In this final short epilogue, I will permit my opinions to emerge. I do not expect to be correct, but hope that I can stimulate research in new directions.

TPL is an enigma. It certainly describes an enormous quantity of data from a score of disparate disciplines. The models purporting to account for it or predict it are legion—I was hardly exhaustive in Chapter 17. Many models overlap in key ways (autocorrelation of populations or characters, for example), but some effectively exclude others (some statistical models effectively exclude process-based models). They cannot all be right. Despite their ability to generate TPLs, I suspect that, as *explanations* of TPL, most, if not all are poor approximations.

While trying to develop models for TPL as a graduate student, I learned one thing and suspected a second. All the models I built to generate TPL that involved a recursion equation, process, or set of rules, iterated in time and/or space produced an acceptable TPL. Some of the models subsequently published and described, I had played with and abandoned. The simplest fractal shapes, like the Koch snowflake or the Sierpinski pyramid (Mandelbrot, 1982), produce a single value for b or a limited range. Other, more complex models such as those based on the logistic equation, produce a broader range, but most stop at $b=2$. It is because of this, I suspect some believe $b>2$ is aberrant. I was looking for a model that could produce the range of slopes then published of $0.5 \leq b \leq 3.5$. The only model to achieve this to my satisfaction was the Δ-model. Although possessing useful descriptive and interpretive value for field biologists, the Δ-model lacked predictive value, and, worse, its parameters did not map to TPL's in any obvious way.

The second insight, then just a suspicion, came from analyzing data and discovering that TPL had limitations. It works poorly for ratios and very poorly for bounded ratios. I had already concluded that TPL works best with unbounded numbers when I discovered it worked well for precipitation, but not for temperature. These two foundational meteorological variables are fundamentally different: one can be reduced to an integer and the other cannot. Ultimately, rainfall can be reduced to an integer number of H_2O molecules and is therefore

Taylor's Power Law. https://doi.org/10.1016/B978-0-12-810987-8.00019-7

countable. Temperature is an irrational number and therefore uncountable. TPL works best with number series that are reducible to positive integers and are unbounded.

The suspicion I had had since 1975 that TPL only applied to integers crystalized when I read about Kendal and Jørgensen's (2015) incredible experiments with the distribution of prime numbers. If TPL works with the unbounded integer sequence of the deviations between the prime counting function and Riemann's approximation, perhaps TPL has nothing to do with ecology. I do not mean that it does not apply or describe features of ecology, but that it is not fundamentally an ecological relationship. This is reinforced by its application to networks, cars passing a point, and the transactions and price of financial instruments. However, it shares with the Fibonacci series, the golden ratio, π, e, and other number theoretic patterns the fact that evolution has adopted it by natural selection, because it confers some advantage to the individuals in a population that adhere to it. The organization in space of populations is a complex process involving birth, death, and movement. While birth and death primarily change number, the rearrangement of the members of a population involves movement under the control of behavior. I include plants here as their spatial behavior operates on a different timescale to animals. The interplay between the individuals in a species population and with the other species in the community determines the success or failure of a genetic line. The behavioral rules selected by evolution to maximize the survival and reproductive probabilities of individuals may be simple or complex, but they are continuously iterated through the life of individuals and the succession of generations.

Outside of ecology and organic evolution, probably nothing is more complex than the evolutionary history of the cosmos. As far back in time as we are able to see, the distribution of luminous matter obeys TPL. The condensation of galaxies and the gravitational dance of galaxies and the stars within them are governed by a few simple rules iterated continuously for billions of years, resulting in the spatial and temporal patterns we see today.

TPL applies to iterated systems that can be resolved to integers, it describes the gaps in prime number predictions, can be generated from feasible sets (Xiao et al., 2015) similar to combinations and permutations of numbers, and is responsive to the proportion of zeros in sets of numbers (RAJT et al., 2017). TPL's connections to Hurst's exponent, to $1/f^{\alpha}$ spectra, and to box-counting algorithms for estimating fractal dimension also suggest the importance of repetition in generating a stream of numbers. Despite their functional differences, all the models generating TPL have in common the generation of a stream of numbers, a stream that if processed systematically, usually produces a linear relationship between $\log(M)$ and $\log(V)$ either in space or time.

Surprisingly, many examples, particularly those in Chapters 7 and 8, showing convincing TPLs of data comprising different taxa, or different segments of populations, or obtained from different habitats, are unified only by a single data collection method. The multitude of ways TPL can be generated by data united by the flimsiest of connections suggests to me that it is the sequence of numbers that matters, not necessarily the realm in which the numbers originated, or even how the numbers were acquired.

That TPL was discovered by an agricultural entomologist and ecologist is perhaps not surprising as it is ubiquitous in agroecology. The fact that for nearly a century, mathematics and statistics had been developed and used at Rothamsted Experimental Station in Harpenden, England, to solve the problems of feeding people added to the likelihood it would be discovered at Rothamsted by one of its scientists. Despite its history in ecology and agriculture and recent history in other disciples, I think the patterns of numbers created by iteration are what TPL characterizes. The patterns so created derive from many different and not necessarily connected rules governing their iteration. No matter the specifics of the processes of cosmic and organic evolution, they are the consequence of a few simple rules iterated both in space and time over billions of years. It is nonlinearity and repetition that all data and models generating TPL have in common. The emergence of complex patterns in cellular automata (Wolfram, 2002) suggests these are number theoretic issues. Perhaps then it is to number theory we should look for a proper understanding of Taylor's power law.

References

Kendal, W.S., Jørgensen, B., 2015. A scale invariant distribution of the prime numbers. Computation 3, 528–540.

Mandelbrot, B.B., 1982. The Fractal Geometry of Nature. W. H. Freeman & Co, San Francisco, CA.

Taylor, R.A.J., Park, S.-J., Grewal, P.S., 2017. Nematode spatial distribution and the frequency of zeros in samples. Nematology 19, 263–270.

Wolfram, S., 2002. A New Kind of Science. Wolfram Media Inc., Champaign, IL.

Xiao, X., Locey, K.J., White, E.P., 2015. A process-independent explanation for the general form of Taylor's law. Am. Nat. 186, E51–E60.

Author index

Note: Page numbers followed by *f* indicate figures and *t* indicate tables.

A

Abdala-Roberts, L., 126–127, 132–140*t*
Abd-Elgawad, M.M., 167, 169–170, 179–180, 213–227*t*, 295*t*
Abdo, Z., 94
Able, K.W., 335–336, 372–378*t*
Abuagob, O., 183, 213–227*t*
Acuña, F.H., 320–321, 324–325*t*
Adamson, M.L., 180–181, 213–227*t*
Adcock, R.J., 53–54
Adler, P.B., 123
Agboka, K., 295*t*
Agboola, O., 21, 389–390
Aguilar, A., 295*t*
Aharony, A., 509
Akhurst, R.J., 173–174
Albieri, A., 295*t*
Alegre, S., 295*t*
Alexander, N., 481–482
Aliakbarpour, H., 295*t*
Alins, G., 295*t*
Allcock, A.L., 311–312, 324–325*t*
Allen, R.B., 524–525
Allsopp, P.G., 37–39, 170, 213–227*t*, 250, 277–279*t*, 295*t*
Alm, S.R., 250–251, 277–279*t*
Alonso, C., 126–127, 132–140*t*
Alston, R.D., 43–44
Altshuler, D.L., 512
Alyokhin, A.V., 295*t*
Amador, J.A., 158, 213–227*t*
Amaral, L.A.N., 400–401, 416–420*t*
Anas, O., 477
Anderson, D.R., 328, 365, 369–370
Anderson, O.R., 69, 70*f*, 97–100*t*
Anderson, R.M., 399–401, 416–420*t*, 449–450, 546–547, 549, 570–571, 592–595
Andow, D.A., 295*t*
Andrew, R.J., 295*t*
Andrews, N., 403–404, 405*f*, 416–420*t*
Andriani, P., 503
Angelini, P., 295*t*
Anscombe, F.J., 31, 160, 213–227*t*, 462–463

Araújo, F.G., 332, 372–378*t*
Arceo-Gómez, G., 126–127, 132–140*t*
Arechavala-Lopez, P., 295*t*
Arenas, A., 434
Arevalo, H.A., 295*t*
Arnold, H., 353–354, 517
Arnold, J.G., 353, 482–483
Arrhenius, O., 515–516
Arroyo, J., 124, 132–140*t*
Arthur, A.L., 295*t*
Atkinson, W.D., 295*t*
Atwood, J.D., 431
Avendano, F., 164, 213–227*t*
Avilla, C., 295*t*
Azam, F., 77, 97–100*t*
Azerefegne, F., 295*t*
Azevedo, R.B.R., 526
Azovskii, A.I., 199, 213–227*t*

B

Bach, L.A., 478–479
Bachlor, C., 239–240, 277–279*t*
Backoulou, G.F., 295*t*
Badenhausser, I., 295*t*
Bader, B., 319–320, 324–325*t*
Badieritakis, E.G., 295*t*
Baffaut, C., 431
Bahlman, J.W., 512
Bai, G., 94
Bai, L., 483–484
Bairden, K., 183, 213–227*t*
Bakaloudis, D.E., 348–349, 372–378*t*
Baker, F.E., 247–248
Baker, R.R., 546
Bal, H.K., 173
Balakrishnan, N., 16–18, 414, 503–504
Balbuena, J.A., 204, 213–227*t*
Baldacchini, F., 295*t*
Balderson, T.K., 249, 277–279*t*
Ballantyne, F., 362–363, 546, 552–553, 570–571, 574, 593–594
Banavar, J.R., 521

Bancroft, T.A., 3, 4f, 18–19, 39, 237, 277–279t, 495
Banks, C.J., 463–464
Barabasi, A.-L., 433–434, 440–441t, 534, 536
Barbercheck, M.E., 173–174, 477
Barnes, H., 328–329
Barnouin, J., 184–185
Barranguet, C., 196, 213–227t
Barrios, L., 178, 213–227t
Barros, E.C., 295t
Bartlett, M.S., 54, 462
Bartos, I., 3, 5, 241–242, 435–437, 533, 556, 572–574, 595–596
Baskerville, G.L., 59
Båtnes, A.S., 295t
Bauer, P.J., 44–45
Baumgärtner, J., 295t
Baumgartner, M.F., 363, 372–378t
Beall, G., 245, 277–279t, 462–463
Beauchamp, J., 344, 372–378t
Becker, E. A., 367, 372–378t, 482–483, 589–590
Bedding, R.A., 173–174
Beddington, J.R., 343–344
Been, T.H., 160–161, 213–227t
Beier, S., 191, 213–227t, 476–477
Bell, G., 349, 560–561
Bell, J.F., 240
Bellini, R., 295t
Beltra, A., 295t
Ben-Hamadou, R., 334, 336, 372–378t
Benke, H., 364–367, 372–378t
Bennetta, P.G., 23
Benothman, M., 183, 213–227t
Benson, V.W., 431
Bentz, J., 295t
Benzing, D.H., 465
Berberet, R.C., 295t
Berg, M., 295t
Berge, J., 295t
Berggren, P., 364–367, 372–378t
Berkson, J., 50
Berlinger, M.J., 295t
Bert, W., 192–193, 213–227t
Berthier, K., 451
Bessière, A., 209, 213–227t
Bethony, J.M., 481–482
Bett, B.J., 447
Bingner, R.L., 353, 482–483
Bini, L.M., 306–307, 324–325t
Binning, L.K., 161, 213–227t
Binns, M.R., 465–467, 471–472

Bishop, J.R.B., 328, 369
Bishop, S.C., 183, 213–227t
Bisseleua, D.H.B., 295t
Bjørn, P.A., 295t
Black, M.J., 510
Blackburn, L.M., 295t
Blackburn, T.M., 353–354, 516–517
Blackshaw, R.P., 40, 207, 213–227t
Blanc, G., 295t
Blanco-Moreno, J.M., 113–114, 132–140t
Blettner, M., 403
Bliss, C.I., 2–3, 32, 36t, 78, 78f, 97–100t, 246–249, 462–463
Boag, B., 144–147, 150–151, 153, 156, 183–186, 207–208, 213–227t
Boag, D.A., 313–314, 324–325t
Boerboom, C.M., 112, 132–140t
Boettner, G., 241
Bonato, O., 295t
Bonecker, C.C., 306–307, 324–325t
Bongers, T., 148, 150–152, 213–227t, 476–477
Bonsdorff, E., 211, 213–227t
Borchers, D.L., 328, 364–367, 369–370, 372–378t
Borges, P.A.V., 275, 295t
Borges, R., 334, 336, 372–378t
Borgonie, G., 192–193, 213–227t
Borgrmeister, C., 295t
Bosch, R.V.D., 267
Bostanian, N.J., 467
Boswell, M.T., 32
Bouix, J., 184–185
Bowles, L.G., 305, 307–309, 324–325t
Boxshall, G.A., 1, 295t, 511–512
Brain, P., 106
Brandt, L.A., 344, 372–378t
Bravo-Mojica, H., 295t
Bretagnolle, F., 115, 132–140t
Brewer, M.J., 295t, 572
Brindley, T.A., 3, 4f, 18–19, 39, 237, 277–279t, 495
Broadbent, S.R., 509
Broccoli, A.J., 431
Brodeur, R.D., 322, 324–325t, 331, 372–378t, 585
Brodeurb, L., 87, 97–100t
Broms, C., 329–330, 372–378t
Brooker, S., 481–482
Brose, U., 515–516
Brøseth, H., 349–350
Brotman, R.M., 94
Brouty, M., 426–427, 440–441t

Brown, D.J.F., 144–147, 156, 213–227t
Brown, J.H., 519–521, 595
Brown, M.W., 3, 18, 295t
Brugel, S., 209, 213–227t
Brunenborg, H., 201, 395, 414, 416–420t, 447, 553, 570–571, 588, 592–593
Buckland, S.T., 328, 365, 369–370
Budha, P., 515
Bujok, M., 434
Bull, R.M., 37–39, 295t
Bulmer, M.G., 19
Bunce, R.G.H., 124–125
Burger, J.R., 354, 358–359, 361–362, 372–378t, 385, 587
Burkness, E.C., 295t
Burlando, B., 526–527
Burnham, K.P., 328, 365, 369–370
Butler, C.D., 295t
Buzas, M.A., 72, 97–100t
Bystrak, D., 351, 353, 538–539

C

Cabaret, J., 184
Cama, A., 481
Cambresy, L., 426–427, 440–441t
Cameron, E.A., 3, 18, 295t
Campbell, C.L., 165–166, 213–227t
Campbell, J.F., 173–174, 176–178, 213–227t
Campbell, R.W., 241
Campo, M.L., 250–251, 277–279t
Campos-Herrera, R., 178, 213–227t
Camps, C., 295t
Cannon, R., 267
Cantú Ros, A.G., 414
Cardina, J., 110–111, 132–140t, 431, 440–441t
Carisse, O., 87, 97–100t
Carrieri, M., 295t
Carroll, B.W., 427–428
Carvalho, M.O., 295t, 585–586
Cascéo, I., 363, 372–378t
Castle, S.J., 269, 295t
Caswell, H., 123–124
Caveness, F.E., 189–190
Cederman, L.E., 528
Celini, L., 295t
Cepák, J., 517
Certain, G., 481
Chambers, S.J., 207–208, 311–312, 324–325t
Chamorro, L., 113–114, 132–140t
Chao, C.-F., 75–76, 97–100t
Charbonnel, N., 451
Charlett, A., 403–404, 405f, 416–420t

Charnov, E.L., 519–521
Chaubey, R., 295t
Chauvel, B., 106, 132–140t
Chaval, Y., 451
Chediak, M., 295t
Chen, G., 198–199, 213–227t
Chen, H., 121, 132–140t
Chen, J., 118, 132–140t, 586
Chen, P., 406–407
Chen, Z.X., 92, 97–100t
Cheng, Z., 148–149, 190
Cherkiss, M.S., 344, 372–378t
Chiang, K.-P., 75–76, 97–100t
Chícharo, A.M., 334, 335f, 336, 372–378t
Chmiel, A., 390–391, 416–420t
Chocron, R., 595
Chołoniewski, J., 390–391, 416–420t
Christel, I., 481
Christianson, D., 369–370, 372–378t
Ciancio, A., 166–167, 213–227t
Clark, P.F., 511–512
Clark, P.J., 28–29
Clark, S.J., 58–59, 72–73, 96, 106, 108, 112, 132–140t, 158, 201–202, 211–212, 249, 252, 259, 267–268, 275–276, 319, 389, 397, 445–446, 472, 494, 524, 581–582
Clarke, R.T., 479–480
Clarke, S., 454–455
Claus, E., 190, 213–227t, 476–477
Cobb, N.A., 143–144
Cochlan, W.P., 77, 97–100t
Cochran, W.G., 13–14, 62
Codeço, C.T., 295t
Cohen, D., 429–430, 553
Cohen, J.E., 93, 97–100t, 201, 209, 274, 395, 414, 416–420t, 429, 440–441t, 447, 454–455, 524–525, 542, 553–558, 570–571, 585–586, 588, 592–593, 595
Coleman, D.C., 150–151, 194, 213–227t
Collett, A., 364–367, 372–378t
Cominelli, S., 369, 372–378t
Cong, B., 198, 213–227t
Coomes, D.A., 524–525
Copes, W.E., 82, 97–100t
Corbett, A.S., 2–3, 30, 586
Corral, A., 509–510
Cosson, J.-F., 451
Costa de Azevedo, M.C., 332, 372–378t
Crawford, J.W., 515, 575–576
Crawley, M.J., 449–450, 546–547, 549, 570–571, 592–595
Creel, S., 369–370, 372–378t

Crespo-Herrera, L.A., 295*t*
Croft, B.A., 295*t*
Croxton, S.D., 295*t*
Cuadra, L., 178, 213–227*t*
Cundill, B., 481–482
Cuperus, G.W., 295*t*
Curnutt, J.L., 21–22, 353, 359, 372–378*t*, 517

D

Dabiré, R.K., 166, 213–227*t*
Daly, E.A., 322, 324–325*t*, 331, 372–378*t*, 585
Dankers, W.H., 150–151, 156
Darbemamieh, M., 295*t*
Darmency, H., 106, 132–140*t*
Dashfield, S.L., 147, 213–227*t*, 580
David, F.N., 18, 30
Dawson, C.G., 250–251, 277–279*t*
de Arcangelis, L., 428, 511
de Bie, T., 192–193, 213–227*t*
de Goede, R.G.M., 151–152, 213–227*t*,
476–477
de Mazancourt, C., 123
de Menezes, M.A., 433–434, 440–441*t*, 534,
536
de Noyelles, F., 516
Dean, G.J., 267
DeAngelis, D.L., 556–557, 592
Dechesne, A., 385–386, 416–420*t*
Dedieu, J.L., 295*t*
Deeds, E.J., 518
DeLong, J.P., 383, 416–420*t*
Delworth, T.L., 431
Demers, S., 209, 213–227*t*
Deming, W.E., 53–54
den Boer, P., 558
Derner, J.D., 94, 119, 122, 132–140*t*, 587
Desbois, M., 509–510
Desrosiers, G., 209, 213–227*t*
Deter, J., 451
Deudero, S., 209, 213–227*t*, 446–447
Dewer, A.J., 267
Di Guilmi, C., 412–413, 416–420*t*
Dichiaro, M.J., 158, 213–227*t*
Dickson, D.W., 92, 97–100*t*, 158, 213–227*t*
Diemert, D., 481–482
Dietrick, E.J., 267
Diggle, P.J., 558–559
Dighton, J., 153–154, 213–227*t*
Dingfelder, H.A., 348, 372–378*t*
Dixon, P.M., 43–44, 481
Dobson, A.P., 181–182, 187, 203–204,
213–227*t*, 481–482

Dong, S.Z., 483–484
Doolittle, R.E., 250
Döring, T.F., 454–455
Dover, M.J., 295*t*
Drakare, S., 515–516
Drake, C.M., 295*t*, 572
Dreisig, H., 560
Duarte, L.F.L., 318, 324–325*t*
Dubois, P., 426–427, 440–441*t*
Duch, J., 434
Dudley, R., 512
Duffey, E., 267
Dun, R.A., 168, 213–227*t*
Duncan, L.W., 168–169, 175–176, 178,
213–227*t*
Duong, V., 405–406, 416–420*t*
Duvel, J.-P., 509–510
Dyer, R.-M., 397–398, 406–407, 416–420*t*
Dyke, G.J., 512

E

Edling, C.R., 400–401, 416–420*t*
Efron, D., 179, 213–227*t*
Eiane, K., 295*t*
Eisler, Z., 3, 5, 241–242, 435–437, 533–534,
556, 572–574, 595–596
Elkinton, J.S., 241–242
Ellegaard, C., 209, 213–227*t*, 295*t*, 313,
324–325*t*
Ellington, C.P., 514
Elliott, J.M., 37, 150–151, 206–207, 213–227*t*,
273, 295*t*, 337–340, 372–378*t*, 571–572, 575,
577, 582
Elliott, N.C., 295*t*
Ellis, J.R., 209, 213–227*t*
El-Morshedy, M.M., 169, 213–227*t*
Elmouttie, D., 295*t*, 585–586
Embody, D.A., 239–240, 277–279*t*
Embree, D.G., 238, 277–279*t*
Emmanouel, N.G., 295*t*
Engen, S., 307–308, 581, 595
Enki, D.G., 403–404, 405*f*, 416–420*t*
Enquist, B.J., 519–521, 595
Escuer, M., 178, 213–227*t*
Espino, L., 295*t*
Ettema, C.H., 150–151, 194, 213–227*t*
Evans, D.A., 31
Evans, F.C., 28–29
Evans, K.L., 517
Evans, N.A., 205, 213–227*t*
Excoffon, A.C., 320–321, 324–325*t*
Eyre, M., 150–151

F

Facca, C., 72, 97–100*t*
Fahay, M.P., 335–336, 372–378*t*
Fall, M.L., 87, 97–100*t*
Falster, D.S., 49, 52–53, 63
Fan, X., 173–174
Fanning, M.A., 109, 132–140*t*
Fantinou, A.A., 295*t*
Farahani, S., 295*t*
Faro, A., 295*t*, 585–586
Farrington, P., 403–404, 405*f*, 416–420*t*
Fathipour, Y., 295*t*
Feder, J., 507, 509, 563–564
Feigenbaum, M.J., 596
Fellous, S., 93, 97–100*t*, 274, 585–586
Fenton, A., 185–186
Ferguson, J.J., 168, 213–227*t*
Fernandez Garcia, A., 44
Ferno, A., 329–330, 372–378*t*
Ferrer, X., 481
Ferris, H., 150–151, 154–155, 158–160, 213–227*t*, 544
Finney, D.J., 59, 243–244, 277–279*t*
Finstad, B., 295*t*
Fisher, R.A., 2–3, 30, 586
Fitton, D.M., 311–312, 324–325*t*
Flather, C.H., 595
Flegg, J.M., 148–149
Fleming, W.E., 247–248
Flesch, A.D., 345–346, 372–378*t*
Flinn, P., 295*t*, 585–586
Flowers, J.D., 431
Fluschnik, T., 414
Foley, J.A., 454–455
Fontana, W., 518
Foord, K.E., 158–160, 544
Forbes, V.E., 383, 416–420*t*
Formentin, M., 558, 592–593
Forsythe, H.Y., 36*t*
Foster, B.L., 516
Fox, H., 247, 277–279*t*
France, R.L., 388–389, 416–420*t*
Frankham, R., 478–479
Franklin, A., 5–8
Freckleton, R.P., 123, 132–140*t*
Freeman, S.M., 317, 324–325*t*
Fretwell, S.D., 197, 349, 560
Friedl, M.A., 354
Friedman, J.A., 528
Friesen, K., 295*t*
Fristoe, T.S., 354, 358–359, 361–362, 372–378*t*, 385, 587

Fronczak, A., 434, 537, 594, 598
Fronczak, P., 434, 537, 594, 598
Frost, P., 21, 389, 416–420*t*
Fu, L., 94
Fukaya, K., 254–256
Fukunaga, G., 411, 416–420*t*
Fumanal, B., 115, 132–140*t*
Furse, M.T., 479–480
Furusawa, C., 504
Futai, K., 171–173, 213–227*t*

G

Gaba, S., 184
Gabryś, E., 507–508
Gaffeo, E., 412–413, 416–420*t*
Gajer, P., 94
Galan, M., 451
Galic, N., 383, 416–420*t*
Galleti, M., 412–413, 416–420*t*
Gallucci, F., 197, 213–227*t*
Gao, D., 188–189, 213–227*t*
Garcia, J.C.C., 451
Garcia-Álvarez, N.C., 295*t*
Garcia-Mari, F., 295*t*
Gardener, B.B.G., 148–149, 190
Garrido, B., 124, 132–140*t*
Garthwaite, P., 403–404, 405*f*, 416–420*t*
Garza-Garcia, R., 295*t*
Gaspar, C., 275, 295*t*
Gasquez, J., 106, 132–140*t*
Gassman, P.W., 431
Gastaminza, G., 295*t*
Gaston, K.J., 275–276, 295*t*, 353–354, 474, 489, 515–517, 595
Gaudot, I., 115, 132–140*t*
Gaugler, R., 173–174, 176–178, 213–227*t*
Geissler, P.H., 351, 353, 538–539
Geluso, K.N., 371, 372–378*t*
Genzano, G.N., 320–321, 324–325*t*
Gerrodette, T., 369
Gettinby, G., 183, 213–227*t*
Gibert, J.P., 383, 416–420*t*
Gibrat, R., 504
Gigon, V., 295*t*
Giles, K.L., 295*t*
Gilioli, G., 37–39, 295*t*
Gilles, A., 367, 372–378*t*, 482–483, 589–590
Gillis, D.M., 349, 560–561
Gilstrap, F.E., 295*t*
Ginot, V., 184
Giometto, A., 558, 592–593
Gladyshev, M.I., 70–71, 97–100*t*

Glazer, I., 179, 213–227t
Godano, C., 428, 511
Godfrey, L.D., 295t
Golizadeh, A., 295t
Gomes da Cruz-Filho, A., 332, 372–378t
Gómez-Ros, J.M., 178, 213–227t
Gonçalves, G.J., 334, 336, 372–378t
Gong, G.-C., 75–76, 97–100t
Gong, J., 121, 132–140t
González-Andójar, J.L., 113–114, 132–140t
González-Carrillo, J.A., 295t
Gonzalez-Zamora, J.E., 295t
Good, P.I., 185–186, 328
Gordon, D.M., 449–450, 546–547, 549, 570–571, 592–595
Gorres, J.H., 158, 213–227t
Gradwell, G.R., 16, 29–30
Graham, J.H., 169
Grant, B.R., 384
Grant, M.C., 103
Grant, P., 449–450
Grant, P.R., 384
Grasso, J.R., 428, 511
Gray, J.S., 189–190
Gray, R.D., 560
Grear, D.A., 187, 213–227t
Green, C.D., 144, 164–165, 213–227t
Green, R.H., 33, 45t, 468
Greene, J.K., 44–45, 295t
Greenewalt, C.S., 512
Greenwood, J.J.D., 516–517
Gregory, R.D., 353–354, 516–517
Grenfell, B.T., 402, 416–420t
Grewal, P.S., 60, 60t, 93–94, 96, 121, 123, 148–149, 151–152, 155–156, 162–163, 167–168, 173–174, 190, 195, 197, 202, 213–227t, 271, 342, 355–356, 362–363, 386–387, 465, 492–493, 500, 585–586, 606
Grieg-Smith, P., 13–14, 35
Griffiths, B.S., 144
Griffiths, G.J.K., 44
Grime, J.P., 118–120, 123
Grof-Tisza, P., 295t
Groom, D.J.E., 512
Grover, J.P., 516
Grundmann, G., 385–386, 416–420t
Gruner, L., 184–185
Guan, Q., 118, 132–140t, 586
Gudgel, R., 431
Guedes, A.P.P., 332, 372–378t
Guedes, R.N.C., 295t

Guegan, J.F., 182, 213–227t
Gunn, R.J.M., 479–480
Guo, H.-R., 75–76, 97–100t
Gutiérrez, C., 178, 213–227t
Gutierrez, J., 295t
Gyrisco, G.G., 36t

H
Hackett, C.A., 150–151, 183–184, 197, 207, 213–227t
Hagan, S.M., 335–336, 372–378t
Hagstrum, D., 295t, 585–586
Haigh, Z., 454–455
Hairston, N.G., 34–35
Haldane, J.B.S., 52–53, 512
Hamid, M.M., 275, 295t, 524
Hamilton, G., 295t, 585–586
Hamilton, W.D., 23, 274, 546
Hamm, P.B., 295t
Hammersley, J.M., 509
Hammond, P.S., 364–367, 372–378t
Hampe, A., 124, 132–140t
Hanley, Q.S., 397–398, 406–407, 416–420t
Hansen, B.W., 209, 213–227t, 295t, 313, 324–325t
Hanski, I., 541, 545–548, 570–571, 592–594
Hapca, S.M., 515
Hardersen, S., 514
Harmel, R.D., 353, 482–483
Harrington, R.H., 43
Harrison, J.M., 144
Harrison, K., 515–516
Hart, K., 344, 372–378t
Harvey, P.H., 63
Harvey, S.C., 187, 213–227t
Hasabo, S.A., 169–170, 213–227t
Hassanpour, M., 295t
Hassell, M.P., 16, 29–30, 449–450, 546–549, 570–571, 592–595
Hauck, L., 431
Hawkes, H.A., 479–480
Hayman, B.I., 463, 475
He, F., 275–276, 295t, 595
Healy, M.J.R., 72–73, 197, 250, 447, 463–464
Hedley, S.L., 328, 369
Heide-Jørgensen, M.P., 364–367, 372–378t
Heimlich, S., 364–367, 372–378t
Heininger, P., 190, 213–227t, 476–477
Heip, C., 198–201, 213–227t
Henderson, P.A., 13–14, 28–29, 189–190, 343–344, 474

Hendrickx, F., 192–193, 213–227*t*
Henne, D.C., 295*t*
Henson, W.R., 31–32
Herms, D.A., 431, 440–441*t*
Hernádez-Fuentes, L.M., 295*t*
Hewitt, M., 42, 45*t*
Hewlett, T.E., 92
Hiby, A.R., 364–367, 372–378*t*
Hilbun, W.S., 295*t*
Hill, M.O., 124–125
Hillebrand, H., 515–516
Hilsenhoff, W.L., 480
Hochberg, M.E., 93, 97–100*t*, 274, 585–586
Hocking, B., 514
Hodasi, J.K.M., 205–206, 213–227*t*
Hoffleit, D., 425, 440–441*t*
Hoffmann, A.A., 295*t*
Hogg, D.B., 295*t*
Holland, J.M., 43–44
Holling, C.S., 174
Holloway, G.J., 348–349, 372–378*t*
Holst, H., 308, 324–325*t*
Holt, R.D., 516
Holyoak, M., 295*t*
Hołyst, J.A., 390–391, 406, 416–420*t*, 435, 440–441*t*, 505, 595–596
Hominick, W.M., 173–174
Honório, N.A., 295*t*
Hooper, D.J., 148–149
Hori, M., 254
Hori, Y., 117–118, 454–455, 523, 586
Höss, S., 190, 213–227*t*, 476–477
Hoste, H., 184–185
Hotez, P., 481–482
Hoy, C.W., 173–174
Hsieh, C.-H., 342–344
Hu, M., 406
Hu, Q., 483–484
Hua, E., 198, 213–227*t*
Huang, L., 399, 416–420*t*
Huang, Y., 121, 132–140*t*
Huang, Z.-G., 399, 416–420*t*
Hudson, P.J., 185–187, 213–227*t*
Hughes, G., 85, 122, 491, 523
Hughes, J.A., 447
Hunter, J.R., 343–344
Hunter, P.J., 315–316, 324–325*t*
Hurst, H.E., 510
Hutchinson, G.E., 515
Hutchison, W.D., 295*t*
Huxley, J.S., 503, 511–512

I

Ibanez, J.J., 526–527
Ims, R.A., 349–350
Innocent, G., 183, 213–227*t*
Isas, M.G., 295*t*
Isbell, F., 123
Ivanova, E.A., 70–71, 97–100*t*
Ives, A.R., 93–94, 362–363, 546, 551, 574, 593–595
Iwao, S., 35, 36*t*, 37–39, 45*t*, 82, 468
Iwasa, Y., 552
Izaurralde, R.C., 431
Izquierdo, J., 113–114, 132–140*t*

J

Jacas, J.A., 295*t*
Jaenike, J., 181, 213–227*t*
James, A.C., 181
James, R., 23
Jaschek, C., 425, 440–441*t*
Jeger, M.J., 146–147, 213–227*t*, 543
Jenkins, E.B., 250–251, 277–279*t*
Jensen, A.S., 295*t*
Jensen, H.J., 189–190
Jeong, J., 431
Jiang, J., 556–557, 592
Jikumaru, S., 172–173, 213–227*t*
Jiménez, J.J., 208, 213–227*t*
Johnson, A.M., 401
Johnson, C.G., 266, 546
Johnson, D., 117, 132–140*t*
Johnson, M.P., 311–312, 324–325*t*
Johnson, N.L., 16–18, 414, 503–504
Johnson, S.J., 295*t*
Jonasson, S., 153–154, 213–227*t*
Jones, C.A., 431
Jones, D.B., 295*t*
Jones, E.W., 243, 277–279*t*
Jones, H., 454–455
Jones, M.M., 295*t*
Jones, V.P., 265, 295*t*, 471–472
Jordan, T.A., 249, 277–279*t*
Jørgensen, B., 21, 437, 538, 553–554, 570–571, 594, 606
Joyeux, J.-C., 336–337, 372–378*t*
Juarez, S., 385–386, 416–420*t*

K

Kaatsch, P., 403
Kadmon, R., 595
Kaiser, G.W., 512

Kalyuzhny, M., 595
Kamali, K., 295*t*
Kaneko, K., 504
Kanno, M., 36*t*
Kappas, A., 390–391, 416–420*t*
Karban, R., 295*t*
Kastdalen, L., 349–350
Kauffman, S., 597–598
Kauscha, H., 308, 324–325*t*
Kawaguchi, A., 81, 97–100*t*
Kaya, H.K., 173–174, 176–179, 213–227*t*
Kazuo, H., 519
Keeling, M.J., 402, 416–420*t*
Kemp, A.W., 16–17, 40–42
Kendal, W.S., 21, 246–247, 389–390,
    416–420*t*, 437, 506–507, 535–536, 538,
    570–571, 578, 594–596, 606
Kenerley, C.M., 146–147, 213–227*t*, 543
Kennedy, M., 560
Kerboeuf, D., 184–185
Kerkhoff, A.J., 546, 552–553, 570–571, 574,
    593–594
Kermack, K.A., 52–53, 512
Kerr, A., 183, 213–227*t*
Kertész, J., 3, 5, 241–242, 435–437, 533–534,
    556, 572–574, 595–596
Kessler, D.A., 595
Kęzdzia, A., 507–508
Khatun, S., 397–398, 406–407, 416–420*t*
Khodayari, S., 295*t*
Kieft, T.L., 517–518
Kiermeier, A., 295*t*, 585–586
Kilpatrick, A.M., 93–94, 362–363, 546, 551,
    574, 593–595
Kim, H., 385–386, 416–420*t*
Kindler, S.D., 295*t*
Kiniry, J.R., 431
Kinzey, D., 369
Kira, T., 519
Kleczkowski, A., 78, 78*f*, 97–100*t*
Klein, M.G., 250, 277–279*t*
Kleinerman, J., 389
Klukowski, Z., 43–44
Knapp, S., 454–455
Knight-Jones, E.W., 258, 277–279*t*
Knutson, A.E., 295*t*
Kobayashi, R., 539
Kobayashi, S., 36*t*
Koenig, S.S.K., 94
Koeschke, V., 478–479
Kolding, J., 525
Kolokotrones, T., 518

Kopittke, R.A., 170, 213–227*t*
Koppenhöfer, A.M., 173–174, 176–179,
    213–227*t*
Koste, W., 308, 324–325*t*
Kotz, S., 16–18, 414, 503–504
Koyama, S., 539
Kramer, D.L., 349, 560–561
Krause, J., 23
Kravchuk, E.S., 70–71, 97–100*t*
Kreiter, S., 295*t*
Kriewald, S., 414
Kropp, J.P., 414
Krzanowski, W.J., 63
Kuhar, T.P., 249, 277–279*t*
Kuklinski, P., 319–320, 324–325*t*
Kumar, P., 432–433
Kummell, C.H., 53–54
Kuno, E., 36*t*, 468
Kuo, T.-C., 342
Küster, D., 390–391, 416–420*t*
Kutner, M., 62
Kvasnes, M.A.J., 349, 372–378*t*

L
Laake, J.L., 328, 365, 369–370
Ladd, T.L., 250, 277–279*t*
Lagerwaard, F.J., 21, 389–390
Lagrue, C., 209, 525
Lai, J., 524–525
Lai, Y.-C., 399, 416–420*t*
Lande, R., 307–308, 581, 595
Langøy, H., 329–330, 372–378*t*
Lansac-Tôha, F.A., 306–307, 324–325*t*
Laport, R.G., 383, 416–420*t*
Larocque, A., 123
Lau, S.S., 150–151
Laub, C.L., 249, 277–279*t*
Lavelle, P., 208, 213–227*t*
Law, R., 525, 595
Lawton, J.H., 474, 516–518, 548–549
Lazarova, S.S., 151–152, 213–227*t*, 476–477
Le Corff, J., 295*t*
Leclerc, Y., 87, 97–100*t*
Lee, K.V., 295*t*
Legg, R., 207–208
Legg, R.K., 150–151, 207, 213–227*t*
Leibold, M.A., 516
Leighty, R.E., 526
Lello, J., 185–186
Lennon, J.L., 515–516
Lentini, A., 37–39, 295*t*
Lentink, D., 512

Leopold, M.F., 364–367, 372–378*t*
Lepš, J., 489
Leroi, A.M., 526
Lewis, E.E., 173, 176–177, 213–227*t*
Lewis, T., 1
Lewontin, R.C., 429–430, 553
Li, B.-D., 85–86, 97–100*t*
Li, B.-H., 85–86, 97–100*t*
Li, B.-L., 574–575
Li, D., 406–407
Li, E., 121, 132–140*t*
Li, J., 198, 213–227*t*
Liang, W.-J., 153, 213–227*t*
Liljeros, F., 400–401, 416–420*t*
Lin, P.-J., 112, 132–140*t*
Lin, Y.-C., 112, 132–140*t*
Lincoln, R.J., 1, 511–512
Lindquist, R.K., 39, 262, 295*t*, 447, 471–472,
    500, 569–570, 572, 590
Linit, M.J., 171, 213–227*t*
Liotta, L.A., 389
Lippiello, E., 428, 511
Lithourgidis, A., 454–455
Liu, H., 399, 416–420*t*
Liu, X., 198, 213–227*t*
Llebot, J.E., 509–510
Lloyd, M.W., 33–34, 45*t*, 125, 132–140*t*
Locey, K.J., 544, 570–571, 606
Loevinsohn, M., 541
Lopes, T., 295*t*
Lopez, C., 295*t*
Lopez-Collado, J., 295*t*
Lordan, J., 295*t*
Loreau, M., 123
Lowe, A.D., 463, 475
Lowrance, R., 150–151, 194, 213–227*t*
Lucas, C., 196, 213–227*t*
Lucas, H.L., 197, 349, 560
Luciano, P., 37–39, 295*t*
Lundberg, P., 478–479
Luschei, E.C., 109, 132–140*t*
Lyons, J.B., 158, 213–227*t*
Lysyk, T.J., 268–269
Lytra, I., 295*t*

M

Ma, Z., 5, 94, 97–100*t*, 211–212, 523, 579, 583,
    586, 595–596
MacArthur, R.H., 515–516
MacCord, F.S., 318, 324–325*t*
Mace, G.M., 63
MacGuidwin, A.E., 161–162, 213–227*t*

Machado, L.A.T., 509–510
Machete, M., 363, 372–378*t*
Macleod, K., 364–367, 365*f*, 372–378*t*
Madansky, A., 49
Madden, L.V., 79, 83, 85–86, 89, 97–100*t*, 122,
    162–164, 213–227*t*, 255–256, 448–449, 491,
    523, 543
Magurran, A.E., 343–344
Mandal, S., 342
Mandelbrot, B.B., 435, 439, 489, 507–508,
    510–511, 519, 592, 605
Mandonnet, N., 184–185
Manhout, J., 192–193, 213–227*t*
Manuel-Pinto, V., 295*t*
Maranon, T., 124, 132–140*t*
Marcum, D.B., 295*t*
Maritan, A., 521, 558, 592–593
Maron, J.L., 295*t*
Marques, T.A., 328, 369
Marshall, C.T., 388–389
Marshall, E.J.P., 44, 104–106, 132–140*t*,
    275–276, 397, 472, 524, 581–582
Martinez, N.D., 515–516
Matielle, T., 166, 213–227*t*
Matthews, K.R., 347, 372–378*t*
Matui, I., 246–247, 535–536
Maurer, B.A., 21–22, 353, 359, 372–378*t*,
    517
Maxwell, T.A.D., 209, 213–227*t*
May, R.M., 29–30, 343–344, 399–400,
    416–420*t*, 515, 546, 548–549, 552,
    592, 595
Mazzotti, F.J., 344, 372–378*t*
McArdle, B.H., 474, 489
McBride, R.S., 335–336, 372–378*t*
McCornack, B.P., 295*t*
McCoy, C.W., 175–176, 178, 213–227*t*
McCoy, E.L., 110–111, 132–140*t*
McCullagh, P., 56–57
McCune, B., 124–125
McGuire, J.A., 512
McGuire, J.U., 3, 4*f*, 18–19, 39, 237, 277–279*t*,
    495
McKellar, Q.A., 183, 213–227*t*
McKelvey, B., 503
McManus, M.L., 240, 265, 474, 497–498
McNeill, S., 518
McSorley, R., 92, 150–151, 156, 158, 169,
    213–227*t*
Mead, R., 30
Meats, A., 295*t*
Meech, K.J., 16

Mefford, M.J., 124–125
Megill, L., 117, 132–140*t*
Melakeberhan, H., 164, 213–227*t*
Merdić, E., 295*t*
Mesquita-Joanes, F., 295*t*
Messiaen, M., 192–193, 213–227*t*
Messing, R.H., 295*t*
Mestre, A., 295*t*
Meyers, M.T., 236, 277–279*t*
Micheli, M., 16
Michelsen, A., 153–154, 213–227*t*
Miller, S.A., 86, 97–100*t*
Milne, A., 268, 295*t*
Milne, B.T., 351, 359, 367–368, 372–378*t*,
    482–483, 589–590
Minarro, M., 295*t*
Mitchell, D.J., 92
Mitchell, S., 183, 213–227*t*
Mitton, J.B., 103
Moens, T., 195, 197, 213–227*t*, 576
Moerkens, R., 295*t*
Mokievskii, V.O., 199, 213–227*t*
Monaghan, K.A., 476, 480, 589
Monro, J., 36*t*
Monros, J.S., 295*t*
Monzo, C., 295*t*
Moodley, L., 198–199, 213–227*t*
Moon, R.D., 295*t*
Moore, D.M., 249, 277–279*t*
Moore, P.G., 18, 30
Moore, R.H., 431, 440–441*t*
Moradi-Vajargah, M., 295*t*
Moraes, W.B., 523
Morais, E.G.F., 295*t*
Morand, S., 182, 204, 213–227*t*, 451
Moreau, G., 87, 97–100*t*
Morgan, G.D., 161, 213–227*t*
Moriasi, D.N., 353, 482–483
Morin, A., 388–389
Morisita, M., 33, 45*t*
Morozov, A., 574–575
Morris, R.D., 449–450
Morris, R.F., 13
Moulins, A., 369, 372–378*t*
Mounport, D., 166, 213–227*t*
Moura, M.F., 295*t*
Mu, F., 198, 213–227*t*
Mukhopadhyay, M.C., 476–477
Mullens, T.A., 158–160, 544
Mulugeta, D., 112, 132–140*t*
Munyaneza, J.E., 295*t*
Murdoch, W.W., 22

Murthy, A.C., 354, 358–359, 361–362,
    372–378*t*, 385, 587
Myers, J.H., 241

N

Nachman, G., 295*t*
Nachtsheim, C., 62
Naiman, R.J., 387, 416–420*t*
Nakaoka, M., 254–256
Naranjo, S.E., 269, 295*t*
Naseri, B., 295*t*
Nash, J.F., 560
Nava-Camberos, U., 295*t*
Navarro-Campos, C., 295*t*
Naveen, N.C., 295*t*
Naylor, A.F., 36*t*
Ndiaye, S., 166, 213–227*t*
Neel, M.C., 125, 132–140*t*
Neher, D.A., 477
Neilson, R., 150–151, 197, 207–208, 213–227*t*
Nelder, J.A., 56–57
Nestel, D., 179, 213–227*t*
Neter, J., 62
Neyman, J., 559
Nie, P., 188–189, 213–227*t*
Nilsen, E.B., 349, 372–378*t*
Nishimura, M., 406, 435, 440–441*t*, 505,
    595–596
Niu, J.T., 483–484
Noble, S., 180–181, 213–227*t*
Noda, T., 254–256
Noe, J.P., 165–166, 213–227*t*
Noling, J.W., 168, 213–227*t*
Norberg, R.A., 514
Norberg, U.M.L., 514
Norris, J.R., 351–352
Nøttestad, L., 329–330, 372–378*t*
Noufaily, A., 403–404, 405*f*, 416–420*t*
Nozais, C., 209, 213–227*t*
Nudds, R.L., 512
Nunan, N., 385–386, 416–420*t*
Nyrop, J.P., 295*t*, 465–466, 471–472, 572

O

Ogato, Y., 428, 440–441*t*, 511
Ogawa, H., 519
Øien, N., 364–367, 372–378*t*
Okuda, T., 254–256
Ord, J.K., 484
Ortin-Angulo, M.C., 295*t*
Orza, G., 176–177, 213–227*t*

Ossó, A., 509–510
Ostlie, D.A., 427–428
Ostling, A., 515–516
Overholt, W.A., 295*t*
Owen, A.R.G., 32, 36*t*, 78, 78*f*, 97–100*t*, 246–247, 462–463

P

Padawer, T., 526
Palmer, L.F., 150–151, 207–208, 213–227*t*
Pan, W., 184, 255
Paoletti, M.G., 465
Park, J.J., 295*t*
Park, M., 183, 213–227*t*
Park, S.-J., 60, 60*t*, 93–94, 96, 121, 123, 148–149, 151–152, 155–156, 162–163, 167–168, 190, 195, 197, 202, 213–227*t*, 271, 342, 355–356, 362–363, 386–387, 465, 492–493, 500, 585–586, 606
Parkinson, L., 44
Parrado, J.L., 150–151, 156
Parra-Tabla, V., 126–127, 132–140*t*
Parrella, M.P., 295*t*
Parrish, J.K., 23
Parry, C.M., 401, 416–420*t*
Patch, L.H., 236, 277–279*t*
Paterson, S., 187, 213–227*t*
Patil, G.P., 32
Pattison, A.B., 170, 213–227*t*
Paturel, G., 426–427, 440–441*t*
Pearce, H., 454–455
Pearson, T.H., 211, 213–227*t*
Pedersen, H.C., 349–350, 372–378*t*
Pelzer, J., 190, 213–227*t*, 476–477
Peña, J.E., 295*t*
Peña-Lomeli, A., 295*t*
Pendleton, B.B., 295*t*
Peneva, V.K., 151–152, 213–227*t*, 476–477
Pérez, M.L.P., 295*t*
Perring, T.M., 295*t*
Perry, J.N., 21–23, 30, 32, 35–37, 39–45, 45*t*, 55–61, 72–73, 90, 96, 106, 108, 112, 132–140*t*, 158, 201–202, 207–208, 211–212, 236, 246, 249, 252, 259, 267–268, 271, 275–276, 295*t*, 319, 356, 357*t*, 359, 361–363, 389, 392, 397, 404, 445–446, 449–450, 472, 481, 494, 516–517, 524, 541, 546–550, 550*f*, 552, 560–561, 563, 569–572, 581–583, 587, 590–592, 594–595
Pertoldi, C., 478–479
Pessanha, A.L.M., 332, 372–378*t*
Peters, A., 174–175, 213–227*t*, 473*t*

Peters, R.H., 388–389, 416–420*t*
Petersen, J.K., 209, 213–227*t*, 295*t*, 313, 324–325*t*
Petit, C., 426–427, 440–441*t*
Petrovskii, S., 574–575
Pfannkuche, O., 189–190
Phillips, J.S., 383, 416–420*t*
Picanço, M.C., 295*t*
Pielou, E.C., 556–557
Pierce, F.J., 164, 213–227*t*
Pimentel, D., 119–120, 123
Pimm, S.L., 21–22, 353, 359, 372–378*t*, 517
Pitts, C.W., 240, 265, 474, 497–498
Plank, M.J., 525, 595
Podoler, H., 295*t*
Poehling, H.-M., 295*t*
Polley, H.W., 94, 119, 122–123, 132–140*t*, 587
Pope, K.L., 347, 372–378*t*
Poswal, M.A., 295*t*
Potere, D., 354
Poulin, R., 209, 525
Powell, B.F., 345–346, 372–378*t*
Powell, W., 275, 295*t*, 524
Powers, D.R., 512
Prager, S.M., 295*t*
Preston, F.W., 356, 515–516, 580–581
Prieto, T., 363, 372–378*t*
Proveaux, A.T., 250
Prugniel, P., 426–427, 440–441*t*
Pye, S.E., 311–312, 324–325*t*

Q

Qi, Y., 121, 132–140*t*
Qi, Z.M., 388
Quénéhervé, P., 166–167, 213–227*t*
Quinn, R.M., 353–354, 516–517
Qureshi, J.A., 295*t*

R

Rabaa, M.A., 401, 416–420*t*
Rafiee-Dastjerdi, H., 295*t*
Rakhshani, E., 295*t*
Ramamurthy, V.V., 295*t*
Ramankutty, N., 454–455
Ramsayer, J., 93, 97–100*t*, 274, 585–586
Rasche, L., 431
Rathbun, S.L., 150–151, 194, 213–227*t*
Ravel, J., 94
Reay-Jones, F.P.F., 44–45, 295*t*
Reddy, G.V.P., 559, 593

Reisig, D.D., 295*t*
Reiss, C.S., 343–344
Rennolls, K., 275, 295*t*, 524
Reusser, D.E., 414
Rexstad, E.A., 328, 369
Reyna-Robles, R., 295*t*
Reynolds, J.S., 150–151, 156
Reynoldson, T.B., 90, 97–100*t*
Rí, P., 334, 336, 372–378*t*
Richardson, C.A., 317, 324–325*t*
Richardson, L.F., 527
Ricker, W.E., 49, 51–53
Riedel, R.M., 79, 97–100*t*, 162–164, 213–227*t*, 543
Rinaldo, A., 521, 558, 592–593
Ristau, K., 192, 213–227*t*, 476–477
Robbins, C.S., 351, 353, 538–539
Rochford, M., 344, 372–378*t*
Rochon, K., 268–269
Rodrigues, L.C., 481–482
Rogers, S.I., 209, 213–227*t*
Room, P.M., 467
Roque-Malo, S., 432–433
Rossi, J.-P., 208, 213–227*t*
Rosso, M., 369, 372–378*t*
Rossow, W.B., 509–510
Roupakias, D., 454–455
Rousseau, J., 426–427, 440–441*t*
Routledge, R.D., 250
Rowe, R.C., 79, 97–100*t*, 162–164, 213–227*t*, 543
Royer, T.A., 295*t*
Ruess, L., 144, 153–154, 213–227*t*
Ruete, A., 74–75, 97–100*t*
Ruiz-Ramos, M., 526–527
Russo, A., 412–413, 416–420*t*
Ruzicka, J.J., 322, 324–325*t*, 331, 372–378*t*, 585
Rybaczuk, M., 507–508
Rybski, D., 414

S

Saachi, R., 514
Sabatelli, L., 481–482
Saeedifar, A., 295*t*
Saether, B.-E., 307–308, 581, 595
Saidel, G.M., 389
Sakamoto, J., 94
Salmah, M.R.C., 295*t*
Salvador, A.R., 451
Salvatore, A.R., 295*t*

Samaniego, H., 351, 359, 367–368, 372–378*t*, 482–483, 589–590
Sanchez, J.A., 295*t*
Sandhu, H.S., 559, 593
Sans, F.X., 113–114, 132–140*t*
Santos, R.S., 363, 372–378*t*
Saphir, M., 249, 277–279*t*
Sarabeev, V., 204, 213–227*t*
Sarasua, M.J., 295*t*
Sarkar, P.K., 476–477
Satake, A., 552
Sato, A.-H., 406, 435, 440–441*t*, 505, 595–596
Savage, V., 518
Sawyer, A.J., 543, 593
Schabenberger, O., 164, 213–227*t*
Schaffer, B., 295*t*
Schembri, P.J., 295*t*
Scherm, H., 82, 97–100*t*
Schiiz, J., 403
Schleicher, D.G., 423, 440–441*t*
Schlinger, E.I., 267
Schmidt, I.K., 153–154, 213–227*t*
Schmiedel, S., 403
Schmitthenner, A.R., 86, 97–100*t*
Schneider, A., 354
Schomaker, C.H., 160–161, 213–227*t*
Schratzberger, M., 209, 213–227*t*
Schreiber, Y., 595
Schroder, G.D., 371, 372–378*t*
Schuette, P., 369–370, 372–378*t*
Schulthess, F., 295*t*
Schumpeter, J.A., 16
Schuster, D.J., 295*t*
Schuster, W.S.F., 201, 395, 414, 416–420*t*, 447, 524, 553, 570–571, 588, 592–593, 595
Schütte, U.M.E., 94
Sciberras, M., 295*t*
Scoles, G.A., 268–269
Scott, E.L., 559
Seabra, M.I., 363, 372–378*t*
Seber, G.A.F., 13–14
Sedaratian, A., 295*t*
Seed, R., 317, 324–325*t*
Segre, L., 295*t*
Segre, P.S., 512
Seiter, N.J., 295*t*
Sérandour, G., 351, 359, 367–368, 372–378*t*, 482–483, 589–590
Serra, G., 37–39, 295*t*
Sétamou, M., 295*t*
Seufert, V., 454–455
Sfriso, A., 72, 97–100*t*

Shaw, D.J., 181–182, 187, 203–204, 213–227*t*, 481–482
Shaw, M.W., 124–125
Shelton, A.M., 572
Shi, P.-J., 559, 593
Shipp, J.L., 39, 262, 295*t*, 447, 471–472, 500, 569–570, 572, 590
Shiyomi, M., 117–118, 122, 132–140*t*, 454–455, 523, 586
Shnerb, N.M., 595
Shorrocks, B., 295*t*
Sienkiewicz, J., 390–391, 416–420*t*
Silva Matos, D.M., 123, 132–140*t*
Silva, M.A., 332, 363, 372–378*t*
Silva-Lima, A.W., 295*t*
Simaika, Y.M., 510
Simeoni, A.E., 250–251, 277–279*t*
Simmons, K.A., 517–518
Šizling, A.L., 515–516
Skandalis, D.A., 512
Skaret, G., 329–330, 372–378*t*
Skellam, J.G., 18
Smith, B., 344, 372–378*t*
Smith, D.C., 77, 97–100*t*
Smith, G.R., 348, 372–378*t*
Smith, H.F., 522, 595
Smith, J.W., 295*t*
Smith, P.G., 481–482
Smith, V.H., 516
Smits, P.H., 473*t*
Snedecor, G.W., 62
Soberón, M.J., 541
Soetaert, K., 200–201, 213–227*t*
Sohlenius, B., 144, 156
Solvang, H., 349, 372–378*t*
Somerfield, P.J., 147, 213–227*t*, 580
Sosnoskie, L.M., 109, 132–140*t*
Soto, A., 295*t*
Southwood, T.R.E., 2, 13–14, 28–29, 189–190, 267–268, 474, 515, 552–553, 595
Sparrow, D.H., 110–111, 132–140*t*
Spivak, M., 295*t*
Spiridonov, S.E., 174, 213–227*t*, 473*t*
Sprugel, D.G., 59
Sriwati, R., 171–173, 213–227*t*
Stanger, B.A., 162, 213–227*t*
Stanley, H.E., 400–401, 416–420*t*
Stansly, P.A., 295*t*
Stanton, J.M., 170, 213–227*t*
Starr, J.L., 146–147, 213–227*t*, 543
Stauffer, D., 509
Stear, M.J., 183, 213–227*t*

Steen, H., 349–350
Steiner, M.Y., 295*t*, 572
Stenalt, E., 209, 213–227*t*, 295*t*, 313, 324–325*t*
Stephen, A.G., 310, 324–325*t*
Stevenson, J.P., 258, 277–279*t*
Steward, G.F., 77, 97–100*t*
Stewart-Oaten, A., 22
Steyaert, M., 196–197, 213–227*t*
Stinner, B.R., 465
Stinner, D.H., 465
Storaas, T., 349, 372–378*t*
Storch, D., 515–517
Stouffer, R.J., 431
Strain, S., 183, 213–227*t*
Strindberg, S., 328, 369
Stuart, A., 484
Stuart, R.J., 173–174
Stubbins, F.L., 295*t*
Subramanyam, B., 295*t*, 585–586
Suenaga-Kanetani, H., 81, 97–100*t*
Sugihara, G., 343–344, 515, 595
Sun, I.-F., 112, 132–140*t*
Svendsen, W., 349–350
Swann, D.E., 345–346, 372–378*t*
Swartz, T.M., 250

T
Takahashi, S., 117–118, 454–455, 523, 586
Takayasu, H., 411, 416–420*t*
Takayasu, M., 411, 416–420*t*
Takemoto, S., 171–173, 213–227*t*
Talebi, A.A., 295*t*
Tamo, M., 295*t*
Tan, C.S., 388
Tao, F.L., 295*t*
Tarquis, A.M., 526–527
Taylor, C.C., 328–329, 372–378*t*
Taylor, D.B., 295*t*
Taylor, L.R. (LRT), 1, 3, 5, 8*f*, 14, 18, 21–23, 27–28, 32, 35–37, 39–42, 45*t*, 56–57, 59, 72–73, 90, 108, 197, 207–208, 236–237, 246, 250, 256, 266–267, 271, 277–279*t*, 340, 355*f*, 356, 357*t*, 359, 361–363, 392, 399, 445–447, 449–450, 463–464, 471–472, 475, 491–492, 491*f*, 498–500, 516–517, 541, 546–547, 550, 560–562, 569–572, 576–577, 579, 583, 587, 590–591
Taylor, R.A.J. (RAJT), 22–23, 27–28, 32, 35–37, 39, 59–60, 60*t*, 93–94, 96, 121, 123, 148, 151–152, 155–156, 162–163, 167–168, 173–175, 190, 195, 197, 202, 213–227*t*, 240–242, 248–249, 251–252, 256, 261–262,

265, 267–268, 271, 277–279*t*, 295*t*, 342, 347, 355–356, 362–363, 386–388, 399, 431, 440–441*t*, 445–448, 450–451, 465, 471–472, 474–475, 492–493, 496–500, 546, 550, 560–562, 569–572, 574–577, 585–586, 590, 592, 594–595, 606
Tepsich, P., 369, 372–378*t*
ter Braak, C.J.F., 481
Terranova, A.C., 175–176, 213–227*t*
Tesfaye, A., 295*t*
Tessier, G., 503, 511–512
Thanh, D.P., 401, 416–420*t*
Thanopoulos, R.C., 295*t*
Theureau, G., 426–427, 440–441*t*
Thiel, H., 189–190
Thomas, C.F.G., 44
Thomas, L., 328, 365, 369–370
Thompson, C.N., 401, 416–420*t*
Thompson, D.W., 503, 507–508
Timmer, L.W., 169
Tippett, M.K., 429, 440–441*t*
Tiwari, S., 249, 277–279*t*
Tixier, M.S., 295*t*
Tobalske, B.W., 512
Tobin, P.C., 295*t*
Togashi, K., 172–173, 213–227*t*
Tokeshi, M., 121
Toledo, M.C.B., 512
Tompkins, D.M., 185–186
Topham, P.B., 145–147, 156, 183–184, 213–227*t*
Tounou, A.K., 295*t*
Townsend, A.M., 295*t*
Traunspurger, W., 190–192, 213–227*t*, 476–477
Tredennick, A.T., 123
Trigal, C., 74–75, 97–100*t*
Trotter, D., 199–200, 213–227*t*
Trumble, J.T., 248–249, 268, 295*t*, 342, 450, 475–476, 572, 577
Trumper, E.V., 295*t*
Tsai, A.-Y., 75–76, 97–100*t*
Tsuiki, M., 117–118, 122, 454–455, 523, 586
Tsutsumi, M., 117–118, 122, 454–455, 523, 586
Tumlinson, J.H., 250
Turechek, W.W., 523
Turner, D.S., 345–346, 372–378*t*

U

Udalov, A.A., 199, 213–227*t*
Uglem, I., 295*t*
Uhlig, G., 189–190

Umina, P.A., 295*t*
Upton, J.E., 295*t*
Urías-López, M.A., 295*t*
Utida, S., 36*t*

V

Vaala, D.A., 348, 372–378*t*
Vaillant, J., 295*t*
van der Heyden, H., 87, 97–100*t*
Van Gaever, S., 447
Van Gansbeke, D., 195, 213–227*t*, 576
van Liew, M.W., 353, 482–483
van Ness, J.W., 439
Vanaclocha, P., 295*t*
Vanaverbeke, J., 196, 213–227*t*
Vanier, C., 117, 132–140*t*
Vanreusel, A., 196, 213–227*t*, 447
Varley, G.C., 16, 29–30
Vavra, J.M., 383, 416–420*t*
Veith, T.L., 353, 482–483
Velidis, G., 150–151, 194, 213–227*t*
Venette, R.C., 150–151
Venturelli, C., 295*t*
Vera-Graziano, J., 295*t*
Vidal, S., 295*t*
Vieites, D.R., 481
Viketoft, M., 151–153, 213–227*t*
Vincx, M., 195–196, 198–201, 209, 213–227*t*, 446–447, 576
Viney, M.E., 187, 213–227*t*
Viquerat, S., 367, 372–378*t*, 482–483, 589–590
Vlachos, C.G., 348–349, 372–378*t*
Vlachostergios, D., 454–455
Vogedes, D., 295*t*
Voronov, D.A., 174, 213–227*t*, 473*t*
Vu Thieu, N.T., 401, 416–420*t*
Vucetich, J.A., 478–479

W

Wagner, R.M., 423, 440–441*t*
Wald, A., 54, 468
Wale, M., 295*t*
Walford, L.A., 328–329
Walker, L.R., 117, 132–140*t*
Wallace, D.S., 183, 213–227*t*
Wallis, J.R., 510–511
Wang, D., 526
Wang, G.T., 188–189, 213–227*t*
Wang, H.-H., 112, 132–140*t*
Wang, X., 388, 431
Wang, Y., 118, 132–140*t*, 405–406, 416–420*t*, 586

Wang, Y.-S., 122, 523
Wang, Z., 388
Ware, G.O., 82, 97–100*t*
Warr, K., 209, 213–227*t*
Warren, J.E., 171, 213–227*t*
Warton, D.I., 49, 52–53, 63
Warwick, R.M., 147, 213–227*t*, 580
Waters, W.E., 31–32
Watkinson, A.R., 123, 132–140*t*
Way, M.O., 295*t*
Webster, J.M., 199–200, 213–227*t*
Webster, R., 145, 156, 183–184
Wedderburn, R.W.M., 56–57
Wei, Z., 118, 132–140*t*, 586
Weiner, J., 383
Weiss, P.W., 114–115, 132–140*t*
Welacky, T., 388
Welch, K.C., 512
Welch, S.M., 295*t*
Weller, D.E., 519
Weryk, R., 16
West, G.B., 519–521, 595
Westoby, M., 49, 52–53, 63
Wheeler, S., 295*t*
Wheeler, T.A., 79, 97–100*t*, 146–147, 162–164, 213–227*t*, 543
White, E.P., 544, 570–571, 606
White, G.F., 173–174
Widmeyer, P.A., 125, 132–140*t*
Wikner, J., 77, 97–100*t*
Williams, C.B., 2–3, 30, 266, 475, 569, 571, 586
Williams, J.R., 431
Wilsey, B.J., 94, 119, 122–123, 132–140*t*, 587
Wilson, B.J., 106
Wilson, E.O., 515–516
Wilson, L.T., 295*t*, 467
Winder, J.M., 479–480
Winder, L., 43–44
Winsor, C.P., 328–329
Wishart, W.D., 313–314, 324–325*t*
Witt, C.C., 512
Witting, D.A., 335–336, 372–378*t*
Woiwod, I.P., 5, 32, 35–37, 39, 56–57, 59, 72, 90, 108, 207–208, 236, 246, 271, 340, 355*f*, 356, 357*t*, 359, 361–363, 392, 445–446, 449–450, 472, 491–492, 491*f*, 516–517, 547, 550, 560–561, 569, 571–572, 579, 583, 587, 590–591
Wolfe, M., 454–455
Wolfram, S., 508, 562, 592, 597, 607
Wright, F.J., 479–480
Wright, I.J., 49, 52–53, 63

Wright, Q., 8–9, 409, 416–420*t*, 527
Wright, R.J., 295*t*
Wu, J., 477
Wu, S.G., 188–189, 213–227*t*

X

Xi, B.W., 188–189, 213–227*t*
Xiao, X., 544, 570–571, 606
Xu, J.M., 483–484
Xu, M., 201, 395, 414, 416–420*t*, 447, 524–526, 542, 553, 570–571, 588, 592–593, 595
Xu, R.-M., 295*t*
Xu, X.-M., 83, 85–86, 97–100*t*, 255–256, 448–449, 523

Y

Yamamoto, T., 254
Yamauchi, A., 342
Yang, F., 121, 132–140*t*
Yang, J.-R., 85–86, 97–100*t*
Yang, J.W., 388
Yang, P., 295*t*
Yang, Y., 118, 132–140*t*, 586
Yasuda, T., 117–118, 454–455, 523, 586
Yates, F., 243–244, 277–279*t*
Yeates, G.W., 144
Yede, 295*t*
Yeh, T., 250–251, 277–279*t*
Yiruhan, Y., 122, 523
Yoccoz, N.G., 349–350
Yoda, K., 519
Yoder, F., 176–177, 213–227*t*
Yosef, A., 397–398, 406–407, 416–420*t*
Yoshimura, J., 117–118, 454–455, 523, 586
Young, I.M., 144, 515, 575–576
Youngman, R.R., 249, 277–279*t*
Yuan, H., 406–407
Yvonne, A., 400–401, 416–420*t*

Z

Zagier, D., 437–438
Zalucki, M.P., 295*t*
Zamponi, M.O., 320–321, 324–325*t*
Zeng, F., 431
Zeo, S.M., 295*t*
Zhang, B., 556–557, 592
Zhang, J., 121, 132–140*t*
Zhang, M., 153, 213–227*t*
Zhang, Q., 405–406, 416–420*t*
Zhang, T., 198, 213–227*t*
Zhang, T.Q., 388

Zhang, X.-K., 153, 213–227*t*
Zhang, Z., 198, 213–227*t*
Zhao, Z.-D., 399, 416–420*t*
Zhong, X., 94
Zhou, B., 414
Zhou, H., 198, 213–227*t*

Zhou, X., 94
Zhu, C., 405–406, 416–420*t*
Zhu, J., 161, 213–227*t*
Zhu, J.J., 295*t*
Zimmermann, H., 308, 324–325*t*
Žitko, T., 295*t*

Index

Note: Page numbers followed by *f* indicate figures and *t* indicate tables.

A

Absolute density, 14
Absolute magnitude, 424–425
Abundance-occupancy curve, 516–517
Acrobeloides, 195
Adès distribution, 39–42, 41*f*, 41*t*, 542, 549
Aedes
  A. aegypti, 265
  A. albopictus, 265
Agent-based models
  arrangement in space, 563–565
  cellular automaton, 562–563
Aggregation measurement, 45*t*
  Green's C_x coefficient, 33
  indices using mean and variance, 29–45
    negative binomial k, 30–32
    variance-mean ratio, 29–30
  Iwao's patchiness index, 35
  Iwao's p index, 35
  Lloyd's mean crowding, 33–35
  Morisita's patchiness index, 33
  nearest-neighbor analysis, 28–29
  variance-mean relationship, 35–45, 41*t*
    Adès distribution, 39–42, 41*f*, 41*t*
    Perry and Hewitt's number of moves index, 42–43
    SADIE, 43–45
    TPL, as index of aggregation, 37–39
Agriotypus armatus, 273
Agroecology, 607
Air traffic control (ATC), 405–406
Akaike Information Criterion (AIC), 255–256
Akinetes, 70–71
Allometric growth, 511–512
Allometry
  biology, power laws of
    allometric growth, 511–512
    binomial power law (BPL), 523–524
    density-size and variance-size laws, 524–526
    flying animals, dimensional relationships for, 512–515
    fractal movement, 515

  genetics and physiology, 526
  Kleiber's law of metabolism, 518
  respiration, 518–519
  self-thinning and space-filling rule, 519–521
  soil fertility and crop yields, 522
  species-area, 515–518
  species size, frequency of, 515
  taxonomy, 526–527
mathematics, power laws of
  DLA, 506–507
  fractals, 507–508
  Pareto distribution, 503–504
  scale-free networks, 505–506
  spectra, 504–505
  Zipf's law, 504
physics, power laws of
  geophysics, 511
  hydrology, 510–511
  inverse square law, 508
  meteorology, 509–510
  percolation, 509
  self-organized criticality, 508
  Stefan–Boltzmann law, 508
sociology, power laws of, 527–528
  Richardson's law of conflict, 527–528
Alopecurus myosuroides. See Black grass
Ambrosia artemisiifolia. See Ragweed
Amphibians
  frogs, in Alpine habitat in California, 347
  larval abundance, in ephemeral ponds, Ohio, 348
Amphisbetia operculata, 320–322, 322*f*
Animal-parasitic nematodes, 180
Annelids, 206–212
Annual crops
  broad beans, 164–165
  eggplant (aubergine), 166
  mixed vegetables, 166–167
  potatoes, 160–164
  soybeans, 164
  tobacco, 165–166
Aphelocoma californica. See Western scrub jay

Apparent magnitude, 424–425
Apples, powdery mildew on, 83–85, 448–449, 449*f*
Aquatic nematodes, 189–190
Arctomecon californica. See Bearpoppy
Arthropods, TPL analysis, 260–279
  changes in scale, effect of, 274–276
  Coleoptera, 242–244
  competition, 273–274
  consistency across space, time, and stage, 261–266
  Crustacea (crustacean), 254–260, 276
  Lepidoptera, 235–242
  parasitism, 273
  predation, 271–273
  sampling methods
    effect of, 268–271
    efficiency and consistency between samplers, 266–268
Astronomy
  cyanogen, in comet Hale-Bopp, 423
  heavenly bodies, distribution of, 424–425
  RA and DC transects
    PGC, galaxies in, 426–428, 427*f*
    SAO star catalog, 425*f*, 426
    YBSC stars, 425–426, 425*f*
Astropecten irregularis, 317, 318*f*
ATC. *See* Air traffic control (ATC)
Aubergine, 166
Autocorrelation matrix, 593–594

B
Bacteria cultures, 93–94
Baermann funnel, 143–144
Balanus balanoides, 257–258
Balearic Islands
  meiobenthos, 446–447, 446*f*
  polychaetes, 209
Banana, 170
Barnacles
  B. balanoides distribution, stratification in, 257–258
  Chthamalus species, in Japan, 254–256
  E. modestus cyprid larvae, settling behavior of, 258–260
Bartlett's 3-group regression method, 54–55, 322–323
Bartlett's method, 398
BBS. *See* Breeding Bird Survey (BBS)
Bearpoppy, 117
Beat cup method, 270
Beetles. *See* Coleoptera

Benthic nematodes, 198–202
Bias, in TPL estimation, 58–59
Binomial power law (BPL), 122, 491, 523–524
Binomial sampling, 466–468
Bioindicators, freshwater nematodes
  farm ponds in Belgium, 192–194
  highland streams in Germany, 191–192
  lakes in Sweden, 192
  restored wetland in Georgia, 194–195
  rivers in Germany, 190–191
Biological control agent (BCA), 91–92
Biological evidence for TPL, 571–572
Biological models, TPL
  agent-based models
    arrangement in space, 563–565
    cellular automaton, 562–563
    dispersal distance, 558–560
    ideal free distribution theory, 560–561
    Lewontin-Cohen model, 553–554
    nonlinear maps
      effect of competition, 551
      models in time, 547–548
      strong density dependence, 548–550
      singularities, in other models, 556–558
      temporal TPL and stability, 552–553
Biological Monitoring Working Party (BMWP) scoring system, 479–480
Biomass, 209–210
Birds
  Audubon Society's Christmas Bird Count (CBC), 359–363
  British Trust for Ornithology (BTO), 356–359, 357*t*
  grassland sparrows, in continental USA, 353–354
  jays, in Western USA
    Breeding Bird Survey (BBS), 351–352
    field data and model output, temporal TPLs of, 351–353, 352*f*
    social systems and behavior, 351
  point transect sampling system, 328
  in urban and nonurban sites
    CBC data, 354
    TPL regressions, 354–356, 355*f*
  willow ptarmigan, in Norway, 349–351
    hybrid spatial TPL analyses, 350–351, 350*f*
    ideal free distribution, 349–350
    line-transect surveys, 349–350
Black grass, 106–108, 107*f*
Boltzmann distribution, 537
Bootstrapping resampling method, 117

Bounded ratios, 451–455
  parasite prevalence, 451–452
  sex ratio, 451, 452*f*
Box counting, 578, 596
Breeding Bird Survey (BBS), 351–353
British Trust for Ornithology (BTO), 353–354,
    356–359, 357*t*
Broad beans, 164–165
"Broken power law" model, 518
Bromus commutatus. See Meadow brome
Brownian motion, 505
Bryozoans, in Greenland Sea, 319–320
BTO. *See* British Trust for Ornithology
    (BTO)
Burlap banding method, 241
Bursaphelenchus
  B. mucronatus, 171–173
  B. xylophilus, 171–173
Butterflies, 235–236

C

Camallanus cotti, 188–189
Capitella capitata, 211
Capsicum annuum. See Peppers
Carp, 188–189
Catastrophic change, 450–451
  pesticide effects, 450–451
CBC. *See* Christmas Bird Count (CBC)
Cellular automaton, 562–563
Censuses, 23
Central limit theorem (CLT), 20
Cestodes, 205–206
Cetaceans
  Azores, 363–364
  British Isles, 364–366
  Ligurian Sea, 369
  North Sea, 366–368
CGF. *See* Cumulant-generating function
    (CGF)
Chenopodium album. See Lambsquarters
Chlamydia trachomatis, 404, 405*f*
Chlorophyll, 387
Christmas Bird Count (CBC), 354, 359–363
Chronogaster, 195
Chthamalus species, in Japan, 254–256
Ciliates
  in East China Sea, 75–76
  on flatworms, 90–91
Citrus, 167–169
City Clustering Algorithm, 414
Clover, 169–170
CLT. *See* Central limit theorem (CLT)

Clupeidae, 337
CN. *See* Cyanogen (CN)
Cockroaches, 180–181
Coefficient of variation (CV), 30, 340–342,
    366, 469–470
Coleoptera, 242
  Colorado potato beetle, 244–247
  Japanese beetle
    adults, 250–252
    in Azores, 252–254
    larvae, 247
    in North America, 247–254
  wireworms, 242–244
Colorado potato beetle (CPB), 244–247
Comet Hale-Bopp, cyanogen in, 423
Commodity crops, 452–455, 453*f*
  maize, 128–129
  sugar cane, 129–130
  potatoes, 131–132
  wheat, 130–131
Common logs, 2
Community TPLs, 204
Compositions approach, 544
Confidence intervals (CI), 328, 366
Convenience store sales, in Japan, 410–412
Corrugator supercilii (COR), 390–391
Cotton, in California, 158–160
Cotton root-knot nematode, 166–167
Countable number series, 591
Cowpeas, in California, 158–160
CPB. *See* Colorado potato beetle (CPB)
Criconemella sphaerocephalus, 146–147
Crime statistics, in England and Wales,
    406–408
Crinoids, in São Paulo State, Brazil,
    318–319
Critical survival time (CT), 339
Crops
  annual crops, 160–167
  commodity crops, 128–131
  in Florida, 158
  perennial crops, 167–173
Crosscorrelations, 546
Crustacea (crustacean), 276
  B. balanoides, 257–258
  Chthamalus species, 254–256
  E. modestus cyprid, 258–260
Cucumber crop, thrips in, 447–448, 448*f*
Culex tritaeniorhynchus, 265
"Curved power law" model, 518
C_x coefficient, 33
Cyanogen (CN), 423

D

Darwin mounds, nematodes on, 446f, 447
Declination (DC) transects
  PGC, galaxies in, 426–428, 427f
  SAO star catalog, 425f, 426
  YBSC stars, 425–426, 425f
Delta technique, 484
Demersal fish, 332–334
Deming regression, 53–54
Density-dependent efficiency, 497–501
Density mass variance allometry (DMVA),
  525–526
Density-size law, 524–526
Depth gradient, benthic nematodes, 198–199
Dermacentor andersoni, 268–269
Devil's thorn, 114–115
Diaphorina citri. See Lemon psyllid
Diatoms, 72
Diffusion-limited aggregation (DLA), 246–247,
  389–390, 506–507, 535–536
Discrete logistic equation, 551
Discrete time models, 556
Dispersal distance, 558–560
Distance sampling
  DISTANCE software, 328
  line transect sampling, 328
  point transect sampling system, 328
  strip sampling, 327–328
DISTANCE software, 328, 369
Ditylenchus dipsaci, 164
DLA. *See* Diffusion-limited aggregation (DLA)
Domestic fowl, 205–206
Drosophilids, 181

E

Earthquakes, in Japan, 428–429
Earthworms
  in Colombia, 208
  in Scotland, 207–208
ECB. *See* European corn borer (ECB)
Echinoderms
  crinoids, in São Paulo State, Brazil, 318–319
  starfish, in North Wales, 317, 318f
Edible palm, in Brazil, 123–124
Eelgrass, 125–126
Eggplant (aubergine), 166
Electrodermal activity (EDA), 390–391
Electromyography (EMG), 390–391
ELISA. *See* Enzyme-linked immunosorbent
  assay (ELISA)
Elminius modestus, 258–260
Emex australis. See Devil's thorn

EMG. *See* Electromyography (EMG)
Enhanced Fujita scale, 429–430
Enoploides longispiculosus, 197
Entomopathogenic nematodes (EPNs),
  143–144, 173–180
  binomial sampling probabilities for,
    467–468, 468t
  power law sampling of, 472, 473t
Environmental assessment
  cost of conservation, 478–479
  nematode community, 476
  perturbations detection, 476–478
  stream water quality, 479–481
  of wind farms, 481
Environmental stochasticity, 551
Enzyme-linked immunosorbent assay (ELISA),
  86–87
EPIC, 431
EPNs. *See* Entomopathogenic nematodes
  (EPNs)
Euclidean spatial dimension, 535
European corn borer (ECB), 3, 4f, 236–237
European wars, 409–410
Euterpe edulis, 123–124
Eutetranychus orientalis, 272
Excitatory impulses, 539–540
Exoenzymes
  bioremediation, 385
  in soil, TPL analysis, 385–387, 386f
Exponential growth model, 395–397

F

Farm ponds, in Belgium, 192–194
Fidelity metrics, 480
Fin whales, in Ligurian Sea, 369
Fish
  Californian commercial fisheries,
    342–344
  demersal fish, 332–334
  haddock and whiting, 329, 330f
  herring and mackerel, 329–331
  larvae entering Pamlico-Albemarle Sound
    lagoon system, 336–337
  pelagic fish larvae
    in New Jersey, 335–336
    in Portugal, 334–335
  salmon, in Northeast Pacific Ocean, 331
  sea trout, in England
    adults, annual rod and commercial catches,
      340–342
    alevins and parr, 337–339
  variation, sources of, 328–329

Fitting Taylor' power law
  Bartlett's 3-group regression, 54–55
  bias, 58–59
  comparison of models, 60–61, 60*t*
  Deming regression, 53–54
  functional regression, 51–55, 51*f*
  geometric mean regression, 52–53
  methods, 55–58, 58*f*, 64
  parallel-line analysis, 61–63
    ordinary dependent regression, 61–62, 62*t*
    residual major axis (geometric mean)
      regression, 62–63
  standard regression model, 50–51
Fixed-precision sampling techniques, 472
Fixed-radius method, 240–241
"Floor-and-ceiling" model, 548
Fluctuation scaling, 3
Flying animals, 512–515
Foraminifera, 72–73
Foreign exchange markets, 435–437
Forest, 151–152
  and old field, 158
Fractal model, 535
Fractal movement, 515
Fractal property, 594–595
Frequency distribution, 3, 4*f*, 462, 475–476,
    537–538, 549
Freshwater nematodes, bioindicators, 190
  farm ponds in Belgium, 192–194
  highland streams in Germany, 191–192
  lakes in Sweden, 192
  restored wetland in Georgia, 194–195
  rivers in Germany, 190–191
Fronczak's model, 537–538
Functional regression, 51–55, 51*f*

G

Galleria-baiting bioassay, 179–180
Gamma distribution, 559
Gastropods
  in Alberta bighorn sheep habitats, 313–315
  re-colonisation in Mariager Fjord, Denmark,
    313
  slugs, in Northumberland, 315–316
Gauss's prime number equation, 437
General Circulation Model (GCM), 431
Generalized linear model (GLM), 56–57
Genetically modified (GM) crops, 481
Geographic Information System (GIS), 360
Geometric mean regression (GMR), 52–53
Geophysical Fluid Dynamics Laboratory
    (GFDL), 431

Geophysics, 511
GILSYM, 431
Glassy-winged sharpshooter, 269
Golden eelworms, 160
Graphidium strigosum, 185–186
Grasslands
  rangeland, in Mongolia, 121–123
  in Shaanxi Province, 118
  spatial heterogeneity, 117–118
  tallgrass prairies, in Texas, 119–121
Grassland sparrows, 353–354
Grassy pasture, in Sweden, 151
Greenhouse whitefly, 263–264
Green oak tortrix, 274
Green Revolution, 128–129, 131
Green's C_x coefficient, 33
Grey mullet, 204
Gutenberg–Richter law, 511
Gymnamoebae, 69–70
Gymnetron pascuorum, 275
Gymnorhinus cyanocephalus. *See* Piñon jays
Gypsy moth, 239–242, 261, 265, 273

H

Habitat heterogeneity, 515–516
Haddock, 329, 330*f*
Harbor porpoise, in North Sea, 366–368
Haser model of gas diffusion, 423
Hausdorff-Besicovitch dimension, 507, 535
Heisenberg uncertainty, 5–8
Herptiles, in Rincón Mountains, 345–347
Herring, 329–331
Heterorhabditis bacteriophora, 175–178
Heterotrophic ciliates, 76
Higher moments, 542–543
Highland streams, in Germany, 191–192
HIV. *See* Human immunodeficiency virus
    (HIV)
Howardula aoronymphium, 181
Human behavior
  crime statistics, in England and Wales,
    406–408
  European wars, 409–410
  stress, in air traffic controllers, 405–406
Human demography
  cyberspace and physical space, movements
    in, 399
  mortality, in England and Wales,
    397–398
  population density, in Norway, 395–397
  and sociology, 587–588
  US decennial census, 392–394

Human health
  disease monitoring, 403–404
  HIV/AIDS, 399–401
  measles and whooping cough, 402–403
  typhoid in Cambodia, 401–402
Human immunodeficiency virus (HIV),
    399–401
Human microbiome, 94–97
Hurst exponent, 510, 533–534
Hydroids, in Argentina, 320–322
Hydrology, 510–511
Hymenolepis tenerrima, 205
HYPERLEDA Galaxy Catalog, 426–427, 427*f*

I

Ideal free distribution (IFD), 349–350, 560–561
Inconsistent sampling, 447–448
  thrips in cucumber crop, 447–448, 448*f*
Infective juveniles (IJs), 173–174
Inhibitory impulses, 539–540
Insectivorous plants, in Morocco and Iberia,
    124–125
Insects
  Coleoptera, 242
    Colorado potato beetle, 244–247
    Japanese beetle (*see* Japanese beetle (JB))
    wireworms, 242–244
  Crustacea (crustacean), 276
    B. balanoides, 257–258
    Chthamalus species, 254–256
    E. modestus cyprid, 258–260
  Lepidoptera
    European corn borer, 236–237
    gypsy moth, 239–242
    life stages, 235–236
    winter moth, 237–238
Intergovernmental Panel on Climate Change
    Fourth Assessment (IPCC-4A), 431
Internet, 536
Intertidal molluscs, Isle of Man, 311–312
Intertidal nematodes, 195–197
Invasive species
  Burmese python, in Florida, 344
  crayfish, in Spain, 276
  Culex tritaeniorhynchus, in Greece, 265
  devil's thorn, in Australia, 114–115
  fire ant, in Louisiana, 271
  flagellate, in Sweden, 73–75
  grey mullet, in the Mediterranean, 204
  gypsy moth, in U.S.A., 239–242
  Japanese beetle, in U.S.A., 247–254
  Japanese beetle in The Azores, 252–254

kudzu bug, in Georgia, 264
ladiesthumb, in U.S.A., 110
mealybug, in Spain, 263–264
Queen Anne's lace, in U.S.A., 242
ragweed, in France, 243
Sally Lightfoot, in Malta, 276
Siphonia rufofascia, in Hawaii, 273
winter moth, in Canada, 237–238
Inverse density, 28
Inverse Gaussian distribution, 462
Inverse square law, 508
Invertebrate hosts, of nematodes
  cockroaches, 180–181
  drosophilids, 181
Invertebrates, TPL analysis, 324–325
  bryozoans, in Greenland Sea, 319–320
  echinoderms, 317–319
  hydroids, in Argentina, 320–322
  jellyfish, in Oregon-Washington coast,
    322–324
  molluscs, 309–316
  rotifers, 305–309
Iwao-Lloyd model, 81–82
Iwao's patchiness index, 35
Iwao's ρ index, 35

J

Japanese beetle (JB), 247
  adult, 250–252
  in Azores, 252–254
  larvae, 247
  in North America, 247–254
Jays, in Western USA
  BBS surveys, 351–352
  field data and model output, temporal TPLs
    of, 351–353, 352*f*
  social systems and behavior, 351
Jellyfish, in Oregon-Washington coast,
    322–324

K

Kangaroo rat mounds, 371–372
Kawaguchi analysis, 81–82
Kelp, nematodes on, 199–200
Kleiber's law of metabolism, 518
Kurtosis, 15

L

Lakes, in Sweden, 192
Lambsquarters, 110–112
Latitudinal gradient, 198

Lattice model, 545–546
Leeches, in Cumbrian stream, 206–207
Lemon psyllid, 264, 270, 273
Lepidoptera
  European corn borer, 236–237
  gypsy moth, 239–242
  winter moth, 237–238
  Leptinotarsa decemlineata. See Colorado
    potato beetle (CPB)
Lewontin-Cohen model, 429–430,
    553–554
Light traps, 474, 496–498
Line transect sampling, 328
Littoral nematodes, 197–198
Liza haematocheilus, 204
Lloyd's mean crowding and patchiness indices,
    112, 317
Logarithmic transformation, 463–464
Logistic equation, 605
Lognormal distribution, 18–19, 19*f*, 504, 543,
    548, 555, 563–564
Longidorus elongatus, 145
Lymantria dispar. See Gypsy moth

M
Mackerel, 329–331
Madden-Hughes power law (BPL), 467
Maize, 128–129
Mammals
  cetaceans
    Azores, 363–364
    British Isles, 364–366
    Ligurian Sea, 369
    North Sea, 366–368
  terrestrial
    kangaroo rat mounds, in New Mexico,
      371–372
    ungulate herds, in Kenya's Rift Valley,
      369–371
Mandelbrot's model, 519
Marine mammals. *See* Cetaceans
Marine bivalves, 313
Marine nematodes, 195–202
Marine viruses, in California and Sweden,
    77–78
Markovian chain model, 558
Matérn cluster process, 559
Mathematical power laws
  DLA, 506–507
  fractals, 507–508
  Pareto distribution, 503–504
  scale-free networks, 505–506

spectra, 504–505
Zipf's law, 504
Meadow brome, 106–108
Mealybug, 263–264
Mean crowding index, 33–35, 36*t*, 317
Mean–variance relationship, 2–3
Measles, 402–404
Mecinus pyraster, 275
Meiobenthos, in Balearic Islands, 446–447,
    446*f*
Meloidogyne
  M. arenaria, 91–93
  M. incognita, 166–167
Menhaden, 336–337
Metabolic footprint, 154–155
Metastatic cancers, 389–390
Meteorology, 509–510
Mice, 187
Microorganisms, TPL analysis, 97–100
  animal hosts
    ciliates, flatworms, 90–91
    *Pasteuria penetrans, Meloidogyne
      arenaria*, 91–93
  free living
    bacteria, Siberian reservoir, 70–71
    ciliates, East China Sea, 75–76
    diatoms, Laguna di Venezia, 72
    foraminifera, Delaware, 72–73
    gymnamoebae, 69–70
    invasive flagellate, Sweden, 73–75
    marine viruses, California and Sweden,
      77–78
  human hosts
    bacteria cultures, 93–94
    human microbiome, 94–97
  plant hosts
    mummy berry disease, blueberries, 82
    Passalora fulva, on tomatoes, 81–82
    pear scab, 85–86
    Phytophthora (see Phytophthora)
    powdery mildew, apples, 83–85
    strawberry anthracnose and rain splashes,
      89
    tobacco mosaic virus, on beans, 78–79
    Verticillium dahliae, in potato fields,
      79–81
Mixed vegetables, 166–167
Molluscs, 309–310
  gastropods
    in Alberta bighorn sheep habitats, 313–315
    re-colonisation in Mariager Fjord,
      Denmark, 313

Molluscs *(Continued)*
  slugs, in Northumberland, 315–316
  intertidal molluscs, Isle of Man, 311–312
  marine bivalves, 313
  Tellina tenuis
    in benthic and littoral zone, 310
    in Firth of Clyde, Scotland, 310, 311*f*
    trocophore larvae, 310
Monochamus
  M. carolinensis, 171
  M. saltuarius, 171–173
Monte Carlo methods, 561–562
Morisita's patchiness index, 33
Mortality, in England and Wales, 397–398
Moths. *See* Lepidoptera
Mountain forest, in China, 153
Mugil cephalus, 204
Mummy berry disease, blueberries, 82

N

Natural logs, 2
Natural Resource Conservation Service
    (NRCS), 360
Nearest-neighbor analysis, 28–29
Negative binomial distribution, 16–17,
    462–463, 467, 543
  common *k*, 32
  fitting, 30–31
  interpretation of, 31–32
  negative binomial *k*, 30–32
Nematodes, 143–144, 213–226
  animal-parasitic nematodes, 180
    aquatic nematodes, 189–190
    benthic, 198–202
    freshwater, 190–195
    intertidal, 195–197
    invertebrate hosts, 180–181
    littoral and sublittoral, 197–198
    marine nematodes, 195–202
    vertebrate hosts, 181–189
  on Darwin mounds, 446*f*, 447
  entomopathogenic nematodes
    baiting effects, 179–180
    habitat, effect of, 178
    Heterorhabditis bacteriophora, 175–178
    S. carpocapsae, 175–178
    S. feltiae and *S. glaseri*, 174–175
  extraction of, 144
  on kelp, 199–200
  plant-parasitic nematodes
    annual crops, 160–167
    perennial crops, 167–173

  sampling of, 144–147
  terrestrial nematodes
    ecological classifications, 148–155
    TPL stability, 156–160
Nerve transmission model, 539
Network model, 536–537
Neyman-Scott process, 559
Nicotiana tabacum, 165–166
Noncountable number series, 591
Nonlinear maps
  effect of competition, 551
  models in time, 547–548
  strong density dependence, 548–550

O

Oak forest, in Bulgaria, 151–152
ODR. *See* Ordinary dependent regression
    (ODR)
Oligochaetes, 207–208
Omori's law, 511
Ooencyrtus kuvanae, 273
Operophtera brumata. See Winter moth
Ordinary dependent regression (ODR), 61–62,
    62*t*
Origins of aggregation, 21–23
Ornstein-Uhlenbeck process, 539
Ostrinia nubilalis. See European corn borer
    (ECB)

P

Parallel-line analysis, 61–63
  ordinary dependent regression, 61–62, 62*t*
  residual major axis (geometric mean)
    regression, 62–63
Parasite prevalence, 451–452
Parasitism, 273
Pareto distribution, 412–414, 503–504, 595
Partitions approach, 544
Passalora fulva, on tomatoes, 81–82
Passalurus ambiguus, 185–186
Pasteuria penetrans, 91–93, 166
Patchiness index
  Iwao's, 35
  Morisita's, 33
PCNs. *See* Potato cyst nematodes (PCNs)
Pea aphid, 263
Pear psyllids, 269–270
Pear scab, 85–86
Pelagic fish larvae
  in New Jersey, 335–336
  in Portugal, 334–335

Peppers, 86–87
Perennial crops
  banana, 170
  citrus, 167–169
  clover, 169–170
  pine trees, 171–173
  sugarcane, 170
Perrine marl soil, in Florida, 156–157
Perry and Hewitt's number of moves index,
    42–43
Perry's lattice model, 593
Perry's spatial analyses, 42–45
Pesticide effects, 450–451
Pezothrips kellyanus, 262
PGC. *See* Principal Galaxy Catalog (PGC)
Phasic skin conductance (PHSC), 390–391
Pheromone traps, 474, 496–498
Phosphorus
  enzyme activity, aggregation, 386–387
  in lakes, 388–389
Physical models, TPL
  biophysical model, 539–541
  diffusion-limited aggregation, 535–536
  ecological sampling, 534
  fractal model, 535
  Hurst exponent, 533–534
  network model, 536–537
  Poisson line, 534
  statistical physics, 537–539
  temporal and ensemble scaling, 533
Physical *vs.* biological sampling, 573–574
Phytophthora
  in air, 87–89
  on peppers and soybeans, 86–87
Phytophthora antigen units (PAU), 87
Pine trees, 171–173
Piñon jays
  BBS surveys, 351–352
  field data and model output, TPLs of,
    351–353, 352*f*
  social systems and behavior, 351
Plant parasitic nematodes (PPNs), 146–147,
    158
  annual crops
    broad beans, 164–165
    eggplant (aubergine), 166
    mixed vegetables, 166–167
    potatoes, 160–164
    soybeans, 164
    tobacco, 165–166
  perennial crops
    banana, 170

  citrus, 167–169
  clover, 169–170
  pine trees, 171–173
  sugarcane, 170
Plants, TPL analysis, 132–140
  commodity crops
    maize, 128–129
    potatoes, 131–132
    sugar cane, 129–130
    wheat, 130–131
  cover, 103–104
  density, 103
  edible palm, in Brazil, 123–124
  eelgrass, in Chesapeake Bay, 125–126
  grasslands
    rangeland in Mongolia, 121–123
    rangeland in Shaanxi Province, 118
    heterogeneity, 117–118
    tallgrass prairie in Texas, 119–121
    temporal heterogenity and stability,
      118–120, 119*f*
  insectivorous plants, in Morocco and Iberia,
    124–125
  pollination success, in Yucatan shrub,
    126–128
  seedbank
    diversity, in Catalonia, 113–114
    endangered bearpoppy, in Nevada, 117
    farmland, in England, 104–109
    field margins, in Wisconsin, 109–110
    invasive devil's thorn, in Australia,
      114–115
    invasive ragweed, France, 115–117
    pest weeds, 104
    in soybean fields, 110–112
    tree seedbank, in Taiwan, 112–113
  spatial distribution, 103
Platyhelminths, 203–204
Plum aphid, 263
Plumularia setacea, 320–322, 322*f*
Point transect sampling system, 328
Poisson distribution, 29–30, 85–86, 117, 464,
    467, 491, 543, 563
  haddock and whiting, 329
  Japanese beetle larvae, 248–249
  kangaroo rats, 371–372
  randomness, 52–53
  snakes and lizards, 349
  wireworms, in England and Wales, 244
Poisson negative binomial distribution,
    535–536
Pollination, in Yucatan shrub, 126–128

Pollution gradient, 197–198
Pólya-Aeppli distribution, 17
Pólya distribution. *See* Negative binomial
distribution
Polycelis tenuis, 90–91
Polychaetes
in Balearic Islands, 209
biomass, 209–210
in Denmark, 209
Popillia japonica. See Japanese beetle (JB)
Population density, 14, 27, 27*f*
Population intensity, 28
Potato cyst nematodes (PCNs), 160
Potatoes, 160–164
Potato leafhopper (PLH), 263
Potato psyllid, 264
Powdery mildew, on apples, 83–85, 448–449,
449*f*
Power dissipation index (PDI), 509–510
Power-law distributions
biology
allometric growth, 511–512
binomial power law (BPL), 523–524
density-size and variance-size laws,
524–526
flying animals, dimensional relationships
for, 512–515
fractal movement, 515
genetics and physiology, 526
Kleiber's law of metabolism, 518
respiration, 518–519
self-thinning and space-filling rule,
519–521
soil fertility and crop yields, 522
species-area, 515–518
species size, frequency of, 515
taxonomy, 526–527
mathematical
DLA, 506–507
fractals, 507–508
Pareto distribution, 503–504
scale-free networks, 505–506
spectra, 504–505
Zipf's law, 504
physical
geophysics, 511
hydrology, 510–511
inverse square law, 508
meteorology, 509–510
percolation, 509
self-organized criticality, 508
Stefan–Boltzmann law, 508

sociological, 527–528
Richardson's law of conflict, 527–528
Power transformation, 463–464
PPNs. *See* Plant parasitic nematodes (PPNs)
Pratylenchus scribneri, 162
Precipitation, 431–433
Predator, intertidal, 197
Presence-absence sampling, 466–467, 467*t*
Prey
Astropecten irregularis, 317
Capitella capitata, 211
eagle prey, in Northern Greece, 348–349
intertidal, 197
starfish, in North Wales, 317
Prime numbers, 437–440, 606
Principal Galaxy Catalog (PGC), 426–428, 427*f*
Prismatolaimus, 195
Probability generating function (PGF),
535–536
Pseudospatial models, 550

R
Rabbits, 185–186
Ragweed, 115–117
Random factors multiplication matrix, 593
Randomness, 16–21
Random sampling, 466
Range of means, sources of, 574–575
Ratio of score per taxon (RSPT), 479–480
Ratios
bounded ratios, 451–452
temperature, 455–456
unbounded ratios, 452–455
Rats, 187
Recursion equation, 605
Relative density, 14
Reliability, 469, 469*t*
Reptiles
eagle prey in Northern Greece, 348–349
in Florida Everglades, 344–345
herptiles, in Rincón Mountains, 345–347
Residual major axis (geometric mean)
regression, 62–63
Restored wetland, in Georgia, 194–195
Return on investment (ROI), 478–479
Rice stinkbug, 270
Richardson's law of conflict, 527–528
Right Ascension (RA) transects
PGC, galaxies in, 426–428, 427*f*
SAO star catalog, 425*f*, 426
YBSC stars, 425–426, 425*f*
Rivers, in Germany, 190–191

Rotifers
  TPL analysis
    in Elbe estuary, Germany, 308–309
    in Lake Eufaula, Oklahoma, 305–306
    in Upper Parana River basin, Brazil,
      306–308

S

SADIE. *See* Spatial Analysis by Distance
  IndicEs (SADIE)
Salmon, 331
Sampling, 2, 445–450, 572–573
  arthropods
    effects, 268–271
    efficiency and consistency between
      samplers, 266–268
  box counting, 578
  distance sampling
    DISTANCE software, 328
    line transect sampling, 328
    point transect sampling system, 328
    strip sampling, 327–328
  efficiency, 472–475, 496–502
  efficiency and consistency between,
    266–268
  effort and efficiency, 576–577
  and feasible sets, 544–545
  inadequate *NQ/NB*, 445–446
  inconsistent (*see* Inconsistent sampling)
  pattern, 572–573
  physical *vs.* biological, 573–574
  plans
    binomial sampling, 466–468
    number of samples, 470–472
    optimum sample size, 469–470
    postscript, 475–476
    random sampling, 466
    sequential sampling, 468–469
    stratified random sampling, 466
    systematic sampling, 466
  range of means, sources of, 574–575
  site, 577
  spatial pattern, 13–15, 15*f*
  time of, 577
  transect sampling, 578
SAOC. *See* Smithsonian Astrophysical
  Observatory Catalog (SAOC)
Satake and Iwasa forest model, 552
Scale effect, 581–583
Scale-free networks, 505–506
Sceloporus clarkii, 346–347, 346*f*
SCN. *See* Soybean cyst nematode (SCN)

Scotland
  earthworms in, 207–208
  forests in, 153
  Tellina tenuis, in Firth of Clyde, 310, 311*f*
Sea trout, in England
  adults, annual rod and commercial catches,
    340–342
  alevins and parr, 337–339
Second law of thermodynamics, 594
Sediment texture, intertidal, 196–197
Seedbank
  diversity, in Catalonia, 113–114
  endangered bearpoppy, in Nevada, 117
  farmland, in England, 104–109
  field margins, in Wisconsin, 109–110
  invasive devil's thorn, in Australia, 114–115
  invasive ragweed, France, 115–117
  pest weeds, 104
  in soybean fields, 110–112
  tree seedbank, in Taiwan, 112–113
Selene setapinnis, 333–334
Self-affine sequences, 505
Self-organized criticality, 508
Self-similarity, 489–491, 507–508
Self-thinning rule, 519–521
Sequential sampling, 82, 468–469
Sertularella mediterranea, 320–322, 322*f*
Sex ratio, 451, 452*f*
Sheep, 183–185
Sheep tick, 268
Shrimps, 205
Sierpinski pyramid, 596
Silo pallipes, 273
Silverleaf whitefly, 263–264
Simple growth model, 548
Simpson's diversity index, 480
Simulated sampling, 543–544
Sitophilus, 270–271
Skewness, 15, 593
Slugs, in Northumberland, 315–316
Small Cetacean Abundance in the North Sea
  (SCANS) survey, 364–366
Smithsonian Astrophysical Observatory
  Catalog (SAOC), 425*f*, 426
Soberon and Levinsohn's model, 541
Soil
  distribution and abundance, 146
  exoenzymes, 385–387, 386*f*
  fertility and crop yields, 522
  nematodes, diversity of, 153
  Perrine marl soil, in Florida, 156–157
Soybean cyst nematode (SCN), 164

Soybeans, 86–87, 164
Space-filling rule, 519–521
Spatial Analysis by Distance IndicEs (SADIE), 43–45, 563
Spatial distributions, 76
Spatial pattern
  randomness, 16–21
    censuses, 23
    inverse Gaussian distribution, 19–20
    lognormal distribution, 18–19, 19f
    negative binomial distribution, 16–17
    origins of aggregation, 21–23
    Poisson distribution, 16
    Polya-Aeppli distribution, 17
    Tweedie family of distributions, 20–21
Spectra, 504–505
Split lines, fitting, 57–58
Square-cube law, 511–512
Stable fly (*Stomoxys calcitrans*), 265
Standard regression model, 50–51
Starfish, in North Wales
Statistical models, TPL
  higher moments, 542–543
  lattice model, 545–546
  reformulation, 541–542
  sampling and feasible sets, 544–545
  simulated sampling, 543–544
Statistical physics model, 594, 597–598
Stefan–Boltzmann law, 508
Steinernema
  S. carpocapsae, 175–178
  S. feltiae, 174–175
  S. glaseri, 174–175
Stepping-stone model, 548–549
Stochastic behavioral model, 399
Stopping rule, 469
Stratified random sampling, 466
Stream water quality, 479–481
Stress, in air traffic controllers, 405–406
Strip sampling, 327–328
Strongyloides ratti, 187
Subarctic, 153–154
Sublittoral nematodes, 197–198
Submersed aquatic vegetation (SAV) beds, 125–126
Suction trap, 497–498
Sugarcane, 129–130, 170
Sugarcane weevil borer, 271
Super-aggregation, 583–584
Sweden
  grassy pasture in, 151
  invasive flagellate, 73–75
  lakes in, 192
  marine viruses, 77–78
Swedish National Monitoring program, 74
Sweet potato whitefly. *See* Silverleaf whitefly
Systematic sampling, 466

T
Tallgrass prairies, in Texas, 119–121
Taxonomy, 526–527
Taylor' power law (TPL), 260–276, 416–420, 440–441, 595–596
  actual and simulated precipitation, 431–433
  amphibians
    frogs, in Alpine habitat in California, 347
    larval abundance, in ephemeral ponds, Ohio, 348
  applications
    environmental assessment and monitoring (*see* Environmental assessment)
    model calibration and validation, 482–483
    quality control, 483–484
    sampling (*see* Sampling)
    to stabilize variance, 463–464
    testing vaccines, 481–482
    transformations, 462–464
  Astropecten irregularis, 317, 318f
  barnacles, 254–260
    Balanus balanoides, 257, 258f
    C. dalli, 255–256, 256t
    E. modestus cyprid larvae, 258–259, 259f
  biological models (*see* Biological models, TPL)
  birds
    Audubon Society's CBC analysis, 359–361, 360f
    BTO's breeding bird survey, 356–359, 358f
    grassland sparrows, 353–354
    piñon and Western scrub jays, 351–353, 352f
    in urban and nonurban sites, 354–356, 355f
    willow ptarmigan, in Norway, 350–351, 350f
  bivalves, 313, 314f
  bryozoans, 320, 321f
  Californian commercial fisheries, 342–344
  cetaceans
    Azores, 364, 365f
    British Isles, 365f, 366
    fin whales, in Ligurian Sea, 369
    harbor porpoise, in North Sea, 367–368, 368f

chlorophyll, intraseasonal concentration of, 387, 388*f*
city size distribution, 414–415
Colorado potato beetle, 245–247, 246*f*
community, mixed species and, 586–587
company size data, 412–413
computation of, 5
convenience store sales, in Japan, 410–412
cyanogen, in comet Hale-Bopp, 423
demersal fish, 332–333, 333*f*
earthquakes, in Japan, 428–429
European corn borer, 236, 237*f*
exoenzymes, in low-input and conventionally managed soils, 385–387, 386*f*
fitting (*see* Fitting Taylor' power law)
forex market transactions, 435–437
gastropods
  in Alberta bighorn sheep habitats, 314–315, 315*f*
  re-colonisation in Mariager Fjord, Denmark, 313, 314*f*
  slugs, in Northumberland, 315–316
gypsy moth
  adult male moths, 241–242, 241*f*
  egg mass, 240–241, 241*f*, 273
  pupae, 241, 241*f*
haddock and whiting, 329, 330*f*
heavenly bodies, distribution of, 424–425
herring and mackerel, 330–331
human behavior
  crime statistics, in England and Wales, 406–408
  European wars, 409–410
  stress, in air traffic controllers, 405–406
human demography
  cyberspace and physical space, movements in, 399
  mortality, in England and Wales, 397–398
  population density, in Norway, 395–397
  US decennial census, 392–394
human health
  disease monitoring, 403–404
  HIV/AIDS, 399–401
  measles and whooping cough, 402–403
  typhoid in Cambodia, 401–402
hydroids, 321–322, 322*f*
as index of aggregation, 37–39
intertidal molluscs, 312, 312*f*
Japanese beetle
  adults, 250–252, 251*f*
  in the Azores, 252–254
  in the United States, 247–254

larvae, 247, 248*f*, 249–250
jellyfish, 322–323, 323*f*
kangaroo rat mounds, in New Mexico, 372
menhaden, 336–337
metastatic melanoma tumors, 389–390, 390*f*
microorganisms (*see* Microorganisms, TPL analysis)
models, 591–596
orthogonal directions, 579–580
pelagic fish larvae
  in New Jersey, 336
  in Portugal, 334–335, 335*f*
phenotypic and population change, rate of, 383–385
phosphorus, in lakes, 388–389
physical models (*see* Physical models, TPL)
physiological responses to stimuli, 391
plants (*see* Plants, TPL analysis)
Poisson line and TPL, intersection of, 580–581
prime numbers, 437–440
RA and DC transects
  PGC, galaxies in, 426–428, 427*f*
  SAO star catalog, 425*f*, 426
  YBSC stars, 425–426, 425*f*
reptiles
  eagle prey in Northern Greece, 349
  in Florida Everglades, 344–345
  herptiles, in Rincón Mountains, 346–347, 346*f*
of rotifer communities
  in Elbe estuary, Germany, 308–309
  in Lake Eufaula, Oklahoma, 305–306
  in Upper Parana River basin, Brazil, 306–308
salmon, 331, 332*f*
scale effect, 581–583
sea trout, in England
  adults, annual rod and commercial catches, 340, 341*f*
  alevins and parr, 338–339, 338*f*
self-similarity, 489–491
small samples, effect of, 491–492
stability
  cowpeas and cotton, in California, 158–160
  crops, in Florida, 158
  forest and old field, 158
  perrine marl soil, in Florida, 156–157
statistical models (*see* Statistical models, TPL)
super-aggregation, 583–584

Taylor' power law (TPL) *(Continued)*
  Tellina tenuis, 310, 311*f*
  temporal *vs.* spatial, 579
  in three dimensions, 580
  timing of failure, 590–591
  tornadoes, in continental USA, 429–431
  traffic, network model, 433–434
  Tropiometra carinata, 319, 319*f*
  ungulate herds, in Kenya's Rift Valley, 370, 371*f*
  universality, 596–599
  uses of, 588–590
  winter moth, 238, 239*f*
  of wireworms
    in England and Wales, 244, 245*f*
    in Washington State, 243, 244*f*
  zeros, effect of
    anatomy, 495–496
    direct effect, 492–495
Tellina tenuis
  in benthic and littoral zone, 310
  in Firth of Clyde, Scotland, 310, 311*f*
  trocophore larvae, 310
Temperature, 455–456, 456*f*
Terrestrial gastropods, in Alberta, 313–315
Terrestrial nematodes
  ecological classifications
    forests, in Scotland, 153
    grassy pasture in Sweden, 151
    metabolic footprint, 154–155
    mountain forest, in China, 151–152
    oak forest, in Bulgaria, 151–152
    subarctic heath, 153–154
    urban turfgrass in Ohio, 148–151
  TPL stability
    cowpeas and cotton, in California, 158–160
    crops, in Florida, 158
    forest and old field, 158
    perrine marl soil, in Florida, 156–157
Thomas cluster process, 559
Tobacco, 165–166
Tobacco mosaic virus (*Tobamovirus* sp.), 78–79
Tomato leafminer, 274–275
Tornadoes, in continental USA, 429–431
TPL. *See* Taylor' power law (TPL)
Traffic, network model, 433–434
Transect sampling, 578
Transmission electron microscope (TEM) imaging, 77
Trap saturation, 448–450, 500–502
  modeling example, 449–450

powdery mildew, on apples, 448–449, 449*f*
Triadic Koch curve, 489–490
Tribolium castaneum, 270–271
Trichostrongylus retortaeformis, 185–186
Trophic interactions
  competition, 585–586
  parasitism, 585
  predation, 584
Tropiometra carinata, 318–319
Tweedie exponential dispersion models, 538, 594
Tweedie family of distributions, 20–21
TWINSPAN, 124–125
Tylenchulus semipenetrans, 167–169
Typhoid, 401–402

U
Unbounded ratios, 452–455
  commodity crops, 452–455, 453*f*
Uncountable numbers, 455–456, 456*f*
Ungulates, in Kenya's Rift Valley, 369–371
Universality, 596–599
Urban turfgrass, in Ohio, 148–151
Urceolaria mitra, 90–91
Urosaurus ornatus, 346–347, 346*f*
US decennial census, 392–394
US Department of Agriculture (USDA), 104

V
Vallisneria americana. See Eelgrass
Variance-mass allometry, 524
Variance-mean ratio, 29–30
Variance-mean relationship, 35–45, 41*t*
  Adès distribution, 39–42, 41*f*, 41*t*
  Perry and Hewitt's number of moves index, 42–43
  SADIE, 43–45
  TPL, as index of aggregation, 37–39
Variance-size law, 524–526
Varroa mite, 273
Vertebrate hosts
  carp, 188–189
  mice, 187
  mixed-species TPLs, 182
  parasite load and aggregation of macroparasites in, 181–182
  rabbits, 185–186
  rats, 187
  sheep, 183–185
Vertebrates, TPL analysis, 372–378
  amphibians and reptiles, 344–349

birds, 349–363
fish, 328–344
mammals, 363–372
Verticillium dahliae, 79–81
Viruses
  HIV/AIDS, 399–401
  oceanic, 77–78
  plant, 78–79

W

Weevils, 275
Weibul distribution, 511
Weighted least squares (WLS), 108, 404
Western flower thrips (WFT), 262, 271–272,
    470–471
Western scrub jay
  BBS surveys, 351–352
  field data and model output, temporal TPLs
      of, 351–353, 352*f*
  social systems and behavior, 351
Wheat, 130–131

Wheel animals. *See* Rotifers
Whiting, 329, 330*f*
Whooping cough, 402–404, 405*f*
Willow ptarmigan
  hybrid spatial TPL analyses, 350–351, 350*f*
  ideal free distribution, 349–350
  line transect survey, 349–350
Wind farms, 481
Winter moth, 237–238
Wireworms, 242–244
  in England and Wales, 243–244
  in Washington State, 243
World's oceans, nematodes in, 199

Y

Yale Bright Star Catalog (YBSC), 425–426

Z

Zipf's law, 414, 504, 595
Zygomaticus major (ZYG), 390–391

Printed in the United States
By Bookmasters